Principles of Signal Detection and Parameter Estimation

Bernard C. Levy

Principles of Signal Detection and Parameter Estimation

Bernard C. Levy
Dept. of Electrical and
Computer Engineering
University of California
1 Shields Avenue
Davis, CA 95616

ISBN: 978-1-4419-4565-5 e-ISBN: 978-0-387-76544-0
DOI: 10.1007/978-0-387-76544-0

Printed on acid-free paper

9 8 7 6 5 4 3 2 1

springer.com

In these matters the only certainty is that nothing is certain.

Pliny the Elder

Preface

As a discipline, signal detection has evolved significantly over the last 40 years. Some changes have been caused by technical advances, like the development of robust detection methods, or the use of the theory of large deviations to characterize the asymptotic performance of tests, but most changes have been caused by transformations in the engineering systems to which detection techniques are applied. While early applications of signal detection focused on radar and sonar signal processing or the design of digital communication receivers, newer areas of application include image analysis and interpretation, document authentification, biometrics, and sensor or actuator failure detection. This expanded scope of application has required some adjustment in standard ways of formulating detection problems. For example, image processing applications typically combine parameter estimation and detection tasks, so the separation of parameter estimation and detection in distinct operations typical of early communication systems, where parameter estimation was accomplished through the use of training signals, needs to be abandoned. Other changes have occured in the design of communication systems which make it increasingly difficult to treat the detection of communications signals and of radar/sonar signals in a unified manner. This common framework assumes implicitly that intersymbol interference is not present and that channel coding and modulation are implemented separately, since in this case modulated signals can be detected one symbol at a time. But modern communication systems are typically designed to operate over bandlimited channels where intersymbol interference is present, and starting with the introduction of trellis coded modulation, modulation and coding have become intertwined. In this context, the detection of modulated signals can no longer be treated on a symbol-by-symbol basis but needs to be viewed as a sequence detection problem, where the sequence is generated by a Markov chain. Another feature of modern radar and communication systems, in particular wireless systems, is that they often need to operate in a rapidly changing environment. So even if training or calibration signals are available to estimate the system parameters, because parameters may change quickly, it is desirable to constantly refresh estimates while at the same time performing detection tasks on received signals. In other words, detection and estimation need to be performed simultaneously and can no longer be viewed as separate tasks. Finally, another feature of modern engineering systems to which detection algorithms are applied is that due to modelling errors, imperfect calibration, changes in the environment, as well as the presence of interfering signals, it is not entirely realistic to assume that accurate models are available, and thus robust detection techniques need to be applied.

The objective of this book is to give a modern presentation of signal detection which incorporates new technical advances, while at the same time addressing issues that reflect the evolution of contemporary detection systems. Recent advances which are covered include the use of the theory of large deviations to characterize the asymptotic performance of detectors, not only for the case of independent identically distributed observations, but also for detection problems involving Gaussian processes or Markov chains. In addition, a chapter discusses robust signal detection, and another the application of the EM algorithm to parameter estimation problems where ML estimates cannot be evaluated in closed form. At the same time, changes in modern communications technology are addressed by examining the detection of partially observed Markov chains, both for the case when the Markov chain model is known, or when the model includes unknown parameters that need to be estimated. To accommodate the need for joint estimation and detection in modern communication systems, particular attention is given to the generalized likelihood ratio test (GLRT), since it explicitly implements detection and estimation as a combined task, and because of its attractive invariance and asymptotic properties.

This book is primarily intended for use in signal detection courses directed at first or second year graduate electrical engineering students. Thus, even though the material presented has been abstracted from actual engineering systems, the emphasis is on fundamental detection principles, rather than on implementation details targeted at specific applications. It is expected that after mastering the concepts discussed here, a student or practicing engineer will be able to analyze a specific detection problem, read the available literature, and design a detector meeting applicable specifications. Since the book is addressed at engineeering students, certain compromises have been made concerning the level of precision applied to mathematical arguments. In particular, no formal exposure to measure theory, modern real analysis, and the theory of operators in Hilbert spaces is assumed. As a consequence, even though derivations are conceptually accurate, they often leave some technical details out. This relatively casual presentation style has for objective to ensure that most students will be able to benefit from the material presented, regardless of preparation. On the other hand, it is expected that readers will have a solid background in the areas of random processes, linear algebra, and convex optimization, which in the aggregate form the common frame of reference of the statistical signal processing community.

Another aspect of this book that may be controversial is that it does not follow a theorem/proof format. To explain this choice, I would like to point out that whereas the hypothesis testing and parameter estimation techniques used in signal detection lend themselves naturally to a formal presentation style, because of its applied nature, signal detection consists primarily of a methodology for converting an observed signal model and some specifications for the detector to be constructed into first a formulation of the problem in hypothesis testing format, followed by a solution meeting the given specifications. In this context, the most important skills needed are first the ability to think geometrically in higher dimensional spaces, and second the capacity to reason in a manner consistent with the assumed observation model. For example, if the parameters appearing in the signal model admit probability distributions, a Bayesian framework needs to be employed to construct a detector, whereas when parameters are unknown but nonrandom, the parameters need to be estimated as part of the detector construction. Slightly different modelling assumptions for the same problem may lead to different detector structures. Accordingly, signal detection cannot really be reduced to a collection of mathematical results. Instead, it is primarily a methodology that can be best explained by employing a continuous presentation flow, without attempting to slice the material into elementary pieces. The continuous flow approach has also the advantage that it makes it easier to connect ideas presented in different parts of the book without having to wait until each analytical derivation is complete.

Obviously, since the field of signal detection covers a vast range of subjects, it has been necessary to leave out certain topics that are either covered elsewhere or that are too advanced or complex to be presented concisely in an introductory text. Accordingly, although Kalman and Wiener filters are employed in the discussion of Gaussian signal detection in Chapter 10, it is assumed that optimal filtering is covered elsewhere as part of a stand-alone course, possibly in combination with adaptive filtering, as is the case at UC Davis. In any case, several excellent presentations of optimal and adaptive filtering are currently available in textbook form, so it makes little sense to duplicate these efforts. Two other topics that have been left out, but for entirely different reasons, are change detection/failure detection, and iterative detection. To explain this choice, let me indicate first that change detection and failure detection represent one of the most interesting and challenging fields of application of the methods presented in this book, since in addition to detecting whether a change occurs, it is necessary to detect when the change occurred, and for safety critical applications, to do so as quickly as possible. However, important advances have occurred in this area over the last 20 years, and it does not appear possible to give a concise presentation of these results in a manner that would do justice to this topic. As for the iterative detection techniques introduced recently for iterative decoding and equalization, it was felt that these results are probably best presented in the context of the communications applications for which they were developed.

The scope of the material presented in this book is sufficiently broad to allow different course organizations depending on length (quarter or semester) and on the intended audience. At UC Davis, within the context of a one quarter course, I usually cover Chapter 2 (hypothesis testing), followed by Chapter 4 (parameter estimation), and the first half of Chapter 5 (composite hypothesis testing). Then I move on to Chapter 7 presenting the Karhunen-Loève decomposition of Gaussian processes, followed by Chapters 8 and 9 discussing the detection of known signals, possibly with unknown parameters, in white and colored Gaussian noise. A semester length presentation directed at a statistical signal processing audience would allow coverage of the second half of Chapter 3 on sequential hypothesis testing, as well as Chapters 10 and 11 on the detection of Gaussian signals, possibly with unknown parameters. On the other hand, a semester course focusing on communications applications would probably add Chapters 12, 13 and parts of Chapter 11 to the one-quarter version of the course outlined above.

The idea of writing this book originated with a lunch conversation I had with a UC Davis colleague, Prof. Zhi Ding about four years ago. I was complaining that available textbooks on signal detection did not include several topics that I thought were essential for a modern presentation of the material, and after listening politely, Zhi pointed out that since I had all these bright ideas, maybe I should write my own book. Against my better judgement, I decided to follow Zhi's suggestion when I became eligible for a sabbatical year in 2004–2005. In spite of the hard work involved, this has been a rewarding experience, since it gave me an opportunity to express my views on signal detection and parameter estimation in a coherent manner. Along the way, I realized how much my understanding of this field had been impacted by teachers, mentors, friends, collaborators, and students. Among the many individuals to whom I am indebted, I would like to start with my teachers Pierre Faurre and Pierre Bernhard at the Ecole des Mines in Paris, who got me interested in optimal filtering and encouraged me to go to Stanford to pursue graduate studies. As soon as I arrived at Stanford, I knew this was the right choice, since in addition to the expert guidance and scientific insights provided by my advisors, Tom Kailath and Martin Morf, I was very fortunate to interact with an unusually talented and lively group of classmates including Sun-Yuan Kung, George Verghese, and Erik Verriest. Later, during my professional life at MIT and UC Davis, I benefited greatly from the mentorship and advice provided by Alan Willsky, Sanjoy Mitter, and Art Krener. I am particularly grateful to Art for showing me through example that good research and fun are not mutually exclusive. In addition, I would like to thank Albert Benveniste and Ramine Nikoukhah for fruitful research collaborations during and after sabbatical visits at INRIA in France. Like most professors, I have learnt a lot from my students, and among those whose research was directly related to the topic of this book, I would like to acknowledge Ahmed Tewfik, Mutlu Koca, Hoang Nguyen and Yongfang Guo. A number of volunteers have helped me in the preparation of this book. Yongfang Guo helped me to get started with MATLAB simulations. Patrick Satarzadeh and Hoang Nguyen read all chapters and made suggestions which improved significantly the presentation of the material. My colleague, Prof. Zhi Ding, used my notes to teach the UCD detection course (EEC264) while I was on sabbatical and provided valuable feedback. Dr. Rami Mangoubi of Draper Laboratory made valuable suggestions concerning the organization of the book, as well as the choice of material presented. I am also grateful to several anonymous referees whose comments helped eliminate a few rough spots in an earlier draft of this book. Finally, I am deeply indebted to my wife, Chuc Thanh, and our two children, Helene and Daniel, for their encouragement and patience during this long project. I am not completely sure that they will ever be persuaded that this book is truly finished.

Bernard C. Levy
Davis, California

Contents

A Note to Instructors

A password protected, solutions manual is available online for qualified instructors utilizing this book in their courses. Please see www.Springer.com for more details or contact Dr Bernard C. Levy directly.

1

Introduction

The field of signal detection and parameter estimation is concerned with the analysis of received signals to determine the presence or absence of signals of interest, to classify the signals present, and to extract information either purposefully or inadvertently included in these signals. For example, in active radar, electromagnetic pulses or pulse trains are transmitted and the reflected signals are analyzed to determine the nature of air traffic (small airplanes, commercial airplanes, or hostile aircraft), to extract information such as distance, speed, and possibly to form an image that would allow the identification of the airplane type. In sonar signal processing, active sonar systems employ sound pulses to probe the ocean and then analyze return echoes by using techniques similar to those employed by radar systems. But active radar and sonar systems have the feature that, since they use probing signals, theses signals give away the location of the transmitter and make it therefore vulnerable to hostile action. Passive sonar systems on the other hand just use ambient ocean noise and other noise sources, such as propeller noise, to analyse ocean traffic without giving away their location. In this context, the information gathered by passive sonar sensors is revealed inadvertently by the sources they track, such as submarines or surface ships.

Signal detection and parameter estimation techniques form also a key element of the design of receivers for both wireline (DSL, cable, fiber) and wireless communications systems. In this context, the digital information extracted from the received signal is inserted intentionally in a transmitted analog signal through the choice of a modulation format. The modulation may be either carrierless, like pulse amplitude modulation (PAM), or rely on a sinusoidal carrier, and in such cases the information may be encoded in the amplitude, the phase or the frequency. In addition to these basic modulation choices, a number of new wrinkles have been introduced over the last 20 years. To increase the information throughput of broadband communication systems without having to design long equalizers, multicarrier modulation schemes [1] such as orthogonal frequency division modulation (OFDM) have been implemented in DSL and broadband wireless communication systems.

B.C. Levy, *Principles of Signal Detection and Parameter Estimation*,
DOI: 10.1007/978-0-387-76544-0_1, © Springer Science+Business Media, LLC 2008

For applications where multiple users share the same spectrum, code division multiple access (CDMA) [2] was introduced to enhance the utilization rate of such systems. Finally, while early communications systems kept coding and modulation separate, it was shown by Ungerboeck [3] that a significant performance gain could be achieved by using trellis coded modulation, which has the effect of merging coding and modulation. These new advances have led to a significant increase in receiver complexity, which typically manifests itself by the need, in parallel with the recovery of the transmitted bits, to estimate a number of system parameters. These parameters include timing, phase and frequency synchronization parameters, channel, and interference noise power, and for the case when an equalizer needs to be implemented, the channel impulse response. Furthermore, it is often of interest to estimate key parameters of interfering signals, such as their direction or arrival or modulation parameters, in order to design operations aimed at reducing their effect on the receiver. So typically, in modern communication systems, estimation and detection need to be performed together as a single task. Note that while it is possible to use training signals to estimate some of the system parameters, such an approach becomes unwieldy for multiuser communication systems since it requires that all users should act cooperatively while parameters are being estimated. Also, while it is ideal to reduce the detection problem to a situation where transmitted symbols can be detected individually, when intersymbol interference is present or for trellis coded modulation systems, an entire sequence of symbols needs to detected. While this problem can be viewed conceptually as a detection problem in a space of large dimensionality, it turns out that the detection complexity can be reduced by exploiting the Markov structure characterizing how information propagates through the communication system formed by the transmitter, channel, and receiver.

In addition to radar, sonar, and communication systems, a number of emerging applications, such as sensor monitoring and failure detection, rely heavily on signal detection techniques. To explain why this type of application has become important observe that automobiles, aircraft, industrial systems such as turbines, or even oil exploration platforms, rely increasingly on embedded electronic sensors to monitor the status of the system and optimize its performance. In airplanes, hardwire controls have been replaced by electronic actuators. All these substitutions are highly beneficial when electronics function properly, since they allow an increased use of automation and ensure that the system functions optimally under normal conditions. However, this leaves the system vulnerable to sensor or actuator failures. To address this issue, it has become customary to put in place failure detection systems which monitor the status of sensors and actuators, in order to allow human intervention when electronics fail. For safety critical applications such as airplane controls, redundant sensors or actuators are usually built in the system, and the purpose of failure detection is to determine when sensing or control functions should be switched from a primary system to its backup.

1.1 Book Organization

The book is divided into three parts. The first part, which is formed by Chapters 2 to 6 focuses on key concepts, and thus is significantly more difficult than the last two parts which apply the general methods of the first part to the detection of deterministic and Gaussian signals, and the detection of Markov chain signals, respectively. The two key chapters of the first part are Chapters 2 and 4, which focus respectively on binary and M-ary hypothesis testing and on parameter estimation. Chapter 2 discusses Bayesian and Neyman-Pearson (NP) tests and characterizes their performance. For binary tests, the optimal decision rule is expressed in terms of the likelihood ratio (LR) statistic and the test performance is analyzed by using the receiver operating characteristic (ROC), whose properties are described. However, a question which is left unanswered by this chapter is: what can be done to ensure that decisions are reached with a high degree of reliability? In signal detection, two different strategies can be employed to reach highly reliable decisions. The first approach consists of ensuring that the detector operates at a sufficiently high signal to noise ratio (SNR). But this is not always possible. The second approach consists of collecting measurements until the reliability of the decision improves. Tests with repeated measurements are discussed in Chapter 3. Tests with a fixed block size are first considered. Cramér's theorem, which is a key result of the theory of large deviations [4, 5], is used to show that the probability of a miss or of false alarm of a test decreases exponentially with the number of observations. The rate of decay is governed by the Kullback-Leibler (KL) divergence of the probability distributions of the two hypotheses for NP tests, and by the Chernoff distance for Bayesian tests. Sequential hypothesis testing represents a more flexible alternative to tests with repeated measurements, since, instead of commiting ahead of time to a test with a fixed size, observations are collected one by one, and depending on the observed values, one decides to either keep collecting observations or to reach a decision based on all collected observations. This can be accomplished optimally by using a sequential probability ratio test (SPRT), which can be viewed in essence as an LR test with two thresholds. As long as the LR statistic stays between the high and low thresholds, it means that enough uncertainty remains and additional observations should be collected, but when either the high or the low threshold is crossed, a high degree of confidence has been reached concerning the validity of one of the two hypotheses, so a decision can be made.

Chapters 2 and 3 assume that the probability distributions of the observations under all hypotheses are known exactly. But in practice, unknown parameters are often present. Chapter 4 examines the estimation of unknown parameters from observations. This problem has a long history in statistics and depending on whether the unknown parameters are viewed as random, or unknown but fixed, different sets of methods can be devised to obtain estimates. Bayesian methods view the parameters as random with a known a-priori probability distribution. This distribution can be derived from physical considerations, it can be based on collected data, or it can just be guessed. Then give a loss function which specifies how estimation errors are weighted, and an estimate is derived. The minimum mean-square error (MMSE) and maximum a-posterori (MAP) estimates are two common Bayesian estimates. Another approach used to derive estimates relies exclusively on the available data and views the parameters as deterministic but unknown. The best known estimate of this type is the maximum likelihood (ML) estimate which, as its name indicates, maximizes the density of the observations over the unknown parameter, so that it represents the parameter value most likely to have produced the observations The asymptotic performance of ML estimates is examined as the number N of measurements increases. Under mild conditions, the ML estimate converges almost surely to the true parameter value, and its standard deviation decreases like $N^{-1/2}$.

The ability to estimate parameters opens the way to the consideration of composite hypothesis testing problems in Chapter 5. In such problems, the probability distribution of the observations includes unknown parameters under at least one hypothesis. When some parameters are unknown, the corresponding hypothesis can be viewed as "composite" since it represents in effect a combination of hypotheses corresponding to each possible parameter value. For such problems, the preferred situation is one where there exists a test which outperforms all other tests independently of the unknown parameter values. Tests of this type are called uniformly most powerful (UMP), but they tend to exist quite rarely. When no UMP exists, if the detection problem admits certain symmetries, it is sometimes possible to identify a test which outperforms all others in the more restricted class of tests that remain invariant under the problem symmetries. A test of this type is called uniformly most powerful invariant (UMPI), but again tests with such a strong property are rare. In the absence of UMP(I) tests, a common strategy to design tests consists of combining estimation and detection by replacing the unknown parameters by their ML estimate under each hypothesis and then forming a likelihood ratio as if the estimates were the true parameter values. A test of this type is called a generalized likelihood ratio test (GLRT). The properties of such tests are examined. In particular, it is shown that the GLRT is invariant under any transformation which leaves the detection problem invariant. Also, for tests with repeated observations and for relatively large classes of observation distributions, it is found that the GLRT is asymptotically optimal in the sense that it maximizes the rate of decay of the probability of a miss or of false alarm. Thus both for ease of implementation and performance reasons, the GLRT is an attractive default test in situations where no UMP test exists. Finally, the first part of the book closes in Chapter 6 with a discussion of robust hypothesis testing. Robust hypothesis testing problems arise whenever the probability distribution of the observations under each hypothesis is known only imperfectly. Unlike composite hypothesis problems, which assume that a complete model of the received signal is available, where some parameters, such as amplitude, phase, or frequency, are unknown, robust hypothesis testing problems occur whenever models are imprecise and may not capture all interferences or noises present in the observations. Problems of this type can be formulated by assuming that under each hypothesis a nominal probability distribution of the observations is given, but the actual probability distribution belongs to a neighborhood of the nominal distribution. What constitutes a neighborhood is based on a notion of proximity between probability distributions. Various measures of proximity exist, such as the Kolmogorov distance, or KL divergence, among others. Then, given neighborhoods representing possible locations of the actual probability distributions under each hypothesis, a robust test is one that optimizes the test performance for the worst possible choice of observation distributions. This type of analysis was pioneered by Huber [6] for binary hypothesis testing and then extended by a number of authors [7, 8] to signal detection. Although robust tests tend to be overly conservative, they highlight an important vulnerability of signal detection methods. Specifically, because detectors are designed to operate with a very small probability of error, relatively small shifts of probability mass for the observation distributions under each hypothesis can result in a catastrophic loss of detector performance. In other words, signal detectors tend to be quite sensitive to model inaccuracies.

The second part of the book focuses on Gaussian signal detection. This class of detection problems includes the case when the received signal is deterministic, possibly with unknown parameters, and is observed in Gaussian noise. It also includes the case where the received signal itself is Gaussian and its model may include unknown parameters. An important aspect of such problems is that whereas the results presented in the first part of the book assumed that the observations formed a finite dimensional vector, for continuous-time (CT) detection problem, the received waveform belongs to an infinite dimensional space. Even if the signal is sampled, the corresponding discrete-time (DT) detection problem may have a large dimensionality. To tackle such problems, it is useful to develop tools that can be used to collapse an infinite dimensional detection problem, or a problem with a large number of dimensions, into one involving only a small number of dimensions. A technique for accomplishing this objective consists of using the Karhunen-Loève expansion [9, Sec. 3.4] of Gaussian processes. This expansion represents Gaussian processes in terms of the eigenfunctions of their covariance. By exploiting the flexibility existing in the choice of eigenfunctions for processes with repeated eigenvalues, it was shown by Middleton [10], Davenport and Root [11], and later by Van Trees [12] and Wozencraft and Jacobs [13] that signal detection and digital communications problems can often be expressed geometrically as detection problems in very low dimensional spaces. This approach has become known as the signal space representation of detection or communication problems. This technique has significant limits. First of all, it is only applicable to Gaussian processes. Also, there is no guarantee that the transformed problem is necessarily finite dimensional. For example, when the Karhunen-Loève decomposition is applied to the detection of a Gaussian signal in white

Gaussian noise, the transformed problem remains infinite dimensional. However, because the Karhunen-Loève expansion is a whitening transformation, the new problem is a lot simpler than the original detection problem. So, on balance, the eigenfunction expansion approach represents a valuable tool for the analysis of Gaussian detection problems. The Karhunen-Loève expansion of Gaussian processes and its properties are examined in Chapter 7. While in the DT case, the computation of the eigenvalues and eigenfunctions can be expressed in matrix form, the computation of the eigenfunctions of the integral operator specified by the covariance kernel of the process of interest is more difficult to accomplish in the CT case. Fortunately, it is observed that for relatively large classes of Gaussian processes, the covariance kernel can be viewed as the Green's function of a self-adjoint Sturm-Liouville differential or difference system, so that the eigenfunction evalution can be expressed as the solution of a self-adjoint differential or difference equation, depending on whether we consider the CT or DT case.

The detection of known signals in white or colored zero-mean Gaussian noise is considered in Chapter 8. Detector structures applicable to binary and M-ary detection are described, which include the well-known matched filter and correlator-integrator structures. These results are illustrated by considering the detection of M-ary phase-shift keyed and frequency-shift-keyed (FSK) signals. Two approaches are presented for the detection of known signals in colored noise. The first method employs a matched filter or correlator integrator structure where the transmitted signal is replaced by a distorted signal which compensates for the coloring of the additive noise. The second approach relies on whitening the additive noise present in the received signalwhich is accomplished by using either Wiener or Kalman filters. Chapter 9 is probably the most important one of the book from the point of view of practical detector implementation, since it considers the detection of deterministic signals with unknown parameters in zero-mean white Gaussian noise (WGN). As such it covers therefore digital communications problems where the amplitude and/or phase have not been acquired, as well as sonar and radar signal processing problems where amplitude, time delay, and possibly a Doppler frequency shift need to be estimated. Depending on the type of application considered, a Bayesian or a GLRT viewpoint are adopted for detector synthesis. In wireless communications, relatively accurate statistical models, such as the Rayleigh or Nakagami/Rice fading channel models [14] have been developed for the amplitude of received signals. So for such applications a Bayesian viewpoint is warranted. On the other hand, for sonar and radar applications, it is essentially impossible to assign meaningful probability distributions to the time delay or Doppler frequency shift of received signals, so GLRT detectors are preferred in this case. In this respect, it is worth noting that in sonar or radar detection problems, the parameter estimates used to implement the GLRT detector provide crucial information about the target, such as its distance and velocity. So proper design of the probing pulse is required to improve the accuracy of the parameter estimates. For the case of range-Doppler radar, it is shown that estimation accuracy is characterized by the ambiguity function of the probing pulse. Its properties are described, and key trade-offs involved in the design of probing pulses are discussed.

While man-made signals are typically deterministic, possibly with unknown parameters, signals formed by the interaction of mechanical systems with their environment are often random. Examples of such signals include random vibrations in turbines or heavy machinery for industrial applications, or ship propeller noise in undersea surveillance. The detection of Gaussian signals in WGN is considered in Chapters 10 and 11. Chapter 10 considers the case where a complete statistical model of the Gaussian signal to be detected is available. Such a model may take the form either of a full set of first- and second-order statistics, or of a Gauss-Markov state-space model. For such problems optimal detectors typically require the implementation of Wiener or Kalman filters. Accordingly, although CT Gaussian detection problems give rise to elegant mathematics [15, 16], we focus our attention primarily on the DT case, since optimal estimation filters are typically implemented in digital form. Both noncausal and causal detector implementations are presented.

Chapter 11 considers the detection of Gaussian signals whose models include unknown parameters. For the case when the observed signal admits a state-space model, this corresponds to a hidden Gauss-Markov model. For such problems, to implement a GLRT detector it is first necessary to evaluate the ML estimates of the unknown parameters. But unlike the case of known signals with unknown parameters, where the ML estimation equations usually admit a closed-form solution, for the case of Gaussian signals with unknown parameters, a closed-form solution of the ML equations is rarely available. Accordingly ML estimates need to be evaluated iteratively. This can be done by using standard iterative methods, such as the Gauss-Newton iteration, for finding the roots of nonlinear equations. However, among iterative techniques, the expectation-maximization (EM) algorithm has the feature that all the quantities that it evaluates have a statistical interpretation. Furthermore, instead of being a precise algorithm, it provides a general framework for simplifying the maximization of likelihood functions. The key idea is as follows: whenever the solution of the likelihood function cannot be performed in closed form, the available observations should be augmented by "missing data" until the "complete data" formed by the combination of the actual and missing data is such that the closed form maximization of the complete data likelihood function can be performed easily. Then, since the missing data is of course not available, it needs to be estimated, so the iteration works in two steps: in the expectation phase (E-phase) the missing data is estimated based on the available data and on the current estimate of the unknown parameters, and in the maximization phase (M-phase), the estimated complete data likelihood function is maximized to obtain a new parameter vector estimate. This iteration has the feature that the likelihood function increases at every step and the parameter estimates converge to a local maximum of the likelihood function. The EM algorithm was proposed by Dempster, Laird and Rubin [17] in 1977, but a version of this algorithm limited to hidden Markov models had been proposed earlier by Baum and Welch [18, 19]. The EM algorithm and its properties are examined in Chapter 11, as well as several extensions, such as the supplemental EM algorithm, which can be used to estimate the error covariance for the parameter vector estimate. For the case of a hidden Gauss Markov model, it is shown that the E-phase of the EM iteration can be implemented by using a double sweep Kalman smoother. Finally, assuming ML estimates of the unknown parameters have been evaluated, the structure of GLRT detectors of Gaussian signals is examined.

The third part of the book concerns the detection of Markov chain signals in WGN. As explained earlier, problems of this type occur in communication systems which exhibit memory effects. Such systems include intersymbol interference channels, as well as modulation systems that rely on memory to produce the transmitted signal, such as trellis coded modulation systems [20], or bandwith efficient modulation schemes of the type described in [21]. As long as memory effects are only of finite duration, such systems can be modelled by finite state Markov chains. If we consider a finite observation block, such problems can be viewed conceptually as M-ary detection problems with as many hypotheses as there exist paths in the Markov chain trellis. But the cardinality of the number of Markov chain paths grows exponentially with the length of the observation block, so the standard M-ary hypothesis technique consisting of evaluating the maximum a-posteriori probability of each hypothesis is not really applicable. Instead, the Markov structure of the model needs to be exploited to eliminate quickly inferior hypotheses. The case when the Markov chain model is completely known is considered first in Chapter 12. It is possible to formulate two different detection problems depending on whether, given the observations, one seeks to maximize the a-posteriori probability of the entire Markov chain state sequence, or the probability of the state at a fixed time instant. It turns out that it is easier to maximize the a-posteriori probability of a state sequence. This is accomplished by the Viterbi algorithm, which is really a special case of Bellman's dynamic programming method. The Viterbi algorithm can be interpreted as a breadth-first pruning technique of the trellis of state trajectories. As such it keeps track of the maximum a-posteriori probability (MAP) state paths, called survivor paths, terminating in each possible Markov chain state at a give time. Then by simple additions and comparisons, the survivor paths are extended from one time instant to the next. When the Viterbi algorithm was introduced in 1967, it was viewed as having a prohibitive complexity compared to the suboptimal sequential decoding algorithms, or depth-first trellis search schemes in common use at the time. With the huge increase in computing power unleashed by Moore's law, the computational load of the Viterbi algorithm is now viewed as quite manageable. In contrast, the forward-backward algorithm developed by Baum, Welch and others [22] for evaluating the pointwise MAP probability for each Markov chain state has a higher complexity. So by itself it is not terribly attractive, but because it evaluates the MAP probability of each state value, it has become recently a key component of iterative decoding algorithms employed by turbo codes and equalizers [23], since at each iteration such algorithms require the exchange of a reliability measure for each transmitted symbol. Chapter 12 is illustrated by examining the equalization problem for ISI channels. Then Chapter 13 examines the detection of signals modelled by a Markov chain with unknown parameters. Models of this type are referred to as hidden Markov models (HMMs). Two approaches are described for solving HMM detection problems. The per survivor processing (PSP) method [24] can be viewed as an approximation of the GLRT, where, instead of fitting a model to each possible Markov chain trajectory, only survivor paths are considered and an adaptive scheme is employed to extend survivor paths while fitting a model to each path. Another approach consists of using the EM algorithm. Two methods can be used to implement the EM iteration in this context. The original approach proposed by Baum and Welch relies on the forward-backward algorithm to implement the E-phase of the EM iteration. However, it was pointed out recently [25] that the Viterbi algorithm could also be employed. The resulting EM Viterbi iteration has a lower computational load, but unlike the forward-backward scheme, which is exact, it involves an approximation of the same type as appearing in the PSP algorithm. Specifically, it involves a truncation of the state trellis consisting of retaining only Viterbi survivors instead of all paths. Both the PSP and EM Viterbi algorithms are illustrated by considering the blind equalization of ISI channels.

1.2 Complementary Readings

Although the intent of this textbook is to provide a relatively complete coverage of key ideas in signal detection, interested readers may find it useful to consult other sources in parallel with the study of this book. The first part of the book on the foundations of hypothesis testing, signal detection, and parameter estimation overlaps significantly with the contents of typical two-semester mathematical statistics courses offered by statistics departments. Such courses tyically cover hypothesis testing, parameter estimation, and large sample theory, based on say, [26–28], or equivalently, a combination of texts among [29–32]. In spite of this overlap, it should be pointed out that detection and estimation problems encountered in signal processing lend themselves to an easier inventory (for examples, parameters to be estimated are typically amplitude, delay, phase, frequency, power, or angle of arrival) than the broader class of problems confronting statisticians, so the examples presented here are likely to be more relevant than those appearing in general statistics texts. Furthermore, due to different traditions and for practical reasons, electrical engineers and statisticians often emphasize different concepts and look at problems in slightly different ways. For example, altough the receiver operating characteristic appears in statistics texts, it plays a central role in detection theory, since it characterizes neatly the realm of achievable detectors. Similarly, statisticians and electrical engineers tend to adopt different perspectives when examining the asymptotic behavior of tests with repeated observations. Statisticians often favor a "local" perspective aimed at characterizing the enhanced sensitivity of tests as more observations are collected. So from this viewpoint, the goal of collecting more observations is to be able to decide between hypotheses with progressively closer parameter values. On the other hand, electrical engineers usually adopt a "nonlocal" viewpoint where the key parameter is the exponential decay rate of the probability of error for tests between two fully distinct hypotheses (with parameter values that are not near each other). This focus on exponential decay rates is influenced by pragmatic considerations. Typical detection problems are nonlocal, so perfect detection can be achieved asymptotically. Thus, it is reasonable to select tests based on their ability to achieve near perfect detection as quickly as possible, i.e., by maximizing the rate of decay of the applicable probability of error.

In addition to statistics sources, the material presented in Parts I and II is also covered with different emphases in a number of signal detection and estimation texts. The three-volume treatise by Van Trees [12, 33, 34] remains an unavoidable starting point, not only due to its completeness, but because of its outstanding exposition of the signal space formulation of signal detection. However, because of its emphasis on CT problems and on Bayesian detection, additional sources are recommended. The book [35] presents a large number of examples where detection problem symmetries can be exploited for designing UMPI tests. The two-volume book by Kay [36, 37] analyses estimation and detection problems from a digital signal processing viewpoint and contains an excellent discussion of the GLRT. Due to its strong emphasis on fundamental principles, Poor's book [38] is probably closest in style to the present text, but with a stronger Bayesian emphasis than adopted here. Finally, Helstrom's book [39], although quite advanced, provides a very balanced treatment of Bayesian and nonBayesian viewpoints.

References

1. J. A. C. Bingham, "Multicarrier modulation for data transmission: an idea whose time has come," *IEEE Communications Magazine*, pp. 5–14, 1990.
2. A. J. Viterbi, *CDMA: Principles of Spread-Spectrum Communication.* Reading, MA: Addison-Wesley, 1995.
3. G. Ungerboeck, "Trellis-coded modulation with redundant signal sets, Parts I and II," *IEEE Communications Magazine*, vol. 25, pp. 5–21, 1987.
4. F. den Hollander, *Large Deviations.* Providence, RI: American Mathematical Soc., 2000.
5. A. Dembo and O. Zeitouni, *Large Deviations Techniques and Applications, Second Edition.* New York: Springer Verlag, 1998.
6. P. J. Huber, *Robust Statistics.* New York: J. Wiley & Sons, 1981.
7. R. D. Martin and S. C. Schwartz, "Robust detection of a known signal in nearly Gaussian noise," *IEEE Trans. Informat. Theory*, vol. 17, pp. 50–56, 1971.

8. S. A. Kassam and H. V. Poor, "Robust techniques for signal processing: a survey," *Proc. IEEE*, vol. 73, pp. 433–480, Mar. 1985.
9. E. Wong and B. Hajek, *Stochastic Processes in Engineering Systems.* New York: Springer-Verlag, 1985.
10. D. Middleton, *An Introduction to Satistical Communication Theory.* New York: McGraw-Hill, 1960. Reprinted by IEEE Press, New York, 1996.
11. W. B. Davenport, Jr. and W. L. Root, *An Introduction to the Theory of Random Signals and Noise.* New York: McGraw-Hill, 1958. Reprinted by IEEE Press, New York, 1987.
12. H. L. Van Trees, *Detection, Estimation and Modulation Theory, Part I: Detection, Estimation and Linear Modulation Theory.* New York: J. Wiley & Sons, 1968. Paperback reprint edition in 2001.
13. J. M. Wozencraft and I. M. Jacobs, *Principles of Communication Engineering.* New York: J. Wiley & Sons, 1965. Reprinted by Waveland Press, Prospect Heights, IL, 1990.
14. M. K. Simon and M.-S. Alouini, *Digital Communication over Fading Channels, Second Edition.* New York: J. Wiley & Sons, 2004.
15. T. Kailath, "The innovations approach to detection and estimation theory," *Proc. IEEE*, vol. 58, pp. 680–695, May 1970.
16. T. Kailath and H. V. Poor, "Detection of stochastic processes," *IEEE Trans. Informat. Theory*, vol. 44, pp. 2230–2259, Oct. 1998.
17. A. P. Dempster, N. M. Laird, and D. B. Rubin, "Maximum likelihood from incomplete data via the EM algorithm," *J. Royal Stat. Society, Series B*, vol. 39, no. 1, pp. 1–38, 1977.
18. L. E. Baum, T. Petrie, G. Soules, and N. Weiss, "A maximization technique occurring in the statistical analysis of probabilistic functions of Markov chains," *Annals Mathematical Statistics*, vol. 41, pp. 164–171, Feb. 1970.
19. L. R. Welch, "The Shannon lecture: Hidden Markov models and the Baum-Welch algorithm," *IEEE Information Theory Soc. Newsletter*, vol. 53, Dec. 2003.
20. E. Biglieri, D. Divsalar, P. J. McLane, and M. K. Simon, *Introduction to Trellis Coded Modulation with Application.* New York: Macmillan, 1991.
21. M. K. Simon, *Bandwidth-Efficient Digital Modulation with Application to Deep-Space Communications.* Wiley-Interscience, 2003.
22. L. R. Bahl, J. Cocke, F. Jelinek, and J. Raviv, "Optimal decoding of linear codes for minimizing symbol error rates," *IEEE Trans. Informat. Theory*, vol. 20, pp. 284–287, Mar. 1974.
23. R. Koetter, A. C. Singer, and M. Tüchler, "Turbo equalization," *IEEE Signal Processing Mag.*, vol. 21, pp. 67–80, Jan. 2004.
24. R. Raheli, A. Polydoros, and C. Tzou, "Per-survivor-processing: A general approach to MLSE in uncertain environments," *IEEE Trans. Commun.*, vol. 43, pp. 354–364, Feb./Apr. 1995.
25. H. Nguyen and B. C. Levy, "Blind and semi-blind equalization of CPM signals with the EMV algorithm," *IEEE Trans. Signal Proc.*, vol. 51, pp. 2650–2664, Oct. 2003.
26. E. L. Lehmann and J. P. Romano, *Testing Statistical Hypotheses, Third Edition.* New York: Springer Verlag, 2005.
27. E. L. Lehmann and G. Casella, *Theory of Point Estimation, Second Edition.* New York: Springer Verlag, 1998.
28. E. L. Lehmann, *Elements of Large-Sample Theory.* New York: Springer Verlag, 1999.

29. T. S. Ferguson, *Mathematical Statistics: A Decision Theoretic Approach.* New York: Academic Press, 1967.
30. P. J. Bickel and K. A. Doksum, *Mathematical Statistics: Basic Ideas and Selected Topics, Second Edition.* Upper Saddle River, NJ: Prentice Hall, 2001.
31. T. S. Ferguson, *A Course in Large Sample Theory.* London: Chapman & Hall, 1996.
32. A. W. van der Vaart, *Asymptotic Statistics.* Cambridge, United Kingdom: Cambridge University Press, 1998.
33. H. L. Van Trees, *Detection, Estimation and Modulation Theory, Part II: Nonlinear Modulation Theory.* New York: J. Wiley & Sons, 1971. Paperback reprint edition in 2001.
34. H. L. Van Trees, *Detection, Estimation and Modulation Theory, Part III: Radar-Sonar Signal Processing and Gaussian Signals in Noise.* New York: J. Wiley & Sons, 1971. Paperback reprint edition in 2001.
35. L. L. Scharf, *Statistical Signal Processing: Detection, Estimation and Time Series Analysis.* Reading, MA: Addison Wesley, 1991.
36. S. M. Kay, *Fundamentals of Statistical Signal Processing: Estimation Theory.* Upper Saddle, NJ: Prentice-Hall, 1993.
37. S. M. Kay, *Fundamentals of Statistical Signal Processing: Detection Theory.* Prentice-Hall, 1998.
38. H. V. Poor, *An Introduction to Signal Detection and Estimation, Second Edition.* New York: Springer Verlag, 1994.
39. C. W. Helstrom, *Elements of Signal Detection & Estimation.* Upper Saddle River, NJ: Prentice-Hall, 1995.

Part I

Foundations

2

Binary and M-ary Hypothesis Testing

2.1 Introduction

Detection problems of the type arising in radar, digital communications, image processing, or failure detection can usually be cast as binary or M-ary hypothesis testing problems. For example, in radar, one has to decide whether a target is present or not based on a noisy return signal that may or may not contain the reflection of a probing pulse. Similarly in digital communications, over each signaling interval $[0, T]$, a pulse $s(t, X)$ is transmitted which encodes the information about a symbol X taking a discrete set of values. This information can be encoded in the amplitude, phase, frequency, or any combination thereof of the transmitted pulse. Then, given noisy observations

$$Y(t) = s(t, X) + V(t) \,,$$

with $0 \leq t \leq T$, where $V(t)$ represents some channel noise, one has to decide which symbol X was transmitted. In the case when X takes only two values, such as for binary phase-shift-keying (BPSK) modulation, this decision gives rise to a binary hypothesis testing problem. But, when X can take $M = 2^k$ values with $k > 1$, the decision problem takes the form of an M-ary hypothesis testing problem.

In this chapter, we restrict our attention to testing between *simple hypotheses.* This means that the probability distribution of the observations under each hypothesis is assumed to be known exactly. For example, for the digital communications problem described above, assuming that $X = x$ is transmitted, the waveform $s(t, x)$ is known exactly, as well as the statistics of the noise $V(t)$. This corresponds to the case of coherent receiver design, for which timing, phase, and frequency synchronization are performed on the received noisy waveform. But obviously, this assumption is not always realistic, as can be seen by considering a simplified failure detection problem. Suppose that when a system is in normal operation, the response of a sensor is

$$Y(t) = s_0(t) + V(t)$$

B.C. Levy, *Principles of Signal Detection and Parameter Estimation*,
DOI: 10.1007/978-0-387-76544-0_2, © Springer Science+Business Media, LLC 2008

for $0 \leq t \leq T$, where $V(t)$ represents some sensor noise. But if a system failure occurs at $t = \tau$, the sensor response takes the form

$$Y(t) = \begin{cases} s_0(t) + V(t) \text{ for } 0 \leq t < \tau \\ s_1(t) + V(t) \text{ for } \tau \leq t \leq T \,, \end{cases}$$

where the failure time τ is usually unknown. In this case, the problem of deciding whether the system has failed or not necessarily involves the failure time τ as an unknown parameter. Hypothesis testing problems involving unknown parameters are called *composite hypothesis testing* problems. The tests applicable to such situations often require the estimation of the unknown parameters, and they will be examined later in Chapter 5.

For the simple binary or M-ary hypothesis testing problems considered in this chapter, our objective is to design tests that are optimal in some appropriate sense. But what constitutes an optimal test can vary depending on the amount of information available to the designer. For the Bayesian tests considered in Section 2.2, it is assumed that a-priori probabilities are known for the different hypotheses, as well as a detailed cost structure for all the possible outcomes of the test. For minimax tests of Section 2.5, a cost structure is known, but priors are not available. Finally, for the Neyman-Pearson tests discussed in Section 2.4, only the probability distribution of the observations under each hypothesis is known. As is customary in engineering texts, two important concepts are developed: the notion of receiver operating characteristic (ROC) which plots the probability of detection P_D versus the probability of false alarm P_F curve for optimum tests. This curve illustrates the inherent trade-off existing between the detection performance and the tolerance of false alarms which is at the heart of detection problems. Secondly, for problems involving known signals in Gaussian noise, we introduce the notion of "distance" between signals, which is the cornerstone of the signal space formulation of digital communications [1,2].

2.2 Bayesian Binary Hypothesis Testing

The goal of binary hypothesis testing is to decide between two hypotheses H_0 and H_1 based on the observation of a random vector $\mathbf{Y}$. We consider the cases where $\mathbf{Y}$ takes continuous or discrete values over a domain $\mathcal{Y}$. In the continuous-valued case, $\mathcal{Y} = \mathbb{R}^n$, and depending on whether H_0 or H_1 holds, $\mathbf{Y}$ admits the probability densities

$$\begin{aligned} H_0 &: \mathbf{Y} \sim f(\mathbf{y}|H_0) \\ H_1 &: \mathbf{Y} \sim f(\mathbf{y}|H_1) \,. \end{aligned} \quad (2.1)$$

In the discrete-valued case, $\mathcal{Y} = \{\mathbf{y}_i, i \in I\}$ is a countable collection of discrete values $\mathbf{y}_i$ indexed by $i \in I$ and depending on whether H_0 or H_1 holds, $\mathbf{Y}$ admits the probability mass distribution functions

$$\begin{aligned} H_0 &: p(\mathbf{y}|H_0) = P[\mathbf{Y} = \mathbf{y}|H_0] \\ H_1 &: p(\mathbf{y}|H_1) = P[\mathbf{Y} = \mathbf{y}|H_1] \,. \end{aligned} \tag{2.2}$$

Then, given $\mathbf{Y}$, we need to decide whether H_0 and H_1 is true. This is accomplished by selecting a decision function $\delta(\mathbf{Y})$ taking values in $\{0, 1\}$, where $\delta(\mathbf{y}) = 1$ if we decide that H_1 holds when $\mathbf{Y} = \mathbf{y}$, and $\delta(\mathbf{y}) = 0$ if we decide that H_0 holds when $\mathbf{Y} = \mathbf{y}$. In effect, the decision function $\delta(\mathbf{y})$ partitions the observation domain $\mathcal{Y}$ into two disjoint sets $\mathcal{Y}_0$ and $\mathcal{Y}_1$ where

$$\begin{aligned} \mathcal{Y}_0 &= \{\mathbf{y} : \delta(\mathbf{y}) = 0\} \\ \mathcal{Y}_1 &= \{\mathbf{y} : \delta(\mathbf{y}) = 1\} \,. \end{aligned} \tag{2.3}$$

Thus, there exist as many decision functions as there are disjoint partitions of $\mathcal{Y}$. Among all decision functions, we seek to obtain decision rules which are "optimal" in some sense. This requires the formulation of an optimization problem whose solution, if it exists, will yield the desired optimal decision rule. The Bayesian formulation of the binary hypothesis testing problem is based on the philosophical viewpoint that all uncertainties are quantifiable, and that the costs and benefits of all outcomes can be measured. This means that the hypotheses H_0 and H_1 admit a-priori probabilities

$$\pi_0 = P[H_0] \quad , \quad \pi_1 = P[H_1] \,, \tag{2.4}$$

with

$$\pi_0 + \pi_1 = 1 \,. \tag{2.5}$$

Furthermore, it is assumed that there exists a cost function C_{ij} with $0 \leq i, j \leq 1$, where C_{ij} represents the cost of deciding that H_i is true when H_j holds. Given this cost function, we can then define the Bayes risk under hypothesis H_j as

$$R(\delta|H_j) = \sum_{i=0}^{1} C_{ij} P[\mathcal{Y}_i|H_j] \,, \tag{2.6}$$

where

$$P[\mathcal{Y}_i|H_j] = \int_{\mathcal{Y}_i} f(\mathbf{y}|H_j) d\mathbf{y} \tag{2.7}$$

when $\mathbf{Y}$ takes continuous values and

$$P[\mathcal{Y}_i|H_j] = \sum_{\mathbf{y} \in \mathcal{Y}_i} p(\mathbf{y}|H_j) \tag{2.8}$$

when it takes discrete values. Consequently, the Bayes risk for decision function δ, which represents the average cost of selecting δ, is given by

$$R(\delta) = \sum_{j=0}^{1} R(\delta|H_j) \pi_j \,. \tag{2.9}$$

The optimum Bayesian decision rule δ_B is obtained by minimizing the risk $R(\delta)$. To find the mimimum we start from the expression

$$R(\delta) = \sum_{i=0}^{1}\sum_{j=0}^{1} C_{ij} P[\mathcal{Y}_i|H_j]\pi_j \,, \tag{2.10}$$

and note that since $\mathcal{Y}_0$ and $\mathcal{Y}_1$ form a disjoint partition of $\mathbb{R}^n$, we have

$$P[\mathcal{Y}_0|H_j] + P[\mathcal{Y}_1|H_j] = 1 \,, \tag{2.11}$$

for $j = 0, \ 1$, so that

$$R(\delta) = \sum_{j=0}^{1} C_{0j}\pi_j + \sum_{j=0}^{1}(C_{1j} - C_{0j}) P[\mathcal{Y}_1|H_j]\pi_j \,. \tag{2.12}$$

When $\mathbf{Y}$ takes continuous values, substituting the expression (2.7) inside (2.12) yields

$$R(\delta) = \sum_{j=0}^{1} C_{0j}\pi_j + \int_{\mathcal{Y}_1} \sum_{j=0}^{1}(C_{1j} - C_{0j})\pi_j f(\mathbf{y}|H_j) d\mathbf{y} \,. \tag{2.13}$$

The risk $R(\delta)$ is minimized if the decision region $\mathbf{Y}_1$ is formed by selecting the values of $\mathbf{y}$ for which the integrand of (2.13) is either negative or zero, i.e.,

$$\mathcal{Y}_1 = \{\mathbf{y} \in \mathbf{R}^n : \sum_{j=0}^{1}(C_{1j} - C_{0j})\pi_j f(\mathbf{y}|H_j) \leq 0\} \,. \tag{2.14}$$

Note that the values of $\mathbf{y}$ for which the integrand of (2.13) is zero do not contribute to the risk, so their inclusion in either $\mathcal{Y}_1$ or $\mathcal{Y}_0$ is discretionary. We have included them here in $\mathcal{Y}_1$, but we could have included them in $\mathcal{Y}_0$, or we could have assigned some to $\mathcal{Y}_1$ and the remainder to $\mathcal{Y}_0$. Similarly, when $\mathbf{Y}$ is discrete-valued, the Bayesian risk $R(\delta)$ can be expressed as

$$R(\delta) = \sum_{j=0}^{1} C_{0j}\pi_j + \sum_{\mathbf{y}\in\mathcal{Y}_1} \Big(\sum_{j=0}^{1}(C_{1j} - C_{0j})\pi_j p(\mathbf{y}|H_j)\Big) \,. \tag{2.15}$$

It is minimized when the summand is either negative or zero, which yields the decision rule

$$\mathcal{Y}_1 = \{\mathbf{y} \in \mathcal{Y} : \sum_{j=0}^{1}(C_{1j} - C_{0j})\pi_j p(\mathbf{y}|H_j) \leq 0\} \,. \tag{2.16}$$

The characterizations (2.14) and (2.16) of the optimum Bayesian decision rule δ_B can be simplified further if we make the reasonable assumption that

$C_{11} < C_{01}$. In other words, selecting H_1 when H_1 holds costs less than making a mistake and selecting H_0 even though H_1 holds. In radar terminology, where H_1 and H_0 indicate respectively the presence or absence of a target, the correct selection of H_1 represents a detection, and the mistaken choice of H_0 when H_1 holds is called a "miss." Obviously a miss can be extremely costly when the target is a hostile airplane, but even for mundane applications of hypothesis testing to digital communications or quality control, making the correct decision is always less costly than making a mistake. In this case, we can introduce the *likelihood ratio*, which depending on whether $\mathbf{Y}$ takes continuous or discrete values, is given by

$$L(\mathbf{y}) = \frac{f(\mathbf{y}|H_1)}{f(\mathbf{y}|H_0)} \tag{2.17}$$

or

$$L(\mathbf{y}) = \frac{p(\mathbf{y}|H_1)}{p(\mathbf{y}|H_0)} \,. \tag{2.18}$$

In the above definition of the likelihood ratio, we note that three cases may arise: the numerator and denominator are both strictly positive, they are both zero, or one is zero and the other is nonzero. When the numerator and denominator are both positive, $L(\mathbf{y})$ is finite and positive. When the numerator and denominator are zero, the observation $\mathbf{y}$ cannot occur under either H_0 or H_1, so it is unnecessary to define the likelihood ratio for this value. When the numerator is zero and the denominator is positive $L(\mathbf{y}) = 0$, indicating that $\mathbf{y}$ cannot be observed under H_1 but can occur under H_0. Conversely, when the numerator is positive and the denominator is zero, we assign the value $L(\mathbf{y}) = +\infty$ to the likelihood ratio, which indicates that $\mathbf{y}$ cannot occur under H_0 but may occur under H_1. In the continuous-valued case, the probability measure corresponding to H_1 is said to be absolutely continuous with respect to the measure under H_0 if the likelihood ratio $L(\mathbf{y})$ is finite almost everywhere, i.e., the set of values for which $L(\mathbf{y})$ is infinite has measure zero.

Consider the threshold

$$\tau = \frac{(C_{10} - C_{00})\pi_0}{(C_{01} - C_{11})\pi_1} \,. \tag{2.19}$$

Then, under the condition $C_{11} < C_{01}$ the optimum Bayesian decision rule given by (2.14) or (2.16) can be rewritten as

$$\delta_{\mathrm{B}}(\mathbf{y}) = \begin{cases} 1 \text{ if } L(\mathbf{y}) \geq \tau \\ 0 \text{ if } L(\mathbf{y}) < \tau \,. \end{cases} \tag{2.20}$$

In the following this decision rule will be written compactly as

$$L(\mathbf{y}) \underset{H_0}{\overset{H_1}{\gtrless}} \tau \,. \tag{2.21}$$

Several special choices of the cost function C_{ij} and a-priori probabilities π_j lead to particular forms of the test.

Minimum probability of error: Assume that the cost function takes the form

$$C_{ij} = 1 - \delta_{ij} \tag{2.22}$$

with

$$\delta_{ij} = \begin{cases} 1 \text{ for } i = j \\ 0 \text{ for } i \neq j \,. \end{cases}$$

Thus a cost is incurred only when an error occurs, and error costs are symmetric, i.e., it costs as much to select H_1 when H_0 holds as it costs to select H_0 when H_1 is true. This is the situation prevailing in digital communications, where hypotheses H_0 and H_1 represent the transmission of a bit equal to 0, or 1, respectively, and since the information value of the bits 0 and 1 is the same, error costs are symmetric. In this case, the Bayesian risk under H_0 and H_1 can be expressed as

$$\begin{aligned} R(\delta|H_0) &= P[\mathcal{Y}_1|H_0] = P[E|H_0] \\ R(\delta|H_1) &= P[\mathcal{Y}_0|H_1] = P[E|H_1] \,, \end{aligned} \tag{2.23}$$

where E denotes an error event, so the Bayesian risk reduces to the error probability

$$P[E] = P[E|H_0]\pi_0 + P[E|H_1]\pi_1 \,. \tag{2.24}$$

Consequently, the optimum Bayesian decision rule minimizes the error probability, and in the likelihood ratio test (LRT) (2.21) the threshold reduces to

$$\tau = \pi_0/\pi_1 \,. \tag{2.25}$$

If we introduce the marginal probability density

$$f(\mathbf{y}) = \sum_{j=0}^{1} f(\mathbf{y}|H_j)\pi_j \tag{2.26}$$

of the observation vector $\mathbf{Y}$, by clearing denominators in (2.21) and dividing both sides by $f(\mathbf{y})$, we obtain

$$\frac{f(\mathbf{y}|H_1)\pi_1}{f(\mathbf{y})} \underset{H_0}{\overset{H_1}{\gtrless}} \frac{f(\mathbf{y}|H_0)\pi_0}{f(\mathbf{y})} \,. \tag{2.27}$$

But according to Bayes' rule, the a-posteriori probabilities of the two hypotheses H_j with $j = 0,\ 1$ can be expressed as

$$P[H_j|\mathbf{Y} = \mathbf{y}] = \frac{f(\mathbf{y}|H_j)\pi_j}{f(\mathbf{y})} \tag{2.28}$$

so the LRT reduces to the *maximum a-posteriori probability* (MAP) decision rule

$$P[H_1|\mathbf{y}] \underset{H_0}{\overset{H_1}{\gtrless}} P[H_0|\mathbf{y}] \,. \tag{2.29}$$

When $\mathbf{Y}$ is discrete-valued, the a-posteriori probabilities appearing in the MAP decision rule (2.29) are given by

$$P[H_j|\mathbf{Y}=\mathbf{y}] = \frac{p(\mathbf{y}|H_j)\pi_j}{p(\mathbf{y})} \,, \tag{2.30}$$

where $p(\mathbf{y})$ is the marginal probability mass function,

$$p(\mathbf{y}) = \sum_{j=0}^{1} p(\mathbf{y}|H_j)\pi_j \,. \tag{2.31}$$

Maximum likelihood decision rule: Suppose that in addition to assuming that the costs are symmetric, i.e., $C_{ij} = 1-\delta_{ij}$, we assume the two hypotheses are equally likely, so

$$\pi_0 = \pi_1 = 1/2$$

This represents the absence of a-priori knowledge about the frequency of occurence of each hypothesis. Then the threshold $\tau = 1$, and the LRT can be expressed as the maximum-likelihood (ML) decision rule given by

$$f(\mathbf{y}|H_1) \underset{H_0}{\overset{H_1}{\gtrless}} f(\mathbf{y}|H_0) \tag{2.32}$$

for continuous-valued observations, and

$$p(\mathbf{y}|H_1) \underset{H_0}{\overset{H_1}{\gtrless}} p(\mathbf{y}|H_0) \tag{2.33}$$

for discrete-valued observations.

To illustrate Bayesian tests, we consider several examples.

Example 2.1: Constant in white Gaussian noise

Let V_k with $1 \leq k \leq N$ be a $N(0, \sigma^2)$ white Gaussian noise (WGN) sequence. Then consider the binary hypothesis testing problem:

$$\begin{aligned} H_0 &: Y_k = m_0 + V_k \\ H_1 &: Y_k = m_1 + V_k \end{aligned}$$

for $1 \leq k \leq N$, where we assume $m_1 > m_0$. The probability density of the vector

$$\mathbf{Y} = \begin{bmatrix} Y_1 \ldots Y_k \ldots Y_N \end{bmatrix}^T$$

under H_j with $j = 0, 1$ is given by

$$f(\mathbf{y}|H_j) = \prod_{k=1}^{N} f(y_k|H_j)$$
$$= \frac{1}{(2\pi\sigma^2)^{N/2}} \exp\big(-\frac{1}{2\sigma^2}\sum_{k=1}^{N}(y_k - m_j)^2\big),$$

and the likelihood ratio function is given by

$$L(\mathbf{y}) = \frac{f(\mathbf{y}|H_1)}{f(\mathbf{y}|H_0)} = \exp\Big(\frac{(m_1 - m_0)}{\sigma^2}\sum_{k=1}^{N} y_k - \frac{N(m_1^2 - m_0^2)}{2\sigma^2}\Big).$$

Taking logarithms on both sides of the LRT (2.21) and observing that the logarithm function is monotonic, i.e., $x_1 < x_2$ if and only if $\ln(x_1) < \ln(x_2)$, we can rewrite the test as

$$\ln(L(\mathbf{Y})) = \frac{(m_1 - m_0)}{\sigma^2}\Big(\sum_{k=1}^{N} Y_k\Big) - \frac{N(m_1^2 - m_0^2)}{2\sigma^2} \underset{H_0}{\overset{H_1}{\gtrless}} \gamma \triangleq \ln(\tau).$$

After simplifications, this gives

$$S \triangleq \frac{1}{N}\sum_{k=1}^{N} Y_k \underset{H_0}{\overset{H_1}{\gtrless}} \eta, \tag{2.34}$$

where

$$\eta = \frac{m_1 + m_0}{2} + \frac{\sigma^2\gamma}{N(m_1 - m_0)}. \tag{2.35}$$

An interesting feature of the above test is that it depends entirely on the sample mean S of the random variables Y_k for $1 \leq k \leq N$, which means that all the other information contained in the Y_k's is unneeded for the purpose of deciding between H_0 and H_1. For this reason, S is called a *sufficient statistic*.

Note that under hypothesis H_j with $j = 0, 1$, S is $N(m_j, \sigma^2/N)$ distributed, and as $N \to \infty$, S converges almost surely to the mean m_j. From this perspective, we see that the threshold η to which we compare S includes two terms: the half sum $(m_1 + m_0)/2$ which is located midway between the means m_0 and m_1 under the two hypotheses, and a correction term $\sigma^2\gamma/N(m_1 - m_0)$ representing the influence of the cost function C_{ij} and the a-priori probabilities π_j of the two hypotheses. Since it varies like N^{-1}, this term decreases as the number of observations becomes larger. Furthermore, when $C_{ij} = 1 - \delta_{ij}$ and the two hypotheses are equally likely, we have $\tau = 1$ and thus $\gamma = 0$, so the correction term vanishes. □

Example 2.2: WGN variance test

Consider the case where we observe a zero-mean WGN over $1 \leq k \leq N$ whose variance differs under H_0 and H_1, i.e.,

$$H_0 : Y_k \sim N(0, \sigma_0^2)$$
$$H_1 : Y_k \sim N(0, \sigma_1^2) \,.$$

For simplicity, we assume $\sigma_1^2 > \sigma_0^2$. The probability density of the vector

$$\mathbf{Y} = \begin{bmatrix} Y_1 \ldots Y_k \ldots Y_N \end{bmatrix}^T$$

under H_j with $j = 0, 1$ is given by

$$f(\mathbf{y}|H_j) = \prod_{k=1}^{N} f(y_k|H_j) = \frac{1}{(2\pi\sigma_j^2)^{N/2}} \exp\big(-\frac{1}{2\sigma_j^2}\sum_{k=1}^{N} y_k^2\big)$$

so the likelihood ratio can be expressed as

$$L(\mathbf{y}) = \frac{f(\mathbf{y}|H_1)}{f(\mathbf{y}|H_0)} = \Big(\frac{\sigma_0^2}{\sigma_1^2}\Big)^{N/2} \exp\Big(\frac{(\sigma_0^{-2} - \sigma_1^{-2})}{2}\sum_{k=1}^{N} y_k^2\Big) \,.$$

Taking logarithms on both sides of the LRT (2.21) gives

$$\ln(L(\mathbf{y})) = \frac{(\sigma_1^2 - \sigma_0^2)}{2\sigma_0^2\sigma_1^2}\Big(\sum_{k=1}^{N} Y_k^2\Big) - N\ln(\sigma_1/\sigma_0) \underset{H_0}{\overset{H_1}{\gtrless}} \gamma \triangleq \ln(\tau) \,,$$

and after rearranging terms, we obtain

$$S \triangleq \frac{1}{N}\sum_{k=1}^{N} Y_k^2 \underset{H_0}{\overset{H_1}{\gtrless}} \eta \tag{2.36}$$

with

$$\eta = \frac{2\sigma_0^2\sigma_1^2}{\sigma_1^2 - \sigma_0^2}[\ln\Big(\frac{\sigma_1}{\sigma_0}\Big) + \frac{\gamma}{N}] \,. \tag{2.37}$$

The test is expressed again in terms of a sufficient statistic S corresponding to the sampled covariance of the observations. Under hypothesis H_j, $Z = NS/\sigma_j^2$ is the sum of N squared $N(0,1)$ random variables, so it admits a chi-squared distribution with N degrees of freedom, i.e., its density is given by

$$f_Z(z|H_j) = \frac{1}{\Gamma(N/2)2^{N/2}} z^{(N/2)-1}\exp(-z/2)u(z)$$

where $\Gamma(\cdot)$ is Euler's gamma function and $u(\cdot)$ denotes the unit step function. Observe also that under H_j, S converges almost surely to σ_j^2 as N becomes large. The test (2.36) compares S to a threshold η which is comprised again of two terms. The first term

$$\frac{\sigma_0^2 \sigma_1^2}{\sigma_1^2 - \sigma_0^2} \ln\Big(\frac{\sigma_1^2}{\sigma_0^2}\Big)$$

is located between σ_0^2 and σ_1^2, as can be seen by applying the inequality

$$1 - \frac{1}{x} \leq \ln(x) \leq x - 1 \tag{2.38}$$

with $x = \sigma_1^2/\sigma_0^2$. The second term represents a correction representing the effect of the cost function and a-priori probabilities. This term varies like N^{-1}, so its influence decreases as more observations are collected. It is also zero if $C_{ij} = 1 - \delta_{ij}$ and $\pi_0 = \pi_1 = 1/2$. □

Example 2.3: Parameter test for Poisson random variables

Consider now the case where we observe a sequence Y_k with $1 \leq k \leq N$ of independent identically distributed (iid) Poisson random variables. Under H_0 and H_1, the parameters of the probability mass distribution of Y_k are λ_0 and λ_1, respectively, i.e.,

$$\begin{aligned} H_0 : P[Y_k = n|H_0] &= \frac{\lambda_0^n}{n!} \exp(-\lambda_0) \\ H_1 : P[Y_k = n|H_1] &= \frac{\lambda_1^n}{n!} \exp(-\lambda_1) \,, \end{aligned}$$

where n denotes a nonnegative integer, and where for simplicity we assume $\lambda_1 > \lambda_0$. Consider now the observation vector

$$\mathbf{Y} = \big[\, Y_1 \dots Y_k \dots Y_N \,\big]^T$$

and the vector of integers

$$\mathbf{n} = \big[\, n_1 \dots n_k \dots n_N \,\big]^T .$$

Since the observations Y_k are iid, the probability mass distribution of $\mathbf{Y}$ under hypothesis H_j with $j = 0,\ 1$ is given by

$$P[\mathbf{Y} = \mathbf{n}|H_j] = \prod_{k=1}^{N} P[Y_k = n_k|H_j] = \frac{\lambda_j^{\sum_{k=1}^N n_k}}{\prod_{k=1}^N n_k!} \exp(-N\lambda_j) \,,$$

so the likelihood ratio can be expressed as

$$L(\mathbf{Y} = \mathbf{n}) = \frac{P[\mathbf{Y} = \mathbf{n}|H_1]}{P[\mathbf{Y} = \mathbf{n}|H_0]} = \Big(\frac{\lambda_1}{\lambda_0}\Big)^{\sum_{k=1}^N n_k} \exp(-N(\lambda_1 - \lambda_0)) \,.$$

Taking logarithms on both sides of the LRT (2.21) gives

$$\ln(L(\mathbf{Y})) = \ln(\lambda_1/\lambda_0)\Big(\sum_{k=1}^{N} Y_k\Big) - N(\lambda_1 - \lambda_0) \underset{H_0}{\overset{H_1}{\gtrless}} \gamma \triangleq \ln(\tau) ,$$

which yields the test

$$S \triangleq \frac{1}{N} \sum_{k=1}^{N} Y_k \underset{H_0}{\overset{H_1}{\gtrless}} \eta \tag{2.39}$$

with

$$\eta = \frac{\lambda_1 - \lambda_0}{\ln(\lambda_1/\lambda_0)} + \frac{\gamma}{N \ln(\lambda_1/\lambda_0)} . \tag{2.40}$$

This test relies on the sampled mean S of the the observations, which tends almost surely to the mean λ_j of the Poisson distribution under H_j. S is compared to a threshold η comprised of two terms. By using the inequality (2.38), it is easy to verify that the first term $(\lambda_1 - \lambda_0)/\ln(\lambda_1/\lambda_0)$ is located between λ_0 and λ_1. The correction term $\gamma/(N \ln(\lambda_1/\lambda_0))$ varies like N^{-1} and represents the influence of the costs and priors. It vanishes when the objective function is the probability of error and the two hypotheses are equally likely. It is worth noting that NS is always an integer, whereas $N\eta$ is real. As a consequence, two thresholds η and η' will correspond to the same test if $N\eta$ and $N\eta'$ have the same integer part. In other words, different thresholds may correspond to the same effective test in practice. □

2.3 Sufficient Statistics

In the previous section, we have seen that likelihood ratio tests can often be expressed in terms of sufficient statistics. Up to this point, the sufficient statistics we have encountered have been scalar, but vector sufficient statistics can also arise. In general, we say that $\mathbf{S}(\mathbf{Y}) \in \mathbb{R}^k$ is a sufficient statistic for the binary hypothesis testing problem if the conditional density $f_{\mathbf{Y}|\mathbf{S}}(\mathbf{y}|\mathbf{s}, H_j)$ is independent of hypothesis H_j. The verification of this condition is simplified if we can exhibit a random vector $\mathbf{A} \in \mathbb{R}^{n-k}$ such that the correspondence

$$\mathbf{Y} \underset{\mathbf{h}}{\overset{\mathbf{g}}{\rightleftarrows}} \begin{bmatrix} \mathbf{S} \\ \mathbf{A} \end{bmatrix} \tag{2.41}$$

is one-to-one, where $\mathbf{g}$ and $\mathbf{h}$ denote respectively the forward and reverse mappings. Then the density of $\mathbf{Y}$ under hypothesis H_j can be expressed as

$$f_{\mathbf{Y}}(\mathbf{y}|H_j) = J(\mathbf{y}) f_{\mathbf{A}|\mathbf{S}}(\mathbf{a}|\mathbf{s}, H_j) f_{\mathbf{S}}(\mathbf{s}|H_j) , \tag{2.42}$$

where $J(\mathbf{y})$ denotes the Jacobian of the transformation h. Note that J is the same for H_0 and H_1, as it depends only on h. The likelihood ratio of $\mathbf{Y}$ can be expressed as

$$\begin{aligned}L_{\mathbf{Y}}(\mathbf{y}) &= \frac{f_{\mathbf{Y}}(\mathbf{y}|H_1)}{f_{\mathbf{Y}}(\mathbf{y}|H_0)} = \frac{f_{\mathbf{S}}(\mathbf{s}|H_1) f_{\mathbf{A}|\mathbf{S}}(\mathbf{a}|\mathbf{s}, H_1)}{f_{\mathbf{S}}(\mathbf{s}|H_0) f_{\mathbf{A}|\mathbf{S}}(\mathbf{a}|\mathbf{s}, H_0)} \\ &= L_{\mathbf{S}}(\mathbf{s}) \frac{f_{\mathbf{A}|\mathbf{S}}(\mathbf{a}|\mathbf{s}, H_1)}{f_{\mathbf{A}|\mathbf{S}}(\mathbf{a}|\mathbf{s}, H_0)} ,\end{aligned} \tag{2.43}$$

which reduces to the likelihood ratio of $\mathbf{S}$ if and only if $f_{\mathbf{A}|\mathbf{S}}(\mathbf{a}|\mathbf{s}, H_j)$ is independent of hypothesis H_j. In this case $\mathbf{S}$ is a sufficient statistic and $\mathbf{A}$ is an auxiliary statistic. Furthermore, $\mathbf{S}$ is a minimal sufficient statistic if it has the smallest dimension among all sufficient statistics. Note that since the likelihood ratio $L = L(\mathbf{Y})$ is itself a sufficient statistic, all minimal sufficient statistics are scalar.

Example 2.1, continued

In Example 2.1, we have already argued that the sampled average $S = \sum_{k=1}^{N} Y_k/N$ is a sufficient statistic. Consider the $N-1$ dimensional vector

$$\mathbf{A} = \begin{bmatrix} Y_2 - Y_1 \dots Y_{k+1} - Y_k \dots Y_N - Y_{N-1} \end{bmatrix}^T .$$

It admits a $N(\mathbf{0}, K_{\mathbf{A}})$ distribution, where the covariance matrix $K_{\mathbf{A}}$ has for entries

$$(K_{\mathbf{A}})_{ij} = \begin{cases} 2\sigma^2 & \text{for } i = j \\ -\sigma^2 & \text{for } i = j \pm 1 \\ 0 & \text{otherwise} , \end{cases}$$

i.e., it is a tridiagonal Toeplitz matrix. Thus the distribution of $\mathbf{A}$ is the same under hypotheses H_0 and H_1, and $\mathbf{A}$ is uncorrelated with S, and therefore independent of S since both random variables are Gaussian. In the case when the probability distribution of $\mathbf{A}$ does not depend on the choice of hypothesis, it is called an *ancillary statistic* for the testing problem. □

Up to this point we have assumed that the statistic $\mathbf{A}$ used in the one-to-one mapping (2.41) takes continuous values, but the previous analysis still holds if A is discrete-valued.

Example 2.4: Discrete auxiliary statistic

Consider the hypothesis testing problem

$$\begin{aligned} H_0 &: Y \sim N(0, \sigma_0^2) \\ H_1 &: Y \sim N(0, \sigma_1^2). \end{aligned}$$

In this case it is easy to verify that $S = |Y|$ is a sufficient statistic. Consider the binary valued random variable $A = \operatorname{sgn}(Y)$. There is a one-to-one correspondence between Y and (S, A). Since the densities $f(y|H_j)$ are symmetric under both hypotheses, we have

$$P[A = \pm 1 | S = s, H_j] = 1/2 ,$$

and because the distribution of $\mathbf{A}$ is independent of both $\mathbf{S}$ and H_j, it is an ancillary statistic, and S is a sufficient statistic. □

A simple method to recognize quickly a sufficient statistic relies on the *factorization criterion*: if the probability density of $\mathbf{Y}$ under the two hypotheses can be factored as

$$f(\mathbf{y}|H_j) = b_j(\mathbf{S}(\mathbf{y}))c(\mathbf{y}) \tag{2.44}$$

then $\mathbf{S}(\mathbf{y})$ is a sufficient statistic for the binary hypothesis problem. To verify this result, one needs only to augment $\mathbf{S}$ with a random vector $\mathbf{A}$ such that the mapping (2.41) is one-to-one, and then check that (2.44) implies that $f_{\mathbf{A}|\mathbf{S}}(\mathbf{a}|\mathbf{s}, H_j)$ is independent of H_j [3, pp. 117–118].

Example 2.5: iid Gaussian sequence

Let the observations Y_k with $1 \leq k \leq N$ be independent identically distributed Gaussian random variables with

$$\begin{aligned} H_0 &: Y_k \sim N(m_0, \sigma_0^2) \\ H_1 &: Y_k \sim N(m_1, \sigma_1^2) \,. \end{aligned}$$

Then the probability density of the vector

$$\mathbf{Y} = \begin{bmatrix} Y_1 \dots Y_k \dots Y_N \end{bmatrix}^T$$

is given by

$$f_{\mathbf{Y}}(\mathbf{y}|H_j) = \prod_{k=1}^{N} f(y_k|H_j) = c_j \exp\left(\boldsymbol{\theta}_j^T \mathbf{S}(\mathbf{y})\right) \tag{2.45}$$

with

$$\begin{aligned} \mathbf{S}(\mathbf{Y}) &= \frac{1}{N} \left[\textstyle\sum_{k=1}^{N} Y_k \; \sum_{k=1}^{N} Y_k^2 \right]^T \\ \boldsymbol{\theta}_j &= \left[N m_j / \sigma_j^2 \;\; -N/2\sigma_j^2 \right]^T , \end{aligned}$$

where c_j is a constant depending on the hypothesis H_j. The density (2.45) is in the form (2.44), so that the vector $\mathbf{S}(\mathbf{y})$ formed by the sampled first and second moments of the observations is a sufficient statistic for the hypothesis testing problem. □

2.4 Receiver Operating Characteristic

Up to this point, we have described only the structure of likelihood ratio tests and the sufficient statistics needed to implement them. In this section, we examine the performance of arbitrary decision rules. As we saw earlier, a decision rule δ specifies a partition of the set $\mathcal{Y}$ into two regions $\mathcal{Y}_0$ and

$\mathcal{Y}_1$ over which we select H_0 and H_1 respectively. The performance of a test will be evaluated in terms of three (really two) quantities: the probability of detection

$$P_D(\delta) = \int_{\mathcal{Y}_1} f(\mathbf{y}|H_1) d\mathbf{y} \,, \tag{2.46}$$

the probability of a miss

$$P_M(\delta) = 1 - P_D(\delta) = \int_{\mathcal{Y}_0} f(\mathbf{y}|H_1) d\mathbf{y} \,, \tag{2.47}$$

and the probability of false alarm

$$P_F(\delta) = \int_{\mathcal{Y}_1} f(\mathbf{y}|H_0) d\mathbf{y} \,. \tag{2.48}$$

In the above expressions, we assume the observations take continous values. When they are discrete-valued, we only need to replace integrals by summations and densities by probability mass functions. Note that since the probability of a miss is $1 - P_D$, it is not really needed to characterize the test performance, but in discussions it will sometimes be handy to refer to it.

The expression (2.6) implies that the conditional Bayes risk under H_0 and H_1 is given by

$$\begin{aligned} R(\delta|H_0) &= C_{10} P_F(\delta) + C_{00}(1 - P_F(\delta)) \\ R(\delta|H_1) &= C_{11} P_D(\delta) + C_{01}(1 - P_D(\delta)) \,, \end{aligned} \tag{2.49}$$

so the overall Bayes risk can be expressed as

$$R(\delta) = C_{00}\pi_0 + C_{01}\pi_1 + (C_{10} - C_{00}) P_F(\delta)\pi_0 + (C_{11} - C_{01}) P_D(\delta)\pi_1 \,. \tag{2.50}$$

This shows that the Bayes risk is parametrized entirely by the pair $(P_F(\delta), P_D(\delta))$.

Ideally, we would like of course to select a test where P_D is as close to one as possible while P_F is as close to zero as possible. However, except for situations where the values taken by $\mathbf{Y}$ under H_0 and H_1 do not overlap, or when the number N of observations tends to infinity, we cannot have both $P_D = 1$ and $P_F = 0$. To separate the domain of achievable pairs (P_F, P_D) from those that cannot be achieved, we can think of looking at the square formed by $0 \leq P_F \leq 1$ and $0 \leq P_D \leq 1$ and divide it into two regions: the region for which we can find a decision rule δ such that $P_F(\delta) = P_F$ and $P_D(\delta) = P_D$, and the region where no decision rule can achieve the pair (P_F, P_D), as shown in Fig. 2.1. In this figure, achievable tests are represented by the shaded region. Nonachievable tests are split into two components formed by the North–West and South–East corners of the square, representing respectively extremely good or extremely poor performances (P_F, P_D). For example, the point $P_F = 0$, $P_D = 1$ at the North–West corner represents an error-free performance, whereas the point $P_F = 1$ and $P_D = 0$ at

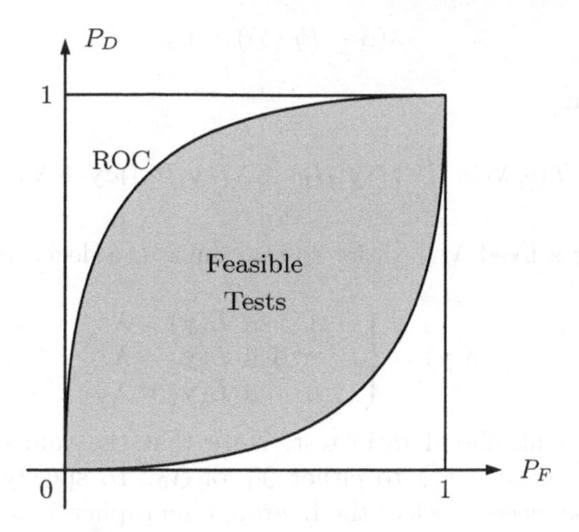

Fig. 2.1. Domain of feasible tests and ROC for a binary hypothesis testing problem.

the South–West corner represents an always incorrect performance. As will be explained later, both cases are unachievable. The upper boundary between achievable and unachievable regions is called the *receiver operating characteristic* (ROC). However, before describing the properties of the ROC, it is convenient to introduce the class of Neyman-Pearson tests.

2.4.1 Neyman-Pearson Tests

These tests play an important historical and practical role in detection theory. A key limitation of the Bayesian tests introduced in Section 4.2 is their reliance on a cost function C_{ij} and priors (π_0, π_1). In [4] Neyman and Pearson formulated the binary hypothesis testing problem pragmatically by selecting the test δ that maximizes $P_D(\delta)$ (or equivalently that minimizes $P_M(\delta)$) while ensuring that the probability of false alarm $P_F(\delta)$ is less than or equal to a number α. Let $\mathcal{D}_\alpha$ denote the domain

$$\mathcal{D}_\alpha = \{\delta \ : \ P_F(\delta) \leq \alpha\}. \tag{2.51}$$

Then the Neyman-Pearson test δ_{NP} solves the constrained optimization problem

$$\delta_{\mathrm{NP}} = \arg \max_{\delta \in \mathcal{D}_\alpha} P_D(\delta) \, . \tag{2.52}$$

This problem can be solved by Lagrangian optimization [5]. Consider the Lagrangian

$$L(\delta, \lambda) = P_D(\delta) + \lambda(\alpha - P_F(\delta)) \tag{2.53}$$

with $\lambda \geq 0$. A test δ will be optimal if it maximizes $L(\delta, \lambda)$ and satisfies the Karush-Kuhn-Tucker (KKT) condition

$$\lambda(\alpha - P_F(\delta)) = 0 . \tag{2.54}$$

From expression

$$L(\delta, \lambda) = \int_{\mathcal{Y}_1} (f(\mathbf{y}|H_1) - \lambda f(\mathbf{y}|H_0))d\mathbf{y} + \lambda\alpha , \tag{2.55}$$

we see that, for a fixed λ, in order to maximize L a decision rule δ needs to satisfy

$$\delta(\mathbf{y}) = \begin{cases} 1 & \text{if } L(\mathbf{y}) > \lambda \\ 0 \text{ or } 1 & \text{if } L(\mathbf{y}) = \lambda \\ 0 & \text{if } L(\mathbf{y}) < \lambda , \end{cases} \tag{2.56}$$

which is again a likelihood ratio test. Note that the values of $\mathbf{y}$ such that $L(\mathbf{y}) = \lambda$ can be allocated to either $\mathcal{Y}_1$ or $\mathcal{Y}_0$. To specify completely the optimum test, we need to select the Lagrange multiplier λ such that the KKT condition (2.54) holds, and we must specify a precise allocation rule for the values of $\mathbf{y}$ such that $L(\mathbf{y}) = \lambda$. To analyze this problem, it is convenient to introduce the cumulative probability distribution

$$F_L(\ell|H_0) = P[L \leq \ell|H_0]$$

of the likelihood ratio $L = L(\mathbf{Y})$ under H_0. Also, for a fixed threshold λ, we denote by $\delta_{L,\lambda}$ and $\delta_{U,\lambda}$ the LRTs with threshold λ such that $\delta_{L,\lambda}(\mathbf{y}) = 0$ and $\delta_{U,\lambda}(\mathbf{y}) = 1$ for all observations $\mathbf{y}$ such that $L(\mathbf{y}) = \lambda$. These two tests represent the two extreme allocation rules for the LRT (2.56): in one case all observations such that $L(\mathbf{y}) = \lambda$ are allocated to $\mathcal{Y}_0$, in the other case they are allocated to $\mathcal{Y}_1$.

Now, if we consider the cumulative probability distribution $F_L(\ell|H_0)$, we denote

$$f_0 = F_L(0|H_0) ,$$

and we recall that $F_L(\ell|H_0)$ is nondecreasing and right continuous, but may include discontinuities. Thus, depending on α, three cases may occur, which are depicted in Fig. 2.2.

(i) When $1 - \alpha < f_0$, we select $\lambda = 0$ and $\delta = \delta_{L,0}$. In this case the KKT condition (2.54) holds and the probability of false alarm of the test is $P_F = 1 - f_0 < \alpha$, so the test is optimal. This case corresponds to the values $1 - \alpha_1$ and $\lambda_1 = 0$ of Fig. 2.2.

(ii) For $1 - \alpha \geq f_0$, suppose there exists a λ such that

$$F_L(\lambda|H_0) = 1 - \alpha , \tag{2.57}$$

i.e., $1 - \alpha$ is in the range of $F_L(\ell|H_0)$. Then by selecting λ as the LRT threshold and $\delta = \delta_{L,\lambda}$ we obtain a test with $P_F = 1 - F_L(\lambda|H_0) = \alpha$, so the KKT condition (2.54) holds and the test is optimal. This case corresponds to the values $1 - \alpha_2$ and λ_2 of Fig. 2.2.

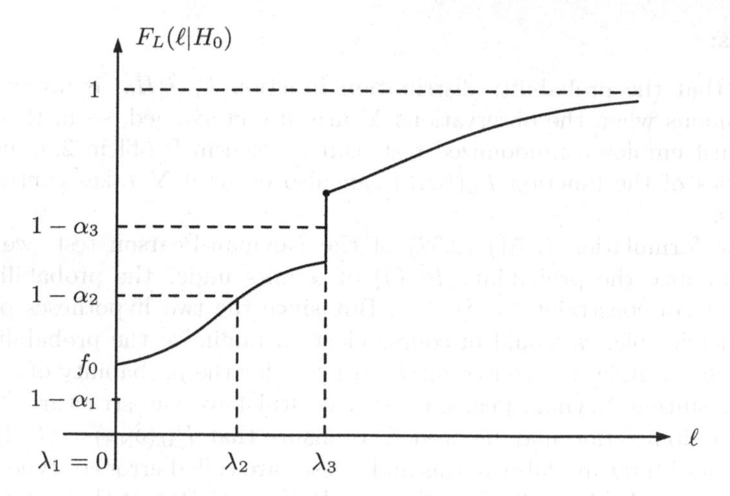

Fig. 2.2. Selection of the Neyman-Pearson threshold λ by using the cumulative probability distribution $F_L(\ell|H_0)$.

(iii) Suppose $1 - \alpha \geq f_0$ is not in the range of $F_L(\ell|H_0)$, i.e., there is a discontinuity point $\lambda > 0$ of $F_L(\ell|H_0)$ such that

$$F_L(\lambda_-|H_0) < 1 - \alpha < F_L(\lambda|H_0) , \tag{2.58}$$

where $F_L(\lambda_-|H_0)$ denotes the left limit of $F_L(\ell|H_0)$ at $\ell = \lambda$. This case corresponds to the values $1-\alpha_3$ and λ_3 of Fig. 2.2. Then if we consider the LRT $\delta_{L,\lambda}$, it has the probability of false alarm $P_F(\delta_{L,\lambda}) = 1-F_L(\lambda|H_0) < \alpha$. Similarly the test $\delta_{U,\lambda}$ has probability of false alarm $P_F(\delta_{U,\lambda}) = 1 - F_L(\lambda_-|H_0) > \alpha$. But according to the KKT condition (2.54) we must have $P_F(\delta) = \alpha$ exactly. The trick to construct an LRT of the form (2.56) that meets this constraint is to use a *randomized test.* Let

$$p = \frac{F_L(\lambda|H_0) - (1-\alpha)}{F_L(\lambda|H_0) - F_L(\lambda_-|H_0)} . \tag{2.59}$$

Clearly $0 < p < 1$. Then let $\delta(p)$ be the test obtained by selecting $\delta_{U,\lambda}$ with probability p and $\delta_{L,\lambda}$ with probability $1-p$. This test has the form (2.56) and its probability of false alarm is

$$\begin{aligned} P_F(\delta(p)) &= pP_F(\delta_{U,\lambda}) + (1-p)P_F(\delta_{L,\lambda}) \\ &= p(1 - F_L(\lambda_-|H_0)) + (1-p)(1 - F_L(\lambda|H_0)) = \alpha , \end{aligned} \tag{2.60}$$

where the last equality was obtained by substituting the value (2.59) of p. This shows that $\delta(p)$ satisfies the KKT condition (2.54), so it is optimal.

Remarks:

(a) Note that the probability distribution function $F_L(\ell|H_0)$ is always discontinuous when the observations $\mathbf{Y}$ are discrete-valued, so in this case we must employ a randomized test. But as seen in Problem 2.6, discontinuities of the function $F_L(\ell|H_0)$ can also occur if $\mathbf{Y}$ takes continuous values.

(b) In the formulation (2.51)–(2.52) of the Neyman-Pearson test, we seek to minimize the probability $P_M(\delta)$ of a miss under the probability of false alarm constraint $P_F(\delta) \leq \alpha$. But since the two hypotheses play a symmetric role, we could of course elect to minimize the probability of false alarm under the constraint $P_M(\delta) \leq \beta$ for the probability of a miss. The resulting Neyman-Pearson test will still have the structure (2.56), and randomization may be needed to ensure that $P_M(\delta_{\mathrm{NP}}) = \beta$. In the statistics literature, false alarms and misses are called errors of type I and II, respectively. Accordingly a Neyman-Pearson (NP) test that minimizes the probability of a miss while placing an upper bound on the probability of false alarm is called an NP test of type I. Conversely, an NP test that minimizes the probability of false alarm while placing an upper bound on the probability of a miss is called an NP test of type II.

2.4.2 ROC Properties

If we return to the problem of characterizing the pairs (P_F, P_D) in the square $[0,1] \times [0,1]$ which are achievable for a test δ, we see from the above analysis that if we fix the probability of false alarm $P_F = \alpha$, the test that maximizes P_D must be an LRT. Thus in the square $[0,1]^2$, the ROC curve separating achievable pairs (P_F, P_D) from nonachievable pairs corresponds to the family of LRTs. In fact, if the likelihood ratio $L(\mathbf{Y})$ admits a probability density $f_L(\ell|H_j)$ under H_j with $j = 0,\ 1$, and if τ denotes the LRT threshold, the probabilities of detection and of false alarm can be expressed as

$$P_D(\tau) = \int_\tau^\infty f_L(\ell|H_1)d\ell$$
$$P_F(\tau) = \int_\tau^\infty f_L(\ell|H_0)d\ell\,, \tag{2.61}$$

so that as τ varies between 0 and ∞, the point $P_D(\tau)$, $P_F(\tau)$ will move continuously along the ROC curve. If we set the threshold $\tau = 0$, we always select hypothesis H_1, so

$$P_D(0) = P_F(0) = 1$$

and thus we achieve a probability of detection of 1, but at the cost of having the false alarm probability equal to 1 as well. Conversely, if we set $\tau = +\infty$, we always select hypothesis H_0, and thus

$$P_D(+\infty) = P_F(+\infty) = 0,$$

so in this case the probability of false alarm is zero, but so is the probability of detection. This establishes the following property.

Property 1: The points $(0,0)$ and $(1,1)$ belong to the ROC.

From the parametrization (2.61) of the ROC, we deduce

$$\begin{aligned} \frac{dP_D}{d\tau} &= -f_L(\tau|H_1) \\ \frac{dP_F}{d\tau} &= -f_L(\tau|H_0) , \end{aligned} \tag{2.62}$$

which implies

$$\frac{dP_D}{dP_F} = \frac{f_L(\tau|H_1)}{f_L(\tau|H_0)} = \tau , \tag{2.63}$$

i.e., we have the following property.

Property 2: The slope of the ROC at point $(P_D(\tau), P_F(\tau))$ is equal to the threshold τ of the corresponding LRT.

In particular, this implies that the slope of the ROC at point $(0,0)$ is $\tau = +\infty$ and the slope at point $(1,1)$ is $\tau = 0$, as shown in Fig. 2.3.

We also have the following important feature.

Property 3: The domain of achievable pairs (P_F, P_D) is convex, which implies the ROC curve is concave.

To establish this result, we use randomization. Let (P_{F1}, P_{D1}) and (P_{F2}, P_{D2}) be two achievable pairs corresponding to the tests δ_1 and δ_2, respectively. Let

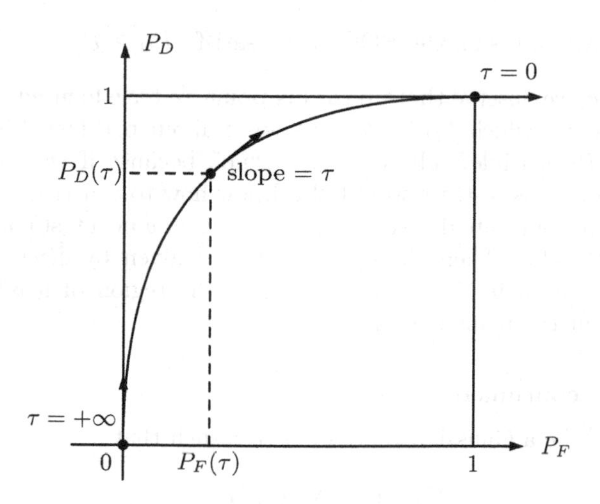

Fig. 2.3. Parametrization of the ROC curve by the LRT threshold τ.

p be an arbitrary number between 0 and 1, i.e., $0 \leq p \leq 1$. The domain of achievable tests will be convex if for any p we can find a test $\delta(p)$ with the probabilities of detection and false alarm

$$\begin{aligned} P_D(p) &= pP_{D1} + (1-p)P_{D_2} \\ P_F(p) &= pP_{F1} + (1-p)P_{F_2} \,. \end{aligned} \tag{2.64}$$

The test $\delta(p)$ we employ is a randomized test: after obtaining the observation vector $\mathbf{Y}$ we generate a binary random variable B taking values in $\{0,1\}$ with probabilities

$$P[B=0] = p \quad , \quad P[B=1] = 1-p .$$

Then if $B = 0$ we select test δ_1 for observations $\mathbf{Y}$ and if $B = 1$ we select test δ_2. Since the binary random variable B is independent of $\mathbf{Y}$, the probabilities of detection and of false alarm for the randomized test are

$$\begin{aligned} P_D &= \sum_{j=0}^{1} P_{D\,j+1}\, P[B=j] = P_D(p) \\ P_F &= \sum_{j=0}^{1} P_{F\,j+1}\, P[B=j] = P_F(p) \,, \end{aligned} \tag{2.65}$$

as desired. Then since the domain of achievable tests is convex, and is located below the ROC curve, the ROC curve is concave. Note that since the points $(0,0)$ and $(1,1)$ are on the ROC, and the ROC curve is concave, the ROC curve must be located above the line $P_D = P_F$, so we have the following property.

Property 4: All points on the ROC curve satisfy $P_D \geq P_F$.

To conclude, we observe that up to this point we have focused our analysis on "good tests" for which $P_D \geq P_F$. But what about bad tests? Is the entire region $P_D \leq P_F$ feasible? The answer is "no" because if we knew how to construct a disastrous test, we would also know how to construct a wonderful test by taking the opposite decisions. Specifically, let δ be a test corresponding to the pair (P_F, P_D). Then the opposite test $\bar{\delta}$ given by $\bar{\delta}(\mathbf{y}) = 1 - \delta(\mathbf{y})$ corresponds to the pair $(1-P_F, 1-P_D)$, so the region of feasible tests is symmetric about the point $(1/2, 1/2)$.

Example 2.1, continued

Let $\mathbf{Y} \in \mathbb{R}^N$ be a Gaussian random vector such that

$$\begin{aligned} H_0 &: \mathbf{Y} \sim N(\mathbf{0}, \sigma^2 \mathbf{I}_N) \\ H_1 &: \mathbf{Y} \sim N(\mathbf{m}, \sigma^2 \mathbf{I}_N) \end{aligned} \tag{2.66}$$

with

$$\mathbf{m} = E^{1/2} \begin{bmatrix} 1 \dots 1 \dots 1 \end{bmatrix}^T ,$$

where $\mathbf{I}_N$ denotes the identity matrix of size N. This problem corresponds to the choice $m_0 = 0$ and $m_1 = E^{1/2}$ in Example 2.1. In this case, according to (2.34)–(2.35), the optimum test can be expressed as

$$S = \frac{1}{N} \sum_{k=1}^{N} Y_k \underset{H_0}{\overset{H_1}{\gtrless}} \eta$$

with

$$\eta = \frac{E^{1/2}}{2} + \frac{\sigma^2 \ln(\tau)}{N E^{1/2}} ,$$

where S admits $N(0, \sigma^2/N)$ and $N(E^{1/2}, \sigma^2/N)$ distributions under H_0 and H_1, respectively. The probabilities of detection and of false alarm are therefore given by

$$\begin{aligned} P_F &= \int_{\eta}^{\infty} \frac{1}{(2\pi\sigma^2/N)^{1/2}} \exp\big(-\frac{(s - E^{1/2})^2}{2\sigma^2/N}\big) ds \\ &= \int_{\frac{(\eta - E^{1/2})N^{1/2}}{\sigma}} \frac{1}{(2\pi)^{1/2}} \exp(-z^2/2) dz \end{aligned} \tag{2.67}$$

$$\begin{aligned} P_F &= \int_{\eta}^{\infty} \frac{1}{(2\pi\sigma^2/N)^{1/2}} \exp\big(-\frac{s^2}{2\sigma^2/N}\big) ds \\ &= \int_{\frac{\eta N^{1/2}}{\sigma}} \frac{1}{(2\pi)^{1/2}} \exp(-z^2/2) dz \,. \end{aligned} \tag{2.68}$$

These probabilities can be expressed in terms of the Gaussian Q function

$$Q(x) = \int_{x}^{\infty} \frac{1}{(2\pi)^{1/2}} \exp(-z^2/2) dz \tag{2.69}$$

and the distance

$$d = (NE)^{1/2}/\sigma$$

between the mean vectors $\mathbf{m}$ and $\mathbf{0}$ of $\mathbf{Y}$ under H_1 and H_0, measured in units of the standard deviation σ. Noting that

$$\frac{(\eta - E^{1/2})N^{1/2}}{\sigma} = -\frac{d}{2} + \frac{\ln(\tau)}{d} ,$$

and using the symmetry property $Q(-x) = 1 - Q(x)$, we find

$$\begin{aligned} P_D(\tau) &= 1 - Q(\frac{d}{2} - \frac{\ln(\tau)}{d}) \\ P_F(\tau) &= Q(\frac{d}{2} + \frac{\ln(\tau)}{d}) \,. \end{aligned} \tag{2.70}$$

The ROC curve is then obtained by letting the threshold τ vary from 0 to ∞ in the parametrization (2.69). By eliminating τ between $P_D(\tau)$ and $P_F(\tau)$, and denoting by $Q^{-1}(\cdot)$ the inverse function of $Q(\cdot)$, we find that the ROC can be expressed in closed form as

$$P_D = 1 - Q(d - Q^{-1}(P_F)) . \tag{2.71}$$

This curve is plotted in Fig. 2.4 for several values of d. For $d = 0$, it just reduces to the diagonal $P_D = P_F$. Then as d increases, the curve gradually approaches the top left hand corner $P_D = 1$, $P_F = 0$ representing the ideal test. As we shall see later, the form (2.70)–(2.71) of the ROC holds for all detection problems involving known signals in zero-mean Gaussian noise. An interesting feature of the ROC curve (2.70) is that it is invariant under the transformation

$$P_M = 1 - P_D \longleftrightarrow P_F ,$$

i.e., it is symmetric with respect to the line $P_D = 1 - P_F$ with slope -1 passing the top left corner $(0, 1)$ of the square $[0, 1]^2$. $\square$

Remark: In the detection literature, d is often referred to as the "deflection coefficient" of the two hypotheses. However, in the Gaussian case, d is a true distance, since it will be shown in the next chapter that the Chernoff distance between two Gaussian densities is proportional to d^2. Thus in this text d will be called a distance whenever densities are Gaussian, and a deflection if the densities are either unknown or nonGaussian.

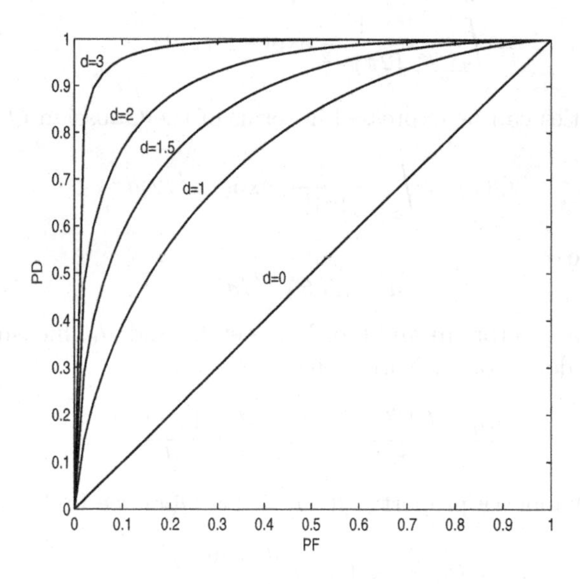

Fig. 2.4. ROC for Gaussian vectors with equal covariance plotted for the values $d = 0$, 1, 1.5, 2 and 3 of the normalized distance between H_0 and H_1.

Example 2.2, continued

We observe a Gaussian vector $\mathbf{Y} \in \mathbb{R}^2$ with the following two possible distributions:

$$\begin{aligned} H_0 &: \mathbf{Y} \sim N(\mathbf{0}, \sigma_0^2 \mathbf{I}_2) \\ H_1 &: \mathbf{Y} \sim N(\mathbf{0}, \sigma_1^2 \mathbf{I}_2) \end{aligned}$$

where $\sigma_1^2 > \sigma_0^2$. This corresponds to the case where $N = 2$ in Example 2.2. Then according to (2.36)–(2.37) the LRT can be expressed as

$$T = Y_1^2 + Y_2^2 \underset{H_0}{\overset{H_1}{\gtrless}} \kappa$$

with

$$\kappa = \frac{2\sigma_0^2\sigma_1^2}{\sigma_1^2 - \sigma_0^2}[\ln(\frac{\sigma_1^2}{\sigma_0^2}) + \ln(\tau)] \, ,$$

and under H_j with $j = 0, 1$, T/σ_j^2 has a chi-squared distribution with two degrees of freedom. The probabilities of detection and false alarm can therefore be expressed as

$$\begin{aligned} P_D(\kappa) &= \frac{1}{2}\int_{\kappa/\sigma_1^2} \exp(-z/2)dz = \exp\big(-\frac{\kappa}{2\sigma_1^2}\big) \\ P_F(\kappa) &= \frac{1}{2}\int_{\kappa/\sigma_0^2} \exp(-z/2)dz = \exp\big(-\frac{\kappa}{2\sigma_0^2}\big) \, , \end{aligned}$$

and by eliminating κ from the above identities we find that the ROC curve admits the closed-form expression

$$P_D = (P_F)^{\sigma_0^2/\sigma_1^2} \, . \tag{2.72}$$

The ROCs corresponding to different values of $r = \sigma_1^2/\sigma_0^2$ are plotted in Fig. 2.5. To verify the concavity of the ROC curve (2.72), note that

$$\frac{d^2 P_D}{d^2 P_F} = \frac{\sigma_0^2}{\sigma_1^2}\Big(\frac{\sigma_0^2}{\sigma_1^2} - 1\Big) P_F^{(\sigma_0^2/\sigma_1^2)-2} < 0 \, ,$$

so the ROC curve is strictly concave. □

Example 2.3, continued

Let Y be a Poisson random variable with parameters λ_0 and λ_1 under H_0 and H_1, i.e.,

$$\begin{aligned} H_0 &: P[Y = n] = \frac{\lambda_0^n}{n!}\exp(-\lambda_0) \\ H_1 &: P[Y = n] = \frac{\lambda_1^n}{n!}\exp(-\lambda_1) \, . \end{aligned}$$

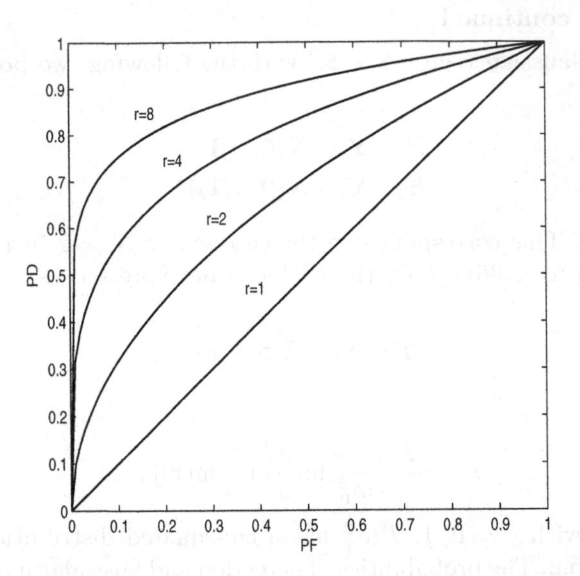

Fig. 2.5. ROC for two zero-mean Gaussian vectors with unequal variances σ_1^2 and σ_0^2. The ROC is plotted for the values $r = 1, 2, 4$ and 8 of the variance ratio $r = \sigma_1^2/\sigma_0^2$.

This corresponds to the case $N = 1$ in Example 2.3. Then according to (2.39)–(2.40) the optimum test is given by

$$Y \underset{H_0}{\overset{H_1}{\gtrless}} \eta$$

with

$$\eta = \frac{\lambda_1 - \lambda_0}{\ln(\lambda_1/\lambda_0)} + \frac{\ln(\tau)}{\ln(\lambda_1/\lambda_0)} .$$

Note that Y takes integer values whereas the threshold η is real. Let $n = \lceil \eta \rceil$ denote the smallest nonnegative integer greater than or equal to η. Then the probabilities of detection and of false alarm can be expressed as

$$P_D(n) = \sum_{m=n}^{+\infty} \frac{\lambda_1^m}{m!} \exp(-\lambda_1)$$

$$P_F(n) = \sum_{m=n}^{+\infty} \frac{\lambda_0^m}{m!} \exp(-\lambda_0) .$$

If $n \geq 1$, all thresholds η such that $n-1 < \eta \leq n$ yield the same pairs $(P_F(n), P_D(n))$, so the tests of the form (2.39)–(2.40) yield only a discrete set of ROC points. To generate a continuous ROC we must connect the points by

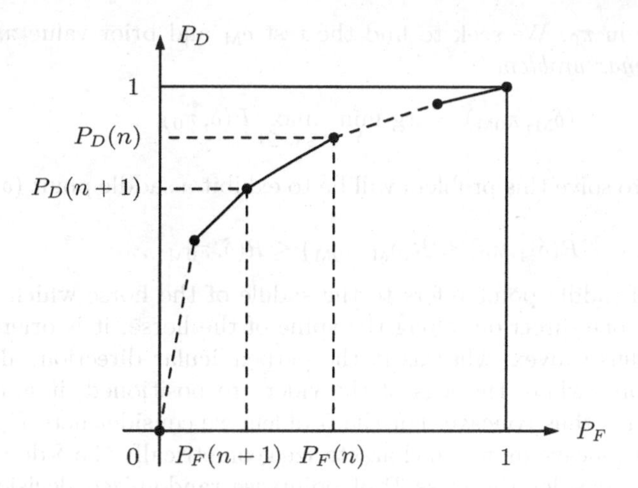

Fig. 2.6. ROC for a Poisson random variable with parameters λ_1 and λ_0. The LR tests correspond to discrete points with coordinates $(P_F(n), P_D(n))$ which can be connected by using randomization.

using randomization. For example, the segment connecting the points $(P_F(n+1), P_D(n+1))$ and $(P_F(n), P_D(n))$ can be obtained by using with probability p, a test with $\eta = n+1$ and with probability $1-p$, a test with $\eta = n$. This means that when $Y > n$ we always select H_1 and when $Y < n$ we pick H_0, but when $Y = n$, we select H_0 with probability p and H_1 with probability $1-p$. The resulting ROC is shown in Fig. 2.6. □

2.5 Minimax Hypothesis Testing

One weakness of the Bayesian formulation of hypothesis testing is its rather optimistic assumption that priors (π_0, π_1) can be assigned to the two hypotheses H_0 and H_1. While for digital communication applications, this is a reasonable assumption as the bits 0 or 1 are equally likely to be transmitted, there exists vast classes of applications, such as failure testing, where it is essentially impossible to assign priors with any degree of certainty. For such applications, instead of guessing, it is preferable to design the test conservatively by assuming the least-favorable choice of priors and selecting the test that minimizes the Bayes risk for this choice of priors. This approach has the advantage of guaranteeing a minimum level of test performance independently of the actual prior values.

Let π_0 be the a-priori probability of H_0. Then $\pi_1 = 1 - \pi_0$ and the Bayes risk corresponding to the decision rule δ and prior π_0 is given by

$$R(\delta, \pi_0) = R(\delta|H_0)\pi_0 + R(\delta|H_1)(1-\pi_0) , \tag{2.73}$$

which is linear in π_0. We seek to find the test δ_M and prior value π_{0M} that solve the *minimax problem*

$$(\delta_M, \pi_{0M}) = \arg \min_{\delta} \max_{0 \leq \pi_0 \leq 1} R(\delta, \pi_0) \,. \tag{2.74}$$

Our approach to solve this problem will be to exhibit a saddle point (δ_M, π_{0M}) satisfying

$$R(\delta_M, \pi_0) \leq R(\delta_M, \pi_{0M}) \leq R(\delta, \pi_{0M}) \,. \tag{2.75}$$

The concept of saddle point refers to the saddle of the horse which has the feature that in one direction, along the spine of the horse, it is oriented upwards and is thus convex, whereas in the perpendicular direction, along the sides of the horse where the legs of the rider are positioned, it is oriented downwards and is thus concave. For the problem we consider here $R(\delta, \pi_0)$ is linear and thus concave in π_0. To characterize geometrically the δ-dependence of $R(\delta, \pi_0)$, we consider the class $\mathcal{D}$ of pointwise randomized decision functions. A function δ of $\mathcal{D}$ takes the value $\delta(\mathbf{y})$ with $0 \leq \delta(\mathbf{y}) \leq 1$ at point $\mathbf{y} \in \mathcal{Y}$ if hypothesis H_1 is chosen with probability $\delta(\mathbf{y})$ and H_0 with probability $1 - \delta(\mathbf{y})$. The set $\mathcal{D}$ is convex since if δ_1 and δ_2 are two functions of $\mathcal{D}$, i.e.,

$$0 \leq \delta_i(\mathbf{y}) \leq 1$$

for all $\mathbf{y} \in \mathcal{Y}$ and $i = 1, 2$, and if α is a real number such that $0 \leq \alpha \leq 1$, the convex combination

$$\delta(\mathbf{y}) = \alpha\delta_1(\mathbf{y}) + (1 - \alpha)\delta_2(\mathbf{y})$$

satisfies $0 \leq \delta(\mathbf{y}) \leq 1$ for all $\mathbf{y} \in \mathcal{Y}$, so it belongs to $\mathcal{D}$. For this class of decision functions, the expression (2.12) for the Bayesian risk can be rewritten as

$$R(\delta, \pi_0) = \sum_{j=0}^{1} C_{0j}\pi_j + \int \Big[\sum_{j=0}^{1} (C_{1j} - C_{0j})\pi_j f(\mathbf{y}|H_j)\Big] \delta(\mathbf{y}) d\mathbf{y} \tag{2.76}$$

which is linear and thus convex in δ. So by extending the definition of the Bayesian risk $R(\delta, \pi_0)$ to the set $\mathcal{D}$ of pointwise randomized decision functions, we have obtained a function which is convex in δ and concave in π_0, so according to von Neumann's minimax theorem [6, p. 319], it necessarily admits a saddle point of the form (2.75). Note that the compactness condition of von Neumann's theorem is satisfied since $\pi_0 \in [0, 1]$, which is compact, and since for any decision function in $\mathcal{D}$, we have

$$||\delta||_\infty = \sup_{\mathbf{y} \in \mathcal{Y}} \delta(\mathbf{y}) \leq 1 \,,$$

so $\mathcal{D}$ is compact for the topology induced by the infinity norm.

Note that defining R over the larger domain $\mathcal{D}$ of pointwise randomized functions instead of the set of binary valued decision functions does not affect the minimization of R. To understand why this is the case, observe that R is a linear function of δ and the constraint $0 \leq \delta(\mathbf{y}) \leq 1$ is also linear, so the minimization of R with respect to δ is a linear programming problem whose solution will necessarily be on the boundary of $\mathcal{D}$, i.e., it will be a binary function. In fact this can be seen directly from expression (2.76) for the risk function, which is clearly minimized by setting $\delta(\mathbf{y})$ equal to 1 or 0 depending on whether the integrand

$$\sum_{j=0}^{1}(C_{1j} - C_{0j})\pi_j f_j(\mathbf{y})$$

is negative or positive, which is of course the usual LR test. When the integrand is zero we could of course use a value of $\delta(\mathbf{y})$ between zero and 1, but it yields the same minimum as a purely binary decision function. So, in conclusion, convexifying R by defining it over the set $\mathcal{D}$ of pointwise randomized decision functions does not affect the minimization process itself.

To demonstrate that if a saddle point exists, it necessarily solves the minimax problem (2.74), we rely on the following result.

Saddle point property: We start by noting that for an arbitrary function $R(\delta, \pi_0)$ we necessarily have

$$\max_{0 \leq \pi_0 \leq 1} \min_{\delta} R(\delta, \pi_0) \leq \min_{\delta} \max_{0 \leq \pi_0 \leq 1} R(\delta, \pi_0) . \tag{2.77}$$

This is due to the fact that, by definition of the maximum,

$$R(\delta, \pi_0) \leq \max_{0 \leq \pi_0 \leq 1} R(\delta, \pi_0) . \tag{2.78}$$

Then minimizing both sides of (2.78) yields

$$\min_{\delta} R(\delta, \pi_0) \leq \min_{\delta} \max_{0 \leq \pi_0 \leq 1} R(\delta, \pi_0)$$

which implies (2.77). Then, if the saddle point identity (2.75) is satisfied, we have

$$\max_{0 \leq \pi_0 \leq 1} R(\delta_{\mathrm{M}}, \pi_0) = R(\delta_{\mathrm{M}}, \pi_{0\mathrm{M}}) = \min_{\delta} R(\delta, \pi_{0\mathrm{M}}) , \tag{2.79}$$

which in turn implies

$$\begin{aligned}\min_{\delta} \max_{0 \leq \pi_0 \leq 1} R(\delta, \pi_0) &\leq \max_{0 \leq \pi_0 \leq 1} R(\delta_{\mathrm{M}}, \pi_0) = R(\delta_{\mathrm{M}}, \pi_{0\mathrm{M}}) \\ &= \min_{\delta} R(\delta, \pi_{0\mathrm{M}}) \leq \max_{0 \leq \pi_0 \leq 1} \min_{\delta} R(\delta, \pi_0) . \end{aligned} \tag{2.80}$$

Combining now (2.77) with (2.80) we obtain the *key property* of saddle points:

$$R(\delta_{\mathrm{M}}, \pi_{0\mathrm{M}}) = \max_{0 \leq \pi_0 \leq 1} \min_{\delta} R(\delta, \pi_0) = \min_{\delta} \max_{0 \leq \pi_0 \leq 1} R(\delta, \pi_0) \,. \tag{2.81}$$

In other words, if $(\delta_{\mathrm{M}}, \pi_{0\mathrm{M}})$ is a saddle point of $R(\delta, \pi_0)$, it solves the minimax problem (2.74) and the solution of the minimax problem is the same as that of the maximin problem, a feature that will be used subsequently in our analysis.

Saddle point construction: To identify the saddle point (2.75), we employ a constructive approach. The first step is to note that the second inequality in (2.75) implies that δ_{M} must be an optimal Bayesian test for the risk function with prior $\pi_{0\mathrm{M}}$, so that

$$\delta_{\mathrm{M}}(\mathbf{y}) = \begin{cases} 1 & \text{if } L(\mathbf{y}) > \tau_{\mathrm{M}} \\ 1 \text{ or } 0 & \text{if } L(\mathbf{y}) = \tau_{\mathrm{M}} \\ 0 & \text{if } L(\mathbf{y}) < \tau_{\mathrm{M}} \,, \end{cases} \tag{2.82}$$

with

$$\tau_{\mathrm{M}} = \frac{(C_{10} - C_{00})\pi_{0\mathrm{M}}}{(C_{01} - C_{11})(1 - \pi_{0\mathrm{M}})} \,, \tag{2.83}$$

where we use the form (2.82) of the LRT instead of (2.20) in case randomization might be needed, which will turn out to be the case. Next, if we consider the first inequality in (2.75), we note that since $R(\delta_{\mathrm{M}}, \pi_0)$ is linear in π_0, three cases can occur:

(i) the slope is positive, i.e.,

$$R(\delta_{\mathrm{M}}|H_0) > R(\delta_{\mathrm{M}}|H_1) \tag{2.84}$$

so the maximum of $R(\delta_{\mathrm{M}}, \pi_0)$ occurs at $\pi_{0\mathrm{M}} = 1$;

(ii) the slope is negative, i.e.,

$$R(\delta_{\mathrm{M}}|H_0) < R(\delta_{\mathrm{M}}|H_1) \tag{2.85}$$

so the maximum of $R(\delta_{\mathrm{M}}, \pi_0)$ occurs at $\pi_{0\mathrm{M}} = 0$; and

(iii) the slope is zero:

$$R(\delta_{\mathrm{M}}|H_0) = R(\delta_{\mathrm{M}}|H_1) \,, \tag{2.86}$$

in which case we have $0 \leq \pi_{0\mathrm{M}} \leq 1$.

Let us consider these three cases:

(i) If $\pi_{0\mathrm{M}} = 1$ we have $\tau_{\mathrm{M}} = +\infty$. Then, if we assume $P[L = +\infty|H_j] = 0$ for $j = 0, 1$ (when $\mathbf{Y}$ takes continuous values, this corresponds to requiring that the probability measure of $\mathbf{Y}$ under H_1 should be absolutely continuous with respect to the measure under H_0), we have $P_F(\delta_{\mathrm{M}}) = P_D(\delta_{\mathrm{M}}) = 0$ and

$$R(\delta_{\mathrm{M}}|H_0) = C_{00} \quad , \quad R(\delta_{\mathrm{M}}|H_1) = C_{01} \,, \tag{2.87}$$

in which case the slope condition (2.84) reduces to

$$C_{00} > C_{01} \,. \tag{2.88}$$

(ii) If $\pi_{0\mathrm{M}} = 0$ we thave $\tau_{\mathrm{M}} = 0$. Then if we assume $P[L = 0|H_j] = 0$ for $j = 0, 1$ (when $\mathbf{Y}$ takes continuous values, this corresponds to requiring that the probability measure of $\mathbf{Y}$ under H_0 should be absolutely continuous with respect to the measure under H_1), we have $P_F(\delta_{\mathrm{M}}) = P_D(\delta_{\mathrm{M}}) = 1$ and

$$R(\delta_{\mathrm{M}}|H_0) = C_{10} \quad , \quad R(\delta_{\mathrm{M}}|H_1) = C_{11} \,, \tag{2.89}$$

in which case the slope condition (2.85) reduces to

$$C_{11} > C_{10} \,. \tag{2.90}$$

(iii) When the slope of $R(\delta_{\mathrm{M}}, \pi_0)$ is zero, the condition (2.86) yields the *minimax receiver equation*

$$C_{11} - C_{00} + (C_{01} - C_{11})P_M(\delta_{\mathrm{M}}) - (C_{10} - C_{00})P_F(\delta_{\mathrm{M}}) = 0 \,. \tag{2.91}$$

But cost functions such that either (2.88) or (2.90) hold tend to be pathological and can be excluded. For example, if we consider a quality control problem where a part is defective under H_1 and meets customer specifications under H_0, the inequality (2.88) would suggest it is more costly to accept a good part than to accept one that is faulty, which is absurd. Similarly the inequality (2.90) implies it is more costly to discard a defective part than a good part, whereas common sense suggests that the cost of discarding a defective part is zero, while the cost of discarding a good part is the value of the part if it was sold to a customer. So, provided the likelihood ratio L does not assign a positive probability to the values 0 or $+\infty$ under either H_0 or H_1, which implies that the points $(0,0)$ and $(1,1)$ belong to the ROC, and provided that the cost function C_{ij} does not satisfy either (2.88) or (2.90), the minimax test is obtained by finding the LRT on the ROC curve that satisfies the linear equation (2.91).

The equation (2.91) takes a simple form when the costs of making correct decisions are symmetric:

$$C_{00} = C_{11} \,, \tag{2.92}$$

which is the case in particular when (2.22) holds. In this case the equation (2.91) can be rewritten as

$$P_D = 1 - \frac{C_{10} - C_{00}}{C_{01} - C_{11}} P_F \tag{2.93}$$

which is a line with negative slope passing through the top left corner $(0, 1)$ of the square $[0, 1]^2$. Since the ROC curve is continuous and connects the points $(0, 0)$ and $(1, 1)$ of this square, the line (2.93) necessarily intersects the ROC, yielding the desired minimax test, as shown in Fig. 2.7, where the intersection point is denoted by M. The ROC slope at point M yields the threshold τ_M. Note, however, that when randomization is needed to generate certain segments of the ROC, the minimax test (2.82) will need to employ

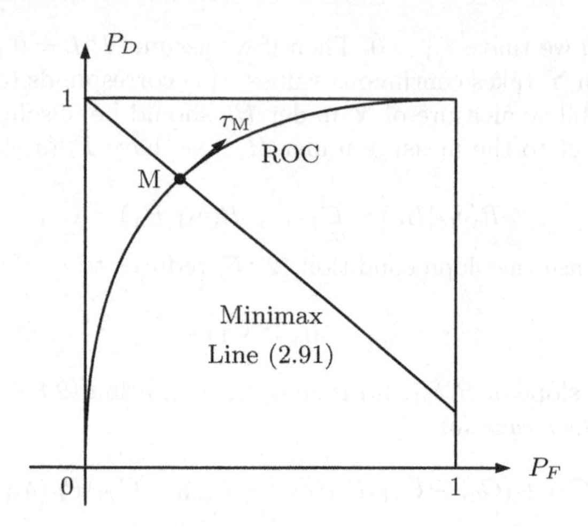

Fig. 2.7. Minimax receiver design procedure.

randomization if the intersection of (2.93) with the ROC occurs on such a segment.

Even if the costs C_{00} and C_{11} do not satisfy the symmetry condition (2.92), it is easy to verify that the line (2.91) will always intersect the ROC as long as L does not assign a positive probability to the points 0 and $+\infty$ under H_0 and H_1 and conditions

$$C_{01} \geq C_{00} \quad , \quad C_{10} \geq C_{11} \tag{2.94}$$

hold. To see this, assume for example that $C_{00} > C_{11}$. Then equation (2.91) can be rewritten as

$$P_D = 1 - \frac{C_{00} - C_{11}}{C_{01} - C_{11}} - \frac{C_{10} - C_{00}}{C_{01} - C_{11}} P_F \tag{2.95}$$

whose intercept v_{M} with the vertical axis $P_F = 0$ satisfies $0 \leq v_{\mathrm{M}} < 1$ as long as $C_{01} \geq C_{00}$. But since the line (2.95) has negative slope and the ROC connects continuously the points $(0,0)$ and $(1,1)$, the line (2.95) necessarily intersects the ROC. Similarly, when $C_{11} > C_{00}$, it is easy to verify that the horizontal intercept h_{M} of the line (2.91) with the line $P_D = 1$ satisfies $0 < h_{\mathrm{M}} \leq 1$, and since the line (2.91) has a negative slope, an intercept with the ROC necessarily occurs.

Once the intersection of line (2.91) with the ROC has been determined, let τ_M be the corresponding threshold. Then the least favorable prior is obtained by solving (2.19) for $\pi_{0\mathrm{M}}$, which gives

$$\pi_{0\mathrm{M}} = \Big[1 + \frac{(C_{10} - C_{00})}{(C_{01} - C_{11})\tau_{\mathrm{M}}}\Big]^{-1} . \tag{2.96}$$

At this point we have exhibited a pair $(\delta_{\mathrm{M}}, \pi_{0\mathrm{M}})$ satisfying the saddle point equation (2.75), so the minimax hypothesis testing problem is solved.

It is worth noting, however, that if a table showing the threshold τ for each ROC point is not available, there exists another procedure for finding the least favorable prior $\pi_{0\mathrm{M}}$. It relies on the observation that according to the property (2.81) of saddle points, we can exchange minimization and maximization operations, so whenever (2.75) holds we have

$$\pi_{0\mathrm{M}} = \arg \max_{0 \leq \pi_0 \leq 1} V(\pi_0) , \tag{2.97}$$

where

$$V(\pi_0) = \min_{\delta} R(\delta, \pi_0) \tag{2.98}$$

is the optimum Bayes risk for the LRT (2.20) with prior π_0. In other words, all we have to do is to find the prior $\pi_{0\mathrm{M}}$ that maximizes the objective function $V(\pi_0)$. In this respect, note that since $R(\delta, \pi_0)$ was linear and thus concave in π_0, and $V(\pi_0)$ is obtained by minimizing $R(\delta, \pi_0)$, the resulting function $V(\pi_0)$ is concave [7, Prop. 1.2.4]. So the maximization problem (2.98) admits a unique solution $\pi_{0\mathrm{M}}$ over $[0, 1]$. This prior specifies the threshold τ_{M} for the minimax test (2.74).

Example 2.1, continued

Consider the binary hypothesis testing problem (2.66) and assume that the costs are symmetric so that $C_{1-i,1-j} = C_{ij}$ for $i, j = 0,\ 1$. Then the ROC takes the form (2.71) and the minimax receiver equation is the line

$$P_D = 1 - P_F \tag{2.99}$$

with slope equal to -1 passing through the top left corner of the square $[0, 1]^2$. Since the ROC is symmetric with respect to this line, we deduce that the tangent to the ROC at its intersection with line (2.99) is perpendicular to this line, so $\tau_{\mathrm{M}} = 1$ and thus $\pi_{0\mathrm{M}} = 1/2$. In other words, the least favorable prior is uniformly distributed. Note that since the ROC takes the form (2.71) for all detection problems involving known signals in zero-mean Gaussian noise, the minimax receiver for this class of problems always uses uniform priors. □

Example 2.6: Binary non-symmetric channel

The following example is adapted from [8, pp. 20–22]. Consider the binary non-symmetric channel (BNSC) depicted in Fig. 2.8. Hypotheses H_0 and H_1 represent the transmission of bit $B = 0$ or bit $B = 1$, respectively. Depending on whether $B = 0$ or $B = 1$ is transmitted, the observation Y takes the binary values $\{0, 1\}$ with the probability distributions

$$\begin{aligned} P[Y = 0|H_0] &= 1 - q_0 \ , \ P[Y = 1|H_0] = q_0 \\ P[Y = 0|H_1] &= q_1 \ , \ P[Y = 1|H_1] = 1 - q_1 \ . \end{aligned}$$

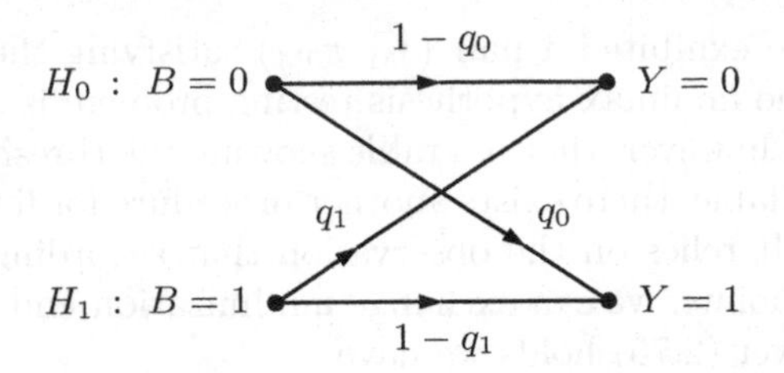

Fig. 2.8. Binary non-symmetric channel model.

In other words, the probability that bit $B = j$ is flipped is q_j. We assume that the bit flipping probability is neither zero nor one, i.e., $0 < q_j < 1$. Since we seek to minimize the probability of error, the cost function is given by $C_{ij} = 1 - \delta_{ij}$ for i, $j = 0$ or 1. Then depending on whether $Y = 0$ or 1, the optimum receiver examines the likelihood ratio values

$$L(0) = \frac{q_1}{1-q_0} \quad , \quad L(1) = \frac{1-q_1}{q_0} \, .$$

To simplify our analysis, we assume $q_0 + q_1 < 1$, which ensures $L(0) < L(1)$. Note that this is a reasonable assumption since in most communication systems the bit flipping probability is small. Then, consider a nonrandomized LRT of the form (2.21). Depending on the location of the threshold τ with respect to $L(0)$ and $L(1)$, three cases may occur:

i) Let δ_L be the test obtained by setting $\tau \leq L(0)$, which means we always select H_1. In this case $P_F(\delta_L) = P_D(\delta_L) = 1$, and expression (2.50) for the Bayes risk yields

$$R(\delta_L, \pi_0) = \pi_0 \, . \tag{2.100}$$

ii) Let δ_I be the test obtained when the threshold satisfies

$$L(0) < \tau \leq L(1) \, .$$

In this case the probabilities of false alarm and of detection are given by

$$P_F(\delta_I) = P[Y = 1|H_0] = q_0 \quad , \quad P_D(\delta_I) = P[Y = 1|H_1] = 1 - q_1 \, ,$$

and the Bayes risk is given by

$$R(\delta_I, \pi_0) = q_0\pi_0 + q_1(1 - \pi_0) \, . \tag{2.101}$$

iii) Let δ_H be the test corresponding to $\tau > L(1)$. In this case we always select H_0, so $P_F(\delta_H) = P_D(\delta_H) = 0$ and the Bayes risk is given by

$$R(\delta_H, \pi_0) = 1 - \pi_0 \, . \tag{2.102}$$

Note that the line (2.101) intersects the lines (2.100) and (2.102) at

$$\pi_L = \frac{q_1}{1 - q_0 + q_1} \quad \text{and} \quad \pi_U = \frac{1 - q_1}{1 + q_0 - q_1} ,$$

respectively, where the condition $q_0 + q_1 < 1$ implies $\pi_L < \pi_U$. The objective function $V(\pi_0)$ defined by (2.98) therefore takes the form

$$V(\pi_0) = \begin{cases} \pi_0 & \text{if } 0 \leq \pi_0 \leq \pi_L \\ (q_0 - q_1)\pi_0 + q_1 & \text{if } \pi_L \leq \pi_0 \leq \pi_U \\ 1 - \pi_0 & \text{if } \pi_U \leq \pi_0 \leq 1 , \end{cases}$$

which is plotted in Fig. 2.9 for the case $q_0 < q_1$. Since $V(\pi_0)$ is piecewise linear, its maximum occurs either at π_U or π_L, depending on whether q_0 is larger or smaller than q_1, so we have

$$\begin{aligned} \pi_{0M} = \pi_U \quad & V(\pi_{0M}) = \frac{q_0}{1 + q_0 - q_1} \quad \text{for } q_0 > q_1 \\ \pi_{0M} = \pi_L \quad & V(\pi_{0M}) = \frac{q_1}{1 - q_0 + q_1} \quad \text{for } q_0 < q_1 \\ \pi_L \leq \pi_{0M} \leq \pi_U \quad & V(\pi_{0M}) = q_0 \quad \text{for } q_0 = q_1 . \end{aligned} \tag{2.103}$$

The ROC is obtained by using randomization to connect the three points $(0, 0)$, $(q_0, 1 - q_1)$ and $(1, 1)$ corresponding to the tests δ_H, δ_I, and δ_L, respectively. This gives

$$P_D = \begin{cases} \dfrac{(1 - q_1)}{q_0} P_F & 0 \leq P_F \leq q_0 \\ \dfrac{q_1}{1 - q_0}(P_F - 1) + 1 & q_0 \leq P_F \leq 1 , \end{cases} \tag{2.104}$$

which is sketched in Fig. 2.10 for the case $q_0 < q_1$.

The minimax test corresponds to the intersection between the ROC and the line $P_D = 1 - P_F$, which is marked by point M in Fig. 2.10. For $q_0 \geq q_1$, we find

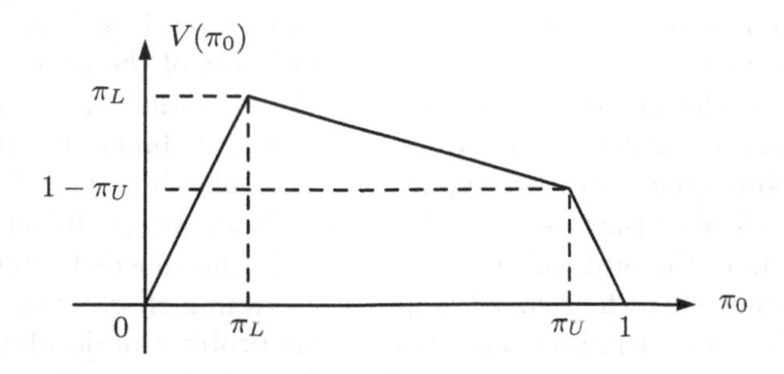

Fig. 2.9. Objective function $V(\pi_0)$ for a binary nonsymmetric channel with $q_0 < q_1$.

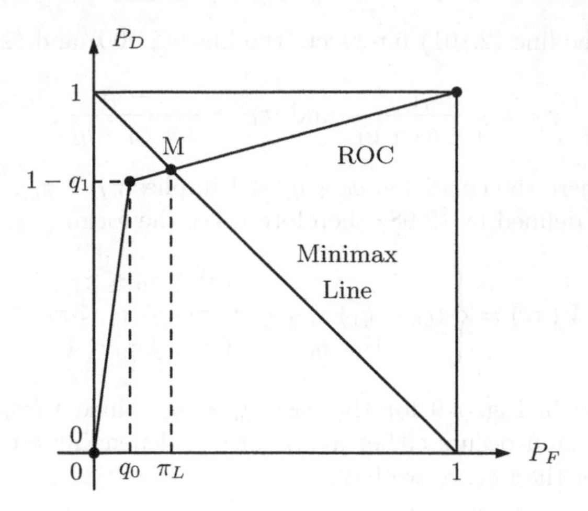

Fig. 2.10. Receiver operating characteristic and minimax receiver selection for a binary nonsymmetric channel with $q_0 < q_1$.

$$P_F(\delta_{\mathrm{M}}) = 1 - \pi_U \quad P_D(\delta_{\mathrm{M}}) = \pi_U \, ,$$

and for $q_0 \leq q_1$ we have

$$P_F(\delta_{\mathrm{M}}) = \pi_L \quad P_D(\delta_{\mathrm{M}}) = 1 - \pi_L \, .$$

To achieve test δ_{M} we must use randomization, except when $q_0 = q_1$. For example, when $q_0 < q_1$, as indicated in Fig. 2.10, the test δ_{M} is obtained by selecting the test δ_I with probability

$$p = \frac{1 - \pi_L}{1 - q_0} = \frac{1}{1 - q_0 + q_1} \, ,$$

and δ_L with probability $1 - p$. This means that we always select H_1 when $Y = 1$, but when $Y = 0$, we select H_0 with probability p and H_1 with probability $1 - p$. □

At this point, it is worth noting that the minimax receiver design technique we have presented above addresses only a small part of the uncertainty that may exist in the formulation of binary hypothesis testing problems, namely the uncertainty regarding priors. However, the probability distributions of the observations under the two hypotheses may also be affected by uncertainties. Theses uncertainties can take two distinct forms. In one case the functional form of the probability distributions is known exactly but depends on some parameters, such as signal amplitudes, timing offsets, noise variance, which are unknown, or imprecisely known. The problem of deciding between H_1 and H_0 then becomes a *composite hypothesis testing* problem, where either we must design tests which do not depend on the unknown parameters, or these parameters must be estimated prior to making a decision. Another form of uncertainty prevails when the functional form itself of the probability distribution of the observations is not known exactly, so only a nominal model is available, together with an uncertainty bound specifying the maximum possible deviation between the actual and the nominal models. Testing between H_1 and H_0 then becomes a *robust hypothesis testing* problem, where the minimax viewpoint can again be adopted to determine the best test for the least-favorable model compatible with the uncertainty affecting the nominal model. Composite and robust tests will be discussed in Chapters 5 and 6, respectively.

2.6 Gaussian Detection

Several of the examples considered earlier involved Gaussian densities with a special structure. In this section we consider the general case where $\mathbf{Y} \in \mathbb{R}^n$ admits the distributions

$$\begin{aligned} H_0 &: \mathbf{Y} \sim N(\mathbf{m}_0, \mathbf{K}_0) \\ H_1 &: \mathbf{Y} \sim N(\mathbf{m}_1, \mathbf{K}_1) \,. \end{aligned} \tag{2.105}$$

By substituting in the LRT (2.21) the expression

$$f(\mathbf{y}|H_j) = \frac{1}{(2\pi)^{n/2}|\mathbf{K}_j|^{1/2}} \exp\Big[-(\mathbf{y}-\mathbf{m}_j)^T \mathbf{K}_j^{-1} (\mathbf{y}-\mathbf{m}_j)/2 \Big] \tag{2.106}$$

with $j = 0,\ 1$, where $|\mathbf{K}|_j$ denotes the determinant of a matrix $\mathbf{K}_j$, and taking logarithms, we obtain

$$\begin{aligned} \ln(f(\mathbf{y}|H_1)) - \ln(f(\mathbf{y}|H_0)) &= \frac{1}{2}\Big[(\mathbf{y}-\mathbf{m}_0)^T \mathbf{K}_0^{-1} (\mathbf{y}-\mathbf{m}_0) \\ -(\mathbf{y}-\mathbf{m}_1)^T \mathbf{K}_1^{-1} (\mathbf{y}-\mathbf{m}_1)\Big] &- \frac{1}{2}\ln(|\mathbf{K}_1|/|\mathbf{K}_0|) \underset{H_0}{\overset{H_1}{\gtrless}} \gamma = \ln(\tau) \,. \end{aligned} \tag{2.107}$$

This reduces to the test

$$S(\mathbf{Y}) \underset{H_0}{\overset{H_1}{\gtrless}} \eta \tag{2.108}$$

where $S(\mathbf{Y})$ denotes the sufficient statistic

$$S(\mathbf{Y}) = \frac{1}{2}\mathbf{Y}^T(\mathbf{K}_0^{-1} - \mathbf{K}_1^{-1})\mathbf{Y} + \mathbf{Y}^T(\mathbf{K}_1^{-1}\mathbf{m}_1 - \mathbf{K}_0^{-1}\mathbf{m}_0) \tag{2.109}$$

and the threshold η is given by

$$\eta = \frac{1}{2}\Big[\mathbf{m}_1^T \mathbf{K}_1^{-1} \mathbf{m}_1 - \mathbf{m}_0^T \mathbf{K}_0^{-1} \mathbf{m}_0 + \ln(|\mathbf{K}_1|/|\mathbf{K}_0|)\Big] + \ln(\tau) \,. \tag{2.110}$$

Note that $S(\mathbf{Y})$ is the sum of a quadratic term and a linear term. Then two important special cases arise for which the detector admits further simplifications.

2.6.1 Known Signals in Gaussian Noise

Suppose the covariances $\mathbf{K}_0 = \mathbf{K}_1 = \mathbf{K}$. This represents a situation where hypothesis H_j can be expressed as

$$\mathbf{Y} = \mathbf{m}_j + \mathbf{V} \tag{2.111}$$

where the noise vector $\mathbf{V} \sim N(\mathbf{0}, \mathbf{K})$. This is the problem we obtain if under H_j we observe a known discrete-time signal $m_j(t)$ in zero-mean colored Gaussian noise $V(t)$ over a finite interval, i.e.,

$$Y(t) = m_j(t) + V(t)$$

over $0 \leq t \leq T$, where the autocorrelation function of $V(t)$ is the same under both hypotheses, and if we regroup all observations in a single vector $\mathbf{Y}$. In this case the quadratic term drops out in $S(\mathbf{Y})$, and if

$$\Delta\mathbf{m} \triangleq \mathbf{m}_1 - \mathbf{m}_0 \,, \tag{2.112}$$

the test (2.108) reduces to

$$S(\mathbf{Y}) = \mathbf{Y}^T \mathbf{K}^{-1} \Delta\mathbf{m} \underset{H_0}{\overset{H_1}{\gtrless}} \eta \,. \tag{2.113}$$

Then if

$$d^2 \triangleq \Delta\mathbf{m}^T \mathbf{K}^{-1} \Delta\mathbf{m} \,, \tag{2.114}$$

the shifted sufficient statistic

$$S_s(\mathbf{Y}) = S(\mathbf{Y}) - \mathbf{m}_0 \mathbf{K}^{-1} \Delta\mathbf{m} \tag{2.115}$$

admits the probability densities

$$\begin{aligned} H_0 &: S_s(\mathbf{Y}) \sim N(0, d^2) \\ H_1 &: S_s(\mathbf{Y}) \sim N(d^2, d^2) \,, \end{aligned} \tag{2.116}$$

which are of the form considered in (2.66) with $N = 1$, $E = \sigma^2 = d^2$. The probabilities of detection and of false alarm for this test are therefore given by (2.70) and the ROC satisfies (2.71). The key quantity in this test is the "distance" d specified by (2.114). To interpret this distance note first that when the additive Gaussian noise is white, so that $\mathbf{K} = \sigma^2 \mathbf{I}_n$, d can be expressed as

$$d = ||\Delta\mathbf{m}||_2 / \sigma \tag{2.117}$$

where $||\mathbf{u}||_2 = (\mathbf{u}^T \mathbf{u})^{1/2}$ denotes the Euclidean norm of a vector $\mathbf{u}$. So in this case, d is just the distance between the two vectors $\mathbf{m}_1$ and $\mathbf{m}_0$ measured with respect to the standard deviation σ of the noise.

Now, let us return to the general case where $\mathbf{K}$ is an arbitrary positive definite covariance matrix. It admits an eigenvalue/eigenvector decomposition of the form

$$\mathbf{K} = \mathbf{P}\mathbf{\Lambda}\mathbf{P}^T \tag{2.118}$$

where the matrices

$$\mathbf{P} = \begin{bmatrix} \mathbf{p}_1 \dots \mathbf{p}_i \dots \mathbf{p}_n \end{bmatrix} \ , \quad \mathbf{\Lambda} = \mathrm{diag}\,\{\lambda_i, 1 \leq i \leq n\} \tag{2.119}$$

are obtained by collecting the eigenvalue/eigenvector pairs $(\lambda_i, \mathbf{p}_i)$ with $1 \leq i \leq n$ satisfying

$$\mathbf{K}\mathbf{p}_i = \lambda_i \mathbf{p}_i \ . \tag{2.120}$$

Note that since $\mathbf{K}$ is symmetric positive definite, the eigenvalues λ_i are real and positive, and by orthogonalizing the eigenvectors corresponding to repeated eigenvalues, we can always ensure that the matrix $\mathbf{P}$ is orthonormal, i.e.,

$$\mathbf{P}^T\mathbf{P} = \mathbf{P}\mathbf{P}^T = \mathbf{I}_n \ . \tag{2.121}$$

Then if

$$\mathbf{\Lambda}^{-1/2} \triangleq \mathrm{diag}\,\{\lambda_i^{-1/2}, 1 \leq i \leq n\} \ , \tag{2.122}$$

by premultiplying the observation equation (2.111) by $\mathbf{\Lambda}^{-1/2}\mathbf{P}^T$, we obtain

$$\bar{\mathbf{Y}} = \bar{\mathbf{m}}_j + \bar{\mathbf{V}} \tag{2.123}$$

with

$$\bar{\mathbf{m}}_j \triangleq \mathbf{\Lambda}^{-1/2}\mathbf{P}^T\mathbf{m}_j \ \text{ and } \ \bar{\mathbf{V}} \triangleq \mathbf{\Lambda}^{-1/2}\mathbf{P}^T\mathbf{V} \ , \tag{2.124}$$

where $\bar{\mathbf{V}} \sim N(\mathbf{0}, \mathbf{I}_m)$, so the noise is now white with unit variance. If

$$\Delta\bar{\mathbf{m}} \triangleq \bar{\mathbf{m}}_1 - \bar{\mathbf{m}}_0 = \mathbf{\Lambda}^{-1/2}\mathbf{P}^T\Delta\mathbf{m} \tag{2.125}$$

the squared distance for the transformed observation is obviously

$$d^2 = ||\Delta\bar{\mathbf{m}}||_2^2 \ , \tag{2.126}$$

which reduces to (2.114) after substitution of (2.125). Thus d in (2.114) should be interpreted as the Euclidean distance between the transformed signals $\bar{\mathbf{m}}_j$ after the noise has been whitened and scaled so its variance is unity.

2.6.2 Detection of a Zero-Mean Gaussian Signal in Noise

Suppose the means $\mathbf{m}_0 = \mathbf{m}_1 = \mathbf{0}$ and $\mathbf{K}_1 > \mathbf{K}_0$. This situation corresponds to an observation model of the form

$$\begin{aligned} H_0 &: \mathbf{Y} = \mathbf{V} \\ H_1 &: \mathbf{Y} = \mathbf{Z} + \mathbf{V} \end{aligned} \tag{2.127}$$

where $\mathbf{Z} \sim N(\mathbf{0}, \mathbf{K}_Z)$ is a zero-mean Gaussian signal independent of the zero-mean Gaussian noise $\mathbf{V} \sim N(\mathbf{0}, \mathbf{K}_V)$. This situation occurs commonly in defense applications where, in order to avoid detection, signals often employ spread spectrum modulation and appear "noise-like" to an eavesdropper. In this case $\mathbf{K}_1 = \mathbf{K}_Z + \mathbf{K}_V > \mathbf{K}_V = \mathbf{K}_0$, and the sufficient statistic $S(\mathbf{Y})$ becomes

$$S(\mathbf{Y}) = \mathbf{Y}^T[\mathbf{K}_V^{-1} - (\mathbf{K}_Z + \mathbf{K}_V)^{-1}]\mathbf{Y}/2 = \mathbf{Y}^T\mathbf{K}_V^{-1}\hat{\mathbf{Z}}/2 \tag{2.128}$$

where, as we shall see in Chapter 4,

$$\hat{\mathbf{Z}} \triangleq \mathbf{K}_Z(\mathbf{K}_Z + \mathbf{K}_V)^{-1}\mathbf{Y} \tag{2.129}$$

is the least-squares estimate of $\mathbf{Z}$ given the observation $\mathbf{Y}$. The expression (2.128) for the sufficient statistic $S(\mathbf{Y})$ has the feature that, except for the scaling factor $1/2$, it is similar to expression (2.113) for the sufficient statistic in the known signal case, with say, $\mathbf{m}_0 = \mathbf{0}$. The only difference is that the known value $\mathbf{m}_1$ of the signal under hypothesis H_1 is now replaced by its estimated value $\hat{\mathbf{Z}}$ based on the received observation vector $\mathbf{Y}$. In the WGN case where $\mathbf{K}_V = \sigma^2\mathbf{I}_n$, the structure (2.128) is called the *estimator-correlator* implementation of the optimum receiver.

2.7 M-ary Hypothesis Testing

While some engineering applications of hypothesis testing, like radar or sonar, typically involve only two hypotheses, others require the use of more than 2 hypotheses. For example, in high-performance digital communication systems, higher order modulation schemes, such as M-PSK or M-QAM with $M = 2^k$ are often employed [2]. In such applications, each received observation Y contains information about k bits which need to be detected simultaneously. Similarly, for image processing applications involving either texture segmentation or texture recognition, the number of possible textures is usually higher than two.

2.7.1 Bayesian M-ary Tests

The Bayesian formulation of such problems assumes there are M hypotheses H_j with $0 \leq j \leq M-1$ with a-priori probabilities $\pi_j = P[H_j]$. The cost of selecting H_i when H_j holds is denoted as C_{ij} with $0 \leq i, j \leq M-1$. Finally, depending on whether the observed random vector $\mathbf{Y} \in \mathcal{Y}$ is continuous-valued or discrete-valued, we assume that under H_j it either admits the probability density $f(\mathbf{y}|H_j)$ or the probability mass function $p(\mathbf{y}|H_j)$ for $0 \leq j \leq M-1$.

The hypothesis testing problem consists of finding an M-ary valued decision function $\delta(\mathbf{y}) \in \{0, \dots, i, \dots, M-1\}$ such that $\delta(\mathbf{y}) = i$ if we decide that hypothesis H_i holds when $\mathbf{Y} = \mathbf{y}$. If

$$\mathcal{Y}_i = \{\mathbf{y} : \delta(\mathbf{y}) = i\} ,$$

the decision function δ specifies therefore an M-fold partition

$$\mathcal{Y} = \cup_{i=0}^{M-1} \mathcal{Y}_i \text{ with } \mathcal{Y}_i \cap \mathcal{Y}_k = \emptyset \text{ for } i \neq k$$

of $\mathcal{Y}$. When $\mathbf{Y}$ is continuous-valued, the Bayes risk for the decision function δ can be expressed as

$$\begin{aligned} R(\delta) &= \sum_{i=0}^{M-1} \sum_{j=0}^{M-1} C_{ij} P[\mathcal{Y}_i | H_j] \pi_j \\ &= \sum_{i=0}^{M-1} \sum_{j=0}^{M-1} C_{ij} \int_{\mathcal{Y}_i} f(\mathbf{y}|H_j) \pi_j d\mathbf{y} \\ &= \sum_{i=0}^{M-1} \int_{\mathcal{Y}_i} \sum_{j=0}^{M-1} C_{ij} P[H_j | \mathbf{Y} = \mathbf{y}] f(\mathbf{y}) d\mathbf{y} \\ &= \sum_{j=0}^{M-1} \int_{\mathcal{Y}_i} C_i(\mathbf{y}) f(\mathbf{y}) d\mathbf{y} \end{aligned} \tag{2.130}$$

where

$$C_i(\mathbf{y}) \triangleq \sum_{j=0}^{M-1} C_{ij} P[H_j | \mathbf{Y} = \mathbf{y}] \tag{2.131}$$

represents the average cost of deciding H_i holds when $\mathbf{Y} = \mathbf{y}$. From expression (2.130), we deduce that the optimum Bayesian decision rule is given by

$$\delta_{\mathrm{B}}(\mathbf{y}) = \arg \min_{0 \leq i \leq M-1} C_i(\mathbf{y}) . \tag{2.132}$$

In other words, when $\mathbf{Y} = \mathbf{y}$, we select hypothesis H_k if

$$C_k(\mathbf{y}) < C_i(\mathbf{y}) \tag{2.133}$$

for all $i \neq k$. When the minimum cost at point $\mathbf{y}$ is achieved by two different hypotheses, say H_k and H_ℓ, $\mathbf{y}$ can be allocated arbitrarily to either $\mathcal{Y}_k$ or $\mathcal{Y}_\ell$. The decision rule (2.132) remains valid when $\mathbf{Y}$ is discrete-valued. The only difference is that the a-posteriori probability $P[H_j|\mathbf{y}]$ of H_j given $\mathbf{Y} = \mathbf{y}$ becomes

$$P[H_j | \mathbf{Y} = \mathbf{y}] = \frac{p(\mathbf{y}|H_j)\pi_j}{p(\mathbf{y})} \tag{2.134}$$

with

$$p(\mathbf{y}) = \sum_{j=0}^{M-1} p(\mathbf{y}|H_j) \pi_j , \tag{2.135}$$

where $p(\mathbf{y}|H_j)$ is the probability mass distribution of $\mathbf{Y}$ under hypothesis H_j.

The decision rule (2.132) can be simplified further if we assume that the cost function has a particular structure.

Minimum error probability/MAP decision rule: If

$$C_{ij} = 1 - \delta_{ij} , \tag{2.136}$$

the average cost function $C_i(\mathbf{y})$ becomes

$$C_i(\mathbf{y}) = \sum_{k \neq i} P[H_k|\mathbf{y}] = 1 - P[H_i|\mathbf{y}] , \tag{2.137}$$

so instead of obtaining δ_{B} by minimizing the costs $C_i(\mathbf{y})$, we can equivalently maximize the a-posteriori probability $P[H_i|\mathbf{y}]$. This gives the MAP decision rule

$$\delta_{\mathrm{MAP}}(\mathbf{y}) = \arg \max_{0 \leq i \leq M-1} P[H_i|\mathbf{Y} = \mathbf{y}] . \tag{2.138}$$

The marginal density $f(\mathbf{y})$ appearing in the denominator of the Bayes expression

$$P[H_i|\mathbf{Y} = \mathbf{y}] = \frac{f(\mathbf{y}|H_i)\pi_i}{f(\mathbf{y})} \tag{2.139}$$

when $\mathbf{Y}$ is continuous-valued is the same for all hypotheses, so the removal of this factor does not affect the maximization (2.133). Consequently, the decision rule can be rewritten as

$$\begin{aligned} \delta_{\mathrm{MAP}}(\mathbf{y}) &= \arg \max_{0 \leq i \leq M-1} f(\mathbf{y}|H_i)\pi_i \\ &= \arg \max_{0 \leq i \leq M-1} [\ln(f(\mathbf{y}|H_i)) + \ln(\pi_i)] , \end{aligned} \tag{2.140}$$

where in (2.140) we have used the fact that the function $\ln(x)$ is monotone increasing and thus preserves ordering.

ML decision rule: When all hypotheses are equally likely, i.e.,

$$\pi_i = 1/M$$

for all i, the decision rule (2.140) reduces to

$$\begin{aligned} \delta_{\mathrm{ML}}(\mathbf{y}) &= \arg \max_{0 \leq i \leq M-1} f(\mathbf{y}|H_i) \\ &= \arg \max_{0 \leq i \leq M-1} \ln(f(\mathbf{y}|H_i)) . \end{aligned} \tag{2.141}$$

As a special case, consider the case when the observation vector $\mathbf{Y} \in \mathbb{R}^n$ admits a Gaussian density

$$H_i \; : \; \mathbf{Y} \sim N(\mathbf{m}_i, \mathbf{K}) , \tag{2.142}$$

for $0 \leq i \leq M-1$, where the covariance $\mathbf{K}$ is the same for all hypotheses. Then

$$\ln(f(\mathbf{y}|H_i)) = -\frac{1}{2}(\mathbf{y} - \mathbf{m}_i)^T \mathbf{K}^{-1}(\mathbf{y} - \mathbf{m}_i) - \frac{n}{2}\ln(2\pi) - \frac{1}{2}\ln(|\mathbf{K}|) \,, \quad (2.143)$$

and by observing that maximizing a function is equivalent to minimizing its negative, and discarding constant additive terms, the ML decision rule (2.141) becomes

$$\delta_{\mathrm{ML}}(\mathbf{y}) = \arg\min\, (\mathbf{y} - \mathbf{m}_i)^T \mathbf{K}^{-1}(\mathbf{y} - \mathbf{m}_i) \quad (2.144)$$

which, in the special case when $\mathbf{K} = \sigma^2 \mathbf{I}_n$ is the covariance of a WGN, reduces to the usual *minimum distance rule*

$$\delta_{\mathrm{ML}}(\mathbf{y}) = \arg\min\, ||\mathbf{y} - \mathbf{m}_i||_2^2 \quad (2.145)$$

for the detection of linearly modulated signals in WGN. Note that when the noise covariance $\mathbf{K}$ is arbitrary, by performing the eigenvalue/eigenvector decomposition (2.118) of $\mathbf{K}$ and premultiplying the observations $\mathbf{Y}$ by $\mathbf{\Lambda}^{-1/2}\mathbf{P}^T$, we can apply the minimum distance rule (2.145) to the transformed observations $\bar{\mathbf{Y}}$ and signals $\bar{\mathbf{m}}_i$. In effect, in the Gaussian case, by performing a linear noise whitening transformation, we can always convert a colored noise problem into an equivalent white noise problem.

Example 2.7: QPSK modulation

In quadrature phase-shift keying modulation (QPSK) [2, pp. 177–178], one of four equally likely signals

$$s_i(t) = (2E/T)^{1/2}\cos(\omega_c t + \theta_i)$$

with phase $\theta_i = \pi/4 + (i-1)\pi/2$ for $0 \leq i \leq 3$ is transmitted over an interval $0 \leq t \leq T$. E denotes here the signal energy over the transmission interval $[0, T]$. We observe

$$Y(t) = s_i(t) + V(t)$$

for $0 \leq t \leq T$, where $V(t)$ denotes an additive white Gaussian noise. Then, as will be shown in Chapter 8, by correlating the observed signal $Y(t)$ with the in-phase and quadrature components of the carrier waveform, the QPSK detection problem can be transformed into a 4-ary hypothesis testing problem in two-dimensional space. In this context, the four hypotheses H_i with $0 \leq i \leq 3$ are equally likely, and the observation vector $\mathbf{Y} \in \mathbb{R}^2$ can be expressed as

$$\mathbf{Y} = \begin{bmatrix} Y_1 \\ Y_2 \end{bmatrix} = \mathbf{m}_i + \mathbf{V}$$

with

$$\mathbf{m}_i = \left(\frac{E}{2}\right)^{1/2} \begin{bmatrix} \pm 1 \\ \pm 1 \end{bmatrix} ,$$

where the noise $\mathbf{V} \sim N(\mathbf{0}, \sigma^2 \mathbf{I}_2)$. The minimimum distance rule for this problem is illustrated in Fig. 2.11. We choose $\mathbf{m}_0$, $\mathbf{m}_1$, $\mathbf{m}_2$, and $\mathbf{m}_3$ depending on whether $\mathbf{Y}$ falls in the first, second, third or fourth quadrant of $\mathbb{R}^2$. □

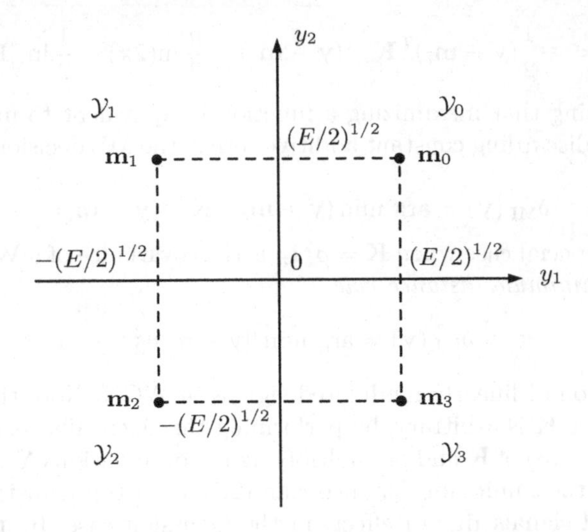

Fig. 2.11. Decision regions for QPSK detection.

2.7.2 Sufficient Statistics for M-ary Tests

By multiplying the costs $C_i(\mathbf{y})$ by the scaling factor $f(\mathbf{y})/f(\mathbf{y}|H_0)$ which is independent of index i, we obtain the reduced costs

$$\begin{aligned}\bar{C}_i(\mathbf{y}) &= C_i(\mathbf{y})\frac{f(\mathbf{y})}{f(\mathbf{y}|H_0)} \\ &= C_{i0}\pi_0 + \sum_{j=1}^{M-1} C_{ij}\pi_j L_j(\mathbf{y}) \,, \end{aligned} \tag{2.146}$$

which are expressed in terms of the likelihood ratios

$$L_j(\mathbf{y}) = \frac{f(\mathbf{y}|H_j)}{f(\mathbf{y}|H_0)} \,, \tag{2.147}$$

with $1 \leq j \leq M-1$, and the Bayesian decision rule (2.132) can be expressed as

$$\delta_{\mathrm{B}}(\mathbf{y}) = \arg \min_{0 \leq i \leq M-1} \bar{C}_i(\mathbf{y}) \,. \tag{2.148}$$

The expression (2.146)–(2.148) of the Bayesian test indicates that the $M-1$ likelihood ratios $L_j(\mathbf{Y})$ form a sufficient statistic for the test, which is consistent with the $M=2$ case for which we saw earlier that $L(\mathbf{Y})$ forms a scalar sufficient statistic.

The expression of the test as a minimization over the costs $\bar{C}_i(\mathbf{Y})$ indicates that the optimum decision rule can be implemented as a sequence of $M-1$ comparisons for which one hypothesis is eliminated after each stage. For example, the comparison

$$\bar{C}_i(\mathbf{Y}) - \bar{C}_k(\mathbf{Y}) = (C_{i0} - C_{k0})\pi_0 + \sum_{j=1}^{M-1}(C_{ij} - C_{kj})\pi_j L_j(\mathbf{Y}) \underset{\text{not } H_k}{\overset{\text{not } H_i}{\gtrless}} 0 \quad (2.149)$$

eliminates either H_i and H_k. Since the cost difference $\bar{C}_i - \bar{C}_k$ depends linearly on the likelihood ratios L_j, the elimination of H_i or H_k is achieved by deciding whether the point $(L_1(\mathbf{Y}), \dots, L_{M-1}(\mathbf{Y}))$ is located on one side or the other of the hyperplane in $(L_1, L_2, \dots, L_{M-1})$ space specified by $\bar{C}_i - \bar{C}_k = 0$.

There are $M(M-1)/2$ possible comparisons $\bar{C}_i - \bar{C}_k$ with $i > k$, but since one hypothesis is eliminated after each comparison, only $M-1$ of these comparisons need to be implemented. However, except for the first comparison, which can be picked arbitrarily, say $\bar{C}_1 - \bar{C}_0$, it is not possible ahead of time to know which of the $M(M-1)/2$ possible comparisons will need to be applied. A simple way to implement the comparison procedure consists in comparing at each stage the two hypotheses with lowest index i that have not been eliminated yet. This method specifies a binary decision tree of depth $M-1$ where each node of the tree correspond to a $\bar{C}_i - \bar{C}_k$ comparison, and the branches emerging from a node correspond to the two possible outcomes of the comparison. Starting from the root node $\bar{C}_1 - \bar{C}_0$, the tree is traversed until a leaf is reached, corresponding to a decision, say $d_B = i$ for some i, as shown in Fig. 2.12 for the case $M = 3$. □

The above discussion assumes that comparisons are performed sequentially. If multiple comparisons can be evaluated simultaneously, we could of course perform all $M(M-1)/2$ possible comparisons at once, but this would be inefficient, since a large fraction of these comparisons is unneeded. If M is a power of 2, an optimum scheme with a latency of only $\log_2(M)$ consists of performing $M/2$ comparisons at the first stage, then for the $M/2$ survivors, $M/4$ comparisons are performed at the second stage, until only one survivor remains after $\log_2(M)$ stages. To illustrate the above discussion, we consider a simple example with $M = 3$.

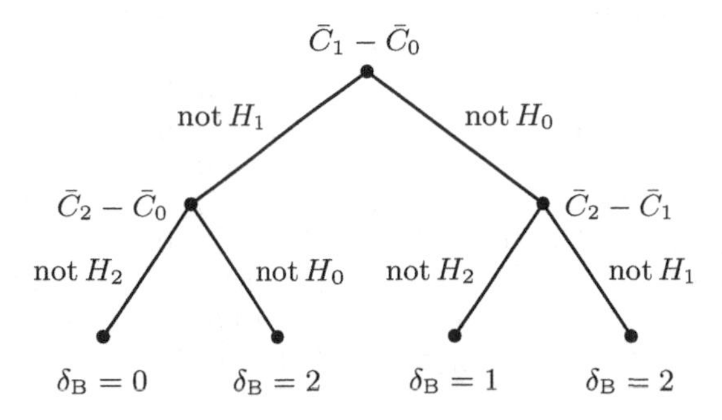

Fig. 2.12. Hypothesis elimination scheme for M-ary hypothesis testing.

Example 2.8: Orthogonal signals in 3-D space

Consider a ternary hypothesis testing problem such that under hypothesis H_i with $i = 0,\ 1,\ 2$ we observe

$$\mathbf{Y} = \mathbf{m}_i + \mathbf{V}$$

where the noise $\mathbf{V} \sim N(\mathbf{0}, \sigma^2 \mathbf{I}_3)$ and the signals

$$\mathbf{m}_0 = E^{1/2} \begin{bmatrix} 1 \\ 0 \\ 0 \end{bmatrix} \quad , \quad \mathbf{m}_1 = E^{1/2} \begin{bmatrix} 0 \\ 1 \\ 0 \end{bmatrix} \quad , \quad \mathbf{m}_2 = E^{1/2} \begin{bmatrix} 0 \\ 0 \\ 1 \end{bmatrix}$$

are orthogonal with energy E. Assume that the hypotheses have a-priori probabilities

$$\pi_0 = \frac{1}{4}\,,\ \pi_1 = \frac{1}{4}\,,\ \text{and}\ \pi_2 = \frac{1}{2}\,,$$

so that hypothesis H_2 is twice as likely as the other two hypotheses. We seek to minimize the probability of error. In this case, by equation (2.147), the two likelihood ratios are given by

$$L_1(\mathbf{y}) = \exp\left(\frac{E^{1/2}}{\sigma^2}(y_2 - y_1)\right) \quad , \quad L_2(\mathbf{y}) = \exp\left(\frac{E^{1/2}}{\sigma^2}(y_3 - y_1)\right)$$

and the reduced costs can be expressed as

$$\begin{aligned}
\bar{C}_0(\mathbf{y}) &= \frac{1}{4}L_1(\mathbf{y}) + \frac{1}{2}L_2(\mathbf{y}) \\
\bar{C}_1(\mathbf{y}) &= \frac{1}{4} + \frac{1}{2}L_2(\mathbf{y}) \\
\bar{C}_2(\mathbf{y}) &= \frac{1}{4} + \frac{1}{4}L_1(\mathbf{y}).
\end{aligned}$$

Then the three comparisons,

$$\begin{aligned}
\bar{C}_1 - \bar{C}_0 &= \frac{1}{4}(1 - L_1) \underset{\text{not } H_0}{\overset{\text{not } H_1}{\gtrless}} 0 \\
\bar{C}_2 - \bar{C}_0 &= \frac{1}{2}(\frac{1}{2} - L_2) \underset{\text{not } H_0}{\overset{\text{not } H_2}{\gtrless}} 0 \\
\bar{C}_2 - \bar{C}_1 &= \frac{1}{2}(\frac{1}{2}L_1 - L_2) \underset{\text{not } H_1}{\overset{\text{not } H_2}{\gtrless}} 0.
\end{aligned}$$

specify three lines in the (L_1, L_2) plane intersecting at point $(L_1 = 1,\ L_2 = 1/2)$, as shown in Fig. 2.13. These three lines specify the boundaries of the

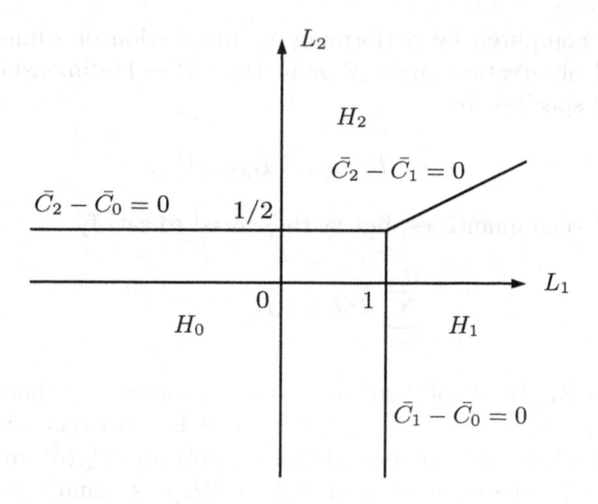

Fig. 2.13. Decision regions in the (L_1, L_2) plane.

decision regions corresponding to the choice of H_0, H_1 and H_2, which are located in the bottom left, bottom right, and top left portions of the plane. It is worth noting that if the prior distribution for all three hypotheses had been uniform, i.e., $\pi_i = 1/3$ for all i, the boundary between H_0 and H_1 would be the same, but the boundaries between H_0 and H_2 would be the line $L_2 = 1$ and the boundary between H_1 and H_2 would be the line $L_1 = L_2$. So a change in priors manifests itself not only in a shift of boundary lines, as is the case in binary hypothesis testing when the threshold is raised or lowered when the priors are changed, but also by a tilt of boundary lines, since the line $L_1 = L_2$ which is valid for the uniform prior case becomes $L_2 = L_1/2$ when hypothesis H_2 is twice as likely as H_1. □

2.7.3 Performance Analysis

To evaluate the performance of a test δ, we express the Bayes risk as

$$R(\delta) = \sum_{j=0}^{M-1} R(\delta|H_j)\pi_j \tag{2.150}$$

with

$$R(\delta|H_j) = \sum_{i=0}^{M-1} C_{ij} P[\delta = i|H_j] \,. \tag{2.151}$$

So the performance of a test δ is characterized entirely by the quantitities

$$P[\delta = i|H_j] = P[\mathcal{Y}_i|H_j] \tag{2.152}$$

which can be computed by performing an integration or summation in the n-dimensional observation space $\mathcal{Y}$ or in the $(M-1)$-dimensional sufficient statistic space specified by

$$\mathbf{L} = \left[L_1 \, L_2 \dots L_{M-1} \right]^T .$$

There are M^2 such quantities, but as they need to satisfy

$$\sum_{i=0}^{M-1} P[\delta = i | H_j] = 1 \tag{2.153}$$

for each j, only $M(M-1)$ of them need to be evaluated, say those with indices $1 \leq i \leq M-1$ and $0 \leq j \leq M-1$. This result is consistent with the binary ($M=2$) case, where only the probabilities $P_F(\delta)$ and $P_D(\delta)$ are required.

Unfortunately, the evaluation of $P[\delta = i|H_j]$ is usually rather difficult to perform since it requires an integration in higher dimensional space over either $f(\mathbf{y}|H_j)$ or over the probability density $f_{\mathbf{L}}(\boldsymbol{\ell}|H_j)$ of the vector sufficient statistic $\mathbf{L}$. Futhermore, the regions of integration are typically polygons or polygonal cones, so symmetries are not always available to reduce the number of variables to be integrated.

Some simplifications arise when the costs admit symmetries. For the minimum probability of error cost function $C_{ij} = 1 - \delta_{ij}$, the risk under hypothesis H_j can be expressed as

$$R(\delta|H_j) = P[\delta \neq j|H_j] = P[E|H_j] \, , \tag{2.154}$$

so in this case only the M probabilities of error $P[E|H_j]$ need to be evaluated. If the hypotheses are equally likely, i.e., $\pi_j = 1/M$ for all j, and the probability densities $f(\mathbf{y}|H_j)$ under the different hypotheses admit symmetries, this number can be reduced even further. Specifically, suppose that there exists a finite group

$$\mathcal{G} = \{(P_i, T_i), \ i \in \mathcal{I}\} \tag{2.155}$$

indexed by $\mathcal{I}$ which is formed by pairs (P_i, T_i), where P_i denotes a permutation of the index set $\{0, \ 1, \dots, \ M-1\}$ and T_i is an orthonormal transformation of $\mathbb{R}^n$, such that for a pair (P, T) in the group we have

$$f(T(\mathbf{y})|H_{P(k)}) = f(\mathbf{y}|H_k) \tag{2.156}$$

for all k. Then using the invariance property (2.156) in the ML decision rule (2.141), we find that the decision regions $\mathcal{Y}_k$ satisfy

$$\mathcal{Y}_{P(k)} = T(\mathcal{Y}_k) \, . \tag{2.157}$$

If $P[C|H_j]$ denotes the probability of a correct decision under hypothesis H_j, the property (2.157) implies

$$\begin{aligned} P[E|H_{P(k)}] &= 1 - P[C|H_{P(k)}] = 1 - P[\mathcal{Y}_{P(k)}|H_{P(k)}] \\ &= 1 - P[T(\mathcal{Y}_k)|H_{P(k)}] = 1 - P[\mathcal{Y}_k|H_k] \\ &= 1 - P[C|H_k] = P[E|H_k] \end{aligned} \tag{2.158}$$

for all k, where the second line uses identity (2.156). When the group of transformations is such that for every index i, there exists a pair (P, T) in the group with $i = P(0)$, which means that all hypotheses belong to the same congruence class, we have

$$P[E|H_i] = P[E|H_0] = P[E] \tag{2.159}$$

for all i, so only one probability of error needs to be evaluated.

Example 2.7, continued

For the QPSK detection problem, an invariance group $\mathcal{G}$ is formed by cyclic permutations of the indices $\{0, 1, 2, 3\}$ and rotations by integer multiples of $\pi/2$. Specifically, for a cyclic permutation P such that $i = P(0)$, the matching T is a rotation by $i\pi/2$ with matrix representation

$$\mathbf{Q} = \begin{bmatrix} \cos(i\pi/2) & -\sin(i\pi/2) \\ \sin(i\pi/2) & \cos(i\pi/2) \end{bmatrix} .$$

Then since $\mathbf{Q}\mathbf{m}_0 = \mathbf{m}_i$ and

$$\mathbf{Q}^T(\sigma^2\mathbf{I}_2)\mathbf{Q} = \sigma^2\mathbf{I}_2 ,$$

we have

$$f(T(\mathbf{y})|H_i) = f(\mathbf{y}|H_0) ,$$

so only one probability of error needs to be evaluted. To evaluate $P[E|H_0]$, we note that it can be expressed as

$$P[E|H_0] = 1 - P[C|H_0] ,$$

where the probability $P[C|H_0]$ of a correct decision is given by

$$\begin{aligned} P[C|H_0] &= P[Y_1 \geq 0, Y_2 \geq 0|H_0] = (P[Y_1 > 0|H_0])^2 \\ &= Q\Big(-\frac{E^{1/2}}{2^{1/2}\sigma}\Big)^2 = \Big(1 - Q\Big(\frac{d}{2}\Big)\Big)^2 . \end{aligned}$$

Here $d = (2E)^{1/2}/\sigma$ denotes the Euclidean distance between two adjacent signals, say $\mathbf{m}_1$ and $\mathbf{m}_0$, scaled by the noise standard deviation. This gives therefore

$$P[E] = 1 - \Big(1 - Q\Big(\frac{d}{2}\Big)\Big)^2 = 2Q\Big(\frac{d}{2}\Big) - Q^2\Big(\frac{d}{2}\Big) . \tag{2.160}$$

□

2.7.4 Bounds Based on Pairwise Error Probability

For the minimum probability of error criterion, we have seen that

$$P[E] = \sum_{j=0}^{M-1} P[E|H_j]\pi_j \tag{2.161}$$

with

$$P[E|H_j] \triangleq P[\mathcal{Y}_j^c|H_j] , \tag{2.162}$$

where $\mathcal{Y}_j^c$ denotes the complement of the decision region $\mathcal{Y}_j$. While integrating over the region $\mathcal{Y}_j^c$ can be difficult, it is useful to note that this region can be constructed as

$$\mathcal{Y}_j^c = \cup_{k \neq j} \mathcal{E}_{kj} , \tag{2.163}$$

where the pairwise error region

$$\begin{aligned} \mathcal{E}_{kj} &= \{\mathbf{y} : P[H_k|\mathbf{y}] > P[H_j|\mathbf{y}]\} \\ &= \{\mathbf{y} : L_{kj}(\mathbf{y}) > \pi_j/\pi_k\} \end{aligned} \tag{2.164}$$

is the domain of observations for which we prefer H_k over H_j, and where

$$L_{kj}(\mathbf{y}) \triangleq \frac{f(\mathbf{y}|H_k)}{f(\mathbf{y}|H_j)}$$

denotes the likelihood ratio for hypotheses H_k and H_j.

In expression (2.163), the error events $\mathcal{E}_{kj}$ usually overlap, and it is often possible to find a set of sufficient pairwise error events such that

$$\mathcal{Y}_j^c = \cup_{k \in N(j)} \mathcal{E}_{kj} , \tag{2.165}$$

with $N(j) \subset \{k : k \neq j, 0 \leq k \leq M-1\}$. In some sense the hypotheses H_k with $k \in N(j)$ are "neighbors" of j, which together cover all the possible error possibilities when we prefer one of them over H_j. Because the error events appearing in either (2.163) or (2.165) may overlap, we have the following error bound

$$\max_{k \neq j} P[\mathcal{E}_{kj}|H_j] \leq P[E|H_j] \leq \sum_{k \in N(j)} P[\mathcal{E}_{kj}|H_j] . \tag{2.166}$$

The right-hand side of this inequality represents an improved form of the *union bound*. When the set $N(j)$ is selected as $\{k : k \neq j\}$, it becomes the usual union bound. The bound (2.166) is very useful to evaluate the performance of detection systems, particularly when the set $N(j)$ of neighbors is selected carefully. The advantage of this bound is that it requires only the evaluation of pairwise error probabilities. But from (2.164), we see that pairwise errors are expressed as likelihood ratio tests, so they can be evaluated in the same way as the probability of false alarm and the probability of a miss of a binary test. In fact, for uniform priors and linearly modulated signals in WGN, we have

$$P[\mathcal{E}_{kj}|H_j] = Q(\frac{d_{kj}}{2}) \tag{2.167}$$

where d_{kj} denotes the distance between signals under hypotheses H_k and H_j.

Example 2.7, continued

Consider again the QPSK detection problem. In this case, it is easy to check that $N(0) = \{1, 3\}$ is a sufficient pairwise error neighborhood of H_0. Specifically, we make an error either when $Y_1 < 0$, in which case we prefer H_1 over H_0, or when $Y_2 < 0$, in which case H_3 is preferred over H_0. So

$$\mathcal{Y}_0^c = \mathcal{E}_{10} \cup \mathcal{E}_{30} .$$

Note that these two sets overlap since the region $\{Y_1 < 0, Y_2 < 0\}$ is included in both sets. Applying the union bound (2.160) yields

$$Q(\frac{d}{2}) \leq P[E|H_0] = P[E] \leq 2Q(\frac{d}{2}) .$$

While the lower bound is rather poor, the upper bound is tight for large d if we compare it to (2.160). □

2.8 Bibliographical Notes

The presentation of binary and M-ary hypothesis in this chapter is rather standard and covers approximately the same material as in [8–10]. Statistics texts on hypothesis testing [3,11] discuss the same topics, but electrical engineers give a greater emphasis to the ROC and, in the Gaussian case, to the notion of distance d between different hypotheses.

2.9 Problems

2.1. In the binary communication system shown in Fig. 2.14, the message values $X = 0$ and $X = 1$ occur with a-priori probabilities $1/4$ and $3/4$, respectively. The random variable V takes the values -1, 0, and 1 with probabilities $1/8$, $3/4$, $1/8$, respectively. The received message is $Y = X + V$.

(a) Given the received signal Y, the receiver must decide if the transmitted message was 0 or 1. The estimated message $\hat{X}$ takes the values 0 or 1. Find the receiver which achieves the maximum probability of a correct decision, i.e., which maximizes $P[\hat{X} = X]$.
(b) Find $P[\hat{X} = X]$ for the receiver of part a).

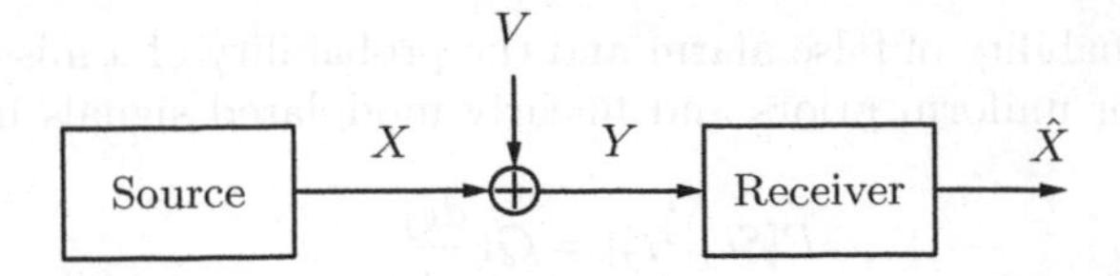

Fig. 2.14. Binary communication system model.

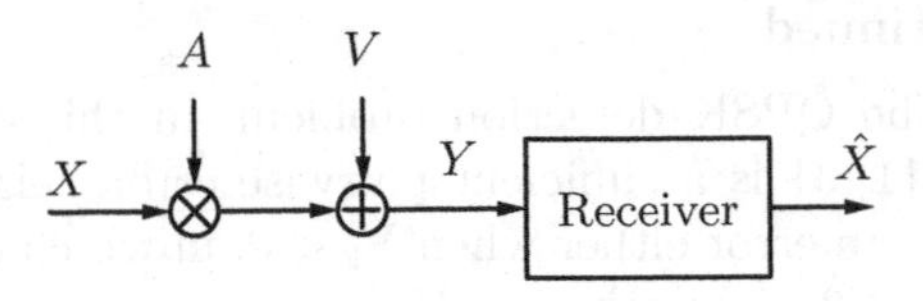

Fig. 2.15. Idealized model of a fading communication channel.

2.2. The system shown of Fig. 2.15 is an idealized model for binary communication through a fading channel. The message X that we want to transmit takes one of two values, 0 or 1, with a-priori probabilities 1/3 and 2/3, respectively. The channel fading A is a $N(1,3)$ Gaussian random variable, i.e., its mean $m_A = 1$ and variance $\sigma_A^2 = 3$. The channel noise V is a $N(0,1)$ Gaussian random variable

(a) Find the minimum probability of error decision rule. Simplify your answer as much as possible.
(b) Sketch the decision regions on a y-axis plot.

2.3. Consider the following binary hypothesis testing problem:

$$H_0 : Y = V$$
$$H_1 : Y = X + V ,$$

where X and V are two independent random variables with densities

$$f_X(x) = \begin{cases} a \exp(-ax) & x \geq 0 \\ 0 & x < 0 , \end{cases}$$

$$f_V(v) = \begin{cases} b \exp(-bv) & v \geq 0 \\ 0 & v < 0 . \end{cases}$$

(a) Prove that the likelihood ratio test reduces to

$$Y \underset{H_0}{\overset{H_1}{\gtrless}} \eta .$$

(b) Find the threshold η for the optimum Bayes test as a function of the costs and a-priori probabilities.

2.4. Consider hypotheses:

$$H_0 : f_Y(y) = \frac{1}{\sqrt{2\pi}} \exp\left(-\frac{1}{2}y^2\right),$$
$$H_1 : f_Y(y) = \frac{1}{2} \exp(-|y|).$$

a) Find the likelihood ratio $L(y)$.
b) The test takes the form

$$L(Y) \underset{H_0}{\overset{H_1}{\gtrless}} \tau.$$

Compute the decision regions for various values of τ.

2.5. Let V_k with $1 \leq k \leq N$ be a sequence of independent identically distributed Laplacian random variables with density

$$f_V(v) = \frac{\lambda}{2} \exp(-\lambda|v|),$$

where the parameter $\lambda > 0$ controls the variance

$$E[V^2] = \frac{2}{\lambda^2}$$

of V. Then consider the binary hypothesis testing problem

$$H_0 : Y_k = V_k$$
$$H_1 : Y_k = A + V_k,$$

for $1 \leq k \leq N$, where A denotes a known positive constant. Consider the symmetric clipping function

$$c(z) = \begin{cases} A & z \geq A/2 \\ 2z & -A/2 \leq z \leq A/2 \\ -A & z \leq -A/2 \end{cases}$$

which is shown in Fig. 2.16. Note that it depends on A.

(a) Show that

$$S = \frac{1}{N} \sum_{k=1}^{N} c(Y_k - A/2)$$

is a sufficient statistic for the LRT, i.e., verify that the LRT can be expressed as

$$S \underset{H_0}{\overset{H_1}{\gtrless}} \eta.$$

(b) Suppose that the two hypotheses are equally likely. Select the threshold η to minimize the probability of error.

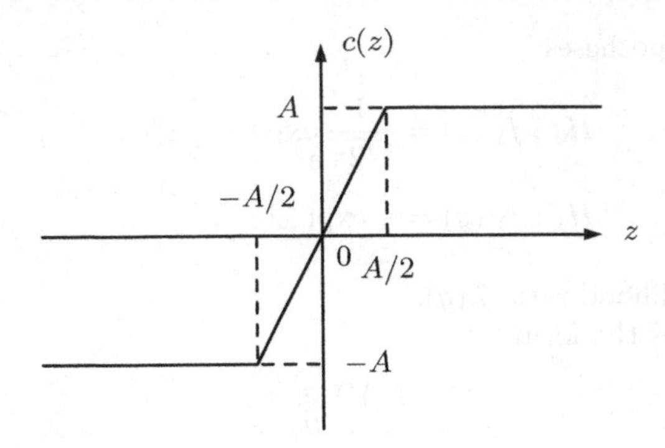

Fig. 2.16. Graph of the clipping function $c(z)$.

(c) Provide an intuitive interpretation of the test of part (b).

2.6. Under hypothesis H_0 an observation Y has the probability density function

$$f(y \mid H_0) = \begin{cases} A_0(a - |y|) & |y| \leq a \\ 0 & |y| > a \,, \end{cases}$$

and under hypothesis H_1, it has the probability density function

$$f(y \mid H_1) = \begin{cases} A_1(b - |y|) & |y| \leq b \\ 0 & |y| > b \,, \end{cases}$$

with $b > a$.

(a) Calculate the constants A_0 and A_1.
(b) Show that the optimum likelihood ratio test for deciding between these two hypotheses takes the form

$$|Y| \underset{H_0}{\overset{H_1}{\gtrless}} \eta \,.$$

(c) Evaluate the probability of detection P_D and the probability of false alarm P_F as a function of the threshold η. For $b = 2a$, sketch the ROC.

2.7. When a part is selected at random from a particular production batch, there is a probability q that the part is defective. Thus, out of n independently selected parts, the number K of defective units is a random variable with the binomial probability distribution

$$P[K = k] = \frac{n!}{k!(n-k)!} q^k (1-q)^{n-k} \,,$$

for $k = 0,\ 1,\ 2,\ \cdots,\ n$.

Suppose we receive a shipment of n parts which either came all from production batch 0, with probability of defect q_0, or came all from batch 1, with probability of defect q_1. Assume that $0 < q_0 < q_1 < 1$, and that the batches 0 and 1 have a priori probabilities π_0 and π_1, respectively. We wish to determine the production batch from which the parts came from by counting the number K of defective parts among the n units received.

(a) Find the minimum probability of error rule for deciding whether the received units were from batch 0 or batch 1.
(b) Suppose $n = 2$. Make a labeled sketch of the ROC for the LRT.

2.8. Consider a binary hypothesis testing problem where we use a scalar observation Y to decide between two hypotheses H_0 and H_1. Under the two hypotheses, Y admits a zero mean Laplacian distribution, but with different variances:

$$f_{Y|H_0}(y|H_0) = \frac{1}{2}\exp(-|y|)$$
$$f_{Y|H_1}(y|H_1) = \exp(-2|y|)\,,$$

with $-\infty < y < \infty$.

(a) Obtain the Bayesian likelihood ratio test for this problem. Express the decision regions in terms of the threshold

$$\tau = \frac{(C_{10} - C_{00})\pi_0}{(C_{01} - C_{11})\pi_1}\,,$$

where C_{ij} denotes the Bayesian cost function and π_i represents the a-priori probability of hypothesis H_i for $i = 0, 1$.
(b) Express the probability of false alarm P_F and the probability of detection P_D in terms of τ.
(c) Evaluate and sketch the receiver operating characteristic (ROC).
(d) Suppose we wish to design a Neyman-Pearson detector with a probability of false alarm less or equal to α. Select the corresponding threshold τ.

2.9. For the binary hypothesis testing problem of Problem 2.8, assume the Bayesian costs $C_{00} = C_{11} = 0$, $C_{10} = 1$ and $C_{01} = 2$.

(a) Obtain a minimax test δ_{M}. Specify the probabilities of false alarm and of detection of this test.
(b) Evaluate and sketch the optimum Bayesian risk function $V(\pi_0)$ and find the least-favorable prior $\pi_{0\mathrm{M}}$.

2.10. Consider a binary hypothesis testing problem where under H_0 we observe a scalar random variable Y which is uniformly distributed over $[0, 1]$, so that its density is given by

$$f_Y(y|H_0) = \begin{cases} 1 & 0 \leq y \leq 1 \\ 0 & \text{otherwise} . \end{cases}$$

Under H_1, Y admits the triangular probability density

$$f_Y(y|H_1) = \begin{cases} 4y & 0 \leq y \leq 1/2 \\ 4(1-y) & 1/2 \leq y \leq 1 \\ 0 & \text{otherwise} , \end{cases}$$

which is plotted in Fig. 2.17.

(a) Obtain the Bayesian likelihood ratio test for this problem. Express the decision regions in terms of the threshold τ. Note that you will need to consider two cases: $\tau \leq 2$, and $\tau > 2$.
(b) Express the probability of false alarm P_F and the probability of detection P_D in terms of τ.
(c) Evaluate and sketch the receiver operating characteristic (ROC).
(d) Suppose we wish to design a Neyman-Pearson detector with a probability of false alarm less or equal to α. Select the corresponding threshold τ.

2.11. For the binary hypothesis testing problem of Problem 2.10, assume the Bayesian costs $C_{00} = C_{11} = 0$, $C_{10} = 1$ and $C_{01} = 2$.

(a) Obtain a minimax test δ_{M}. Specify the probabilities of false alarm and of detection of this test.
(b) Evaluate and sketch the optimum Bayesian risk function $V(\pi_0)$ and find the least-favorable prior $\pi_{0\text{M}}$.

2.12. Consider the binary hypothesis testing problem

$$H_0 : Y = V$$
$$H_1 : Y = A + V ,$$

where $0 < A < 1$ is known and the noise V is uniformly distributed over interval $[-1/2, 1/2]$, so that its density

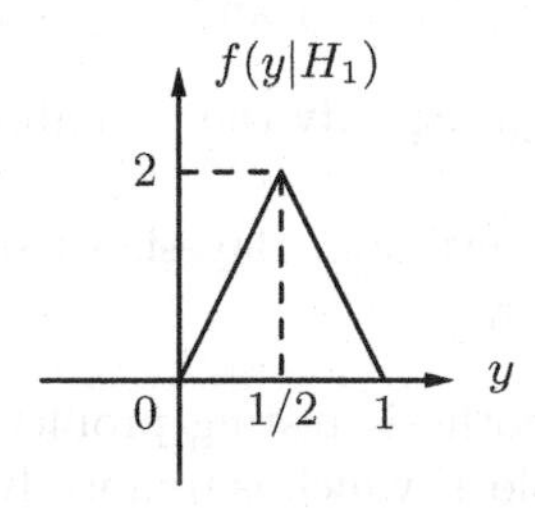

Fig. 2.17. Triangular probability density of Y under H_1.

$$f_V(v) = \begin{cases} 1 & -1/2 \leq v \leq 1/2 \\ 0 & \text{otherwise} . \end{cases}$$

(a) Design a Neyman-Pearson test such that the probability of false alarm is $P_F = \alpha$. Observe that, as the likelihood ratio takes on only a finite number of possible values, randomization is necessary. Evaluate and sketch the ROC for this test.
(b) Let $A = 1/2$. Assume a Bayesian cost function $C_{00} = C_{11} = 0$, $C_{10} = 1$ and $C_{01} = 2$. Plot the optimum Bayesian risk function $V(\pi_0)$ and design a minimax test. Specify the least-favorable prior π_{0M}, and if randomization is required, describe precisely the randomization rule needed to implement the minimax test δ_M.

2.13. Consider the ternary hypothesis testing problem

$$H_0 : \mathbf{Y} = \mathbf{s}_0 + \mathbf{V}$$
$$H_1 : \mathbf{Y} = \mathbf{s}_1 + \mathbf{V}$$
$$H_2 : \mathbf{Y} = \mathbf{s}_3 + \mathbf{V} ,$$

where

$$\mathbf{Y} = \begin{bmatrix} Y_1 \\ Y_2 \end{bmatrix}$$

is a two-dimensional observation vector,

$$\mathbf{s}_0 = E^{1/2} \begin{bmatrix} 1 \\ 0 \end{bmatrix} \quad , \quad \mathbf{s}_1 = \frac{E^{1/2}}{2} \begin{bmatrix} -1 \\ \sqrt{3} \end{bmatrix} \quad , \quad \mathbf{s}_2 = -\frac{E^{1/2}}{2} \begin{bmatrix} 1 \\ \sqrt{3} \end{bmatrix} ,$$

and $\mathbf{V}$ is a two-dimensional zero-mean Gaussian random vector with covariance matrix $\sigma^2 I_2$. For this problem we seek to minimize the probability of error, so that the Bayesian costs take the form $C_{ij} = 1 - \delta_{ij}$ for $0 \leq i, j \leq 2$.

(a) Assume first that the three hypotheses are equally likely, i.e., $\pi_i = 1/3$ for $i = 0, 1, 2$. Draw the decision regions in the Y_1-Y_2 plane.
(b) Obtain an expression for the probability of error. To do so, you may want to evaluate first the probability of a correct decision.
(c) Assume now that the a-priori probabilities for the three hypotheses are given by

$$\pi_0 = \frac{1}{2} \quad , \quad \pi_1 = \pi_2 = \frac{1}{4} .$$

Draw the decision regions in the L_1-L_2 plane, where

$$L_i(\mathbf{Y}) = \frac{f(\mathbf{Y}|H_i)}{f(\mathbf{Y}|H_0)}$$

denotes the likelihood ratio between the i-th and the zeroth hypothesis.

2.14. Consider the ternary hypothesis testing problem

$$H_0 : \mathbf{Y} = \mathbf{s}_0 + \mathbf{V}$$
$$H_1 : \mathbf{Y} = \mathbf{s}_1 + \mathbf{V}$$
$$H_2 : \mathbf{Y} = \mathbf{s}_3 + \mathbf{V} ,$$

where

$$\mathbf{Y} = \begin{bmatrix} Y_1 \; Y_2 \; Y_3 \end{bmatrix}^T$$

is a three-dimensional observation vector,

$$\mathbf{s}_0 = E^{1/2} \begin{bmatrix} 1 \\ 0 \\ 0 \end{bmatrix} \quad , \quad \mathbf{s}_1 = E^{1/2} \begin{bmatrix} 0 \\ 1 \\ 0 \end{bmatrix} \quad , \quad \mathbf{s}_2 = E^{1/2} \begin{bmatrix} 0 \\ 0 \\ 1 \end{bmatrix} \quad ,$$

and $\mathbf{V}$ is a three-dimensional zero-mean Gaussian random vector with covariance matrix $\sigma^2 I_3$. The Bayes costs are given by

$$C_{ij} = 1 - \delta_{ij}$$

for $0 \leq i, \, j \leq 2$.

(a) Let

$$\mathbf{p}(\mathbf{y}) = \begin{bmatrix} P[H_0 \text{ is true } | \mathbf{Y} = \mathbf{y}] \\ P[H_1 \text{ is true } | \mathbf{Y} = \mathbf{y}] \\ P[H_2 \text{ is true } | \mathbf{Y} = \mathbf{y}] \end{bmatrix} = \begin{bmatrix} p_0 \\ p_1 \\ p_2 \end{bmatrix} .$$

(i) Specify the optimum decision rules in terms of p_0, p_1, and p_2.
(ii) Given that $p_0 + p_1 + p_2 = 1$, express these rules completely in terms of p_0 and p_1, and sketch the decision regions in the p_0, p_1 plane.

(b) Suppose that the three hypotheses are equally likely a-priori, i.e.,

$$\pi_0 = \pi_1 = \pi_2 = 1/3 .$$

Show that the optimum decision rules can be specified in terms of the pair of sufficient statistics

$$S_1(\mathbf{Y}) = Y_2 - Y_1 \quad , \quad S_2(\mathbf{Y}) = Y_3 - Y_1 .$$

To do so, you may want to use the form (2.148) of the optimum decision rule, which is expressed in terms of the likelihood ratios

$$L_i(\mathbf{Y}) = \frac{f(\mathbf{Y}|H_i)}{f(\mathbf{Y}|H_0)}$$

with $i = 1, \, 2$.

References

1. J. M. Wozencraft and I. M. Jacobs, *Principles of Communication Engineering.* New York: J. Wiley & Sons, 1965. Reprinted by Waveland Press, Prospect Heights, IL, 1990.
2. J. G. Proakis, *Digital Communications, Fourth Edition.* New York: McGraw-Hill, 2000.
3. T. S. Ferguson, *Mathematical Statistics: A Decision Theoretic Approach.* New York: Academic Press, 1967.
4. J. Neyman and E. Pearson, "On the problem of the most efficient tests of statistical hypotheses," *Phil. Trans. Royal Soc. London*, vol. A 231, pp. 289–337, Feb. 1933.
5. D. Bertsekas, *Nonlinear Programming, Second Edition.* Belmont, MA: Athena Scientific, 1999.
6. J.-P. Aubin and I. Ekland, *Applied Nonlinear Analysis.* New York: J. Wiley, 1984.
7. D. Bertsekas, A. Nedic, and A. E. Ozdaglar, *Convex Analsis and Optimization.* Belmont, MA: Athena Scientific, 2003.
8. H. V. Poor, *An Introduction to Signal Detection and Estimation, Second Edition.* New York: Springer Verlag, 1994.
9. H. L. Van Trees, *Detection, Estimation and Modulation Theory, Part I: Detection, Estimation and Linear Modulation Theory.* New York: J. Wiley & Sons, 1968. paperback reprint edition in 2001.
10. S. M. Kay, *Fundamentals of Statistical Signal Processing: Detection Theory.* Prentice-Hall, 1998.
11. E. L. Lehmann, *Elements of Large-Sample Theory.* New York: Springer Verlag, 1999.

3

Tests with Repeated Observations

3.1 Introduction

In Chapter 2, we described how to design optimal Bayesian and Neyman-Pearson tests by implementing likelihood ratio tests (LRTs) with appropriately selected thresholds. The performance of the test was described by the pair (P_D, P_F) formed by the probabilities of detection and of false alarm. However, except in the case of known signals in zero-mean Gaussian noise, for which the distance d between hypotheses was directly related to (P_D, P_F), we were not yet able to quantify how long the observation window has to be in order to achieve some target values for (P_D, P_F), or equivalently for (P_M, P_F). In the first part of this chapter, we show that for LRTs with identically distributed observations, the probability P_M of a miss and the probability P_F of false alarm decay exponentially with the number N of observations and we describe how to evaluate the exponents governing this decay. This allows, of course, the selection of the length N of a fixed size test to ensure that both P_M and P_F fall below certain target values. In the second part of this chapter, we describe the sequential hypothesis testing approach proposed in 1943 by statistician, Abraham Wald, which is presented in detail in his 1947 monograph [1]. In this formulation, instead of deciding ahead of time how many observations will be collected, samples are examined one by one until the observer has enough information to be able to reach a decision with a high degree of confidence. In many ways this hypothesis testing approach represents an accurate model of human behavior, since, when confronted with uncertain situations, most individuals collect information until they become reasonably sure about the situation they are facing, and then take immediate action, without waiting for a preset amount of time. Thus, given a set of observations, the sequential hypothesis testing approach adds the option of taking another sample to the usual binary choice between two hypotheses. Of course, taking an extra sample has a cost, and the performance of different sampling and decision strategies can be characterized by a Bayesian risk function which is minimized by the class of *sequential probability ratio tests*

B.C. Levy, *Principles of Signal Detection and Parameter Estimation*,
DOI: 10.1007/978-0-387-76544-0_3, © Springer Science+Business Media, LLC 2008

(SPRTs). These tests, which can be viewed as extending LRTs, evaluate the likelihood ratios of all observations up to the current time, and compare the likelihood ratio (LR) to two thresholds: a high threshold and a low threshold. As long as the LR stays between the two thresholds, another observation is collected, but as soon as the LR exceeds the high threshold or falls below the low threshold, hypotheses H_1 or H_0 are selected, respectively. In essence this test applies the principle of "when in doubt, collect another observation," as long as the expected gain from the additional observation exceeds its cost. An elegant feature of the test is that the high and low thresholds can be selected to achieve prespecified probabilities of false alarm and of a miss, P_F and P_M. So the number N of observations that need to be collected is random, but the test performance is guaranteed. Also, although the number of samples is random, it can be shown that under both hypotheses, the expected number of samples until a decision is reached is smaller than the corresponding number of samples for a fixed size test with the same performance.

3.2 Asymptotic Performance of Likelihood Ratio Tests

Consider a binary hypothesis testing problem where a sequence of independent identically distributed random vectors $\mathbf{Y}_k \in \mathbb{R}^n$ is observed for $1 \leq k \leq N$. Under hypothesis H_j with $j = 0, 1$, depending on whether $\mathbf{Y}_k$ takes continuous or discrete values, it has either the probability density $f(\mathbf{y}|H_j)$ or the probability mass distribution $p(\mathbf{y}|H_j)$. Then if

$$L(\mathbf{y}) = \frac{f(\mathbf{y}|H_1)}{f(\mathbf{y}|H_0)} \tag{3.1}$$

denotes the likelihood ratio function for a single observation, an LRT for this problem takes the form

$$\prod_{k=1}^{N} L(\mathbf{Y}_k) \underset{H_0}{\overset{H_1}{\gtrless}} \tau(N) , \tag{3.2}$$

where $\tau(N)$ denotes a threshold that may depend on the number of observations. For example, for the Bayesian tests considered in Section 2.2, it is reasonable to assume the costs C_{ij} vary linearly with N, i.e.,

$$C_{ij}(N) = \bar{D}_{ij} N + \bar{C}_{ij}$$

where $\bar{C}_{ij}$ and $\bar{D}_{ij}$ represent respectively the fixed and incremental components of the cost. Assuming that $\bar{D}_{ij}$ is strictly positive for $0 \leq i, j \leq 1$, we then see that the threshold $\tau(N)$ given by (2.19) depends on N but tends to a constant as $N \to \infty$. Taking logarithms on both sides of (3.2), and denoting $Z_k = \ln(L(\mathbf{Y}_k))$, we find that the LRT reduces to

$$S_N = \frac{1}{N} \sum_{k=1}^{N} Z_k \underset{H_0}{\overset{H_1}{\gtrless}} \gamma(N) = \frac{\ln(\tau(N))}{N} . \tag{3.3}$$

When $\tau(N)$ tends to a constant as $N \rightarrow \infty$, $\gamma \triangleq \lim_{N\rightarrow\infty} \gamma(N) = 0$. In the following we consider the more general situation where the threshold $\gamma(N)$ on the right-hand side tends to a constant γ as $N \rightarrow \infty$. In this respect, observe that the test (3.3) is expressed in terms of the sufficient statistic S_N, which is the sampled mean of the log-likelihood functions Z_k. Under H_1, S_N converges almost surely to the mean

$$m_1 = E[Z_k|H_1] = \int \ln\Big(\frac{f(\mathbf{y}|H_1)}{f(\mathbf{y}|H_0)}\Big) f(\mathbf{y}|H_1)d\mathbf{y} \\ = D(f_1|f_0)\,, \tag{3.4}$$

where $D(f_1|f_0)$ denotes the Kullback-Leibler (KL) divergence, also known as the relative entropy, of the densities $f_1 = f(\cdot|H_1)$ and $f_0 = f(\cdot|H_0)$. Similarly, under H_0, S_N converges almost surely to the mean

$$m_0 = E[Z_k|H_0] = \int \ln\Big(\frac{f(\mathbf{y}|H_1)}{f(\mathbf{y}|H_0)}\Big) f(\mathbf{y}|H_0)d\mathbf{y} \\ = -D(f_0|f_1)\,. \tag{3.5}$$

The KL divergence has the property that $D(f|g) \geq 0$ with equality if and only if $f = g$ almost everywhere [2]. However it is not a conventional distance since it is not symmetric, i.e., $D(f|g) \neq D(g|f)$ and it does not satisfy the triangle inequality $D(f|g) + D(g|h) \geq D(f|h)$ for three arbitrary densities f, g, and h. Yet it has been shown to be a useful measure of model mismatch in statistical modelling [3], and by adopting a differential geometric viewpoint, it is possible to demonstrate the KL divergence obeys Pythagoras' theorem in an appropriate sense, so it is argued in [4] that it represents the natural notion of distance between random systems.

When the observations $\mathbf{Y}$ take discrete values and admit the probability mass distributions $p_j = p(\cdot|H_j)$, the means $m_j = E[Z_k|H_j]$ can be expressed as

$$m_1 = D(p_1|p_0) \quad , \quad m_0 = -D(p_0|p_1) \tag{3.6}$$

where the discrete KL divergence [5] of two probability mass distributions is defined as

$$D(p|q) \triangleq \sum_{\mathbf{y}\in\mathcal{Y}} \ln\Big(\frac{p(\mathbf{y})}{q(\mathbf{y})}\Big)p(\mathbf{y})\,. \tag{3.7}$$

For limited purposes, we observe from (3.4) and (3.5) that as the number N of observations tends to infinity, S_N tends to a positive number m_1 under H_1 and a negative number m_0 under H_0. Thus, as long as we are willing to collect an arbitrarily large number of observations, it is possible to separate perfectly the hypotheses H_1 and H_0. In other words, for a Bayesian test, the probability $P_F(N)$ of false alarm and the probability $P_M(N)$ of a miss will simultaneously go to zero as $N \rightarrow \infty$. From the above analysis we also see how to design an optimal Neyman-Pearson (NP) test for an infinite number of observations. If our goal is to minimize the probability of a miss while keeping the probability of false alarm below α, in the test (3.3) we should select the threshold

$$\gamma_{\mathrm{NP}}^{\mathrm{I}} = m_0 + \epsilon \tag{3.8}$$

where $\epsilon > 0$ is arbitrarily close to zero. Then under H_0, as $N \to \infty$, S_N falls below $\gamma_{\mathrm{NP}}^{\mathrm{I}}$ almost surely ensuring that the probability of false alarm is less than or equal to α. Similarly, if we seek to minimize the probability of false alarm while keeping the probability of a miss below β, we should select the threshold

$$\gamma_{\mathrm{NP}}^{\mathrm{II}} = m_1 - \epsilon \tag{3.9}$$

where $\epsilon > 0$ can be arbitrarily close to zero. The superscripts I and II for the thresholds (3.8) and (3.9) refer to the fact that they correspond to NP tests of types I and II, respectively.

Then for a test of the form (3.3) with threshold γ, we would like to determine how quickly the probability of false alarm $P_F(N)$ and the probability of a miss $P_M(N)$ tend to zero as N becomes large. This problem was first examined by Chernoff [6], with subsequent elaborations by Kullback [2] and others. Our discussion is based on the following result due to the statistician, Harald Cramér [7], which is at the origin of a form of analysis called the *theory of large deviations* [8,9] for evaluating the probabilities of rare random events.

Cramér's theorem: Let $\{Z_k,\ k \geq 1\}$ be a sequence of i.i.d. random variables with cumulant generating function

$$G(u) = E[\exp(uZ)]\,, \tag{3.10}$$

and let

$$S_N = \frac{1}{N}\sum_{k=1}^{N} Z_k \tag{3.11}$$

denote the sample average of this sequence. Then for all $z > m = E[Z]$,

$$\lim_{N\to\infty} \frac{1}{N} \ln P[S_N \geq z] = -I(z)\,, \tag{3.12}$$

where the rate function $I(z)$ is defined by

$$I(z) \triangleq \max_{u \in \mathbb{R}} (zu - \ln(G(u)))\,. \tag{3.13}$$

This theorem is proved in Appendix 3.A.. Before applying this result to the asymptotic analysis of LRTs, it is worth spending some time understanding all quantities appearing in expressions (3.12)–(3.13). As a first step, note that both the cumulant generating function $G(u)$ and the log-generating function $\ln(G(u))$ are convex.

For $G(u)$, this is just a consequence of the fact that

$$\frac{d^2}{du^2} G(u) = E[Z^2 \exp(uZ)] \geq 0 \tag{3.14}$$

where the last inequality follows from the fact that the quantity in the expectation is nonnegative for all u. In fact $G(u)$ is strictly convex, i.e., its second derivative is strictly positive, as long as Z is not equal to zero almost surely. The convexity of $\ln(G(u))$ follows from Hölder's inequality, which can be viewed as a generalization of the Cauchy-Schwartz inequality: if X and Y are two random variables and $1 \leq p, q \leq \infty$ satisfy $p^{-1} + q^{-1} = 1$, then [10, p. 47]

$$|E[XY]| \leq E[|XY|] \leq (E[|X|^p])^{1/p} E[|Y|^q]^{1/q} . \tag{3.15}$$

The Cauchy-Schwartz inequality corresponds to the choice $p = q = 2$. For the case at hand, let u and v be two real numbers for which $G(u)$ and $G(v)$ are finite, and let $0 \leq \lambda \leq 1$. Then consider

$$G(\lambda u + (1 - \lambda)v) = E[\exp(\lambda u + (1 - \lambda)v)Z)] . \tag{3.16}$$

By applying Hölder's inequality with

$$X = \exp(\lambda u Z) \quad , \quad Y = \exp((1 - \lambda)vZ) ,$$

$p = \lambda^{-1}$ and $q = (1 - \lambda)^{-1}$, we obtain

$$G(\lambda u + (1 - \lambda)v) \leq G(u)^{\lambda} G(v)^{1-\lambda} , \tag{3.17}$$

which after taking logarithms on both sides yields

$$\ln(G(\lambda u + (1 - \lambda)v)) \leq \lambda \ln(G(u)) + (1 - \lambda) \ln(G(v)) , \tag{3.18}$$

so $\ln(G(u))$ is convex.

Once we know that $\ln(G(u))$ is convex, the rate function $I(z)$ becomes easy to interpret: it is the *Legendre transform* of the log-generating function $\ln(G(u))$. This transform, which arises in optimization, classical mechanics, and thermodynamics, has the feature that it transforms convex functions into convex functions and is its own inverse [11, Section 7.1]. Specifically, if $\mathcal{L}$ denotes the Legendre transform, and $I = \mathcal{L}(\ln G)$, then $\ln G = \mathcal{L}(I)$. In other words, Legendre transforming the rate function I yields back the log-generating function $\ln(G(u))$, so $I(z)$ and $G(u)$ contain exactly the same information about the random variable Z. The Legendre transform $I(z)$ of the convex function $\ln G(u)$ has a simple geometric interpretation: consider the line with slope z which is tangent to $\ln(G(u))$. Because of the convexity of $\ln(G(u))$, there can be at the most one such tangent. Then the intercept of the tangent with the vertical axis is $-I(z)$, as shown in Fig. 3.1, where T denotes the point of tangency and $u(z)$ represents its horizontal coordinate.

Now that we have discussed the geometric picture behind Cramér's theorem, let us return to the hypothesis testing problem, for which

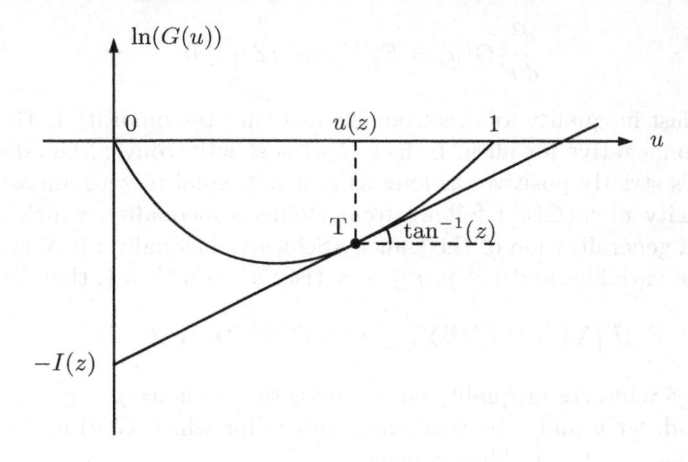

Fig. 3.1. Geometric illustration of the Legendre transform of the convex function $\ln(G(u))$.

$$Z_k = \ln L(\mathbf{Y}_k)$$

is the log-likelihood function of the observations. Substituting this expression in the definition of the cumulant generating function yields

$$\begin{aligned} G_0(u) &= E[\exp(uZ)|H_0] = \int f(\mathbf{y}|H_1)^u f(\mathbf{y}|H_0)^{1-u} d\mathbf{y} \\ G_1(u) &= E[\exp(uZ)|H_1] = \int f(\mathbf{y}|H_1)^{u+1} f(\mathbf{y}|H_0)^{-u} d\mathbf{y} \\ &= G_0(u+1) \,. \end{aligned} \tag{3.19}$$

From expressions (3.19) we find

$$G_0(0) = G_0(1) = 1 \,, \tag{3.20}$$

and

$$\begin{aligned} \frac{dG_0}{du}(0) &= m_0 = -D(f_0|f_1) \\ \frac{dG_0}{du}(1) &= m_1 = D(f_1|f_0) \,. \end{aligned} \tag{3.21}$$

For $0 \leq u \leq 1$, by observing that $u = (1-u)0 + u1$ and using the convexity of $G_0(u)$, we find

$$G_0(u) \leq (1-u)G_0(0) + uG_0(1) = 1 \,. \tag{3.22}$$

The above properties imply that the log-generating function $\ln(G_0(u))$ is zero at $u = 0$ and $u = 1$, has derivatives

$$\frac{d\ln(G_0)}{du}(0) = m_0 = -D(f_0|f_1)$$
$$\frac{d\ln(G_0)}{du}(1) = m_1 = D(f_1|f_0)\,. \tag{3.23}$$

at these points and is nonpositive over the interval $[0,1]$, as shown in part (a) of Fig. 3.2. Since the derivative of a convex function is monotone nondecreasing, the above properties indicate that, as long as we are interested only in slopes z between m_0 and m_1, the point of tangency in the Legendre transform

$$I_0(z) = \max_{u \in \mathbb{R}}(zu - \ln(G_0(u))) \tag{3.24}$$

will occur for $0 \leq u \leq 1$, so the maximization can just be performed over $[0,1]$, and the intercept with the vertical axis will be such that $I_0(z) > 0$,

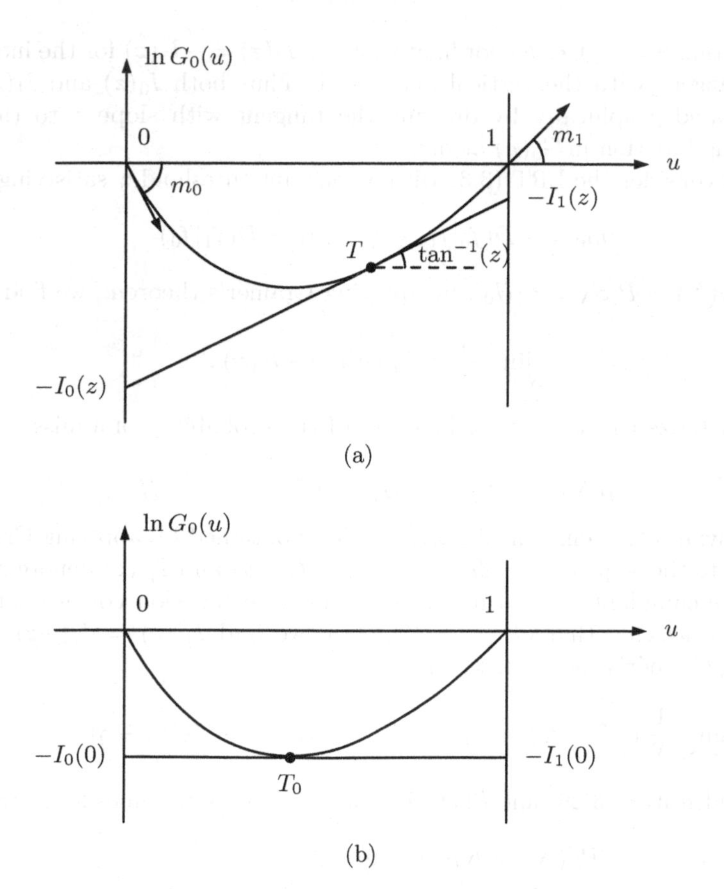

Fig. 3.2. a) Log-generating function $\ln(G_0(u))$ of the log-likelihood ratio under hypothesis H_0 and rate functions $I_0(z)$ and $I_1(z)$; (b) equalized rate functions $I_0(0) = I_1(0)$ for a tangent with slope $z = 0$.

except at $z = m_0$, where $I_0(m_0) = 0$. Also the property $G_1(u) = G_0(u+1)$ implies that the rate function

$$I_1(z) = \max_{u \in \mathbb{R}}(zu - \ln G_1(u)) \tag{3.25}$$

corresponding to $G_1(u)$ satisfies

$$I_1(z) = I_0(z) - z\,. \tag{3.26}$$

As indicated in part (a) of Fig. 3.2, $-I_1(z)$ is the intercept of the tangent with slope z to $\ln G_0(u)$ and the vertical line passing through $u = 1$. To see why this is the case, observe that since its intercept with the vertical axis is $-I_0(z)$, the tangent line to $\ln G_0(u)$ with slope z has for equation

$$y = zu - I_0(z)\,.$$

Then setting $u = 1$ gives a coordinate $y = z - I_0(z) = -I_1(z)$ for the intercept of the tangent with the vertical line $u = 1$. Thus both $I_0(z)$ and $I_1(z)$ can be evaluated graphically by drawing the tangent with slope z to the log-generating function $\ln G_0(u)$ at point T.

Then consider the LRT (3.3) with a constant threshold γ satisfying

$$m_0 = -D(f_0|f_1) < \gamma < m_1 = D(f_1|f_0)\,. \tag{3.27}$$

Since $P_F(N) = P[S_N \geq \gamma|H_0]$, by applying Cramér's theorem, we find

$$\lim_{N\to\infty} \frac{1}{N} \ln P_F(N) = -I_0(\gamma)\,. \tag{3.28}$$

To characterize the asymptotic behavior of the probability of a miss, we note that

$$P_M(N) = P[S_N < \gamma|H_1] = P[-S_N > -\gamma|H_1]\,, \tag{3.29}$$

so the asymptotic behavior of $P_M(N)$ can be obtained by applying Cramér's theorem to the sequence $\{-Z_k,\ k \geq 0\}$. Let $G_1^-(u)$ and $I_1^-(z)$ denote respectively the cumulant generating function and rate function corresponding to $-Z$. By observing that $G_1^-(u) = G_1(-u)$, we find $I_1^-(z) = I_1(-z)$. So by applying Cramér's theorem, we find

$$\lim_{N\to\infty} \frac{1}{N} \ln P_M(N) = -I_1^-(-\gamma) = -I_1(\gamma) = -[I_0(\gamma) - \gamma]\,. \tag{3.30}$$

The identities (3.29) and (3.30) indicate that as N becomes large, we have

$$\begin{aligned} P_F(N) &\approx K_F(N)\exp(-NI_0(\gamma)) \\ P_M(N) &\approx K_M(N)\exp(-N(I_0(\gamma) - \gamma))\,, \end{aligned} \tag{3.31}$$

where the functions $K_F(N)$ and $K_M(N)$ depend on the probability densities/distributions of the observations under hypotheses H_1 and H_0 and satisfy

$$\lim_{N\to\infty} \frac{1}{N} \ln(K_F(N)) = \lim_{N\to\infty} \frac{1}{N} \ln(K_M(N)) = 0 \,. \tag{3.32}$$

This last condition indicates that the exponential decay terms dominate in the asymptotic expressions (3.31) for P_F and P_M. So, just as we suspected, for any threshold γ located between m_0 and m_1, the probability of false alarm and probability of a miss converge simultaneously to zero as the number N of observations becomes infinite. But we now know that the rate of decay is *exponential* and is governed by the rate function I_0.

Let us consider now some special choices of the threshold γ, starting with the two Neyman-Pearson tests corresponding to (3.8) and (3.9). For $\gamma_{\text{NP}}^{I} = m_0 + \epsilon$ with $\epsilon > 0$ vanishingly small, we have $I_0(\gamma_{\text{NP}}^{I}) \approx I_0(m_0) = 0$, so

$$P_M(N) \approx K_M(N) \exp(-ND(f_0|f_1)) \,, \tag{3.33}$$

i.e., the decay exponent of $P_M(N)$ is the KL divergence $D(f_0|f_1)$. Similarly, for of $\gamma_{\text{NP}}^{II} = m_1 - \epsilon$ with ϵ vanishingly small, we have $I_0(\gamma_{\text{NP}}^{II}) \approx I_0(m_1) = m_1$, so

$$P_F(N) \approx K_F(N) \exp(-ND(f_1|f_0)) \,. \tag{3.34}$$

Next, suppose we consider the probability of error

$$\begin{aligned} P_E(N) &= P_F(N)\pi_0 + P_M(N)(1-\pi_0) \\ &= K_F(N)\pi_0 \exp(-NI_0(\gamma)) + K_M(N)(1-\pi_0)\exp(-NI_1(\gamma)) \,, \end{aligned} \tag{3.35}$$

and we ask where the threshold γ should be located to make the exponential decay of $P_E(N)$ as rapid as possible. This question was resolved by Chernoff [6] and it requires a careful analysis, but the answer can be guessed easily. The minimization of the probability of error corresponds to the choice $C_{ij} = 1 - \delta_{ij}$ and thus $\tau(N) = 1$ and $\gamma(N) = 0$ in the Bayesian test (3.3). So we can guess right away that the decay maximizing threshold is $\gamma = 0$. For this choice, we note that the decay rates $I_0(0) = I_1(0)$ of the probability of false alarm and of a miss are equalized, which makes sense, since as we shall see below, any attempt at improving one of these two rates will come at the expense of making the other worse. But electrical engineers familiar with circuit analysis know that it is always the slowest decaying exponential that dominates. So starting from two equal rates of decay, any attempt at improving one rate of decay by making the other worse is necessarily counterproductive, since the worst rate becomes dominant.

From equation (3.35), we see that to maximize the decay rate of $P_E(N)$ we must maximize $\min(I_0(\gamma), I_1(\gamma)$. This is equivalent of course to minimizing

$$-\min(I_0(\gamma), I_1(\gamma)) = \max(-I_0(\gamma), -I_1(\gamma)) \,.$$

If we return to Fig. 3.2, we see that since the tangent with slope z to the log-generating function has intercepts $-I_0(z)$ and $-I_1(z)$ with vertical lines $u = 0$ and $u = 1$, the only way to make sure that the maximum of these two intercepts is as small as possible is to select a horizontal tangent with slope $z = 0$, in which case $-I_0(0) = -I_1(0)$ as shown in part (b) of Fig. 3.2. Suppose that instead of selecting a zero slope, we select a slope $z > 0$ as shown in part (a) of Fig. 3.2, then the right intercept $-I_1(z)$ is above $-I_1(0) = -I_0(0)$, so the rate of decay of the probability of a miss has become worse. Similarly, if we tip the tangent line the other way by selecting a negative slope $z < 0$, then the left intercept $-I_0(z)$ is above $-I_0(0)$, so the probability of false alarm is now worse.

Since the optimum slope is $z = 0$, we deduce therefore that the intercept $-I_0(0) = -I_1(0)$ represents the minimum of $\ln G_0(u)$ for $0 \leq u \leq 1$. So we have found that

$$\begin{aligned} I_0(0) &= - \min_{0 \leq u \leq 1} \ln G_0(u) \\ &= - \min_{0 \leq u \leq 1} \ln \int f_1^u(\mathbf{y}) f_0^{1-u}(\mathbf{y}) d\mathbf{y} = C(f_1, f_0) \end{aligned} \tag{3.36}$$

where $C(f_1, f_0)$ denotes the Chernoff distance between densities f_1 and f_0. It is nonnegative, i.e., $C(f_1, f_0) \geq 0$ with equality if and only if $f_1 = f_0$. Furthermore, unlike the KL divergence, it is symmetric in its two arguments, i.e., $C(f_1, f_0) = C(f_0, f_1)$. Selecting the threshold optimally as $\gamma = 0$ yields

$$P_E(N) = (K_F(N)\pi_0 + K_M(N)(1 - \pi_0)) \exp(-NC(f_1, f_0)) \,, \tag{3.37}$$

so the rate of decay of the probability of error is governed by the Chernoff distance.

Example 3.1: Poisson distributed random variables

Consider a test where we observe a sequence $\{Y_k,\ k \geq 1\}$ of i.i.d. Poisson random variables, where under hypothesis H_j, Y_k has parameter λ_j with $j = 0,\ 1$, i.e.,

$$\begin{aligned} H_0 : P[Y_k = n|H_0] &= \frac{\lambda_0^n}{n!} \exp(-\lambda_0) \\ H_1 : P[Y_k = n|H_1] &= \frac{\lambda_1^n}{n!} \exp(-\lambda_1) \,, \end{aligned}$$

where we assume $\lambda_1 > \lambda_0$. Then

$$Z_k = \ln L(Y_k) = \ln(\lambda_1/\lambda_0) Y_k - (\lambda_1 - \lambda_0)$$

and

$$\begin{aligned} G_0(u) &= \Big[\sum_{n=0}^{\infty} \frac{(\lambda_1^u \lambda_0^{1-u})^n}{n!} \Big] \exp(-(u\lambda_1 + (1-u)\lambda_0)) \\ &= \exp[\lambda_1^u \lambda_0^{1-u} - (u\lambda_1 + (1-u)\lambda_0)] \,. \end{aligned}$$

This implies

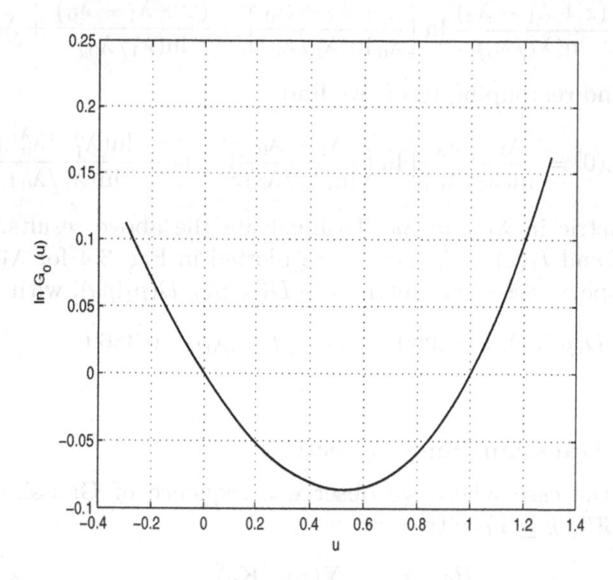

Fig. 3.3. Log generating function $\ln G_0(u)$ for testing Poisson distributions with parameters $\lambda_1 = 2$ and $\lambda_0 = 1$.

$$\ln(G_0(u)) = \lambda_0 \exp(u \ln(\lambda_1/\lambda_0)) - (u\lambda_1 + (1-u)\lambda_0) \,, \tag{3.38}$$

which is plotted in Fig. 3.3 for $\lambda_1 = 2$ and $\lambda_0 = 1$.

By differentiating we find

$$\frac{d}{du} \ln(G_0(u)) = \lambda_0 \ln(\lambda_1/\lambda_0) \exp(u \ln(\lambda_1/\lambda_0)) - (\lambda_1 - \lambda_0) \,, \tag{3.39}$$

so that

$$\begin{aligned} \Big(\frac{d}{du} \ln G_0\Big)(0) &= -D(p_0|p_1) = \lambda_0 \ln(\lambda_1/\lambda_0) - (\lambda_1 - \lambda_0) \\ \Big(\frac{d}{du} \ln G_0\Big)(1) &= D(p_1|p_0) = \lambda_1 \ln(\lambda_1/\lambda_0) - (\lambda_1 - \lambda_0) \,. \end{aligned} \tag{3.40}$$

To find the rate function $I_0(z)$, we note that since $zu - \ln(G_0(u))$ is concave in u, its maximum is found by setting its derivative with respect to u equal to zero, which requires solving

$$z = \frac{d}{du} \ln(G_0(u)) \,. \tag{3.41}$$

This gives

$$u = \ln \Big[\frac{z + \lambda_1 - \lambda_0}{\lambda_0 \ln(\lambda_1/\lambda_0)}\Big] \Big/ \ln(\lambda_1/\lambda_0) \,,$$

which after substitution in expression (3.24) for $I_0(z)$ gives

$$I_0(z) = \frac{(z + \lambda_1 - \lambda_0)}{\ln(\lambda_1/\lambda_0)} \ln\Big[\frac{z + \lambda_1 - \lambda_0}{\lambda_0 \ln(\lambda_1/\lambda_0)}\Big] - \frac{(z + \lambda_1 - \lambda_0)}{\ln(\lambda_1/\lambda_0)} + \lambda_0 \,.$$

Setting $z = 0$ and regrouping terms we find

$$C(p_1, p_0) = I_0(0) = \frac{\lambda_1 - \lambda_0}{\ln(\lambda_1/\lambda_0)} \Big[\ln\Big(\frac{\lambda_1 - \lambda_0}{\ln(\lambda_1/\lambda_0)}\Big) - 1\Big] + \frac{\ln(\lambda_1^{\lambda_0}/\lambda_0^{\lambda_1})}{\ln(\lambda_1/\lambda_0)} \,, \tag{3.42}$$

which is symmetric in λ_1 and λ_0. To illustrate the above results, the rate functions $I_0(z)$ and $I_1(z) = I_0(z) - z$ are plotted in Fig. 3.4 for $\lambda_1 = 2$ and $\lambda_0 = 1$. The slope z varies over interval $[-D(p_0|p_1), D(p_1|p_0)]$ with

$$D(p_0|p_1) = 0.3069 \quad \text{and} \quad D(p_1|p_0) = 0.3869 \,.$$

□

Example 3.2: Gaussian random vectors

Consider now the case where we observe a sequence of Gaussian random vectors $\{\mathbf{Y}_k \in \mathbb{R}^n,\ k \geq 1\}$ with

$$H_0 : \mathbf{Y}_k \sim N(\mathbf{m}_0, \mathbf{K}_0)$$
$$H_1 : \mathbf{Y}_k \sim N(\mathbf{m}_1, \mathbf{K}_1) \,.$$

Then by performing the integration (3.19) we obtain

$$G_0(u) = \frac{|K(u)|^{1/2}}{|K_1|^{u/2}|K_0|^{(1-u)/2}} \exp\Big[\frac{1}{2}\Big(\mathbf{m}(u)^T\mathbf{K}^{-1}(u)\mathbf{m}(u) - u\mathbf{m}_1^T\mathbf{K}_1^{-1}\mathbf{m}_1 - (1-u)\mathbf{m}_0^T\mathbf{K}_0^{-1}\mathbf{m}_0\Big)\Big] \,,$$

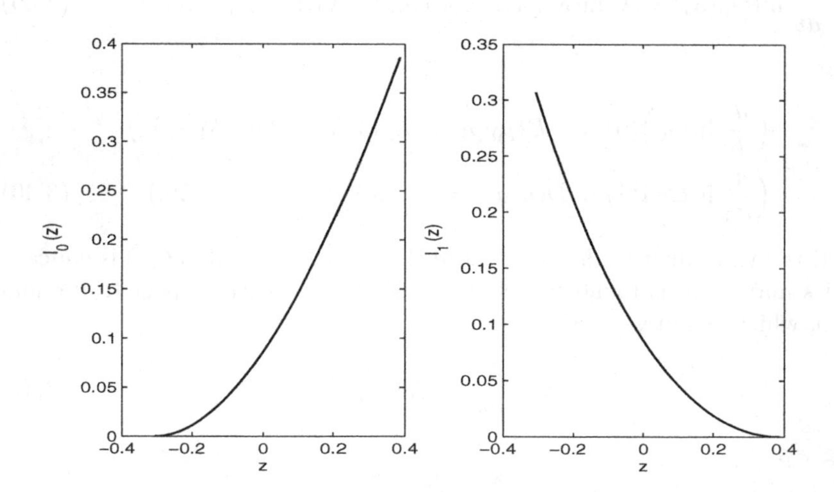

Fig. 3.4. Rate functions $I_0(z)$ and $I_1(z)$ for testing Poisson distributions with parameters $\lambda_1 = 2$ and $\lambda_0 = 1$.

where $\mathbf{K}(u)$ and $\mathbf{m}(u)$ satisfy

$$\begin{aligned}\mathbf{K}^{-1}(u) &= u\mathbf{K}_1^{-1} + (1-u)\mathbf{K}_0^{-1} \\ \mathbf{K}^{-1}(u)\mathbf{m}(u) &= u\mathbf{K}_1^{-1}\mathbf{m}_1 + (1-u)\mathbf{K}_0^{-1}\mathbf{m}_0 \,.\end{aligned}$$

Taking logarithms and regrouping terms, we obtain

$$\begin{aligned}\ln(G_0(u)) = \frac{1}{2}\Big[& \ln(|\mathbf{K}|(u)) - u\ln(|\mathbf{K}_1|) - (1-u)\ln(|\mathbf{K}_0|) \\ & -u(1-u)\Delta\mathbf{m}^T[u\mathbf{K}_0 + (1-u)\mathbf{K}_1]^{-1}\Delta\mathbf{m}\Big] \,, \quad (3.43)\end{aligned}$$

where

$$\Delta\mathbf{m} = \mathbf{m}_1 - \mathbf{m}_0 \,.$$

Taking derivatives and using the matrix differentiation identities [12, Chap. 8]

$$\begin{aligned}\frac{d}{du}\ln(|\mathbf{M}(u)|) &= \mathrm{tr}\big[\mathbf{M}^{-1}(u)\frac{d\mathbf{M}}{du}(u)\big] \\ \frac{d}{du}\mathbf{M}^{-1}(u) &= -\mathbf{M}^{-1}(u)\frac{d\mathbf{M}}{du}(u)\mathbf{M}^{-1}(u) \,,\end{aligned}$$

where $\mathbf{M}(u)$ denotes an arbitrary square invertible matrix, we obtain

$$\begin{aligned}\frac{d}{du}\ln(G_0(u)) = \frac{1}{2}\Big[& -\mathrm{tr}\big(\mathbf{K}(u)(\mathbf{K}_1^{-1} - \mathbf{K}_0^{-1})\big) - \ln(|\mathbf{K}_1|/|\mathbf{K}_0|) \\ & +(2u-1)\Delta\mathbf{m}^T[u\mathbf{K}_0 + (1-u)\mathbf{K}_1]^{-1}\Delta\mathbf{m} \\ & +u(1-u)\Delta\mathbf{m}^T[u\mathbf{K}_0 + (1-u)\mathbf{K}_1]^{-1} \\ & .(\mathbf{K}_0 - \mathbf{K}_1)[u\mathbf{K}_0 + (1-u)\mathbf{K}_1]^{-1}\Delta\mathbf{m}\Big] \,. \quad (3.44)\end{aligned}$$

Setting $u = 0$ and $u = 1$ yields

$$\begin{aligned}\Big(\frac{d}{du}\ln G_0\Big)(0) &= -D(f_0|f_1) \\ &= \frac{1}{2}\big[-\mathrm{tr}(\mathbf{K}_0\mathbf{K}_1^{-1} - \mathbf{I}_n) + \ln(|\mathbf{K}_0|/|\mathbf{K}_1|) - \Delta\mathbf{m}^T\mathbf{K}_1^{-1}\Delta\mathbf{m}\big] \\ \Big(\frac{d}{du}\ln G_0\Big)(1) &= D(f_1|f_0) \\ &= \frac{1}{2}\big[\mathrm{tr}(\mathbf{K}_1\mathbf{K}_0^{-1} - \mathbf{I}_n) - \ln(|\mathbf{K}_1|/|\mathbf{K}_0|) + \Delta\mathbf{m}^T\mathbf{K}_0^{-1}\Delta\mathbf{m}\big] \,, \quad (3.45)\end{aligned}$$

which coincides with the expression given in [13] for the KL divergence of Gaussian densities.

Unfortunately, in this case it is not possible to evaluate the rate function in closed form, except when $\mathbf{K}_0 = \mathbf{K}_1 = \mathbf{K}$. For this special choice

$$\ln G_0(u) = -\frac{u(1-u)}{2}d^2$$

is quadratic in u, where

$$d^2 = \Delta\mathbf{m}^T\mathbf{K}^{-1}\Delta\mathbf{m}$$

denotes the "distance" between hypotheses H_1 and H_0. Then to compute $I_0(z)$, we solve equation (3.41), which gives

$$u = (z + \frac{d^2}{2})/d^2 ,$$

and after substitution in expression (3.24) for $I_0(z)$ we obtain

$$I_0(z) = \frac{(z + d^2/2)^2}{2d^2} ,$$

which is again quadratic in z. Setting $z = 0$ gives

$$C(f_1, f_0) = \frac{d^2}{8} . \tag{3.46}$$

Note also that in this case the expressions (3.45) for the KL divergence reduce to

$$D(f_0|f_1) = D(f_1|f_0) = \frac{d^2}{2} ,$$

so the KL divergence and Chernoff distance are both proportional to d^2. To illustrate the above results, the rate functions $I_0(z)$ and $I_1(z) = I_0(z) - z$ are plotted in Fig. 3.5 for Gaussian distributions with unequal means such that $d = 1$.

To verify the asymptotic expression (3.37) for the probability of error of a LRT with threshold $\gamma = \ln(\tau) = 0$, we note from expressions (2.70) that for

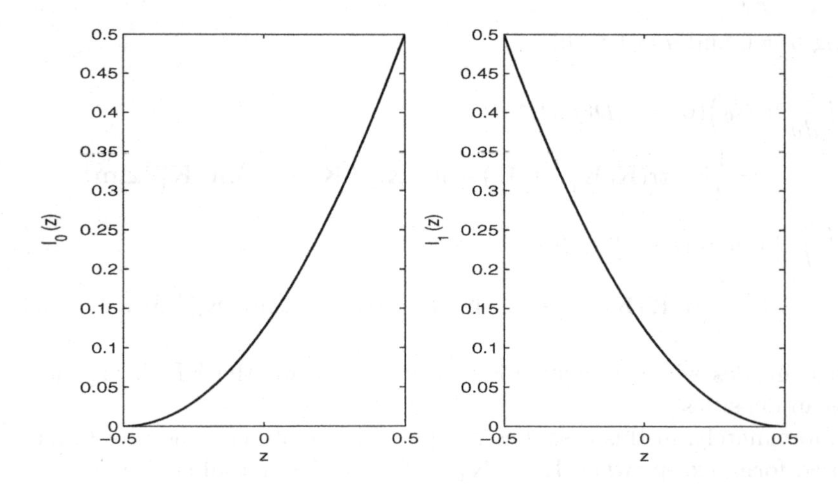

Fig. 3.5. Rate functions $I_0(z)$ and $I_1(z)$ for testing Gaussian distributions with $d = 1$.

a test with N i.i.d. Gaussian observations where the distance per observation is d, the probabilities of false alarm and of a miss are given by

$$P_F(N) = P_M(N) = Q(\frac{N^{1/2}d}{2}) .$$

But as x become large, $Q(x)$ admits the asymptotic approximation [14, p. 39]

$$Q(x) \approx \frac{1}{(2\pi)^{1/2}x} \exp\big(-\frac{x^2}{2}\big) .$$

This implies that for large N,

$$P_F(N) = P_M(N) \approx \frac{2^{1/2}}{(\pi N)^{1/2}d} \exp\big(-\frac{Nd^2}{8}\big) ,$$

which matches expressions (3.37) and (3.46) for the probability of error and the Chernoff distance. □

Finally, as noted in [8, p. 95], the rate functions $I_0(z)$ and $I_1(z)$ admit an interesting geometric interpretation. For the case when $\mathbf{Y}$ is continuous-valued with densities $f_0(\mathbf{y})$ and $f_1(\mathbf{y})$ under H_0 and H_1, consider the function

$$f_u(\mathbf{y}) = \frac{f_1^u(\mathbf{y}) f_0^{1-u}(\mathbf{y})}{G_0(u)} \tag{3.47}$$

Since it is nonnegative for all $\mathbf{y}$ and its integral is one, it is a probability density for $0 \leq u \leq 1$. As u varies from 0 to 1, this density connects continuously the densities f_0 and f_1, and it is shown in [15,16] that in a differential geometric sense, it forms a geodesic (a curve with zero acceleration) linking f_0 to f_1. The KL divergences between f_u and f_j with $j = 0,\ 1$ satisfy

$$\begin{aligned} D(f_u|f_0) &= \frac{u}{G_0(u)} \int f_1^u f_0^{1-u} \ln\big(\frac{f_1}{f_0}\big) d\mathbf{y} - \ln(G_0(u)) \\ D(f_u|f_1) &= -\frac{(1-u)}{G_0(u)} \int f_1^u f_0^{1-u} \ln\big(\frac{f_1}{f_0}\big) d\mathbf{y} - \ln(G_0(u)) , \end{aligned} \tag{3.48}$$

so that we have

$$u(D(f_u|f_0) - D(f_u|f_1)) - \ln(G_0(u)) = D(f_u|f_0) \tag{3.49}$$

and

$$D(f_u|f_0) - D(f_u|f_1) = \int f_1^u f_0^{1-u} \ln\big(\frac{f_1}{f_0}\big) d\mathbf{y} / G_0(u) = \frac{d}{du} \ln(G_0(u)) . \tag{3.50}$$

From (3.50) and (3.49) we see that if $u = u(z)$ is selected such that

$$z = D(f_u|f_0) - D(f_u|f_1) , \tag{3.51}$$

then
$$I_0(z) = D(f_u|f_0) , \tag{3.52}$$
which in the light of (3.51) implies
$$I_1(z) = D(f_u|f_1) . \tag{3.53}$$
Thus the rate functions $I_0(z)$ and $I_1(z)$ are the "distances" in the KL divergence sense between $f_{u(z)}$ and f_0 and f_1, respectively, where $u(z)$ denotes the horizontal coordinate of the point of tangency of the tangent with slope z to $\ln(G_0(u))$, as indicated in Fig. 3.1. Furthermore, when $z = 0$, the density $f_{u(0)}$ is such that
$$D(f_{u(0)}|f_0) = D(f_{u(0)}|f_1) = C(f_1, f_0) ,$$
i.e., it is located midway (in the KL sense) between f_0 and f_1 on the geodesic linking these two densities, and the Chernoff distance $C(f_1, f_0)$ can be interpreted geometrically as the KL divergence between $f_{u(0)}$ and f_j with $j = 0, 1$.

3.3 Bayesian Sequential Hypothesis Testing

All tests discussed up to this point employ a fixed number of observations. In sequential hypothesis testing, the number of observations is not fixed a priori, and depending on the observed sample values, a decision may be taken after just a few samples, or if the observed samples are inconclusive, a large number of samples may be collected until a decision is reached. We assume that an infinite sequence of independent identically distributed observations $\{\mathbf{Y}_k \ ; \ k \geq 1\}$ is available, but for each realization $\omega \in \Omega$, where Ω denotes the set of all possible realizations, a random number $N(\omega)$ of observations is examined until a decision is reached. Under hypothesis H_j with $j = 0, 1$, depending on whether $\mathbf{Y}_k$ takes continuous or discrete values, it has either the probability density $f(\mathbf{y}|H_j)$ or the probability mass distribution function $p(\mathbf{y}|H_j)$. Each observation takes value in domain $\mathcal{Y}$.

A sequential decision rule is formed by a pair $(\boldsymbol{\phi}, \boldsymbol{\delta})$, where $\boldsymbol{\phi} = \{\phi_n, n \in \mathbb{N}\}$ is a *sampling plan* or *stopping rule*, and $\boldsymbol{\delta} = \{\delta_n, n \in \mathbb{N}\}$ denotes the *terminal decision rule.* The function $\phi_n(\mathbf{y}_1, \dots, \mathbf{y}_n)$ maps $\mathcal{Y}^n$ into $\{0, 1\}$. After observing $\mathbf{Y}_k = \mathbf{y}_k$ for $1 \leq k \leq n$, the choice $\phi_n(\mathbf{y}_1, \dots, \mathbf{y}_n) = 0$ indicates that we should take one more sample, whereas for $\phi_n(\mathbf{y}_1, \dots, \mathbf{y}_n) = 1$ we should stop sampling and make a decision. The random number of observations
$$N(\omega) = \min\{n : \phi_n(\mathbf{Y}_1, \dots, \mathbf{Y}_n) = 1\} \tag{3.54}$$
is called the *stopping time* of the decision rule. The second element of a sequential decision rule is the terminal decision function $\delta_n(\mathbf{y}_1, \dots, \mathbf{y}_n)$ taking values in $\{0, 1\}$, where $\delta_n(\mathbf{y}_1, \dots, \mathbf{y}_n) = 0$ or 1 depending on whether we decide that H_0 holds or H_1 holds based on the observations $Y_k = \mathbf{y}_k$ for $1 \leq k \leq n$. Note that the terminal decision function $\delta_n(\mathbf{y}_1, \dots, \mathbf{y}_n)$ needs to be defined only

for the values $(\mathbf{y}_1, \ldots, \mathbf{y}_n)$ such that $\phi_n(\mathbf{y}_1, \ldots, \mathbf{y}_n) = 1$, i.e., a decision needs to be reached only if we have decided to stop sampling.

From this perspective, a decision rule δ for a test with a fixed number N of observations can be expressed as a sequential decision rule $(\boldsymbol{\phi}, \boldsymbol{\delta})$ with

$$\phi_n(\mathbf{y}_1, \ldots, \mathbf{y}_n) = \begin{cases} 0 \text{ for } n \neq N \\ 1 \text{ for } n = N \end{cases}$$

and

$$\delta_n(\mathbf{y}_1, \ldots, \mathbf{y}_n) = \begin{cases} \delta(\mathbf{y}_1, \ldots, \mathbf{y}_n) \text{ for } n = N \\ \quad \text{undefined} \quad \text{for } n \neq N \, . \end{cases}$$

To express the Bayes risk associated to a sequential decision rule $(\boldsymbol{\phi}, \boldsymbol{\delta})$ we assume that the hypotheses H_0 and H_1 admit the a-priori probabilities

$$\pi_0 = P[H_0] \ , \quad \pi_1 = 1 - \pi_0 = P[H_1] \, . \tag{3.55}$$

For simplicity, we assume that no cost is incurred for a correct decision. The cost of a miss, i.e., the cost of choosing H_0 when H_1 is true, is denoted by C_M, and the cost of a false alarm, i.e., the cost of choosing H_1 when H_0 holds, is written as C_F. We also assume that each observation has a cost $D > 0$. Note that D needs to be strictly positive. Otherwise, it would be optimal to collect an infinite number of observations, which would ensure that the decision is error-free. Then the Bayes risk under hypothesis H_0 is given by

$$R(\boldsymbol{\phi}, \boldsymbol{\delta}|H_0) = C_F P[\delta_N(\mathbf{Y}_1, \ldots, \mathbf{Y}_N) = 1|H_0] + DE[N|H_0] \, , \tag{3.56}$$

where N denotes the stopping time, and under hypothesis H_1, it can be expressed as

$$R(\boldsymbol{\phi}, \boldsymbol{\delta}|H_1) = C_M P[\delta_N(\mathbf{Y}_1, \ldots, \mathbf{Y}_N) = 0|H_1] + DE[N|H_1] \, . \tag{3.57}$$

The overall Bayes risk is therefore given by

$$R(\boldsymbol{\phi}, \boldsymbol{\delta}) = \sum_{j=0}^{1} R(\boldsymbol{\phi}, \boldsymbol{\delta}|H_j)\pi_j \, , \tag{3.58}$$

and an optimum Bayesian sequential decision rule $(\boldsymbol{\phi}_{\mathrm{B}}, \boldsymbol{\delta}_{\mathrm{B}})$ is one that minimizes $R(\boldsymbol{\phi}, \boldsymbol{\delta})$. Its cost viewed as a function of the a-priori probability π_0 is denoted as

$$V(\pi_0) = \min_{\boldsymbol{\phi}, \boldsymbol{\delta}} R(\boldsymbol{\phi}, \boldsymbol{\delta}) \, . \tag{3.59}$$

To characterize the structure of the optimum Bayesian test, following the classical paper by Arrow, Blackwell and Girshick [17], it is convenient to perform a first sample analysis. Specifically, we divide the set of all sequential decision rules into the set

$$\mathcal{S} = \{(\boldsymbol{\phi}, \boldsymbol{\delta}) : \phi_0 = 0\}$$

of decision rules which take at least one sample, so that $\phi_0 = 0$, and those that do not take any sample, in which case $\phi_0 = 1$. The minimum Bayes risk, viewed as a function of π_0, for strategies in $\mathcal{S}$ is denoted as

$$J(\pi_0) = \min_{(\boldsymbol{\phi},\boldsymbol{\delta})\in\mathcal{S}} R(\boldsymbol{\phi},\boldsymbol{\delta}) \tag{3.60}$$

Because $N \geq 1$ for all strategies in $\mathcal{S}$, we have $J(\pi_0) \geq D$ and in fact

$$J(0) = J(1) = D\,,$$

because no error occurs when making a decision if the correct hypothesis is known a-priori. Also, since $R(\boldsymbol{\phi},\boldsymbol{\delta})$ is linear and thus concave in π_0, according to Proposition 1.2.4 [11], the function $J(\pi_0)$ obtained by minimizing $R(\boldsymbol{\phi},\boldsymbol{\delta})$ over $\mathcal{S}$ must be concave with respect to π_0. Now, if we consider the decision rules that do not take any sample, only two decisions are possible $\delta_0 = 1$ or $\delta_0 = 0$. The Bayes risks for these decisions are

$$\begin{aligned} R(\phi_0 = 1, \delta_0 = 1) &= C_F\pi_0 \\ R(\phi_0 = 1, \delta_0 = 0) &= C_M(1-\pi_0)\,. \end{aligned} \tag{3.61}$$

The minimum Bayes risk for strategies that do not take any sample is therefore given by the piecewise linear function

$$\begin{aligned} T(\pi_0) &= \min\{C_F\pi_0, C_M(1-\pi_0)\} \\ &= \begin{cases} C_F\pi_0 & \text{for } \pi_0 < \dfrac{C_M}{C_F+C_M} \\ C_M(1-\pi_0) & \text{for } \pi_0 > \dfrac{C_M}{C_F+C_M} \end{cases} \end{aligned} \tag{3.62}$$

which is plotted in Fig. 3.6. We can therefore conclude that the Bayes risk $V(\pi_0)$ satisfies

$$V(\pi_0) = \min\{T(\pi_0), J(\pi_0)\}\,. \tag{3.63}$$

Then, two cases are possible depending on whether $J(\pi_0)$ always remains above the piecewise linear function $T(\pi_0)$ or whether it intersects it at two points π_L and π_U with $\pi_L < \pi_U$, as shown in Fig. 3.6. In the first case, observations are so costly that it is always better to make a decision immediately, without taking any observation. The optimum strategy in this case is

$$\delta_0 = \begin{cases} 1 \text{ if } \pi_0 \leq \dfrac{C_M}{C_M+C_F} \\ 0 \text{ if } \pi_0 > \dfrac{C_M}{C_M+C_F}\,. \end{cases} \tag{3.64}$$

In the second case, which is more interesting, the optimum strategy is

$$\begin{aligned} \phi_0 = 1\ ,\ \ \delta_0 = 1 \text{ for } \pi_0 \leq \pi_L \\ \phi_0 = 1\ ,\ \ \delta_0 = 0 \text{ for } \pi_0 \geq \pi_U \end{aligned} \tag{3.65}$$

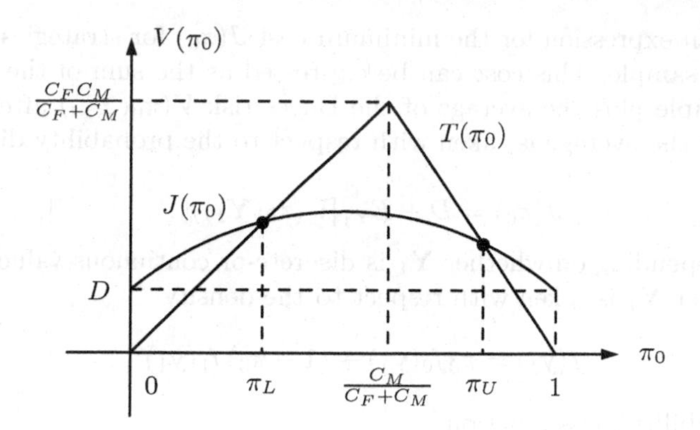

Fig. 3.6. Plot of the Bayes risk $V(\pi_0)$ obtained as the minimum of the nonsampling risk $T(\pi_0)$ and the risk function $J(\pi_0)$ for strategies in $\mathcal{S}$.

and

$$\phi_0 = 0 \quad \text{for} \quad \pi_L < \pi_0 < \pi_U \,. \tag{3.66}$$

So we decide H_1 or H_0 if π_0 falls below a low threshold π_L or above a high threshold π_U, respectively, and we take at least one sample in the intermediate range $\pi_L \leq \pi_0 \leq \pi_U$.

When $\pi_L < \pi_0 < \pi_U$, we must decide what to do after obtaining a sample $\mathbf{Y}_1 = \mathbf{y}_1$. At this point, we observe that the problem takes the same form as before, since the costs C_M, C_F and D are the same, except that the a-priori probability π_0 is replaced by the conditional probability

$$\pi_0(\mathbf{y}_1) = P[H_0|\mathbf{Y}_1 = \mathbf{y}_1] \,. \tag{3.67}$$

Depending on whether $\mathbf{Y}_1$ is continuous-or discrete-valued, by applying Bayes' rule, $\pi_0(\mathbf{y}_1)$ can be expressed as

$$\pi_0(\mathbf{y}_1) = \frac{\pi_0 f_0(\mathbf{y}_1)}{\pi_0 f_0(\mathbf{y}_1) + (1 - \pi_0) f_1(\mathbf{y}_1)} \,, \tag{3.68}$$

or as

$$\pi_0(\mathbf{y}_1) = \frac{\pi_0 p_0(\mathbf{y}_1)}{\pi_0 p_0(\mathbf{y}_1) + (1 - \pi_0) p_1(\mathbf{y}_1)} \,, \tag{3.69}$$

where $f_j(\mathbf{y}_1)$ and $p_j(\mathbf{y}_1)$ denote respectively the probability density or the probability mass function of $\mathbf{Y}_1$ under H_j with $j = 0,\ 1$. So the optimum decision rule is to choose H_1 if $\pi_0(\mathbf{y}_1) \leq \pi_L$, to choose H_0 if $\pi_0(\mathbf{y}_1) \geq \pi_U$, and to take another sample if if $\pi_L < \pi_0(\mathbf{y}_1) < \pi_U$. From this analysis, we can obtain an expression for the minimum cost $J(\pi_0)$ for strategies that take at least one sample. This cost can be expressed as the sum of the cost D of the first sample plus the average of the Bayes risk $V(\pi_0(\mathbf{Y}_1))$ after the first stage, where the average is taken with respect to the probability distribution of $\mathbf{Y}_1$, i.e.,

$$J(\pi_0) = D + E_{\mathbf{Y}_1}[V(\pi_0(\mathbf{Y}_1))] \,. \tag{3.70}$$

In (3.70), depending on whether $\mathbf{Y}_1$ is discrete-or continuous-valued, the expectation over $\mathbf{Y}_1$ is taken with respect to the density

$$f(\mathbf{y}_1) = \pi_0 f_0(\mathbf{y}_1) + (1 - \pi_0) f_1(\mathbf{y}_1)$$

or the probability mass function

$$p(\mathbf{y}_1) = \pi_0 p_0(\mathbf{y}_1) + (1 - \pi_0) p_1(\mathbf{y}_1) \,.$$

Combining (3.63) and (3.70), we find that $V(\pi_0)$ satisfies the functional equation

$$V(\pi_0) = \min\{T(\pi_0), D + E_{\mathbf{Y}_1}[V(\pi_0(\mathbf{Y}_1))]\} \,. \tag{3.71}$$

Let us continue the induction argument and assume that n samples $\mathbf{Y}_k = \mathbf{y}_k$ with $1 \leq k \leq n$ have been obtained. Let $\pi_0(\mathbf{y}_1, \ldots, \mathbf{y}_n)$ denote the conditional probability of H_0 given $\mathbf{Y}_k = \mathbf{y}_k$ for $1 \leq k \leq n$. It can be computed by using Bayes's rule, in the same manner as in (3.68) and (3.69). For example, when the observations are continuous-valued we have

$$\begin{aligned} \pi_0(\mathbf{y}_1, \ldots, \mathbf{y}_n) &= \frac{\pi_0 \prod_{k=1}^n f_0(\mathbf{y}_k)}{\pi_0 \prod_{k=1}^n f_0(\mathbf{y}_k) + (1 - \pi_0) \prod_{k=1}^n f_1(\mathbf{y}_k)} \\ &= \frac{\pi_0}{\pi_0 + (1 - \pi_0) L_n(\mathbf{y}_1, \ldots, \mathbf{y}_n)} \end{aligned} \tag{3.72}$$

where

$$L_n(\mathbf{y}_1, \ldots, \mathbf{y}_n) = \prod_{k=1}^n f_1(\mathbf{y}_n)/f_0(\mathbf{y}_n) \tag{3.73}$$

denotes the likelihood ratio of the two hypotheses based on the first n observations. Then by applying the same reasoning as for the first stage, we conclude that we choose H_1 for $\pi_0(\mathbf{y}_1, \ldots, \mathbf{y}_n) \leq \pi_L$, we choose H_0 for $\pi_0(\mathbf{Y}_1, \ldots, \mathbf{y}_n) \geq \pi_U$ and we take another sample in the intermediate range $\pi_L < \pi_0(\mathbf{y}_1, \ldots, \mathbf{y}_n) < \pi_U$. Thus we have found that the optimal Bayesian decision rule is given by the sampling plan

$$\phi_{\mathrm{B}n}(\mathbf{y}_1, \ldots, \mathbf{y}_n) = \begin{cases} 0 \text{ if } \pi_L < \pi_0(\mathbf{y}_1, \ldots, \mathbf{y}_n) < \pi_U \\ 1 \text{ otherwise}, \end{cases} \tag{3.74}$$

and the terminal decision rule is

$$\delta_{\mathrm{B}n}(\mathbf{y}_1, \ldots, \mathbf{y}_n) = \begin{cases} 1 \text{ for } \pi_0(\mathbf{y}_1, \ldots, \mathbf{y}_n) \leq \pi_L \\ 0 \text{ for } \pi_0(\mathbf{y}_1, \ldots, \mathbf{y}_n) \geq \pi_U \,. \end{cases} \tag{3.75}$$

By using expression (3.72), this decision rule can be expressed equivalently in terms of the likelihood function L_n as

$$\phi_{\mathrm{B}n}(\mathbf{y}_1, \dots, \mathbf{y}_n) = \begin{cases} 0 \text{ if } A < L_n(\mathbf{y}_1, \dots, \mathbf{y}_n) < B \\ 1 \text{ otherwise}, \end{cases} \tag{3.76}$$

and

$$\delta_{\mathrm{B}n}(\mathbf{y}_1, \dots, \mathbf{y}_n) = \begin{cases} 1 \text{ for } L_n(\mathbf{y}_1, \dots, \mathbf{y}_n) \geq A \\ 0 \text{ for } L_n(\mathbf{y}_1, \dots, \mathbf{y}_n) \leq B, \end{cases} \tag{3.77}$$

where the lower and upper thresholds A and B are given by

$$A = \frac{\pi_0(1 - \pi_U)}{(1 - \pi_0)\pi_U} \qquad B = \frac{\pi_0(1 - \pi_L)}{(1 - \pi_0)\pi_L}. \tag{3.78}$$

Thus the optimum Bayesian rule is expressed as a *sequential probability ratio test* (SPRT) where we keep sampling as long as the likelihood ratio L_n stays between A and B, and we select H_0 or H_1 as soon as L_n falls below A, or exceeds B, respectively. From the form (3.78) of the thresholds, note that since $\pi_L < \pi_0 < \pi_U$ is required to ensure that we pick at least one sample, we have $A < 1 < B$.

So at this point we have completely characterized the structure of the optimum Bayesian decision rule, but the two most important quantities needed for its implementation, namely the values of π_L and π_U, or equivalently A and B, are still unspecified. Ideally, it would also be useful to evaluate the Bayesian risk function $V(\pi_0)$, since this function can be used to find the minimax Bayesian decision rule by selecting

$$\pi_{0\mathrm{M}} = \arg\max_{\pi_0} V(\pi_0) \tag{3.79}$$

as the least favorable a-priori probability for H_0. Conceptually, $V(\pi_0)$ and thus π_L and π_U can be obtained by solving the functional equation (3.71), but this equation cannot usually be solved in closed form. So, in practice, $V(\pi_0)$, π_L and π_U need to be evaluated numerically in terms of the costs C_F, C_M, D and the probability densities $f_j(\mathbf{y})$ (resp. the probability mass functions $p_j(\mathbf{y})$) for $j = 0,\ 1$. This can be accomplished by using a backward dynamic programming method which is described in [18] and [19]. In this method, the infinite horizon sequential hypothesis testing problem is truncated to a finite horizon, so that at the most a fixed number K of observations can be collected. Then by proceeding backward in time and letting K increase, it is possible to compute a sequence of functions $V_K(\pi_0)$ and thresholds $\pi_L(K)$, $\pi_U(K)$ which converge to $V(\pi_0)$, π_L and π_U as $K \to \infty$. In this respect, it is worth noting that a significant defect of sequential hypothesis testing is that, even though a decision is almost surely reached in finite time as will be proved in the next section, we cannot guarantee that a decision is reached by a fixed time. While this aspect is of little consequence in applications such as quality control, it is unacceptable for safety related hypothesis testing problems. To fix this problem, it is customary to consider a truncated sequential hypothesis testing problem where the maximum number K of observations is fixed.

3.4 Sequential Probability Ratio Tests

We have just shown that the optimal sequential Bayesian decision rule takes the form of an SPRT. It is therefore of interest to examine the properties of such tests. Consider an SPRT with lower and upper thresholds A and B such that

$$0 < A < 1 < B < \infty \,. \tag{3.80}$$

We select H_0 whenever

$$\prod_{k=1}^{n} L(\mathbf{Y}_k) \leq A \,,$$

H_1 whenever

$$\prod_{k=1}^{n} L(\mathbf{Y}_k) \geq B \,,$$

and we take another sample if

$$A < \prod_{k=1}^{n} L(\mathbf{Y}_k) < B \,.$$

It is convenient to express this test in logarithmic form. Thus let

$$\Lambda_n = \sum_{k=1}^{n} Z_k \tag{3.81}$$

with $Z_k = \ln(L(\mathbf{Y}_k))$, and let $a = \ln(A)$, $b = \ln(B)$. Note that the random variables Z_k are independent identically distributed with mean $m_0 < 0$ and $m_1 > 0$ under H_0 and H_1, respectively. Depending on whether $\mathbf{Y}_k$ is continuous-valued or discrete-valued, $m_0 = -D(f_0|f_1)$ or $-D(p_0|p_1)$, and $m_1 = D(f_1|f_0)$ or $D(p_1|p_0)$. Also, the assumption (3.80) ensures that b and a are finite with $a < 0 < b$. Then, as shown in Fig. 3.7, the SPRT(a,b) is such that we select H_0 whenever $\Lambda_n \leq a$, H_1 whenever $\Lambda_n \geq b$, and we keep sampling if

$$a < \Lambda_n < b \,. \tag{3.82}$$

Since the random walk Λ_n is the sum of n i.i.d. random variables with means $m_0 < 0$ and $m_1 > 0$ under H_0 and H_1, it can be viewed as obtained by adding zero-mean fluctuations to the mean trajectories nm_0 or nm_1, depending on whether H_0 or H_1 holds. The mean nm_0 (resp. nm_1) drags Λ_n towards the lower (resp. upper) threshold, so when H_0 holds Λ_n will in general cross the lower threshold a, resulting in the correct decision. Under H_0, an incorrect decision occurs only when the fluctuations are so large that they overwhelm the drift and create an excursion above b before Λ_n has had the chance to cross the threshold a. Obviously, the lower and higher we select the thresholds a and b, respectively, the less likely it is that such an excursion will take place.

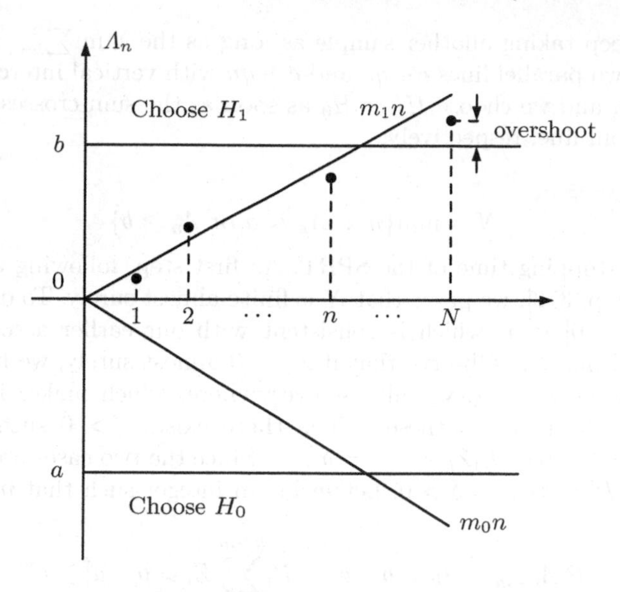

Fig. 3.7. Illustration of a realization of the SPRT(a,b). The lines $m_0 n$ and $m_1 n$ represent the means of the random walk Λ_n under H_0 and H_1.

Example 3.1, continued

Consider a sequence $\{Y_k, k \geq 1\}$ of i.i.d. Poisson random variables, where under hypothesis H_j, Y_k has parameter λ_j, so that

$$P[Y_k = n|H_j] = \frac{\lambda_j^n}{n!} \exp(-\lambda_j)$$

with $n \in \mathbb{N}$. Then

$$\Lambda_n = \sum_{k=1}^{n} \ln \left[(\lambda_1/\lambda_0)^{Y_k} \exp(-(\lambda_1 - \lambda_0)) \right]$$

$$= \ln(\lambda_1/\lambda_0) \sum_{k=1}^{n} Y_k - n(\lambda_1 - \lambda_0)$$

and, if we assume that $\lambda_1 > \lambda_0$, the SPRT rule for taking another sample can be expressed as

$$c + \eta n < \sum_{k=1}^{n} Y_k < d + \eta n$$

with

$$c \triangleq \frac{a}{\ln(\lambda_1/\lambda_0)} \quad , \quad d \triangleq \frac{b}{\ln(\lambda_1/\lambda_0)} \, ,$$

and

$$\eta \triangleq \frac{\lambda_1 - \lambda_0}{\ln(\lambda_1/\lambda_0)} > 0 \, .$$

Thus, we keep taking another sample as long as the sum $\sum_{k=1}^{n} Y_k$ stays between the two parallel lines $c + \eta n$ and $d + \eta n$ with vertical intercepts c and d and slope η, and we choose H_1 or H_0 as soon as the sum crosses the top line or the bottom line, respectively. □

Let

$$N = \min\{n \ : \ \Lambda_n \leq a \text{ or } \Lambda_n \geq b\} \tag{3.83}$$

denote the stopping time of the SPRT. As first step, following an argument given in [20, p. 372], we prove that N is finite almost surely. To do so, assume that $P[Z_k = 0] < 1$, which is consistent with our earlier assumption that $m_1 > 0$ and $m_0 < 0$. Observe that if $Z_k = 0$ almost surely, we have $f_1(\mathbf{y}) = f_0(\mathbf{y})$ (resp. $p_1(\mathbf{y}) = p_0(\mathbf{y})$) almost everywhere, which makes it impossible to separate the two hypotheses. Then there exists $\epsilon > 0$ such that either $P[Z_k > \epsilon] = \delta > 0$ or $P[Z_k < -\epsilon] = \delta > 0$. Since the two cases are symmetric, we assume $P[Z_k > \epsilon] = \delta > 0$. Let m be an integer such that $m\epsilon > (b - a)$. Then

$$P[\Lambda_{k+m} - \Lambda_k > b - a] = P[\sum_{k+1}^{k+m} Z_i > b - a] \geq \delta^m$$

for all k. This implies that for all integers ℓ

$$P[N > \ell m] = P[a < S_k < b \text{ for } k = 1, \ldots, \ell m] < (1 - \delta^m)^\ell .$$

Let $c = (1 - \delta^m)^{-1}$ and $r = (1 - \delta^m)^{1/m} < 1$. For an arbitrary n, we can find ℓ such that $\ell m < n < (\ell + 1)m$ and

$$P[N > n] \leq P[N > \ell m] \leq (1 - \delta^m)^\ell = cr^{(\ell+1)m} \leq cr^n .$$

This implies $P[N < \infty] = 1$. Also the geometric decay of $P[N > n]$ implies that the generating function $E[\exp(uN)]$ is defined for $u < -\ln(r)$, so that N has finite moments.

Next, we introduce two concepts that will be needed in the discussion below. We say that a random process M_n with $n \geq 0$ is a *martingale* if it satisfies the following two properties: (i) $E[|M_n|] < \infty$ and (ii)

$$E[M_{n+1}|M_k \ , \ 0 \leq k \leq n] = M_n . \tag{3.84}$$

We also say that a nonnegative random variable N is a *stopping time* adapted to the martingale M_n if the event $N = n$ can be expressed in terms of the values M_k for $0 \leq k \leq n$. In other words, the stopping rule $N = n$ must depend on the past of the process M_n up to time n. For our purposes we restrict our attention to stopping times which are finite, i.e., $P[N < \infty] = 1$.

To see how these definitions are related to the SPRT, let $\{Z_k \ ; \ k \geq 1\}$ be an infinite sequence of independent identically distributed random variables whose generating function $G(u) = E[\exp(uZ)]$ is finite over an interval (u_-, u_+) with $u_- < 0 < u_+$. Then, as indicated in (3.81), let Λ_n be the sum of the random variables Z_k up to time n, and consider the process

$$M_n = \frac{\exp(u\Lambda_n)}{G(u)^n} \tag{3.85}$$

with $M_0 = 1$, where u is in the interval (u_-, u_+). This process is a martingale. To see this, note that since the random variables Z_k are independent

$$E[M_{n+1}|M_k \ , \ 0 \leq k \leq n] = M_n E[\exp(uZ_{n+1})]/G(u) = M_n \ . \tag{3.86}$$

Note also that since M_n is positive, we have

$$E[|M_n|] = E[M_n] = E[M_0] = 1 \ ,$$

where the second equality is a consequence of the martingale property. Also, if N is defined as in (3.83), it is a stopping time adapted to Λ_n and thus M_n, since $N = n$ whenever $a < \Lambda_k < b$ for $1 \leq k \leq n-1$ and $\Lambda_n \leq a$ or $\Lambda_n \geq b$. In other words, $N = n$ is expressible entirely as a condition on the sample path of the process Λ_k up to time n. However it is possible to construct other adapted stopping times for the process M_n.

We are now in a position to establish the main result from which all properties of the SPRT follow.

Wald's identity: Let $\{Z_k \ ; k \geq 1\}$ be a sequence of i.i.d. random variables whose generating function $G(u)$ is defined over (u_-, u_+) with $u_- < 0 < u_+$ and let M_n be the martingale (3.85) with $u_- < u < u_+$. Let N be an adapted stopping time such that

$$\lim_{n \to \infty} E[M_n|N > n]P[N > n] = 0 \, . \tag{3.87}$$

Then

$$E[M_N] = E[\frac{\exp(u\Lambda_N)}{G(u)^N}] = 1 \, . \tag{3.88}$$

This result is a consequence of Doob's optional sampling theorem [21, p. 261], which states that if M_n is a martingale and N is a finite stopping time satisfying (3.87), then $E[M_N] = E[M_0]$. Note that the condition (3.87) cannot be omitted. We refer the reader to [22, pp. 328–339] for a proof that the stopping time N of an SPRT satisfies this condition. Wald's identity contains all the information we need to perform a complete analysis of the SPRT. In analyzing the SPRT, we rely on the "zero-overshoot" approximation introduced by Wald [1] in his original study of the SPRT. Whenever $\Lambda_N \leq b$ or $\Lambda_N \geq a$, it consists in approximating Λ_N by b and a respectively. To develop an intuitive understanding of this approximation consider Fig. 3.7. From the figure, we see that when $N = n$, the random walk Λ_k has stayed between b and a for $1 \leq k \leq n-1$, and the addition of the last random variable Z_n was just enough to make Λ_n cross either the lower threshold a or the upper threshold b. But if the increments Z_k are small compared to the size of the thresholds a and b, it is reasonable to assume that Λ_n cannot have strayed too far from the threshold it just crossed, so the overshoot can be neglected.

For the case when $Z_k = \ln(L(\mathbf{Y}_k))$ is the log-likelihood ratio of the probability distributions of $\mathbf{Y}_k$ under H_1 and H_0, we recall from (3.20) that the generating functions $G_0(u)$ and $G_1(u)$ of Z_k under H_0 and H_1 satisfy

$$\begin{aligned} G_0(0) &= G_0(1) = 1 \\ G_1(-1) &= G_1(0) = 1 \,. \end{aligned} \tag{3.89}$$

We note also that

$$P_F = P[\Lambda_N \geq b|H_0] \quad , \quad P_M = P[\Lambda_N \leq a|H_1] \,. \tag{3.90}$$

Then we evaluate (3.88) under hypothesis H_0 with $u = 1$. We obtain

$$1 = BP_F + A(1 - P_F) \tag{3.91}$$

where we have used the zero-overshoot approximation in setting

$$\begin{aligned} E[\exp(\Lambda_N)|\Lambda_N \geq b, H_0] &\approx \exp(b) = B \\ E[\exp(\Lambda_N)|\Lambda_N \leq a, H_0] &\approx \exp(a) = A \,. \end{aligned} \tag{3.92}$$

From (3.91) we find

$$P_F = \frac{1 - A}{B - A} \,. \tag{3.93}$$

Similarly, if we evaluate (3.88) under H_1 with $u = -1$, we find

$$1 = B^{-1}(1 - P_M) + A^{-1} P_M \tag{3.94}$$

which yields

$$P_M = \frac{1 - B^{-1}}{A^{-1} - B^{-1}} = \frac{A(B-1)}{B - A} \,. \tag{3.95}$$

This shows that for an SPRT, under the zero-overshoot approximation, the probabilities of false alarm and of a miss can be expressed directly in terms of the thresholds A and B. Conversely, we have

$$A = \frac{P_M}{1 - P_F} \quad , \quad B = \frac{1 - P_M}{P_F} = \frac{P_D}{P_F} \,, \tag{3.96}$$

so that given any desired probabilities of false alarm and of a miss P_F and P_M, we can set the thresholds accordingly.

Next, if we differentiate (3.88) once with respect to u, we obtain

$$0 = E[(\Lambda_N - N\frac{d\ln(G)}{du}(u))M_N(u)] \,, \tag{3.97}$$

where the dependence of M_n on u is denoted explicitly. We evaluate this identity at $u = 0$ under H_0 and H_1, where as indicated in (3.23) we have

$$\frac{d\ln(G_0)}{du}(0) = m_0 \quad , \quad \frac{d\ln(G_1)}{du}(0) = m_1 \,. \tag{3.98}$$

We recall that m_0 and m_1 are assumed to be nonzero, since the probability distributions of $\mathbf{Y}_k$ under H_0 and H_1 are undistinguishable otherwise. Under H_0, this gives

$$\begin{aligned} E[N|H_0]m_0 &= E[\Lambda_N|H_0] = bP_F + a(1-P_F) \\ &= \frac{(1-A)\ln(B) + (B-1)\ln(A)}{B-A} \,, \end{aligned} \tag{3.99}$$

and under H_1 we obtain

$$\begin{aligned} E[N|H_1]m_1 &= E[\Lambda_N|H_1] = aP_M + b(1-P_M) \\ &= \frac{(B-1)A\ln(A) + (1-A)B\ln(B)}{B-A} \,. \end{aligned} \tag{3.100}$$

Equivalently, the expected number of samples under the two hypotheses can be expressed in terms of P_F and P_M as

$$\begin{aligned} E[N|H_0] &= \frac{1}{m_0}[P_F\ln(\frac{1-P_M}{P_F}) + (1-P_F)\ln(\frac{P_M}{1-P_F})] \\ E[N|H_1] &= \frac{1}{m_1}[P_M\ln(\frac{P_M}{1-P_F}) + (1-P_M)\ln(\frac{1-P_M}{P_F})] \,. \end{aligned} \tag{3.101}$$

By differentiating (3.88) a second time, we obtain also

$$0 = E[(\Lambda_N - N\frac{d\ln(G)}{du}(u))^2 M_N(u)] - E[N\frac{d^2\ln(G)}{du^2}M_N(u)] \,. \tag{3.102}$$

Setting $u = 0$ and evaluating this identity under H_j we obtain

$$E[(\Lambda_N - Nm_j)^2|H_j] = \sigma_j^2 E[N|H_j] \tag{3.103}$$

with

$$\sigma_j^2 = \frac{d^2\ln(G_j)}{du^2}(0) = E[(Z_k - m_j)^2|H_j] \,,$$

which can be used to characterize the second order statistics of N under H_j.

Comparison between sequential and fixed size tests: The expressions (3.101) for the expected number of samples in a sequential test under H_0 and H_1 can be used to show that, on average, sequential tests require fewer samples than fixed size tests. Specifically, assume that we select $P_F \ll 1$ and $P_M \ll 1$. Then by using $x\ln(x) \approx 0$ for small x, the expressions (3.101) yield the approximations:

$$\begin{aligned} E[N|H_0] &\approx -\frac{\ln(P_M)}{D(f_0|f_1)} \\ E[N|H_1] &\approx -\frac{\ln(P_F)}{D(f_1|f_0)} \,. \end{aligned} \tag{3.104}$$

Consider first the case $P_F = P_M = P_E$. By comparison, for a fixed size test, it was shown in Section 3.2 that the threshold maximizing the decay of the probability of error is $\gamma = 0$ and in this case, according to (3.37), the size N of the test needed to achieve a probability of error P_E is given by

$$N \approx -\frac{\ln(P_E)}{C(f_1, f_0)} . \tag{3.105}$$

But the Chernoff distance $C(f_1, f_0)$ is always less than the KL divergences $D(f_0|f_1)$ and $D(f_1|f_0)$. In fact in the Gaussian case where $\mathbf{Y}_k \sim N(\mathbf{m}_j, \mathbf{K})$ under hypothesis H_j with $j = 0,\ 1$, it was shown in Example 3.2 that

$$D(f_0|f_1) = D(f_1|f_0) = \frac{d^2}{2}$$

and

$$C(f_1, f_0) = \frac{d^2}{8} ,$$

where d denotes the "distance" between the two hypotheses. So in this case the SPRT requires on average only 25% of the observations needed by a fixed size test to achieve the same probability of error.

It is also instructive to compare expressions (3.104) for the expected number of samples of an SPRT to asymptotic approximations (3.33) and (3.34) for the probabilities of a miss $P_M(N)$ and of false alarm $P_F(N)$ of the Neyman-Pearson tests of types I and II as the size N of the test becomes large. The Neyman-Pearson test of type I minimizes P_M while keeping P_F below a fixed value α. From (3.33) we see that the size N of an NP test of type I required to achieve P_M is given by

$$N \approx -\frac{\ln(P_M)}{D(f_0|f_1)} .$$

Similarly, the NP test of type II minimizes P_F while keeping P_M below a fixed value β. The size needed by such a test to achieve P_F is

$$N \approx -\frac{\ln(P_F)}{D(f_1|f_0)} .$$

Comparing (3.104) to these expressions, we see that, on average, the SPRT achieves the decay, maximizing rates of both NP tests simultaneously. But since identities (3.104) concern the expected stopping time N under H_0 and H_1, they only characterize the average behavior of the SPRT; so for any given test the number of collected samples may be higher or lower than for NP tests of types I and II.

3.5 Optimality of SPRTs

It turns out that SPRTs have a strong optimality property in the class of sequential tests, which, as noted earlier, includes fixed-size tests. Specifically, let (ϕ, δ) denote an SPRT with thresholds A and B. Let (ϕ', δ') be an arbitrary sequential decision rule such that

$$P_F(\phi', \delta') \leq P_F(\phi, \delta) \quad \text{and} \quad P_M(\phi', \delta') \leq P_M(\phi, \delta) . \tag{3.106}$$

Then

$$E[N|H_0, \phi] \leq E[N|H_0, \phi'] \tag{3.107}$$

and

$$E[N|H_1, \phi] \leq E[N|H_1, \phi'] , \tag{3.108}$$

so that among all sequential decision rules which achieve certain probabilities of false alarm and of a miss, the SPRT requires on average the smallest number of samples under hypotheses H_0 and H_1.

This result is remarkable, since among sequential tests (ϕ', δ') satisfying (3.106), it would have been reasonable to expect that it might be possible to trade off $E[N|H_0, \phi']$ against $E[N|H_1, \phi']$ and to be able to slip one below the value attained by SPRT test (ϕ, δ) while keeping the other above the corresponding SPRT value. But it turns out that this is not the case and, under both hypotheses, the SPRT minimizes the expected number of observations required to achieve a certain performance level. The derivation of (3.107)–(3.108) is rather intricate and only the main line of the argument is provided. The reader is referred to [20, Sec. 7.6] and [23] for a detailed proof. As starting point, observe that it was shown in Section 3.3 that the optimum decision rule for a sequential Bayesian decision problem is an SPRT. To establish the optimality of SPRT (ϕ, δ), we perform a reverse mapping and associate to it not one, but two different Bayesian cost structures for which (ϕ, δ) is the optimum decision rule. Note that the SPRT (ϕ, δ) is specified by thresholds A and B with $0 < A < 1 < B < \infty$. On the other hand, the Bayesian risk $R(\phi, \delta)$ is specified by decision costs C_M, C_F, sampling cost D, and the a-priori probability π_0. However, scaling the Bayesian risk by an arbitrary constant does not change the underlying optimization problem, so we can always apply a normalization such that

$$C_M + C_F = 1$$

and thus only parameters C_M, D and π_0 are needed to specify the Bayesian risk. Hence, based on a degrees of freedom argument, we see that starting with thresholds A and B, there should be more than one way to select matchings C_M, D and π_0. It turns out that this is the case and for a vanishingly small $\epsilon > 0$, it is shown in [20] that it is possible to find costs $C_M(\epsilon)$, $D(\epsilon)$ and a-priori probability π_0 with $0 < \pi_0 < \epsilon$ for which the SPRT is optimal. It is also possible to find costs $\bar{C}_M(\epsilon)$, $\bar{D}(\epsilon)$ and a-priori probability π_0 with $1 - \epsilon < \pi_0 < 1$ for which the SPRT is also optimal.

We consider the first set of costs. Due to the optimality of the SPRT (ϕ, δ), for this choice of Bayesian risk, any other sequential test (ϕ', δ') satisfying constraints (3.106) must be such that

$$\begin{aligned}
0 &\leq R(\boldsymbol{\phi}', \boldsymbol{\delta}') - R(\boldsymbol{\phi}, \boldsymbol{\delta}) \\
&= \pi_0 C_F (P_F(\boldsymbol{\phi}', \boldsymbol{\delta}') - P_F(\boldsymbol{\phi}, \boldsymbol{\delta})) \\
&\quad + (1-\pi_0) C_M (P_M(\boldsymbol{\phi}', \boldsymbol{\delta}') - P_M(\boldsymbol{\phi}, \boldsymbol{\delta})) \\
&\quad + \pi_0 D(E[N|H_0, \boldsymbol{\phi}'] - E[N|H_0, \boldsymbol{\phi}]) \\
&\quad + (1-\pi_0) D(E[N|H_1, \boldsymbol{\phi}'] - E[N|H_1, \boldsymbol{\phi}]) \\
&\leq \epsilon D(E[N|H_0, \boldsymbol{\phi}'] - E[N|H_0, \boldsymbol{\phi}]) \\
&\quad + D(E[N|H_1, \boldsymbol{\phi}'] - E[N|H_1, \boldsymbol{\phi}]) \,,
\end{aligned} \tag{3.109}$$

where the last inequality uses inequalities (3.106). But since ϵ is infinitesimally small, this implies (3.108). Proceeding in a similar manner with cost structure $\bar{C}_M$, $\bar{D}$ and $1 - \epsilon < \pi_0 < 1$, we obtain inequality (3.107).

3.6 Bibliographical Notes

With the exception of [5, 24], the asymptotic behavior of LRTs is rarely discussed in engineering texts. This is somewhat unfortunate, as this study provides a strong justification for the use of information theoretic quantities such as the KL divergence or Chernoff distance as measures of separation between statistical hypotheses. Quite surpringly, Cramér's theorem, which is applicable to independent identically distributed random variables, admits an extension to dependent random variables called the Gärtner-Ellis theorem, which is one of the key tools of the theory of large deviations [8, 9]. This result will be introduced later in the text to analyze the asymptotic performance of detectors for stationary Gaussian processes and stationary Markov chains. In addition to Wald's monograph [1], detailed presentations of sequential hypothesis testing can be found in the books by De Groot [18] and Ferguson [20]. See also [25, Chap. 40] and [19, Sec. 5.5] for a stochastic dynamic programming viewpoint of this material.

3.7 Problems

3.1. Consider a sequence $\{Y_k, \, k \geq 1\}$ of i.i.d. observations, where under hypothesis H_j, the probability density of Y_k is exponential with parameter λ_j, i.e.,

$$f_j(y) = \lambda_j \exp(-\lambda_j y) u(y)$$

for $j = 0, 1$. We seek to characterize the rate of decay of the probability of false alarm and of a miss for the LR test

$$\frac{1}{N} \sum_{k=1}^{N} Z_k \underset{H_0}{\overset{H_1}{\gtrless}} \gamma$$

with
$$Z_k = \ln L(Y_k) = \ln \left(\frac{f_1(Y_k)}{f_0(Y_k)} \right)$$
as the length N of the observation block becomes large.

(a) Verify that generating function
$$G_0(u) = E[\exp(uZ_k)|H_0] = \int_0^\infty f_1(y)^u f_0(y)^{1-u} dy$$
of Z_k can be expressed as
$$G_0(u) = \frac{\lambda_1^u \lambda_0^{1-u}}{u\lambda_1 + (1-u)\lambda_0} .$$
(b) Use the expression for $G_0(u)$ obtained in (a) to evaluate the means
$$m_0 = -D(f_0|f_1) \quad , \quad m_1 = D(f_1|f_0)$$
of Z_k under H_0 and H_1, respectively.
(c) Obtain a closed form expression for the rate function $I_0(z)$ obtained by Legendre transformation of $\ln G_0(u)$.
(d) Evaluate the Chernoff distance
$$C(f_1, f_0) = I_0(0) .$$

3.2. Consider a binary hypothesis testing problem where we observe a sequence $\{Y_k, k \geq 1\}$ of i.i.d. $N(m_j, \sigma^2)$ distributed random variables under H_j with $j = 0, 1$.

(a) Use the results of Example 3.2 to evaluate the log-generating function $\ln G_0(u)$ of $Z_k = \ln L(Y_k)$.
(b) Use the expression for $\ln G_0(u)$ obtained in (a) to evaluate the means m_0 and m_1 of Z_k under H_0 and H_1, respectively.
(c) Obtain a closed form expression for the rate function $I_0(z)$ obtained by Legendre transformation of $\ln G_0(u)$.
(d) Evaluate the Chernoff distance
$$C(f_1, f_0) = I_0(0) .$$

3.3. Consider a binary hypothesis testing problem where we observe a sequence $\{Y_k, k \geq 1\}$ of i.i.d. $N(0, \sigma_j^2)$ distributed random variables under H_j with $j = 0, 1$.

(a) Use the results of Example 3.2 to show that the log-generating function of $Z_k = \ln L(Y_k)$ can be expressed as
$$\ln G_0(u) = \frac{1}{2}\Big[-\ln \Big(\frac{u}{\sigma_1^2} + \frac{(1-u)}{\sigma_0^2} \Big) - u \ln(\sigma_1^2) - (1-u) \ln(\sigma_0^2) \Big] .$$

(b) Use the expression for $\ln G_0(u)$ obtained in (a) to evaluate the means m_0 and m_1 of Z_k under H_0 and H_1, respectively.
(c) Obtain a closed form expression for the rate function $I_0(z)$ obtained by Legendre transformation of $\ln G_0(u)$.
(d) Evaluate the Chernoff distance

$$C(f_1, f_0) = I_0(0) .$$

3.4. Consider a binary hypothesis testing problem where we observe a sequence $\{Y_k, k \geq 1\}$ of i.i.d. observations, which are uniformly distributed over interval $[-w_j/2, w_j/2]$ under hypothesis H_j, so that

$$f_j(y) = \begin{cases} 1/w_j & -w_j/2 \leq y \leq w_j/2 \\ 0 & \text{otherwise} , \end{cases}$$

for $j = 0, 1$. Assume $w_1 > w_0$. This problem arises whenever we wish to determine whether the step size of a roundoff quantizer matches its nominal design value.

(a) Evaluate the log-generating function $\ln G_0(u)$ of $Z_k = \ln L(Y_k)$. Note that $G_0(u)$ is discontinuous at $u = 1$.
(b) Use the expression for $G_0(u)$ obtained in (a) to evaluate

$$m_0 = -D(f_0|f_1) .$$

What is the value of $m_1 = D(f_1|f_0)$? Note that you can perform this evaluation directly by using definition (3.4) of the KL divergence.
(c) Evaluate the Legendre transform $I_0(z)$ of $\ln G_0(u)$.

3.5. Consider a sequence $\{Y_k, k \geq 1\}$ of i.i.d. observations where under H_0, the observations are uniformly distributed over $[0, 1]$, so they admit the density

$$f_0(y) = \begin{cases} 1 & 0 \leq y \leq 1 \\ 0 & \text{otherwise} , \end{cases}$$

and under H_1, they admit the probability density

$$f_1(y) = \begin{cases} 4y & 0 \leq y \leq 1/2 \\ 4(1-y) & 1/2 \leq y \leq 1 \\ 0 & \text{otherwise} , \end{cases}$$

which is sketched in Fig. 2.17.

(a) Evaluate the generating function $G_0(u)$ of $Z_k = \ln L(Y_k)$.
(b) Use the expression for $\ln G_0(u)$ obtained in (a) to evaluate the means m_0 and m_1 of Z_k under H_0 and H_1, respectively.
(c) Obtain a closed form expression for the rate function $I_0(z)$ obtained by Legendre transformation of $\ln G_0(u)$.

(d) Evaluate the Chernoff distance

$$C(f_1, f_0) = I_0(0) .$$

3.6. Consider a sequence of i.i.d. observations $\{Y_k, k \geq 1\}$ where under hypothesis H_j, Y_k is a binary random variable with

$$p_j = P[Y_k = 1|H_j] \quad , \quad q_j = 1 - p_j = P[Y_k = 0|H_j]$$

for $j = 0, 1$.

(a) Evaluate the generating function $G_0(u)$ of $Z_k = \ln L(Y_k)$.
(b) Use the expression for $\ln G_0(u)$ obtained in (a) to evaluate the means m_0 and m_1 of Z_k under H_0 and H_1, respectively.
(c) Obtain a closed form expression for the rate function $I_0(z)$ obtained by Legendre transformation of $\ln G_0(u)$.
(d) Evaluate the Chernoff distance

$$C(p_1, p_0) = I_0(0) .$$

3.7. Consider Problem 2.5 where the goal is to detect of a constant $A > 0$ in independent identically distributed Laplacian noise with parameter λ. Thus under the two hypotheses, we have $f_1(y) = f_V(y - A)$ and $f_0(y) = f_V(y)$ with

$$f_V(v) = \frac{\lambda}{2} \exp(-\lambda|v|) .$$

(a) Evaluate the generating function $G_0(u)$ of $Z_k = \ln L(Y_k)$.
(b) Suppose that for a test with N observations for N large, we decide to implement a type I NP test. Evaluate the rate of decay $-\ln(P_M)/N$ of the probability of a miss as N increases. This rate should be expressed as a function of A and λ.

3.8. Consider a sequential hypothesis testing problem where we observe a sequence $\{Y_k, k \geq 1\}$ of i.i.d. $N(m_j, \sigma^2)$ distributed random variables under H_j with $j = 0, 1$. Assume that $m_1 > m_0$.

(a) Show that the SPRT rule for taking another sample can be expressed as

$$c + \eta n < S_n = \sum_{k=1}^{n} Y_k < d + \eta n ,$$

so that we keep taking additional samples as long as the sum S_n stays between two parallel lines with vertical intercepts c and d and slope η. Express the slope and intercepts in terms of m_1, m_0, σ^2, $a = \ln(A)$ and $b = \ln(B)$.
(b) Let $\sigma^2 = 1$, $m_1 = 2$, $m_0 = 1$. Use Wald's zero-overshoot approximation to select A and B such that $P_F = P_M = 10^{-4}$. Evaluate the slope η and intercepts c and d.

(c) For the values of part (b), evaluate the expected number of samples $E[N|H_j]$ for $j = 0, 1$.
(d) Compare the values you have obtained in part (c) with the estimate (3.105) for the number of samples of a fixed sized test achieving a probability of error P_E. What is the reduction in the number of samples required by the SPRT, compared to a fixed size test?

3.9. Consider a sequential hypothesis testing problem where we observe a sequence $\{Y_k, k \geq 1\}$ of i.i.d. $N(0, \sigma_j^2)$ distributed random variables under H_j with $j = 0$, 1. Assume that $\sigma_1 > \sigma_0$.

(a) Show that the SPRT rule for taking another sample can be expressed as

$$c + \eta n < S_n = \sum_{k=1}^{n} Y_k^2 < d + \eta n ,$$

so that we keep taking additional samples as long as the sum S_n stays between two parallel lines with vertical intercepts c and d and slope η. Express the slope and intercepts in terms of σ_1, σ_0, $a = \ln(A)$ and $b = \ln(B)$.
(b) Let $\sigma_0^2 = 1$ and $\sigma_1^2 = 2$. Use the zero-overshoot approximation to select A and B such that $P_F = P_M = 10^{-4}$. Evaluate the slope η and intercepts c and d.
(c) For the values of part (b), evaluate the expected number of samples $E[N|H_j]$ for $j = 0$, 1.
(d) Compare the values you have obtained in part (c) with the estimate (3.105) for the number of samples of a fixed sized test achieving a probability of error P_E. What is the reduction in the number of samples required by the SPRT, compared to a fixed size test? For this problem you will need to use the expression for the Chernoff distance of $N(0, \sigma_j^2)$, $j = 0$, 1 random variables derived in Problem 3.3.

3.10. Consider the sequential hypothesis testing problem

$$H_0 : Y_k = V_k$$
$$H_1 : Y_k = L + V_k$$

for $k \geq 1$, where L with $0 < L < 1$ is a known constant and the noise V_k is white (i.i.d.) uniformly distributed over $[-1/2, 1/2]$.

(a) Suppose that we implement a SPRT test with thresholds $A < 1 < B$. Describe the rule for selecting an additional sample in function of the values of successive observations.
(b) Indicate whether the total number of samples that need to be examined has a fixed (nonrandom) upper bound.
(c) Evaluate the expected number of samples $E[N|H_j]$ for $j = 0$, 1.

3.11. Consider now a Bayesian formulation of the sequential hypothesis testing Problem 3.10. We set $C_F = 1$, $C_M = 2$, $D = 1/10$, and the constant level $L = 1/2$.

(a) Evaluate the Bayes risk function $V(\pi_0)$ satisfying the functional equation (3.71). To do so, observe that the a-posteriori probability $\pi_0(y_1)$ of H_0 after taking the first sample is either 0, π_0, or 1. Sketch $V(\pi_0)$ and specify π_L and π_U.

(b) Verify that the optimum Bayesian decision rule (δ, ϕ) forms an SPRT test. Interpret the value of $V(\pi_0)$ for $\pi_L < \pi_0 < \pi_U$ by using the result obtained in part (c) of Problem 3.10 for the expected number of samples $E[N|H_j]$, $j = 0, 1$.

3.12. Consider the sequential version of the binary test of Problem 3.4. We observe a sequence $\{Y_k,\, k \geq 1\}$ of i.i.d. observations which are uniformly distributed over interval $[-w_1/2, w_1/2]$ under H_1 and over interval $[-w_0/2, w_0/2]$ under H_0. Assume $w_1 > w_0$.

(a) Suppose that we implement an SPRT test with thresholds $A < 1 < B$. Describe the rule for selecting an additional sample in function of the values of successive observations.

(b) Evaluate the probabilities of false alarm and of a miss for the SPRT rule of part (a). Note that these probabilities will depend on A and B, but they will be exact, i.e., you do not need to use the zero-overshoot approximation.

(c) Evaluate the expected number of samples $E[N|H_j]$ for $j = 0, 1]$. Note again that the expressions you obtain do not require any approximation.

3.13. Consider a sequential hypothesis testing problem where we observe a sequence of i.i.d. binomial random variables $\{Y_k,\, k \geq 1\}$ such that

$$p_j = P[Y_k = 1|H_j] \quad , \quad q_j = 1 - p_j = P[Y_k = 0|H_j]$$

for $j = 0,\ 1$. Assume that $p_1 > p_0$.

(a) Let $N_1(n)$ denote the number of observations Y_k which are equal to one, among the first n observations. Show that the SPRT rule for taking another sample can be expressed as

$$c + \eta n < N_1(n) < d + \eta n \ ,$$

so that we keep taking additional samples as long as $N_1(n)$ stays between two parallel lines with vertical intercepts c and d and slope η. Express the slope and intercepts in terms of p_1, p_0, $a = \ln(A)$ and $b = \ln(B)$.

(b) Let $p_1 = 4/5$, $p_0 = 1/5$. Use Wald's zero-overshoot approximation to select A and B such that $P_F = P_M = 10^{-4}$. Evaluate the slope η and intercepts c and d.

(c) For the values of part (b), evaluate the expected number of samples $E[N|H_j]$ for $j = 0, 1$.

(d) Compare the values you have obtained in part (c) with the estimate (3.105) for the number of samples of a fixed sized test achieving a probability of error P_E. What is the reduction in the number of samples achieved by the SPRT, compared to a fixed size test?

3.14. To evaluate the Bayes risk function $V(\pi_0)$ obeying the functional equation (3.71), a common strategy consists of considering a finite horizon version of the sequential hypothesis testing problem. For a Bayesian sequential test with costs C_F, C_M, and D, let $V_k(\pi_0)$ be the risk corresponding to the optimum strategy based on collecting at most k observations. For an optimum sequential test where at most $k+1$ samples are collected, note that if a first observation is collected, then the optimum sequential testing procedure from this point on must be optimal among strategies which allow at most k samples.

(a) Use this last observation to prove that $V_k(\pi_0)$ satisfies the backward recursion

$$V_{k+1}(\pi_0) = \min\{T(\pi_0), D + E_{\mathbf{Y}_1}[V_k(\pi_0(\mathbf{Y}_1))]\}$$

for $k \geq 0$, with initial condition $V_0(\pi_0) = T(\pi_0)$. Hence the Bayesian infinite horizon risk function

$$V(\pi_0) = \lim_{k\to\infty} V_k(\pi_0).$$

(b) Consider a Bayesian formulation of sequential hypothesis testing Problem 3.13. We set $C_F = 1$, $C_M = 1$, $D = 1/10$, $p_1 = 4/5$ and $p_0 = 1/5$. Evaluate risk functions $V_k(\pi_0)$ for $k \geq 0$ and obtain $V(\pi_0)$. Specify π_L and π_U.

3.A. Proof of Cramér's Theorem

To keep things simple, we assume that the generating function $G(u)$ is defined for all $u \in \mathbb{R}$. Without loss of generality, we can restrict our attention to the case $z = 0$ and $m = E[Z_k] < 0$. To see why this is the case, observe that under the random variable transformation $Z'_k = Z_k - z$, the mean and generating function of Z'_k can be expressed in terms of those of Z_k as $m' = m - z < 0$ and $G'(u) = \exp(-uz)G(u)$, and the transfomed rate function satisfies $I'(0) = I(z)$, so proving the result with $z = 0$ for the Z'_k's is the same as proving it at z for the Z_k's.

To simplify the notation, we denote

$$\rho = \min_{u\geq 0} G(u)\,,$$

so that $I(0) = -\ln(\rho)$. We need to show that

$$\lim_{N\to\infty} \frac{1}{N} \ln P[S_N \geq 0] = \ln(\rho) . \tag{3A.1}$$

Depending on where the probability mass is located, three cases need to be considered:

(i) $P[Z_k < 0] = 1$. Then for $u \geq 0$, the generating function $G_0(u)$ is a strictly decreasing function of u which converges to 0 from above as $u \to \infty$. According to definition (3.13), this implies $I(0) = +\infty$. Since $P[S_N \geq 0] = 0$ (the random variables Z_k are strictly negative), the left-hand side of (3A.1) equals $-\infty$, so the result holds.

(ii) $P[Z_k \leq 0] = 1$ and $P[Z_k = 0] > 0$. The generating function $G(u)$ is strictly decreasing for $u \geq 0$ and $\lim_{u\to\infty} G(u) = P[Z_k = 0] = \rho$. Since

$$P[S_N \geq 0] = P[Z_k = 0;\ 1 \leq k \leq N] = \rho^N ,$$

the result holds.

(iii) $P[Z_k < 0] > 0$ and $P[Z_k > 0] > 0$. Then $\lim_{u\to\infty} G(u) = +\infty$. Since $G(u)$ is stricly convex, its first derivative is a strictly increasing function. At $u = 0$, we have $dG/du(0) = m < 0$. So for $G(u)$ to become infinite for u large, there must be a unique point $u_0 > 0$ such that

$$\frac{dG}{du}(u_0) = 0 \quad , \quad G(u_0) = \rho .$$

For the remainder of our proof, we focus on Case (iii). We prove the equality (3A.1) by establishing two inequalities going in opposite directions. First, by applying the Markov inequality for random variables [22, p. 12], we have

$$P[S_N \geq 0] = P[\exp(u_0 N S_N) \geq 1] \leq E[\exp(u_0 N S_N)] = G^N(u_0) = \rho^N .$$

Taking logarithms, scaling by $1/N$, and letting N tend to infinity gives the upper bound

$$\lim_{N\to\infty} \frac{1}{N} \ln P[S_N \geq 0] \leq \ln(\rho) . \tag{3A.2}$$

which represents the "easy part" of the proof.

To obtain the lower bound, following Cramér's original proof, we introduce a sequence $\{Y_k,\ k \geq 1\}$ of i.i.d. random variables whose probability density is expressed in terms of the density f_Z of the Z_k's as

$$f_Y(y) = \exp(u_0 y) \frac{f_Z(y)}{\rho} . \tag{3A.3}$$

This density is often referred to as a "tilted density" in the context of the theory of large deviations. It has the important property that the Y_k's have zero mean, since

$$E[Y_k] = \frac{1}{\rho} \frac{dG}{du}(u_0) = 0 \,,$$

and

$$\sigma_Y^2 = E[Y_k^2] = \frac{1}{\rho} \frac{d^2G}{du^2}(u_0)$$

is finite. Then let

$$\Sigma_N = \frac{1}{N} \sum_{k=1}^{N} Y_k$$

be the sampled mean of the sequence Y_k, $k \geq 1$, and denote by $\mathcal{P}$ the subset of $\mathbb{R}^N$ formed by the n-tuples $(z_k,\ 1 \leq k \leq N)$ such that $\sum_{k=1}^{N} z_k \geq 0$. By applying the transformation (3A.3), we find

$$\begin{aligned} P[S_N \geq 0] &= \int_{\mathcal{P}} \prod_{k=1}^{N} f_Z(z_k) dz_k \\ &= \rho^N \int_{\mathcal{P}} \exp(-u_0 \sum_{k=1}^{N} y_k) \prod_{k=1}^{N} f_Y(y_k) dy_k \\ &= \rho^N E[\exp(-u_0 N \Sigma_N) 1_{\{\Sigma_N \geq 0\}}] \end{aligned} \tag{3A.4}$$

where

$$1_{\{\Sigma_N \geq 0\}} = \begin{cases} 1 & \Sigma_N \geq 0 \\ 0 & \text{otherwise}\,. \end{cases}$$

Because the random variables Y_k have zero-mean and have finite variance, the central limit theorem is applicable to Σ_N, so as N becomes large $N^{1/2}\Sigma_N/\sigma_Y$ admits a $N(0,1)$ distribution. Let $C > 0$ be such that

$$\frac{1}{(2\pi)^{1/2}} \int_0^C \exp(-x^2/2) dx = \frac{1}{4} \,.$$

The expectation on the right hand side of (3A.4) admits the lower bound

$$\begin{aligned} E[\exp(-u_0 N \Sigma_N) 1_{\{\Sigma_N \geq 0\}}] &\geq \exp(-u_0 C \sigma_Y N^{1/2}) P[0 \leq N^{1/2}\Sigma_N/\sigma_Y < C] \\ &\approx \frac{1}{4} \exp(-u_0 C \sigma_Y N^{1/2}) \,. \end{aligned} \tag{3A.5}$$

Combining this inequality with (3A.4), taking logarithms, and scaling by $1/N$ we find

$$\frac{1}{N} \ln P[S_N \geq 0] \geq \ln(\rho) - \frac{u_0 C \sigma_Y}{N^{1/2}} - \frac{\ln(4)}{N}$$

for N large. Then by letting N tend to ∞, we obtain the lower bound

$$\lim_{N\to\infty} \frac{1}{N} \ln P[S_N \geq 0] \geq \ln(\rho) \,, \tag{3A.6}$$

which together with the opposite inequality proves (3A.1).

Roughly speaking, Cramér's theorem characterizes large deviations, i.e., the tail of probability distributions, whereas the central limit theorem characterizes moderate deviations or the bulk of probability distributions.

References

1. A. Wald, *Sequential Analysis.* New York: J. Wiley & Sons, 1947. Reprinted by Dover Publ., Mineola, NY, 2004.
2. S. Kullback, *Information Theory and Statistics.* New York: J. Wiley & Sons, 1959. Reprinted by Dover Publ., Mineola, NY, 1968.
3. N. Seshadri and C.-E. W. Sundberg, "List Viterbi decoding algorithms with applications," *IEEE Trans. Commun.*, vol. 42, pp. 313–323, Feb./Apr. 1994.
4. S.-I. Amari and H. Nagaoka, *Methods of Information Geometry.* Providence, RI: American Mathematical Soc., 2000.
5. T. M. Cover and J. A. Thomas, *Elements of Information Theory, 2nd edition.* New York: J. Wiley & Sons, 2006.
6. H. Chernoff, "A measure of asymptotic efficiency for tests of a hypothesis based on the sum of observations," *Annals Math. Statist.*, vol. 23, pp. 493–507, 1952.
7. H. Cramér, "Sur un nouveau théorème-limite de la théorie des probabilités," in *Actalités Scientifiques et Industrielles*, vol. 736, Paris: Hermann, 1938.
8. A. Dembo and O. Zeitouni, *Large Deviations Techniques and Applications, Second Edition.* New York: Springer Verlag, 1998.
9. F. den Hollander, *Large Deviations.* Providence, RI: American Mathematical Soc., 2000.
10. K. L. Chung, *A Course in Probability Theory, Second Edition.* New York: Academic Press, 1968.
11. D. Bertsekas, A. Nedic, and A. E. Ozdaglar, *Convex Analsis and Optimization.* Belmont, MA: Athena Scientific, 2003.
12. J. R. Magnus and H. Neudecker, *Matrix Differential Calculus with Applications in Statistics and Econometrics.* Chichester, England: J. Wiley & Sons, 1988.
13. M. Basseville, "Information: Entropies, divergences et moyennes," Tech. Rep. 1020, Institut de Recherche en Informatique et Systèmes Aléatoires, Rennes, France, May 1996.
14. H. L. Van Trees, *Detection, Estimation and Modulation Theory, Part I: Detection, Estimation and Linear Modulation Theory.* New York: J. Wiley & Sons, 1968. paperback reprint edition in 2001.
15. A. G. Dabak, *A Geometry for Detection Theory.* PhD thesis, Electrical and Computer Engineering Dept., Rice University, Houston, TX, 1992.
16. A. G. Dabak and D. H. Johnson, "Geometrically based robust detection," in *Proc. Conf. Information Sciences and Systems*, (Baltimore, MD), pp. 73–77, The Johns Hopkins Univ., Mar. 1993.
17. K. J. Arrow, D. Blackwell, and M. A. Girshick, "Bayes and minimax solutions of sequential decision problems," *Econometrica*, vol. 17, pp. 213–244, Jul.-Oct. 1949.
18. M. H. DeGroot, *Optimal Statistical Decisions.* New York: McGraw-Hill, 1970. Reprinted by Wiley-Interscience, New York, 2004.

19. D. Bertsekas, *Dynamic Programming and Optimal Control, Vol. I.* Belmont, MA: Athena Scientific, 1995.
20. T. S. Ferguson, *Mathematical Statistics: A Decision Theoretic Approach.* New York: Academic Press, 1967.
21. S. Karlin and H. M. Taylor, *A First Course in Stochastic Processes.* New York: Academic Press, 1975.
22. R. G. Gallager, *Discrete Stochastic Processes.* Boston: Kluwer Acad. Publ., 1996.
23. D. L. Burkholder and R. A. Wijsman, "Optimum properties and admissibility of sequential tests," *Annals. Math. Statistics*, vol. 34, pp. 1–17, Mar. 1963.
24. D. H. Johnson, "Notes for ELEC 530: Detection Theory." Dept. Elec. Comp. Eng., Rice University, 2003.
25. P. Whittle, *Optimization Over Time, Vol. II.* Chichester, England: J. Wiley & Sons, 1983.

4

Parameter Estimation Theory

4.1 Introduction

As noted earlier, the signals to be detected typically contain parameters, such as amplitude, frequency, and phase, which must be estimated before any detection can take place. To estimate these parameters, a number of methods can be applied, which differ either by their design philosophy or by the metric used to evaluate their performance. For the Bayesian estimation methods described in Section 4.2, all parameters are viewed as random, and a probability distribution is specified for the vector formed by all parameters. This probability distribution may be obtained either from physical modelling considerations, or empirically, by collecting data over a large number of experiments. Bayesian methods require also the specification of a loss function for the estimation error. This loss function essentially specifies what type of error needs to be avoided. When the loss function is quadratic, we obtain the class of minimum mean-square error (MMSE) estimators. These estimators have the attractive feature that in the Gaussian case, they are linear, and thus easy to implement. Another choice of loss function that penalizes equally all errors beyond a certain threshold, leads to maximum a posteriori (MAP) estimators. In the frequentist approach to estimation problems, which is described in Section 4.4, all parameters are viewed as nonrandom and must be estimated exclusively from the observations. The performance of estimators designed in this manner can be characterized by their bias, and their mean-square error, which in the unbiased case is bounded from below by the Cramér-Rao lower bound. Among estimators of this type, the maximum likelihood (ML) estimator maximizes the probability of the observations given the unknown parameters. It is shown in Section 4.5 that this estimator has highly desirable properties, since under mild conditions, for independent identically distributed observations, the ML estimate asymptotically admits a Gaussian distribution, reaches the Cramér-Rao lower bound, and converges almost surely to the true parameter vector as the number of observations becomes large.

B.C. Levy, *Principles of Signal Detection and Parameter Estimation*,
DOI: 10.1007/978-0-387-76544-0_4,

4.2 Bayesian Estimation

In this section we describe the Bayesian formulation of estimation problems. As we shall see, this formulation takes a form similar to the Bayesian formulation of detection problems. Given a random observation vector $\mathbf{Y} \in \mathbb{R}^n$ and a parameter vector to be estimated $\mathbf{X} \in \mathbb{R}^m$, the difference between detection and estimation is that, in the detection case, the decision rule $\delta(\mathbf{Y})$ takes discrete values, whereas for estimation problems, the estimator $\hat{\mathbf{X}}(\mathbf{Y})$, which plays the role of the decision rule, takes values in $\mathbf{R}^m$. Otherwise, as will soon become obvious, Bayesian estimation and Bayesian detection proceed along very similar lines. For this reason it is, in fact, possible to treat Bayesian detection and estimation in a unified manner, as demonstrated in [1]. However, this viewpoint leads to an unnecessary level of abstraction, and we follow here a more conventional approach where estimation and detection are kept separate.

The Bayesian formulation of estimation problems relies on the following elements:

(i) An *observation model* specifying the probability density $f_{\mathbf{Y}|\mathbf{X}}(\mathbf{y}|\mathbf{x})$ of the observations $\mathbf{Y} \in \mathbb{R}^n$ in function of the value $\mathbf{x} \in \mathbb{R}^m$ taken by the parameter vector $\mathbf{X}$ to be estimated. Note that the conditional probability density $f_{\mathbf{Y}|\mathbf{X}}(\mathbf{y}|\mathbf{x})$ can be specified either explicitly, or indirectly in terms of a measurement model. For example, suppose that $\mathbf{Y}$ is a noisy observation

$$\mathbf{Y} = \mathbf{h}(\mathbf{X}) + \mathbf{V} \tag{4.1}$$

of a vector function $\mathbf{h}$ of $\mathbf{X}$, where the additive noise $\mathbf{V}$ has density $f_{\mathbf{V}}(\mathbf{v})$. Then the conditional density of $\mathbf{Y}$ given $\mathbf{X}$ can be expressed as

$$f_{\mathbf{Y}|\mathbf{X}}(\mathbf{y}|\mathbf{x}) = f_{\mathbf{V}}(\mathbf{y} - \mathbf{h}(\mathbf{x})) \,. \tag{4.2}$$

(ii) A *prior density* $f_{\mathbf{X}}(\mathbf{x})$ for the vector $\mathbf{X}$ of unknown parameters. Note that this means that the parameter vector $\mathbf{X}$ is viewed as random. The density $f_{\mathbf{X}}(\mathbf{x})$ need not be informative. For example, if Θ denotes a random phase, it is customary to assume that Θ is uniformly distributed over $[0, 2\pi]$, i.e.,

$$f_\Theta(\theta) = \begin{cases} \frac{1}{2\pi} & 0 \leq \theta < 2\pi \\ 0 & \text{otherwise} \,, \end{cases}$$

which amounts to saying that we do not know anything about Θ. Unfortunately, this approach cannot be applied to situations where the range of possible values for the parameter vector $\mathbf{X}$ is unbounded. In this case, the choice of a density necessarily biases the estimator towards values of $\mathbf{x}$ for which $f_{\mathbf{X}}(\mathbf{x})$ is larger.

(iii) A *cost function* $C(\mathbf{x}, \hat{\mathbf{x}})$ expressing the cost of estimating $\mathbf{x}$ as $\hat{\mathbf{x}}$.

Comparing the quantities described above to the formulation of Bayesian detection, we see that $f(\mathbf{y}|\mathbf{x})$ plays the role of $f(\mathbf{y}|H_j)$, $f(\mathbf{x})$ corresponds to the prior distribution $\pi_j = P[H_j]$ for the different hypotheses, and $C(\mathbf{x}, \hat{\mathbf{x}})$ replaces C_{ij}, so Bayesian estimation and detection are indeed quite similar.

Loss function: Let

$$\mathbf{E} = \mathbf{X} - \hat{\mathbf{X}} \tag{4.3}$$

denote the estimation error vector. Tyically, the cost function $C(\mathbf{x}, \hat{\mathbf{x}})$ is expressed in terms of a nondecreasing loss function $L(\mathbf{e})$ for the error $\mathbf{e} = \mathbf{x} - \hat{\mathbf{x}}$ as

$$C(\mathbf{x}, \hat{\mathbf{x}}) = L(\mathbf{e}) \,. \tag{4.4}$$

Several types of loss functions will be considered in the discussions below.

(a) **Mean-square error:** In this case we select

$$L_{\text{MSE}}(\mathbf{e}) = ||\mathbf{e}||_2^2 \tag{4.5}$$

where the Euclidean norm (or 2-norm) of the vector

$$\mathbf{e} = \begin{bmatrix} e_1 \dots e_i \dots e_m \end{bmatrix}^T$$

can be expressed in terms of its coordinates e_i as

$$||\mathbf{e}||_2 = \Big(\sum_{i=1}^{m} e_i^2\Big)^{1/2} \,. \tag{4.6}$$

For the case when $m = 1$, so that the error is scalar, the loss function $L(\mathbf{e})$ is sketched in part (a) of Fig. 4.1. Because it is quadratic, it progressively penalizes the errors more severely as they become large. While this is reasonable, this choice has the effect of placing a larger weight on the tails of probability distributions. Since these tails are often known only approximately, mean-square error (MSE) estimators tend to be sensitive to modelling errors.

(b) **Mean absolute error:** Let

$$L_{\text{MAE}}(\mathbf{e}) = ||\mathbf{e}||_1 \tag{4.7}$$

where the sum norm (or 1-norm) of the error vector $\mathbf{e}$ is defined as

$$||\mathbf{e}||_1 = \sum_{i=1}^{m} |e_i| \,. \tag{4.8}$$

For the case of a scalar error $\mathbf{e}$, the function $L(\mathbf{e})$ is plotted in part (b) of Fig. 4.1. Because it just rectifies the error, $L(\mathbf{e})$ weighs equally the magnitude of all error components. So estimators designed with this loss function will be more prone to large errors than MSE estimators based on (4.5), since those place a more severe penalty on large errors.

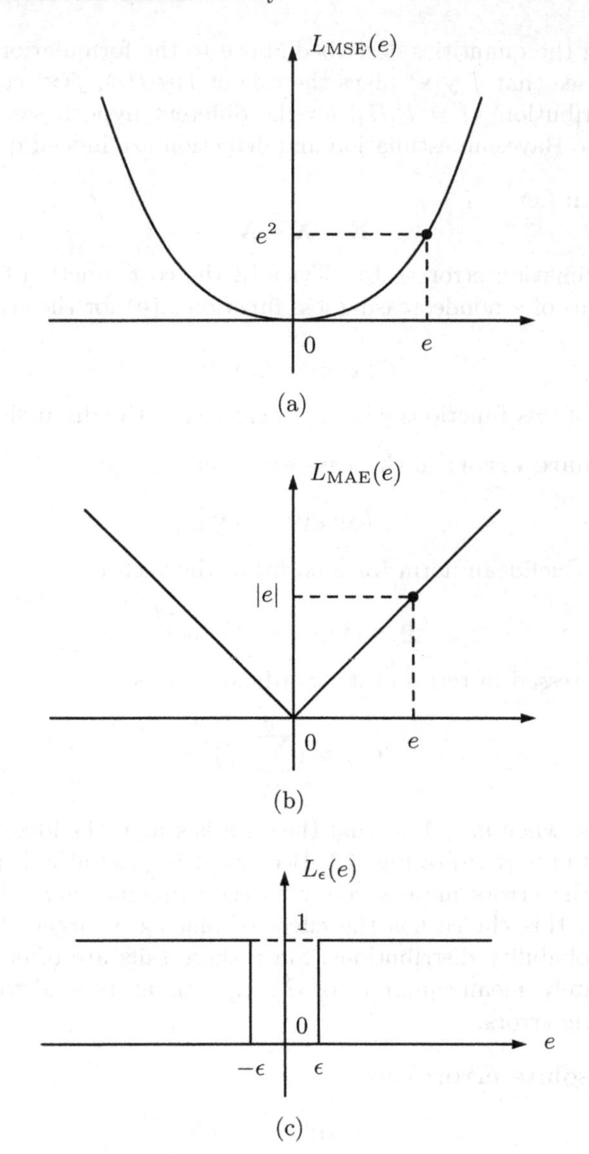

Fig. 4.1. Plots of the (a) MSE, (b) MAE, and (c) notch loss functions.

(c) **Notch function:** Let

$$L_\epsilon(\mathbf{e}) = \begin{cases} 0 \text{ if } ||\mathbf{e}||_\infty < \epsilon \\ 1 \text{ otherwise}, \end{cases} \tag{4.9}$$

where the infinity norm of the error vector $\mathbf{e}$ is defined as

$$||\mathbf{e}||_\infty = \max_{1 \leq i \leq m} |e_i| \,. \tag{4.10}$$

The function $L_\epsilon(\mathbf{e})$ is plotted in part (c) of Fig. 4.1 for a scalar error. This function privileges very small errors and does not care about the size of errors beyond ϵ. It is worth noting that while the mean-square error (MSE) and mean-absolute error (MAE) loss functions are convex in $\mathbf{e}$, the notch function $L_\epsilon(\mathbf{e})$ is not convex.

4.2.1 Optimum Bayesian Estimator

The objective of Bayesian estimation is to find a vector estimator

$$\hat{\mathbf{X}}(\mathbf{Y}) = \left[\hat{X}_1(\mathbf{Y}) \ldots \hat{X}_i(\mathbf{Y}) \ldots \hat{X}_m(\mathbf{Y}) \right]^T \tag{4.11}$$

taking values in $\mathbb{R}^m$ which minimizes the expected cost $E[C(\mathbf{X}, \hat{\mathbf{X}}(\mathbf{Y}))]$. To characterize this estimator, we observe that the expected cost can be expressed as

$$\begin{aligned} E[C(\mathbf{X}, \hat{\mathbf{X}}(\mathbf{Y}))] &= \int\int C(\mathbf{x}, \hat{\mathbf{X}}(\mathbf{y}) f_{\mathbf{X},\mathbf{Y}}(\mathbf{x}, \mathbf{y}) d\mathbf{x} d\mathbf{y} \\ &= \int \Big[\int C(\mathbf{x}, \hat{\mathbf{X}}(\mathbf{y})) f_{\mathbf{X}|\mathbf{Y}}(\mathbf{x}|\mathbf{y}) d\mathbf{x} \Big] f_{\mathbf{Y}}(\mathbf{y}) d\mathbf{y} \end{aligned} \tag{4.12}$$

where the posterior density of the parameter vector $\mathbf{X}$ given the observations $\mathbf{Y}$ is evaluated by applying Bayes's rule

$$f_{\mathbf{X}|\mathbf{Y}}(\mathbf{x}|\mathbf{y}) = \frac{f_{\mathbf{Y}|\mathbf{X}}(\mathbf{y}|\mathbf{x}) f_{\mathbf{X}}(\mathbf{x})}{f_{\mathbf{Y}}(\mathbf{y})} \,. \tag{4.13}$$

Note that in (4.13) the marginal density $f_{\mathbf{Y}}(\mathbf{y})$ is obtained integrating the joint density of $\mathbf{X}$ and $\mathbf{Y}$ with respect to $\mathbf{x}$, i.e.,

$$f_{\mathbf{Y}}(\mathbf{y}) = \int f_{\mathbf{Y}|\mathbf{X}}(\mathbf{y}|\mathbf{x}) f_{\mathbf{X}}(\mathbf{x}) d\mathbf{x} \,. \tag{4.14}$$

From expression (4.12), we see that since $f_{\mathbf{Y}}(\mathbf{y}) \geq 0$, the expected cost will be minimized if the term between brackets is minimized for each $\mathbf{y}$. This gives

$$\hat{\mathbf{X}}(\mathbf{y}) = \arg \min_{\hat{\mathbf{x}} \in \mathbb{R}^m} \int C(\mathbf{x}, \hat{\mathbf{x}}) f_{\mathbf{X}|\mathbf{Y}}(\mathbf{x}|\mathbf{y}) d\mathbf{x} \,. \tag{4.15}$$

Furthermore, by observing that for a fixed $\mathbf{y}$, the marginal density $f_{\mathbf{Y}}(\mathbf{y})$ appearing in the denominator of the conditional density $f_{\mathbf{X}|\mathbf{Y}}(\mathbf{x}|\mathbf{y})$ in (4.13) is an arbitrary scaling factor that does not affect the outcome of the minimization (4.15), we have also

$$\hat{\mathbf{X}}(\mathbf{y}) = \arg \min_{\hat{\mathbf{x}} \in \mathbb{R}^m} \int C(\mathbf{x}, \hat{\mathbf{x}}) f_{\mathbf{X},\mathbf{Y}}(\mathbf{x}, \mathbf{y}) d\mathbf{x} \,. \tag{4.16}$$

MSE Estimate: For

$$C(\mathbf{x}, \hat{\mathbf{x}}) = ||\mathbf{x} - \hat{\mathbf{x}}||_2^2 ,$$

the minimum mean-square error (MMSE) estimate $\hat{\mathbf{X}}_{\text{MSE}}(\mathbf{y})$ minimizes the conditional mean-square error

$$J(\hat{\mathbf{x}}|\mathbf{y}) = \int ||\mathbf{x} - \hat{\mathbf{x}}||^2 f_{\mathbf{X}|\mathbf{Y}}(\mathbf{x}|\mathbf{y}) d\mathbf{x} \tag{4.17}$$

To minimize this function, we set its gradient

$$\nabla_{\hat{\mathbf{x}}} J(\hat{\mathbf{x}}|\mathbf{y}) = 2 \int (\hat{\mathbf{x}} - \mathbf{x}) f_{\mathbf{X}|\mathbf{Y}}(\mathbf{x}|\mathbf{y}) d\mathbf{x} \tag{4.18}$$

equal to zero, where the gradient is defined as

$$\nabla_{\hat{\mathbf{x}}} J = \left[\frac{\partial J}{\partial \hat{x}_1} \cdots \frac{\partial J}{\partial \hat{x}_i} \cdots \frac{\partial J}{\partial \hat{x}_m} \right]^T .$$

Solving $\nabla_{\hat{\mathbf{x}}} J(\hat{\mathbf{x}}|\mathbf{y}) = 0$ and taking into account

$$\int f_{\mathbf{X}|\mathbf{Y}}(\mathbf{x}|\mathbf{y}) d\mathbf{x} = 1$$

gives

$$\hat{\mathbf{X}}_{\text{MSE}}(\mathbf{y}) = \int \mathbf{x} f_{\mathbf{X}|\mathbf{Y}}(\mathbf{x}|\mathbf{y}) d\mathbf{x} = E[\mathbf{X}|\mathbf{Y} = \mathbf{y}] , \tag{4.19}$$

so the mean-square error estimate $\mathbf{X}_{\text{MSE}}(\mathbf{Y})$ is just the *conditional mean* of $\mathbf{X}$ given $\mathbf{Y}$.

The performance of the estimator can then be expressed as

$$J(\hat{\mathbf{X}}_{\text{MSE}}(\mathbf{y})|\mathbf{y}) = \text{tr}(\mathbf{K}_{\mathbf{X}|\mathbf{Y}}(\mathbf{y})) \tag{4.20}$$

where the conditional covariance matrix $\mathbf{K}_{\mathbf{X}|\mathbf{Y}}(\mathbf{y})$ of $\mathbf{X}$ given $\mathbf{Y} = \mathbf{y}$ takes the form

$$\begin{aligned} \mathbf{K}_{\mathbf{X}|\mathbf{Y}}(\mathbf{y}) &= E[(\mathbf{X} - E[\mathbf{X}|\mathbf{y}])(\mathbf{X} - E[\mathbf{X}|\mathbf{y}])^T|\mathbf{Y} = \mathbf{y}] \\ &= \int (\mathbf{x} - E[\mathbf{X}|\mathbf{y}])(\mathbf{x} - E[\mathbf{X}|\mathbf{y}])^T f_{\mathbf{X}|\mathbf{Y}}(\mathbf{x}|\mathbf{y}) d\mathbf{x} . \end{aligned} \tag{4.21}$$

Averaging with respect to $\mathbf{Y}$, we find that the minimum mean-square error (MMSE) can be expressed as

$$\text{MMSE} = E[||\mathbf{X} - E[\mathbf{X}|\mathbf{Y}]||_2^2] = \text{tr}\,(\mathbf{K}_{\text{E}}) , \tag{4.22}$$

where the error covariance matrix $\mathbf{K}_{\text{E}}$ is given by

$$\begin{aligned} \mathbf{K}_{\text{E}} &= E[(\mathbf{X} - E[\mathbf{X}|\mathbf{Y}])(\mathbf{X} - E[\mathbf{X}|\mathbf{Y}])^T] \\ &= \int\int (\mathbf{x} - E[\mathbf{X}|\mathbf{y}])(\mathbf{x} - E[\mathbf{X}|\mathbf{y}])^T f_{\mathbf{X},\mathbf{Y}}(\mathbf{x}, \mathbf{y}) d\mathbf{x} d\mathbf{y} . \end{aligned} \tag{4.23}$$

MAE Estimate: For

$$C(\mathbf{x}, \hat{\mathbf{x}}) = ||\mathbf{x} - \mathbf{x}||_1 ,$$

the minimum mean absolute error estimate (MMAE) $\hat{\mathbf{X}}_{\mathrm{MAE}}(\mathbf{y})$ minimizes the objective function

$$J(\hat{\mathbf{x}}|\mathbf{y}) = \int ||\mathbf{x} - \hat{\mathbf{x}}||_1 f_{\mathbf{X}|\mathbf{Y}}(\mathbf{x}|\mathbf{y}) d\mathbf{x} .$$

Taking the partial derivative of J with respect to $\hat{x}_i$ for $1 \leq i \leq m$ gives

$$\frac{\partial J}{\partial \hat{x}_i} = - \int \operatorname{sgn}(x_i - \hat{x}_i) f_{\mathbf{X}_i|\mathbf{Y}}(x_i|\mathbf{y}) dx_i , \tag{4.24}$$

where the sign function is defined as

$$\operatorname{sgn}(z) = \begin{cases} 1 \text{ for } z \geq 0 \\ -1 \text{ for } z < 0 . \end{cases}$$

Setting $\frac{\partial J}{\partial \hat{x}_i} = 0$ and noting that the total probability mass of the density $f_{X_i|\mathbf{Y}}(x_i|\mathbf{y})$ equals one, we find

$$\int_{-\infty}^{\hat{x}_i} f_{X_i|\mathbf{Y}}(x_i|\mathbf{y}) d\mathbf{y} = \int_{\hat{x}_i}^{\infty} f_{X_i|\mathbf{Y}}(x_i|\mathbf{y}) d\mathbf{y} = \frac{1}{2} , \tag{4.25}$$

so that for each i, the i-th entry of $\mathbf{X}_{\mathrm{MAE}}(\mathbf{y})$ is the *median* of the conditional density $f_{X_i|\mathbf{Y}}(x_i|\mathbf{y})$. We recall that the median of the probability density of a random variable is the point on the real axis where half of the probability mass is located on one side of the point, and the other half on the other side.

MAP Estimate: For

$$C(\mathbf{x}, \hat{\mathbf{x}}) = L_\epsilon(\mathbf{x} - \hat{\mathbf{x}}) ,$$

the maximum a posteriori (MAP) estimate $\mathbf{X}_{\mathrm{MAP}}(\mathbf{y})$ minimizes

$$\begin{aligned} J(\hat{\mathbf{x}}|\mathbf{y}) &= \int L_\epsilon(\mathbf{x} - \hat{\mathbf{x}}) f_{\mathbf{X}|\mathbf{Y}}(\mathbf{x}|\mathbf{y}) d\mathbf{x} \\ &= 1 - \int_{||\hat{\mathbf{x}} - \mathbf{x}||_\infty < \epsilon} f_{\mathbf{X}|\mathbf{Y}}(\mathbf{x}|\mathbf{y}) d\mathbf{x} \end{aligned} \tag{4.26}$$

for ϵ vanishingly small. In the light of (4.26), this is equivalent to maximizing

$$\int_{||\mathbf{x} - \hat{\mathbf{x}}||_\infty < \epsilon} f_{\mathbf{X}|\mathbf{Y}}(\mathbf{x}|\mathbf{y}) d\mathbf{x} \approx f_{\mathbf{X}|\mathbf{Y}}(\hat{\mathbf{x}}|\mathbf{y})(2\epsilon)^m$$

for small ϵ, so that

$$\hat{\mathbf{X}}_{\mathrm{MAP}}(\mathbf{y}) = \arg \max_{\mathbf{x} \in \mathbb{R}^m} f_{\mathbf{X}|\mathbf{Y}}(\mathbf{x}|\mathbf{y}) \tag{4.27}$$

corresponds to the maximum, which is also called the *mode*, of the a posteriori density $f_{\mathbf{X}|\mathbf{Y}}(\mathbf{x}|\mathbf{y})$. Note that this choice makes sense only if the conditional density $f_{\mathbf{X}|\mathbf{Y}}(\mathbf{x}|\mathbf{y})$ has a dominant peak. Otherwise, when the posterior density has several peaks of similar size, selecting the largest one and ignoring the others may lead to unacceptably large errors.

Example 4.1: Jointly Gaussian random vectors

Let $\mathbf{X} \in \mathbb{R}^m$ and $\mathbf{Y} \in \mathbb{R}^n$ be jointly Gaussian, so that

$$\begin{bmatrix} \mathbf{X} \\ \mathbf{Y} \end{bmatrix} = N(\mathbf{m}, \mathbf{K})$$

with

$$\mathbf{m} = \begin{bmatrix} \mathbf{m_X} \\ \mathbf{m_Y} \end{bmatrix} = \begin{bmatrix} E[\mathbf{X}] \\ E[\mathbf{Y}] \end{bmatrix}$$

and

$$\mathbf{K} = \begin{bmatrix} \mathbf{K_X} & \mathbf{K_{XY}} \\ \mathbf{K_{YX}} & \mathbf{K_Y} \end{bmatrix} = E\left[\begin{bmatrix} \mathbf{X} - \mathbf{m_X} \\ \mathbf{Y} - \mathbf{m_Y} \end{bmatrix} \left[(\mathbf{X} - \mathbf{m_X})^T \; (\mathbf{Y} - \mathbf{m_Y})^T \right] \right].$$

Then the conditional density of $\mathbf{X}$ given $\mathbf{Y}$ is also Gaussian, i.e.,

$$f_{\mathbf{X}|\mathbf{Y}}(\mathbf{x}|\mathbf{y}) = \frac{1}{(2\pi)^{m/2}|\mathbf{K_{X|Y}}|^{1/2}} \cdot \exp\left(-\frac{1}{2}(\mathbf{x} - \mathbf{m_{X|Y}})^T \mathbf{K}_{\mathbf{X}|\mathbf{Y}}^{-1} (\mathbf{x} - \mathbf{m_{X|Y}}) \right) \tag{4.28}$$

where

$$\mathbf{m_{X|Y}} = \mathbf{m_X} + \mathbf{K_{XY}} \mathbf{K_Y}^{-1} (\mathbf{Y} - \mathbf{m_Y}) \tag{4.29}$$

and

$$\mathbf{K_{X|Y}} = \mathbf{K_X} - \mathbf{K_{XY}} \mathbf{K_Y}^{-1} \mathbf{K_{YX}} \tag{4.30}$$

denote respectively the conditional mean vector and the conditional covariance matrix of $\mathbf{X}$ given $\mathbf{Y}$. In the following we will write (4.28) compactly as

$$f_{\mathbf{X}|\mathbf{Y}}(\mathbf{x}|\mathbf{y}) \sim N(\mathbf{m_{X|Y}}, \mathbf{K_{X|Y}}) .$$

Then

$$\hat{\mathbf{X}}_{\text{MSE}}(\mathbf{Y}) = \mathbf{m_{X|Y}} = \mathbf{m_X} - \mathbf{K_{XY}} \mathbf{K_Y}^{-1} (\mathbf{Y} - \mathbf{m_Y})$$

depends *linearly* on the observation vector $\mathbf{Y}$, and the conditional error covariance matrix $K_{\mathbf{X}|\mathbf{Y}}$ is given by (4.30). Note that this matrix does not depend on the observation vector $\mathbf{Y}$, so that the error covariance matrix K_{E} takes the form

$$\mathbf{K_E} = \mathbf{K_X} - \mathbf{K_{XY}} \mathbf{K_Y}^{-1} \mathbf{K_{YX}} .$$

Since the median of a Gaussian distribution equals its mean, and since the maximum of a Gaussian density is achieved at its mean, we have also

$$\hat{\mathbf{X}}_{\text{MAE}}(\mathbf{Y}) = \hat{\mathbf{X}}_{\text{MAP}}(\mathbf{Y}) = \mathbf{m}_{\mathbf{X}|\mathbf{Y}} ,$$

so in the Gaussian case the MSE, MAE and MAP estimates coincide. □

However, this last property does not hold in general as can be seen from the following example:

Example 4.2: Exponential observation of an exponential parameter

Assume that Y is an exponential random variable with parameter X, so that

$$f_{Y|X}(y|x) = \begin{cases} x\exp(-xy) & y \geq 0 \\ 0 & \text{otherwise} , \end{cases}$$

and the parameter X is itself exponentially distributed with parameter $a > 0$, i.e.,

$$f_X(x) = \begin{cases} a\exp(-ax) & x \geq 0 \\ 0 & x < 0 . \end{cases}$$

Then the joint density of X and Y can be expressed as

$$f_{X,Y}(x,y) = ax\exp(-(y+a)x)u(x)u(y) ,$$

where $u(\cdot)$ denotes the unit step function, and by using integration by parts we find that the marginal density

$$\begin{aligned} f_Y(y) &= \int_0^\infty ax\exp(-(y+a)x)dx u(y) \\ &= \frac{a}{(y+a)^2}u(y) . \end{aligned}$$

The conditional density of X given $Y = y$ is therefore given by

$$f_{X|Y}(x|y) = \frac{f_{X,Y}(x,y)}{f_Y(y)} = (y+a)^2 x\exp(-(y+a)x)u(x)$$

which is sketched in Fig. 4.2. By integration by parts, we find

$$\hat{X}_{\text{MSE}}(y) = \int_{-\infty}^{\infty} x f_{X|Y}(x|y)dx = \frac{2}{y+a} .$$

To obtain the MAE estimate, we need to solve

$$\begin{aligned} \frac{1}{2} = \int_{-\infty}^{\hat{x}} f_{X|Y}(x|y)dx &= (y+a)^2 \int_0^{\hat{x}} x\exp(-(y+a)x)dx \\ &= [1+(a+y)\hat{x}]\exp(-(y+a)\hat{x}) \end{aligned}$$

for $\hat{x}$. Replacing $(y+a)\hat{x}$ by c in this equation, we obtain

$$(1+c)\exp(-c) = \frac{1}{2} . \tag{4.31}$$

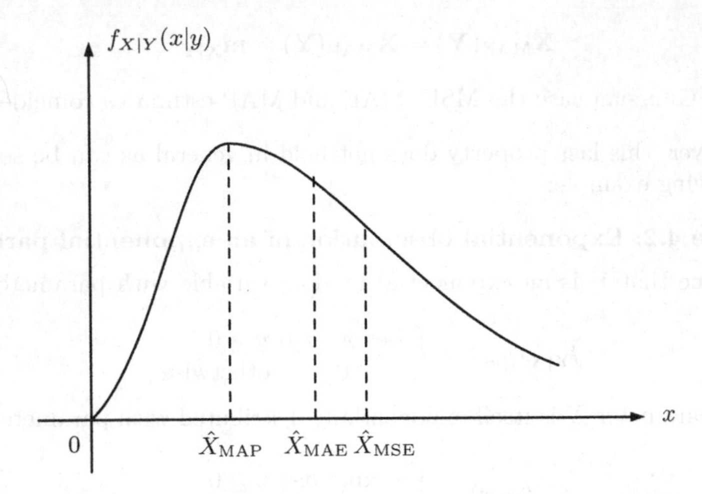

Fig. 4.2. Conditional probability density $f_{X|Y}(x|y)$ and MAP, MAE and MSE estimates.

The MAE estimate is therefore given by

$$\hat{X}_{\rm MAE}(y) = \frac{c}{y+a}$$

where $c \approx 1.68$ is the unique solution of (4.31). Finally, to find the MAP estimate, i.e., the maximum of the conditional density, we set the derivative

$$\frac{\partial}{\partial x} f_{X|Y}(x|y) = (y+a)^2 \exp(-(y+a)x)[1-(y+a)x]$$

equal to zero, which yields

$$\hat{X}_{\rm MAP}(y) = \frac{1}{y+a} .$$

So for this example, the MAP, MAE and MSE estimates take different values.
□

For scalar estimation, it is possible in fact to identify a set of conditions under which all estimates coincide [2, pp. 60–61]. Suppose the loss function $L(e)$ is convex and even in e. Then, if the conditional density $f_{X|\mathbf{Y}}(x|\mathbf{y})$ is symmetric about its mode $\hat{X}_{\rm MAP}(\mathbf{y})$, so that

$$f_{X|\mathbf{Y}}(\hat{X}_{\rm MAP} + z) = f_{X|\mathbf{Y}}(\hat{X}_{\rm MAP} - z) \tag{4.32}$$

for all z, and if $\hat{X}(\mathbf{Y})$ denotes the optimum Bayesian estimate corresponding to the loss function L, we have

$$\hat{X}(\mathbf{Y}) = \hat{X}_{\rm MAP}(\mathbf{Y}) . \tag{4.33}$$

Since the loss functions L_{MSE} and L_{MAE} are convex and even, this implies that in this case the MSE, MAE and MAP estimates coincide.

To prove this result, consider the conditional Bayesian cost

$$J(\hat{x}|\mathbf{y}) = \int L(x - \hat{x}) f_{X|\mathbf{Y}}(x|\mathbf{y}) dx$$

and perform the change of variable $Z = X - \hat{X}_{\text{MAP}}$. Then the cost can be rewritten as

$$J(\hat{x}|\mathbf{y}) = \int L(z + \hat{X}_{\text{MAP}} - \hat{x}) f_{Z|\mathbf{Y}}(z|\mathbf{y}) dz \tag{4.34}$$

$$= \int L(z - (\hat{X}_{\text{MAP}} - \hat{x})) f_{Z|\mathbf{Y}}(z|\mathbf{y}) dz \,, \tag{4.35}$$

where in going from (4.34) to (4.35) we have used the evenness of both L and $f_{Z|\mathbf{Y}}$. Averaging expressions (4.34) and (4.35) for J and using the convexity of L gives

$$J(\hat{x}|\mathbf{y}) = \int \frac{1}{2} \left[L(z + \hat{X}_{\text{MAP}} - \hat{x}) + L(z - (\hat{X}_{\text{MAP}} - \hat{x})) \right] f_{Z|\mathbf{Y}}(z|\mathbf{y}) dz$$

$$\geq \int L(z) f_{Z|\mathbf{Y}}(z|\mathbf{y}) dz = J(\hat{X}_{\text{MAP}}|\mathbf{y}) \,, \tag{4.36}$$

so that in order to minimize J we need to select $\hat{x} = \hat{X}_{\text{MAP}}$.

4.2.2 Properties of the MSE Estimator

In this section we describe briefly two properties of the MSE estimator $\hat{\mathbf{X}}_{\text{MSE}}(\mathbf{Y}) = E[\mathbf{X}|\mathbf{Y}]$.

Orthogonality Property: The conditional mean $E[\mathbf{X}|\mathbf{Y}]$ is the unique function of $\mathbf{Y}$ such that the error

$$\mathbf{E} = \mathbf{X} - E[\mathbf{X}|\mathbf{Y}] \tag{4.37}$$

is *uncorrelated* with any function $g(\mathbf{Y})$ of the observations, i.e.,

$$E[(\mathbf{X} - E[\mathbf{X}|\mathbf{Y}])g(\mathbf{Y})] = 0 \tag{4.38}$$

for all g. This property is just an expression of the fact that the conditional mean $E[\mathbf{X}|\mathbf{Y}]$ extracts all the information contained in $\mathbf{Y}$ that can be used to reduce the MSE.

To prove the identity (4.38), we employ two properties of conditional expectations:

(i) For an arbitrary scalar function g, we have

$$E[g(\mathbf{Y})\mathbf{X}|\mathbf{Y}] = g(\mathbf{Y})E[\mathbf{X}|\mathbf{Y}] \, , \tag{4.39}$$

since after conditioning with respect to $\mathbf{Y}$, the function $g(\mathbf{Y})$ can be viewed as a constant and pulled out of the expectation.

(ii) Given an arbitrary scalar function h

$$E_{\mathbf{Y}}[E_{\mathbf{X}}[h(\mathbf{X},\mathbf{Y})|\mathbf{Y}]] = E_{\mathbf{X},\mathbf{Y}}[h(\mathbf{X},\mathbf{Y})] \, , \tag{4.40}$$

i.e., to take the expectation of $h(\mathbf{X},\mathbf{Y})$, we can first take its expectation conditioned on $\mathbf{Y}$, and then average with respect to $\mathbf{Y}$, or we can take the expectation with respect to $\mathbf{X}$ and $\mathbf{Y}$ in one step. This is known as the *repeated expectation* property. It can be verified directly by noting that

$$E_{\mathbf{X},\mathbf{Y}}[h(\mathbf{X},\mathbf{Y})] = \int \Big[\int h(\mathbf{x},\mathbf{y}) f_{\mathbf{X}|\mathbf{Y}}(\mathbf{x},\mathbf{y}) d\mathbf{x} \Big] f_{\mathbf{Y}}(\mathbf{y}) d\mathbf{y} \, .$$

To verify that the conditional expectation satisfies (4.38) we note that

$$E[E[\mathbf{X}|\mathbf{Y}]g(\mathbf{Y})] = E[E[\mathbf{X}g(\mathbf{Y})|\mathbf{Y}] = E[\mathbf{X}g(\mathbf{Y})]$$

where the first and second equalities are due respectively to properties (i) and (ii) of the conditional expectation. Next, we need to show that the conditional expectation $E[\mathbf{X}|\mathbf{Y}]$ is the unique vector function of $\mathbf{Y}$ taking values in $\mathbb{R}^m$ such that (4.38) holds. Suppose that $\mathbf{h}(\mathbf{Y})$ is another function taking values in $\mathbb{R}^m$ with the same property, so that

$$E[(\mathbf{X} - \mathbf{h}(\mathbf{Y}))g(\mathbf{Y})] = 0$$

for all functions g. Then

$$E[||E[\mathbf{X}|\mathbf{Y}] - \mathbf{h}(\mathbf{Y})||_2^2] = E[\mathbf{g}^T(\mathbf{Y})(E[\mathbf{X}|\mathbf{Y}] - \mathbf{X} + \mathbf{X} - \mathbf{h}(\mathbf{Y}))] = 0 \tag{4.41}$$

with

$$\mathbf{g}(\mathbf{Y}) \triangleq E[\mathbf{X}|\mathbf{Y}] - \mathbf{h}(\mathbf{Y}) \, ,$$

where the last equality in (4.41) is due to the property (4.38) satisfied by both $E[\mathbf{X}|\mathbf{Y}]$ and $\mathbf{h}(\mathbf{Y})$. This implies $\mathbf{h}(\mathbf{Y}) = E[\mathbf{X}|\mathbf{Y}]$ almost surely, so the conditional mean is the unique vector function satisfying (4.38).

Variance Reduction: If

$$\mathbf{K}_{\mathbf{X}} = E[(\mathbf{X} - \mathbf{m}_{\mathbf{X}})(\mathbf{X} - \mathbf{m}_{\mathbf{X}})^T] \tag{4.42}$$

denotes the a-priori covariance matrix of $\mathbf{X}$ and

$$\mathbf{K}_{\mathbf{E}} = E[(\mathbf{X} - E[\mathbf{X}|\mathbf{Y}])(\mathbf{X} - E[\mathbf{X}|\mathbf{Y}])^T] \tag{4.43}$$

denotes the covariance matrix of the mean-square estimate of $\mathbf{X}$ given $\mathbf{Y}$, we have

$$\mathbf{K_E} \leq \mathbf{K_X} \tag{4.44}$$

where the inequality (4.44) indicates that the matrix $\mathbf{K_X} - \mathbf{K_E}$ is nonnegative definite. Furthermore, the inequality (4.44) is an equality if and only if

$$E[\mathbf{X}|\mathbf{Y}] = \mathbf{m_X} \tag{4.45}$$

i.e., if the knowledge of the observations $\mathbf{Y}$ does not improve the estimate of $\mathbf{X}$.

To verify the property (4.44), note that

$$\mathbf{X} - \mathbf{m_X} = \mathbf{X} - E[\mathbf{X}|\mathbf{Y}] + E[\mathbf{X}|\mathbf{Y}] - \mathbf{m_X} ,$$

so that

$$\mathbf{K_X} = \mathbf{K_E} + E[(\mathbf{X} - E[\mathbf{X}|\mathbf{Y}])\boldsymbol{\Delta}^T(\mathbf{Y})] + \mathbf{K_\Delta} , \tag{4.46}$$

where

$$\boldsymbol{\Delta}(\mathbf{Y}) \stackrel{\triangle}{=} E[\mathbf{X}|\mathbf{Y}] - \mathbf{m_X}$$

depends on $\mathbf{Y}$ only, and

$$\mathbf{K_\Delta} = E[\boldsymbol{\Delta}(\mathbf{Y})\boldsymbol{\Delta}^T(\mathbf{Y})] .$$

Since the error $\mathbf{X} - E[\mathbf{X}|\mathbf{Y}]$ is uncorrelated with $\boldsymbol{\Delta}(\mathbf{Y})$, the second term on the right-hand side of (4.46) is zero, which gives

$$\mathbf{K_X} = \mathbf{K_E} + \mathbf{K_\Delta} . \tag{4.47}$$

This implies (4.44) since $\mathbf{K_\Delta}$ is a covariance matrix. Furthermore, $\mathbf{K_\Delta}$ is zero if and only if its trace is zero, i.e.,

$$\mathrm{tr}\,(\mathbf{K_\Delta}) = \mathrm{tr}\, E[\boldsymbol{\Delta}\boldsymbol{\Delta}^T] = E[||\boldsymbol{\Delta}||_2^2] = 0 , \tag{4.48}$$

where the second equality is due to the identity

$$||\mathbf{v}||_2^2 = \mathbf{v}^T\mathbf{v} = \mathrm{tr}\,(\mathbf{v}\mathbf{v}^T) \tag{4.49}$$

for an arbitray vector $\mathbf{v}$. From (4.47) and (4.48) we see therefore that $\mathbf{K_X} = \mathbf{K_E}$ if and only if $\boldsymbol{\Delta}$ is zero almost surely, or equivalently if (4.45) holds.

4.3 Linear Least-squares Estimation

In situations where not enough statistical information is available about the observation vector $\mathbf{Y} \in \mathbb{R}^n$ and parameter vector $\mathbf{X} \in \mathbb{R}^m$ to allow the estimation of the densities $f_{\mathbf{Y}|\mathbf{X}}$ and $f_{\mathbf{X}}$, it is often of interest to consider the class of Bayesian *linear estimators*, since such estimators do not need a complete statistical description of the vectors $\mathbf{X}$ and $\mathbf{Y}$ and require only the knowledge of their first and second order statistics. Thus, assume that $\mathbf{X}$ and $\mathbf{Y}$ have mean vectors

$$\mathbf{m_X} = E[\mathbf{X}] \ , \ \ \mathbf{m_Y} = E[\mathbf{Y}]$$

and joint covariance matrix

$$\begin{aligned}\mathbf{K} &= E\left[\begin{bmatrix}\mathbf{X}-\mathbf{m_X}\\ \mathbf{Y}-\mathbf{m_Y}\end{bmatrix}\begin{bmatrix}(\mathbf{X}-\mathbf{m_X})^T & (\mathbf{Y}-\mathbf{m_Y})^T\end{bmatrix}\right] \\ &= \begin{bmatrix}\mathbf{K_X} & \mathbf{K_{XY}}\\ \mathbf{K_{YX}} & \mathbf{K_Y}\end{bmatrix} .\end{aligned}$$

In our discussion we assume that the matrix $\mathbf{K_Y}$ is positive definite and thus invertible. Otherwise, there exists a nontrivial linear combination of the entries of $\mathbf{Y}-\mathbf{m_Y}$ which is zero almost surely, so that we could replace the observation vector $\mathbf{y}$ by a vector of lower dimension. Then consider the class $\mathcal{L}$ of linear estimators of the form

$$\hat{\mathbf{X}}_{\mathrm{L}}(\mathbf{Y}) = \mathbf{AY} + \mathbf{b} \tag{4.50}$$

where $\mathbf{A}$ is a matrix of size $m \times n$ and $\mathbf{b}$ is a vector of $\mathbb{R}^m$. The linear least-squares estimator $\hat{\mathbf{X}}_{\mathrm{LLS}}(\mathbf{Y})$ is the estimator in $\mathcal{L}$ that minimizes the MSE

$$J = E[||\mathbf{E}||_2^2] \tag{4.51}$$

where $\mathbf{E}$ denotes the estimation error

$$\mathbf{E} = \mathbf{X} - \hat{\mathbf{X}}_{\mathrm{L}}(\mathbf{Y}) = \mathbf{X} - \mathbf{AY} - \mathbf{b} . \tag{4.52}$$

The mean of the error vector $\mathbf{E}$ is given by

$$\mathbf{m_E} = E[\mathbf{E}] = \mathbf{m_X} - \mathbf{Am_Y} - \mathbf{b} \tag{4.53}$$

and by observing that

$$\begin{aligned}\mathbf{E} - \mathbf{m_E} &= (\mathbf{X} - \mathbf{AY} - \mathbf{b}) - (\mathbf{m_X} - \mathbf{Am_Y} - \mathbf{b}) \\ &= \begin{bmatrix}\mathbf{I}_m & -\mathbf{A}\end{bmatrix}\begin{bmatrix}\mathbf{X}-\mathbf{m_X}\\ \mathbf{Y}-\mathbf{m_Y}\end{bmatrix} ,\end{aligned} \tag{4.54}$$

we find that the covariance matrix of the error can be expressed as

$$\begin{aligned}\mathbf{K_E} &= E[(\mathbf{E}-\mathbf{m_E})(\mathbf{E}-\mathbf{m_E})^T] \\ &= \begin{bmatrix}\mathbf{I}_m & -\mathbf{A}\end{bmatrix}\begin{bmatrix}\mathbf{K_X} & \mathbf{K_{XY}}\\ \mathbf{K_{YX}} & \mathbf{K_Y}\end{bmatrix}\begin{bmatrix}\mathbf{I}_m\\ -\mathbf{A}^T\end{bmatrix} .\end{aligned} \tag{4.55}$$

To find the linear least-squares estimate (LLSE), we need to find the pair ($\mathbf{A}$, $\mathbf{b}$) that minimizes

$$J = E[\mathbf{E}^T\mathbf{E}] = E[(\mathbf{E} - \mathbf{m_E})^T(\mathbf{E} - \mathbf{m_E})] + ||\mathbf{m_E}||_2^2 . \tag{4.56}$$

Using again the trace identity (4.48), we find

$$J = \text{tr}\,(\mathbf{K_E}) + ||\mathbf{m_E}||_2^2 . \tag{4.57}$$

An important feature of this identity is that the covariance $\mathbf{K_E}$ depends on $\mathbf{A}$ only, whereas the mean $\mathbf{m_E}$ depends on both $\mathbf{A}$ and $\mathbf{b}$. In fact, assuming that $\mathbf{A}$ has already been selected, we can always ensure that $\mathbf{m_E} = 0$ by selecting

$$\mathbf{b} = \mathbf{m_X} - \mathbf{A}\mathbf{m_Y} . \tag{4.58}$$

So all we need to do is to minimize $\text{tr}\,(\mathbf{K_E})$ with respect to $\mathbf{A}$. To do so, observe that by completing the square $\mathbf{K_E}$ can be decomposed as

$$\begin{aligned} \mathbf{K_E} &= \begin{bmatrix} \mathbf{I}_m & -\mathbf{A} \end{bmatrix} \begin{bmatrix} \mathbf{K_X} & \mathbf{K_{XY}} \\ \mathbf{K_{YX}} & \mathbf{K_Y} \end{bmatrix} \begin{bmatrix} \mathbf{I_m} \\ -\mathbf{A}^T \end{bmatrix} \\ &= (\mathbf{K_{XY}}\mathbf{K_Y}^{-1} - \mathbf{A})\mathbf{K_Y}(\mathbf{K_{XY}}\mathbf{K_Y}^{-1} - \mathbf{A})^T + \mathbf{S} , \end{aligned} \tag{4.59}$$

where

$$\mathbf{S} = \mathbf{K_X} - \mathbf{K_{XY}}\mathbf{K_Y}^{-1}\mathbf{K_{YX}} \tag{4.60}$$

denotes the Schur complement $\mathbf{K_Y}$ of the matrix block $\mathbf{K_Y}$ inside $\mathbf{K}$. See [3, pp. 725–726] for a definition of the Schur complement and its role in evaluating the determinant and inverse of block matrices.

Since the covariance $\mathbf{K_Y}$ has been assumed positive definite, the decomposition (4.59) indicates that $\text{tr}\,(\mathbf{K_E})$ is minimized if and only if we choose

$$\mathbf{A} = \mathbf{K_{XY}}\mathbf{K_Y}^{-1} . \tag{4.61}$$

Combining (4.61) and (4.58), this implies that the linear least-squares estimate can be expressed as

$$\hat{\mathbf{X}}_{\text{LLS}}(\mathbf{Y}) = \mathbf{m_X} + \mathbf{K_{XY}}\mathbf{K_Y}^{-1}(\mathbf{Y} - \mathbf{m_Y}) . \tag{4.62}$$

To interpret this identity, note that $\mathbf{m_X}$ represents the a-priori mean of $\mathbf{X}$ and $\mathbf{Y} - \mathbf{m_Y}$ represents the deviation between the observation vector $\mathbf{Y}$ and its a-priori mean, which in some sense represents the new information or "innovation" provided by the observation. So the identity (4.62) adds to the a-priori mean $\mathbf{m_X}$ a correction term proportional to the the new information $\mathbf{Y} - \mathbf{m_Y}$ provided by the observations, where the matrix $\mathbf{A}$ given by (4.61) represents the gain or weighting applied to this new information.

The mean of the error $\mathbf{E} = \mathbf{X} - \hat{\mathbf{X}}_{\text{LLS}}(\mathbf{Y})$ is zero and from (4.59) we find that its covariance is given by

$$\mathbf{K}_{\text{LLS}} = \mathbf{K_X} - \mathbf{K_{XY}}\mathbf{K_Y}^{-1}\mathbf{K_{YX}} . \tag{4.63}$$

Remark: It is interesting to compare the expressions (4.62) and (4.63) for the LLS estimate and its covariance with those obtained in Example 4.1 for the MSE estimation of Gaussian random vectors. Thus, assume that $\mathbf{X}$ and $\mathbf{Y}$ are jointly Gaussian with

$$\begin{bmatrix} \mathbf{X} \\ \mathbf{Y} \end{bmatrix} \sim N\left(\begin{bmatrix} \mathbf{m_X} \\ \mathbf{m_Y} \end{bmatrix}, \begin{bmatrix} \mathbf{K_X} & \mathbf{K_{XY}} \\ \mathbf{K_{YX}} & \mathbf{K_Y} \end{bmatrix} \right).$$

Then it was shown in (4.29) that the mean-square error estimate of $\mathbf{X}$ given $\mathbf{y}$ takes the form

$$\hat{\mathbf{X}}_{\mathrm{MSE}}(\mathbf{Y}) = E[\mathbf{X}|\mathbf{Y}] = \mathbf{m_X} + \mathbf{K_{XY}} \mathbf{K_Y}^{-1} (\mathbf{Y} - \mathbf{m_Y}) ,$$

so that we see that in the Gaussian case, the MSE and LLS estimates coincide! This provides in some sense some underlying justification for the use of linear estimates. We just need to assume that all random variables of interest are jointly Gaussian. Then all MSE estimates will necessarily be linear.

Remark: As desired, the expressions (4.62) and (4.63) for the LLS estimate and its error covariance matrix depend only on the means $\mathbf{m_X}$ and $\mathbf{m_Y}$ and the blocks of the joint covariance matrix $\mathbf{K}$ of $\mathbf{X}$ and $\mathbf{Y}$. So the LLS estimate requires only the knowledge of the first-and second-order statistics of $\mathbf{X}$ and $\mathbf{Y}$, which are generally much easier to estimate than the densities $f_{\mathbf{X}|\mathbf{Y}}(\mathbf{x}|\mathbf{y})$ and $f_{\mathbf{X}}(\mathbf{x})$ required by general Bayesian estimation methods. Note, however, that linear-least squares estimation can still be viewed as a Bayesian method, albeit simplified, since it treats $\mathbf{X}$ as random and requires the knowledge of its mean $\mathbf{m_X}$ and covariance $\mathbf{K_X}$.

Geometric characterization: Like the MSE estimate, the LLS estimate can be characterized in terms of an *orthogonality property*, which is weaker than the similar property satisfied by MSE estimates. Specifically,

$$\hat{\mathbf{X}}_{\mathrm{LLS}}(\mathbf{Y}) = \mathbf{m_X} + \mathbf{K_{XY}} \mathbf{K_Y}^{-1} (\mathbf{Y} - \mathbf{m_Y})$$

is the unique linear function of $\mathbf{Y}$ taking values in $\mathbb{R}^m$ such that

$$E[(\mathbf{X} - \hat{\mathbf{X}}_{\mathrm{LLS}}(\mathbf{Y}))g(\mathbf{Y})] = 0 \tag{4.64}$$

for all *linear* functions g of the form

$$g(\mathbf{Y}) = g_0 + \mathbf{Y}^T \mathbf{g}_1 , \tag{4.65}$$

where g_0 and $\mathbf{g}_1$ denote respectively an arbitrary scalar and a vector of $\mathbb{R}^n$. If we compare the orthogonality relations (4.38) and (4.64) satisfied by $\hat{\mathbf{X}}_{\mathrm{MSE}}$ and $\hat{\mathbf{X}}_{\mathrm{LLS}}$ we see that the difference lies in the fact that (4.64) holds only for the class $\mathcal{L}$ of linear functions of $\mathbf{Y}$, whereas (4.38) is valid for all nonlinear functions of $\mathbf{Y}$.

To verify (4.64) we write the estimation error and $g(\mathbf{Y})$ as

$$\mathbf{X} - \hat{\mathbf{X}}_{\mathrm{LLS}}(\mathbf{Y}) = \begin{bmatrix} \mathbf{I}_m & -\mathbf{K_{XY}} \mathbf{K_Y}^{-1} \end{bmatrix} \begin{bmatrix} \mathbf{X} - \mathbf{m_X} \\ \mathbf{Y} - \mathbf{m_Y} \end{bmatrix}$$

and

$$g(\mathbf{Y}) = g_0 + \mathbf{m}_{\mathbf{Y}}^T \mathbf{g}_1 + (\mathbf{Y} - \mathbf{m}_{\mathbf{y}})^T \mathbf{g}_1 \,.$$

Since the error $\mathbf{X} - \hat{\mathbf{X}}_{\text{LLS}}$ is a zero-mean random vector, it is clearly uncorrelated with the constant term $g_0 + \mathbf{m}_{\mathbf{Y}}^T \mathbf{g}_1$ of $g(\mathbf{Y})$, so that

$$E[(\mathbf{X} - \hat{\mathbf{X}}_{\text{LLS}})g(\mathbf{Y})] = \begin{bmatrix} \mathbf{I}_m & -\mathbf{K}_{\mathbf{XY}}\mathbf{K}_{\mathbf{Y}}^{-1} \end{bmatrix} E\left[\begin{bmatrix} \mathbf{X} - \mathbf{m}_{\mathbf{X}} \\ \mathbf{Y} - \mathbf{m}_{\mathbf{Y}} \end{bmatrix} (\mathbf{Y} - \mathbf{m}_{\mathbf{Y}})^T \right] \mathbf{g}_1$$

$$= \begin{bmatrix} \mathbf{I}_m & -\mathbf{K}_{\mathbf{XY}}\mathbf{K}_{\mathbf{Y}}^{-1} \end{bmatrix} \begin{bmatrix} \mathbf{K}_{\mathbf{XY}} \\ \mathbf{K}_{\mathbf{Y}} \end{bmatrix} \mathbf{g}_1 = 0 \,,$$

which proves the identity.

To verify that $\hat{\mathbf{X}}_{\text{LLS}}(\mathbf{Y})$ is the only linear estimator with the orthogonality property (4.64), suppose that $\mathbf{h}(\mathbf{Y})$ is another linear estimator with the same property. Then

$$E[||\hat{\mathbf{X}}_{\text{LLS}}(\mathbf{Y}) - \mathbf{h}(\mathbf{Y})||_2^2] = E[\mathbf{g}^T(\mathbf{Y})(\hat{\mathbf{X}}_{\text{LLS}} - \mathbf{X} + \mathbf{X} - \mathbf{h}(\mathbf{Y}))] = 0 \quad (4.66)$$

with

$$\mathbf{g}(\mathbf{Y}) \triangleq \hat{\mathbf{X}}_{\text{LLS}}(\mathbf{Y}) - \mathbf{h}(\mathbf{Y})$$

where the last equality in (4.66) is due to the orthogonality property of $\hat{\mathbf{X}}_{\text{LLS}}(\mathbf{Y})$ and $\mathbf{h}(\mathbf{Y})$ and the observation that since both of these estimators are linear, $\mathbf{g}(\mathbf{Y})$ is linear.

Example 4.3: Linear measurement in additive noise

Consider the case where the observation vector $\mathbf{Y} \in \mathbb{R}^n$ is a linear transformation of $\mathbf{X} \in \mathbb{R}^m$ observed in the presence of an additive noise vector $\mathbf{V} \in \mathbb{R}^n$, so that we have the measurement model

$$\mathbf{Y} = \mathbf{H}\mathbf{X} + \mathbf{V}$$

which is depicted in Fig. 4.3. This model arises, for example, when linearly modulated signals are transmitted through a multiple-input multiple-output (MIMO) flat-fading communication channel with channel matrix $\mathbf{H}$ in the presence of additive noise [4].

We assume here that the noise $\mathbf{V}$ has zero-mean, is uncorrelated with $\mathbf{X}$, and that its covariance

$$\mathbf{R} = E[\mathbf{V}\mathbf{V}^T]$$

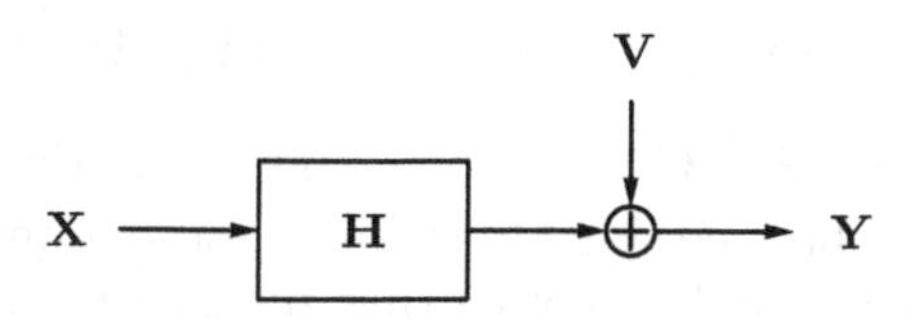

Fig. 4.3. Linear measurement in additive noise.

is positive definite. Then to be able to construct the LLS estimate of $\mathbf{X}$ given $\mathbf{Y}$ we must first evaluate the mean and covariance of $\mathbf{Y}$ and its cross-covariance with $\mathbf{X}$. Since $\mathbf{V}$ has zero-mean, we find

$$\mathbf{m_Y} = \mathbf{H}\mathbf{m_X} \tag{4.67}$$

and

$$\mathbf{Y} - \mathbf{m_Y} = \mathbf{H}(\mathbf{X} - \mathbf{m_X}) + \mathbf{V} \,. \tag{4.68}$$

Since $\mathbf{V}$ is uncorrelated with $\mathbf{X}$, the identity (4.68) implies

$$\begin{aligned} K_{\mathbf{YX}} &= E[(\mathbf{Y} - \mathbf{m_Y})(\mathbf{X} - \mathbf{m_X})^T] \\ &= \mathbf{H}E[(\mathbf{X} - \mathbf{m_X})(\mathbf{X} - \mathbf{m_X})^T] = \mathbf{H}\mathbf{K_X} \end{aligned} \tag{4.69}$$

and

$$\begin{aligned} K_{\mathbf{Y}} &= E[(\mathbf{Y} - \mathbf{m_Y})(\mathbf{Y} - \mathbf{m_Y})^T] \\ &= \mathbf{H}E[(\mathbf{X} - \mathbf{m_X})(\mathbf{X} - \mathbf{m_X})^T]\mathbf{H}^T + E[\mathbf{V}\mathbf{V}^T] \\ &= \mathbf{H}\mathbf{K_X}\mathbf{H}^T + \mathbf{R} \,. \end{aligned} \tag{4.70}$$

Plugging in the expressions (4.67), (4.69) and (4.70) for $\mathbf{m_Y}$, $\mathbf{K_{YX}}$ and $\mathbf{K_Y}$ inside the expressions (4.62) and (4.63) for the LLS estimate and its covariance, we obtain

$$\hat{\mathbf{X}}_{\mathrm{LLS}}(\mathbf{Y}) = \mathbf{m_X} + \mathbf{K_X}\mathbf{H}^T(\mathbf{H}\mathbf{K_X}\mathbf{H}^T + \mathbf{R})^{-1}(\mathbf{Y} - \mathbf{H}\mathbf{m_X}) \tag{4.71}$$

and

$$\mathbf{K}_{\mathrm{LLS}} = \mathbf{K_X} - \mathbf{K_X}\mathbf{H}^T(\mathbf{H}\mathbf{K_X}\mathbf{H}^T + \mathbf{R})^{-1}\mathbf{H}\mathbf{K_X} \,. \tag{4.72}$$

These two expressions form the key to the derivation of Kalman filters [3]. It turns out that when $\mathbf{K_X}$ is invertible, the identities (4.71) and (4.72) can be written more compactly as

$$\mathbf{K}_{\mathrm{LLS}}^{-1}\hat{\mathbf{X}}_{\mathrm{LLS}} = \mathbf{K_X}^{-1}\mathbf{m_X} + \mathbf{H}^T\mathbf{R}^{-1}\mathbf{Y} \tag{4.73}$$

and

$$\mathbf{K}_{\mathrm{LLS}}^{-1} = \mathbf{K_X}^{-1} + \mathbf{H}^T\mathbf{R}^{-1}\mathbf{H} \,, \tag{4.74}$$

which are called the "information forms" of the linear estimate and its covariance matrix. The expression (4.74) can be obtained directly from (4.72) by using the Sherman-Morrison-Woodbury matrix inversion identity [5, p. 48]

$$(\mathbf{A} + \mathbf{BCD})^{-1} = \mathbf{A}^{-1} - \mathbf{A}^{-1}\mathbf{B}(\mathbf{C}^{-1} + \mathbf{D}\mathbf{A}^{-1}\mathbf{B})^{-1}\mathbf{D}\mathbf{A}^{-1} \tag{4.75}$$

with $\mathbf{A} = \mathbf{K_X}^{-1}$, $\mathbf{B} = \mathbf{H}^T$, $\mathbf{C} = \mathbf{R}^{-1}$ and $\mathbf{D} = \mathbf{H}$. Expression (4.73) for the estimate $\hat{\mathbf{X}}_{\mathrm{LLS}}$ is obtained directly from the estimation formula (4.71) by using the identity (4.74), and by noting that the gain matrix

$$\mathbf{G} = \mathbf{K_X}\mathbf{H}^T(\mathbf{H}\mathbf{K_X}\mathbf{H}^T + \mathbf{R})^{-1} \tag{4.76}$$

which multiplies the vector $\mathbf{Y} - \mathbf{H}\mathbf{m}_\mathbf{X}$ in (4.71) can be rewritten as

$$\mathbf{G} = \mathbf{K}_{\mathrm{LLS}}\mathbf{H}^T\mathbf{R}\,. \tag{4.77}$$

To verify that expressions (4.76) and (4.77) are equivalent, note that by multiplication on the left by $\mathbf{K}_\mathbf{E}^{-1}$ and on the right by $\mathbf{H}\mathbf{K}_\mathbf{X}\mathbf{H}^T + \mathbf{R}$, these two expressions are equal if and only if

$$(\mathbf{K}_\mathbf{X}^{-1} + \mathbf{H}^T\mathbf{R}^{-1}\mathbf{H})\mathbf{K}_\mathbf{X}\mathbf{H}^T = \mathbf{H}^T\mathbf{R}^{-1}(\mathbf{H}\mathbf{K}_\mathbf{X}\mathbf{H}^T + \mathbf{R})\,,$$

which can be verified by inspection. □

4.4 Estimation of Nonrandom Parameters

In the previous two sections we have assumed that some a-priori statistical information was available for the parameter vector $\mathbf{X}$ in the form of either the density $f_\mathbf{X}(\mathbf{x})$ for general Bayesian estimation, or the mean vector $\mathbf{m}_\mathbf{X}$ and covariance matrix $\mathbf{K}_\mathbf{X}$ for linear least-squares estimation. An alternate perspective consists of viewing the parameter vector $\mathbf{X}$ as unknown but nonrandom. In this approach, the only available information is the measurement vector $\mathbf{Y} \in \mathbb{R}^n$ and the observation model specified by the density $f_{\mathbf{Y}|\mathbf{X}}(\mathbf{y}|\mathbf{x})$. The density $f_{\mathbf{Y}|\mathbf{X}}(\mathbf{y}|\mathbf{x})$, when viewed as a function of $\mathbf{x}$, is usually called the likelihood function since it indicates how likely we are to observe $\mathbf{Y} = \mathbf{y}$ when the parameter vector is $\mathbf{x}$.

Any function $\mathbf{g}(\mathbf{Y})$ of the observations taking values in $\mathbb{R}^m$ can be viewed as an estimator of $\mathbf{X}$. Among such estimators, the *maximum likelihood* estimator $\mathbf{X}_{\mathrm{ML}}(\mathbf{Y})$ maximizes the likelihood function $f_{\mathbf{Y}|\mathbf{X}}(\mathbf{y}|\mathbf{x})$, i.e.,

$$\hat{\mathbf{X}}_{\mathrm{ML}}(\mathbf{y}) = \arg \max_{\mathbf{x}\in\mathbb{R}^m} f_{\mathbf{Y}|\mathbf{X}}(\mathbf{y}|\mathbf{x})\,. \tag{4.78}$$

Since the logarithm is a monotone function, i.e. $z_1 < z_2$ if and only if $\ln(z_1) < \ln(z_2)$, we can equivalently obtain the ML estimate by maximizing the log-likelihood function:

$$\hat{\mathbf{X}}_{\mathrm{ML}}(\mathbf{y}) = \arg \max_{\mathbf{x}\in\mathbf{R}^m} \ln(f_{\mathbf{Y}|\mathbf{X}}(\mathbf{y}|\mathbf{x}))\,. \tag{4.79}$$

This choice is convenient if $f_{\mathbf{Y}|\mathbf{X}}(\mathbf{y}|\mathbf{x})$ belongs to the exponential class of densities of the form

$$f_{\mathbf{Y}|\mathbf{X}}(\mathbf{y}|\mathbf{x}) = u(\mathbf{y})\exp(\mathbf{x}^T\mathbf{S}(\mathbf{y}) - t(\mathbf{x}))\,, \tag{4.80}$$

which includes as special cases the Poisson, exponential, and Gaussian distributions.

Although the MAP estimate (4.27) was obtained by using a different approach, it is closely related to the ML estimate. Specifically, by observing that the marginal density $f_{\mathbf{Y}}(\mathbf{y})$ appearing in the denominator of expression (4.13) for the conditional density $f_{\mathbf{X}|\mathbf{Y}}$ is a scaling factor that does not affect the outcome of the maximization (4.27), we can rewrite the MAP estimate as

$$\begin{aligned}\hat{\mathbf{X}}_{\mathrm{MAP}}(\mathbf{y}) &= \arg \max_{\mathbf{x}\in\mathbb{R}^m} f_{\mathbf{Y},\mathbf{X}}(\mathbf{y},\mathbf{x}) \\ &= \arg \max_{\mathbf{x}\in\mathbb{R}^m} \left[\ln(f_{\mathbf{Y}|\mathbf{X}}(\mathbf{y}|\mathbf{x})) + \ln(f_{\mathbf{X}}(\mathbf{x}))\right] . \qquad (4.81)\end{aligned}$$

Comparing (4.81) and (4.79), we see that the only difference between the MAP and ML estimates is that the objective function minimized by the MAP estimate is formed by adding to the log-likelihood function a term $\ln f_{\mathbf{X}}(\mathbf{x})$ representing the a-priori information about $\mathbf{X}$. When $\mathbf{X}$ admits a uniform distribution, so that $f_{\mathbf{X}}(\mathbf{x})$ is constant, the two estimates coincide. Thus, when the range of values of $\mathbf{X}$ is bounded, we can interpret the ML estimate as the MAP estimate obtained for a noninformative prior distribution. This indicates that the dichotomy between the random and nonrandom formulations of parameter estimation is not completely strict, and it is often possible to switch from one viewpoint to the other.

Example 4.4: Signal with unknown amplitude

We observe an N-dimensional Gaussian vector

$$\mathbf{Y} \sim N(A\mathbf{s}, \sigma^2\mathbf{I}_N) ,$$

where the noise variance σ^2 and the signal $\mathbf{s} \in \mathbb{R}^N$ are known, but the amplitude A is unknown, so the density of $\mathbf{Y}$ can be expressed as

$$f_{\mathbf{Y}|A}(\mathbf{y}|A) = \frac{1}{(2\pi\sigma^2)^{N/2}} \exp\left(- \frac{1}{2\sigma^2}||\mathbf{y} - A\mathbf{s}||_2^2\right) .$$

Then $\hat{A}_{\mathrm{ML}}$ is obtained by maximizing

$$\ln f_{\mathbf{Y}|A}(\mathbf{y}|A) = -\frac{1}{2\sigma^2}||\mathbf{y} - A\mathbf{s}||_2^2 + c$$

where c does not depend on A and $\mathbf{y}$, or equivalently, by minimizing $||\mathbf{y}-A\mathbf{s}||_2^2$ over A, which gives

$$\hat{A}_{\mathrm{ML}}(\mathbf{y}) = \frac{\mathbf{s}^T\mathbf{y}}{||\mathbf{s}||_2^2} .$$

□

An important feature of the ML estimate is that it is *parametrization independent.* Thus, instead of using the vector $\mathbf{x}$ to parametrize the observation density $f_{\mathbf{Y}|\mathbf{X}}(\mathbf{y}|\mathbf{x})$, suppose we choose the parametrization $\mathbf{z} = \mathbf{g}(\mathbf{x})$ where the transformation $\mathbf{g}(\cdot)$ is *one-to-one.* If $\mathbf{g}^{-1}$ denotes the inverse transformation, the density of the observations can be expressed in terms of the parameter vector $\mathbf{z}$ as

$$f_{\mathbf{Y}|\mathbf{Z}}(\mathbf{y}|\mathbf{z}) = f_{\mathbf{Y}|\mathbf{X}}(\mathbf{y}|\mathbf{g}^{-1}(\mathbf{z}))\,.$$

Then if $\hat{\mathbf{X}}_{\mathrm{ML}}(\mathbf{Y})$ denotes the ML estimate of $\mathbf{x}$, the ML estimate of $\mathbf{z}$ is just the image of $\hat{\mathbf{X}}_{\mathrm{ML}}$ under $\mathbf{g}$, i.e.

$$\hat{\mathbf{Z}}_{\mathrm{ML}}(\mathbf{Y}) = \mathbf{g}(\hat{\mathbf{X}}_{\mathrm{ML}}(\mathbf{Y}))\,. \tag{4.82}$$

Example 4.5: Parametrization of Gaussian densities

Consider a sequence $\{Y_k,\ 1 \leq k \leq N\}$ of independent identically distributed $N(m,v)$ random variables. The density of the random vector

$$\mathbf{Y} = \begin{bmatrix} Y_1\, Y_2 \ldots Y_N \end{bmatrix}^T$$

can be expressed in terms of the (m,v) parametrization as

$$f_{\mathbf{Y}}(\mathbf{y}|m,v) = \frac{1}{(2\pi v)^{N/2}} \exp\Big(\frac{-1}{2v}\sum_{k=1}^{N}(y_k - m)^2\Big)\,. \tag{4.83}$$

However, this density can also be expressed in the canonical exponential form (4.80) by selecting

$$\mathbf{x} = \begin{bmatrix} x_1 \\ x_2 \end{bmatrix} = N \begin{bmatrix} m/v \\ -1/2v \end{bmatrix}$$

with the vector sufficient statistic

$$\mathbf{S}(\mathbf{y}) = \frac{1}{N}\begin{bmatrix} \sum_{k=1}^{N} y_k \\ \sum_{k=1}^{N} y_k^2 \end{bmatrix},$$

and the constant

$$t(\mathbf{x}) = -\frac{N}{2}(\frac{m^2}{v} + \ln(v)) = -\frac{N}{2}\Big(- \frac{x_1^2}{2Nx_2} + \ln(-N/2x_2)\Big)\,.$$

Then, independently of what parametrization we choose, the ML estimate for one choice of parameter vector immediately gives the ML estimate for all other parametrizations. □

4.4.1 Bias

To evaluate the performance of estimators, we rely on several quantities. The *bias* $\mathbf{b}(\mathbf{x})$ of an estimator $\hat{\mathbf{X}}(\mathbf{Y})$ is the expectation of its error, i.e.,

$$\mathbf{b}(\mathbf{x}) = \mathbf{x} - E[\hat{\mathbf{X}}(\mathbf{Y})]\,. \tag{4.84}$$

An estimator is said to be *unbiased* if its bias is zero, i.e. if $\mathbf{b}(\mathbf{x}) = \mathbf{0}$. This property just indicates that when the estimator is averaged over a large number of realizations, it gives the correct parameter vector value. This property is rather weak since it does not ensure that for a single realization, the estimator $\hat{\mathbf{X}}(\mathbf{Y})$ is close to the true parameter vector.

Example 4.6: ML estimates of the mean and variance of i.i.d. Gaussian random variables

Assume that we observe a sequence $\{Y_k,\ 1 \leq k \leq N\}$ of i.i.d. $N(m, v)$ random variables. The probability density of vector

$$\mathbf{Y} = \begin{bmatrix} Y_1 \, Y_2 \dots Y_N \end{bmatrix}^T$$

is given by (4.83). We consider three estimation cases:

Case 1: m unknown, v known. In this case the observation vector density is denoted as $f_{\mathbf{Y}}(\mathbf{y}|m)$, and we observe it has the form considered in Example 4.4, with $A = m$ and

$$\mathbf{s} = \mathbf{u} \triangleq \begin{bmatrix} 1 \, 1 \dots 1 \end{bmatrix}^T , \tag{4.85}$$

so that

$$\widehat{m}_{\mathrm{ML}}(\mathbf{Y}) = \frac{1}{N} \sum_{k=1}^{N} Y_k , \tag{4.86}$$

which is the *sampled mean* of the observations. We have

$$E[\widehat{m}_{\mathrm{ML}}(\mathbf{Y})] = \frac{1}{N} \sum_{k=1}^{N} E[Y_k] = m ,$$

so the estimator is unbiased.

Case 2: m known, v unknown. In this case the observation vector density is denoted as $f_{\mathbf{Y}}(\mathbf{y}|v)$, and the log-likelihood is given by

$$\ln(f_{\mathbf{Y}}(\mathbf{y}|v)) = -\frac{N}{2} \ln(2\pi v) - \frac{1}{2v} \sum_{k=1}^{N} (y_k - m)^2 .$$

Taking its derivative with respect to v yields

$$\frac{\partial}{\partial v} \ln(f_{\mathbf{Y}}(\mathbf{y}|v)) = -\frac{N}{2v} + \frac{1}{2v^2} \sum_{k=1}^{N} (y_k - m)^2 .$$

So by setting this derivative equal to zero, we obtain the ML estimate

$$\hat{v}_{\mathrm{ML}}(\mathbf{Y}) = \frac{1}{N} \sum_{k=1}^{N} (Y_k - m)^2 , \tag{4.87}$$

which is the *sampled variance* of the observations. To ensure that the estimate (4.87) represents a maximum of the log-likelihood function, we need to verify that the second derivative

$$\frac{\partial^2}{\partial v^2} \ln(f_{\mathbf{Y}}(\mathbf{y}|v)) = \frac{N}{2v^2} - \frac{1}{v^3}\sum_{k=1}^{N}(y_k - m)^2$$

is negative at this point. This is the case since

$$\frac{\partial^2}{\partial v^2} \ln(f_{\mathbf{Y}}(\mathbf{y}|\hat{v}_{\mathrm{ML}})) = -\frac{N}{2(\hat{v}_{\mathrm{ML}})^2} < 0 .$$

By observing that

$$E[\hat{v}_{\mathrm{ML}}(\mathbf{Y})] = \frac{1}{N}\sum_{k=1}^{N} E[(Y_k - m)^2] = v ,$$

we conclude that the ML estimate is unbiased.

Case 3: m and v unknown. In this case, the derivatives of the log-likelihood function

$$\begin{aligned} L(\mathbf{y}|m, v) &\triangleq \ln(f_{\mathbf{Y}}(\mathbf{y}|m, v) \\ &= -\frac{N}{2}\ln(2\pi v) - \frac{1}{2v}\sum_{k=1}^{N}(y_k - m)^2 \end{aligned}$$

with respect to m and v are given by

$$\frac{\partial}{\partial m} L(\mathbf{y}|m, v) = \frac{1}{v}\sum_{k=1}^{N}(y_k - m) , \tag{4.88}$$

and

$$\frac{\partial}{\partial v} L(\mathbf{y}|m, v) = -\frac{N}{2v} + \frac{1}{2v^2}\sum_{k=1}^{N}(y_k - m)^2 . \tag{4.89}$$

Setting these two derivatives equal to zero and solving the resulting coupled equations, we find that the ML estimate of m is given by (4.86), so it is the same as in the case where the variance is known, but the ML estimate of v is now given by

$$\hat{v}_{\mathrm{ML}}(\mathbf{Y}) = \frac{1}{N}\sum_{k=1}^{N}(Y_k - \widehat{m}_{\mathrm{ML}})^2 . \tag{4.90}$$

This estimate can be viewed as obtained by replacing the unknown mean m in the estimator (4.86) by the sampled mean $\widehat{m}_{\mathrm{ML}}$. Even though this is a reasonable choice, it affects the properties of the resulting estimator, and in particular its bias. To see this, let $\mathbf{u}$ be the vector of dimension N with all its entries equal to 1 defined in (4.85), and introduce the matrix

$$\mathbf{P} = \mathbf{I}_N - \frac{1}{N}\mathbf{u}\mathbf{u}^T .$$

Since $\mathbf{Pu} = \mathbf{0}$ and $\mathbf{P}^2 = \mathbf{P}$, it represents a projection on the $N-1$ dimensional subspace perpendicular to $\mathbf{u}$. Then if

$$\mathbf{Z} = \mathbf{PY} = \mathbf{P}(\mathbf{Y} - m\mathbf{u}) \tag{4.91}$$

the estimator (4.90) can be written compactly as

$$\hat{v}_{\mathrm{ML}}(\mathbf{Y}) = \frac{||\mathbf{Z}||_2^2}{N} . \tag{4.92}$$

Since the vector $\mathbf{Y} - m\mathbf{u}$ admits a $N(0, v\mathbf{I}_N)$ distribution and $\mathbf{Z}$ represents its projection on an $N-1$ dimensional subspace, its decomposition

$$\mathbf{Z} = \sum_{k=1}^{N-1} Z_k \mathbf{e}_k$$

with respect to an arbitrary orthonormal basis $\{\mathbf{e}_k, 1 \leq k \leq N-1\}$ of the subspace orthogonal to $\mathbf{u}$ is such that the coordinates Z_k with $1 \leq k \leq N-1$ are independent $N(0, v)$ distributed. Accordingly, the expected value of the estimator (4.92) is given by

$$E[\hat{v}_{\mathrm{ML}}(\mathbf{Y})] = \frac{1}{N} \sum_{k=1}^{N-1} E[Z_k^2] = \frac{(N-1)}{N} v ,$$

which indicates the estimator is *biased*. Again, to ensure that the ML estimates $(\widehat{m}_{\mathrm{ML}}, \hat{v}_{\mathrm{ML}})$ represent a maximum of the likelihood function, it is necessary to verify that the Hessian matrix formed by the second order derivatives of L with respect to m and v is negative definite at the point specified by (4.86) and (4.90). This is left as an exercise for the reader. □

4.4.2 Sufficient Statistic

As in the detection case, a statistic $\mathbf{S}(\mathbf{Y}) \in \mathbb{R}^k$ is said to be sufficient for $\mathbf{x}$ if it contains all the information about the observation vector $\mathbf{Y}$ necessary to estimate $\mathbf{x}$. Formally, $\mathbf{S}(\mathbf{Y})$ is sufficient for $\mathbf{x}$ if the conditional density of $\mathbf{Y}$ given $\mathbf{S}(\mathbf{Y})$ is independent of $\mathbf{x}$. This independence property indicates that all the information about $\mathbf{x}$ has been "squeezed" in $f_{\mathbf{S}}(\mathbf{s}|\mathbf{x})$ and there is no leftover information about $\mathbf{x}$ that could be extracted from $f_{\mathbf{Y}|\mathbf{S}}(\mathbf{y}|\mathbf{s})$, which means this density must be independent of $\mathbf{x}$. We also say that the sufficient statistic $\mathbf{S}(\mathbf{Y})$ is minimal if it has the smallest dimension among all sufficient statistics. Equivalently, a sufficient statistic is minimal if it represents a maximal compression of the observations. Thus if $\mathbf{T}(\mathbf{Y})$ is another sufficient statistic, $\mathbf{S}(\mathbf{Y})$ is necessarily expressible as a compression of $\mathbf{T}(\mathbf{y})$:

$$\mathbf{S}(\mathbf{Y}) = c(\mathbf{T}(\mathbf{Y})) .$$

Consequently, if two sufficient statistics are minimal, they are related by a one-to-one mapping, so they have the same dimension.

The definition given above is not convenient since it requires guessing a sufficient statistic $\mathbf{S}(\mathbf{Y})$ and then verifying it is sufficient by evaluating the conditional density $f_{\mathbf{Y}|\mathbf{S}}(\mathbf{y}|\mathbf{s})$. A method that takes the guessing out of finding sufficient statistics is provided by the *Neyman-Fisher factorization criterion*, according to which a statistic $\mathbf{S}(\mathbf{Y})$ is sufficient if and only if the density $f_{\mathbf{Y}}$ admits a factorization of the form

$$f_{\mathbf{Y}}(\mathbf{y}|\mathbf{x}) = b(\mathbf{S}(\mathbf{y}), \mathbf{x})c(\mathbf{y}) . \tag{4.93}$$

To prove this result, we proceed as in Section 2.3. Specifically, we select an arbitrary auxiliary statistic $\mathbf{A}$ such that the mapping

$$\mathbf{Y} \underset{\mathbf{h}}{\overset{\mathbf{g}}{\rightleftarrows}} \begin{bmatrix} \mathbf{S} \\ \mathbf{A} \end{bmatrix} \tag{4.94}$$

is one-to-one. Then if $J(\mathbf{s}, \mathbf{a})$ denotes the Jacobian of $\mathbf{h}$, the joint density of $\mathbf{S}$ and $\mathbf{A}$ is given by

$$\begin{aligned} f_{\mathbf{S},\mathbf{A}}(\mathbf{s}, \mathbf{a}|\mathbf{x}) &= J(\mathbf{s}, \mathbf{a}) f_{\mathbf{Y}}(\mathbf{h}(\mathbf{s}, \mathbf{a})|\mathbf{x}) \\ &= J(\mathbf{s}, \mathbf{a}) b(\mathbf{s}, \mathbf{x}) c(\mathbf{h}(\mathbf{s}, \mathbf{a})) , \end{aligned} \tag{4.95}$$

from which we immediately see that

$$f_{\mathbf{A}|\mathbf{S}}(\mathbf{a}|\mathbf{s}) = \frac{J(\mathbf{s}, \mathbf{a})c(\mathbf{h}(\mathbf{s}, \mathbf{a}))}{\int J(\mathbf{s}, \mathbf{a})c(\mathbf{h}(\mathbf{s}, \mathbf{a}))d\mathbf{a}} \tag{4.96}$$

is independent of $\mathbf{x}$.

Example 4.5, continued

Consider a sequence $\{Y_k,\ 1 \le k \le N\}$ of i.i.d. $N(m, v)$ random variables with density (4.83). We have seen in Example 4.5 that it can be rewritten in the canonical exponential form (4.80) which is in the factored form (4.93) with

$$b(\mathbf{S}, \mathbf{x}) = \exp(\mathbf{x}^T \mathbf{S} - v(\mathbf{x}))$$

and $c(\mathbf{y}) = u(\mathbf{y})$. This implies that

$$\mathbf{S}(\mathbf{Y}) = \begin{bmatrix} S_1(\mathbf{Y}) \\ S_2(\mathbf{Y}) \end{bmatrix} = \frac{1}{N} \begin{bmatrix} \sum_{k=1}^N Y_k \\ \sum_{k=1}^N Y_k^2 \end{bmatrix} \tag{4.97}$$

is a sufficient statistic for estimating

$$\mathbf{x} = N \begin{bmatrix} m/v \\ -1/2v \end{bmatrix}$$

or equivalently, for estimating the pair (m, v). This is confirmed by noting that the ML estimator obtained for the mean and variance in Case 3 of Example 4.6 can be expressed as

$$\begin{bmatrix} \widehat{m}_{\rm ML}(\mathbf{Y}) \\ \hat{v}_{\rm ML}(\mathbf{Y}) \end{bmatrix} = \begin{bmatrix} S_1(\mathbf{Y}) \\ S_2(\mathbf{Y}) - S_1^2(\mathbf{Y}) \end{bmatrix} . \tag{4.98}$$

Note however that a sufficient statistic is always defined in relation to a parameter vector $\mathbf{x}$ to be estimated. For example, for the Case 1 of Example 4.6, where only the mean m needs to be estimated and the variance v is known, the previous argument can be modified to show that $S_1(\mathbf{Y})$ is sufficient. Similarly for Case 2 where m is known and v is unknown, $S_2(\mathbf{Y})$ is sufficient. □

4.4.3 Cramér-Rao Lower Bound

In addition to its bias, an arbitrary estimator $\hat{X}(\mathbf{Y})$ is usually characterized by its error correlation matrix

$$\mathbf{C_E}(\mathbf{x}) = E[(\mathbf{x} - \hat{\mathbf{X}}(\mathbf{Y}))(\mathbf{x} - \hat{\mathbf{X}}(\mathbf{Y}))^T] \tag{4.99}$$

and its mean-square error

$$\mathrm{MSE}(\mathbf{x}) = \mathrm{tr}\,(\mathbf{C_E}(\mathbf{x})) = E[||\mathbf{x} - \hat{\mathbf{X}}(\mathbf{Y})||_2^2] . \tag{4.100}$$

In this context, for the class of unbiased estimators, a useful metric for evaluating estimator performance is provided by the Cramér-Rao lower bound. We start by introducing the Fisher information matrix $\mathbf{J}(\mathbf{x})$ which characterizes the information in $\mathbf{Y}$ about the parameter vector $\mathbf{x}$. If $\nabla_{\mathbf{x}}$ denotes the vector gradient

$$\nabla_{\mathbf{x}} = \left[\frac{\partial}{\partial x_1} \ \frac{\partial}{\partial x_2} \cdots \frac{\partial}{\partial x_m} \right]^T , \tag{4.101}$$

the Fisher information is the $m \times m$ matrix given by

$$\mathbf{J}(\mathbf{x}) = E_{\mathbf{Y}}[\nabla_{\mathbf{x}} \ln(f_{\mathbf{Y}}(\mathbf{y}|\mathbf{x}))\big(\nabla_{\mathbf{x}} \ln(f_{\mathbf{Y}}(\mathbf{y}|\mathbf{x}))\big)^T] , \tag{4.102}$$

where the expectation represents an average with respect to $f_{\mathbf{Y}}(\mathbf{y}|\mathbf{x})$. It turns out that this matrix can also be expressed in terms of the negative of the Hessian of the log-likelihood function as

$$\mathbf{J}(\mathbf{x}) = -E_{\mathbf{Y}}[\nabla_{\mathbf{x}} \nabla_{\mathbf{x}}^T \ln(f_{\mathbf{Y}}(\mathbf{y}|\mathbf{x}))] . \tag{4.103}$$

To see how (4.103) follows from (4.102), we start from

$$\int f_{\mathbf{Y}}(\mathbf{y}|\mathbf{x}) d\mathbf{y} = 1 , \tag{4.104}$$

which expresses that the total probability mass of the density $f_{\mathbf{Y}}(\mathbf{y}|\mathbf{x})$ is equal to one. Applying $\nabla_{\mathbf{x}}^T$ to (4.104) and noting that

$$\nabla_{\mathbf{x}} \ln(f_{\mathbf{Y}}(\mathbf{y}|\mathbf{x})) = \frac{1}{f_{\mathbf{Y}}(\mathbf{y}|\mathbf{x})} \nabla_{\mathbf{x}} f_{\mathbf{Y}}(\mathbf{y}|\mathbf{x}) , \tag{4.105}$$

we find

$$\int [\nabla_{\mathbf{x}}^T \ln(f_{\mathbf{Y}}(\mathbf{y}|\mathbf{x}))] f_{\mathbf{Y}}(\mathbf{y}|\mathbf{x}) d\mathbf{y} = \mathbf{0}^T , \tag{4.106}$$

where $\mathbf{O}^T$ denotes an all-zero row vector of length m. Then applying $\nabla_{\mathbf{x}}$ to (4.106) and taking into account (4.105) gives

$$\begin{aligned} &\int [\nabla_{\mathbf{x}} \nabla_{\mathbf{x}}^T \ln(f_{\mathbf{Y}}(\mathbf{y}|\mathbf{x}))] f_{\mathbf{Y}}(\mathbf{y}|\mathbf{x}) d\mathbf{y} \\ &+ \int [\nabla_{\mathbf{x}} \ln(f_{\mathbf{Y}}(\mathbf{y}|\mathbf{x}))][\nabla_{\mathbf{x}} \ln(f_{\mathbf{Y}}(\mathbf{y}|\mathbf{x}))]^T f_{\mathbf{Y}}(\mathbf{y}|\mathbf{x}) d\mathbf{y} = \mathbf{0} , \end{aligned}$$

which is just a statement of the equality of (4.103) and (4.102).

Thus the (i, j)-th entry of the Fisher information matrix can be written as

$$\begin{aligned} \mathbf{J}_{ij}(\mathbf{x}) &= E_{\mathbf{Y}}[\frac{\partial}{\partial x_i} \ln(f_{\mathbf{Y}}(\mathbf{Y}|\mathbf{x})) \frac{\partial}{\partial x_j} \ln(f_{\mathbf{Y}}(\mathbf{Y}|\mathbf{x}))] \\ &= -E_{\mathbf{Y}}[\frac{\partial^2}{\partial x_i \partial x_j} \ln(f_{\mathbf{Y}}(\mathbf{Y}|\mathbf{x}))]. \end{aligned}$$

It does not matter which one of these two expressions we choose to evaluate, but in general it is easier to take the expectation of the second-order derivative of the log-likelihood function with respect to x_i and x_j than to compute the expectation of the cross product of the first-order derivatives of the log-likelihood function with respect to x_i and x_j.

Next, given an arbitrary estimator $\hat{\mathbf{X}}(\mathbf{Y})$ with bias $\mathbf{b}(\mathbf{x})$, we form the $2m$-dimensional random vector

$$\mathbf{Z} = \begin{bmatrix} \mathbf{x} - \hat{\mathbf{X}}(\mathbf{Y}) - \mathbf{b}(\mathbf{x}) \\ \nabla_{\mathbf{x}} \ln(f_{\mathbf{Y}}(\mathbf{Y}|\mathbf{x})) \end{bmatrix} \tag{4.107}$$

and let

$$\mathbf{C}_{\mathbf{Z}} = \begin{bmatrix} \mathbf{C}_{11} & \mathbf{C}_{12} \\ \mathbf{C}_{21} & \mathbf{C}_{22} \end{bmatrix} = E[\mathbf{Z}\mathbf{Z}^T] \tag{4.108}$$

denote its correlation matrix, where the blocks $\mathbf{C}_{ij}(\mathbf{x})$ have dimension $m \times m$. From the definitions (4.99) and (4.102) of the error correlation matrix and Fisher information matrix, we can immediately identify

$$\begin{aligned} \mathbf{C}_{11}(\mathbf{x}) &= \mathbf{C}_{\mathbf{E}}(\mathbf{x}) - \mathbf{b}(\mathbf{x})\mathbf{b}^T(\mathbf{x}) \\ \mathbf{C}_{22}(\mathbf{x}) &= \mathbf{J}(\mathbf{x}) . \end{aligned} \tag{4.109}$$

It remains to evaluate

$$C_{21}^T(\mathbf{x}) = C_{12}(\mathbf{x}) = E_{\mathbf{Y}}[(\mathbf{x} - \hat{\mathbf{X}}(\mathbf{Y}) - \mathbf{b}(\mathbf{x}))\nabla_{\mathbf{x}}^T \ln(f_{\mathbf{Y}}(\mathbf{y}|\mathbf{x}))]$$
$$= \int (\mathbf{x} - \hat{\mathbf{X}}(\mathbf{y}) - \mathbf{b}(\mathbf{x}))\nabla_{\mathbf{x}}^T f_{\mathbf{Y}}(\mathbf{y}|\mathbf{x})d\mathbf{y} \,. \tag{4.110}$$

To do so, we integrate the identity

$$\nabla_{\mathbf{x}}^T\{[\mathbf{x} - \hat{\mathbf{X}}(\mathbf{y}) - \mathbf{b}(\mathbf{x})]f_{\mathbf{Y}}(\mathbf{y}|\mathbf{x})\}$$
$$= f_{\mathbf{Y}}(\mathbf{y}|\mathbf{x})\nabla_{\mathbf{x}}^T[\mathbf{x} - \hat{\mathbf{X}}(\mathbf{y}) - \mathbf{b}(\mathbf{x})] + [\mathbf{x} - \hat{\mathbf{X}}(\mathbf{y}) - \mathbf{b}(\mathbf{x})]\nabla_{\mathbf{x}}^T f_{\mathbf{Y}}(\mathbf{y}|\mathbf{x})$$
$$= f_{\mathbf{Y}}(\mathbf{y}|\mathbf{x})[\mathbf{I}_m - \nabla_{\mathbf{x}}^T\mathbf{b}(\mathbf{x})] + [\mathbf{x} - \hat{\mathbf{X}}(\mathbf{y}) - \mathbf{b}(\mathbf{x})]\nabla_{\mathbf{x}}^T f_{\mathbf{Y}}(\mathbf{y}|\mathbf{x})$$

with respect to $\mathbf{y}$. By taking into account the definition (4.84) of the bias, we find that the left-hand side of the resulting expression is zero. Accordingly, this gives

$$\mathbf{C}_{12}(\mathbf{x}) = -\mathbf{I}_m + \nabla_{\mathbf{x}}^T\mathbf{b}(\mathbf{x}) \,. \tag{4.111}$$

Because $\mathbf{C_Z}(\mathbf{x})$ is a correlation matrix, it is necessarily non-negative definite. The information matrix $\mathbf{J}(\mathbf{x})$ appearing as its (2,2) block is also a correlation matrix, and we assume that it is positive definite at point $\mathbf{x}$. Then since $\mathbf{C_Z}(\mathbf{x})$ is non-negative, the Schur complement

$$\mathbf{S} = \mathbf{C}_{11} - \mathbf{C}_{12}\mathbf{C}_{22}^{-1}\mathbf{C}_{21}$$
$$= \mathbf{C_E} - \mathbf{b}\mathbf{b}^T - (\mathbf{I}_m - \nabla_{\mathbf{x}}^T\mathbf{b})\mathbf{J}^{-1}(\mathbf{I}_m - \nabla_{\mathbf{x}}^T\mathbf{b})^T \tag{4.112}$$

of the (2,2) block must be non-negative definite. This gives the following *Cramér-Rao lower bound* for the error correlation matrix

$$\mathbf{C_E}(\mathbf{x}) \geq \mathbf{b}(\mathbf{x})\mathbf{b}^T(\mathbf{x}) + (\mathbf{I}_m - \nabla_{\mathbf{x}}^T\mathbf{b}(\mathbf{x}))\mathbf{J}^{-1}(\mathbf{x})(\mathbf{I}_m - \nabla_{\mathbf{x}}^T\mathbf{b}(\mathbf{x}))^T \tag{4.113}$$

where we recall that if $\mathbf{A}$ and $\mathbf{B}$ are two symmetric matrices, the inequality $\mathbf{A} \geq \mathbf{B}$ indicates the matrix $\mathbf{A} - \mathbf{B}$ is non-negative definite.

If the estimator $\hat{\mathbf{X}}(\mathbf{Y})$ is unbiased, so that $\mathbf{b}(\mathbf{x}) = 0$, the error correlation matrix $\mathbf{C_E}(\mathbf{x})$ coincides with the error covariance matrix, and the Cramér-Rao lower bound (CRLB) takes the simple form

$$\mathbf{C_E}(\mathbf{x}) \geq \mathbf{J}^{-1}(\mathbf{x}) \tag{4.114}$$

relating the error covariance matrix to the inverse of the Fisher information matrix. By noting that if a matrix is non-negative definite, its diagonal elements must be non-negative, the inequality (4.114) implies

$$E[(x_i - \hat{X}_i(\mathbf{Y}))^2] \geq [\mathbf{J}^{-1}]_{ii}(\mathbf{x}) \,, \tag{4.115}$$

so the diagonal elements of the inverse of the Fisher information matrix provide a lower-bound for the mean-square error performance of *any* unbiased estimator.

Example 4.7: i.i.d. Gaussian sequence

Consider again the case where we observe a sequence $\{Y_k,\ 1 \leq k \leq N\}$ of i.i.d. $N(m, v)$ random variables. The first order derivatives of the log-likelihood function

$$L(\mathbf{y}|m,v) = \ln(f_{\mathbf{Y}}(\mathbf{y}|m,v)) = -\frac{N}{2}\ln(2\pi v) - \frac{1}{2v}\sum_{k=1}^{N}(y_k - m)^2 .$$

are given by (3.88) and (3.89), and its second-order derivatives can be expressed as

$$\begin{aligned}
\frac{\partial^2}{\partial m^2}L(\mathbf{y}|m,v) &= -\frac{N}{v} \\
\frac{\partial^2}{\partial m \partial v}L(\mathbf{y}|m,v) &= -\frac{1}{v^2}\sum_{k-1}^{N}(y_k - m) \\
\frac{\partial^2}{\partial v^2}L(\mathbf{y}|m,v) &= \frac{N}{2v^2} - \frac{1}{v^3}\sum_{k=1}^{N}(y_k - m)^2 .
\end{aligned} \tag{4.116}$$

Evaluating these derivatives at $\mathbf{Y}$ and taking expectations with respect to $\mathbf{Y}$ we find that the Fisher information matrix

$$J(m,v) = N\begin{bmatrix} v^{-1} & 0 \\ 0 & (2v^2)^{-1} \end{bmatrix}$$

has a diagonal structure. Then its inverse is also diagonal, and the error variances of the entries of any unbiased estimator must satisfy the lower bounds

$$\begin{aligned}
E[(m - \widehat{m}(\mathbf{Y}))^2] &\geq \frac{v}{N} \\
E[(v - \hat{v}(\mathbf{Y}))^2] &\geq \frac{2v^2}{N}
\end{aligned} \tag{4.117}$$

□

Efficiency: In the following we will say that an unbiased estimator $\hat{\mathbf{X}}(\mathbf{Y})$ is efficient if it reaches its CRLB. Note that since we consider here vector estimators, an estimator $\hat{\mathbf{X}}(\mathbf{Y})$ can be partially efficient if one or more of the eigenvalues of $\mathbf{C_E}(\mathbf{x}) - \mathbf{J}^{-1}(\mathbf{x})$ is zero, but it is completely efficient only if this matrix is identically zero. For simplicity, we restrict our attention to completely efficient estimators. The first question we need to ask is whether such estimators exist, and if so what is their structure?

To answer this question, we note that $\hat{\mathbf{X}}(\mathbf{Y})$ is an efficient estimator if and only if the Schur complement matrix $\mathbf{S}$ defined in (4.112) is zero, which means that the correlation matrix $\mathbf{C_Z}$ has rank m. So in the definition (4.107) of the vector $\mathbf{Z}$, the first m-dimensional subvector $\mathbf{x} - \hat{\mathbf{X}}(\mathbf{Y})$ must be expressible as a linear transformation of the second m-dimensional subvector $\nabla_{\mathbf{x}} \ln(f_{\mathbf{Y}}(\mathbf{y}|\mathbf{x}))$, which means we have

$$\mathbf{x} - \hat{\mathbf{X}}(\mathbf{Y}) = \mathbf{M}(\mathbf{x}) \nabla_{\mathbf{x}} \ln(f_{\mathbf{Y}}(\mathbf{Y}|\mathbf{x})) , \tag{4.118}$$

where $\mathbf{M}(\mathbf{x})$ is an $m \times m$ transformation matrix. Note that in the above analysis, we have set $\mathbf{b}(\mathbf{x}) = \mathbf{0}$ since we consider an unbiased estimator. To identify the matrix $\mathbf{M}(\mathbf{x})$, observe that by multiplying identity (4.118) by $(\nabla_{\mathbf{x}} \ln(f_{\mathbf{Y}}(\mathbf{Y}|\mathbf{x})))^T$ and taking expectations, we get

$$\mathbf{C}_{12}(\mathbf{x}) = \mathbf{M}(\mathbf{x})\mathbf{J}(\mathbf{x}) , \tag{4.119}$$

which in the light of (4.111) implies

$$\mathbf{M}(\mathbf{x}) = -\mathbf{J}^{-1}(\mathbf{x}) . \tag{4.120}$$

The relation (4.118) can be rewritten as

$$\hat{\mathbf{X}}(\mathbf{Y}) = \mathbf{x} + \mathbf{J}^{-1}(\mathbf{x}) \nabla_{\mathbf{x}} \ln(f_{\mathbf{Y}}(\mathbf{Y}|\mathbf{x})) , \tag{4.121}$$

where the left-hand side does not depend on $\mathbf{x}$. This indicates that an unbiased efficient estimator exists if and only if the right-hand side of (4.121) does not depend on $\mathbf{x}$.

By setting $\mathbf{x} = \hat{\mathbf{X}}(\mathbf{Y})$ in this identity, we also find

$$\nabla_{\mathbf{x}} \ln(f_{\mathbf{Y}}(\mathbf{Y}|\hat{\mathbf{X}}(\mathbf{Y}))) = \mathbf{0} , \tag{4.122}$$

so that if an unbiased efficient estimator $\hat{\mathbf{X}}(\mathbf{y})$ exists, it must be a stationary point of the likelihood function and assuming there is only one such point, it must be the ML estimator.

So under the assumption that the likelihood function has a single maximum, we have discovered that if an estimator is efficient, it must be the ML estimator. However, this does not mean all ML estimators are efficient, since, as we have found in Example 4.6, an ML estimator can be biased.

Example 4.8: i.i.d. exponential sequence

Consider a sequence $\{Y_k,\ 1 \leq k \leq N\}$ of i.i.d. exponential random variables with unknown parameter $1/\theta$, so that

$$f_{Y_k}(y_k|\theta) = \frac{1}{\theta} \exp(-y_k/\theta) u(y_k) .$$

The joint probability density of the observations can be expressed as

$$f_{\mathbf{Y}}(\mathbf{y}|\theta) = \frac{1}{\theta^N} \exp(-S(\mathbf{y})/\theta) \prod_{k=1}^{N} u(y_k) ,$$

with

$$S(\mathbf{y}) = \sum_{k=1}^{N} y_k \, ,$$

which is in the factored form (4.93), so that $S(\mathbf{Y})$ is a sufficient statistic for estimating θ. Then the log-likelihood function $L(\mathbf{y}|\theta) = \ln(f_{\mathbf{Y}}(\mathbf{y}|\theta))$ is given by

$$L(\mathbf{y}|\theta) = -N\ln(\theta) - S(\mathbf{y})/\theta$$

when $y_k \geq 0$ for all k, and $L(\mathbf{y}|\theta) = -\infty$ otherwise. The first and second derivatives of L with respect to θ are given by

$$\frac{\partial}{\partial\theta} L(\mathbf{y}|\theta) = -\frac{N}{\theta} + \frac{S(\mathbf{y})}{\theta^2}$$
$$\frac{\partial^2}{\partial\theta^2} L(\mathbf{y}|\theta) = \frac{N}{\theta^2} - 2\frac{S(\mathbf{y})}{\theta^3} \, .$$

Setting the first derivative equal to zero, we find that the ML estimate of θ is just the sampled mean of the observations, i.e.,

$$\hat{\theta}_{\mathrm{ML}}(\mathbf{Y}) = \frac{S(\mathbf{Y})}{N} \, .$$

Since $E[Y_k] = \theta$, this estimate is unbiased. Also, taking the expectation of the second derivative gives the Fisher information

$$J(\theta) = E_{\mathbf{Y}}[-\frac{\partial^2}{\partial\theta^2} L(\mathbf{Y}|\theta)] = \frac{N}{\theta^2} \, . \tag{4.123}$$

To determine whether the ML estimate is efficient, we evaluate the right-hand side of (4.121), which yields

$$\theta + \frac{\theta^2}{N}\Big(- \frac{N}{\theta} + \frac{S(\mathbf{Y})}{\theta^2}\Big) = \frac{S(\mathbf{Y})}{N} = \hat{\theta}_{\mathrm{ML}}(\mathbf{Y})$$

as desired, so that the ML estimator is efficient. Its variance is therefore given by

$$E[(\theta - \theta_{\mathrm{ML}})^2] = J^{-1}(\theta) = \frac{\theta^2}{N} \, .$$

Unlike the ML estimate itself, which is preserved under a nonlinear transformation of the parameter vector, the unbiasedness and efficiency of an ML estimator are affected by parameter transformations. For the bias, this is due to the fact that if $\mathbf{g}$ denotes an arbitrary nonlinear parameter vector transformation and if $\hat{\mathbf{X}}(\mathbf{Y})$ denotes an arbitrary estimate of $\mathbf{x}$, we have

$$E[\mathbf{g}(\hat{\mathbf{X}}(\mathbf{Y}))] \neq \mathbf{g}(E[\hat{\mathbf{X}}(\mathbf{Y})]) \, .$$

To illustrate this observation, assume that the parametrization $\lambda = 1/\theta$ is employed for the exponentially distributed observations $\{Y_k, \; 1 \leq k \leq N\}$. Then the ML estimate of λ is given by

$$\hat{\lambda}_{\mathrm{ML}}(\mathbf{Y}) = \frac{N}{S(\mathbf{Y})} = \frac{1}{\hat{\theta}_{\mathrm{ML}}(\mathbf{Y})} .$$

To evaluate the mean of the ML estimate of λ, we note that the sum S of N independent exponential random variables with parameter λ admits the Erlang/gamma density

$$f_S(s|\lambda) = \frac{\lambda^N}{(N-1)!} s^{N-1} \exp(-\lambda s) u(s)$$

This gives

$$E[\hat{\lambda}_{\mathrm{ML}}] = N E[S^{-1}] = \frac{N}{N-1} \lambda ,$$

so $\hat{\lambda}_{\mathrm{ML}}$ is biased even though $\hat{\theta}_{\mathrm{ML}}$ was unbiased. To evaluate the Cramér-Rao lower bound for unbiased estimates of λ, we note that if $\theta(\lambda) = 1/\lambda$, by using the chain rule of differentiation, the derivative of L with respect to λ can be expressed as

$$\frac{\partial}{\partial \lambda} L(\mathbf{y}|\lambda) = \frac{\partial}{\partial \theta} L(\mathbf{y}|\theta(\lambda)) \frac{d\theta}{d\lambda}$$

so that by using the squared form (4.102) of the Fisher information matrix we find

$$J(\lambda) = \left(\frac{d\theta}{d\lambda} \right)^2 J(\theta(\lambda)) .$$

Taking into account (4.123) and $d\theta/d\lambda = -1/\lambda^2$, this gives

$$J(\lambda) = \frac{N}{\lambda^2} ,$$

and thus the CRLB for the variance of an abitrary unbiased estimator of λ is given by

$$E[(\lambda - \hat{\lambda})^2] \geq J^{-1}(\lambda) = \frac{\lambda^2}{N} .$$

Even though the ML estimator is biased, it is of interest to evaluate its MSE. By using the Erlang distribution of S we find that

$$E[(\hat{\lambda}_{\mathrm{ML}})^2] = N^2 E[S^{-2}] = \frac{N^2}{(N-1)(N-2)} \lambda^2 ,$$

which in turn gives

$$E[(\lambda - \hat{\lambda}_{\mathrm{ML}})^2] = \frac{N+2}{(N-1)(N-2)} \lambda^2 .$$

It is worth noting that as N becomes large, the bias term

$$b = \lambda - E[\hat{\lambda}_{\mathrm{ML}}] = -\lambda/(N-1)$$

tends to zero, so that the ML estimator is *asymptotically unbiased.* Also, the leading term of the ML error variance is λ^2/N, which coincides with the CRLB, and hence the estimator is *asymptotically efficient.* As we shall see in next section, it turns out that all ML estimators have these two properties. □

Example 4.7, continued

We return to the case of an i.i.d. $N(m, v)$ sequence $\{Y_k,\ 1 \leq k \leq N\}$. In this case the right-hand side of (4.121) can be expressed as

$$\begin{bmatrix} m \\ v \end{bmatrix} + \frac{1}{N} \begin{bmatrix} v & 0 \\ 0 & 2v^2 \end{bmatrix} \begin{bmatrix} \frac{1}{v} \sum_{k=1}^{N} (Y_k - m) \\ -\frac{N}{2v} + \frac{1}{2v^2} \sum_{k=1}^{N} (Y_k - m)^2 \end{bmatrix} = \frac{1}{N} \begin{bmatrix} \sum_{k=1}^{N} Y_k \\ \sum_{k=1}^{N} (Y_k - m)^2 \end{bmatrix} . \quad (4.124)$$

The first entry of the vector on the right-hand side of (4.124) is $\hat{m}_{\text{ML}}(\mathbf{Y})$, which depends on $\mathbf{Y}$ only, but the second entry depends on m and differs from $\hat{v}_{\text{ML}}$ by the fact that it includes m instead of its estimate. So we know that the ML estimator will not be efficient, which is not a surprise, since we have seen that $\hat{v}_{\text{ML}}$ is biased. Also, if

$$\mathbf{x} \triangleq \begin{bmatrix} m \ v \end{bmatrix}^T \quad , \quad \text{and} \ \hat{\mathbf{X}} = \begin{bmatrix} \hat{m} \ \hat{v} \end{bmatrix}^T$$

denote the vector of unknown parameters and an arbitrary unbiased estimator, the error covariance matrix of $\hat{\mathbf{X}}(\mathbf{Y})$ must satisfy the CRLB

$$E[(\mathbf{x} - \hat{\mathbf{X}})(\mathbf{x} - \hat{\mathbf{X}})^T] \geq \mathbf{J}^{-1}(\mathbf{x}) = \frac{1}{N} \begin{bmatrix} v & 0 \\ 0 & 2v^2 \end{bmatrix} .$$

The ML estimator is in fact partly efficient, since the mean estimate is unbiased and its variance

$$E[(m - \widehat{m}_{\text{ML}})^2] = \frac{1}{N^2} E[\Big(\sum_{k=1}^{N} (m - Y_k)\Big)^2] = \frac{v}{N}$$

equals the (1,1) entry of $J^{-1}(\mathbf{x})$. This matches our earlier observation that the first entry of the vector on the right-hand side of (4.124) is $\hat{\mathbf{m}}_{\text{ML}}$. Next, let $\mathbf{u}$ be the vector of $\mathbb{R}^N$ with all one entries defined in (4.85). We can write the mean estimation error as

$$m - \widehat{m}_{\text{ML}} = -\mathbf{u}^T(\mathbf{Y} - m\mathbf{u})/N ,$$

where the random vector $\mathbf{Y} - m\mathbf{u}$ has zero mean. By using the expression (4.91)–(4.92) for the estimated variance and recalling that the expectation of odd monomials of zero-mean Gaussian random variables is always zero, we can conclude that

$$E[(m - \widehat{m}_{\mathrm{ML}})(v - \hat{v}_{\mathrm{ML}})] = 0 ,$$

so the mean and variance errors are uncorrelated. By representing the $N-1$ dimensional vector $\mathbf{Z}$ in (4.91) in terms of its coordinates Z_k, we also find

$$E[(\hat{v}_{\mathrm{ML}})^2] = \frac{1}{N^2} E[(\sum_{k=1}^{N-1} Z_k^2)^2] = \frac{N^2-1}{N^2} v^2 , \tag{4.125}$$

where the expectation of products of squares for the $N(0, v)$ random variables Z_k is evaluated by using

$$E[Z_k^2 Z_l^2] = \begin{cases} 3v^2 \text{ for } k = l \\ v^2 \text{ for } k \neq l . \end{cases}$$

This implies that the MSE of the variance estimator is

$$E[(v - \hat{v}_{\mathrm{ML}})^2] = \frac{(2N-1)}{N^2} v^2 . \tag{4.126}$$

An interesting feature of this expression is that it is smaller than the CRLB, which is given by

$$(\mathbf{J}^{-1})_{22} = \frac{2v^2}{N} .$$

This is due to the fact that the estimator $\hat{v}_{\mathrm{ML}}$ is biased. So allowing a small bias can sometimes be beneficial. Note again that as N becomes large, the bias term goes to zero and the MSE (4.126) approaches the CRLB, so the estimator $\hat{v}_{\mathrm{ML}}$ is asymptotically unbiased and efficient. □

Example 4.9: Phase estimation

Now consider a two-dimensional observation vector

$$\mathbf{Y} = \begin{bmatrix} Y_c \\ Y_s \end{bmatrix} = A \begin{bmatrix} \cos(\theta) \\ \sin(\theta) \end{bmatrix} + \mathbf{V} ,$$

where the 2-dimensional vector $\mathbf{V}$ has a $N(\mathbf{0}, \sigma^2 \mathbf{I}_2)$ distribution. We assume that that amplitude A and noise variance σ^2 are known, but the phase θ is unknown and needs to be estimated. As will be seen later in this text, this problem arises in the phase synchronization of a modulated signal with known envelope observed in WGN. The log-likelihood function of the observations is given by

$$L(\mathbf{y}|\theta) = \ln(f_{\mathbf{Y}}(\mathbf{y}|\theta)) = \frac{A}{\sigma^2}(y_c \cos(\theta) + y_s \sin(\theta)) + c(\mathbf{y})$$

where $c(\mathbf{y})$ does not depend on θ. Its first- and second-order derivatives with respect to θ are given by

$$\frac{\partial}{\partial\theta}L(\mathbf{y}|\theta) = \frac{A}{\sigma^2}(-y_c\sin(\theta) + y_s\cos(\theta))$$
$$\frac{\partial^2}{\partial\theta^2}L(\mathbf{y}|\theta) = -\frac{A}{\sigma^2}(y_c\cos(\theta) + y_s\sin(\theta))\,.$$

By setting the first-order derivative to zero, we find that the ML estimate is given by

$$\hat{\theta}_{\mathrm{ML}}(\mathbf{Y}) = \tan^{-1}(Y_s/Y_c)\,.$$

Also, taking the expectation of the second derivative yields the Fisher information

$$J(\theta) = \frac{A}{\sigma^2}E[Y_c\cos(\theta) + Y_s\sin(\theta)] = \frac{A^2}{\sigma^2}\,.$$

Note that with respect to the polar representation

$$Y_c = R\cos(\Phi) \quad , \quad Y_s = R\sin(\Phi)$$

of the observations, we have $\hat{\theta}_{\mathrm{ML}}(\mathbf{Y}) = \Phi$. In this coordinate system, the joint density of R and Φ can be expressed as

$$f_{R,\Phi}(r,\phi|\theta) = \frac{r}{2\pi\sigma^2}\exp(-\frac{1}{2\sigma^2}(A^2 + r^2 - 2Ar\cos(\phi-\theta))\,,$$

which is an even function of $\phi - \theta$. Then

$$E[\Phi] = \theta + E[(\Phi - \theta)] = \theta\,,$$

where the expectation of $\Phi - \theta$ is zero since the integral of an odd function is always zero. This shows that $\hat{\theta}_{\mathrm{ML}}$ is unbiased. To determine whether this estimate is efficient, we note that the right-hand side of equation (4.121) can be expressed as

$$\begin{aligned}\theta + \frac{\sigma^2}{A^2}\frac{A}{\sigma^2}(-Y_c\sin(\theta) + Y_s\cos(\theta) &= \theta + \frac{R}{A}\sin(\Phi - \theta)\\ &\approx \Phi = \hat{\theta}_{\mathrm{ML}}\,.\end{aligned}$$

Thus as long as $R \approx A$ and $\Phi \approx \theta$, the ML estimate is efficient, i.e.

$$E[(\Phi - \theta)^2] \approx J^{-1} = \frac{\sigma^2}{A^2} = \frac{1}{\mathrm{SNR}}\,,$$

which indicates that the phase variance is inversely proportional to the SNR. This result admits a simple geometric interpretation. As shown in Fig. 4.4, let $\mathbf{S}$ be the two dimensional vector OP representing a signal with amplitude A and phase θ, and let the noise $\mathbf{V}$ be represented by the vector PQ, so that $\mathbf{Y} = \mathbf{S} + \mathbf{V}$ is represented by the vector OQ with magnitude R and phase Φ. Assume that the SNR is large, so that $A \gg ||\mathbf{V}||_2$. Then the angular error $\Phi - \theta$ can be approximated as $V^{\perp}/A$, where where $V^{\perp}$ denotes the projection of $\mathbf{V}$ perpendicular to $\mathbf{S}$. Since the two coordinates of $\mathbf{V}$ are independent

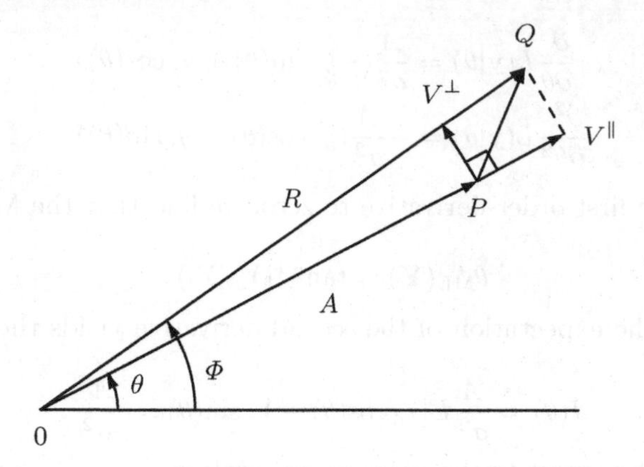

Fig. 4.4. Approximation of the phase estimation error for high SNR.

$N(0, \sigma^2)$, this property is preserved by any coordinate system in which we choose to represent $\mathbf{V}$. This implies that $V^\perp$ is a $N(0, \sigma^2)$ and thus

$$\Phi - \theta \approx \frac{V^\perp}{A} \sim N(0, \sigma^2/A) .$$

The variation of the MSE $E[(\Phi - \theta)^2]$ with the SNR is plotted in Fig. 4.5. In this simulation $A = 1$, $\theta = \pi/6$, and the noise variance is selected to

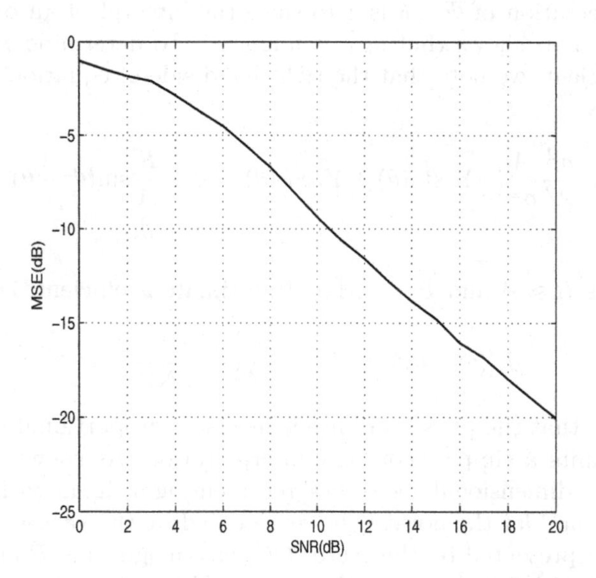

Fig. 4.5. Plot of the sampled angle MSE $E[(\Phi - \theta)^2]$ in dB versus SNR in dB obtained from 10,000 trials with $A = 1$ and $\theta = \pi/6$.

achieve the desired SNR. The curve shows the sampled MSE averaged over 10,000 trials. Both the MSE and SNR are plotted in dBs. For high SNR, the curve is a straight line with slope -1, which shows that the MSE is inversely proportional to the SNR, but as the SNR decreases, the curve becomes flatter. This threshold effect is due to the fact that the high SNR assumption required to establish the efficiency of the ML estimate no longer holds. □

Example 4.10: i.i.d. Poisson sequence

In this chapter we have focused our attention primarily on continuous-valued random variables, but all the results presented can be adapted easily to the discrete-valued case. Consider, for example, a sequence $\{Y_k,\ k \geq 1\}$ of i.i.d. Poisson random variables with unknown parameter λ. The probability distribution of the vector

$$\mathbf{Y} = \begin{bmatrix} Y_1\ Y_2 \ldots Y_N \end{bmatrix}^T$$

formed by the first N elements of the sequence is

$$P[\mathbf{Y} = \mathbf{y}|\lambda] = p(\mathbf{y}|\lambda) = \frac{1}{\prod_{k=1}^{N} y_k!} \exp(S(\mathbf{y})\ln(\lambda) - N\lambda)\,,$$

where

$$S(\mathbf{Y}) = \sum_{k=1}^{N} Y_k$$

is a sufficient statistic, since $p(\mathbf{y}|\lambda)$ is in the factored form (4.93). The log-likelihood function can be expressed as

$$L(\mathbf{y}|\lambda) = S(\mathbf{y})\ln(\lambda) - N\lambda + c(\mathbf{y})$$

where the constant $c(\mathbf{y})$ does not depend on λ. The first two derivatives of L with respect to λ are given by

$$\begin{aligned} \frac{\partial}{\partial\lambda} L(\mathbf{y}|\lambda) &= \frac{S(\mathbf{y})}{\lambda} - N \\ \frac{\partial^2}{\partial\lambda^2} L(\mathbf{y}|\lambda) &= -\frac{S(\mathbf{y})}{\lambda^2}\,. \end{aligned}$$

Then, since the Poisson random variables Y_k have mean λ, the ML estimate

$$\lambda_{\mathrm{ML}}(\mathbf{Y}) = \frac{S(\mathbf{Y})}{N}\,.$$

is unbiased, and the Fisher information is given by

$$J(\lambda) = E\big[-\frac{\partial^2}{\partial\lambda^2} L(\mathbf{Y}|\lambda)\big] = \frac{N}{\lambda}\,.$$

The CRLB for an abitrary unbiased estimate $\hat{\lambda}(\mathbf{Y})$ of λ is therefore given by

$$E[(\lambda - \hat{\lambda}(\mathbf{b}))^2] \geq J^{-1}(\lambda) = \frac{\lambda}{N} \,.$$

It is attained by the ML estimate, which is therefore efficient, since its variance is given by

$$E[(\lambda - \hat{\lambda}_{\mathrm{ML}})^2] = E\Big[\Big(\frac{1}{N}\sum_{k=1}^{N}(\lambda - Y_k)\Big)^2\Big] = \frac{1}{N}E[(\lambda - Y_1)^2] = \frac{\lambda}{N} \,,$$

where we have used the fact that the variance of a Poisson random variable equals its parameter λ. Alternatively, we note that the right-hand side of equation (4.121) can be expressed as

$$\lambda + \frac{\lambda}{N}\Big(\frac{S(\mathbf{y})}{\lambda} - N\Big) = \frac{S(\mathbf{Y})}{N} \,,$$

which is independent of λ, so $\hat{\lambda}_{\mathrm{ML}}$ is efficient. □

4.4.4 Uniform Minimum Variance Unbiased Estimates

We have seen that ML estimates are not always unbiased. Since it is often desirable to focus on estimators that are unbiased, this leads us to the question of whether we can find an unbiased estimator which is optimal in the MSE sense. Thus if $\hat{\mathbf{X}}(\mathbf{Y})$ denotes an arbitrary unbiased estimator of $\mathbf{x}$, consider the MSE

$$J(\hat{\mathbf{X}}, \mathbf{x}) = E[||\mathbf{x} - \hat{\mathbf{X}}(\mathbf{Y})||_2^2] \,. \tag{4.127}$$

We say that an estimate $\hat{\mathbf{X}}_{\mathrm{UMVU}}(\mathbf{Y})$ is a uniform minimum variance unbiased estimate (UMVUE) if it satisfies

$$J(\hat{\mathbf{X}}_{\mathrm{UMVU}}, \mathbf{x}) \leq J(\hat{\mathbf{X}}, \mathbf{x}) \tag{4.128}$$

for all other unbiased estimators $\hat{\mathbf{X}}$ and all $\mathbf{x}$. This is a very strong property as it indicates that the UMVU estimator outperforms any other estimator for all values of $\mathbf{x}$. Accordingly, there is no guarantee that such an estimator exists. However, if it does, how do we find it?

There are two ways of finding UMVUEs. The first one is obviously to compute the ML estimate and check whether it is unbiased and reaches the CRLB. If so, we are done. But if this approach fails, a second technique is available. It relies on the concept of a *complete* sufficient statistic. Let $\mathbf{S}(\mathbf{Y})$ be a sufficient statistic. We say $\mathbf{S}$ is complete if any function $\mathbf{h}(\mathbf{S})$ that satisfies

$$E[\mathbf{h}(\mathbf{S})] = \mathbf{0} \tag{4.129}$$

for all $\mathbf{x}$ must necessarily be identically zero. Equivalently, the sufficient statistic $\mathbf{S}$ is complete if there is at the most one unbiased estimator $\hat{\mathbf{X}}(\mathbf{S})$ of $\mathbf{x}$

depending on $\mathbf{S}$ only. Note indeed that if there are two estimators $\hat{\mathbf{X}}_i(\mathbf{S})$ with $i = 1,\ 2$ such that

$$E[\hat{\mathbf{X}}_i(\mathbf{S})] = \mathbf{x}\,, \tag{4.130}$$

for $i = 1,\ 2$ then their difference $\mathbf{h}(\mathbf{S}) = \hat{\mathbf{X}}_1(\mathbf{S}) - \hat{\mathbf{X}}_2(\mathbf{S})$ satisfies (4.129) for all $\mathbf{x}$. Also, if there exists a nonzero function $\mathbf{h}(\mathbf{S})$ such that (4.130) holds, and if $\hat{\mathbf{X}}_1(\mathbf{S})$ is a first unbiased estimate of $\mathbf{x}$, then $\hat{\mathbf{X}}_1(\mathbf{S}) + \mathbf{h}(\mathbf{S})$ would be a second such estimate. In general, determining whether a sufficient statistic is complete is a difficult task and we refer the reader to [6, Sec. 1.6] for a discussion of this issue. For our purposes, we note that for the canonical exponential class of densities of the form (4.80), $\mathbf{S}(\mathbf{Y})$ is a complete sufficient statistic. Since this class contains Gaussian, exponential, and Poisson (for the discrete version of the class) random variables, it gives us enough to work with.

To find a UMVU estimate, we can employ the following result due to Rao, Blackwell, Lehmann and Scheffe, which will therefore be referred to as the RBLS theorem/method.

RBLS Theorem: Let $\check{\mathbf{X}}(\mathbf{Y})$ denote an unbiased estimate of $\mathbf{x}$, and let $\mathbf{S}(\mathbf{Y})$ denote a sufficient statistic for $\mathbf{x}$. Then the estimate

$$\hat{\mathbf{X}}(\mathbf{S}) = E[\check{\mathbf{X}}(\mathbf{Y})|\mathbf{S}] \tag{4.131}$$

is unbiased and if

$$\begin{aligned} \check{\mathbf{K}}(\mathbf{x}) &= E[(\mathbf{x} - \check{\mathbf{X}})(\mathbf{x} - \check{\mathbf{X}})^T] \\ \hat{\mathbf{K}}(\mathbf{x}) &= E[(\mathbf{x} - \hat{\mathbf{X}})(\mathbf{x} - \hat{\mathbf{X}})^T] \end{aligned} \tag{4.132}$$

denote the covariance matrices of estimators $\check{\mathbf{X}}$ and $\hat{\mathbf{X}}$, we have

$$\hat{\mathbf{K}}(\mathbf{x}) \leq \check{\mathbf{K}}(\mathbf{x})\,, \tag{4.133}$$

where the inequality is interpreted again in the sense of nonnegative definite matrices. Furthermore, if the sufficient statistic $\mathbf{S}$ is complete, the estimate $\hat{\mathbf{X}}(\mathbf{S})$ is a UMVU estimate of $\mathbf{x}$.

A proof of this result is given in Appendix 4.A.. To see how this scheme works out in practice, we consider some examples.

Example 4.11: UMVUE for an independent exponential sequence

Consider an i.i.d. sequence of exponential random variables $\{Y_k,\ 1 \leq k \leq N\}$ with parameter $1/\theta$ of the type examined in Example 4.8. Since the joint density

$$f_{\mathbf{Y}}(\mathbf{y}|\theta) = \frac{1}{\theta^N} \exp(-S(\mathbf{y})/\theta) \prod_{k=1}^{N} u(y_k)$$

with

$$S(\mathbf{y}) = \sum_{k=1}^{N} y_k$$

is of exponential type, $S(\mathbf{Y})$ is a complete sufficient statistic. Also, because the Y_k's have mean θ, the estimate

$$\breve{\theta} = Y_1$$

is unbiased. To evaluate the conditional expectation (4.131), we must first find the joint probability density of Y_1 and S. By performing a linear transformation, we find that the joint density of the Y_k's with $1 \leq k \leq N-1$ and S is given by

$$f_{Y_1,\dots,Y_{N-1},S}(y_1,\dots,y_{N-1},s|\theta) = \frac{1}{\theta^N} \exp(-s/\theta) \prod_{k=1}^{N-1} u(y_k) u(s - \sum_{k=1}^{N-1} y_k) \,.$$

Then computing the marginal density with respect to Y_1 and S, we obtain

$$f_{Y_1,S}(y_1,s|\theta) = \frac{1}{\theta^N} \frac{(s-y_1)^{N-2}}{(N-2)!} \exp(-s/\theta) u(y_1) u(s-y_1) \,.$$

Integrating with respect to y_1, we find S satisfies an Erlang distribution of order N with parameter $1/\theta$, so that

$$f_{Y_1|S}(y_1|s,\theta) = \frac{(N-1)}{s^{(N-1)}} (s-y_1)^{N-2} u(y_1) u(s-y_1) \,.$$

Then

$$\begin{aligned} E[Y_1|S=s] &= \int y_1 f_{Y_1,S}(y_1|s,\theta) dy_1 \\ &= s - \frac{N-1}{s^{N-1}} \int_0^s (s-y_1)^{N-1} dy_1 = s - \frac{(N-1)}{N} s = \frac{s}{N} \,, \end{aligned}$$

so the RBLS procedure yields the ML estimator

$$\hat{\theta}_{\mathrm{ML}} = \frac{S}{N} \,,$$

which was shown in Example 4.8 to be unbiased and efficient. □

As the above example indicates, applying the RBLS procedure to an arbitrary unbiased estimate of $\mathbf{x}$ often involves tedious computations. But there is a shortcut. Note that the final estimate is unbiased and is a function of the complete sufficient statistic $\mathbf{S}$. Furthermore, because $\mathbf{S}$ is complete we know that there is at the most one unbiased estimate that is a function of $\mathbf{S}$. So once a complete sufficient statistic $\mathbf{S}$ has been identified all we need to do to obtain a UMVU estimate is to find an unbiased estimate of $\mathbf{X}$ that depends on $\mathbf{S}$ only. As the following examples indicate, this is often easy to accomplish.

Example 4.8, continued

For the i.i.d. exponential sequence $\{Y_k,\ 1 \leq k \leq N$ with parameter λ considered in the second half of Example 4.8, we saw that the ML estimate is given by

$$\hat{\lambda}_{\text{ML}} = \frac{N}{S(\mathbf{Y})} ,$$

where $S(\mathbf{Y}) = \sum_{k=1}^{N} Y_k$ is a sufficient statistic, which is biased since its expectation is given by

$$E[\hat{\lambda}_{\text{ML}}] = \frac{N}{N-1} \lambda .$$

By rescaling the ML estimate, we immediately find that

$$\hat{\lambda}_{\text{UMVU}} = \frac{N-1}{N} \hat{\lambda}_{\text{ML}} = \frac{N-1}{S}$$

is unbiased and must be the UMVU estimate since it depends on S only. In this case we have

$$E[(\hat{\lambda}_{\text{UMVU}})^2] = \frac{(N-1)^2}{N^2} E[(\hat{\lambda}_{\text{ML}})^2] = \frac{(N-1)}{N-2} \lambda^2$$

so that the MSE of the UMVU estimate is given by

$$E[(\lambda - \hat{\lambda}_{\text{UMVU}})^2] = \frac{\lambda^2}{N-2} ,$$

which, as expected, is above the CRLB equal to λ^2/N computed in Example 4.8, but is lower than the MSE of the ML estimate. $\square$

Example 4.7, continued

Consider now the case of an i.i.d. sequence $\{Y_k,\ 1 \leq k \leq N\}$ of $N(m, v)$ random variables. In this case, the ML estimate $\widehat{m}_{\text{ML}}$ coincides with the sampled mean and is unbiased, but the ML variance estimate $\hat{v}_{\text{ML}}$ given by (4.90) is biased since

$$E[\hat{v}_{\text{ML}}] = \frac{(N-1)}{N} v .$$

But since as indicated in (4.98), the estimate $\hat{v}_{\text{ML}}$ is a function of the sufficient statistic $\mathbf{S}(\mathbf{Y})$ defined in (4.97), and since this statistic is complete as Gaussians belong to the exponential class, all we need to obtain a UMVU estimate of the variance is to rescale the ML estimate to obtain an unbiased estimate. This gives

$$\hat{v}_{\text{UMVU}} = \frac{N}{N-1} \hat{v}_{\text{ML}} = \frac{1}{N-1} \sum_{k=1}^{N} (Y_k - \widehat{m}_{\text{ML}})^2 .$$

We have therefore

$$E[(\hat{v}_{\mathrm{UMVU}})^2] = \frac{N^2}{(N-1)^2} E[(\hat{v}_{\mathrm{ML}})^2] = \frac{(N+1)}{N-1} v^2$$

and the MSE of the MVU estimator is given by

$$E[(v - \hat{v}_{\mathrm{UMVU}})^2] = \frac{2v^2}{N-1} .$$

It is larger than the CRLB, which is equal to $2v^2/N$, and consequently, it is also larger than the MSE of the ML estimator $\hat{v}_{\mathrm{ML}}$, which as we noted earlier is smaller than the CRLB. So in this case, insisting on an unbiased estimator results in a loss of performance in the MSE sense. □

4.5 Asymptotic Behavior of ML Estimates

It is reasonable to expect that the estimation accuracy of an estimator will improve as the number N of observations increases. Ideally, independently of whether the parameter vector is random or not, if $\hat{\mathbf{X}}_N$ denotes the estimate produced by an estimation scheme based on the first N observations, we would like $\hat{\mathbf{X}}_N$ to converge to the true value as N tends to infinity, and to do so as quickly as possible within the underlying limits imposed by the estimation problem. To keep our discussion short, since the ML estimator plays a key role in implementing detectors with unknown parameters, we focus here on ML estimates, so that the parameter vector to be estimated is nonrandom.

4.5.1 Consistency

Given a sequence $\{\mathbf{Y}_k,\ k \geq 1\}$ of observations $\mathbf{Y}_k \in \mathbb{R}^n$ and an unknown parameter vector $\mathbf{x} \in \mathcal{X}$, if $\hat{\mathbf{X}}_N$ denotes an estimate of $\mathbf{x}$ based on the first N observations, we say that $\hat{X}_N$ is *consistent* almost surely if it converges to $\mathbf{x}$ with probability one as $N \to \infty$.

To keep the analysis simple, we consider the case when the observations $\mathbf{Y}_k$ are independent, identically distributed with density $f(\mathbf{y}|\mathbf{x})$, and we denote the true parameter vector by $\mathbf{x}_0$. We assume that the parametrization of the density $f(\mathbf{y}|\mathbf{x})$ is such that two different vectors $\mathbf{x}$ cannot correspond to the same density. Then the ML estimate $\hat{\mathbf{X}}_N$ based on the observations $\{\mathbf{Y}_k,\ 1 \leq k \leq N\}$ is given by

$$\hat{\mathbf{X}}_N = \arg\max_{\mathbf{x} \in \mathcal{X}} S_N(\mathbf{x}) , \tag{4.134}$$

where

$$S_N(\mathbf{x}) = \frac{1}{N} \sum_{k=1}^{N} \ln(f(\mathbf{Y}_k|\mathbf{x})) . \tag{4.135}$$

Let $f_0 = f(\cdot|\mathbf{x}_0)$, $f_{\mathbf{x}} = f(\cdot|\mathbf{x})$ and denote

$$\ell_{\mathbf{x}}(\mathbf{y}) \triangleq \ln(f_0(\mathbf{y})/f_{\mathbf{x}}(\mathbf{y})) . \tag{4.136}$$

For $k \geq 1$, the random variables $\ell_{\mathbf{x}}(\mathbf{Y}_k)$ are i.i.d. with mean

$$\begin{aligned} E_0[\ell_{\mathbf{x}}(\mathbf{Y}_k)] &= \int \ln\left(\frac{f_0(\mathbf{y})}{f_{\mathbf{x}}(\mathbf{y})}\right) f_0(\mathbf{y}) d\mathbf{y} \\ &= D(f_0|f_{\mathbf{x}}) \geq 0 . \end{aligned} \tag{4.137}$$

Thus, according to the strong law of large numbers, as N tends to ∞ we have

$$\left| \frac{1}{N} \sum_{k=1}^{N} \ell_{\mathbf{x}}(\mathbf{Y}_k) - D(f_0|f_{\mathbf{x}}) \right| \to 0 \tag{4.138}$$

almost surely for every *fixed* $\mathbf{x} \in \mathcal{X}$. Since $\hat{\mathbf{X}}_N$ maximizes $S_N(\mathbf{x})$, the inequality

$$\begin{aligned} 0 \geq S_N(\mathbf{x}_0) - S_N(\hat{\mathbf{X}}_N) &= \frac{1}{N} \sum_{k=1}^{N} \ell_{\hat{\mathbf{X}}_N}(\mathbf{Y}_k) \\ &= \frac{1}{N} \sum_{k=1}^{N} \ell_{\hat{\mathbf{X}}_N}(\mathbf{Y}_k) - D(f_0, f_{\hat{\mathbf{X}}_N}) + D(f_0, f_{\hat{\mathbf{X}}_N}) , \end{aligned} \tag{4.139}$$

holds. This implies

$$0 \leq D(f_0, f_{\hat{\mathbf{X}}_N}) \leq \left| \frac{1}{N} \sum_{k=1}^{N} \ell_{\hat{\mathbf{X}}_N}(\mathbf{Y}_k) - D(f_0, f_{\hat{\mathbf{X}}_N}) \right| . \tag{4.140}$$

Consequently, provided that

$$\left| \frac{1}{N} \sum_{k=1}^{N} \ell_{\hat{\mathbf{X}}_N}(\mathbf{Y}_k) - D(f_0|f_{\hat{\mathbf{X}}_N}) \right| \to 0 , \tag{4.141}$$

we can conclude that

$$\lim_{N\to\infty} D(f_0, f_{\hat{\mathbf{X}}_N}) = 0 ,$$

which implies

$$\lim_{N\to\infty} f_{\hat{\mathbf{X}}_N} = f_0 ,$$

and thus

$$\lim_{N\to\infty} \hat{\mathbf{X}}_N = \mathbf{x}_0 \tag{4.142}$$

almost surely as $N \to \infty$, so that the ML estimate is consistent almost surely.

If we compare the condition (4.141) with the strong law of large numbers (4.138) it is tempting to think that (4.141) is implied by (4.138), since it is obtained by setting $\mathbf{x} = \hat{\mathbf{X}}_N$ in (4.138). Unfortunately this is not true as (4.138) holds only pointwise. What is needed is a stronger set of conditions on the random variables $\ell_{\mathbf{x}}(\mathbf{Y}_k)$ so that a *uniform law of large numbers* (ULLN) holds:

$$\sup_{\mathbf{x}\in\mathbf{X}} \left| \frac{1}{N} \sum_{k=1}^{N} \ell_{\mathbf{x}}(\mathbf{Y}_k) - D(f_0|f_{\mathbf{x}}) \right| \to 0 \tag{4.143}$$

almost surely as $N \to \infty$. Finding a set of conditions such that (4.143) holds is a rather technical issue, but it is shown in [7, p. 108] that (4.143) is satisfied if the following conditions hold:

(i) The domain $\mathcal{X}$ is compact.
(ii) The function $\ell_{\mathbf{x}}(\mathbf{y})$ is continuous in $\mathbf{x}$ for all $\mathbf{y}$.
(iii) There exists a function $C(\mathbf{Y})$ such that $E_0[C(\mathbf{Y})] < \infty$ and $|\ell_{\mathbf{x}}(\mathbf{y})| \leq C(\mathbf{y})$ for all $\mathbf{x}$ and $\mathbf{y}$.

This shows that under relatively mild smoothness conditions for the density $f(\mathbf{y}|\mathbf{x})$ the ML estimate is consistent almost surely. Note that since the ML estimate is usually expressed in terms of sufficient statistics, instead of checking whether the density $f(\mathbf{y}|\mathbf{x})$ satisfies conditions which ensure that the ULLN holds, it is often easier to verify the almost sure consistency property by inspection.

Example 4:12: Consistency for an i.i.d. Gaussian sequence

Let $\{Y_k, k \geq 1\}$ denote an i.i.d. sequence of $N(m, v)$ random variables. Then the ML estimates for the mean and variance can be expressed as

$$\begin{aligned} \widehat{m}_N &= \frac{1}{N} \sum_{k=1}^{N} Y_k \\ \hat{v}_N &= \frac{1}{N} \sum_{k=1}^{N} (Y_k - \widehat{m}_N)^2 = \frac{1}{N} \sum_{k=1}^{N} (Y_k - m)^2 - (m - \widehat{m}_N)^2 , \end{aligned}$$

and by invoking the strong law of large numbers for the sequences $\{Y_k,\ k \geq 1\}$ and $\{(Y_k - m)^2,\ k \geq 1\}$, we deduce immediately that the ML estimates are consistent almost surely. □

Remark: Consider the Kullback-Leibler divergence

$$D(\mathbf{x}_0|\mathbf{x}) = D(f_0|f_{\mathbf{x}}) . \tag{4.144}$$

The ML estimation method can be viewed as an indirect way of maximizing $-D(\mathbf{x}_0|\mathbf{x})$ by maximizing instead the sampled average of random variables with mean $-D(\mathbf{x}_0|\mathbf{x})$. In addition to imposing a uniform law of large numbers condition which just guarantees that the method works correctly, to ensure that the method yields an accurate estimate of the parameter vector $\mathbf{x}_0$, the maximum of $-D(\mathbf{x}_0|\mathbf{x})$ (or equivalently the minimum of $D(\mathbf{x}_0|\mathbf{x})$) when viewed as a function of $\mathbf{x}$ needs to stand out. If we consider $D(\mathbf{x}_0|\mathbf{x})$, this means that its surface in the vicinity of $\mathbf{x}_0$ should not be shallow. Observing that $D(\mathbf{x}_0|\mathbf{x}_0) = 0$, by performing a Taylor series expansion of $D(\mathbf{x}_0|\mathbf{x})$ with respect to $\mathbf{x}$ in the vicinity of $\mathbf{x}_0$, we get

$$D(\mathbf{x}_0|\mathbf{x}) \approx \frac{1}{2}(\mathbf{x}-\mathbf{x}_0)^T \mathbf{J}(\mathbf{x}_0)(\mathbf{x}-\mathbf{x}_0) . \tag{4.145}$$

To obtain this expression, we have used the fact that since $D(\mathbf{x}_0|\mathbf{x})$ is maximized at $\mathbf{x} = \mathbf{x}_0$, its gradient with respect to $\mathbf{x}$ is zero at $\mathbf{x}_0$, so that the Taylor series does not include a linear term. To evaluate the Hessian, note that

$$\nabla_{\mathbf{x}} \nabla_{\mathbf{x}}^T D(\mathbf{x}_0|\mathbf{x}) = -\int \nabla_{\mathbf{x}} \nabla_{\mathbf{x}}^T \ln(f(\mathbf{y}|\mathbf{x})) f(\mathbf{y}|\mathbf{x}_0) d\mathbf{y} , \tag{4.146}$$

which coincides with the Fisher information matrix $\mathbf{J}(\mathbf{x}_0)$ at $\mathbf{x} = \mathbf{x}_0$. Thus the Fisher information matrix can be interpreted as the Hessian of the Kullback-Leibler divergence $D(\mathbf{x}_0|\mathbf{x})$ at $\mathbf{x}_0$. The larger the eigenvalues of $J(\mathbf{x}_0)$, the more curved the surface $D(\mathbf{x}_0|\mathbf{x})$ will be at $\mathbf{x} = \mathbf{x}_0$, and the easier it will be to find its minimum. This also explains why the Cramér-Rao bound is expressed in terms of the inverse of $\mathbf{J}(\mathbf{x}_0)$, since when $\mathbf{J}(\mathbf{x}_0)$ is large, estimation errors will be smaller.

4.5.2 Asymptotic Distribution of the ML Estimate

This leads us to our next topic, namely the asymptotic distribution of the ML estimate $\hat{\mathbf{X}}_N$ as N becomes large. Since we already know that the ML estimate converges to $\mathbf{x}_0$, we are interested in its deviations away from $\mathbf{x}_0$. We show now that as N becomes large

$$N^{1/2}(\hat{\mathbf{X}}_N - \mathbf{x}_0) \sim N(\mathbf{0}, \mathbf{J}^{-1}(\mathbf{x}_0)) . \tag{4.147}$$

So the deviations are Gaussian and since

$$E[(\hat{\mathbf{X}}_N - \mathbf{x}_0)(\hat{\mathbf{X}}_N - \mathbf{x}_0)^T] = \frac{\mathbf{J}^{-1}(\mathbf{x}_0)}{N} \tag{4.148}$$

for N large, where the right-hand side corresponds to the CRLB for N independent observations, we conclude that the ML estimate is asymptotically efficient.

To prove that the distribution (4.147) holds, we write the log-likelihood function for observations $\{\mathbf{Y}_k, 1 \leq k \leq N\}$ as

$$\Lambda_N(\mathbf{x}) = \sum_{k=1}^{N} \ln(f(\mathbf{Y}_k|\mathbf{x})) \tag{4.149}$$

and we introduce the *score function*

$$\mathbf{U}_N(\mathbf{x}) \triangleq \nabla_{\mathbf{x}} \Lambda_N(\mathbf{x}) = \sum_{k=1}^{N} \nabla_{\mathbf{x}} \ln(f(\mathbf{Y}_k|\mathbf{x})) . \tag{4.150}$$

Since the ML estimate $\hat{\mathbf{X}}_N$ maximizes the likelihood function, it satisfies $U_N(\hat{\mathbf{X}}_N) = \mathbf{0}$. As N becomes large, the ML estimate approaches $\mathbf{x}_0$. Accordingly, we can perform a Taylor series expansion of the score function in the vicinity of $\mathbf{x}_0$ where we neglect quadratic and higher order terms. We find

$$0 = \mathbf{U}_N(\hat{\mathbf{X}}_N) \approx \mathbf{U}_N(\mathbf{x}_0) - \mathbf{H}_N(\mathbf{x}_0)(\hat{\mathbf{X}}_N - \mathbf{x}_0) , \tag{4.151}$$

where

$$\mathbf{H}_N(\mathbf{x}) \triangleq -\nabla_{\mathbf{x}} \nabla_{\mathbf{x}}^T \Lambda_N(\mathbf{x}) = -\sum_{k=1}^{N} \nabla_{\mathbf{x}} \nabla_{\mathbf{x}}^T \ln(f(\mathbf{Y}_k|\mathbf{x})) \tag{4.152}$$

denotes the Hessian of $-\Lambda_N(\mathbf{x})$. Rescaling terms, this gives

$$\frac{1}{N} \mathbf{H}_N(\mathbf{x}_0)[N^{1/2}(\hat{\mathbf{X}}_N - \mathbf{x}_0)] = \frac{\mathbf{U}_N(\mathbf{x}_0)}{N^{1/2}} . \tag{4.153}$$

In this expression, by applying the strong law of large numbers, the matrix sequence $\mathbf{H}_N(\mathbf{x}_0)/N$ converges almost surely to the Fisher information matrix $\mathbf{J}(\mathbf{x}_0)$, since

$$E_0[-\nabla_{\mathbf{x}} \nabla_{\mathbf{x}}^T \ln(f(\mathbf{Y}_k|\mathbf{x}_0))] = \mathbf{J}(\mathbf{x}_0) ,$$

where $E_0[\cdot]$ denotes the expectation with respect to $f(\mathbf{y}|\mathbf{x}_0)$. Consequently for N sufficiently large, the matrix $\mathbf{H}_N(\mathbf{x}_0)/N$ will be invertible, so that we can rewrite (4.153) as

$$N^{1/2}(\hat{\mathbf{X}}_N - \mathbf{x}_0) = \left[\mathbf{H}_N(\mathbf{x}_0)/N\right]^{-1} \frac{\mathbf{U}_N(\mathbf{x}_0)}{N^{1/2}} . \tag{4.154}$$

Consider the term

$$\frac{\mathbf{U}_N(\mathbf{x}_0)}{N^{1/2}} = \frac{1}{N^{1/2}} \sum_{k=1}^{N} \nabla_{\mathbf{x}} \ln(f(\mathbf{Y}_k|\mathbf{x}_0)) .$$

It is obtained by scaling by $N^{1/2}$ the sum of the i.i.d. random vectors $\mathbf{Z}_k = \nabla_{\mathbf{x}} \ln(f(\mathbf{Y}_k|\mathbf{x}_0))$, where $\mathbf{Z}_k$ has zero mean and covariance matrix

$$E[\mathbf{Z}_k \mathbf{Z}_k^T] = E_0[\nabla_{\mathbf{x}} \ln(f(\mathbf{Y}_k|\mathbf{x}_0))(\nabla_{\mathbf{x}} \ln(f(\mathbf{Y}_k|\mathbf{x}_0)))^T] = \mathbf{J}(\mathbf{x}_0) .$$

According to the central limit theorem, it converges in law to a $N(\mathbf{0}, \mathbf{J}(\mathbf{x}_0))$ distributed random vector. If we consider the term on the right-hand side of (4.154), it is the product of the matrix $[\mathbf{H}_N(\mathbf{x}_0)/N]^{-1}$ which converges almost surely to $\mathbf{J}^{-1}(\mathbf{x}_0)$ and the vector $\mathbf{U}_N(\mathbf{x}_0)/N^{1/2}$ which converges in law to a $N(\mathbf{0}, \mathbf{J}(\mathbf{x}_0))$ distributed random vector. Employing a theorem due to Slutzky [7, Chap. 6] to combine these two modes of convergence, we then conclude that the vector sequence $N^{1/2}(\hat{\mathbf{X}}_N - \mathbf{x}_0)$ converges in law to a $N(\mathbf{0}, \mathbf{J}^{-1}(\mathbf{x}_0))$ random vector, as desired.

4.6 Bibliographical Notes

The Bayesian estimation of random parameters and estimation theory of nonrandom parameters are discussed in detail in several electrical engineering or statistics textbooks [6, 8–10]. In addition to Ferguson's concise but easy-to-follow introduction to asymptotic statistics [7], several recent texts [11, 12] cover in a unified manner the asymptotic behavior of the class of M-estimators, which in addition to the ML estimator includes least-squares estimators and variants thereof. The approach that was employed in Section 4.5 to prove the consistency of ML estimates was first proposed by Wald [13]. In a larger context, the uniform law of large numbers condition that we have placed on the log-likelihood ratio between the density of the observations under the true parameter $\mathbf{x}_0$ and under an arbitrary parameter vector is called a Glivenko-Cantelli condition. This type of condition plays an important role in machine learning theory [14] since, like the ML estimation technique, machine learning theory replaces the minimization of an objective function by a corresponding minimization on a sampled average.

4.7 Problems

4.1. Let X and Y be two independent identically distributed random variables with the uniform probability density $f(u) = 1$ for $0 \leq u \leq 1$. Let $Z = X + Y$.

(a) Find $\hat{X}_{\text{LLS}}(Z)$ and K_{LLS}, the linear least-squares estimate of X based on the observation Z, and its mean-square error variance.
(b) Find $\hat{X}_{\text{MSE}}(Z)$ and K_{MSE}, the least-squares estimate of X based on the observation Z, and its mean-square error variance.

4.2. When a light beam of known intensity I is shined on a photomultiplier, the number of photocounts observed in a T-seconds interval is a Poisson random variable with mean αI, where $\alpha > 0$ is a known constant. The number of photocounts observed in nonoverlapping T-seconds intervals are statistically independent random variables. Suppose that we shine a light beam of intensity I, where I is now a random variable, on a photomultiplier and measure

$$\mathbf{Y} = \begin{bmatrix} Y_1 \\ Y_2 \\ \vdots \\ Y_P \end{bmatrix},$$

where Y_i is the number of counts in the i-th of a set of P T-seconds nonoverlapping intervals. Then

$$P[Y_i = k \mid I] = \frac{(\alpha I)^k}{k!} \exp(-\alpha I),$$

for $k = 0, 1, 2, \cdots$

Suppose that the intensity I admits the probability density function

$$f_I(i) = (\bar{I})^{-1} \exp(-i/\bar{I}) u(i) ,$$

where $\bar{I}$ is a constant, and $u(i)$ is the unit step function. Find the Bayes least-squares estimate $\hat{I}_{\text{MSE}}(Y)$ of I given Y, and the corresponding mean-square error variance.

4.3. Let

$$Y = X + V ,$$

where X, and V are independent random variables. X takes the values $+1$ and -1 with equal probability, and V is a zero-mean Gaussian random variable with variance σ^2, respectively.

(a) Evaluate the a-posteriori probabilities $P[X = 1 \mid Y]$ and $P[X = -1 \mid Y]$.
(b) Find the maximum a-posteriori estimate $\hat{X}_{\text{MAP}}(Y)$ of X given the observation Y.
(c) Find the Bayes least-squares estimate $\hat{X}_{\text{MSE}}(Y)$ of X given Y.

4.4. Let X be a random parameter with the exponential density

$$f_X(x) = \begin{cases} a \exp(-ax) & \text{for } x > 0 \\ 0 & \text{otherwise} , \end{cases}$$

where $a > 0$ is known. Suppose we observe a Poisson random variable N with rate X, i.e.,

$$P[N = n \mid X] = \frac{X^n}{n!} \exp(-X) .$$

(a) Find the a posteriori conditional density of X given N.
(b) Find the least-squares estimate $\hat{X}_{\text{MSE}}$.
(c) Find the a maximum a-posteriori estimate $\hat{X}_{\text{MAP}}$.

For this problem, you may need to use the integral

$$\int_0^\infty \frac{x^n}{n!} \exp(-bx) dx = \frac{1}{b^{n+1}} .$$

4.5. Consider the observation

$$Y = X + V ,$$

where X is a random variable uniformly distributed on the interval $[0, 1]$ and the noise V is independent of X, with the exponential probability density

$$f_V(v) = \begin{cases} \exp(-v) & v \geq 0 \\ 0 & v < 0 . \end{cases}$$

(a) Find the MAP estimate $\hat{X}_{\text{MAP}}(Y)$.
(b) Find the estimate $\hat{X}_{\text{MAE}}(Y)$ which minimizes the expected absolute error $E[|X - \hat{X}|]$.
(c) Find the least-squares estimate $\hat{X}_{\text{MSE}}(Y)$ of X.

4.6. Let $X > 0$ be a random variable with probability density

$$f_X(x) = \begin{cases} x\exp(-x) & \text{for } x > 0 \\ 0 & \text{otherwise} \end{cases},$$

and let Y denote an observation which is uniformly distributed over $[0, X]$, so its density is given by

$$f_{Y|X}(y|x) = \begin{cases} x^{-1} & 0 \leq y \leq x \\ 0 & \text{otherwise} \end{cases}.$$

(a) Find the MSE estimate $\hat{X}_{\text{MSE}}(Y)$ of X given Y.
(b) Find the MAE estimate $\hat{X}_{\text{MAE}}(Y)$.
(c) Find the MAP estimate $\hat{X}_{\text{MAP}}(Y)$.

4.7. Consider the communication system shown in Fig. 4.6. The message X is a zero-mean Gaussian random variable with variance σ_X^2. The transmitter output is HX, and the receiver input is $Y = HX + V$, where V is an $N(0, r)$ random variable which is statistically independent of X.
Suppose that the transmitter is subject to intermittent failure, so that H is a random variable taking on values 1 or 0 with probabilities p and $1 - p$, respectively. Assume that H is statistically independent of both X and V.

(a) Find $\hat{X}_{\text{LLS}}(Y)$, the linear least-squares estimate of X based on the observation Y, and K_{LLS}, the corresponding error variance.
(b) Show that

$$E[X|Y] = \sum_{k=0}^{1} E[X|Y, H = k]P[H = k|Y].$$

(c) Find $\hat{X}_{\text{MSE}}(Y)$, the Bayes mean-squares estimate of X based on the observation Y.

4.8. You have been asked to determine whether a binary random number generator is working properly. The number generator produces a stream of

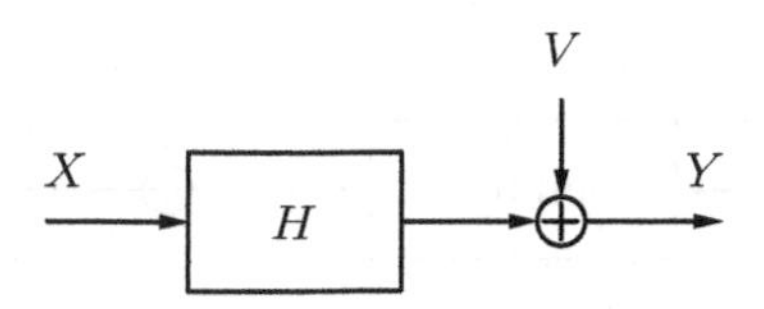

Fig. 4.6. Model of an intermittent communication system.

bits equal to either 0 or 1. As a first step, you decide to estimate the unknown probability p that a zero is generated. Obviously, the generator is not working properly unless p equals 1/2.

(a) Suppose that you observe a stream of n bits, and each time that a bit is generated, it has probability p of being equal to 0. Let Z be the total number of zero bits in the observed sequence. Find the maximum likelihood estimate of of p based on the knowledge of Z.
(b) Evaluate the bias and mean-square error of the ML estimate.
(c) Is the ML estimate efficient? Is it consistent? Briefly explain your reasoning.

4.9. A highly simplified speech communication system that uses a logarithmic compressor is shown in Fig. 4.7 below. In this block diagram,

$$h(X) = \begin{cases} -\ln(1+|X|) & X < 0 \\ 0 & X = 0 \\ \ln(1+|X|) & X > 0\,, \end{cases}$$

X is an unknown input, V is a zero-mean Gaussian random variable with variance σ^2, and the output $Y = h(X) + V$.

(a) Find the maximum likelihood estimate of X based on the observation of Y.
(b) Find an explicit lower bound on the mean-square estimation error of an arbitrary unbiased estimate of X based on observation Y.

4.10. Let

$$\mathbf{Y} = \begin{bmatrix} Y_1 \\ Y_2 \end{bmatrix}$$

be a two-dimensional Gaussian random vector with zero-mean and covariance matrix

$$\mathbf{K}_Y = v \begin{bmatrix} 1 & \rho \\ \rho & 1 \end{bmatrix}.$$

So Y_1 and Y_2 have variance v and correlation coefficient ρ.

(a) Find the ML estimates $\hat{v}_{\mathrm{ML}}$ and $\hat{\rho}_{\mathrm{ML}}$ of v and ρ given Y_1 and Y_2.
(b) Are these estimates unbiased?

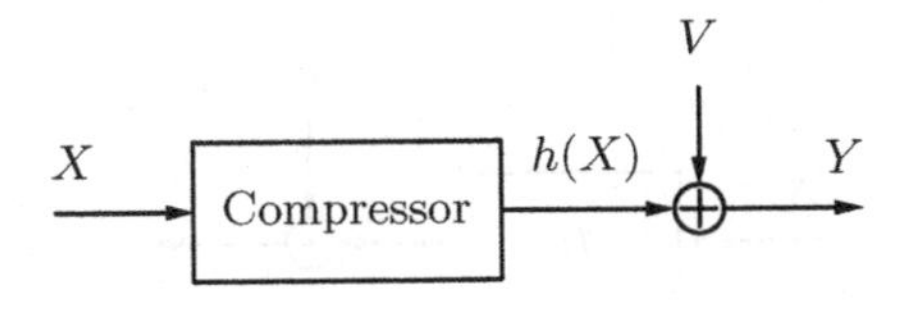

Fig. 4.7. Model of a speech communication system with logarithmic compression.

(c) Evaluate the Fisher information matrix $\mathbf{J}(v, \rho)$ and determine whether the ML estimates are efficient.

4.11. Consider observations

$$Y(t) = Z_c \cos(\omega_0 t) + Z_s \sin(\omega_0 t) + V(t)$$

for $0 \leq t \leq N-1$ of a sinusoidal signal with frequency ω_0 in the presence of a zero mean WGN $V(t)$ with known variance σ^2. If

$$\begin{aligned}
\mathbf{Y} &= \begin{bmatrix} Y(0) \cdots Y(t) \cdots Y(N-1) \end{bmatrix}^T \\
\mathbf{V} &= \begin{bmatrix} V(0) \cdots V(t) \cdots V(N-1) \end{bmatrix}^T \\
\mathbf{H}(\omega_0) &= \begin{bmatrix} \cos(0) & \cdots & \cos(\omega_0 t) & \cdots & \cos(\omega_0(N-1)) \\ \sin(0) & \cdots & \sin(\omega_0 t) & \cdots & \sin(\omega_0(N-1)) \end{bmatrix}^T ,
\end{aligned}$$

are the vectors and matrix representing the observations, noise and sinusoidal signal space, respectively, the observed signal can be expressed in vector form as

$$\mathbf{Y} = \mathbf{H}(\omega_0) \begin{bmatrix} Z_c \\ Z_s \end{bmatrix} + \mathbf{V} .$$

In this expression, we seek to estimate amplitudes Z_c, Z_s, and frequency ω_0, so the unknown parameter vector is given by

$$\mathbf{x} = \begin{bmatrix} Z_c \\ Z_s \\ \omega_0 \end{bmatrix} .$$

In the analysis below, it is assumed that $0 < \omega_0 < \pi$. It is also assumed that the number N of observations is sufficiently large to ensure that the columns of $\mathbf{H}(\omega_0)$ are approximately orthogonal, i.e.,

$$\frac{2}{N} \mathbf{H}^T(\omega_0) \mathbf{H}(\omega_0) \approx \mathbf{I}_2 . \tag{4.155}$$

(a) Find the ML estimate of $\mathbf{x}$ by maximizing the log-likelihood function, $\ln f_{\mathbf{Y}}(\mathbf{y}|\mathbf{x})$. This can be achieved in two stages by maximizing first with respect to

$$\mathbf{Z} = \begin{bmatrix} Z_c \; Z_s \end{bmatrix}^T ,$$

and then with respect to ω_0. Express the ML estimate $\hat{\omega}_{0,\mathrm{ML}}$ in terms of the projection $\mathbf{P_H}(\omega_0)\mathbf{Y}$ of the observation vector $\mathbf{Y}$ onto the space spanned by the columns of $\mathbf{H}(\omega_0)$, where

$$\mathbf{P_H}(\omega_0) \triangleq \mathbf{H}(\omega_0)(\mathbf{H}^T(\omega_0)\mathbf{H}(\omega_0))^{-1}\mathbf{H}^T(\omega_0) .$$

Then, use the approximation (4.155) to verify that $\hat{\omega}_{0,\mathrm{ML}}$ maximizes the periodogram

$$\frac{2}{N}|\sum_{t=0}^{N-1} Y(t)\exp(-j\omega t)|^2 ,$$

which is the squared discrete-time Fourier transform (DTFT) of observation sequence $Y(t)$, $0 \leq t \leq T$.

(b) Evaluate the Fisher information matrix $\mathbf{J}(\mathbf{x})$ for the parameter vector $\mathbf{x}$. To do so, use approximations

$$\frac{1}{N^{k+1}} \sum_{t=0}^{N-1} t^k \cos(2\omega_0 t) \approx 0$$

$$\frac{1}{N^{k+1}} \sum_{t=0}^{N-1} t^k \sin(2\omega_0 t) \approx 0$$

for $k = 1,\ 2$. Obtain the matrix CRLB for the MSE of unbiased estimates of $\mathbf{x}$.

(c) Verify that if the ML estimate $\hat{\omega}_{0,\text{ML}} \approx \omega_0$, the ML estimates

$$\begin{bmatrix} \hat{Z}_{c\text{ML}} \\ \hat{Z}_{s\text{ML}} \end{bmatrix}$$

of the cosine and sine components of the sinusoidal signal are unbiased.

(d) Evaluate

$$\mathbf{x} + \mathbf{J}^{-1}(\mathbf{x}) \nabla_{\mathbf{x}} \ln f_{\mathbf{Y}}(\mathbf{y}|\mathbf{x})$$

and determine if all or some of the components of the ML vector estimate $\hat{\mathbf{x}}_{\text{ML}}$ are efficient. See [15] for a complete analysis of frequency estimation.

4.12. Let $\{Y_k,\ 1 \leq k \leq N\}$ be N independent uniformly distributed random variables over $[0, X]$, so the density of each Y_k takes the form

$$f(y|X) = \begin{cases} 1/X & 0 \leq y \leq X \\ 0 & \text{otherwise} . \end{cases}$$

We seek to estimate the range X of the density from the N observations $\{Y_k,\ 1 \leq k \leq N\}$.

(a) Verify that the joint density of the Y_k's can be expressed as

$$f(\mathbf{y}|X) = \frac{1}{X^N} u(X - z_1(\mathbf{y})) u(z_2(\mathbf{y}))$$

where

$$z_1(\mathbf{Y}) = \max_{1 \leq k \leq N} Y_k \quad , \quad z_2(\mathbf{Y}) = \min_{1 \leq k \leq N} Y_k$$

and $u(\cdot)$ denotes the unit step function. Then use the Neyman-Fisher factorization criterion to verify that $z_1(\mathbf{Y})$ is a sufficient statistic for estimating X based on observations Y_k, $1 \leq k \leq N$.

(b) Evaluate the ML estimate $\hat{X}_{\text{ML}}(\mathbf{Y})$ of X based on the Y_k's, $1 \leq k \leq N$.
(c) Evaluate the bias and MSE of the ML estimate. To answer this question, you will need to use the fact that if $\{Y_k, 1 \leq k \leq N\}$ are N i.i.d. random variables with probability density $f_Y(y)$ and cumulative probability distribution $F_Y(y)$, the probability density of $Z = \max_{1 \leq k \leq N} Y_k$ is given by [16, p. 246]

$$f_Z(z) = N F_Y^{N-1}(z) f_Y(z) .$$

(d) Find the UMVU estimate $\hat{X}_{\text{UMVU}}(\mathbf{Y})$ of X based on the observations Y_k, $1 \leq k \leq N$. Evaluate its MSE and compare it to the MSE of the ML estimate. Draw some conclusions based on this comparison.

4.13. A sinusoid of known frequency ω_0 is observed in the presence of an additive WGN $V(n)$ with zero mean and variance σ^2. The observations can be expressed as

$$Y(t) = A \cos(\omega_0 t) + V(n)$$

for $0 \leq t \leq N - 1$.

(a) Find the uniform minimum variance unbiased estimate of amplitude A, assuming σ^2 is known.
(b) Assuming now that *both* A and σ^2 are unknown, find the minimum variance unbiased estimates of A and σ^2.
(c) Repeat parts (a) and (b) for the ML estimates. Are the ML and UMVU estimates identical?
(d) Evaluate the matrix Cramér-Rao bound for the joint estimation of A and σ^2.
(e) Evaluate the bias and variances of the joint ML and UMVU estimates of A and σ^2 and compare them to the Cramér-Rao bound.

4.14. Let $\{Y_k, 1 \leq k \leq N\}$ be N independent random variables with the geometric distribution

$$P[Y = m] = p q^m ,$$

where $q = 1 - p$, for $m = 0, 1, 2, \ldots$. Recall [16] that the generating function of Y takes the form

$$G_Y(z) = E[z^Y] = \frac{p}{1 - qz}$$

and its mean and variance are given by

$$E[Y] = \frac{q}{p} \quad , \quad K_Y = \frac{q}{p^2} ,$$

We seek to estimate $p = P[Y = 0]$.

(a) Verify that $S = \sum_{k=1}^{N} Y_k$ is a sufficient statistic for estimating p. By observing that the generating function $G_S(z)$ of S can be expressed as

$$G_S(z) = (G_Y(z))^N ,$$

show that S admits the probability distribution

$$P[S = s|p] = \frac{\prod_{i=1}^{N-1}(s+i)}{(N-1)!} p^N q^s$$

for $s = 0, 1, 2 \ldots$.

(b) Obtain the ML estimate $\hat{p}_{\mathrm{ML}}$ of p.

(c) Evaluate the Fisher information $J(p)$ and the CRLB for unbiased estimates of p.

(d) Evaluate

$$p + J^{-1}(p)\frac{d}{dp}\ln P[s|p]$$

and determine whether an unbiased efficient estimate exists.

4.15. Consider Problem 4.14. We seek to construct an UMVUE of p by employing the RBLS procedure. Recall that $S = \sum_{k=1}^{N} Y_k$ is a sufficient statistic, which is complete, since geometric random variables belong to the exponential class of distributions.

(a) Given a single observation Y_1, let

$$T(Y_1) = \begin{cases} 1 & \text{if } Y_1 = 0 \\ 0 & \text{otherwise} . \end{cases}$$

This estimator is rather crude, but it is unbiased since

$$E[T] = P[Y_1 = 0] = p\,.$$

(b) Next, apply the RBLS procedure to $T(Y_1)$ to find the UMVU estimate as

$$\hat{p}_{\mathrm{UMVU}} = E[T(Y_1)|S]\,.$$

To do so, observe that

$$\begin{aligned} E[T|S=s] &= P[Y_1 = 0|S = s] \\ &= \frac{P[Y_1 = 0]P[\sum_{k=2}^{N} Y_k = s]}{P[S=s]} \,. \end{aligned}$$

Compare this estimate to the ML estimate $\hat{p}_{\mathrm{ML}}$ derived in Problem 4.14.

4.A. Derivation of the RBLS Theorem

Consider the estimate $\hat{\mathbf{X}}(\mathbf{S}) = E[\check{\mathbf{X}}(\mathbf{Y})|\mathbf{S}]$ obtained by applying the RBLS method to an unbiased estimate $\check{\mathbf{X}}(\mathbf{Y})$. Let $\mathbf{A}$ denote an auxiliary statistic such that the mapping

$$\mathbf{Y} \to \begin{bmatrix} \mathbf{S} \\ \mathbf{A} \end{bmatrix}$$

is one-to-one. By repeated expectations, we find

$$\begin{aligned} E_{\mathbf{S}}[\hat{\mathbf{X}}(\mathbf{S})] &= E_{\mathbf{S}}[E_{\mathbf{A}}[\check{\mathbf{X}}(\mathbf{Y})|\mathbf{S}]] \\ &= E_{\mathbf{Y}}[\check{\mathbf{X}}(\mathbf{Y})] = \mathbf{x} \end{aligned} \tag{4A.1}$$

so that the RBLS estimate is *unbiased*. Next, we rewrite the covariance matrix $\check{\mathbf{K}}(\mathbf{x})$ of $\check{\mathbf{X}}$ given by (4.132) as

$$\begin{aligned} \check{\mathbf{K}}(\mathbf{x}) &= E[(\mathbf{x} - \hat{\mathbf{X}} + \hat{\mathbf{X}} - \check{\mathbf{X}})(\mathbf{x} - \hat{\mathbf{X}} + \hat{\mathbf{X}} - \check{\mathbf{X}})^T] \\ &= \hat{\mathbf{K}}(\mathbf{x}) + E[(\mathbf{x} - \hat{\mathbf{X}})(\hat{\mathbf{X}} - \check{\mathbf{X}})^T] \\ &\quad + E[(\hat{\mathbf{X}} - \check{\mathbf{X}})(\mathbf{x} - \hat{\mathbf{X}})^T] + E[(\hat{\mathbf{X}} - \check{\mathbf{X}})(\hat{\mathbf{X}} - \check{\mathbf{X}})^T] \,. \end{aligned} \tag{4A.2}$$

The second and third terms are transposes of one another, so we only need to evaluate the second term. We employ the orthogonality property of mean-square estimates. By construction, $\hat{\mathbf{X}}(\mathbf{S})$ is the MSE estimate of $\check{\mathbf{X}}(\mathbf{Y})$ given $\mathbf{S}$. The error $\hat{\mathbf{X}}(\mathbf{S}) - \check{\mathbf{X}}(\mathbf{Y})$ must therefore be orthogonal to any function of $\mathbf{S}$ only. But $\mathbf{x} - \hat{\mathbf{X}}(\mathbf{S})$ depends on $\mathbf{S}$ only. So by the orthogonality property of MSEs, the second and third terms drop out of (4A.2). This leaves us with

$$\check{\mathbf{K}}(\mathbf{x}) = \hat{\mathbf{K}}(\mathbf{x}) + E[(\hat{\mathbf{X}} - \check{\mathbf{X}})(\hat{\mathbf{X}} - \check{\mathbf{X}})^T] \,, \tag{4A.3}$$

where the second matrix on the right-hand side is a nonnegative correlation matrix. This proves the covariance reduction property (4.133) of the RBLS method.

The RBLS method produces an unbiased estimate of $\mathbf{x}$ that depends on the sufficient statistic $\mathbf{S}$. If the statistic $\mathbf{S}$ is complete, there is only one unbiased estimate of $\mathbf{x}$ depending on $\mathbf{S}$ only. So by applying the RBLS scheme to any unbiased estimate $\check{\mathbf{X}}(\mathbf{Y})$ of $\mathbf{x}$, the resulting estimate $\hat{\mathbf{X}}(\mathbf{S})$ is always the same. This estimate must be a UMVU estimate of $\mathbf{x}$. Otherwise, if there was an unbiased estimator $\hat{\mathbf{X}}'$ of $\mathbf{x}$ that outperformed $\hat{\mathbf{X}}(\mathbf{S})$ for a value $\mathbf{x}_0$ of the parameter vector, by applying the RBLS method to $\hat{\mathbf{X}}'$, we obtain $\hat{\mathbf{X}}(\mathbf{S})$, which according to (4.133) outperforms $\mathbf{X}'$ at $\mathbf{x}_0$, a contradiction.

References

1. J. O. Berger, *Statistical Decision Theory and Bayesian Analysis, Second Edition.* Springer, 1985.
2. H. L. Van Trees, *Detection, Estimation and Modulation Theory, Part I: Detection, Estimation and Linear Modulation Theory.* New York: J. Wiley & Sons, 1968. Paperback reprint edition in 2001.
3. T. Kailath, A. H. Sayed, and B. Hassibi, *Linear Estimation.* Upper Saddle River, NJ: Prentice Hall, 2000.

4. A. Paulraj, R. Nabar, and D. Gore, *Introduction to Space-Time Wireless Communications.* Cambridge, UK: Cambridge University Press, 2003.
5. A. J. Laub, *Matrix Analysis for Scientists and Engineers.* Philadelphia, PA: Soc. for Industrial and Applied Math., 2005.
6. E. L. Lehmann and G. Casella, *Theory of Point Estimation, Second Edition.* New York: Springer Verlag, 1998.
7. T. S. Ferguson, *A Course in Large Sample Theory.* London: Chapman & Hall, 1996.
8. S. M. Kay, *Fundamentals of Statistical Signal Processing: Estimation Theory.* Prentice-Hall, 1993.
9. H. V. Poor, *An Introduction to Signal Detection and Estimation, Second Edition.* New York: Springer Verlag, 1994.
10. P. J. Bickel and K. A. Doksum, *Mathematical Statistics: Basic Ideas and Selected Topics, Second Edition.* Upper Saddle River, NJ: Prentice Hall, 2001.
11. A. W. van der Vaart, *Asymptotic Statistics.* Cambridge, UK: Cambridge University Press, 1998.
12. S. van de Geer, *Empirical Processes in M-Estimation.* Cambridge, UK: Cambridge University Press, 2000.
13. A. Wald, "Note on the consistency of the maximum likelihood estimate," *Annals Math. Statistics*, vol. 20, pp. 595–601, 1949.
14. V. N. Vapnik, *Statistical Learning Theory.* New York: Wiley & Sons, 1998.
15. B. G. Quinn and E. J. Hannan, *The Estimation and Tracking of Frequency.* Cambridge, UK: Cambridge University Press, 2001.
16. A. Papoulis and S. U. Pillai, *Probability, Random Variables and Stochastic Processes, Fourth Edition.* New York: McGraw Hill, 2002.

5

Composite Hypothesis Testing

5.1 Introduction

For all hypothesis testing problems considered up to this point, it has been assumed that a complete description of the probability distribution of observations is available under each hypothesis. However, for practical signal detection problems, even though an exact model of the received signal may be available, it often includes some unknown parameters, such as amplitude, phase or time delay, frequency, or an additive noise variance. Problems of this type are called composite hypothesis testing problems since each hypothesis is really a family of sub-hypotheses parametrized by the unknown parameter vector. For problems of this kind, in some rare cases the presence of unknown parameters does not affect the detection problem in the sense that there exists a detector which is best for all values of the signal parameters. Tests with this property are called *uniformly most powerful* (UMP). Needless to say, such situations are rather rare and are examined in Section 5.2. An important feature of composite hypothesis testing problems is that the unknown parameter vector may admit a group of transformations that leaves the problem invariant. For example, for a signal of unknown amplitude received in additive white noise of unknown variance, the detection problem is obviously scale invariant. To tackle problems with this property, it is natural to require that the detector should be invariant under such transformations. The resulting class of invariant detectors is smaller than the entire class of detectors, and one may then search for a UMP detector in this subclass. Detectors of this type are called *uniformly most powerful invariant* (UMPI) and are discussed in Section 5.3. Another important feature of composite hypothesis problems is that some of the parameters appearing in a hypothesis may be unique to the hypothesis, such as the amplitude or phase of a signal to be detected, whereas other parameters, such as those of interference signals, may be common to all hypotheses. Parameters which are common to all hypotheses are usually called *nuisance parameters*. As their name indicates, parameters of this type are an inconvenience since they do not play any role in determining which

B.C. Levy, *Principles of Signal Detection and Parameter Estimation*,
DOI: 10.1007/978-0-387-76544-0_5, © Springer Science+Business Media, LLC 2008

hypothesis is true. So it is in general a good idea to remove as many of these parameters as possible, and in Section 5.4 we describe a general method to do so for a class of linear detection problems. Yet, for many detection problems, no UMP(I) test exists, and in such situations one must resort to joint estimation and detection techniques. The best known joint estimation and detection method is the generalized likelihood ratio test (GLRT) described in Section 5.5. Its principle is quite straightforward: it consists of finding the maximum likelihood (ML) estimate of the parameter vector under each hypothesis, and then plugging the estimate in the probability distribution of the corresponding hypothesis and and treating the detection problem as if the estimated values were correct. This common sense test gives good results in general, and it can be shown that the GLRT is invariant with respect to transformations whenever the detection problem is itself invariant under the same transformations [1,2]. Furthermore, as explained in Section 5.6, for certain classes of distributions, the GLRT has an asymptotically optimal performance in the sense that it outperforms all other tests that do not have any knowledge of the parameter vector. In other words, for these distributions, the GLRT forms what is called a *universal test*.

5.2 Uniformly Most Powerful Tests

We consider a binary hypothesis testing problem where we observe a random vector $\mathbf{Y}$ in $\mathbb{R}^n$ which under the two hypotheses admits probability densities

$$\begin{aligned} H_0 &: \mathbf{Y} \sim f(\mathbf{y}|\mathbf{x}, H_0) \ , \ \mathbf{x} \in \mathcal{X}_0 \\ H_1 &: \mathbf{Y} \sim f(\mathbf{y}|\mathbf{x}, H_1) \ , \ \mathbf{x} \in \mathcal{X}_1 \end{aligned} \tag{5.1}$$

which are parametrized by an unknown parameter vector $\mathbf{x}$. This corresponds to a *composite hypothesis testing* problem, since we are testing a family of densities against another. For the case of discrete-valued observations, $\mathbf{Y}$ admits the probability mass distribution function $p(\mathbf{y}|\mathbf{x}, H_j)$ with $\mathbf{x} \in \mathcal{X}_j$ under hypothesis H_j with $j = 0,\ 1$.

In the above formulation, we assume that the probability densities or distributions are different under each hypothesis. To illustrate this situation, consider the following example:

Example 5.1: Test of distribution type

We observe a sequence $\{Y_k,\ 1 \leq k \leq N\}$ of independent identically distributed random variables, where under hypothesis H_0, the Y_k's are Gaussian distributed with unknown mean m and variance σ^2, and under H_1, they admit a Laplacian distribution with unknown mean m and parameter λ, so that

$$H_0 : Y_k \sim f(y|m, \sigma^2, H_0) = \frac{1}{(2\pi\sigma^2)^{1/2}} \exp(-\frac{1}{2\sigma^2}(y-m)^2)$$

$$H_1 : Y_k \sim f(y|m, \lambda, H_1) = \frac{\lambda}{2} \exp(-\lambda|y-m|) .$$

So this problem is one of distribution classification. In a communications context, a problem of the same nature arises whenever given a random signal of unknown modulation type, we need to determine the modulation employed (QAM, PSK, continuous-phase modulation, etc ...). □

However, in many detection problems, the distribution of the observations is the same under both hypotheses, and it is only the range of values of the parameters that differs. In such cases, we will drop the H_j dependence when writing the observation density, which becomes $f(\mathbf{y}|\mathbf{x})$, and it is only the ranges $\mathcal{X}_0$ and $\mathcal{X}_1$ of allowed parameter values that distinguish the two hypotheses. When the observation density is the same under both hypotheses, we will denote by $\mathbf{x}_j$ with $j = 0,\ 1$ the parameter vector under hypothesis H_j to keep track of which hypothesis is being considered.

Example 5.2: Detection of a signal with unknown amplitude in noise

Let

$$Y(k) = As(k) + V(k) \tag{5.2}$$

with $1 \leq k \leq N$ be a set of discrete-time observations of a known signal $s(k)$ with unknown amplitude A corrupted by a zero-mean WGN $V(k)$ of unknown variance σ^2. Under H_0, the signal is absent ($A = 0$) and under H_1 it is present ($A \neq 0$). So if

$$\mathbf{Y} = \begin{bmatrix} Y(1)\ Y(2) \ldots Y(N) \end{bmatrix}^T$$

and

$$\mathbf{s} = \begin{bmatrix} s(1)\ s(2) \ldots s(N) \end{bmatrix}^T$$

denote respectively the vectors formed by the observations and the signal to be detected, $\mathbf{Y}$ admits the probability distribution

$$f(\mathbf{y}|A, \sigma^2) \sim N(A\mathbf{s}, \sigma^2 \mathbf{I}_N) \tag{5.3}$$

and H_0 and H_1 differ only by the domains

$$\mathcal{X}_0 = \{(A, \sigma^2) :\ A = 0\}$$

and

$$\mathcal{X}_1 = \{(A, \sigma^2) :\ A \neq 0\}$$

of the parameter vector (A, σ^2). Note that since the noise variance σ^2 does not play a role in determining which hypothesis holds, it is a nuisance parameter.

□

Obviously, when the observations obey the same distribution under both hypotheses, the vector parameter domains $\mathcal{X}_0$ and $\mathcal{X}_1$ corresponding to the two hypotheses need to satisfy $\mathcal{X}_0 \cap \mathcal{X}_1 = \emptyset$. Otherwise the detection problem cannot be solved, since the same observation distribution could arise under both hypotheses. An important special case occurs when among the two hypotheses, only H_1 is composite, but hypothesis H_0, which is often called the "null hypothesis," is simple. This means that the domain $\mathcal{X}_0$ reduces to a single point $\mathbf{x}_0$. In this case, as we shall see, the analysis of the detection problem is usually easier than in the general case where both hypotheses are composite.

Let $\delta(\mathbf{y})$ denote a decision function taking values 0 or 1 depending on whether H_0 or H_1 are selected. This function specifies a disjoint partition $\mathcal{Y} = \mathcal{Y}_0 \cup \mathcal{Y}_1$ of the observations, where

$$\mathcal{Y}_j = \{\mathbf{y} : \delta(\mathbf{y}) = j\}$$

with $j = 0, 1$. Then to each decision rule δ corresponds a probability of detection and a probability of false alarm

$$\begin{aligned} P_D(\delta, \mathbf{x}) &= P[\delta = 1|\mathbf{x}, H_1] = \int_{\mathcal{Y}_1} f(\mathbf{y}|\mathbf{x}, H_1) d\mathbf{y} \\ P_F(\delta, \mathbf{x}) &= P[\delta = 1|\mathbf{x}, H_0] = \int_{\mathcal{Y}_1} f(\mathbf{y}|\mathbf{x}, H_0) d\mathbf{y} \,. \end{aligned} \tag{5.4}$$

Note that both P_D and P_F are functions of the parameter vector $\mathbf{x}$, and when viewed as a function of $\mathbf{x}$, $P_D(\delta, \mathbf{x})$ is called the *power* of the test. Although it is possible to discuss the design of a test in a Bayesian setting, we consider here Neyman-Pearson (NP) tests. Specifically, given an upper bound α for the probability of false alarm, we consider all tests δ such that

$$\max_{\mathbf{x} \in \mathcal{X}_0} P_F(\delta, \mathbf{x}) \leq \alpha \,. \tag{5.5}$$

In this context, α is usually called the "size" of the test. The bound (5.5) specifies an NP test of type I, but we can of course define an NP of type II in a similar manner. Then, among all tests δ satisfying (5.5), we say that δ_{UMP} is a *uniformly most powerful test* if it satisfies

$$P_D(\delta, \mathbf{x}) \leq P_D(\delta_{\mathrm{UMP}}, \mathbf{x}) \tag{5.6}$$

for all $\mathbf{x} \in \mathcal{X}_1$ and for all tests δ obeying (5.5). So from a conceptual viewpoint, a UMP test is similar to a UMVU estimator: it is a test that outperforms all other tests for all possible parameter vectors in $\mathcal{X}_1$. Needless to say, the UMP property is very strong and only a few hypothesis testing problems admit UMP tests. An important consequence of definition (5.6) is that a UMP test cannot depend on the parameter vector $\mathbf{x}$. Also, if we view $\mathbf{x}$ as being fixed, δ_{UMP} must be the optimum test in the sense of the Neyman-Pearson (NP) tests discussed in Chapter 2, so it must take the form of a likelihood-ratio test (LRT), possibly involving randomization. From these two observations, we immediately deduce a relatively simple approach for finding UMP tests: we need to find the LRT and then try to transform it in such a way that the parameter vector $\mathbf{x}$ disappears from the test statistic. When such a transformation can be found, a UMP test exists, provided a threshold can be selected such that the upper bound (5.5) is satisfied. To illustrate this methodology, we consider an example:

Example 5.2, continued

Consider the noisy observations $Y(k)$ with $1 \leq k \leq N$ of a known signal $s(k)$ with unknown amplitude A in a WGN $V(k)$ of variance σ^2 given by (5.2). In this example, we assume first that the noise variance σ^2 is *known*, and that we seek to test

$$H_1 \; : \; \mathcal{X}_1 = \{A : \; A > 0\}$$

against

$$H_0 \; : \; \mathcal{X}_0 = \{A = 0\}$$

so that H_0 is *simple*. Note that the range of values $\{A > 0\}$ of the parameter A under H_1 is located on one side of the value $A = 0$ under H_0. Such a test is said to be *one-sided*. On the other hand, if H_1 was such that $\mathcal{X}_1 = \{A \neq 0\}$, the test would be *two-sided*, since the range of possible values of A under H_1 is located on both sides of the value $A = 0$ corresponding to H_0. It turns out that a UMP detector does not exist when $A \neq 0$ under H_1, but as we shall see below, such a detector exists in the one-sided case with $A > 0$ (or $A < 0$). Consider the likelihood ratio test

$$L(\mathbf{y}|A) = \frac{f(\mathbf{y}|A)}{f(\mathbf{y}|A=0)} \underset{H_0}{\overset{H_1}{\gtrless}} \tau \tag{5.7}$$

corresponding to the two hypotheses for an arbitrary value $A > 0$ of the parameter under H_1, where the threshold τ is yet to be specified. If we denote the energy of the signal $s(k)$ by

$$E = ||\mathbf{s}||_2^2 = \sum_{k=1}^{N} s^2(k) ,$$

by using expression (5.3) for the observations density, and taking logarithms, (5.7) can be rewritten as

$$\ln(L(\mathbf{y}|A)) = A\frac{\mathbf{s}^T\mathbf{y}}{\sigma^2} - \frac{A^2 E}{2\sigma^2} \underset{H_0}{\overset{H_1}{\gtrless}} \ln(\tau) . \tag{5.8}$$

Then consider the sufficient statistic

$$S(\mathbf{Y}) = \frac{\mathbf{s}^T\mathbf{Y}}{E^{1/2}} \sim N(AE^{1/2}, \sigma^2) ,$$

which can be evaluated without knowing the parameter A. Noting that under H_1, the amplitude A is strictly positive, by multiplying (5.8) by $\sigma^2/(AE^{1/2})$, we can rewrite the LR test as

$$S(\mathbf{Y}) \underset{H_0}{\overset{H_1}{\gtrless}} \eta \tag{5.9}$$

with

$$\eta = \frac{AE^{1/2}}{2} + \frac{\sigma^2}{AE^{1/2}} \ln(\tau) .$$

Observe that the test (5.9) is now completely independent of A. The threshold η needs to be selected such that the size of the test is α. But the hypothesis H_0 is simple and corresponds to $A = 0$. Under H_0, the statistic $S(\mathbf{Y})$ is $N(0, \sigma^2)$ distributed, so that the probability of false alarm of the test is $\alpha = Q(\eta/\sigma)$, which gives

$$\eta = \sigma Q^{-1}(\alpha) . \tag{5.10}$$

Since the test (5.9) and threshold η given by (5.10) are independent of the value $A > 0$ under H_1, a UMP test exists. The power of the test is

$$\begin{aligned} P_D(A) &= P[S(\mathbf{Y}) \geq \eta | A] = 1 - P[S(\mathbf{Y}) < \eta | A] \\ &= 1 - Q(\frac{AE^{1/2} - \eta}{\sigma}) = 1 - Q(\frac{AE^{1/2}}{\sigma} - Q^{-1}(\alpha)) , \end{aligned} \tag{5.11}$$

which is a monotone increasing function of $A > 0$.

Up to this point we have assumed that $A > 0$ and σ is known. When the domain of A under H_1 is $\{A : A \neq 0\}$, so that H_1 is two-sided, the transformation employed above to convert the LRT test (5.7) into a test independent of A no longer works. The key step is the division of (5.8) by $AE^{1/2}$. When $A > 0$ the direction of the inequality (5.8) is preserved, but when $A < 0$, it is reversed. So, if all we know is $A \neq 0$, we cannot transform the inequality (5.8) to make the left-hand side independent of A. Now suppose that σ^2 is unknown, so that under H_1 we have $A > 0$ and σ^2 unknown, and under H_0 $A = 0$ and σ^2 unknown. In this case H_0 is a composite hypothesis. Then the LRT transformation to the form (5.9) still works. The difficulty arises when we seek to select a threshold η which ensures that the test has size α. If σ^2 is completely unknown, we must select $\eta = \infty$, so that in this case we never choose H_1, and the power of the test is zero. However, if we know that σ^2 belongs to the interval $[\sigma_{\mathrm{L}}^2, \sigma_{\mathrm{U}}^2]$, by selecting

$$\eta = \sigma_{\mathrm{U}} Q^{-1}(\alpha)$$

the constraint (5.5) on the size of the test is satisfied, and a UMP test still exists. The power of the test is still given by (5.11), which is now a function of both A and σ. □

A class of distributions that admits a UMP test was identified by Karlin and Rubin [3] who proved the following result:

Monotone Likelihood-ratio Criterion: Consider a random vector $\mathbf{Y}$ whose probability density $f(\mathbf{y}|x)$ (resp. probability mass distribution $p(\mathbf{y}|x)$) is parametrized by a scalar parameter x. We say that the density f (resp. distribution p) specifies a monotone likelihood-ratio (MLR) family if the densities/distributions corresponding to different values of x are distinct and if there is a scalar statistic $S(\mathbf{Y})$ such that for $x_0 < x_1$, the likelihood-ratio function

$$L(\mathbf{y}|x_1, x_0) = \frac{f(\mathbf{y}|x_1)}{f(\mathbf{y}|x_0)} \tag{5.12}$$

is a monotone increasing function of $S(\mathbf{y})$. In the case of discrete-valued random vectors, the densities need to be replaced by probability distributions when evaluating the likelihood ratio (5.12). Then consider the decision function

$$\delta_\eta(\mathbf{y}) = \begin{cases} 1 \text{ for } S(\mathbf{y}) \geq \eta \\ 0 \text{ for } S(\mathbf{y}) < \eta \,, \end{cases} \tag{5.13}$$

which does not depend on the parameter x of the density $f(\mathbf{y}|x)$ (resp. $p(\mathbf{y}|x)$) and let

$$\alpha = P_F(\delta_\eta, x_0) = P[S(\mathbf{Y}) \geq \eta | x_0] \,. \tag{5.14}$$

This test has the following properties:

(i) With α given by (5.14), δ_η is a UMP test of size α for testing $H_1 : x > x_0$ against $H_0 : x \leq x_0$.
(ii) For each η, the power function

$$P_D(\delta_\eta, x) = P[S(\mathbf{Y}) \geq \eta | x] \tag{5.15}$$

is a monotone increasing function of x.

To prove (i), we first consider the simple hypothesis $H'_0 : x = x_0$. Then let $x_1 > x_0$ and consider an NP test for deciding between the simple hypothesis $H'_1 : x = x_1$ against H'_0. It takes the form

$$L(\mathbf{y}; x_1, x_0) = g(S(\mathbf{y})) \underset{H'_0}{\overset{H'_1}{\gtrless}} \tau \,, \tag{5.16}$$

where the threshold τ is yet to be determined and $g(S)$ is monotone increasing. Since $g(S)$ is monotone increasing, it admits an inverse function $g^{-1}(S)$ which, when applied to inequality (5.16) yields the test (5.13), where the left-hand side is now independent of x_1 and $\eta = g^{-1}(\tau)$. Since the threshold η is given by (5.14), the size of the test is α. Thus by transforming the optimal LR test for testing H'_1 against H'_0 we have obtained a test independent of x_1. This test is therefore UMP for testing the composite hypothesis H_1 against the simple hypothesis H'_0.

Next, consider the power of the test δ_η. At $x = x_0$ we have

$$P_D(\delta_\eta, x_0) = \alpha \,. \tag{5.17}$$

For $x_1 > x_0$, since the δ_η is an optimal NP test of size α for testing the simple hypothesis $H_1' : x = x_1$ against $H_0' : x = x_0$, and since the probability of detection of an NP test is always greater than the probability of false alarm, we have

$$P_D(\delta_\eta, x_1) > \alpha \,.$$

Now consider $x_2 < x_0$. In this case δ_η is still an optimal NP test with size

$$\breve{\alpha} = P[S(\mathbf{y}) > \eta | \mathbf{x}_2]$$

for testing the simple hypothesis $\breve{H}_0 : x = x_2$ against $\breve{H}_1 : x = x_0$. Using again the fact that for an NP test, the probability of detection is necessarily higher than the probability of false alarm, we find

$$P_D(\delta_\eta, x_2) = \breve{\alpha} < P_D(\delta_\eta, x_0) \,. \tag{5.18}$$

This shows that the power function is monotone increasing in x. Also, in the light of (5.17), we have $\breve{\alpha} < \alpha$, so the test δ_η is UMP for testing $H_1 : x > x_0$ against the composite hypothesis $H_0 : x \leq x_0$.

Remark: Consider the one parameter class of exponential densities given by

$$f(\mathbf{y}|x) = u(\mathbf{y}) \exp(\pi(x) S(\mathbf{y}) - t(x)) \,. \tag{5.19}$$

Then, if $\pi(x)$ is a strictly monotone increasing function of x, the class of densities given by (5.19) satisfies the MLR criterion, so that the Karlin-Rubin theorem is applicable.

Example 5.3: Parameter test for exponential random variables

Consider a sequence $\{Y_k,\ 1 \leq k \leq N\}$ of i.i.d. exponential random variables with parameter $1/\theta$. The joint probability density of the observations can be expressed as

$$f(\mathbf{y}|\theta) = \frac{1}{\theta^N} \exp(-S(\mathbf{y})/\theta) \prod_{k=1}^{N} u(y_k) \,,$$

with

$$S(\mathbf{Y}) = \sum_{k=1}^{N} Y_k \,,$$

which is in the form (5.19), where $\pi(\theta) = -1/\theta$ is a monotone increasing function of θ, so that the test

$$S(\mathbf{Y}) \underset{H_0}{\overset{H_1}{\gtrless}} \eta$$

is UMP for testing $H_1 : \theta > \theta_0$ against $H_0 : \theta \leq \theta_0$. To find the size of the test in terms of η (or vice-versa), note that $S(\mathbf{Y})$ admits the Erlang density

$$f_S(s|\theta) = \frac{1}{\theta^N} \frac{s^{N-1}}{(N-1)!} \exp(-s/\theta) u(s) .$$

The size of the test can therefore be expressed as

$$\begin{aligned} \alpha &= P[S(\mathbf{Y}) \geq \eta | \theta_0] \\ &= \frac{1}{\theta_0^N} \int_\eta^\infty \frac{s^{N-1}}{(N-1)!} \exp(-s/\theta_0) ds \\ &= \left[1 + \frac{\eta}{\theta_0} + \ldots + \frac{1}{(N-1)!} \left(\frac{\eta}{\theta_0} \right)^{N-1} \right] \exp(-\eta/\theta_0) , \end{aligned}$$

where we have used integration by parts to evaluate the integral. Unfortunately, it is not possible to express the threshold η in closed form in a function of α, except when $N = 1$, in which case

$$\eta = \theta_0 \ln(1/\alpha) .$$

□

5.3 Invariant Tests

An important feature of many detection problems is that they may admit groups of transformations that leave the problem invariant. We recall that $\mathcal{G}$ defines a group if the composition $g_2 \circ g_1$ of two group elements belongs to $\mathcal{G}$ and if every group element g admits an inverse g^{-1}. Consider an observation vector $\mathbf{Y}$ whose probability distribution is parametrized by a vector $\mathbf{x} \in \mathcal{X}$. A transformation $\mathbf{Z} = \mathbf{g}(\mathbf{Y})$ where $\mathbf{g}$ belongs to a transformation group $\mathcal{G}$ defines a random vector $\mathbf{Z}$ with probability distribution

$$P[\mathbf{Z} \in A|\mathbf{x}] = P[\mathbf{g}(\mathbf{Y}) \in A|\mathbf{x}] , \tag{5.20}$$

where A denotes an arbitrary set. We say that the probability distribution of the observations is invariant under the group $\mathcal{G}$ if there exists an induced group $\tilde{\mathcal{G}}$ of transformations of the parameter set $\mathcal{X}$ such that for each transformation $\mathbf{Z} = \mathbf{g}(\mathbf{Y})$ in $\mathcal{G}$ there is a matching transformation $\boldsymbol{\xi} = \tilde{\mathbf{g}}(\mathbf{x})$ in $\tilde{\mathcal{G}}$ such that

$$P[\mathbf{g}(\mathbf{Y}) \in A|\mathbf{x}] = P[\mathbf{Y} \in \mathcal{A}|\tilde{\mathbf{g}}(\mathbf{x})] . \tag{5.21}$$

Furthermore, we say that the detection problem is invariant under $\mathcal{G}$ if the domains $\mathcal{X}_0$ and $\mathcal{X}_1$ of the parameter vector $\mathbf{x}$ under H_0 and H_1 are invariant under the transformations $\tilde{\mathbf{g}}$ of $\tilde{\mathcal{G}}$, i.e.,

$$\tilde{\mathbf{g}}(\mathcal{X}_j) = \mathcal{X}_j$$

for $j = 0,\ 1$. To fix ideas, let us consider some examples.

Example 5.4: Scale transformations

Suppose we are given observations

$$Y(k) = As(k) + V(k), \ 1 \leq k \leq N$$

of a known signal $s(k)$ with unknown amplitude A in a zero-mean WGN $V(k)$ with unknown variance σ^2 and we wish to test $H_1 : A \neq 0$ against $H_0 : A = 0$. Then, if we multiply the received observations by a scale factor $c > 0$, i.e.,

$$Z(k) = cY(k)$$

for $1 \leq k \leq N$, the group of scale transformations leaves the problem invariant, since to each such transformation corresponds a matching scale transformation

$$A' = cA \quad , \quad \sigma' = c\sigma$$

in parameter space where $A' \neq 0$ whenever $A \neq 0$. This transformation is illustrated in Fig. 5.1. Let

$$\begin{aligned}
\mathbf{s} &= \begin{bmatrix} s(1) \ s(2) \ \dots \ s(N) \end{bmatrix} \\
\mathbf{V} &= \begin{bmatrix} V(1) \ V(2) \ \dots \ V(N) \end{bmatrix} \\
\mathbf{Y} &= \begin{bmatrix} Y(1) \ Y(2) \ \dots \ Y(N) \end{bmatrix}
\end{aligned} \tag{5.22}$$

denote the N-dimensional vectors representing the signal, noise, and observations over the interval $[1, N]$. Then, as indicated in Fig. 5.1, the scale factor c defines a similarity transformation between the triangle specified by the vector relation

$$\mathbf{Y} = A\mathbf{s} + \mathbf{V}$$

and the one obtained by scaling the signal, noise and observations by c. Note that we are allowed to scale freely $A\mathbf{s}$ and $\mathbf{V}$ since the signal amplitude A is unknown and the noise standard deviation is unspecified. □

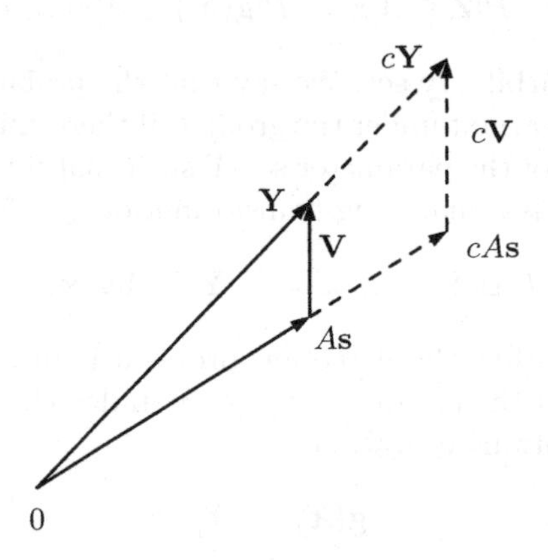

Fig. 5.1. Similarity transformation defined by a scale change by a factor c.

Example 5.5: Permutation transformation

Consider a sequence of observations

$$Y(k) = A\delta(k-\ell) + V(k)$$

with $1 \leq k \leq N$, where $\delta(\cdot)$ is the Kronecker delta function and ℓ denotes an unknown integer parameter between 1 and N. The amplitude A is unknown and $V(k)$ is a zero-mean WGN with known variance σ^2. We wish to test $H_1 : A \neq 0$ against $H_0 : A = 0$. Because of the presence of integer ℓ, the parameter sets under H_1 and H_0 are

$$\mathcal{X}_1 = \{(A,\ell) : A \neq 0, \ 1 \leq \ell \leq N\}$$

and

$$\mathcal{X}_0 = \{(A,\ell) : A = 0, \ 1 \leq \ell \leq N\}.$$

Roughly speaking, the detection problem consists of finding an impulse of size A at an unknown location in an observation segment of length N. Clearly, the location ℓ of the blip is a nuisance parameter, since it does not affect the outcome of the detection problem. To see this, we rewrite the observations in vector form as

$$\mathbf{Y} = A\mathbf{e}_\ell + \mathbf{V},$$

where the vectors $\mathbf{Y}$ and $\mathbf{V}$ are defined as in (5.22), and $\mathbf{e}_\ell$ denotes the ℓ-th unit vector in $\mathbb{R}^N$. Then let

$$\mathbf{Z}_\mathrm{C} = \begin{bmatrix} 0\,0 \ldots 0\,1 \\ 1\,0 \ldots 0\,0 \\ 0\,1 \ldots 0\,0 \\ \vdots \qquad \vdots \\ 0\,0 \ldots 1\,0 \end{bmatrix}$$

denote the cyclic shift matrix and consider the N-element group

$$\mathcal{G} = \{\mathbf{Z}_\mathrm{C}^m, \ 0 \leq m \leq N-1\}$$

formed by all possible cyclic shifts. Under the action of $\mathbf{Z}_\mathrm{C}^m$, the observation equation becomes

$$\mathbf{Y}' = \mathbf{Z}_\mathrm{C}^m \mathbf{Y} = A\mathbf{e}_{(\ell+m) \bmod N} + \mathbf{V}'$$

where $\mathbf{V}' = \mathbf{Z}_\mathrm{C}^m \mathbf{V}$ is $N(\mathbf{0}, \sigma^2 \mathbf{I}_N)$ distributed. So the action of $\mathbf{Z}_\mathrm{C}^m$ induces the parameter space transformation

$$A' = A \quad , \quad \ell' = (\ell + m) \bmod N$$

which leaves the detection problem invariant since it affects only the nuisance parameter ℓ. □

Example 5.6: Rotation transformation

Consider a two-dimensional observation vector

$$\mathbf{Y} = \begin{bmatrix} Y_c \\ Y_s \end{bmatrix} = A \begin{bmatrix} \cos(\theta) \\ \sin(\theta) \end{bmatrix} + \mathbf{V} , \tag{5.23}$$

where the amplitude A and phase θ are unknown, and $\mathbf{V}$ is a $N(\mathbf{0}, \sigma^2 \mathbf{I}_2)$ distributed vector in $\mathbb{R}^2$. We would like to test $H_1 : A \neq 0$ against $H_0 : A = 0$. The parameter sets under the two hypotheses are

$$\mathcal{X}_1 = \{(A, \theta) : \; A \neq 0, 0 \leq \theta < 2\pi\}$$

and

$$\mathcal{X}_0 = \{(A, \theta) : \; A = 0, 0 \leq \theta < 2\pi\} .$$

In this setup, the phase θ is a nuisance parameter since it does not affect the outcome of the detection problem. Consider the action of the group of rotation matrices represented by

$$\mathbf{Q}(\phi) = \begin{bmatrix} \cos(\phi) & -\sin(\phi) \\ \sin(\phi) & \cos(\phi) \end{bmatrix} , \; 0 \leq \phi < 2\pi$$

on the observation vector $\mathbf{Y}$. Let

$$\mathbf{S}(\theta) \triangleq \begin{bmatrix} \cos(\theta) \\ \sin(\theta) \end{bmatrix} .$$

As depicted in Fig. 5.2, the rotation $\mathbf{Q}(\phi)$ takes the signal vector $A\mathbf{S}(\theta)$ with angular coordinate θ located on a circle of radius A and moves it to the point

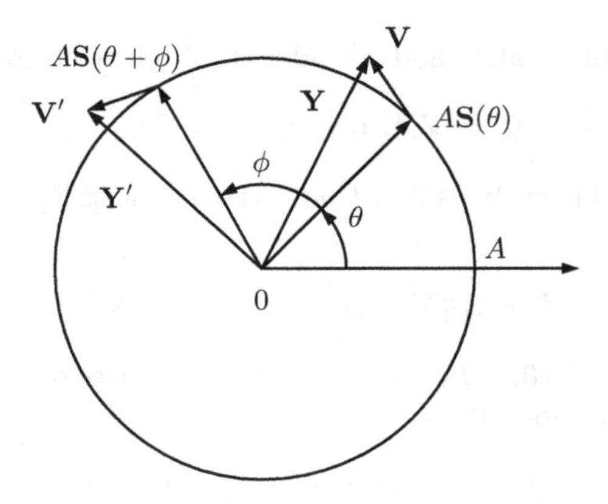

Fig. 5.2. Transformation of observation equation (5.23) under the action of a rotation by an angle ϕ.

$A\mathbf{S}(\theta+\phi)$ located on the same circle, but with angle $\theta+\phi$. Under $\mathbf{Q}(\phi)$, the noise and observation vectors become

$$\mathbf{V}' = \mathbf{Q}(\phi)\mathbf{V}$$

and

$$\mathbf{Y}' = \mathbf{Q}(\phi)\mathbf{Y} = A \begin{bmatrix} \cos(\theta+\phi) \\ \sin(\theta+\phi) \end{bmatrix} + \mathbf{V}' . \tag{5.24}$$

The covariance matrix of the transformed noise $\mathbf{V}'$ is given by

$$E[\mathbf{V}'\mathbf{V}'^T] = \mathbf{Q}^T(\phi)E[\mathbf{V}\mathbf{V}^T]\mathbf{Q}(\phi) = \sigma^2\mathbf{I}_2 ,$$

where we have used the orthonormality of $\mathbf{Q}(\phi)$. Thus, $\mathbf{V}'$ admits the same $N(\mathbf{0}, \sigma^2\mathbf{I}_2)$ distribution as $\mathbf{V}$, so the transformed observation equation (5.24) has exactly the same structure as the original observation (5.23). This shows that the group of rotations leaves the observations invariant and equation (5.24) indicates that the parameter space transformation corresponding to $\mathbf{Q}(\phi)$ is given by

$$A' = A \quad , \quad \theta' = (\theta+\phi) \mod 2\pi .$$

This transformation leaves $\mathcal{X}_j$ invariant, hence rotations form an invariant group for the detection problem. □

It is worth noting that groups of invariant transformations are not always apparent and usually require looking carefully at the problem at hand until they can be identified. There is no truly systematic way of guessing what transformations will leave the problem invariant. So finding groups of invariant transformations is usually a skill that is developed more by experience than by deep mathematical insight. To perform the search for invariant transformations, it is usually recommended to represent the signal, noise, and observations in vector form and then check if the detection problem is left unchanged by simple geometric transformations such as similarities, translations, rotations, or symmetries. This search can be simplified somewhat if the noise vector $\mathbf{V} \in \mathbb{R}^N$ is additive and $N(\mathbf{0}, \sigma^2\mathbf{I}_N)$ distributed, like in Example 5.6. This class of noise vectors is important, since it corresponds to the representation of zero-mean WGN with a constant variance. Then $\mathbf{V}$ has the feature that its probability distribution remains the same in any orthonormal coordinate system. Specifically, if

$$\mathbf{V}' = \mathbf{Q}\mathbf{V}$$

where $\mathbf{Q}$ denotes an abitrary orthonormal matrix, then $\mathbf{V}'$ is still $N(\mathbf{0}, \sigma^2\mathbf{I}_N)$ distributed. Accordingly, if we consider a detection problem with additive $N(\mathbf{0}, \sigma^2\mathbf{I}_N)$ noise, when the variance σ^2 is known, the search for invariant transformations can be restricted to orthonormal transformations. When σ^2 is unknown, one should look for the composition of a scale transformation and an orthonormal transformation.

Having characterized the group invariance property, we must now try to take advantage of it for detection purposes. To do so, we will rely on the following concept:

Maximal invariant statistics and invariant tests: When $\mathbf{Y}$ admits an invariant distribution under $\mathcal{G}$, we say that a statistic $S(\mathbf{Y})$ is $\mathcal{G}$-invariant if

$$\mathbf{S}(\mathbf{g}(\mathbf{Y})) = \mathbf{S}(\mathbf{Y}) \tag{5.25}$$

for all $\mathbf{g} \in \mathcal{G}$. The statistic $S(\mathbf{Y})$ is a *maximal invariant* if it satisfies (5.25) and if

$$\mathbf{S}(\mathbf{Y}_1) = \mathbf{S}(\mathbf{Y}_2) \Longrightarrow \mathbf{Y}_2 = \mathbf{g}(\mathbf{Y}_1) \text{ for some } \mathbf{g} \in \mathcal{G} \,. \tag{5.26}$$

Then a test $\delta(\mathbf{Y})$ is invariant if it can be expressed as a function of a maximal invariant statistic, i.e.,

$$\delta(\mathbf{Y}) = \delta(\mathbf{S}(\mathbf{Y}))\,. \tag{5.27}$$

The purpose of considering invariant tests for detection problems that admit a group of invariant transformations is that in situations where no UMP test exists, it is sometimes possible to find an optimal test in the smaller class of invariant tests.

UMP Invariant Test: An invariant test δ_{UMPI} such that

$$\max_{\mathbf{x} \in \mathcal{X}_0} P_F(\delta_{\mathrm{UMPI}}, \mathbf{x}) \leq \alpha \tag{5.28}$$

is a uniformly most powerful invariant (UMPI) test of size α for testing $H_0 : \mathbf{x} \in \mathcal{X}_0$ against $H_1 : \mathbf{X} \in \mathcal{X}_1$ if for all invariant tests δ of size α it satisfies

$$P_D(\delta_{\mathrm{UMPI}}, \mathbf{x}) \geq P_D(\delta, \mathbf{x}) \tag{5.29}$$

for all $\mathbf{x} \in \mathcal{X}_1$.

So a UMPI test has uniformly higher power over the parameter set $\mathcal{X}_1$ than any other *invariant* test. In other words, by restricting our comparison to the smaller class of invariant tests, it is sometimes possible to exhibit a UMPI test in situations where no UMP test exists. This approach is very reasonable, since it is very natural to seek detectors that respect the invariances of the problem, and it turns out that there are important classes of detection problems for which a UMPI exists even though no UMP test is available. We consider two examples below:

Example 5.7: Signal with unknown amplitude in noise

Consider the case where we observe a signal

$$Y(k) = As(k) + V(k) \tag{5.30}$$

for $1 \leq k \leq N$ where the signal $s(k)$ is known, but the amplitude A is unknown, and $V(k)$ is a zero-mean WGN with variance σ^2. It was shown in

Example 5.2 that when σ^2 is known, (5.9) is a UMP test for choosing between $H_1 : A > 0$ and $H_0 : A \leq 0$. However, there is no UMP test when either the test is two-sided, or σ^2 is unknown. We prove here that UMPI tests exist for the following three cases:

(i) σ^2 known, $H_1 : A \neq 0$, $H_0 : A = 0$.
(ii) σ^2 unknown, $H_1 : A > 0$, $H_0 : A \leq 0$.
(iii) σ^2 unknown, $H_1 : A \neq 0$, $H_0 : A = 0$.

As starting point, note that if $\mathbf{Y}$ and $\mathbf{s}$ denote the N-dimensional vectors formed by writing the observations and signal in vector form, the probability density of the observations is given by (5.3), which can be rewritten as

$$\begin{aligned} f(\mathbf{Y}|A,\sigma^2) &= \frac{1}{(2\pi\sigma^2)^{N/2}} \exp\big(-\frac{1}{2\sigma^2}||\mathbf{Y} - A\mathbf{s}||_2^2\big) \\ &= c(A,\sigma^2)\exp\big(-\frac{1}{2\sigma^2}||\mathbf{Y}||_2^2 + \frac{A}{\sigma^2}\mathbf{s}^T\mathbf{Y}\big)\,, \end{aligned} \tag{5.31}$$

where $c(A, \sigma^2)$ is a constant depending only on A and σ^2. So when both A and σ^2 are unknown,

$$\mathbf{T}(\mathbf{Y}) = \begin{bmatrix} \mathbf{s}^T\mathbf{Y}/E^{1/2} \\ ||\mathbf{Y}||_2 \end{bmatrix} \tag{5.32}$$

is a sufficient for the detection problem, where $E = ||\mathbf{s}||_2^2$ denotes the energy of signal $\mathbf{s}$. But when σ^2 is known, only the first component

$$T_1(\mathbf{Y}) = \mathbf{s}^T\mathbf{Y}/E^{1/2} \tag{5.33}$$

of $\mathbf{T}(\mathbf{Y})$ is a sufficient statistic.

Case 1: Assume that σ^2 is known and we need to decide between $H_1 : A \neq 0$ and $H_0 : A = 0$. Consider the two element group of transformations $\mathcal{G}_m$ formed by the identity (no transformation) and the symmetry with respect to the $n-1$ subspace $\mathbf{S}^\perp$ orthogonal to the vector $\mathbf{s}$. This symmetry is represented by the orthonormal symmetric matrix

$$\mathbf{Q} = \mathbf{I}_N - \frac{2}{E}\mathbf{s}\mathbf{s}^T\,.$$

It is easy to verify that $\mathbf{Q}$ maps $\mathbf{s}$ into $-\mathbf{s}$, but leaves all vectors orthogonal to $\mathbf{s}$ invariant. The effect of $\mathbf{Q}$ on an arbitary vector $\mathbf{Y}$ is illustrated in part (a) of Fig. 5.3. The component

$$\mathbf{Y}^{\|} = \frac{\mathbf{s}^T\mathbf{Y}}{E}\mathbf{s}$$

of $\mathbf{Y}$ which is colinear with $\mathbf{s}$ is mapped into mirror image $-\mathbf{Y}^{\|}$, whereas the component

$$\mathbf{Y}^{\perp} = \mathbf{Y} - \mathbf{Y}^{\|}$$

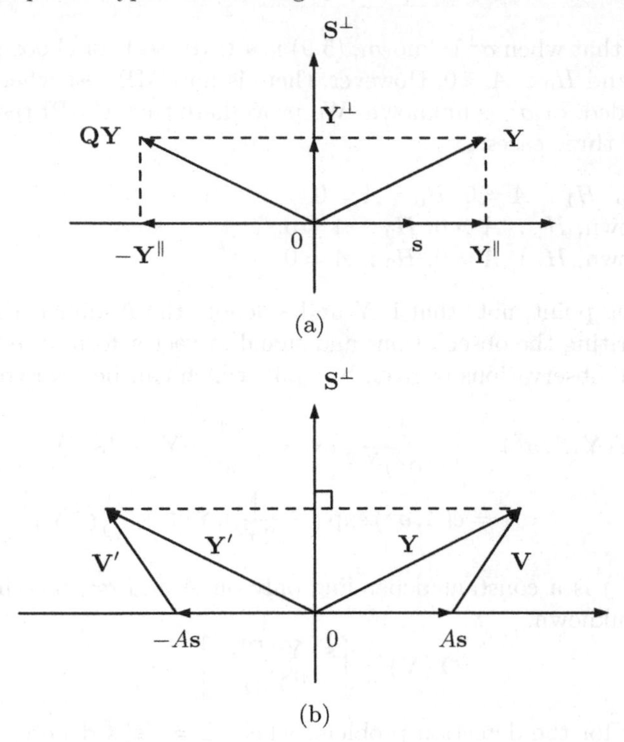

Fig. 5.3. Effect of the symmetry transformation $\mathbf{Q}$ on: (a) an arbitray vector $\mathbf{Y}$; (b) the signal plus noise representation of the observation vector.

located in $\mathbf{S}^{\perp}$ is unchanged. As shown in part (b) of Fig. 5.3, under the transformation $\mathbf{Q}$, the observation vector becomes

$$\mathbf{Y}' = \mathbf{QY} = -A\mathbf{s} + \mathbf{V}' \tag{5.34}$$

where $\mathbf{V}' = \mathbf{QV}$ is $N(\mathbf{0}, \sigma^2 \mathbf{I}_N)$ distributed, so the parameter space transformation induced by (5.32) takes the form

$$A' = -A \ , \ \ \sigma' = \sigma \ ,$$

which leaves the detection problem invariant.
Under the transformation (5.34), the sufficient statistic $T_1(\mathbf{Y})$ becomes

$$T_1(\mathbf{Y}') = \mathbf{s}^T \mathbf{Q}_s \mathbf{Y}/E^{1/2} = -\mathbf{s}^T \mathbf{Y}/E^{1/2} \ ,$$

so that

$$Z(\mathbf{Y}) = |\mathbf{s}^T \mathbf{Y}|/E^{1/2} \tag{5.35}$$

is a maximal invariant statistic under $\mathcal{G}_m$. To determine whether a UMPI test exists, we need to evaluate the probability densities of $T_1(\mathbf{Y})$ and $Z(\mathbf{Y})$. Since

$\mathbf{Y} \sim N(A\mathbf{s}, \sigma^2 \mathbf{I}_N)$ and $T_1(\mathbf{Y})$ is just a linear combination of the components of $\mathbf{Y}$, $T_1(\mathbf{Y})$ admits the $N(AE^{1/2}, \sigma^2)$ probability density

$$f_{T_1}(t_1|A) = \frac{1}{(2\pi\sigma^2)^{1/2}} \exp\big(- \frac{1}{2\sigma^2}(t_1 - AE^{1/2})^2 \big) .$$

The density of the maximal invariant statistic $Z(\mathbf{Y})$ is therefore given by

$$\begin{aligned} f_Z(z|A) &= f_{T_1}(z|A) + f_{T_1}(-z|A) \\ &= \big(\frac{2}{\pi\sigma^2}\big)^{1/2} \exp\big(- \frac{z^2 + A^2 E}{2\sigma^2}\big) \cosh\big(\frac{|A|Ez}{\sigma^2}\big) u(z) , \end{aligned} \tag{5.36}$$

where $u(\cdot)$ denotes the unit step function. With $A \neq 0$, the likelihood ratio

$$\begin{aligned} L(z|A) &= \frac{f_Z(z|A)}{f_Z(z|A=0)} \\ &= \exp\big(- \frac{A^2 E}{2\sigma^2}\big) \cosh\big(\frac{|A|Ez}{\sigma^2}\big) \end{aligned}$$

is a monotone increasing function of $z \geq 0$, so by applying the monotone likelihood ratio criterion, we conclude that the test

$$Z(\mathbf{Y}) = |\mathbf{s}^T \mathbf{Y}|/E^{1/2} \underset{H_0}{\overset{H_1}{\gtrless}} \eta \tag{5.37}$$

is UMPI under the group of transformations $\mathcal{G}_m$. To select the threshold η, we note that under $H_0: \; A = 0$, $Z(\mathbf{Y})$ admits the density

$$f_Z(z|A=0) = \big(\frac{2}{\pi\sigma^2}\big)^{1/2} \exp\big(- \frac{z^2}{2\sigma^2}\big) u(z) ,$$

so that

$$\alpha = P[Z > \eta | A = 0] = 2Q(\eta/\sigma) , \tag{5.38}$$

or equivalently

$$\eta = \sigma Q^{-1}(\alpha/2) . \tag{5.39}$$

With this choice of threshold, the power of the test is given by

$$\begin{aligned} P_D^{\mathrm{UMPI}}(A) &= P[Z > \eta | A] = \int_\eta^\infty f_Z(z|A) dz \\ &= \frac{1}{(2\pi\sigma^2)} \int_\eta^\infty \exp\left(\frac{(z - |A|E^{1/2})^2}{2\sigma^2} \right) dz \\ &\quad + \frac{1}{(2\pi\sigma^2)} \int_\eta^\infty \exp\left(\frac{(z + |A|E^{1/2})^2}{2\sigma^2} \right) dz \\ &= Q\Big(\frac{\eta - |A|E^{1/2}}{\sigma} \Big) + Q\Big(\frac{\eta + |A|E^{1/2}}{\sigma} \Big) \\ &= Q\Big(Q^{-1}\big(\frac{\alpha}{2}\big) - \frac{|A|E^{1/2}}{\sigma} \Big) + Q\Big(Q^{-1}\big(\frac{\alpha}{2}\big) + \frac{|A|E^{1/2}}{\sigma} \Big) . \end{aligned} \tag{5.40}$$

Case 2: Assume now that σ^2 is unknown and we want to test $H_1 : A > 0$ against $H_0 : A \leq 0$. From Example 5.4, we know that the problem is scale invariant. So starting from the sufficient statistic $\mathbf{T}(\mathbf{Y})$ given by (5.32), we need to find a maximal invariant statistic under the group of scale transformations $\mathcal{G}_s$. The statistic

$$S(\mathbf{Y}) = \frac{\mathbf{s}^T\mathbf{Y}}{E^{1/2}||\mathbf{Y}||_2} \tag{5.41}$$

is what we are looking for, since it is unchanged when we replace $\mathbf{Y}$ by $c\mathbf{Y}$ (or equivalently, $\mathbf{T}$ by $c\mathbf{T}$) with $c > 0$, and since $S_1 = S_2$ implies

$$\mathbf{T}_1 = c\mathbf{T}_2$$

with

$$c = ||\mathbf{Y}_2||_2/||\mathbf{Y}_1||_2 \,.$$

To determine whether a UMPI test exists we need to evaluate the probability densities of $\mathbf{T}(\mathbf{Y})$ and $S(\mathbf{Y})$. To do so, it is convenient to change the coordinate system so as to make the vector $\mathbf{s}$ colinear with the first unit vector $\mathbf{e}_1$ of $\mathbb{R}^N$. The observation

$$\mathbf{Y} = A\mathbf{s} + \mathbf{V}$$

becomes

$$\mathbf{Y}' = AE^{1/2}\mathbf{e}_1 + \mathbf{V}'$$

where $\mathbf{V}'$ is $N(\mathbf{0}, \sigma^2\mathbf{I}_N)$ since, as noted earlier, the distribution of a zero-mean Gaussian vector whose covariance matrix is proportional to the identity is not affected by the choice of coordinate system. Note that the first coordinate of $\mathbf{Y}'$ can be expressed in terms of the original observations as

$$Y_1' = \mathbf{s}^T\mathbf{Y}/E^{1/2} = T_1(\mathbf{Y}) \,.$$

We can then express $\mathbf{Y}'$ in a system of hyperspherical coordinates as

$$\mathbf{Y}' = \begin{bmatrix} Y_1' \\ Y_2' \\ \vdots \\ Y_k' \\ \vdots \\ Y_{N-1}' \\ Y_N' \end{bmatrix} = R \begin{bmatrix} \cos(\Phi_1) \\ \sin(\Phi_1)\cos(\Phi_2) \\ \vdots \\ (\prod_{j=1}^{k-1}\sin(\Phi_j))\cos(\Phi_k) \\ \vdots \\ (\prod_{j=1}^{N-2}\sin(\Phi_j))\cos(\Theta) \\ (\prod_{j=1}^{N-2}\sin(\Phi_j))\sin(\Theta) \end{bmatrix} , \tag{5.42}$$

where the polar angles $\Phi_k \in [0, \pi]$ and the azimuthal angle $\Theta \in [0, 2\pi)$. Note that in this coordinate system $R = ||\mathbf{Y}||_2$ and

$$\cos(\Phi_1) = \frac{Y_1'}{||\mathbf{Y}||_2} = S(\mathbf{Y}) \,,$$

where $S(\mathbf{Y})$ is the statistic (5.41). The Jacobian of the transformation (5.42) is given by

$$J = r^{N-1} \prod_{k=1}^{N-1} (\sin(\phi_k))^{N-1-k}$$

so that in hyperspherical coordinates, the probability density of the observations can be expressed as

$$\begin{aligned} & f(r, \theta, \phi_k,\ 1 \leq k \leq N-2 | A, \sigma^2) \\ & = \frac{r^{N-1}}{(2\pi\sigma^2)^{N/2}} \prod_{k=1}^{N-1} (\sin(\phi_k))^{N-1-k} \exp\big(-\frac{r^2 + A^2 E}{2\sigma^2}\big) \exp\big(\frac{A E^{1/2} r \cos(\phi_1)}{\sigma^2}\big) . \end{aligned}$$

Just as expected

$$\mathbf{T}(\mathbf{Y}) = \begin{bmatrix} R\cos(\Phi_1) \\ R \end{bmatrix}$$

is a sufficient statistic for this density. Integrating out all the angular variables except the first, the joint density of R and Φ_1 takes the form

$$\begin{aligned} & f(r, \phi_1 | A, \sigma^2) = \\ & B r^{N-1} (\sin(\phi_1))^{N-2} \exp\big(-\frac{r^2 + A^2 E}{2\sigma^2}\big) \exp\big(\frac{A E^{1/2} r \cos(\phi_1)}{\sigma^2}\big) , \end{aligned} \tag{5.43}$$

where B denotes a constant, and the marginal density of angle Φ_1 is given by

$$\begin{aligned} f(\phi_1 | A, \sigma^2) &= B (\sin(\phi_1))^{N-2} \\ & \cdot \int_0^\infty r^{N-1} \exp\big(-\frac{r^2 + A^2 E}{2\sigma^2}\big) \exp\big(\frac{A E^{1/2} r \cos(\phi_1)}{\sigma^2}\big) dr . \end{aligned} \tag{5.44}$$

For $A_1 > 0$ and $A_0 \leq 0$, the likelihood ratio

$$L(\phi_1 | A_1, \sigma_1^2, A_0, \sigma_0^2) = \frac{f(\phi_1 | A_1, \sigma_1^2)}{f(\phi_1 | A_0, \sigma_0^2)}$$

is a monotone increasing function of $\cos(\phi_1)$, so by applying the monotone likelihood ratio criterion, we conclude that under the group of scale transformations, there exists a UMPI test for testing $H_0 : A > 0$ against $H_1 :\ A \leq 0$. This test takes the form

$$S(\mathbf{Y}) = \cos(\Phi_1) \underset{H_0}{\overset{H_1}{\gtrless}} \eta , \tag{5.45}$$

where $\cos(\Phi_1)$ is a maximal invariant statistic under scale transformations after the detection problem has been expressed in terms of the sufficient statistic $\mathbf{T}$. To compute the size of the test, observe that for $A = 0$, Φ_1 admits the density

$$f(\phi_1|A=0,\sigma^2) = \begin{cases} \dfrac{\Gamma(N/2)}{\pi^{1/2}\Gamma((N-1)/2)}(\sin(\phi_1))^{N-2} & \text{for } 0 \leq \phi_1 \leq \pi \\ 0 & \text{otherwise} , \end{cases} \tag{5.46}$$

which is independent of σ^2. This implies

$$\begin{aligned} \alpha &= P[\cos(\Phi_1) > \eta | A = 0] \\ &= \frac{\Gamma(N/2)}{\pi^{1/2}\Gamma((N-1)/2)} \int_0^{\cos^{-1}(\eta)} (\sin(\phi_1))^{N-2} d\phi_1 , \end{aligned} \tag{5.47}$$

which can be evaluated by integration by parts. The power of the test can be expressed as

$$P_D^{\mathrm{UMPI}}(A,\sigma^2) = \int_0^{\cos^{-1}(\eta)} f(\phi_1|A,\sigma^2) d\phi_1 ,$$

where the density of Φ_1 is given by (5.44). Unfortunately, $P_D^{\mathrm{UMPI}}(A,\sigma^2)$ does not admit a simple closed form expression, so it must usually be evaluated by Monte-Carlo simulation.

Case 3: Suppose now that σ^2 is unknown and that we want to test $H_1 : A \neq 0$ against $H_0 : A = 0$. Then the detection problem is invariant under the group $\mathcal{G}_m \times \mathcal{G}_s$ obtained by composing symmetries with respect to the hyperplane orthonormal to $\mathbf{s}$ and scale transformations. Under this larger group of transformations, a maximal invariant statistic is now given by

$$|\cos(\Phi_1)| = |S(\mathbf{Y})| = \frac{|\mathbf{s}^T\mathbf{Y}|}{E^{1/2}||\mathbf{Y}||_2} . \tag{5.48}$$

Let

$$\check{\Phi}_1 = \begin{cases} \Phi_1 & \text{for } 0 \leq \Phi_1 \leq \pi/2 \\ \pi - \Phi_1 & \text{for } \pi/2 < \Phi_1 < \pi . \end{cases}$$

Then $\cos(\check{\Phi}_1) = |\cos(\Phi_1)|$ and the density of $\check{\Phi}_1$ is given by

$$\begin{aligned} f(\check{\phi}_1|A,\sigma^2) &= f(\phi_1|A,\sigma^2) + f(\pi - \phi_1|A,\sigma^2) = 2B(\sin(\check{\phi}_1))^{N-2} \\ &\cdot \int_0^\infty r^{N-1} \exp\Big(-\frac{r^2 + A^2E}{2\sigma^2}\Big) \cosh\Big(\frac{|A|E^{1/2} r\cos(\check{\phi}_1)}{\sigma^2}\Big) dr \end{aligned} \tag{5.49}$$

for $0 \leq \check{\phi}_1 \leq \pi/2$. Since it is monotone increasing in $|\cos(\phi_1)|$, by applying the monotone likelihood ratio criterion, we find that under the product group $\mathcal{G}_m \times \mathcal{G}_s$, the test

$$|S(\mathbf{Y})| = |\cos(\Phi_1)| \underset{H_0}{\overset{H_1}{\gtrless}} \eta \tag{5.50}$$

is UMPI for testing H_1 against H_0. □

Remark: The tests (5.45) and (5.50) obtained in Cases 2 and 3 for detecting a signal with unknown amplitude in a noise of unknown variance σ^2 constitute examples of *constant false-alarm rate* (CFAR) detectors [4, Chap. 8], [5]. It was shown in Example 5.2 that the matched filter statistic $\mathbf{s}^T\mathbf{Y}/E^{1/2}$ used to implement the optimum Neyman-Pearson test (5.9) has a variance proportional to σ^2 under the null hypothesis, so when σ^2 is unknown we cannot adjust the test threshold η to ensure that the false alarm rate stays below α. The angular statistic $\cos(\Phi_1)$ used to implement the CFAR detector has the feature that under the null hypothesis, its probability density (5.44) is independent of the noise variance σ^2, which allows the selection of a threshold that guarantees that the false alarm rate will be α (and thus constant) independently of the noise variance.

Example 5.7, continued

To illustrate more concretely Example 5.7, consider the case where in model (5.30) we have

$$s(k) = \frac{1}{2}\cos(k\pi/4)$$

and $N = 8$. Thus the sinusoidal signal $s(k)$ is observed over a single period, and it has been normalized so that $E = 1$. We examine Case 1 where A is unknown, but σ is known, with the value $\sigma = 1$. The UMPI test specified by (5.37) and (5.39) takes the form

$$Z = \frac{1}{2}|\sum_{k=1}^{8} Y(k)\cos(k\pi/4)| \underset{H_0}{\overset{H_1}{\gtrless}} Q^{-1}(\alpha/2)\,.$$

We compare it to the NP test (5.9), (5.10) which can be expressed here as

$$S = \frac{\mathrm{sgn}(A)}{2}\sum_{k=1}^{8} Y(k)\cos(k\pi/4) \underset{H_0}{\overset{H_1}{\gtrless}} Q^{-1}(\alpha)\,,$$

where sign(A) denotes the sign of A. Note that the only piece of information that is needed to be able to implement the NP test instead of the UMPI test is the sign of A, not its actual value. The power of the UMPI test is given by (5.40) with $E = \sigma = 1$, whereas for the NP test it is given by setting $\sigma = 1$ in

$$P_D^{\mathrm{NP}}(A) = Q\big(Q^{-1}(\alpha) - \frac{|A|}{\sigma}\big) = 1 - Q\big(\frac{|A|}{\sigma} - Q^{-1}(\alpha)\big)\,. \tag{5.51}$$

The receiver operating characteristic (ROC) of the UMPI and NP tests is shown in Fig. 5.4 for $A = 2.5$. This indicates that the apparently small difference in information corresponding to the knowledge of the sign of A results in a significant performance difference between the two tests. Note, of course, that since $\sigma = 1$, for larger values of A, the performance difference between the two tests decreases. This is confirmed by Fig. 5.5 where the power of the NP and UMPI tests is plotted for $\alpha = P_F = 0.1$ as A varies over the interval $[-4, 4]$. For large values of A, the power of the two tests approaches 1, but

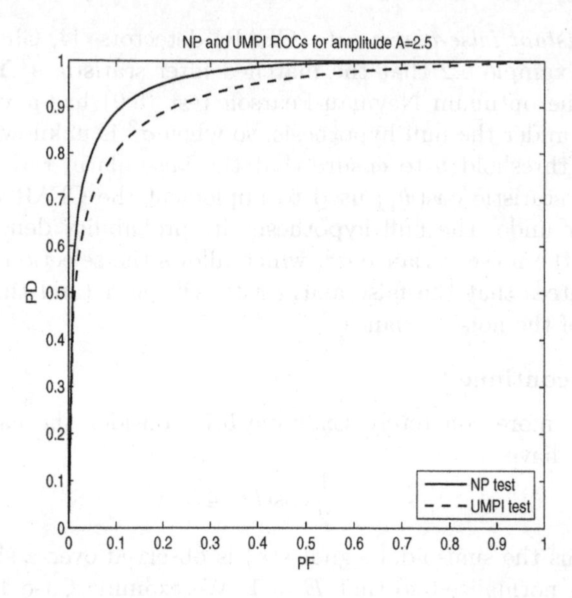

Fig. 5.4. Receiver operating characteristic for the NP and UMPI tests for a signal of amplitude $A = 2.5$.

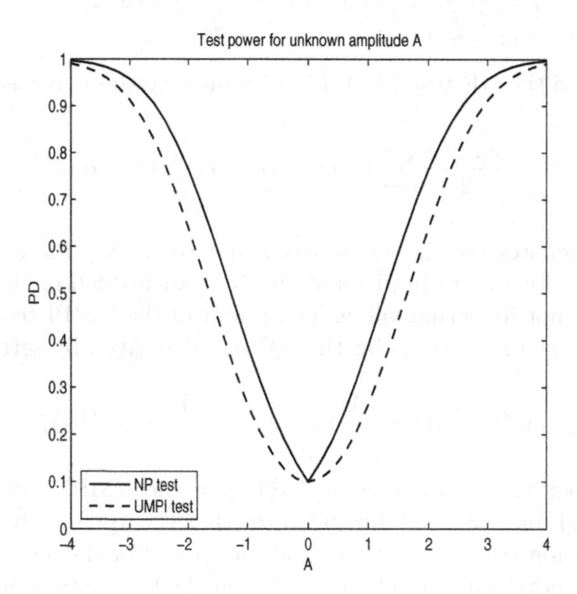

Fig. 5.5. Power of the NP and UMPI tests designed for $P_F = 0.1$, as the signal amplitude varies over $[-4, 4]$.

for small values of A the NP test significantly outperforms the UMPI test. In the literature, the NP test is often referred to as a clairvoyant test, since it assumes all the unknown parameters are known. Although this test cannot be implemented, it establishes a benchmark for all other tests and it is helpful in quantifying the value of knowing certain parameters for test design purposes.

□

Example 5.6, continued

Consider an observation vector of the form (5.23) with A and θ unknown, for which we wish to test $H_1 : A > 0$ against $H_0 : A = 0$. The variance σ^2 of the noise is known. We have seen that this detection problem is invariant under the group of rotations. Then

$$R = ||\mathbf{Y}||_2 = (Y_c^2 + Y_s^2)^{1/2} \tag{5.52}$$

is an invariant statistic since a rotation does not affect the Euclidean norm. It is also maximal since when two observation vectors $\mathbf{Y}_1$ and $\mathbf{Y}_2$ satisfy $||\mathbf{Y}_1||_2 = ||\mathbf{Y}_2||_2$, they are located on the same circle centered at the origin and can be obtained from each other by rotation. The observation components Y_c and Y_s can be expressed in polar coordinates as

$$Y_c = R\cos(\Phi) \quad , \quad Y_s = R\sin(\Phi) \, ,$$

with $\Phi = \tan^{-1}(Y_s/Y_c) \in [0, 2\pi)$. In this system of coordinates the joint density of R and Φ can be expressed as

$$f(r, \phi|A, \theta) = \frac{r}{2\pi\sigma^2} \exp\big(-\frac{r^2 + A^2}{2\sigma^2}\big) \exp\big(\frac{Ar}{\sigma^2}\cos(\phi - \theta)\big) \, .$$

The marginal density of the invariant statistic R is therefore given by

$$\begin{aligned} f(r|A) &= \int_0^{2\pi} f(r, \phi|A, \theta) d\phi \\ &= \frac{r}{\sigma^2} \exp\big(-\frac{r^2 + A^2}{2\sigma^2}\big) I_0(Ar/\sigma^2) \, , \end{aligned} \tag{5.53}$$

where the modified Bessel function of zero-th order $I_0(\cdot)$ is defined by

$$I_0(z) = \frac{1}{2\pi} \int_0^{2\pi} \exp(z\cos(\psi)) d\psi \, .$$

Note that $I_0(z)$ is a monotone increasing function of $z > 0$. As expected, since R is rotation invariant, its density $f(r|A)$ is independent of the unknown parameter θ. Since the likelihood ratio

$$L(r|A) = \frac{f(r|A)}{f(r|A = 0)} = \exp(-\frac{A^2}{2\sigma^2}) I_0(Ar/\sigma^2)$$

is a monotone increasing function of r for $A > 0$, we conclude a UMPI test exists. This test is given by

$$R = ||\mathbf{Y}||_2 \underset{H_0}{\overset{H_1}{\gtrless}} \eta \,, \tag{5.54}$$

where η is selected such that

$$\begin{aligned}\alpha &= P[R \geq \eta | A = 0] \\ &= \int_\eta^\infty \frac{r}{\sigma^2} \exp\big(-\frac{r^2}{2\sigma^2}\big) dr = \exp\big(-\frac{\eta^2}{2\sigma^2}\big) \,,\end{aligned}$$

which gives

$$\eta = \sigma(2\ln(1/\alpha))^{1/2} \,. \tag{5.55}$$

The power of the test is then given by

$$\begin{aligned}P_D^{\text{UMPI}}(A) &= P[R > \eta | A] \\ &= \int \frac{r}{\sigma^2} \exp\big(-\frac{r^2 + A^2}{2\sigma^2}\big) I_0(rA/\sigma^2) dr \,.\end{aligned} \tag{5.56}$$

It cannot be evaluated in closed form, but it can be expressed in terms of Marcum's Q function

$$Q(\alpha, \beta) \triangleq \int_\beta^\infty z \exp\big(-\frac{z^2 + \alpha^2}{2}\big) I_0(\alpha z) dz \,, \tag{5.57}$$

where α^2 is called the noncentrality parameter. Then, by performing the change of variable $z = r/\sigma$ in (5.56), we find

$$P_D(A) = Q\big(\frac{A}{\sigma}, \frac{\eta}{\sigma}\big) \,. \tag{5.58}$$

Note that even though the phase θ is unknown, the power of the UMPI test is independent of θ, since as noted earlier, the density (5.53) of the test statistic R is independent of θ.

To evaluate the performance of the UMPI test (5.53), we compare it to the Neyman-Pearson test

$$\text{sign}(A)[Y_c \cos(\theta) + Y_s \sin(\theta)] \underset{H_0}{\overset{H_1}{\gtrless}} \eta$$

with $\eta = \sigma Q^{-1}(\alpha)$ for the testing problem where under H_1 the observation vector $\mathbf{Y}$ is given by (5.23) with A and θ known, and under H_0, $A = 0$. The power of this test takes again the form (5.51). For $A = 3$ and $\sigma = 1$, the ROCs of the NP and UMPI tests are plotted in Fig. 5.6. Note again that there is an appreciable performance difference between the two tests. The power of the NP and UMPI tests is shown in Fig. 5.7 for $\sigma = 1$ and

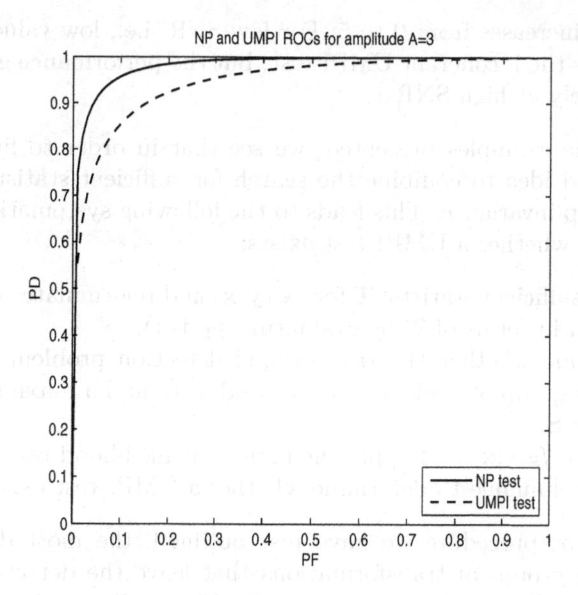

Fig. 5.6. Receiver operating characteristic for the NP and UMPI tests with $A = 3$ and θ arbitrary.

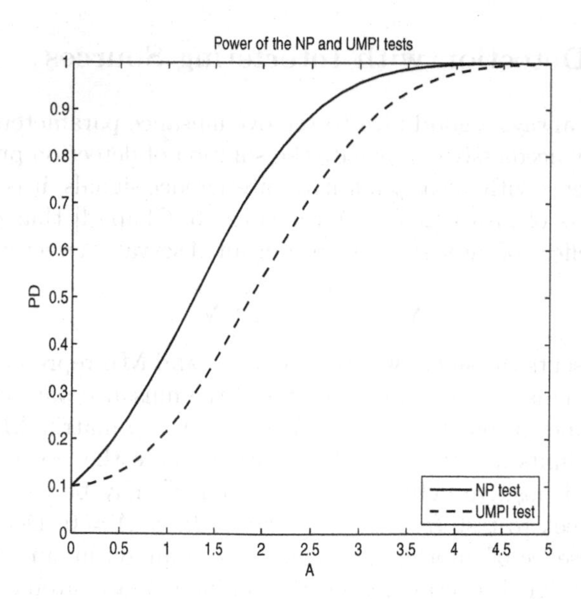

Fig. 5.7. Power of the NP and UMPI tests designed for $P_F = 0.1$, as the amplitude A varies over 0 to 5.

$P_F = 0.1$ as A increases from 0 to 5. For low SNR, i.e., low values of A, the NP outperforms the incoherent UMPI test, but the performance gap vanishes almost completely at high SNR. □

Based on the examples presented, we see that in order to find a UMPI test, it is a good idea to combine the search for sufficient statistics with an analysis of group invariance. This leads to the following systematic procedure for determining whether a UMPI test exists:

Step 1. Find a sufficient statistic $\mathbf{T}$ for $f_{\mathbf{Y}}(\mathbf{y}|\mathbf{x})$ and reformulate the detection problem in terms of $\mathbf{T}$ by evaluating $f_{\mathbf{T}}(\mathbf{t}|\mathbf{x})$.
Step 2. Determine whether the transformed detection problem is invariant under a group of transformations, and if so, find a maximal invariant statistic $\mathbf{S}$.
Step 3. Evaluate $f_{\mathbf{S}}(\mathbf{s}|\mathbf{x})$ and apply the monotone likelihood ratio test or any other technnique to determine whether a UMPI test exists.

In the 3-steps procedure we have just outlined, the most difficult part involves finding groups of transformations that leave the detection problem invariant. As indicated by Examples 5.4–5.6, this usually involves writing the observations in vector form and then identifying a combination of classical geometric transformations that preserve the structure of detection problem.

5.4 Linear Detection with Interfering Sources

In general, it is always a good idea to remove nuisance parameters as soon as possible, as they needlessly complicate the solution of detection problems. For detection problems with linearly additive interference signals, it is not difficult to exhibit a class of invariant transformations [6, Chap. 4] that can be used to remove the effect of interferers. Consider an observation vector $\mathbf{Y} \in \mathbb{R}^N$ of the form

$$\mathbf{Y} = A\mathbf{s} + \mathbf{M}\mathbf{n} + \mathbf{V} \tag{5.59}$$

where $A\mathbf{s}$ represents the signal we seek to detect and $\mathbf{M}\mathbf{n}$ represents the effect of interference signals. The amplitude A and the nuisance parameter vector $\mathbf{n} \in \mathbb{R}^p$ are unknown, but the vector $\mathbf{s}$ and the $N \times p$ matrix $\mathbf{M}$ are known. The noise $\mathbf{V}$ admits a $N(\mathbf{0}, \sigma^2\mathbf{K})$ distribution where the covariance matrix $\mathbf{K}$ is known and positive definite, but the gain σ^2 may be either known or unknown. We seek to test $H_1 : A \neq 0$ against $H_0 : A = 0$. Detection problems in the presence of interference signals are common in surveillance since electromagnetic signals containing a source of interest are often contaminated by a variety of other users sharing the the same bandwidth. Without loss of generality we can assume that $\mathbf{M}$ has full column rank, since otherwise we can always compress its column space and reparametrize $\mathbf{n}$ accordingly.

To study this detection problem, we first apply a noise whitening transformation to bring the problem to a form with $\mathbf{K} = \mathbf{I}_N$. To do so, we perform an eigenvalue decomposition

$$\mathbf{K} = \mathbf{P}\boldsymbol{\Lambda}\mathbf{P}^T \tag{5.60}$$

of the covariance matrix $\mathbf{K}$, where $\mathbf{P}$ is orthonormal. Then, by premultiplying (5.59) by $\boldsymbol{\Lambda}^{-1/2}\mathbf{P}^T$, we obtain

$$\bar{\mathbf{Y}} = A\bar{\mathbf{s}} + \bar{\mathbf{M}}\mathbf{n} + \bar{\mathbf{V}} \tag{5.61}$$

with

$$\bar{\mathbf{Y}} = \boldsymbol{\Lambda}^{-1/2}\mathbf{P}^T\mathbf{Y} \quad , \quad \bar{\mathbf{V}} = \boldsymbol{\Lambda}^{-1/2}\mathbf{P}^T\mathbf{V}$$

and

$$\bar{\mathbf{s}} = \boldsymbol{\Lambda}^{-1/2}\mathbf{P}^T\mathbf{s} \quad , \quad \bar{\mathbf{M}} = \boldsymbol{\Lambda}^{-1/2}\mathbf{P}^T\mathbf{M} \, .$$

The transformed observation equation (5.61) is such that the noise $\bar{\mathbf{V}}$ is $N(\mathbf{0}, \sigma^2\mathbf{I}_N)$ distributed, as desired. So in the following we will assume that $\mathbf{K} = \mathbf{I}_N$.

The form (5.59) of the observations suggests that any component of the observation vector in the subspace $\mathcal{M}$ spanned by the columns of $\mathbf{M}$ cannot be trusted since it can as easily come from the interferer as from the signal to be detected. This means that two observation vectors $\mathbf{Y}$ and $\mathbf{Y}'$ related to each other by the transformation

$$\mathbf{Y}' = \mathbf{Y} + \mathbf{M}\mathbf{d} \tag{5.62}$$

with $\mathbf{d} \in \mathbb{R}^p$ contain the same information. The transformations (5.62) form a group $\mathcal{G}_i$, and since the parameter space transformation

$$A' = A \quad , \quad \mathbf{n}' = \mathbf{n} + \mathbf{d} \tag{5.63}$$

corresponding to (5.62) does not affect A, it leaves the detection problem invariant. Consider the singular value decomposition (SVD)

$$\mathbf{M} = \mathbf{U}\boldsymbol{\Sigma}\mathbf{W}^T \tag{5.64}$$

of $\mathbf{M}$, where $\mathbf{U}$ and $\mathbf{W}$ are orthonormal matrices of size $N \times N$ and $p \times p$ respectively. Since we have assumed that $\mathbf{M}$ has full column rank, the matrices $\mathbf{U}$ and $\boldsymbol{\Sigma}$ can be partitioned as

$$\mathbf{U} = \begin{bmatrix} \mathbf{U}^{\|} \; \mathbf{U}^{\perp} \end{bmatrix} \quad , \quad \boldsymbol{\Sigma} = \begin{bmatrix} \boldsymbol{\Gamma} \\ \mathbf{0} \end{bmatrix} \tag{5.65}$$

where the $p \times p$ invertible diagonal matrix $\boldsymbol{\Gamma}$ regroups the singular values of $\mathbf{M}$ and the $N \times p$ and $N \times (N-p)$ matrices $\mathbf{U}^{\|}$ and $\mathbf{U}^{\perp}$ span the column space of $\mathbf{M}$ and its orthogonal complement, respectively. We refer the reader to [7, Chap. 5] for a discussion of the SVD and its properties. Then the statistic

$$\mathbf{S}(\mathbf{Y}) = (\mathbf{U}^{\perp})^T \mathbf{Y} \tag{5.66}$$

is a maximal invariant under the group $\mathcal{G}_i$ of transformations of the form (5.62) since it is invariant, and if

$$\mathbf{0} = \mathbf{S}(\mathbf{Y}_1) - \mathbf{S}(\mathbf{Y}_2) = (\mathbf{U}^{\perp})^T (\mathbf{Y}_1 - \mathbf{Y}_2) ,$$

we must have

$$\mathbf{Y}_1 = \mathbf{Y}_2 + \mathbf{M}\mathbf{d}$$

for some vector $\mathbf{d}$ in $\mathbb{R}^p$. In other words, if the statistic $\mathbf{S}$ corresponds to two different observation vectors $\mathbf{Y}_1$ and $\mathbf{Y}_2$, they are necessarily related by a transformation of the form (5.62).

The statistic $\mathbf{S}(\mathbf{Y})$ can be expressed as

$$\mathbf{S}(\mathbf{Y}) = A\mathbf{s}' + \mathbf{V}' \tag{5.67}$$

where $\mathbf{s}' = (\mathbf{U}^{\perp})^T \mathbf{s}$ and $\mathbf{V}' = (\mathbf{U}^{\perp})^T \mathbf{V}$ is $N(\mathbf{0}, \sigma^2 \mathbf{I}_{N-p})$ distributed, so that the detection problem is of the form considered in Example 5.7. When $\mathbf{s}' = \mathbf{0}$, the detection problem is degenerate and cannot be solved. This corresponds to a situation where the signal $\mathbf{s}$ is contained in the interference space $\mathcal{M}$, in which case the signal cannot be detected. To interpret the observation (5.67) geometrically, observe that

$$\mathbf{Q}^{\perp} = \mathbf{U}^{\perp}(\mathbf{U}^{\perp})^T = \mathbf{I}_N - \mathbf{U}^{\|}(\mathbf{U}^{\|})^T \tag{5.68}$$

is the projection onto the space orthogonal to $\mathcal{M}$, so (5.67) corresponds to the coordinate representation of the projection

$$\mathbf{Q}^{\perp}\mathbf{Y} = A\mathbf{Q}^{\perp}\mathbf{s} + \mathbf{Q}^{\perp}\mathbf{V} \tag{5.69}$$

of the observations onto the $N - p$ dimensional space orthogonal to $\mathcal{M}$.

So by performing successive transformations, we have whitened the noise $\mathbf{V}$ and removed the interference component from the observation equation (5.59). Then, depending on whether σ^2 is known or unknown, the reduced equation (5.69) specifies a detection problem of the type considered in Cases 1 and 3 of Example 5.7, respectively. The tests obtained for these two cases will be UMPI under the composition of the interference equivalence group $\mathcal{G}_i$ with the groups $\mathcal{G}_m$ or $\mathcal{G}_m \times \mathcal{G}_s$ considered in these two cases.

Remarks:

(a) The noise whitening transformation $\mathbf{\Lambda}^{-1/2}\mathbf{P}^T$ which was employed above to whiten the noise $\mathbf{V}$ is not unique, and in fact if $\mathbf{K}^{1/2}$ denotes any matrix square root of $\mathbf{K}$, i.e., if it satisfies

$$\mathbf{K} = \mathbf{K}^{1/2}\mathbf{K}^{T/2} ,$$

then premultiplying (5.59) by $\mathbf{K}^{-1/2}$ will yield a $N(\mathbf{0}, \sigma^2 \mathbf{I}_N)$ distributed noise vector $\bar{\mathbf{V}} = \mathbf{K}^{-1/2}\mathbf{V}$. For example, if

$$\mathbf{K} = \mathbf{L}\mathbf{L}^T$$

denotes a lower times upper Cholesky factorization of $\mathbf{K}$ [7, p. 101], $\mathbf{L}$ is a matrix square root of $\mathbf{K}$.

(b) The idea of projecting observations either onto a signal subspace, or onto a subspace orthogonal to interference signals is central not only to detection problems, but also to modern eigenstructure-based array processing methods used for source location or tracking [8].

5.5 Generalized Likelihood Ratio Tests

The generalized likelihood ratio test is a systematic joint estimation and detection procedure which is applicable under all circumstances, independently of whether a UMP(I) test exists or not. Consider the composite hypothesis testing problem specified by

$$\begin{aligned} H_0 &: \mathbf{Y} \sim f(\mathbf{y}|\mathbf{x}, H_0), \ \mathbf{x} \in \mathcal{X}_0 \\ H_1 &: \mathbf{Y} \sim f(\mathbf{y}|\mathbf{x}, H_1), \ \mathbf{x} \in \mathcal{X}_1 \,. \end{aligned} \tag{5.70}$$

The generalized likelihood ratio (GLR) statistic associated to this problem is given by

$$L_{\mathrm{G}}(\mathbf{y}) = \frac{\max_{\mathbf{x} \in \mathcal{X}_1} f(\mathbf{y}|\mathbf{x}, H_1)}{\max_{\mathbf{x} \in \mathcal{X}_0} f(\mathbf{y}|\mathbf{x}, H_0)} = \frac{f(\mathbf{y}|\hat{\mathbf{x}}_1, H_1)}{f(\mathbf{y}|\hat{\mathbf{x}}_0, H_0)} \,, \tag{5.71}$$

where

$$\hat{\mathbf{x}}_j = \arg \max_{\mathbf{x} \in \mathcal{X}_j} f(\mathbf{y}|\mathbf{x}, H_j) \tag{5.72}$$

denotes the maximum likelihood estimate of the parameter vector $\mathbf{x}$ under H_j with $j = 0,\ 1$. Thus the GLR statistic is obtained by replacing the unknown parameter vector by its ML estimate under each hypothesis. In the special case where hypothesis H_0 is simple, which is often the case, the GLR statistic reduces to

$$L_{\mathrm{G}}(\mathbf{y}) = \frac{\max_{\mathbf{x} \in \mathcal{X}_1} f(\mathbf{y}|\mathbf{x}, H_1)}{f(\mathbf{y}|\mathbf{x}_0, H_0)} = \frac{f(\mathbf{y}|\hat{\mathbf{x}}_1, H_1)}{f(\mathbf{y}|\mathbf{x}_0, H_0)} \,, \tag{5.73}$$

where $\mathbf{x}_0$ denotes the value of the parameter vector under H_0. Then the generalized likelihood ratio test (GLRT) takes the form

$$L_{\mathrm{G}}(\mathbf{Y}) \underset{H_0}{\overset{H_1}{\gtrless}} \tau \,, \tag{5.74}$$

or equivalently

$$\ln(L_{\rm G}(\mathbf{Y})) \underset{H_0}{\overset{H_1}{\gtrless}} \gamma \triangleq \ln(\tau) , \tag{5.75}$$

where the threshold τ (or γ) is selected such that

$$\max_{\mathbf{x}\in\mathcal{X}_0} P[L_{\rm G}(\mathbf{Y}) > \tau|\mathbf{x}, H_0] = \alpha . \tag{5.76}$$

To illustrate the above discussion we consider several examples.

Example 5.8: Signal with unknown amplitude in noise

Assume again that we observe

$$Y(k) = As(k) + V(k)$$

for $1 \leq k \leq N$, where the signal $s(k)$ is known, but the amplitude A is unknown, and $V(k)$ is a zero-mean WGN with variance σ^2. We seek to decide between $H_1 : \ A \neq 0$ and $H_0 : \ A = 0$. We will consider two cases: (i) σ^2 known, and (ii) σ^2 unknown.

Case 1: If $\mathbf{Y}$ and $\mathbf{s}$ denote again the N-dimensional vector representations of the observations $Y(k)$ and the signal $s(k)$, the probability density of $\mathbf{Y}$ under H_1 is given by

$$f(\mathbf{y}|A) = \frac{1}{(2\pi\sigma^2)^{N/2}} \exp\big(-\frac{||\mathbf{y} - A\mathbf{s}||_2^2}{2\sigma^2}\big) ,$$

and under H_0, it is given by

$$f(\mathbf{y}|A = 0) = \frac{1}{(2\pi\sigma^2)^{N/2}} \exp\big(-\frac{||\mathbf{y}||_2^2}{2\sigma^2}\big) ,$$

where since σ^2 is known, H_0 is a simple hypothesis. The ML estimate of the amplitude A under H_1 is given by

$$\hat{A}_1 = \mathbf{s}^T\mathbf{Y}/E \tag{5.77}$$

with $E = ||\mathbf{s}||_2^2$, so that the GLR statistic takes the form

$$\begin{aligned} L_{\rm G}(\mathbf{Y}) &= \frac{f(\mathbf{Y}|\hat{A}_1)}{f(\mathbf{Y}|A = 0)} \\ &= \exp\big(\frac{(\mathbf{s}^T\mathbf{Y})^2}{2\sigma^2 E}\big) . \end{aligned} \tag{5.78}$$

Taking logarithms, the GLRT can be expressed as

$$\frac{(\mathbf{s}^T\mathbf{Y})^2}{2\sigma^2 E} \underset{H_0}{\overset{H_1}{\gtrless}} \gamma . \tag{5.79}$$

Multiplying both sides by $2\sigma^2$, and taking square roots, this test is equivalent to

$$|\mathbf{s}^T\mathbf{Y}|/E^{1/2} \underset{H_0}{\overset{H_1}{\gtrless}} \eta = \sigma(2\gamma)^{1/2} , \tag{5.80}$$

which coincides with the UMPI test (5.37).

Case 2: When σ^2 is unknown, under H_1, the density of $\mathbf{Y}$ admits the parametrization

$$f(\mathbf{y}|A,\sigma^2) = \frac{1}{(2\pi\sigma^2)^{N/2}} \exp\big(-\frac{||\mathbf{y}-A\mathbf{s}||_2^2}{2\sigma^2}\big) ,$$

and under H_0 it is given by

$$f(\mathbf{y}|A=0,\sigma^2) = \frac{1}{(2\pi\sigma^2)^{N/2}} \exp\big(-\frac{||\mathbf{y}||_2^2}{2\sigma^2}\big) .$$

Hypothesis H_0 is no longer simple since σ^2 is unknown. Then under H_1, the ML estimate of the amplitude A is still given by (5.77), and the ML variance estimates under H_1 and H_0 are given respectively by

$$\widehat{\sigma_1^2} = \frac{1}{N}||\mathbf{Y}-\hat{A}_1\mathbf{s}||_2^2 = \frac{1}{N}(||\mathbf{Y}||_2^2 - \frac{|\mathbf{s}^T\mathbf{Y}|^2}{E}) \tag{5.81}$$

and

$$\widehat{\sigma_0^2} = \frac{1}{N}||\mathbf{Y}||_2^2 . \tag{5.82}$$

The GLR statistic can then be expressed as

$$\begin{aligned} L_G(\mathbf{Y}) &= \frac{f(\mathbf{Y}|\hat{A}_1,\widehat{\sigma_1^2})}{f(\mathbf{Y}|A=0,\widehat{\sigma_0^2})} \\ &= \Big(\frac{\widehat{\sigma_0^2}}{\widehat{\sigma_1^2}}\Big)^{N/2} . \end{aligned} \tag{5.83}$$

Next consider the GLRT (5.74). By noting that

$$\frac{\widehat{\sigma_0^2}}{\widehat{\sigma_1^2}} = \Big(1 - \frac{|\mathbf{s}^T\mathbf{Y}|^2}{E||\mathbf{Y}||_2^2}\Big)^{-1} ,$$

we see that in (5.74) the threshold $\tau > 1$, since otherwise we always select H_1. Then, inverting terms on both sides of (5.74) and reversing the inequality, passing terms from one side to the other, and taking square roots, we find that the GLRT can be expressed as

$$\frac{|\mathbf{s}^T\mathbf{Y}|}{E^{1/2}||\mathbf{Y}||_2} \underset{H_0}{\overset{H_1}{\gtrless}} \eta = (1-\tau^{-2/N})^{1/2} , \tag{5.84}$$

which coincides with the UMPI test (5.50). □

Example 5.9: Incoherent detection

Consider the situation already examined in Example 5.6, where we observe a two-dimensional vector

$$\mathbf{Y} = \begin{bmatrix} Y_c \\ Y_s \end{bmatrix} = A \begin{bmatrix} \cos(\theta) \\ \sin(\theta) \end{bmatrix} + \mathbf{V}$$

where $A \geq 0$ and $\theta \in [0, 2\pi)$ are unknown and $\mathbf{V}$ is $N(\mathbf{0}, \sigma^2 \mathbf{I}_2)$ distributed. The noise variance σ^2 is known, and we seek to test $H_1 : \ A \neq 0$ against $H_0 : \ A = 0$. In the polar coordinate system

$$Y_c = R\cos(\Phi) \ \ , \ \ Y_s = R\sin(\Phi)$$

corresponding to

$$R = (Y_c^2 + Y_s^2)^{1/2} \ \ , \ \ \Phi = \tan^{-1}(Y_s/Y_c) \, ,$$

the density of the observations under H_1 can be expressed as

$$f(r, \phi | A, \theta) = \frac{r}{2\pi\sigma^2} \exp\big(- \frac{(r^2 + A^2)}{2\sigma^2} \big) \exp\big(\frac{Ar}{\sigma^2} \cos(\phi - \theta) \big)$$

and the ML estimates of θ and A are given by

$$\hat{\theta} = \Phi \ \ , \ \ \hat{A} = R \, . \tag{5.85}$$

Under the simple hypothesis H_0, the observation density is given by

$$f(r, \phi | A = 0) = \frac{r}{2\pi\sigma^2} \exp\big(- \frac{r^2}{2\sigma^2} \big) \, ,$$

so the GLR statistic can be expressed as

$$\begin{aligned} L_{\mathrm{G}}(R) &= \frac{f(R, \Phi | \hat{A}, \hat{\theta})}{f(R, \Phi | A = 0)} \\ &= \exp\big(\frac{R^2}{2\sigma^2} \big) \end{aligned} \tag{5.86}$$

Taking logarithms, the GLRT can be expressed as

$$\frac{R^2}{2\sigma^2} \mathop{\gtrless}_{H_0}^{H_1} \gamma = \ln(\tau) \, ,$$

or equivalently, as

$$R \mathop{\gtrless}_{H_0}^{H_1} \eta = \sigma(2\ln(\tau))^{1/2} \tag{5.87}$$

which coincides with the UMPI test (5.54). □

Up to this point, all GLR tests have coincided with UMPI tests, but we now describe an example where, in one case, the GLR test does not coincide with a UMPI test, and in another, there is no UMPI test, so in this situation the GLRT becomes our best option after the search for an optimal test has failed.

Example 5.10: Impulse detector

Consider the impulse detection problem of Example 5.5, which is taken from [9] and [2]. As noted in Example 5.5, it can be written in vector form as

$$\mathbf{Y} = A\mathbf{e}_\ell + \mathbf{V}$$

where $\mathbf{e}_\ell$ denotes the ℓ-th basis vector of $\mathbb{R}^N$ (all its entries are zero, except the ℓ-th, which is equal to one) and the noise $\mathbf{V}$ is $N(\mathbf{0}, \sigma^2 \mathbf{I}_N)$ distributed where σ^2 is known. We want to decide between $H_1 : \ A \neq 0$ and $H_0 : \ A = 0$. We will examine two cases: (i) the amplitude A is known and positive, and (ii) it is unknown.

Case 1: When $A > 0$ is known, the density of the observations under H_1 can be expressed as

$$f(\mathbf{y}|A, \ell) = \frac{1}{(2\pi\sigma^2)^{N/2}} \exp\big(- \frac{(||\mathbf{y}||_2^2 + A^2)}{2\sigma^2}\big) \exp\big(\frac{A y_\ell}{\sigma^2}\big) , \tag{5.88}$$

so that the ML estimate of ℓ is given by

$$\hat{\ell} = \arg \max_{1 \leq \ell \leq N} Y_\ell , \tag{5.89}$$

where y_ℓ denotes the ℓ-th entry of $\mathbf{y}$. Under the simple hypothesis H_0 the observations admit the distribution

$$f(\mathbf{y}|A = 0) = \frac{1}{2\pi\sigma^2} \exp\big(- \frac{||\mathbf{y}||_2^2}{2\sigma^2}\big) ,$$

so that the GLR statistic is given by

$$\begin{aligned} L_{\mathrm{G}}(\mathbf{Y}) &= \frac{f(\mathbf{Y}|A, \hat{\ell})}{f(\mathbf{Y}|A = 0)} \\ &= \exp\big((Y_{\max} - \frac{A}{2})\frac{A}{\sigma^2}\big) , \end{aligned} \tag{5.90}$$

where

$$Y_{\max} \stackrel{\triangle}{=} Y_{\hat{\ell}} = \max_{1 \leq \ell \leq N} Y_\ell .$$

Taking logarithms the GLRT can be expressed as

$$Y_{\max} \underset{H_0}{\overset{H_1}{\gtrless}} \eta = \frac{A}{2} + \frac{\sigma^2}{A} \ln(\tau) . \tag{5.91}$$

To select the threshold η so that the probability of false alarm is α, observe that under H_0 the entries Y_k of $\mathbf{Y}$ are independent $N(0, \sigma^2)$ distributed. But given N i.i.d. random variables Y_k with density $f(y)$ and cumulative probability distribution $F(y)$, the density of $Y_{\max}$ is given by [10, p. 246]

$$f_{Y_{\max}}(y) = NF^{N-1}(y)f(y) .$$

Applying this result, we find

$$\begin{aligned} \alpha &= P[Y_{\max} > \eta | A = 0] \\ &= \frac{N}{(2\pi)^{1/2}} \int_{\eta/\sigma}^{\infty} (1 - Q(z))^{N-1} \exp\big(-\frac{z^2}{2}\big) dz , \end{aligned} \tag{5.92}$$

where the $Q(\cdot)$ function is defined in (2.69).

To determine whether a UMPI exists for the impulse detection problem, note that by applying a random cyclic shift to $\mathbf{Y}$, we can always randomize the impulse location. Specifically, let M be a random integer with

$$P[M = m] = \frac{1}{N} \text{ for } 0 \leq m \leq N-1 .$$

Then, let

$$\bar{\mathbf{Y}} = \mathbf{Z}_C^M \mathbf{Y} = A\mathbf{e}_{(\ell+M) \bmod N} + \bar{\mathbf{V}}$$

be the vector obtained by cyclically shifting $\mathbf{Y}$ by M places, where $\bar{\mathbf{V}} = \mathbf{Z}_C^M \mathbf{V}$ is $N(\mathbf{0}, \mathbf{I}_N)$ distributed. Its density is given by

$$\begin{aligned} f_{\bar{\mathbf{Y}}}(\bar{\mathbf{y}}|A) &= \sum_{m=0}^{N-1} f_{\bar{\mathbf{Y}}}(\bar{\mathbf{y}}|A, m) P[M = m] \\ &= \frac{1}{N} \sum_{m=0}^{N-1} \frac{1}{(2\pi\sigma^2)^{N/2}} \exp\big(-\frac{1}{2\sigma^2} ||\bar{\mathbf{y}} - A\mathbf{e}_{(\ell+m) \bmod N}||_2^2\big) \\ &= \frac{1}{N(2\pi\sigma^2)^{N/2}} \exp\big(-\frac{(||\bar{\mathbf{y}}||_2^2 + A^2)}{2\sigma^2}\big) \sum_{k=0}^{N-1} \exp\big(\frac{A\bar{y}_k}{\sigma^2}\big) , \end{aligned} \tag{5.93}$$

where $k = (\ell + m) \bmod N$ and $\bar{Y}_k$ denotes the k-th entry of $\bar{\mathbf{Y}}$. Note that since $\bar{\mathbf{Y}}$ is just a cyclically shifted version of $\mathbf{Y}$, we have $||\bar{\mathbf{Y}}||_2^2 = ||\mathbf{Y}||_2^2$ and

$$\sum_{k=0}^{N-1} \exp\big(\frac{A\bar{Y}_k}{\sigma^2}\big) = \sum_{k=0}^{N-1} \exp\big(\frac{AY_k}{\sigma^2}\big) .$$

Then the LR for the randomly shifted observation vector $\bar{\mathbf{Y}}$ is given by

$$\begin{aligned} L(\bar{\mathbf{Y}}) &= \frac{f_{\bar{\mathbf{Y}}}(\bar{\mathbf{Y}}|A)}{f_{\bar{\mathbf{Y}}}(\bar{\mathbf{Y}}|A=0)} \\ &= \frac{1}{N} \sum_{k=0}^{N-1} \exp\big(\frac{AY_k}{\sigma^2}\big) , \end{aligned} \tag{5.94}$$

which is an invariant statistic of $\mathbf{Y}$. Consequently if

$$S(\mathbf{Y}) \triangleq \frac{\sigma^2}{A} \ln \Big(\sum_{k=0}^{N-1} \exp \big(\frac{AY_k}{\sigma^2} \big) \Big) , \tag{5.95}$$

the test

$$S(\mathbf{Y}) \underset{H_0}{\overset{H_1}{\gtrless}} \eta \tag{5.96}$$

is UMPI under the action of the group $\mathcal{G}$ of cyclic shifts. This test is different from the GLRT (5.91), even though the GLR statistic $Y_{\max}$ is invariant under $\mathcal{G}$. Note however that for high SNR when $A/\sigma \gg 1$, the sum of exponentials in (5.95) is dominated by its largest term $\exp(AY_{\max}/\sigma^2)$, so that

$$S(\mathbf{Y}) \approx Y_{\max} ,$$

and thus in this case the UMPI test reduces to the GLRT.

Case 2: When $A \neq 0$ is unknown under H_1, there no longer exists a UMPI test since the invariant statistic (5.95) depends on A. Then the ML estimates of ℓ and A are given by

$$\begin{aligned} \hat{\ell} &= \arg \max_{1 \leq \ell \leq N} |Y_\ell| \\ \hat{A} &= |Y|_{\max} \triangleq \max_{1 \leq \ell \leq N} |Y_\ell| , \end{aligned} \tag{5.97}$$

so the GLR statistic can be expressed as

$$\begin{aligned} L_{\mathrm{G}}(\mathbf{Y}) &= \frac{f(\mathbf{Y}|\hat{A}, \hat{\ell})}{f(\mathbf{Y}|A=0)} \\ &= \exp \big(\frac{|Y|^2_{\max}}{2\sigma^2} \big) . \end{aligned} \tag{5.98}$$

Taking logarithms, multiplying both sides by $2\sigma^2$ and then taking square roots, the GLRT can be written as

$$|Y|_{\max} \underset{H_0}{\overset{H_1}{\gtrless}} \eta . \tag{5.99}$$

□

Invariance of the GLR statistic: The GLR statistic has the important property that it is invariant under any group of transformations that leaves the detection problem invariant [2]. To derive this result, note that if $\mathbf{Z} = \mathbf{g}(\mathbf{Y})$

denotes a one-to-one transformation, then the density of $\mathbf{Z}$ can be expressed in terms of the density of $\mathbf{Y}$ as [10, p. 201]

$$f_{\mathbf{Z}}(\mathbf{g}(\mathbf{y})) = J^{-1}(\mathbf{y}) f_{\mathbf{Y}}(\mathbf{y}) \tag{5.100}$$

where the Jacobian J of the transformation $g(\mathbf{y})$ is defined as

$$J(\mathbf{y}) = |\det(\nabla^T \mathbf{g}(\mathbf{y}))| \,. \tag{5.101}$$

Taking this transformation into account in the definition (5.25) of invariance, we find that if the transformation $\mathbf{Z} = \mathbf{g}(\mathbf{Y})$ with induced parameter transformation $\tilde{\mathbf{g}}$ leaves the detection problem invariant, the observation vector density satisfies

$$f(\mathbf{y}|\mathbf{x}) = J(\mathbf{y}) f(\mathbf{g}(\mathbf{y})|\tilde{\mathbf{g}}(\mathbf{x})) \,. \tag{5.102}$$

Substituting this identity in the definition (5.71) of the GLR statistic gives

$$L_{\mathrm{G}}(\mathbf{y}) = \frac{\max_{\mathbf{x} \in \mathcal{X}_1} f(\mathbf{g}(\mathbf{y})|\tilde{\mathbf{g}}(\mathbf{x}))}{\max_{\mathbf{x} \in \mathcal{X}_0} f(\mathbf{g}(\mathbf{y})|\tilde{\mathbf{g}}(\mathbf{x}))} \,. \tag{5.103}$$

But, if we replace the dummy maximization variable $\mathbf{x}$ by $\boldsymbol{\xi} = \tilde{\mathbf{g}}(\mathbf{x})$ and note that the transformation $\tilde{\mathbf{g}}$ leaves $\mathcal{X}_j$ invariant, we obtain

$$L_{\mathrm{G}}(\mathbf{y}) = \frac{\max_{\boldsymbol{\xi} \in \mathcal{X}_1} f(\mathbf{g}(\mathbf{y})|\boldsymbol{\xi})}{\max_{\boldsymbol{\xi} \in \mathcal{X}_0} f(\mathbf{g}(\mathbf{y})|\boldsymbol{\xi})} = L_{\mathrm{G}}(\mathbf{g}(\mathbf{y})) \,, \tag{5.104}$$

which proves the invariance of the GLR statistic.

5.6 Asymptotic Optimality of the GLRT

As starting point we observe that since the GLRT is obtained in a heuristic manner by replacing the unknown parameter vector $\mathbf{x}$ by its ML estimates $\hat{\mathbf{x}}_0$ and $\hat{\mathbf{x}}_1$ under H_0 and H_1, it does not have a "built-in" optimality property like UMP or UMPI tests. Furthermore, if the GLRT has any optimality property, it must be of an asymptotic nature, as it is applied to a sequence of i.i.d. observations $\{Y_k, 1 \leq k \leq N\}$. This is due to the fact that the ML estimate does not have any specific accuracy property for a finite sample size N, but as the number of observations N becomes large, it has the feature that under mild conditions, it converges almost surely to the true parameter value $\mathbf{x}$, and the estimation error covariance matrix decreases like $1/N$. Thus suppose we consider a binary hypothesis testing problem where the null hypothesis is simple, so we need to decide between $H_0 : \mathbf{x} = \mathbf{x}_0$ and $H_1 : \mathbf{x} \neq \mathbf{x}_0$. A relatively naive but yet effective strategy consists of comparing the ML estimate $\hat{\mathbf{x}}_{\mathrm{ML}}$ of $\mathbf{x}$ to $\mathbf{x}_0$. When $\hat{\mathbf{x}}_{\mathrm{ML}}$ is close to $\mathbf{x}_0$ in an appropriate sense, we choose H_0, whereas when the distance is beyond a certain threshold, we select H_1. Furthermore, note that the neighborhood of x_0 where we select H_0 can be shrunk progressively as N increases. This shrinking can be accomplished in such a way that the probability of false alarm goes to zero as N tends to infinity, while the power of the test goes to 1 for parameter vectors $\mathbf{x}$ at a fixed positive distance of $\mathbf{x}_0$. So we know ahead of time that the GLRT is likely to be highly effective in an asymptotic sense, since it benefits from the strong convergence properties of ML estimates, but if we wish to establish optimality in some sense, proving that both the probability of false alarm and the probability of a miss go to zero is not good enough, since most reasonable tests will have this property.

It is also worth noting that although Chernoff [11] and other researchers were able to show that as N tends to infinity, the GLR statistic converges in distribution to a chi-squared distribution under H_0, and a noncentered chi-squared distribution under H_1 (for the case when the test is "local," i.e., when the deviation between $\mathbf{x}_0$ and parameter vector $\mathbf{x}$ under H_1 is proportional to $N^{-1/2}$ as N increases), these results are not as helpful as one might expect for characterizing the asymptotic performance of the GLRT. This is due to the fact that convergence in distribution results for the GLR statistic focus on regions where most of the probability distribution is concentrated, i.e., on small deviations. In contrast, the asymptotic probability of false alarm and the probability of a miss are affected by the tails of the GLR distribution, i.e., by large deviations. Thus, if the chi-squared GLR model is inaccurate in the tail region for large but finite N, convergence in distribution is not affected, but the evaluation of the test performance becomes incorrect.

5.6.1 Multinomial Distributions

So an evaluation of the asymptotic behavior of the GLR test needs to focus primarily on the test performance, not on the GLR statistic itself. The first important result in this area was derived by Hoeffding [12] for the case of multinomial distributions. Specifically, consider a sequence of i.i.d. random variables $\{Y_k,\ 1 \leq k \leq N\}$, where each Y_k takes one of K possible discrete values, say $\{1, 2 \ldots, K\}$ with probability

$$P[Y_k = i] = p_i \,. \tag{5.105}$$

The probability distribution of Y_k can be represented by a row vector

$$\mathbf{p} = \begin{bmatrix} p_1 \cdots p_i \cdots p_K \end{bmatrix}$$

which belongs to the simplex $\mathcal{P}$ formed by vectors of $\mathbb{R}^K$ such that $0 \leq p_i \leq 1$ and

$$\sum_{i=1}^{K} p_i = 1 \,. \tag{5.106}$$

Consider a partition

$$\mathcal{P} = \mathcal{P}_0 \cup \mathcal{P}_1 \quad , \quad \mathcal{P}_0 \cap \mathcal{P}_1 \tag{5.107}$$

of the simplex $\mathcal{P}$. We wish to test $H_0 : \mathbf{p} \in \mathcal{P}_0$ against $H_1 : \mathbf{p} \in \mathcal{P}_1$. For the special case where $\mathcal{P}_0$ reduces to a point $\mathbf{p}_0$ and $\mathcal{P}_1 = \mathcal{P} - \mathbf{p}_0$, the hypothesis H_0 is simple and H_1 corresponds to vectors $\mathbf{p} \neq \mathbf{p}_0$.

Likelihood ratio test: For the case of two simple hypotheses where the Y_k's have distribution $\mathbf{p}_i$ under H_i with $i = 0, 1$, the likelihood ratio for the observations $\{Y_k, 1 \leq k \leq N\}$ can be expressed as

$$L(\mathbf{Y}) = \prod_{k=1}^{N} \frac{p^1_{Y_k}}{p^0_{Y_k}} = \prod_{i=1}^{K} \left(\frac{p^1_i}{p^0_i}\right)^{N_i} \tag{5.108}$$

where N_i with $1 \leq i \leq K$ denotes the number of observations $\{Y_k, 1 \leq k \leq N\}$ that are equal to i. This indicates that the K-tuple

$$\mathbf{N} = \left[N_1 \cdots N_i \cdots N_K \right],$$

or equivalently, the empirical probability distribution

$$\mathbf{q}_N \triangleq \left[\frac{N_1}{N} \cdots \frac{N_i}{N} \cdots \frac{N_K}{N} \right], \tag{5.109}$$

is a sufficient statistic for testing H_1 against H_0. If the Y_k's have distribution $\mathbf{p}$ and if

$$\mathbf{n} = \left[n_1 \cdots n_i \cdots n_K \right]$$

represents an arbitrary integer K-tuple with

$$\sum_{i=1}^{K} n_i = N, \tag{5.110}$$

the random vector $\mathbf{N}$ admits the multinomial probability distribution

$$P[\mathbf{N} = \mathbf{n}|\mathbf{p}] = \frac{N!}{n_1! n_2! \dots n_K!} \prod_{i=1}^{K} p_i^{n_i}. \tag{5.111}$$

Accordingly, the log-likelihood ratio for two simple hypotheses can be expressed as

$$\begin{aligned}
\ln L(\mathbf{Y}) = \ln L(\mathbf{N}) &= N \sum_{i=1}^{K} \frac{N_i}{N} \ln(p^1_i / p^0_i) \\
&= N \Big[\sum_{i=1}^{K} q^N_i \ln(q^N_i / p^0_i) - \sum_{i=1}^{K} q^N_i \ln(q^N_i / p^1_i) \Big] \\
&= N [D(\mathbf{q}_N | \mathbf{p}_0) - D(\mathbf{q}_N | \mathbf{p}_1)],
\end{aligned} \tag{5.112}$$

where $D(\mathbf{q}|\mathbf{p})$ denotes the discrete Kullback-Leibler (KL) divergence defined in (3.7), and $q^N_i = n_i/N$ is the i-th element of the empirical probability distribution vector $\mathbf{q}_N$. Note that q_i^N can be expressed in terms of observations $\{Y_k, 1 \leq k \leq N\}$ as

$$q_i^N = \frac{1}{N} \sum_{k=1}^{N} \delta(Y_k - i) \tag{5.113}$$

where $\delta(\cdot)$ denotes the Kronecker delta function.

From (5.112) we find that the LRT admits the simple geometric expression

$$D(\mathbf{q}_N|\mathbf{p}_0) - D(\mathbf{q}_N|\mathbf{p}_1) \underset{H_0}{\overset{H_1}{\gtrless}} \gamma \tag{5.114}$$

where γ can be expressed in terms of the Bayesian threshold of (2.18) as

$$\gamma = \ln(\tau)/N \, .$$

For the case of equally likely hypotheses and a cost structure of the form $C_{ij} = 1 - \delta_{ij}$, the threshold $\gamma = 0$, in which case the optimal decision rule can be expressed as the minimum KL metric rule

$$\delta(\mathbf{q}_N) = \arg \min_{i=0,\,1} D(\mathbf{q}_N|\mathbf{p}_i) \, . \tag{5.115}$$

GLR Test: Unfortunately the testing problem that we really need to solve consists of choosing between sets $\mathcal{P}_0$ and $\mathcal{P}_1$ satisfying (5.107). The GLR statistic is given by

$$L_{\mathrm{G}}(\mathbf{Y}) = L_{\mathrm{G}}(\mathbf{N}) = \frac{\max_{\mathbf{p}_1 \in \mathcal{P}_1} P[\mathbf{N}|\mathbf{p}_1]}{\max_{\mathbf{p}_0 \in \mathcal{P}_0} P[\mathbf{N}|\mathbf{p}_0]} \tag{5.116}$$

where $P[\mathbf{N}|\mathbf{p}]$ is the multinomial distribution (5.111). Note that, strictly speaking, when the set $\mathcal{P}_i$ is not closed, the maximum over $\mathcal{P}_i$ with $i = 0, 1$ should be replaced by a supremum, where we recall that the supremum of a function f over a domain $\mathcal{D}$ is defined as the smallest upper bound for the values of f over $\mathcal{D}$. Since the logarithm is monotone increasing, the maximization and logarithm operations can be exchanged, so the logarithm of the GLR statistic can be expressed as

$$\begin{aligned}
\ln L_{\mathrm{G}}(\mathbf{Y}) &= \max_{\mathbf{p}_1 \in \mathcal{P}_1} \ln P[\mathbf{N}|\mathbf{p}_1] - \max_{\mathbf{p}_0 \in \mathcal{P}_0} \ln P[\mathbf{N}|\mathbf{p}_0] \\
&= N\Big[\min_{\mathbf{p}_0 \in \mathcal{P}_0} D(\mathbf{q}_N|\mathbf{p}_0) - \min_{\mathbf{p}_1 \in \mathcal{P}_1} D(\mathbf{q}_N|\mathbf{p}_1)\Big] \\
&= N[D(\mathbf{q}_N|\hat{\mathbf{p}}_0^{\mathrm{ML}}) - D(\mathbf{q}_N|\hat{\mathbf{p}}_1^{\mathrm{ML}}) \, ,
\end{aligned} \tag{5.117}$$

where

$$\hat{\mathbf{p}}_i^{\mathrm{ML}} = \arg \min_{\mathbf{p} \in \mathcal{P}_i} D(\mathbf{q}_N|\mathbf{p}) \tag{5.118}$$

denotes the ML estimate of $\mathbf{p}$ under H_i with $i = 0, 1$. Thus the ML estimate of $\mathbf{p}$ under H_i is the point of $\mathcal{P}_i$ closest to the empirical distribution $\mathbf{q}_N$ in the sense of the KL metric.

Let $\mathcal{Q}_N$ denote the set of all possible empirical probability distributions of the form (5.109), where the integers N_i with $1 \leq i \leq N$ satisfy the constraint (5.110). Clearly, $\mathcal{Q}_N$ is included in the simplex $\mathcal{P}$, but because $\mathbf{q}_N$ is obtained by partitioning the integer N into K integers N_i, $1 \leq i \leq K$, when N is fixed $\mathcal{Q}_N$ contains only a finite number of points. The total number of points of $\mathcal{Q}_N$ is $\binom{N+K}{K-1}$. Obviously as N becomes large, $\mathcal{Q}_N$ is dense in $\mathcal{P}$. In fact, by the strong law of large numbers, the empirical probability distribution $\mathbf{q}_N$ converges almost surely to the true probability distribution $\mathbf{p}_1$ under H_1, or to $\mathbf{p}_0$ under H_0.

Since the sets $\mathcal{P}_0$ and $\mathcal{P}_1$ form a partition of $\mathcal{P}$, the empirical distribution $\mathbf{q}_N$ necessarily falls in one of these sets. Accordingly, the ML estimates $\mathbf{p}_i^{\mathrm{ML}}$ satisfy

$$\begin{aligned} \hat{\mathbf{p}}_0^{\mathrm{ML}} &= \mathbf{q}_N \quad \text{if} \quad \mathbf{q}_N \in \mathcal{P}_0 \\ \hat{\mathbf{p}}_1^{\mathrm{ML}} &= \mathbf{q}_N \quad \text{if} \quad \mathbf{q}_N \in \mathcal{P}_1 . \end{aligned} \tag{5.119}$$

The GLR statistic can be therefore expressed as

$$\frac{1}{N} \ln L_G(\mathbf{q}_N) = \begin{cases} D(\mathbf{q}_N|\mathcal{P}_0) & \text{for } \mathbf{q}_N \in \mathcal{P}_1 \\ -D(\mathbf{q}_N|\mathcal{P}_1) & \text{for } \mathbf{q}_N \in \mathcal{P}_0 , \end{cases} \tag{5.120}$$

where if $\mathbf{p}$ denotes a simplex point outside a set $\mathcal{R} \subset \mathcal{P}$, the KL "distance" between $\mathbf{p}$ and $\mathcal{R}$ is defined as

$$D(\mathbf{p}|\mathcal{R}) \triangleq \inf_{\mathbf{r} \in \mathcal{R}} D(\mathbf{p}|\mathbf{r}) , \tag{5.121}$$

where we recall that the infimum of a function f over a domain $\mathcal{D}$ is the largest lower bound for the values of f over this domain.

The expression (5.120) indicates that the GLRT takes two distinct forms depending on whether we select a positive or a negative threshold η. If $\eta > 0$, by observing that all points $\mathbf{q}_N$ inside $\mathcal{P}_0$ necessarily fall below the threshold, the GLRT can be expressed as

$$D(\mathbf{q}_N|\mathcal{P}_0) \underset{H_0}{\overset{H_1}{\gtrless}} \eta , \tag{5.122}$$

which is often referred to as *Hoeffding's test.* In other words the GLRT selects H_1 if the KL distance between $\mathbf{q}_N$ and $\mathcal{P}_0$ is greater than η, and it selects H_0 otherwise. So the test partitions $\mathcal{P}$ into two disjoint regions

$$\mathcal{R}_0 = \{\mathbf{p} : D(\mathbf{p}|\mathcal{P}_0) < \eta\} \tag{5.123}$$

and

$$\mathcal{R}_1 = \{\mathbf{p} : D(\mathbf{p}|\mathcal{P}_0) \geq \eta\} \tag{5.124}$$

and selects H_0 or H_1 depending on whether $\mathbf{q}_N$ is in $\mathcal{R}_0$ or $\mathcal{R}_1$. The definitions (5.123)–(5.124) imply that regions $\mathcal{R}_i$ satisfy

$$\mathcal{P}_0 \subset \mathcal{R}_0 \quad \text{and} \quad \mathcal{R}_1 \subset \mathcal{P}_1 \,. \tag{5.125}$$

As we shall see below, for all points of $\mathcal{P}_0$, the probability of false alarm has an asymptotic exponential decay rate greater or equal to η, but as shown in Fig. 5.8, the test power at points of $\mathcal{P}_1$ located inside $\mathcal{R}_0$ will be asymptotically zero, since we always declare H_0 for empirical distributions $\mathbf{q}_N$ in this region even though the actual probability distribution belongs to $\mathcal{P}_1$. This test is biased towards ensuring that the size of the test has a guaranteed exponential decay rate, so it can be called a GLRT of type I.

Suppose instead that we select a negative threshold η . Then the GLRT can be rewritten as

$$D(\mathbf{q}_N|\mathcal{P}_1) \underset{H_1}{\overset{H_0}{\gtrless}} -\eta \,, \tag{5.126}$$

where the set $\mathcal{P}_1$ plays the same role as $\mathcal{P}_0$ in (5.122) and the two hypotheses are exchanged. This test performs a partition of $\mathcal{P}$ into two disjoint regions $\mathcal{S}_0$ and $\mathcal{S}_1$ and selects H_0 and H_1 depending on whether $\mathbf{q}_N$ belongs to $\mathcal{S}_0$ or $\mathcal{S}_1$. In this case the sets $\mathcal{S}_i$ satisfy

$$\mathcal{S}_0 \subset \mathcal{P}_0 \quad \text{and} \quad \mathcal{P}_1 \subset \mathcal{S}_1 \,, \tag{5.127}$$

and the test if biased towards ensuring that for the probability of a miss has a guaranteed exponential decay rate, so it forms a GLRT test of type II.

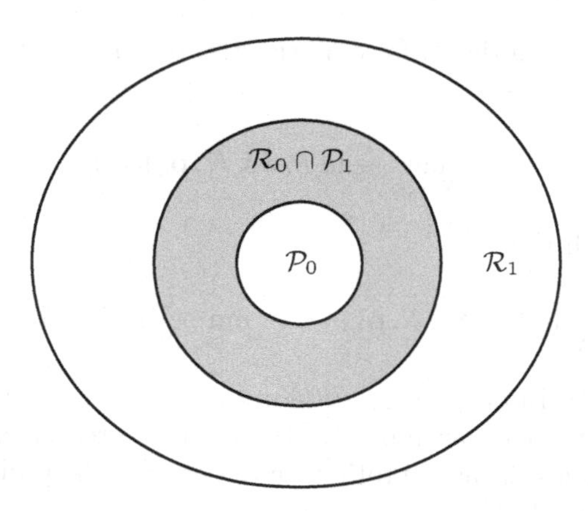

Fig. 5.8. The GLRT of type I assigns points in the guard band marked in grey to the decision region $\mathcal{R}_0$ even though they correspond to probability distributions in $\mathcal{P}_1$. Points in the outer band located further away from $\mathcal{P}_0$ are assigned to $\mathcal{R}_1$.

GLRT Properties: Since the GLRTs of types I and II are obviously identical after exchanging the roles of hypotheses H_0 and H_1, we consider only the GLRT of type I (5.122), which is denoted as δ_{G}. Let

$$P_F(\delta_{\text{G}}, N, \mathbf{p}_0) = P[\mathbf{q}_N \in \mathcal{R}_1 | \mathbf{p}_0] \tag{5.128}$$

be the probability of false alarm for a point $\mathbf{p}_0 \in \mathcal{P}_0$ and let

$$P_M(\delta_{\text{G}}, N, \mathbf{p}_1) = P[\mathbf{q}_N \in \mathcal{R}_0 | \mathbf{p}_1] \tag{5.129}$$

denote the probability of a miss for a point $\mathbf{p}_1 \in \mathcal{P}_1$. The test δ_{G} has the following two properties:

(a) The probability of false alarm has an exponential decay rate

$$-\lim_{N\to\infty} \frac{1}{N} \ln P_F(\delta_G, N, \mathbf{p}_0) \geq \eta \tag{5.130}$$

for all points $\mathbf{p}_0$ of $\mathcal{P}_0$. In other words, the size of the test has a guaranteed asymptotic exponential decay rate of η.

(b) Let δ be any other test specified by

$$\delta(\mathbf{q}_N) = \begin{cases} 1 & \mathbf{q}_N \in \mathcal{R}'_1 \\ 0 & \mathbf{q}_N \in \mathcal{R}'_0 \,, \end{cases} \tag{5.131}$$

where the sets $\mathcal{R}'_0$ and $\mathcal{R}'_1$ are disjoint and form a partition of the simplex $\mathcal{P}$. Let also

$$\begin{aligned} P_F(\delta, N, \mathbf{p}_0) &= P[\mathbf{q}_N \in \mathcal{R}'_1 | \mathbf{p}_0] \quad \text{for } \mathbf{p}_0 \in \mathcal{P}_0 \\ P_M(\delta, N, \mathbf{p}_1) &= P[\mathbf{q}_N \in \mathcal{R}'_0 | \mathbf{p}_1] \quad \text{for } \mathbf{p}_1 \in \mathcal{P}_1 \end{aligned} \tag{5.132}$$

denote the probabilities of false alarm and of a miss for the test δ. Then if the test δ satisfies

$$-\lim_{N\to\infty} \frac{1}{N} \ln P_F(\delta, N, \mathbf{p}_0) \geq \eta \,, \tag{5.133}$$

it must be such that

$$-\lim_{N\to\infty} \frac{1}{N} \ln P_M(\delta_{\text{G}}, N, \mathbf{p}_1) \geq -\lim_{N\to\infty} \frac{1}{N} \ln P_M(\delta, N, \mathbf{p}_1) \tag{5.134}$$

for all $\mathbf{p}_1 \in \mathcal{P}_1$. In other words, among all tests that guarantee that the size of the test decays asymptotically at a rate greater or equal to η, the GLRT maximizes the asymptotic rate of decay of the probability of a miss at all points $\mathbf{p}_1$ of $\mathcal{P}_1$.

To establish the above properties of the GLRT, we shall employ an important result of the theory of large deviations called Sanov's theorem. Whereas

Cramér's theorem introduced in Chapter 3 characterizes deviations of the empirical mean of a sequence of i.i.d. random variables away from their statistical mean, Sanov's theorem characterizes deviations of the empirical probability distribution (5.109) away from the actual probability distribution. Note however that while Cramér's theorem applies equally to continuous- or discrete-valued random variables, Sanov's theorem is restricted to discrete-valued random variables with only a finite number of possible values. The version of the theorem that we employ is proved in Appendix 5.A. and can be stated as follows:

Sanov's Theorem: Consider a sequence of i.i.d. random variables $\{Y_k\,,\, k \geq 1\}$ with probability distribution vector $\mathbf{p}$, and let $\{\mathbf{q}_N\}$ be the corresponding sequence of empirical probability measures. Let $\mathcal{R}$ be a subset of the simplex $\mathcal{P}$ such that all points $\mathbf{r}$ of $\mathcal{R}$ are at least at distance $\epsilon > 0$ of $\mathbf{p}$ for some appropriate distance between probability distribution vectors. Then

$$\lim_{N \to \infty} \frac{1}{N} \ln P[\mathbf{q}_N \in \mathcal{R} | \mathbf{p}] = -D(\mathcal{R}|\mathbf{p}) \tag{5.135}$$

where

$$D(\mathcal{R}|\mathbf{p}) \triangleq \inf_{\mathbf{r} \in \mathcal{R}} D(\mathbf{r}|\mathbf{p})\,. \tag{5.136}$$

To clarify the above theorem statement, let us point out that the condition that the set $\mathcal{R}$ must be at a fixed positive distance ϵ of $\mathbf{p}$ ensures that the deviations considered are large. Specifically, the empirical measure $\mathbf{q}_N$ converges almost surely to $\mathbf{p}$ as N tends to infinity, so normally $\mathbf{q}_N$ will be located in the immmediate vicinity of $\mathbf{p}$ as N becomes large. By considering sets $\mathcal{R}$ that are not close to $\mathbf{p}$, we are ensuring that the empirical probability distribution $\mathbf{q}_N$ undergoes a large deviation. The choice of metric that can be selected to measure distances between points of $\mathcal{P}$ is somewhat arbitrary [13, pp. 13–15], but a common choice is the total variation

$$d(\mathbf{r}, \mathbf{p}) = \frac{1}{2} \sum_{i=1}^{K} |r_i - p_i|\,,$$

where r_i and p_i represent the i-th entries of vectors $\mathbf{r}$ and $\mathbf{p}$, respectively.

To evaluate the probability of false alarm of the GLRT (5.122) we set $\mathcal{R} = \mathcal{R}_1$ and $\mathbf{p} = \mathbf{p}_0 \in \mathcal{P}_0$ in Sanov's theorem. This gives

$$\lim_{N \to \infty} \frac{1}{N} \ln P_F(\delta_{\mathrm{G}}, N, \mathbf{p}_0) = -D(\mathcal{R}_1 | \mathbf{p}_0) \leq -\eta\,, \tag{5.137}$$

where the last inequality is due to the fact that the construction (5.124) of the set $\mathcal{R}_1$ ensures all its points are at a KL "distance" of at least η from all points of $\mathcal{P}_0$. Thus property (a) of the GLRT is proved.

To evaluate the probability of a miss, we can distinguish two cases, depending on whether $\mathbf{p}_1 \in \mathcal{P}_1$ is in the guard band $\mathcal{P}_1 \cap \mathcal{R}_0$ surrounding $\mathcal{P}_0$

which is used by the GLRT to ensure that the size has the proper exponential decay rate, or whether it belongs to $\mathcal{R}_1$.

Case 1: When $\mathbf{p}_1$ is in the interior of the guard band $\mathcal{P}_1 \cap \mathcal{R}_0$, by selecting $\mathcal{R} = \mathcal{R}_1$ and $\mathbf{p} = \mathbf{p}_1$ in Sanov's theorem we find

$$\lim_{N \to \infty} \frac{1}{N} \ln(1 - P_M(\delta_G, N, \mathbf{p}_1)) = -D(\mathcal{R}_1|\mathbf{p}_1) < 0 \,. \tag{5.138}$$

This implies

$$\lim_{N \to \infty} P_M(\delta_G, N, \mathbf{p}_1) = 1 \,,$$

and thus

$$\lim_{N \to \infty} \frac{1}{N} \ln P_M(\delta_G, N, \mathbf{p}_1) = 0 \,. \tag{5.139}$$

In this case, the points $\mathbf{p}_1$ are just too close to $\mathcal{P}_0$ to be detectable under the given exponential decay constraint for the size of the test.

Case 2: When $\mathbf{p}_1$ is in the interior of $\mathcal{R}_1$, by applying Sanov's theorem with $\mathcal{R} = \mathcal{R}_0$ and $\mathbf{p} = \mathbf{p}_1$, we obtain

$$\lim_{N \to \infty} \frac{1}{N} \ln P_M(\delta_G, N, \mathbf{p}_1) = -D(\mathcal{R}_0|\mathbf{p}_1) < 0 \,, \tag{5.140}$$

so that the probability of a miss decreases exponentially.

Let us turn now to proving the asymptotic optimality of the GLRT. If δ is an arbitrary test statisfying the asymptotic size decay constraint (5.133), by applying Sanov's theorem with $\mathcal{R} = \mathcal{R}_1'$ and $\mathbf{p} = \mathbf{p}_0 \in \mathcal{P}_1$, we have

$$-\lim_{N \to \infty} \frac{1}{N} \ln P_F(\delta, N, \mathbf{p}_0) = D(\mathcal{R}_1'|\mathbf{p}_0) \geq \eta \,. \tag{5.141}$$

Thus all points of $\mathcal{R}_1'$ must be at a KL distance η of all points of $\mathcal{P}_0$. This implies that the decision region $\mathcal{R}_1'$ must be contained in $\mathcal{R}_1$, i.e.

$$\mathcal{R}_1' \subseteq \mathcal{R}_1 \,, \tag{5.142}$$

or equivalently

$$\mathcal{R}_0 \subseteq \mathcal{R}_0' \,, \tag{5.143}$$

as illustrated by Fig. 5.9. Then, by using an approach similar to the one used in (5.139) and (5.140) to evaluate the asymptotic rates of decay of the probability of a miss of the GLRT, we find that when $\mathbf{p}_1 \in \mathcal{P}_1$ is located inside $\mathcal{R}_0' \cap \mathcal{P}_1$, the probability of a miss is asymptotically equal to 1, so that

$$\lim_{N \to \infty} \frac{1}{N} \ln P_M(\delta, N, \mathbf{p}_1) = 0 \,, \tag{5.144}$$

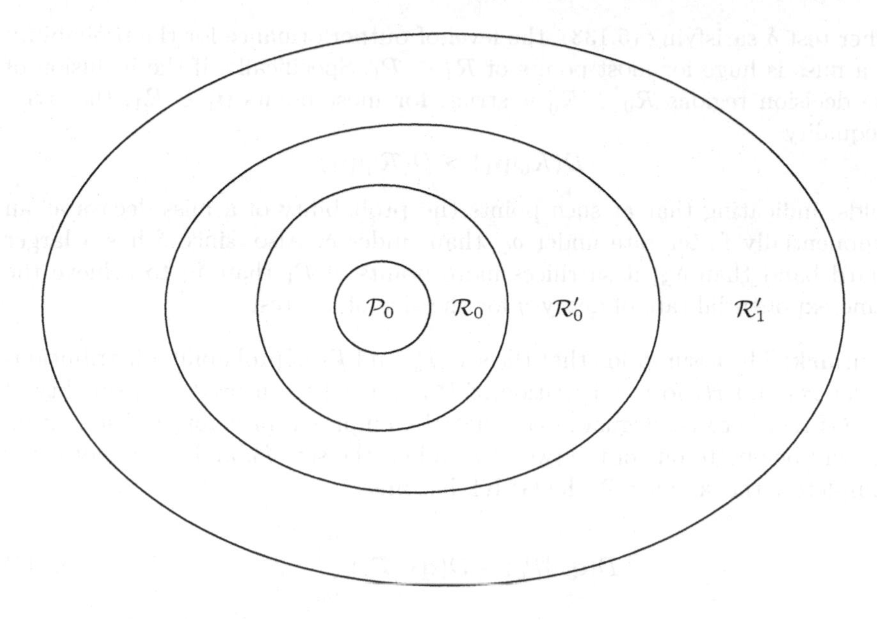

Fig. 5.9. Inclusion relation for the decision regions $\mathcal{R}_0$ and $\mathcal{R}'_0$ of tests δ_G and δ satisfying (5.130).

and when $\mathbf{p}_1$ is located inside $\mathcal{R}'_1$, we have

$$\lim_{N \to \infty} \frac{1}{N} \ln P_M(\delta, N, \mathbf{p}_1) = -D(\mathcal{R}'_0 | \mathbf{p}_1) . \tag{5.145}$$

But since $\mathcal{R}_0$ is contained in $\mathcal{R}'_0$ we necessarily have

$$D(\mathcal{R}_0 | \mathbf{p}_1) \geq D(\mathcal{R}'_0 | \mathbf{p}_1) \tag{5.146}$$

for all $\mathbf{p}_1 \in \mathcal{R}_1$. Taking this observation into account, by combining (5.139)–(5.140) with (5.144)–(5.145) we obtain (5.134) for all $\mathbf{p}_1 \in \mathcal{P}_1$.

The optimality property (5.134) of the GLRT of type I suggests that for composite hypotheses and an infinite sequence of observations, the GLRT plays the same role as the NP test for simple hypotheses and a finite block of observations. In [14, Chap. 7] and [15] a test satisfying Properties (a) and (b) is called a *universal test* since under the exponential decay constraint for the size of the test, it outperforms any other test that does not have any knowledge of the true probability distributions $\mathbf{p}_0$ and $\mathbf{p}_1$ of the Y_k's under H_0 and H_1. It is also worth noting that, in addition to ensuring that the probability of a miss decays exponentially at a rate greater or equal to η for points of $\mathcal{P}_0$, the probability of a miss also decays exponentially at rate $D(\mathcal{R}_0 | \mathbf{p}_1)$ for all points $\mathbf{p}_1$ in the interior of $\mathcal{R}_1$. The only region for which the test is ineffective is the guard region $\mathcal{R}_0 \cap \mathcal{P}_1$, where the probability of a miss tends asymptotically to one. Further, not only does δ_G outperform any other test δ satisfying (5.133), the level of outperformance for the probability of a miss is huge for most points of $\mathcal{R}_1 \subset \mathcal{P}_1$. Specifically, if the inclusion of the decision regions $\mathcal{R}_0 \subset \mathcal{R}'_0$ is strict, for most points $\mathbf{p}_1 \in \mathcal{R}_1$, the strict inequality

$$D(\mathcal{R}_0|\mathbf{p}_1) > D(\mathcal{R}'_0|\mathbf{p}_1)$$

holds, indicating that at such points the probability of a miss decays at an exponentially faster rate under δ_G than under δ. Also, since δ has a larger guard band than δ_G, it sacrifices more points of $\mathcal{P}_1$ than δ_G to achieve the same exponential rate of decay η for the size of the test.

Remark: The assumption that the sets $\mathcal{P}_0$ and $\mathcal{P}_1$ of probability distributions under H_0 and H_1 form a partition of $\mathcal{P}$ is crucial to ensure the optimality of the GLRT, because it guarantees that the empirical probability distribution $\mathbf{q}_N$ will belong to one of the two sets. When the sets $\mathcal{P}_0$ and $\mathcal{P}_1$ do not cover completely the simplex $\mathcal{P}$, the GLRT becomes

$$D(\mathbf{q}_N|\mathcal{P}_0) - D(\mathbf{q}_N|\mathcal{P}_1) \underset{H_0}{\overset{H_1}{\gtrless}} \gamma \,. \tag{5.147}$$

In this case, a simple counterexample was constructed in [16] showing that the GLRT is not necessarily optimal. The main problem is that for a target rate of decay η of the size, it is not always possible to adjust the threshold $\gamma(\eta)$ in such a way that the GLRT becomes asymptotically optimal.

Simple null hypothesis: Consider the special case where $\mathcal{P}_0$ reduces to a single point $\mathbf{p}_0$ and $\mathcal{P}_1$ is the set obtained by removing $\mathbf{p}_0$ from the simplex $\mathcal{P}$. In this case

$$\mathcal{R}_0 = \{\mathbf{r} \in \mathcal{P} : D(\mathbf{r}|\mathbf{p}_0) < \eta\}$$

is the KL "ball" of radius η centered at $\mathbf{p}_0$, and the GLRT can be expressed as

$$D(\mathbf{q}_N|\mathbf{p}_0) \underset{H_0}{\overset{H_1}{\gtrless}} \eta \,. \tag{5.148}$$

This test fits exactly the template for an asymptotic test of a simple hypothesis against the alternative proposed at the beginning of this section, since it just involves comparing the ML estimate $\hat{\mathbf{p}} = \mathbf{q}_N$ of the probability distribution to the probability distribution $\mathbf{p}_0$ under H_0 and then selecting H_1 or H_0 depending on how close $\mathbf{q}_N$ is to $\mathbf{p}_0$. The only new information is that the metric used to determine whether $\mathbf{q}_N$ is close to $\mathbf{p}_0$ is the KL metric.

In the previous discussion of the asymptotic optimality of the GLRT, the decay rate η for the probability of false alarm is arbitrary. But suppose that for the case of a simple null hypothesis we select η arbitrarily small. The decision region $\mathcal{R}_0$ shrinks to a small neighborhood of $\mathbf{p}_0$, so for an arbitrary point $\mathbf{p} \neq \mathbf{p}_0$, the asymptotic decay rate of the probability of a miss is given by

$$
\begin{aligned}
-\lim_{N\to\infty} \frac{1}{N} \ln P_M(\delta_{\mathrm{G}}, N, \mathbf{p}) &= D(\mathcal{R}_0|\mathbf{p}) \\
&= D(\mathbf{p}_0|\mathbf{p}) - \epsilon \,, \qquad (5.149)
\end{aligned}
$$

where ϵ can be made arbitrarily small. If we compare this result to expression (3.33) of Section 3.2 for the asymptotic decay rate of the probability of a miss of an NP test of type I for two simple hypotheses, we see that the GLRT (5.148) achieves the same level of performance as an NP test where the probability distributions of the observations under H_0 and H_1 are known at the outset. Thus, for a vanishingy small η, the GLRT of type I is an *asymptotically clairvoyant* test in the sense that even though it does not know $\mathbf{p}$ under H_1, it achieves the same level of performance as the NP test of type I based on the knowledge of $\mathbf{p}$. Note, of course, that while the GLRT has the same asymptotic performance as the NP test of type I, it is suboptimal for a finite number of observations, so the "clairvoyance" is only of an asymptotic nature.

Example 5.11: Random number generator

A random number generator produces a sequence $\{Y_k,\ k \geq 1\}$ of i.i.d. binary random variables. If the random number generator is working properly, the Y_k's have for probability distribution

$$
P[Y_k = 0|H_0] = P[Y_k = 1|H_0] = \frac{1}{2}
$$

whereas if the algorithm used to generate the Y_k's is incorrect, we have

$$
P[Y_k = 0|H_1] = p \quad , \quad P[Y_k = 1|H_1] = 1 - p
$$

with $p \neq 1/2$. Thus we need to decide whether the sequence of observations has probability distribution

$$
\mathbf{p}_0 = \begin{bmatrix} 1/2 \ 1/2 \end{bmatrix}
$$

or

$$
\mathbf{p} = \begin{bmatrix} p \ 1-p \end{bmatrix} .
$$

Among the first N observations, let N_0 and N_1 denote the number of observations which are equal to zero or one, respectively. Thus $N_0 + N_1 = N$. Then the empirical distribution takes the form

$$
\mathbf{q}_N = \begin{bmatrix} q_1^N \ q_2^N \end{bmatrix} = \begin{bmatrix} N_0/N \ N_1/N \end{bmatrix}
$$

and the GLRT takes the form

$$
\begin{aligned}
D(\mathbf{q}_N|\mathbf{p}_0) &= q_1^N \ln(2q_1^N) + q_2^N \ln(2q_2^N) \\
&= \ln(2) - H(\mathbf{q}_N) \underset{H_0}{\overset{H_1}{\gtrless}} \eta \,,
\end{aligned}
$$

where

$$H(\mathbf{p}) \triangleq -\left[p\ln(p) + (1-p)\ln(1-p)\right]$$

denotes the entropy of probability distribution $\mathbf{p}$. □

5.6.2 Exponential Families

The class of multinomial distributions is not the only one for which the GLRT is asymptotically optimal and, in fact soon after Hoeffding's work, his results were extended first to Gaussian distributions by Herr [17], and then to exponential families of probability distributions by Efron and Truax [18] for the case of a scalar parameter, and by Kallenberg [19, Chap 3] (see also [20]) for a vector parameter. Since the derivation of these results involves advanced mathematical techniques, we will only describe the results and refer the reader to the original sources for proofs.

Consider the exponential family of densities of the form

$$f(\mathbf{y}|\mathbf{x}) = u(\mathbf{y})\exp(\mathbf{x}^T\mathbf{S}(\mathbf{y}) - t(\mathbf{x})) \tag{5.150}$$

introduced in (4.80). In this expression, the observation vector is $\mathbf{y} \in \mathbb{R}^n$ and the parameter vector $\mathbf{x} \in \mathbb{R}^m$. The domain $\mathcal{X}$ of $\mathbf{x}$ is formed by the values of $\mathbf{x}$ such that the integral

$$Z(\mathbf{x}) \triangleq \int u(\mathbf{y})\exp(\mathbf{x}^T\mathbf{S}(\mathbf{y}))d\mathbf{y} \tag{5.151}$$

converges. Then

$$t(\mathbf{x}) = \ln Z(\mathbf{x}) \tag{5.152}$$

represents the log-generating function of the exponential density. To understand its role, observe that in (5.150) the vector $\mathbf{S}(\mathbf{Y}) \in \mathbb{R}^m$ is a sufficient statistic for the density. Let $S_i(\mathbf{Y})$ denote the i-th entry of $\mathbf{S}(\mathbf{Y})$. Then

$$\begin{aligned} m_i(\mathbf{x}) &= E[S_i(\mathbf{Y})|\mathbf{x}] \\ &= \int S_i(\mathbf{y})f(\mathbf{y}|\mathbf{x})d\mathbf{y} = \frac{\partial}{\partial x_i}t(\mathbf{x}) \, , \end{aligned}$$

and for $p \geq 2$, if i_k with $1 \leq k \leq p$ denotes an arbitrary sequence of indices between 1 and m, some of them possibly repeated, we have

$$\begin{aligned} E[\prod_{k=1}^{p}(S_{i_k}(\mathbf{Y}) - m_{i_k}(\mathbf{x}))|\mathbf{x}] &= \int \prod_{k=1}^{p}(S_{i_k}(\mathbf{y}) - m_{i_k}(\mathbf{x}))f(\mathbf{y}|\mathbf{x})d\mathbf{y} \\ &= \frac{\partial^p t(\mathbf{x})}{\partial x_{i_1}\cdots\partial x_{i_p}} \, . \end{aligned} \tag{5.153}$$

In other words all cumulants of the entries of the sufficient statistic vector can be obtained by taking partial derivatives of $t(\mathbf{x})$. As was noted earlier, the exponential family (5.150) includes exponential distributions, Gaussians, and a number of other distributions. In the context of statistical mechanics, (5.150) forms a Gibbs distribution, and $Z(\mathbf{x})$ represents the partition function.

ML estimate: If $\{\mathbf{Y}_k, \ 1 \leq k \leq N\}$ is a block of N i.i.d. observations vectors with density (5.150), the logarithm of their joint density can be expressed as

$$\begin{aligned} \ln(f(\mathbf{Y}|\mathbf{x})) &= \sum_{k=1}^{N} \ln(f(\mathbf{Y}_k|\mathbf{x})) \\ &= N[\mathbf{x}^T\bar{\mathbf{S}} - t(\mathbf{x}) + \overline{\ln(u)}] \,, \end{aligned} \tag{5.154}$$

where

$$\bar{\mathbf{S}} \stackrel{\triangle}{=} \frac{1}{N} \sum_{k=1}^{N} S(\mathbf{Y}_k) \tag{5.155}$$

and

$$\overline{\ln(u)} \stackrel{\triangle}{=} \frac{1}{N} \sum_{k=1}^{N} \ln\left(u(\mathbf{Y}_k)\right)$$

denote respectively the sampled means of the sufficient statistic and of $\ln(u(\mathbf{Y}_k)$ for $1 \leq k \leq N$. The ML estimate of the parameter vector $\mathbf{x}$ is obtained by maximizing the log-likelihood function (5.154). Setting its derivative equal to zero yields

$$\begin{aligned} \mathbf{0} &= \nabla_{\mathbf{x}} \ln f(\mathbf{Y}|\mathbf{x}) \\ &= \bar{S} - \nabla_{\mathbf{x}} t(\mathbf{x}) \,. \end{aligned} \tag{5.156}$$

Thus the ML estimate $\hat{\mathbf{x}}_{\mathrm{ML}}$ is obtained by solving

$$\bar{S} = E[\mathbf{S}(\mathbf{Y})|\mathbf{x}] \,, \tag{5.157}$$

i.e., by equating the sampled mean to the statistical mean of the sufficient statistic vector. To evaluate the maximized value of the likelihood function, it is convenient to introduce the m-dimensional Legendre transform

$$I(\mathbf{s}) \stackrel{\triangle}{=} \max_{\mathbf{x} \in \mathcal{X}} (\mathbf{x}^T \mathbf{s} - t(\mathbf{x})) \tag{5.158}$$

of the log-generating function $t(\mathbf{x})$. In this case

$$\ln(f(\mathbf{Y}|\hat{\mathbf{x}}_{\mathrm{ML}}) = N(I(\bar{S}) + \overline{\ln(u)}) \,. \tag{5.159}$$

From this perspective we alse see that the ML estimate

$$\hat{\mathbf{x}}_{\mathrm{ML}} = \arg \max_{\mathbf{x} \in \mathcal{X}} (\mathbf{x}^T \bar{S} - t(\mathbf{x})) \tag{5.160}$$

is the tangency point of the hyperplane perpendicular to the vector $(\bar{S}, -1)$ and tangent to the surface $(\mathbf{x}, t(\mathbf{x}))$.

GLR statistic: For the distribution (5.150) consider a composite hyothesis testing problem where we must decide between hypotheses

$$H_i : \mathbf{x} \in \mathcal{X}_i$$

for $i = 0, 1$, where the sets $\mathcal{X}_0$ and $\mathcal{X}_1$ form a partitition of $\mathcal{X}$, i.e., $\mathcal{X} = \mathcal{X}_0 \cup \mathcal{X}_1$ and $\mathcal{X}_0 \cap \mathcal{X}_1 = \emptyset$. Then if $\mathbf{Y}$ is the vector representing all observations $\mathbf{Y}_k$ with $1 \leq k \leq N$, the GLR statistic can be expressed as

$$L_{\mathrm{G}}(\mathbf{Y}) = \frac{\max_{\mathbf{x}_1 \in \mathcal{X}_1} f(\mathbf{Y}|\mathbf{x})}{\max_{\mathbf{x}_0 \in \mathcal{X}_0} f(\mathbf{Y}|\mathbf{x})} . \quad (5.161)$$

Taking logarithms, exchanging maximizations with the logarithm operation, and substituting the form (5.154) of the log-density gives

$$\begin{aligned} H(\mathbf{Y}) &\stackrel{\triangle}{=} \frac{1}{N} \ln L_{\mathrm{G}}(\mathbf{Y}) \\ &= \max_{\mathbf{x}_1 \in \mathcal{X}_1} (\mathbf{x_1}^T \bar{S}(\mathbf{Y}) - t(\mathbf{x}_1)) - \max_{\mathbf{x}_0 \in \mathcal{X}_0} (\mathbf{x_0}^T \bar{S}(\mathbf{Y}) - t(\mathbf{x}_0)) . \quad (5.162) \end{aligned}$$

Let now

$$\hat{\mathbf{x}}_i \stackrel{\triangle}{=} \arg \max_{\mathbf{x}_i \in \mathcal{X}_i} (\mathbf{x}_i^T \bar{\mathbf{S}}(\mathbf{Y}) - t(\mathbf{x}_i)) \quad (5.163)$$

for $i = 0, 1$. Observe that

$$\begin{aligned} \hat{\mathbf{x}}_0 &= \hat{\mathbf{x}}_{\mathrm{ML}} \text{ for } \hat{\mathbf{x}}_{\mathrm{ML}} \in \mathcal{X}_0 \\ \hat{\mathbf{x}}_1 &= \hat{\mathbf{x}}_{\mathrm{ML}} \text{ for } \hat{\mathbf{x}}_{\mathrm{ML}} \in \mathcal{X}_1 \quad (5.164) \end{aligned}$$

so that by taking into account the definition (5.158) of the Legendre transform, we obtain

$$\begin{aligned} H(\mathbf{Y}) &= I(\bar{S}) - \max_{\mathbf{x}_0 \in \mathcal{X}_0} (\mathbf{x}_0^T \bar{\mathbf{S}}(\mathbf{Y}) - t(\mathbf{x}_0)) \\ &= \min_{\mathbf{x}_0 \in \mathcal{X}_0} \Big[I(\bar{S}) - (\mathbf{x}_0^T \bar{\mathbf{S}}(\mathbf{Y}) - t(\mathbf{x}_0)) \Big] \quad (5.165) \end{aligned}$$

for $\hat{\mathbf{x}}_{\mathrm{ML}} \in \mathcal{X}_1$ and

$$= - \min_{\mathbf{x}_1 \in \mathcal{X}_1} \Big[I(\bar{S}) - (\mathbf{x}_1^T \bar{\mathbf{S}}(\mathbf{Y}) - t(\mathbf{x}_1)) \Big] \quad (5.166)$$

for $\hat{\mathbf{x}}_{\mathrm{ML}} \in \mathcal{X}_0$.

To interpret expressions (5.165) and (5.166) for the scaled log-GLR statistic, note that the KL divergence of $f(\cdot|\mathbf{x}_1)$ and $f(\cdot|\mathbf{x}_0)$ can be expressed as

$$\begin{aligned} D(\mathbf{x}_1|\mathbf{x}_0) &= \int f(\mathbf{y}|\mathbf{x}_1) \ln \Big(\frac{f(\mathbf{y}|\mathbf{x}_1)}{f(\mathbf{y}|\mathbf{x}_0)} \Big) d\mathbf{y} \\ &= (\mathbf{x}_1 - \mathbf{x}_0)^T E[\mathbf{S}(\mathbf{y})|\mathbf{x}_1] - (t(\mathbf{x}_1) - t(\mathbf{x}_0)) . \quad (5.167) \end{aligned}$$

Evaluating this expression at $\mathbf{x} = \hat{\mathbf{x}}_{\mathrm{ML}}$ and taking into account (5.157) yields

$$\begin{aligned} D(\hat{\mathbf{x}}_{\mathrm{ML}}|\mathbf{x}_0) &= (\hat{\mathbf{x}}_{\mathrm{ML}} - \mathbf{x}_0)^T \bar{S} - (t(\hat{\mathbf{x}}_{\mathrm{ML}}) - t(\mathbf{x}_0)) \\ &= I(\bar{S}) - (\mathbf{x}_0^T \bar{S} - t(\mathbf{x}_0)) \end{aligned} \tag{5.168}$$

Then, denote the KL "distance" between a point $\mathbf{x}_1 \in \mathcal{X}_1$ and set $\mathcal{X}_0$ as

$$D(\mathbf{x}_1|\mathcal{X}_0) = \inf_{\mathbf{x}_0 \in \mathcal{X}_0} D(\mathbf{x}_1|\mathbf{x}_0) . \tag{5.169}$$

Comparing identities (5.168)–(5.169) with expressions (5.165)–(5.166) for the normalized log-GLR statistic, we find

$$H(\mathbf{Y}) = \begin{cases} D(\hat{\mathbf{x}}_{\mathrm{ML}}|\mathcal{X}_0) & \text{for } \hat{\mathbf{x}}_{\mathrm{ML}} \in \mathcal{X}_1 \\ -D(\hat{\mathbf{x}}_{\mathrm{ML}}|\mathcal{X}_1) & \text{for } \hat{\mathbf{x}}_{\mathrm{ML}} \in \mathcal{X}_0 , \end{cases} \tag{5.170}$$

which has a form similar to Hoeffding's statistic (5.120) for the case of multinomial distributions.

GLRT: If we select a positive threshold $\eta > 0$, we obtain a GLRT of type I:

$$D(\hat{\mathbf{x}}_{\mathrm{ML}}|\mathcal{X}_0) \underset{H_0}{\overset{H_1}{\gtrless}} \eta . \tag{5.171}$$

This test selects H_0 if the ML estimate $\hat{\mathbf{x}}_{\mathrm{ML}}$ is at a KL distance less than η of the set $\mathcal{X}_0$, and it selects H_1 otherwise. As we shall see below, this test is biased towards ensuring that the size of the test decays with rate η. Similarly for a negative threshold $\eta < 0$, we obtain the GLRT of type II:

$$D(\hat{\mathbf{x}}_{\mathrm{ML}}|\mathcal{X}_1) \underset{H_1}{\overset{H_0}{\gtrless}} -\eta . \tag{5.172}$$

which is biased towards ensuring that the probability of a miss decays with rate $-\eta$ for all points of $\mathcal{X}_1$. Comparing tests (5.171) and (5.172) to the multinomial tests (5.122) and (5.126), we see that the tests have exactly the same form, except that the empirical distribution vector $\mathbf{q}_N$ is replaced by the ML estimate $\hat{\mathbf{x}}_{\mathrm{ML}}$ of the parameter vector $\mathbf{x}$, and sets $\mathcal{P}_i$ are replaced by $\mathcal{X}_i$ with $i = 0, 1$. This analogy is not just formal, since the GLRT admits almost exactly the same properties for exponential families of distributions as in the multinomial case. The only significant difference is that while the simplex $\mathcal{P}$ is bounded, the domain $\mathcal{X}$ of the exponential density (5.150) may be unbounded. Thus, let δ_{G} denote test (5.171). Then under the assumption that $\mathcal{X}_0$ is contained in a compact set in interior of $\mathcal{X}$, the following results are proved in Theorems 3.2.1 and 3.3.2 of [19]:

(a) The probability of false alarm has the exponential decay rate

$$-\lim_{N \to \infty} \frac{1}{N} \ln P_F(\delta_G, N, \mathbf{x}_0) \geq \eta \tag{5.173}$$

for all points $\mathbf{x}_0$ of $\mathcal{X}_0$.

(b) Let δ be any other test specified by

$$\delta(\hat{\mathbf{x}}_{\mathrm{ML}}) = \begin{cases} 1 & \hat{\mathbf{x}}_{\mathrm{ML}} \in \mathcal{R}'_1 \\ 0 & \hat{\mathbf{x}}_{\mathrm{ML}} \in \mathcal{R}'_0 \,, \end{cases} \tag{5.174}$$

where the sets $\mathcal{R}'_0$ and $\mathcal{R}'_1$ are disjoint and form a partition of $\mathcal{X}$. Then if the test δ satisfies

$$-\lim_{N\to\infty} \frac{1}{N} \ln P_F(\delta, N, \mathbf{x}_0) \geq \eta \,, \tag{5.175}$$

for all $\mathbf{x}_0 \in \mathcal{X}_0$, it must be such that

$$-\lim_{N\to\infty} \frac{1}{N} \ln P_M(\delta_{\mathrm{G}}, N, \mathbf{x}_1) \geq -\lim_{N\to\infty} \frac{1}{N} \ln P_M(\delta, N, \mathbf{x}_1) \tag{5.176}$$

for all $\mathbf{x}_1 \in \mathcal{X}_1$.

Example 5.12: Zero-mean Gaussian densities

Consider a sequence $\{Y_k,\ k \geq 1\}$ of $N(0,v)$ i.i.d. random variables for which we wish to test between $H_0 : v = v_0 > 0$ and $H_1 : v \neq v_0$. The density

$$f(Y|v) = \frac{1}{(2\pi v)^{1/2}} \exp(-\frac{y^2}{2v})$$

can be written in the exponential form (5.145) by setting $x = (2v)^{-1}$, $S(Y) = -Y^2$ and

$$t(x) = \ln((\pi/x)^{1/2}) \,.$$

So the parameter x is scalar and its domain is $\mathcal{X} = \mathbb{R}^+$ where $\mathbb{R}^+$ denotes the set of strictly positive real numbers. With this parametrization, the ML equation (5.156) takes the form

$$-\overline{Y^2} = -\frac{1}{N} \sum_{k=1}^{N} Y_k^2 = E[-Y^2|x] = -\frac{1}{2x} \,,$$

so that

$$\hat{x}_{\mathrm{ML}} = \frac{1}{2\overline{Y^2}} \,.$$

Setting $x_0 = (2v_0)^{-1}$, the KL divergence between $f(y|x)$ and $f(y|x_0)$ can be expressed as

$$\begin{aligned} D(x|x_0) &= (x - x_0)E[-Y^2|x] - (t(x) - t(x_0)) \\ &= \frac{1}{2}\Big[(\frac{x_0}{x} - 1) - \ln\big(\frac{x_0}{x}\big)\,.\Big] \end{aligned}$$

Then the GLRT

$$2D(\hat{x}_{\mathrm{ML}}|x_0) = (x_0/\hat{x}_{\mathrm{ML}}) - 1 - \ln(x_0/\hat{x}_{\mathrm{ML}}) \underset{H_0}{\overset{H_1}{\gtrless}} 2\eta$$

is asymptotically optimal among all tests that do not know x and satisfy (5.175). □

Finally, it is worth noting that an approach was proposed by Brown [21] for extending the asymptotic optimality properties of the GLRT to the case of arbitrary densities, but at the cost of a slight enlargement of the parameter space $\mathcal{X}$ and of its $\mathcal{X}_0$, $\mathcal{X}_1$ components.

5.7 Bibliographical Notes

While the topic of composite hypothesis testing plays a prominent role in statistics texts [1, 22], the coverage devoted to this subject in signal detection textbooks varies greatly depending on whether a Bayesian or a frequentist viewpoint is adopted. From a Bayesian viewpoint, composite hypothesis testing is completely unnecessary, since the unknown parameter vector $\mathbf{x}$ can be assigned a probability density $f(\mathbf{x}|H_j)$ under hypothesis H_j with $j = 0,\ 1$, which can then be used to evaluate the marginal density

$$f_{\mathbf{Y}}(\mathbf{y}|H_j) = \int f_{\mathbf{Y}}(\mathbf{y}|\mathbf{x}, H_j) f(\mathbf{x}|H_j) d\mathbf{x}$$

for $j = 0,\ 1$, thereby reducting the hypothesis testing problem to a binary hypothesis testing problem of the type discussed in Chapter 2. In-depth treatments of composite hypothesis testing in a detection context are given in [6, 23]. The focus in [6] is primarily on the derivation of UMPI tests and the presentation we have adopted in Section 5.3 has been influenced significantly by this work. On the other hand, [23] provides an in-depth treatment of the GLRT and its properties and applications. The discussion of the asymptotic performance of the GLRT for multinomial distributions in Section 5.6 is based both on Hoeffding's paper [12] and the analysis of the GLRT given in [14, Section 3.5]. Over the last 15 years, a significant amount of attention has been addressed at the development of universal tests for composite hypothesis testing. For the case when H_0 is simple, an abstract extension of Hoeffding's result is proposed in [15], and a competitive approach to the construction of universal tests is described in [24, 25]. In this chapter, we have adopted an agnostic viewpoint with respect to the selection of Bayesian, or UMPI/GLRT approaches for detector design. In practice, the prior distribution for the parameter vector used in the Bayesian approach is obtained either from physical considerations by studying the interaction of the transmitted signal with the propagation channel, or from empirically obtained channel models. In general all approaches tend to produce similar, if not identical, detectors. The GLRT is rather attractive as a default methodology, since it does not require the selection of prior distributions on $\mathbf{x}$, or an analysis of the symmetries satisfied by the detection problem. In fact, it finds the symmetries automatically, since the GLR is always invariant under all groups of transformations that leave the detection problem invariant. However, it is worth noting that the ML estimate of the parameter vector, which is needed to implement the GLRT, is not always easy to evaluate. This suggests that no single detector design technique is likely to work all the time, and consequently a pragmatic approach is recommended when tackling complex detection problems.

5.8 Problems

5.1. Let

$$Y = A + V$$

where the random variable V is uniformly distributed over $[-1/2, 1/2]$ and A is an unknown amplitude. Given Y, we seek to test $H_1 : 0 < A \leq 1$ against the simple hypothesis $H_0 : A = 0$. The probability P_F of false alarm must be less than or equal to α.

(a) Determine whether the test has the monotone likelihood ratio property.
(b) Does a UMP test exist? If so, express its threshold in function of α.
(c) If a UMP test δ_{UMP} exists, evaluate its power $P_D(\delta_{\text{UMP}}, A)$.

5.2. Given an $N(0, v)$ distributed observation Y, we wish to test the composite hypothesis $H_1 : v > v_0$ against the simple hypothesis $H_0 : v = v_0$. The probability P_F of false alarm of the test must be less than or equal to α.

(a) Does a uniformly most powerful test exist? If so, express the threshold in function of α.
(b) If the answer to part (a) is affirmative, evaluate the power $P(\delta_{\text{UMP}}, v)$ of the test in function of the unknown variance v.
(c) Suppose now that you wish to test H_1 against the composite hypothesis $H_0' : 0 < v \leq v_0$. It is required that the size of the test should be α, i.e.,

$$\max_{0 < v \leq v_0} P_F(\delta, v) \leq \alpha \,.$$

Does a UMP test exist? If so, explain how its threshold is selected.
(d) Does the answer to part (c) change if the range of values for v under H_0' is $v_0/2 \leq v \leq v_0$ with $v_0 > 0$?

5.3. Consider the composite binary hypothesis testing Problem 5.2, where, given an $N(0, v)$ random variable Y, we seek to test $H_1 : v > v_0$ against the simple hypothesis $H_0 : v = v_0$.

(a) Design a GLRT test δ_G. Select its threshold so that the probability of false alarm is α.
(b) Compare the GLRT to the UMP test derived in Problem 5.2.

5.4. Let $\{Y_k, 1 \leq k \leq N\}$ denote N independent identically distributed Poisson random variables with parameter λ, i.e., the random variables Y_k admit the probability distribution

$$P[Y_k = n] = \frac{\lambda^n}{n!} \exp(-\lambda)$$

for $n \geq 0$.

(a) Show that

$$S = \sum_{k=1}^{N} Y_k$$

is a sufficient statistic for estimating λ from observations of the random variables Y_k with $1 \leq k \leq N$.
(b) Show that S is a Poisson random variable with parameter $N\lambda$.
(c) Consider the binary hypothesis testing problem where, given the observations $Y_k, 1 \leq k \leq N$, we seek to decide between H_0: $\lambda = \lambda_0$ and H_1: $\lambda = \lambda_1$. Obtain the likelihood ratio test.
(d) Determine whether there exists a UMP test of size α for testing H_0: $\lambda \leq 1$ versus H_1: $\lambda > 1$. Recall that if $P_F(\delta, \lambda)$ denotes the probability of false alarm of a test δ as a function of the unknown parameter λ, the test will have size α if

$$\max_{\lambda \leq 1} P_F(\delta, \lambda) = \alpha \,.$$

Explain how the threshold is selected.

5.5. Let Y be a random variable with the Laplace probability density

$$f_Y(y) = \frac{1}{2} \exp(-|y - x|)$$

whose mean x is unknown. We seek to test $H_1 : x > 0$ against $H_0 : x = 0$. The probability of false alarm should not exceed α, with $\alpha < 1/2$.

(a) Determine whether a UMP test δ_{UMP} exists. If so, evaluate its threshold in function of α.
(b) Evaluate the power of test δ_{UMP}.
(c) Design a GLR test δ_G. Express its threshold in function of α.
(d) Evaluate the power of test δ_G.
(e) Compare the UMP and GLR tests.

5.6. Consider the DT detection problem

$$\begin{aligned} H_0 &: Y(t) = V(t) \\ H_1 &: Y(t) = A + Bt + V(t) \end{aligned}$$

for $0 \leq t \leq T$, which consists of detecting a straight line in the presence of a zero-mean WGN $V(t)$ with known variance σ^2. Let $N = T+1$ denote the number of observations. The observations under H_1 can be written conveniently in vector form as

$$\mathbf{Y} = \mathbf{H} \begin{bmatrix} A \\ B \end{bmatrix} + \mathbf{V} ,$$

with

$$\mathbf{Y} = \begin{bmatrix} Y(0) \\ \vdots \\ Y(t) \\ \vdots \\ Y(T) \end{bmatrix} , \quad \mathbf{H} = \begin{bmatrix} 1 & 0 \\ \vdots & \vdots \\ 1 & t \\ \vdots & \vdots \\ 1 & T \end{bmatrix} , \quad \mathbf{V} = \begin{bmatrix} V(0) \\ \vdots \\ V(t) \\ \vdots \\ V(T) \end{bmatrix} ,$$

where the noise vector $\mathbf{V}$ is $N(\mathbf{0}, \sigma^2 \mathbf{I}_N)$ distributed.

(a) Assume that A and B are known. Obtain an NP test for the detection problem. Select the threshold η to ensure the probability P_F of false alarm is less than or equal to α.
(b) Now let A and B be unknown. Explain why no UMP test exists in this case.
(c) Design a GLRT test. This will require finding the ML estimates of A and B under H_1. Select the test threshold to ensure the probability of false alarm does not exceed α.

5.7. Consider the detection of a discrete-time signal $As(t)$ with unknown amplitude $A > 0$ in the presence of an unknown burst interference signal $i(t)$ which is nonzero over a subinterval $I = [t_0, t_1]$ strictly included inside the interval of observation $[0, T]$. So the detection problem takes the form

$$\begin{aligned} H_0 &: Y(t) = i(t) + V(t) \\ H_1 &: Y(t) = As(t) + i(t) + V(t) , \end{aligned} \tag{5.177}$$

for $0 \leq t \leq T$, where $V(t)$ is a zero-mean WGN with variance σ^2.

(a) Write the detection problem in matrix form and verify that under H_1, the observation vector $\mathbf{Y}$ can be written in the form (5.59). What is the matrix $\mathbf{M}$? This implies that the problem is invariant under the group $\mathcal{G}$ of transformations of the form

$$\mathbf{Y}' = \mathbf{Y} + \mathbf{M}\mathbf{d} , \tag{5.178}$$

where $\mathbf{d}$ denotes an arbitrary vector.

(b) Use the approach of Section 5.4 to construct a statistic $\mathbf{S}(\mathbf{Y})$ which is a maximal invariant under $\mathcal{G}$. Then use the results of Example 5.2 to verify that a UMPI test exists for problem (5.177). Set the test threshold to ensure that P_F stays below α.
(c) Evaluate the power $P_D(\delta_{\text{UMPI}}, A)$ of the UMPI test of part (b).

5.8. Consider a GLRT formulation of detection problem (5.177).

(a) Find the ML estimate of amplitude A and burst signal $i(t)$, $t_0 \leq t \leq t_1$ under H_1. Similarly, find the ML estimate of $i(t)$, $t_0 \leq t \leq t_1$ under H_0.
(b) Evaluate the GLR statistic for problem (5.177). Verify that it is invariant under the group $\mathcal{G}$ of transformations specified by (5.178).
(c) Is the GLRT identical to the UMPI test obtained in Problem 5.7? Discuss differences, if any are present.

5.9. Consider the detection of a discrete-time signal $s(t)$ with unknown amplitude $A > 0$ in the presence of an interfering sinusoidal signal with known frequency ω_i and unknown amplitude B and phase Θ in WGN:

$$\begin{aligned} H_0 &: Y(t) = B\cos(\omega_i t + \Theta) + V(t) \\ H_1 &: Y(t) = As(t) + B\cos(\omega_i t + \Theta) + V(t)\,, \end{aligned}$$

$0 \leq t \leq T$. For simplicity, we assume that the length $N = T + 1$ of the observation interval is even, i.e $N = 2p$, and that it is an integer multiple of the period of the interfering sinusoidal signal, so that $\omega_i = 2\pi m/N$, where m is an integer satisfying $0 < m < N/2$. The noise $V(t)$ is a zero-mean WGN with variance σ^2. This problem can be written in the vector form

$$\begin{aligned} H_0 &: \mathbf{Y} = \mathbf{Mn} + \mathbf{V} \\ H_1 &: \mathbf{Y} = A\mathbf{s} + \mathbf{Mn} + \mathbf{V}\,, \end{aligned}$$

where the N-dimensional vectors $\mathbf{Y}$, $\mathbf{s}$, and $\mathbf{V}$ are obtained by stacking successive samples of $Y(t)$, and

$$\mathbf{M}^T = \begin{bmatrix} 1 \cdots \cos(\omega_i t) \cdots \cos(\omega_i T) \\ 0 \cdots \sin(\omega_i t) \cdots \sin(\omega_i T) \end{bmatrix}$$

with

$$\mathbf{n}^T = \begin{bmatrix} B\cos(\Theta) & -B\sin(\Theta) \end{bmatrix}\,.$$

This problem is now in the form of a linear detection problem with interfering sources of the type considered in Section 5.4.

(a) Verify that the problem is invariant under the group $\mathcal{G}$ of transformations of the form

$$\mathbf{Y}' = \mathbf{Y} + \mathbf{Md}\,.$$

(b) To obtain a statistic $\mathbf{T}(\mathbf{Y})$ that is a maximal invariant under $\mathcal{G}$, consider the orthonormal matrix

$$\mathbf{Q} = \begin{bmatrix} \mathbf{C} \\ \mathbf{S} \end{bmatrix}$$

where $\mathbf{C}$ is the $(N/2+1) \times N$ discrete cosine transform (DCT) matrix with entries

$$c_{k\ell} = \begin{cases} (1/N)^{1/2} & k = 1 \\ (2/N)^{1/2} \cos((k-1)(\ell-1)2\pi/N) & 2 \leq k \leq N/2 \\ (1/N)^{1/2} \cos((\ell-1)\pi) & \end{cases}$$

for $1 \leq \ell \leq N$, and $\mathbf{S}$ is the $(N/2-1) \times N$ discrete sine transform (DST) matrix with entries

$$s_{k\ell} = (2/N)^{1/2} \sin(k\ell 2\pi/N)$$

for $1 \leq k \leq N/2 - 1$ and $1 \leq \ell \leq N$. Then denote by

$$\tilde{\mathbf{Y}} = \mathbf{Q}\mathbf{Y} \quad , \quad \tilde{\mathbf{s}} = \mathbf{Q}\mathbf{s} \quad , \quad \tilde{\mathbf{V}} = \mathbf{Q}\mathbf{V}$$

the vectors obtained by applying the transformation $\mathbf{Q}$ to the observations, signal and noise. Note that since $\mathbf{Q}$ is orthonormal, the transformed noise $\tilde{\mathbf{V}}$ remains $N(\mathbf{0}, \sigma^2 \mathbf{I}_N)$ distributed. Since the interfering signal has frequency $\omega_i = m2\pi/N$, all the rows of the transformed matrix

$$\tilde{\mathbf{M}} = \mathbf{Q}\mathbf{M}$$

are zero, except rows $m+1$ and $N/2+m+1$. Let $\check{\mathbf{Y}}$, $\check{\mathbf{s}}$, and $\check{\mathbf{V}}$ be the vectors obtained by deleting the $m+1$-th and $N/2+m+1$-th entries of $\tilde{\mathbf{Y}}$, $\tilde{\mathbf{s}}$ and $\tilde{\mathbf{V}}$, respectively. Show that

$$\mathbf{T}(\mathbf{Y}) = \check{\mathbf{Y}}$$

is a maximal invariant statistic under $\mathcal{G}$, and verify that under the two hypotheses we have

$$\begin{aligned} H_0 &: \check{\mathbf{Y}} = \check{\mathbf{V}} \\ H_1 &: \check{\mathbf{Y}} = A\check{\mathbf{s}} + \check{\mathbf{V}} \end{aligned} \qquad (5.179)$$

(c) Apply the analysis of Example 5.2 to the transformed detection problem (5.179) to obtain a UMPI test for the original detection problem. Set the test threshold to ensure that P_F stays below α.

(d) Evaluate the power $P_D(\delta_{\mathrm{UMPI}}, A)$ of the UMPI test of part c).

5.10. In digital communications, it is sometimes necessary to determine whether a communications link is active or not [26]. Suppose that we are trying to decide whether a BPSK link is active. To simplify the analysis, assume that phase and timing synchronization are perfect. Under H_1, the link is active and after demodulation and sampling, the received discrete-time signal can be expressed as

$$Y(t) = AI(t) + V(t)$$

for $1 \leq t \leq N$, where the binary symbols $I(t)$ are independent and take values ± 1 with probability $1/2$ each. The amplitude A is positive, and $V(t)$ is a zero-mean WGN independent of $I(t)$ with variance σ^2. Under H_0, the link is inactive and the received signal is given by

$$Y(t) = V(t)$$

for $1 \leq t \leq N$. Again, the composite hypothesis testing problem can be written in vector form as

$$\begin{aligned} H_0 &: \mathbf{Y} = \mathbf{V} \\ H_1 &: \mathbf{Y} = A\mathbf{I} + \mathbf{V} \end{aligned}$$

where $\mathbf{Y}$, $\mathbf{I}$ and $\mathbf{V}$ are the N-dimensional vectors with t-th entry $Y(t)$, $I(t)$ and $V(t)$, respectively.

(a) Assume that A and σ^2 are known. The probability density $f_{\mathbf{Y}}(\mathbf{y}|\mathbf{I})$ for the observations under H_1 is parametrized by the unknown symbol vector $\mathbf{I}$. To design a GLR test, the transmitted symbols $I(t)$, $1 \leq t \leq N$ need to be detected under H_1. By solving this detection problem, show that the GLR statistic is given by

$$S(\mathbf{Y}) = \frac{1}{N} \sum_{t=1}^{N} |Y(t)| \,. \tag{5.180}$$

(b) Repeat part (a) for the case where $A > 0$ is unknown but σ^2 is known. Is the resulting GLR statistic different from the one obtained in (a)?

(c) For the GLRT obtained in parts (a) and (b) select the threshold η such that the probability of false alarm does not exceed α. To do so, you may assume that N is large.

(d) Assume now that $\mathbf{I}$, $A > 0$ and σ^2 are all unknown. Show that the GLR test can be expressed as

$$\frac{\frac{1}{N}\sum_{t=1}^{N} |Y(t)|}{(\frac{1}{N}\sum_{t=1}^{N} Y^2(t))^{1/2}} \underset{H_0}{\overset{H_1}{\gtrless}} \eta \,.$$

For the case when N is large, use the strong law of large numbers and the result of part (c) to select the test threshold.

5.11. Consider a complex observation vector of the form

$$\mathbf{Y} = \mathbf{M}\mathbf{X} + \mathbf{V}\,, \tag{5.181}$$

where $\mathbf{X}$ is an unknown complex vector of dimension n, $\mathbf{Y}$ has dimension $p \geq n$ and the complex noise vector $\mathbf{V} = \mathbf{V}_R + j\mathbf{V}_I$ has dimension p and is such that its real and imaginary components $\mathbf{V}_R$ and $\mathbf{V}_I$ are independent zero-mean Gaussian vectors with covariance matrix $\mathbf{K}/2$. This ensures that the complex Gaussian vector $\mathbf{V}$ has the circular zero-mean Gaussian density

$$f_{\mathbf{V}}(\mathbf{v}) = \frac{1}{\pi^p |\mathbf{K}|} \exp\big(-\mathbf{v}^H \mathbf{K}^{-1} \mathbf{v}\big)\,,$$

where $\mathbf{v}^H$ denotes the Hermitian transpose (complex conjugate transpace) of $\mathbf{v}$. We assume that $\mathbf{K}$ is positive definite and the complex matrix $\mathbf{M}$ has full rank, i.e., its rank equals its column size n.

(a) Verify that the maximum likelihood estimate of $\mathbf{X}$ given $\mathbf{Y}$ is given by

$$\hat{\mathbf{X}}_{\text{ML}}(\mathbf{Y}) = (\mathbf{M}^H \mathbf{K}^{-1} \mathbf{M})^{-1} \mathbf{M}^H \mathbf{K}^{-1} \mathbf{Y}\,.$$

Is this estimate unbiased? You are encouraged to perform your analysis in the complex domain, but if you do not feel comfortable doing so, you can always separate the real and imaginary parts of equation (5.181) and then examine the ML estimation in the real domain.

(b) Find the Fisher information matrix for the given observation, and verify that the Cramér-Rao lower bound for any unbiased vector estimate $\hat{\mathbf{X}}(\mathbf{Y})$ takes the form

$$K_E = E[(\mathbf{X} - \hat{\mathbf{X}})(\mathbf{X} - \hat{\mathbf{X}})^H] \geq (\mathbf{M}^H \mathbf{K}^{-1} \mathbf{M})^{-1}\,.$$

Indicate whether the ML estimate of part (a) is efficient.

Consider now the linear uniform antenna array of Fig. 5.10 which consists of p antenna elements with spacing d. For a signal arriving at an angle θ with respect to the normal to the array, a delay occurs between its time of arrival at the first array element and at other elements. Specifically, if the source generating the signal is located in the far field, the received signal takes the form of a plane wave. The relative delay measuring the difference in time at which the plane wave reaches the first and k-th element is given by $\tau_k = (k-1)d\sin(\theta)/c$, where c denotes the speed of light, and where $2 \leq k \leq p$. Assume that over a time snapshot $0 \leq t \leq T$, the signal arriving at the first array element has approximately a constant amplitude and phase and takes the form

$$s(t) = A\cos(\omega_c t + \Phi)$$

where the carrier frequency ω_c is known, but the amplitude A and phase Φ are unknown. The signal received by the k-th array element is therefore

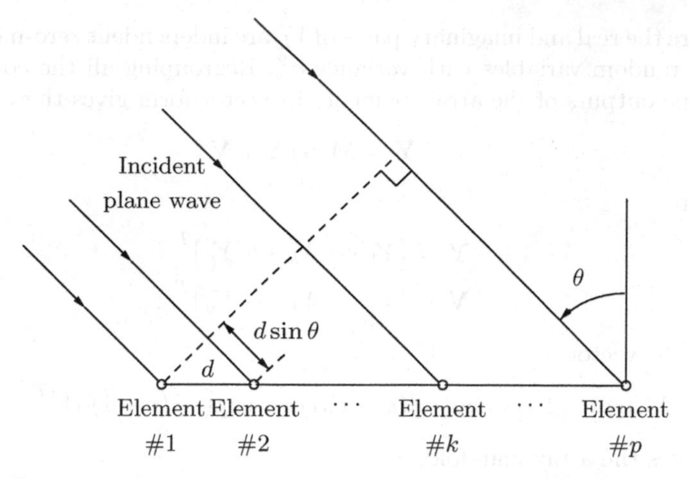

Fig. 5.10. Plane wave incident on a linear antenna array.

$s_k(t) = s(t - \tau_k)$ with $\tau_1 = 0$. The signal is measured in the presence of WGN with intensity σ^2, so for $0 \leq t \leq T$, the signal measured by the k-th antenna element takes the form

$$Y_k(t) = s_k(t) + V_k(t)$$

with $1 \leq k \leq p$. The WGNs $V_k(t)$ at different antenna elements are independent.

(c) Verify that the signal $s_k(t)$ admits the bandpass representation

$$s_k(t) = \Re\Big\{ \exp(-j(k-1)\alpha) X \exp(j\omega_c t) \Big\}$$

where if $\lambda = 2\pi c/\omega_c$ denotes the wavelength of the received plane wave,

$$\alpha = 2\pi d \sin(\theta)/\lambda$$

represents the phase shift created by the longer distance needed for the plane wave to travel between two successive antenna elements. Here $X = A\exp(j\Phi)$ denotes the complex wave amplitude. Then, by using the bandpass representations

$$V_k(t) = \Re\Big\{ V_k \exp(j\omega_c t) \Big\}$$

$$Y_k(t) = \Re\Big\{ Y_k \exp(j\omega_c t) \Big\}$$

for narrowband random signals centered at the frequency ω_c, show that the observation at the k-th array element can be written in the complex envelope domain as

$$Y_k = \exp(-j(k-1)\alpha) X + V_k \tag{5.182}$$

where the real and imaginary parts of V_k are independent zero-mean Gaussian random variables with variance σ^2. Regrouping all the complex envelope outputs of the array elements in vector form gives therefore

$$\mathbf{Y} = \mathbf{M}(\alpha)X + \mathbf{V} \tag{5.183}$$

where

$$\mathbf{Y} = \begin{bmatrix} Y_1 \cdots Y_k \cdots Y_p \end{bmatrix}^T$$
$$\mathbf{V} = \begin{bmatrix} V_1 \cdots V_k \cdots V_p \end{bmatrix}^T ,$$

and the vector

$$\mathbf{M}(\alpha) = \begin{bmatrix} 1 \cdots \exp(-j(k-1)\alpha) \cdots \exp(-j(p-1)\alpha) \end{bmatrix}^T \tag{5.184}$$

denotes the array manifold.

(d) In the vector observation (5.183), both the complex number X and α (or equivalently the incidence angle θ) are unknown. Use the results of part (a) to evaluate the joint maximum likelihood estimate of X and α.

(e) Find the 2×2 Fisher information matrix $\mathbf{J}(X, \alpha)$ for the joint estimation of X and α and use it to obtain a matrix Cramér-Rao lower bound for the error covariance matrix of joint unbiased estimates of X and α.

(f) Design a GLRT test for the detection problem where under hypothesis H_1 a signal with incidence angle θ is impinging upon the array in the presence of sensor noise, i.e.,

$$H_1 \;:\; Y_k(t) = s_k(t) + V_k(t) \;\; 0 \leq t \leq T ,$$

with $1 \leq k \leq p$, and under hypothesis H_0 no signal is present, so that

$$H_0 \;:\; Y_k(t) = V_k(t) \;\; 0 \leq t \leq T ,$$

with $1 \leq k \leq p$.

(g) Assume now that in addition to a signal $s(t)$ of unknown magnitude, phase and incidence angle, an interfering signal $i(t)$ with known angle of incidence ψ, but unknown magnitude and phase, is present so that the vector formed by the complex envelopes of the array elements takes the form

$$\mathbf{Y} = \mathbf{M}(\alpha)X + \mathbf{M}(\beta)Z + \mathbf{V} ,$$

where $\beta = 2\pi d \sin(\psi)/\lambda$ and the complex number Z represents the unknown amplitude and phase of $i(t)$. Note that β is *known*, but Z is unknown. By first estimating X and Z and then α obtain a procedure for the joint ML estimation of X, Z and α. Obtain a condition involving β and α under which the estimates for X and α reduce to those obtained in part (d) without the interfering signal $i(t)$. In other words, find a condition for α and β such that the presence of the interfering signal $i(t)$ does not affect the estimation of X and α. When expressed in terms of θ and ψ, this condition specifies the *resolution* of the array.

5.12. Let Y_k, $1 \leq k \leq N$ be independent identically distributed exponential random variables with parameter λ.

(a) Verify that their density

$$f_Y(y|\lambda) = \lambda \exp(-\lambda y) u(y) \, ,$$

where $u(\cdot)$ denotes the unit step function, can be written in the form (5.150). Identify x, $S(Y)$ and $t(x)$.

(b) If $f(\cdot|x)$ denotes the parametrization of the exponential density obtained in (a), use expression (5.167) to evaluate the KL divergence $D(x_1|x_0)$ of exponential densities $f(\cdot|x_1)$ and $f(\cdot|x_0)$. Recall that if Y is exponential with parameter λ, its mean is $1/\lambda$.

(c) Assume that the parameter λ, or equivalently x, is unknown. Given observations Y_k, $1 \leq k \leq N$, solve equation (5.157) to obtain the ML estimate $\hat{x}_{\text{ML}}$ of the unknown parameter x.

(d) Design a GLR test for choosing between $H_0 : x = x_0$ and $H_1 : x \neq x_0$. Verify that the GLRT can be expressed as

$$D(\hat{x}_{\text{ML}}|x_0) \underset{H_0}{\overset{H_1}{\gtrless}} \eta \, ,$$

where $D(x_1|x_0)$ is the KL divergence obtained in part (b), and $\hat{x}_{\text{ML}}$ is the ML estimate of part (c).

5.A. Proof of Sanov's Theorem

We follow the proof given in [13, Sec. 2.1]. Let $\mathbf{q}_N$ be the empirical probability distribution (5.109). As noted earlier, the set $\mathcal{Q}_N$ of empirical probability distributions based on N samples has $|\mathcal{Q}_N| = \binom{N+K}{K-1}$ points. So if we let

$$\begin{aligned} M(\mathcal{R}) &\triangleq \max_{\mathbf{n}/N \in \mathcal{R}} P[\mathbf{q}_N = \frac{\mathbf{n}}{N}|\mathbf{p}] \\ &= \max_{\mathbf{n}/N \in \mathcal{R}} \Big(N! \prod_{i=1}^{K} \frac{p_i^{n_i}}{n_i!} \Big) \end{aligned} \qquad (5A.1)$$

denote the maximum of the multinomial distribution over $\mathcal{Q}_N \cap \mathcal{R}$ where in (5A.1) $\mathbf{n}$ represents an integer partition of N, we have

$$M(\mathcal{R}) \leq P[\mathbf{q}_N \in \mathcal{R}|\mathbf{p}] \leq |\mathcal{Q}_N| M(\mathcal{R}) \, . \qquad (5A.2)$$

Next, Stirling's formula implies

$$\begin{aligned} \frac{1}{N} \ln \Big(N! \prod_{i=1}^{K} \frac{p_i^{n_i}}{n_i!} \Big) &= - \sum_{i=1}^{N} \frac{n_1}{N} \Big(\ln p_i - \ln(n_i/N) \Big) + O\Big(\frac{\ln(N)}{N}\Big) \\ &= -D(\mathbf{q}_N|\mathbf{p}) + O\Big(\frac{\ln(N)}{N}\Big) \, , \end{aligned} \qquad (5A.3)$$

where (5A.3) holds uniformly over $\mathcal{Q}_N \cap \mathcal{R}$. Since

$$\frac{1}{N} \ln(|\mathcal{Q}_N|) = O\Big(\frac{\ln(N)}{N}\Big) ,$$

from (5A.2) and (5A.3), we find

$$\begin{aligned} \frac{1}{N} \ln P[\mathbf{q}_N \in \mathcal{R}|\mathbf{p}] &= \frac{1}{N} \ln M(\mathcal{R}) + O\Big(\frac{\ln(N)}{N}\Big) \\ &= - \min_{\mathbf{q}_N \in \mathcal{Q}_N \cap \mathcal{R}} D(\mathbf{q}_N|\mathbf{p}) + O\Big(\frac{\ln(N)}{N}\Big) . \quad (5A.4) \end{aligned}$$

By letting N tend to infinity, we obtain (5.120) where the minimum in (5A.4) is replaced by an infimum since the limit of the points $\mathbf{q}_N \in \mathcal{Q}_N \cap \mathcal{R}$ minimizing $D(\cdot|\mathbf{p})$ may belong to the closure of $\mathcal{R}$.

References

1. E. L. Lehmann, *Elements of Large-Sample Theory.* New York: Springer Verlag, 1999.
2. S. M. Kay and J. R. Gabriel, "An invariance property of the generalized likelihood ratio test," *IEEE Signal Proc. Letters*, vol. 10, pp. 352–355, Dec. 2003.
3. S. Karlin and H. Rubin, "The theory of decision procedures for distributions with monotone likelihood ratios," *Annals Math. Statistics*, vol. 27, pp. 272–299, 1956.
4. C. W. Helstrom, *Elements of Signal Detection & Estimation.* Upper Saddle River, NJ: Prentice-Hall, 1995.
5. S. Bose and A. O. Steinhardt, "Optimum array detector for a weak signal in unknown noise," *IEEE Trans. Aerospace Electronic Systems*, vol. 32, pp. 911–922, July 1996.
6. L. L. Scharf, *Statistical Signal Processing: Detection, Estimation and Time Series Analysis.* Reading, MA: Addison Wesley, 1991.
7. A. J. Laub, *Matrix Analysis for Scientists and Engineers.* Philadelphia, PA: Soc. for Industrial and Applied Math., 2005.
8. H. L. Van Trees, *Optimum Array Processing.* New York: J. Wiley & Sons, 2002.
9. F. Nicolls and G. de Jager, "Uniformly most powerful cyclic permutation invariant detection for discrete-time signals," in *Proc. 2001 IEEE Internat. Conf. on Acoustics, Speech, Signal Proc. (ICASSP '01)*, vol. 5, (Salt Lake City, UT), pp. 3165–3168, May 2001.
10. A. Papoulis and S. U. Pillai, *Probability, Random Variables and Stochastic Processes, Fourth Edition.* New York: McGraw Hill, 2002.
11. H. Chernoff, "On the distribution of the likelihood ratio," *Annals Math. Statist.*, vol. 25, pp. 573–578, Sept. 1954.
12. W. Hoeffding, "Asymptotically optimal tests for multinomial distributions," *Annals Math. Statistics*, vol. 36, pp. 369–401, Apr. 1965.
13. F. den Hollander, *Large Deviations.* Providence, RI: American Mathematical Soc., 2000.

14. A. Dembo and O. Zeitouni, *Large Deviations Techniques and Applications, Second Edition.* New York: Springer Verlag, 1998.
15. O. Zeitouni and M. Gutman, "On universal hypothesis testing with large deviations," *IEEE Trans. Informat. Theory*, vol. 37, pp. 285–290, Mar. 1991.
16. O. Zeitouni, J. Ziv, and N. Merhav, "When is the generalized likelihood ratio test optimal?," *IEEE Trans. Informat. Theory*, vol. 38, pp. 1597–1602, 1992.
17. D. G. Herr, "Asymptotically optimal tests for multivariate normal distributions," *Annals Math. Statistics*, vol. 38, pp. 1829–1844, Dec. 1967.
18. B. Efron and D. Truax, "Large deviations theory in exponential families," *Annals Math. Statistics*, vol. 39, pp. 1402–1424, Oct. 1968.
19. W. C. M. Kallenberg, *Asymptotic Optimality of Likelihood Ratio Tests in Exponential Families.* Amsterdam: Mathematical Centre, 1978.
20. S. Kourouklis, "A large deviations result for the likelihood ratio statistic in exponential families," *Annals of Statistics*, vol. 12, pp. 1510–1521, Dec. 1984.
21. L. D. Brown, "Non-local asymptotic optimality of appropriate likelihood ratio tests," *Annals Math. Statistics*, vol. 42, pp. 1206–1240, 1971.
22. P. J. Bickel and K. A. Doksum, *Mathematical Statistics: Basic Ideas and Selected Topics, Second Edition.* Upper Saddle River, NJ: Prentice Hall, 2001.
23. S. M. Kay, *Fundamentals of Statistical Signal Processing: Detection Theory.* Prentice-Hall, 1998.
24. M. Feder and N. Merhav, "Universal composite hypothesis testing: a composite minimax approach," *IEEE Trans. Informat. Theory*, vol. 48, pp. 1504–1517, June 2002.
25. E. Levitan and N. Merhav, "A competitive Neyman-Pearson approach to universal hypothesis testing with applications," *IEEE Trans. Informat. Theory*, vol. 48, pp. 2215–2229, Aug. 2002.
26. A. A. Tadaion, M. Derakhtian, S. Gazor, M. M. Nayebi, and M. A. Aref, "Signal activity detection of phase-shift keying signals," *IEEE Trans. Commun.*, vol. 54, pp. 1439–1445, Aug. 2006.

6

Robust Detection

6.1 Introduction

In our discussion of composite binary hypothesis testing in the last chapter, it was assumed that the form of the probability density $f(\mathbf{y}|\mathbf{x}, H_j)$ of the observations $\mathbf{Y}$ under hypothesis H_j, $j = 0, 1$ is known exactly, and that only the parameter vector $\mathbf{x}$ is unknown. But this formulation is rather optimistic since in practice, secondary physical effects affecting the observed signal often go unmodelled. To handle such situations, we need to develop tests that recognize the fact that the probability densities of the observations under the two hypotheses are known only imprecisely. In other words, the tests that we consider must have a built-in tolerance which ensures that the test performs appropriately not only for the given model, but for an entire class of models in the vicinity of the assumed model. How this tolerance gets factored in the testing procedure depends not only on the nature of the modelling errors, but also on the target application. Specifically, for quality control applications, one is typically interested in applying a test that will perform well "on average" for the uncertainty class of probabilistic models centered about the nominal model. In other words, we are only interested in average performance, since the goal is to reject bad parts while retaining good parts so as to maximize the income generated by a product line. However, for safety oriented applications, such as radar, sonar, or earthquake detection, one is interested in worst-case outcomes, since the consequences of incorrect decisions can be rather severe. In such situations, it is natural to seek to design tests that are optimal for the worst-case model compatible with the nominal model and the allowed mismodelling tolerance. This minimax approach is similar to the one that was employed in Chapter 2 for designing a test that is insensitive to the choice of priors for hypotheses H_0 and H_1. Of course, the problem is far more difficult, since we need now to select least-favorable probability densities for the observations. The resulting test is said to be robust since it guarantees a minimum level of performance for all models in the vicinity of the nominal model. This test design approach tends to be rather conservative since the minimum level

B.C. Levy, *Principles of Signal Detection and Parameter Estimation*,
DOI: 10.1007/978-0-387-76544-0_6, © Springer Science+Business Media, LLC 2008

of performance is based on a worst-case analysis and actual performance for models selected at random in the vicinity of the nominal model is usually far better than predicted by a worst-case analysis.

This chapter is organized as follows: Several metrics measuring the proximity of probabilistic models are described in Section 6.2. Section 6.3 describes the methodology developed by Huber [1] for designing robust tests based on a single observation vector. An alternative design methodlogy which is applicable asymptotically to tests designed for repeated observations is described in Section 6.4. Finally, Section 6.5 describes how robust binary hypothesis testing techniques can be adapted to signal detection problems involving a sequence of observations over a finite interval.

6.2 Measures of Model Proximity

In order to formulate the robust hypothesis testing problem we need first to specify when two probabilistic models are close to one another. Since the degree of proximity of two stochastic models depends on the modelling process itself, the concept of closeness can be defined in several ways [1–3]. We describe briefly several measures of proximity. In our discussion, we consider a random vector $\mathbf{Y}$ of $\mathbb{R}^n$ and assume that a nominal cumulative probability distribution $F(\mathbf{y})$ or a density $f(\mathbf{y})$ is given for $\mathbf{Y}$. This specifies implicitly a probability measure P for all sets A in the Borel field $\mathcal{B}$ of $\mathbb{R}^n$, so in the following $F(\mathbf{y})$, $f(\mathbf{y})$ and P will be used interchangeably when we refer to the nominal probabilistic description of $\mathbf{Y}$. The actual or true model of $\mathbf{Y}$ is different from this model. It is specified by a cumulative probability distribution $G(\mathbf{y})$, or a density $g(\mathbf{y})$ or a probablity measure Q defined over the Borel field $\mathcal{B}$. So we are interested in characterizing the neighborhood $\mathcal{F}$ of actual models that are close to the nominal model. Such a neighborhood can be defined in several ways.

Contamination model: Given the nominal probability distribution $F(\mathbf{y})$ and a tolerance ϵ with $0 \leq \epsilon \leq 1$, we define the set $\mathcal{F}$ of ϵ-contaminated probability distributions centered about F as

$$\mathcal{F} = \{G \;:\; G = (1-\epsilon)F + \epsilon H\} \tag{6.1}$$

where $H(\mathbf{y})$ denotes an arbitrary cumulative probability distribution function. Equivalently, if the probability distributions G, F and H have densities g, f and h, $g(\mathbf{y})$ is an ϵ-contaminated version of the nominal density $f(\mathbf{y})$ if it can be expressed as

$$g(\mathbf{y}) = (1-\epsilon)f(\mathbf{y}) + \epsilon h(\mathbf{y})\,. \tag{6.2}$$

This indicates that the samples of the true observation density $g(\mathbf{y})$ are drawn with probability $1-\epsilon$ from the nominal density $f(\mathbf{y})$, and with probability ϵ from an unknown probability density $h(\mathbf{y})$. In other words, this model of closeness allows for the possibility that outliers may contaminate the observations.

This type of situation arises often in signal processing applications. For example, in underwater acoustics, while ambient noise models usually take into account the effect of wind-induced surface waves, other types of unexpected noise sources, such as fish, often go unmodelled. The use of contamination models aims at addressing this modelling limitation.

Total variation: Given two probability measures P and Q over the Borel field $\mathcal{B}$ of $\mathbb{R}^n$, and an arbitrary real function $h(\mathbf{y})$ defined over $\mathbb{R}^n$, the total variation distance between P and Q is defined as

$$d_{\mathrm{TV}}(P, Q) = \max_{||h||_\infty \leq 1} |E_P[h(\mathbf{Y})] - E_Q[h(\mathbf{Y})]| \,, \tag{6.3}$$

where

$$||h||_\infty = \sup_{\mathbf{y} \in \mathbb{R}^n} |h(\mathbf{y})|$$

denotes the infinity norm of h, and $E_P[\cdot]$ represents the expectation taken with respect to the measure P. Let A be an arbitrary set of $\mathcal{B}$, and let A^c denote its complement. By considering the function

$$h(\mathbf{y}) = 1_A(\mathbf{y}) - 1_{A^c}(\mathbf{y})$$

where

$$1_A(\mathbf{y}) = \begin{cases} 1 & \mathbf{y} \in A \\ 0 & \text{otherwise} \end{cases}$$

is the indicator function of the set A, and observing that

$$E_P[h(\mathbf{Y})] = 2P[A] - 1 \,,$$

the total variation distance can be written as

$$d_{\mathrm{TV}}(P, Q) = 2 \sup_{A \in \mathcal{B}} |P[A] - Q[A]| \,. \tag{6.4}$$

Let $F(\mathbf{y}) = P[\mathbf{Y} \leq \mathbf{y}]$ and $G(\mathbf{y}) = Q[\mathbf{Y} \leq \mathbf{y}]$ denote the cumulative probability distributions corresponding to P and Q, and assume they admit the densities $f(\mathbf{y})$ and $g(\mathbf{y})$. Then the function $h(\cdot)$ that attains the maximum in (6.3) is obviously $h(\mathbf{y}) = \mathrm{sign}((f - g)(\mathbf{y}))$, and hence

$$d_{\mathrm{TV}}(f, g) = \int |f(\mathbf{y}) - g(\mathbf{y})| d\mathbf{y} \tag{6.5}$$

is the L_1 norm of $f - g$. Note that expression (6.5) for the total variation is written as a metric over probability densities. The equivalence of expressions (6.3)–(6.5) for the total variation is established in Problem 6.1. Given the metric (6.5) and a tolerance ϵ, we can then define the neighborhood of probability densities located at a distance less than ϵ of the nominal density f as

$$\mathcal{F} = \{g \; : \; d_{\mathrm{TV}}(f, g) < \epsilon\} \,. \tag{6.6}$$

Kolmogorov distance: Given two cumulative probability distributions $F(\mathbf{y})$ and $G(\mathbf{y})$, the Kolmogorov or uniform distance is defined as

$$d_{\mathrm{K}}(F,G) = \sup_{\mathbf{y}\in\mathbb{R}^n} |F(\mathbf{y}) - G(\mathbf{y})| \,. \tag{6.7}$$

By restricting our attention in (6.4) to the subclass of the Borel field formed by $\{\mathbf{Y} \leq \mathbf{y}\}$, we conclude that

$$d_{\mathrm{TV}}(F,G) \geq 2 d_{\mathrm{K}}(F,G) \,. \tag{6.8}$$

Lévy distance: For the case of a scalar random variable, i.e., when $n = 1$, the Lévy distance is defined as

$$d_{\mathrm{L}}(F,G) = \inf\{\epsilon > 0 \; : \; G(y-\epsilon) - \epsilon < F(y) < G(y+\epsilon) + \epsilon \text{ for all } y\} \,. \tag{6.9}$$

This distance has the feature that it metrizes the weak convergence of probability measures, or equivalently, the convergence in distribution/law of random variables. Specifically if, for $k \geq 1$, Y_k denotes a sequence of random variables with probability distribution $F_k(y)$, Y_k converges in distribution to a random variable Y with probability distribution $F(y)$ if and only if

$$\lim_{k\to\infty} d_{\mathrm{L}}(F_k, F) = 0 \,.$$

Its extension to random vectors is given by the Prohorov distance between probability measures. If P and Q are two probability measures defined over the Borel field $\mathcal{B}$ and A denotes an arbitrary element of $\mathcal{B}$, the Prohorov distance is defined as

$$d_{\mathrm{P}}(P,Q) = \inf\{\epsilon > 0 \; : \; Q(A) < P(A^{\epsilon}) + \epsilon \text{ for all } A \in \mathcal{B}\} \,, \tag{6.10}$$

where

$$A^{\epsilon} = \{\mathbf{y} \in \mathbb{R}^n \; : \; d(\mathbf{y}, A) \leq \epsilon\}$$

is the set of points located at a distance less than or equal to ϵ of A. This more abstract definition reduces to the Lévy distance for a scalar random variable if we restrict our attention to sets A of the form $(-\infty, y]$ or $[y, \infty)$. Like the Lévy distance, the Prohorov distance metrizes the weak convergence of probability measures.

Then, if d represents the Kolmogorov or Lévy distance, given a nominal probability distribution F, if ϵ denotes an arbitrary tolerance level, the neigborhood

$$\mathcal{F} = \{G \; : \; d(G,F) \leq \epsilon\}$$

is formed by the set of actual probability distributions located at a distance less than ϵ of F.

Kullback-Leibler divergence: Assume the nominal and actual models have probability densities $f(\mathbf{y})$ and $g(\mathbf{y})$, respectively. Although the KL divergence

$$D(g|f) = \int \ln\Big(\frac{g(\mathbf{y})}{f(\mathbf{y})}\Big) g(\mathbf{y}) d\mathbf{y} \tag{6.11}$$

is not a distance, it is often used in statistics [4] as a measure of model mismatch. For example, as explained in [5], it provides the underlying metric for establishing the convergence of the Expectation-Maximization method of mathematical statistics. Because the function $x \log(x)$ is convex, $D(g|f)$ is a convex function of g, and accordingly, if ϵ is a prespecified uncertainty tolerance, the neighborhood

$$\mathcal{F} = \{g \ : \ D(g|f) \leq \epsilon\}$$

is convex.

6.3 Robust Hypothesis Testing

6.3.1 Robust Bayesian and NP Tests

We now have at our disposal all the tools necessary to formulate robust hypothesis testing. As in Section 2.5 of Chapter 2, we consider the convex set $\mathcal{D}$ of pointwise randomized decision functions. Recall that a decision function takes the value $\delta(\mathbf{y})$ at point $\mathbf{y} \in \mathbb{R}^n$ if, when observation $\mathbf{Y} = \mathbf{y}$, we select hypothesis H_1 with probability $\delta(\mathbf{y})$, and H_0 with probability $1 - \delta(\mathbf{y})$. Assume that the nominal probability density of the observations under H_j is denoted as $f_j(\mathbf{y})$ with $j = 0,\ 1$ and that the actual density $g_j(\mathbf{y})$ belongs to a convex neighborhood $\mathcal{F}_j$, $j = 0,\ 1$. Then the probabilities of false alarm and of a miss (types I and II errors) can be expressed as

$$\begin{aligned} P_F(\delta, g_0) &= \int \delta(\mathbf{y}) g_0(\mathbf{y}) d\mathbf{y} \\ P_M(\delta, g_1) &= \int (1 - \delta(\mathbf{y})) g_1(\mathbf{y}) d\mathbf{y} \, , \end{aligned} \tag{6.12}$$

which are linear in δ and g_j, $j = \ 0,\ 1$. The Bayes risk takes the form

$$\begin{aligned} R(\delta, g_0, g_1) = & \ C_{00}\pi_0 + (C_{10} - C_{00})\pi_0 P_F(\delta, g_0) \\ & + C_{11}\pi_1 + (C_{01} - C_{11})\pi_1 P_M(\delta, g_1) \end{aligned} \tag{6.13}$$

which is of course linear in δ, g_0 and g_1. So the functions $P_F(\delta, g_0)$, $P_M(\delta, g_1)$ and $R(\delta, g_0, g_1)$ are convex in δ and concave in g_0 and g_1.

Let $\mathcal{F} = \mathcal{F}_0 \times \mathcal{F}_1$. Following Huber [1,6,7], we consider the robust Bayesian hypothesis testing problem

$$\min_{\delta \in \mathcal{D}} \max_{(g_0, g_1) \in \mathcal{F}} R(\delta, g_0, g_1) \tag{6.14}$$

and the robust Neyman-Pearson testing problem

$$\min_{\delta \in \mathcal{D}_\alpha} \max_{g_1 \in \mathcal{F}_1} P_M(\delta, g_1) \tag{6.15}$$

where the set $\mathcal{D}_\alpha$ is defined as

$$\mathcal{D}_\alpha = \{\delta \in \mathcal{D} : \max_{g_0 \in \mathcal{F}_0} P_F(\delta, g_0) \leq \alpha\} . \tag{6.16}$$

In this respect, note that since $P_F(\delta, g_0)$ is a convex function of δ for each $g_0 \in \mathcal{F}_0$, according to Proposition 1.2.4 of [8], the function

$$P_F(\delta, \mathcal{F}_0) \triangleq \max_{g_0 \in \mathcal{F}_0} P_F(\delta, g_0) \tag{6.17}$$

is also convex, which in turn implies that the set $\mathcal{D}_\alpha$ is convex. Thus, both the minimax Bayesian problem (6.14) and the minimax Neyman-Pearson problem (6.15) are of convex-concave type with arguments defined over convex domains. In these problems, the set $\mathcal{D}$ of randomized decision functions is compact with respect to the infinity norm, and by construction, the neighborhoods $\mathcal{F}_0$ and $\mathcal{F}_1$ are compact for some appropriate proximity metric. The von Neumann minimax theorem on p. 319 of [9] is therefore applicable, so for both problems there exists a saddle point $(\delta_{\rm R}, g_0^{\rm L}, g_1^{\rm L})$, where $\delta_{\rm R}$ represents the robust test, and the pair $(g_0^{\rm L}, g_1^{\rm L})$ specifies the associated least-favorable densities. For the robust Bayesian hypothesis testing problem (6.14), the saddle point satisfies

$$R(\delta_{\rm R}, g_0, g_1) \leq R(\delta_{\rm R}, g_0^{\rm L}, g_1^{\rm L}) \leq R(\delta, g_0^{\rm L}, g_1^{\rm L}) \tag{6.18}$$

for all $(g_0, g_1) \in \mathcal{F}$ and $\delta \in \mathcal{D}$. The right-hand side of this inequality implies that $\delta_{\rm R}$ is the optimum likelihood-ratio test for the least-favorable pair of densities $(g_0^{\rm L}, g_1^{\rm L})$, i.e., it takes the form

$$L_{\rm L}(\mathbf{y}) \triangleq \frac{g_1^{\rm L}(\mathbf{y})}{g_0^{\rm L}(\mathbf{y})} \underset{H_0}{\overset{H_1}{\gtrless}} \tau = \frac{(C_{10} - C_{00})\pi_0}{(C_{01} - C_{11})\pi_1} . \tag{6.19}$$

Note that this test is binary-valued, so $\delta_{\rm R}$ solves also the minimax problem (6.14) where δ is restricted to belong to the smaller set $\mathcal{D}_b$ of binary-valued decision functions. In other words, like the minimax test of Chapter 2, convexifying the set of decision functions does not affect the solution of the minimax Bayesian hypothesis testing problem.

Next, if we consider the left-hand side of (6.18) and observe that the risk function is a linear functional of $P_F(\delta, g_0)$ and $P_M(\delta, g_1)$ we find that (6.18) is equivalent to the simultaneous inequalities

$$\begin{aligned} P_F(\delta_{\rm R}, g_0) &\leq P_F(\delta_{\rm R}, g_0^{\rm L}) \\ P_M(\delta_{\rm R}, g_1) &\leq P_M(\delta_{\rm R}, g_1^{\rm L}) \end{aligned} \tag{6.20}$$

for all $g_0 \in \mathcal{F}_0$ and $g_1 \in \mathcal{F}_1$, respectively. Thus, given the test δ_R, g_0^L and g_1^L are obtained respectively by maximizing the probability of false alarm P_F and the miss probability P_M.

So finding a saddle point of the risk function $R(\delta, g_0, g_1)$ reduces to finding an LR test δ_R and two densities g_0^L and g_1^L satisfying inequalities (6.20). It turns out that this is not quite as simple as it looks, and the most effective approach is often to guess the solution and then verify that it works.

6.3.2 Clipped LR Tests

To illustrate the guess and verify approach to identifying saddle points, we consider an example described in [1, 7] which includes as a special case the result of [6]. Although this result is not as general as one obtained later by Huber and Strassen [10] for neighborhoods described by alternating 2-capacities, the saddle point obtained for this example has a simple and intuitive structure which is ultimately applicable to several detection problems. So we assume that the observation Y is scalar and that the nominal likelihood ratio $L(y) = f_1(y)/f_0(y)$ is a monotone increasing function of y. In the following, we denote the inverse function of $\ell = L(y)$ as $y = K(\ell)$, which is also monotone increasing. Then for some numbers $0 \leq \epsilon_0, \epsilon_1, \nu_0, \nu_1 < 1$, consider the neighborhoods specified by

$$\begin{aligned} \mathcal{F}_0 &= \{g_0 \ : \ G_0(y) \geq (1-\epsilon_0)F_0(y) - \nu_0 \text{ for all } y \in \mathbb{R}\} \\ \mathcal{F}_1 &= \{g_1 \ : \ 1 - G_1(y) \geq (1-\epsilon_1)(1 - F_1(y)) - \nu_1 \text{ for all } y \in \mathbb{R}\}\,. \end{aligned} \quad (6.21)$$

Altough the constraints are expressed in terms of the cumulative probability distributions G_j, they specify implicitly a linear set of constraints on the densities, since

$$G_j(y) = \int_{-\infty}^{y} g_j(u)du\,.$$

Since the constraints are linear, the sets $\mathcal{F}_0$ and $\mathcal{F}_1$ are convex. Intuitively, we see that a density belongs to $\mathcal{F}_0$ if its cumulative probability distribution stays above the curve specified by $(1-\epsilon_0)F_0(y) - \nu_0$ for all y. Similarly, a density belongs to $\mathcal{F}_1$ if its cumulative probability distribution stays below the curve $(1-\epsilon_1)F_1(y) + \epsilon_1 + \nu_1$ for all y.

Geometrically, since $P_F(\delta_R, g_0)$ and $P_M(\delta_R, g_1)$ depend linearly on g_0 and g_1, the least-favorable densities g_0^L and g_1^L will solve the maximization problems (6.20) only if g_0^L and g_1^L are located on the boundary of $\mathcal{F}_0$ and $\mathcal{F}_1$. Accordingly, the trial least-favorable densities are such that the the inequalities specifying $G_0(y)$ and $G_1(y)$ are equalities over an interval $[y_L, y_U]$. Specifically, we select

$$\begin{aligned} G_0^L(y) &= (1-\epsilon_0)F_0(y) - \nu_0 \\ G_1^L(y) &= (1-\epsilon_1)F_1(y) + \epsilon_1 + \nu_1\,, \end{aligned} \quad (6.22)$$

which implies

$$g_j^{\mathrm{L}}(y) = (1-\epsilon_j)f_j(y) , \tag{6.23}$$

for $y_L \leq y \leq y_U$ and $j = 0,\ 1$. The limits $y_L < y_U$ of interval $[y_L, y_U]$ have yet to be selected. This implies that over $[y_L, y_U]$, the least-favorable likelihood ratio

$$L_{\mathrm{L}}(y) = \frac{1-\epsilon_1}{1-\epsilon_0} L(y) \tag{6.24}$$

is proportional to the nominal LR, and, in fact, $L_{\mathrm{L}}(y) = L(y)$ if the tolerance coefficients ϵ_1 and ϵ_0 are equal. Over the intervals $(-\infty, y_L)$ and (y_U, ∞), the least-favorable densities are selected as linear combinations of the two nominal densities:

$$g_j^{\mathrm{L}}(y) = a_j f_0(y) + b_j f_1(y) ,$$

with $j = 0,\ 1$, in such a way that the least-favorable likelihood ratio is constant over these intervals, i.e., it is a censored or clipped version of the nominal LR function. To express the least-favorable densities, let

$$\begin{aligned} v' &= \frac{\epsilon_1 + \nu_1}{1-\epsilon_1} \ , \ v'' = \frac{\epsilon_0 + \nu_0}{1-\epsilon_0} \\ w' &= \frac{\nu_0}{1-\epsilon_0} \ , \ w'' = \frac{\nu_1}{1-\epsilon_1} \ , \end{aligned} \tag{6.25}$$

and let $\ell_L = L(y_L)$ and $\ell_U = L(y_U)$. Then, if we assume $v' > 0$ and $v'' > 0$, the trial least-favorable densities are selected as

$$g_0^{\mathrm{L}}(y) = \begin{cases} \dfrac{1-\epsilon_0}{v' + w'\ell_L}(v' f_0(y) + w' f_1(y)) & y < y_L \\ (1-\epsilon_0) f_0(y) & y_L \leq y \leq y_U \\ \dfrac{1-\epsilon_0}{w'' + v''\ell_U}(w'' f_0(y) + v'' f_1(y)) & y_U < y \end{cases} \tag{6.26}$$

and

$$g_1^{\mathrm{L}}(y) = \begin{cases} \dfrac{(1-\epsilon_1)\ell_L}{v' + w'\ell_L}(v' f_0(y) + w' f_1(y)) & y < y_L \\ (1-\epsilon_1) f_1(y) & y_L \leq y \leq y_U \\ \dfrac{(1-\epsilon_1)\ell_U}{w'' + v''\ell_U}(w'' f_0(y) + v'' f_1(y)) & y_U < y \ . \end{cases} \tag{6.27}$$

For this choice, we find that

$$L_{\mathrm{L}}(y) = \frac{1-\epsilon_1}{1-\epsilon_0} C(L(y)) , \tag{6.28}$$

where $C(\cdot)$ denotes the clipping nonlinearity

$$C(L) = \begin{cases} \ell_L & L < \ell_L \\ L & \ell_L \leq L \leq \ell_U \\ \ell_U & L > \ell_U \end{cases} \tag{6.29}$$

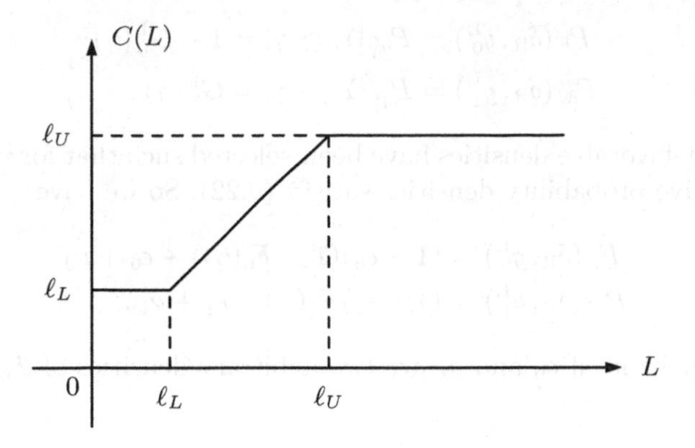

Fig. 6.1. Clipping transformation of the nominal likelihood ratio function.

which is sketched in Fig. 6.1.

Next, consider the robust decision rule $\delta_{\mathrm{R}}(y)$ specified by

$$L_{\mathrm{L}}(y) \underset{H_0}{\overset{H_1}{\gtrless}} \tau \tag{6.30}$$

where τ is the Bayesian threshold appearing in (6.19). If η denotes the renormalized threshold

$$\eta = \frac{1-\epsilon_0}{1-\epsilon_1}\tau\,, \tag{6.31}$$

it is equivalent to

$$C(L(y)) \underset{H_0}{\overset{H_1}{\gtrless}} \eta, \tag{6.32}$$

and three cases may occur.

Case 1: $\eta < \ell_L$. Then, the clipped LR $C(L(y))$ is always above ℓ_L, so $\delta_{\mathrm{R}}(\mathrm{y}) = 1$ for all y, i.e., we always choose hypothesis H_1. For this decision rule, we have

$$P_F(\delta_{\mathrm{R}}, g_0) = 1 \qquad P_M(\delta_{\mathrm{R}}, g_1) = 0 \tag{6.33}$$

for all choices of densities $g_0 \in \mathcal{F}_0$ and $g_1 \in \mathcal{F}_1$, so the saddle point condition (6.20) is satisfied.

Case 2: $\ell_L \leq \eta \leq \ell_U$. In this case, the decision rule (6.32) is equivalent to the test

$$L(y) \underset{H_0}{\overset{H_1}{\gtrless}} \eta \tag{6.34}$$

for the nominal LR. Let $\gamma = K(\eta)$, where $K(\cdot)$ is the inverse function of $L(\cdot)$. Note that since L is monotone increasing, we have $y_L \leq \gamma \leq y_U$. The probabilities of false alarm and of a miss for the test (6.34) are given by

$$P_F(\delta_R, g_0^L) = P_{g_0^L}[Y \geq \gamma] = 1 - G_0^L(\gamma)$$
$$P_M(\delta_R, g_1^L) = P_{g_1^L}[Y < \gamma] = G_1^L(\gamma) \,. \quad (6.35)$$

But the least-favorable densities have been selected such that for $y_L \leq y \leq y_U$, the cumulative probability densities satisfy (6.22). So we have

$$P_F(\delta_R, g_0^L) = (1-\epsilon_0)(1 - F_0(\gamma)) + \epsilon_0 + \nu_0$$
$$P_M(\delta_R, g_1^L) = (1-\epsilon_1)F_1(\gamma) + \epsilon_1 + \nu_1 \,. \quad (6.36)$$

On the other hand, if g_0 and g_1 are two arbitrary densities of $\mathcal{F}_0$ and $\mathcal{F}_1$, we have

$$P_F(\delta_R, g_0) = 1 - G_0(\gamma) \leq (1-\epsilon_0)(1 - F_0(\gamma)) + \epsilon_0 + \nu_0$$
$$P_M(\delta_R, g_1) = G_1(\gamma) \leq (1-\epsilon_1)F_1(\gamma) + \epsilon_1 + \nu_1 \,, \quad (6.37)$$

and thus the least-favorable densities g_0^L and g_1^L satisfy the saddle point condition (6.20).

Case 3: $\eta > \ell_U$. Then, the clipped LR $C(L(y))$ is always below ℓ_U, so $\delta_R(y) = 0$ for all y, i.e., we always choose hypothesis H_0. For this decision rule, we have

$$P_F(\delta_R, g_0) = 0 \qquad P_M(\delta_R, g_1) = 1 \quad (6.38)$$

for all choices of densities $g_0 \in \mathcal{F}_0$ and $g_1 \in \mathcal{F}_1$, so the saddle point condition (6.20) holds.

The last issue that needs to be settled is whether we can find some values $y_L < y_U$ such that the least-favorable densities g_0^L and g_1^L specified by (6.26) and (6.27) satisfy

$$G_0^L(y) - (1-\epsilon_0)F_0(y) \begin{cases} \geq -\nu_0 & y < y_L \\ = -\nu_0 & y_L \leq y \leq y_U \\ \geq -\nu_0 & y > y_U \end{cases} \quad (6.39)$$

and

$$(1 - G_1^L(y)) - (1-\epsilon_1)(1 - F_1(y)) \begin{cases} \geq -\nu_1 & y < y_L \\ = -\nu_1 & y_L \leq y \leq y_U \\ \geq -\nu_1 & y > y_U \,. \end{cases} \quad (6.40)$$

We consider first the condition (6.39) for G_0^L. Note that $G_0^L(-\infty) - (1-\epsilon_0)F_0(-\infty) = 0$ and for $y < y_L$, we have

$$\frac{d}{dy}\Big[G_0^L(y) - (1-\epsilon_0)F_0(y)\Big] = g_0^L(y) - (1-\epsilon_0)f_0(y)$$
$$= \frac{(1-\epsilon_0)w'}{v' + w'\ell_L}[f_1(y) - \ell_L f_0(y)] < 0 \,, \quad (6.41)$$

where the last inequality is due to the fact that the nominal LR is monotone increasing and $\ell_L = L(y_L)$, and where we note that $(1-\epsilon_0)w' = \nu_0$. So over the interval $(-\infty, y_L)$, the function $G_0^{\rm L}(y) - (1-\epsilon_0)F_0(y)$ is monotone decreasing and stays above $-\nu_0$ before being equal to $-\nu_0$ over $[y_L, y_U]$ as long as it hits $-\nu_0$ at $y = y_L$:

$$G_0^{\rm L}(y_L) - (1-\epsilon_0)F_0(y_L) = \frac{\nu_0}{v' + w'\ell_L} \int_{-\infty}^{y_L} [f_1(y) - \ell_L f_0(y)]dy$$
$$= \frac{\nu_0}{v' + w'L(y_L)} [F_1(y_L) - L(y_L)F_0(y_L)] = -\nu_0 \,. \tag{6.42}$$

For $y > y_U$, we have

$$\frac{d}{dy}\Big[G_0^{\rm L}(y) - (1-\epsilon_0)F_0(y)\Big] = g_0^{\rm L}(y) - (1-\epsilon_0)f_0(y)$$
$$= \frac{(1-\epsilon_0)v''}{w'' + v''\ell_U}[f_1(y) - \ell_U f_0(y)] > 0 \,, \tag{6.43}$$

where the last inequality is due to the fact that $L(y) \geq \ell_U = L(y_U)$ over (y_L, ∞), and where we have $(1-\epsilon_0)v'' = \epsilon_0 + \nu_0$. This indicates that the least favorable cumulative probability distribution $G_0^{\rm L}(y)$ satisfies (6.39) for $y > y_U$. However, we still need to ensure that the cumulative probability distribution admits the normalization $G_0^{\rm L}(\infty) = 1$. This gives the condition

$$\epsilon_0 + \nu_0 = [G_0^{\rm L}(\infty) - (1-\epsilon_0)F_0(\infty)] - [G_0^{\rm L}(y_U) - (1-\epsilon_0)F_0(y_U)]$$
$$= \frac{\epsilon_0 + \nu_0}{w'' + v''\ell_U} \int_{y_U}^{\infty} [f_1(y) - \ell_U f_0(y)]dy$$
$$= \frac{\epsilon_0 + \nu_0}{w'' + v''\ell_U} [(1 - F_1(y_U)) - L(y_U)(1 - F_0(y_U))] \,. \tag{6.44}$$

By analyzing in a similar manner the condition (6.40) for $G_1^{\rm L}(y)$, we obtain constraints which are identical to (6.42) and (6.44).

So we need to find y_L and y_U with $y_L < y_U$ such that

$$\Gamma(y_L) = \Xi(y_U) = 1 \tag{6.45}$$

with

$$\Gamma(y) \triangleq \frac{L(y)F_0(y) - F_1(y)}{v' + w'L(y)} \tag{6.46}$$

$$\Xi(y) \triangleq \frac{(1 - F_1(y)) - L(y)(1 - F_0(y))}{w'' + v''L(y)} \,. \tag{6.47}$$

Since the two equations are similar, we focus on the evaluation of y_L. Since

$$\frac{d\Gamma}{dy} = \frac{v'F_0(y) + w'F_1(y)}{(v' + w'L(y))^2} \frac{dL}{dy} > 0 \tag{6.48}$$

the function $\Gamma(\cdot)$ is monotone increasing, with $\Gamma(-\infty) = 0$ and

$$\Gamma(\infty) = \frac{L(\infty) - 1}{v' + w' L(\infty)} \,. \tag{6.49}$$

This implies that the equation (6.45) for y_L admits a unique solution as long as $\Gamma(\infty) > 1$, or equivalently,

$$L(\infty) > \frac{1 + v'}{1 - w'} = \frac{(1 - \epsilon_0)(1 + \nu_1)}{(1 - \epsilon_1)(1 - \epsilon_0 - \nu_0)} \,. \tag{6.50}$$

The value y_U is obtained similarly, and the condition $y_L < y_U$ holds provided the neighborhoods $\mathcal{F}_0$ and $\mathcal{F}_1$ do not intersect [1, pp. 270–271], since otherwise it is obviously impossible to test H_0 versus H_1.

This concludes the construction of the least-favorable densities g_0^{L} and g_1^{L} and the verification of the saddle point condition (6.18) for the minimax Bayesian hypothesis testing problem (6.14). Let us now turn our attention to the minimax Neyman-Pearson problem (6.15). Observe that the least-favorable densities g_0^{L} and g_1^{L} depend on the nominal densities f_j and the parameters ϵ_j, ν_j with $j = 0,\ 1$, but are completely independent of the threshold τ. Accordingly, if we denote by $\delta_{\mathrm{R}}(\tau)$ the robust test (6.30) with threshold τ, to solve the minimax Neyman-Pearson problem (6.15), all we need to do is to select τ such that

$$P_F(\delta_{\mathrm{R}}(\tau), g_0^{\mathrm{L}}) = \alpha \,, \tag{6.51}$$

i.e., τ specifies an NP test with probability of false alarm α for the least-favorable pair $(g_0^{\mathrm{L}}, g_1^{\mathrm{L}})$. Then if δ_{R} denotes the resulting test, the inequalities (6.20) ensure that δ_{R} solves the minimax NP testing problem (6.15).

One final issue that needs to be addressed is how the neighborhoods $\mathcal{F}_0$ and $\mathcal{F}_1$ specified by (6.21) relate to the measures of model proximity examined in Section 6.2. By selecting appropriately the parameters ϵ_j and ν_j for $j = 0,\ 1$, we show that the neighborhoods $\mathcal{F}_0$ and $\mathcal{F}_1$ contain several of the proximity neighborhoods discussed in Section 6.2, so the result presented above is applicable to several of the proximity metrics discussed earlier.

Contamination model: Set $\nu_j = 0$ with $j = 0,\ 1$ in (6.21). Then, if $H(y)$ denotes an arbitrary probability distribution function, the contamination neighborhood

$$\mathcal{N}_j = \{g_j \ : \ G_j(y) = (1 - \epsilon_j) F_j(y) + \epsilon_j H(y) \text{ for all } y\} \tag{6.52}$$

is contained in the neighborhood $\mathcal{G}_j$ of (6.21) and the least-favorable density g_j^{L} belongs to $\mathcal{N}_j$ for $j = 0,\ 1$. Accordingly, the test δ_{R} and the densities $(g_0^{\mathrm{L}}, g_1^{\mathrm{L}})$ solve the minimax problems obtained by replacing $\mathcal{F}$ by $\mathcal{N} = \mathcal{N}_0 \times \mathcal{N}_1$ in (6.14), and $\mathcal{F}_1$ by $\mathcal{N}_1$ in (6.15).

Kolmogorov/total variation metric: Set $\epsilon_j = 0$ with $j = 0,\ 1$ in (6.21). Then the ball

$$\mathcal{N}_j^{\mathrm{K}} = \{g_j \ : \ |G_j(y) - F_j(y)| \leq \nu_j \text{ for all } y\} \tag{6.53}$$

of functions located at a Kolmogorov distance less than or equal to ν_j of F_j is contained in the neighborhood $\mathcal{F}_j$ of (6.21), and g_j^{L} belongs to $\mathcal{N}_j^{\mathrm{K}}$ for $j = 0,\ 1$. So again, the test δ_{R} and least-favorable densities $(g_0^{\mathrm{L}}, g_1^{\mathrm{L}})$ solve the minimax Bayesian and NP testing problems with respect to the Kolmogorov metric. For $j = 0,\ 1$, consider also the ball

$$\mathcal{N}_j^{\mathrm{TV}} = \{g_j \ : \ d_{\mathrm{TV}}(g_j, f_j) \leq 2\nu_j\} \tag{6.54}$$

specified by the total variation metric. The inequality (6.8) implies that it is contained in the ball $\mathcal{N}_j^{\mathrm{K}}$. Furthermore, setting $\epsilon_j = 0$ in identities (6.42) and (6.44) yields

$$\begin{aligned} d_{\mathrm{TV}}(g_0^{\mathrm{L}}, f_0) &= \int_{-\infty}^{\infty} |g_0^{\mathrm{L}} - f_0| dy \\ &= \int_{-\infty}^{y_L} (f_0 - g_0^{\mathrm{L}}) dy + \int_{y_U}^{\infty} (g_0^{\mathrm{L}} - f_0) dy = 2\nu_0 \ , \end{aligned} \tag{6.55}$$

so g_0^{L} belongs to $\mathcal{N}_0^{\mathrm{TV}}$. Similarly, g_1^{L} belongs to $\mathcal{N}_1^{\mathrm{TV}}$. This establishes that the test δ_{R} and least favorable densities $(g_0^{\mathrm{L}}, g_1^{\mathrm{L}})$ solve the minimax Bayesian and NP testing problems for the total variation metric.

Lévy metric: Set $\epsilon_j = 0$ with $j = 0,\ 1$ and replace $F_0(y)$ by $F_0(y - \nu_0)$ and $F_1(y)$ by $F_1(y + \nu_1)$ in the definition (6.21) of $\mathcal{F}_0$ and $\mathcal{F}_1$. This means in particular that the densities $f_0(y)$ and $f_1(y)$ appearing in expressions (6.26) and (6.27) for the least-favorable densities must be replaced by $f_0(y - \nu_0)$ and $f_1(y + \nu_1)$, respectively. Then the Lévy metric ball

$$\mathcal{N}_j^{\mathrm{L}} = \{g_j \ : \ F_j(y - \nu_j) - \nu_j \leq G_j(y) \leq F_j(y + \nu_j) + \nu_j \text{ for all } y\} \tag{6.56}$$

is contained in $\mathcal{F}_j$ and g_j^{L} belongs to $\mathcal{N}_j^{\mathrm{L}}$ for $j = 0,\ 1$. This implies that the robust test δ_{R} and the least-favorable densities $(g_0^{\mathrm{L}}, g_1^{\mathrm{L}})$ solve the minimax Bayesian and NP test problems corresponding to the Lévy metric.

In general, Huber's formulation of robust hypothesis testing relies on the observation that the tails of probability densities tend to be undermodelled, since, when observations are collected to estimate probability densities for each hypothesis, few observations fall in the tail region, so density estimates tend to be less accurate in the tail area than in regions where the probability mass is concentrated. Also, tests are usually designed to produce small probabilities of false alarm or of a miss. Accordingly, by shifting a small amount of probability from the main body of a distribution to its tails, or from one tail to another, it is possible to leave the main part of the distribution almost unchanged, while affecting significantly the tails in such a way that the test performance can be changed dramatically. In effect, for a nominal test designed to produce a probability of false alarm equal to 10^{-3}, if only 1% of the the total probability mass of f_0 can be shifted to an area where we always choose H_1 (the region where $L = f_1/f_0$ is large), we have in effect increased the probability of false alarm by an order of magnitude. This explains why relatively minor modelling errors can impact significantly the performance of tests, thus necessitating the recourse to robust testing techniques that take explicitly into account the presence of model uncertainties. To illustrate the impact of tail distribution changes on test performance, we consider a simple example.

Example 6.1: Gaussian densities with opposite means

Consider a binary hypothesis testing problem where we observe a nominally Gaussian distributed random variable Y with means of opposite sign under each hypothesis:

$$\begin{aligned} f_0(y) &\sim N(-m, \sigma^2) \\ f_1(y) &\sim N(m, \sigma^2) \,. \end{aligned} \tag{6.57}$$

The modelling uncertainty is fixed by a total variation tolerance, so that in the specification (6.21) of the uncertainty balls $\mathcal{F}_j$ with $j = 0,\ 1$, we have $\epsilon_0 = \epsilon_1 = 0$ and

$$\nu_0 = \nu_1 = \nu \,.$$

Then the nominal likelihood ratio function

$$L(y) = \frac{f_1(y)}{f_0(y)} = \exp(2my/\sigma^2) \,,$$

is monotone increasing, so the procedure outlined above for constructing the least-favorable densities is applicable. By symmetry, in the specification (6.26)–(6.27) of the least-favorable densities, we have $y_L = -y_U$ and

$$\ell_L = \exp(2my_L/\sigma^2) = (\ell_U)^{-1} \,.$$

By performing simple algebraic manipulations on expressions (6.26)–(6.27) for the least-favorable densities, if

$$d_j(y) = g_j^{\mathrm{L}}(y) - f_j(y)$$

with $j = 0,\ 1$ denotes the deviation between the least-favorable and nominal density for each hypothesis, we find

$$d_0(y) = \begin{cases} \exp\left(-\frac{(y^2+m^2)}{2\sigma^2}\right) \dfrac{\sinh(m(y-y_L)/\sigma^2)}{(2\pi\sigma^2)^{1/2}\cosh(my_L/\sigma^2)} & y < y_L \\ 0 & y_L \le y \le y_U \\ \exp\left(-\frac{(y^2+m^2)}{2\sigma^2}\right) \dfrac{\sinh(m(y-y_U)/\sigma^2)}{(2\pi\sigma^2)^{1/2}\cosh(my_U/\sigma^2)} & y > y_U \end{cases} \tag{6.58}$$

and

$$d_1(y) = -d_0(y)\,. \tag{6.59}$$

Note from expression (6.58) that $d_0(y)$ is an odd function, so the least-favorable density $g_0^{\mathrm{L}}(y)$ is obtained by shifting a small portion of the probability mass of $f_0(y)$ from its left tail, where $L(y) < 1$ and H_0 is favored, to its right tail, where $L(y) > 1$ so H_1 is preferred. This has the effect of increasing the probability of false alarm. Similarly, $g_1^{\mathrm{L}}(y)$ is obtained by shifting a small portion of its probability mass from its right tail to its left tail, thereby increasing the probability of a miss. The value of y_L is obtained by solving equation (6.45), where if

$$\Phi(x) = \frac{1}{(2\pi)^{1/2}} \int_{-\infty}^{x} \exp(-u^2/2)du$$

denotes the cumulative probability distribution of a $N(0,1)$ random variable, we have

$$\Gamma(y) = \frac{\exp(my/\sigma^2)\Phi((y+m)/\sigma) - \exp(-my/\sigma^2)\Phi((y-m)/\sigma)}{2\nu\cosh(my/\sigma^2)}\,.$$

In this case, by symmetry, the condition $y_L < y_U$ is equivalent to $y_L < 0$, which implies

$$\Gamma(0) = \frac{1}{2\nu}(\Phi(m/\sigma) - \Phi(-m/\sigma)) > 1\,.$$

Noting $\Phi(x) = 1 - Q(x)$, this gives

$$\frac{1}{2} - \nu > Q(m/\sigma)$$

or equivalently, since $Q(x)$ is monotone decreasing,

$$\frac{m}{\sigma} > Q^{-1}(\frac{1}{2} - \nu)\,.$$

In other words, the mean m measured in units of the standard deviation, must be sufficiently large to exceed the modelling uncertainty of the probability distributions. To illustrate these results, the deviation $d_0(y)$ is plotted in Fig. 6.2 for the case where $m = 2$, $\sigma = 1$, and $\nu = 10^{-2}$. We find $y_U = 1.075$. Also, if we select $\tau = \eta = 1$ as threshold, the nominal and robust tests δ and δ_{R} are identical and reduce to

$$Y \underset{H_0}{\overset{H_1}{\gtrless}} 0\,.$$

However, the nominal probability of false alarm (or of a miss since the test is symmetric)

$$P_F(\delta, f_0) = Q(2) = 0.0228$$

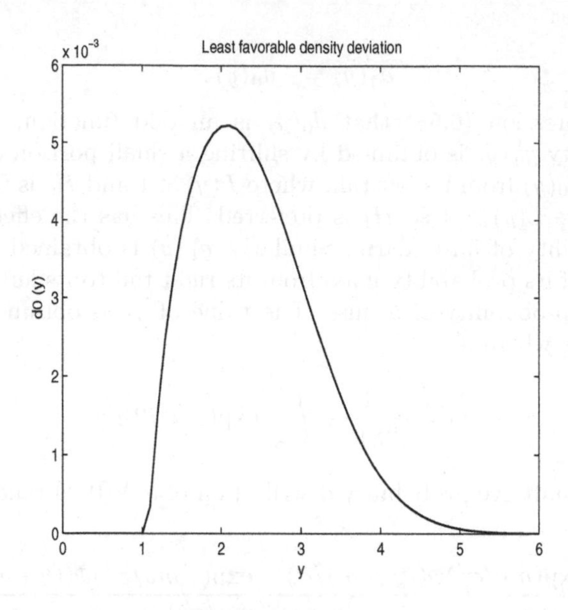

Fig. 6.2. Least-favorably probability density deviation $d_0(y)$ for a test between two Gaussian densities with $\sigma = 1$ and opposite means of amplitude $m = 2$.

is increased significantly since we have

$$P_F(\delta, g_0^{\mathrm{L}}) = Q(2) + \nu = 0.0328 \,.$$

□

The above example indicates clearly that since detectors are designed to operate with a small probability of error, relatively trivial modelling errors can have a big impact on test performance. Of course it is worth noting that the worst-case analysis performed above presumes that nature always conspires to produce the worst probabilistic models for the test we select. This viewpoint is very conservative, and in actual situations, outcomes tend to fall somewhere in between the optimistic results predicted by a nominal model analysis and the dire outcomes envisioned by the minimax approach.

6.4 Asymptotic Robustness

Instead of attempting to design binary tests that are robust for a single observation, it is possible to define robustness in an asymptotic sense. From this perspective, the goal is to find robust tests whose performance is optimal as the number of observations goes to infinity. A design methodology for achieving this objective was proposed by Dabak and Johnson [11]. It relies on the observation that, as shown in Section 3.2, for repeated i.i.d. observations with probability density $f_j(\mathbf{y})$ under H_j for $j = 0,\ 1$, as the number N of observations tends to infinity, the probability of a miss of a type I Neyman-Pearson test decays exponentially with rate

$$\lim_{N\to\infty} \frac{1}{N} \ln P_M(N) = -D(f_0|f_1)\,. \tag{6.60}$$

Similarly, the probability of false alarm of a type II Neyman-Pearson test decays exponentially with the number N of observations according to

$$\lim_{N\to\infty} \frac{1}{N} \ln P_F(N) = -D(f_1|f_0)\,. \tag{6.61}$$

Assume that actual probability densities g_0 and g_1 differ from the nominal densities f_0 and f_1 and are located in the convex balls

$$\begin{aligned} \mathcal{F}_0 &= \{g_0 :\ D(g_0|f_0) \le \epsilon_0\} \\ \mathcal{F}_1 &= \{g_1 :\ D(g_1|f_1) \le \epsilon_1\}\,, \end{aligned} \tag{6.62}$$

where the constants ϵ_j with $j = 0,\ 1$ represent the mismodelling tolerance in the sense of the KL divergence.

Given that for any pair $(g_0, g_1) \in \mathcal{F}_0 \times \mathcal{F}_1$, the optimum NP tests of types I and II have exponential rates of decay of the form (6.60)–(6.61) with f_0 and f_1 replaced by g_0 and g_1, Dabak and Johnson concluded that a least-favorable pair $(g_0^{\mathrm{L}}, g_1^{\mathrm{L}})$ in the sense of these two tests must have the properties

$$g_0^{\mathrm{L}} = \arg \min_{g_0\in\mathcal{F}_0} D(g_0|g_1^{\mathrm{L}}) \tag{6.63}$$

$$g_1^{\mathrm{L}} = \arg \min_{g_1\in\mathcal{F}_1} D(g_1|g_0^{\mathrm{L}})\,. \tag{6.64}$$

In other words, given g_1^{L}, g_0^{L} is the density that minimizes the rate of decay of the probability of a miss for a type I NP test between g_1^{L} and $g_0 \in \mathcal{F}_0$, and given g_0^{L}, g_1^{L} is the density that minimizes the rate of decay of the probability of false alarm for a type II NP test between $g_1 \in \mathcal{F}_1$ and g_0^{L}.

6.4.1 Least Favorable Densities

Note that expressions (6.63) and (6.64) involve the minimization of a convex function over a convex domain, and thus the minimum exists and can be obtained by the method of Lagrange multipliers. For the problem (6.62), we consider the Lagrangian

$$\begin{aligned} L(g_0, \lambda, \mu) &= D(g_0|g_1^{\mathrm{L}}) + \lambda(D(g_0|f_0) - \epsilon_0) + \mu(\int g_0 d\mathbf{y} - 1) \\ &= \int [\ln\big(\frac{g_0}{g_1^{\mathrm{L}}}\big) + \lambda \ln\big(\frac{g_0}{f_0}\big) + \mu] g_0 d\mathbf{y} - \lambda\epsilon_0 - \mu\,, \end{aligned} \tag{6.65}$$

where the Lagrange multiplier $\lambda \geq 0$ is associated to the constraint that g_0 belongs to $\mathcal{F}_0$, and the multiplier μ corresponds to the requirement that the total probability mass should equal one. The nonnegativity constraint $g_0(\mathbf{y}) \geq 0$ for all $\mathbf{y}$ does not need to be imposed explicity, since, as we shall see below, the function that minimizes L satisfies this condition automatically. To minimize L, we set to zero its Gateaux derivative with respect to g_0 in the direction of an arbitrary function z. This derivative is defined as [8, p. 17]

$$\nabla_{g_0,z} L(g,\lambda,\mu) \triangleq \lim_{h \to 0} \frac{L(g_0 + hz, \lambda, \mu) - L(g_0, \lambda, \mu)}{h} .$$

We find

$$\nabla_{g_0,z} L(g,\lambda,\mu) = \int [\ln\big(\frac{g_0}{g_1^{\mathrm{L}}}\big) + \lambda \ln\big(\frac{g_0}{f_0}\big) + \lambda + \mu] z d\mathbf{y} = 0 , \tag{6.66}$$

and since z is arbitrary, this gives

$$\ln\big(\frac{g_0^{1+\lambda}}{g_1^{\mathrm{L}} f_0^{\lambda}}\big) + \lambda + \mu = 0 . \tag{6.67}$$

Consequently,

$$g_0 = \frac{(g_1^{\mathrm{L}})^r (f_0)^{1-r}}{Z_0(r)} , \tag{6.68}$$

where $r = 1/(1+\lambda)$ satisfies $0 \leq r \leq 1$ and

$$Z_0(r) \triangleq \int (g_1^{\mathrm{L}})^r (f_0)^{1-r} d\mathbf{y} . \tag{6.69}$$

Since g_0 depends on the parameter r, in the following discussion, it is denoted as $g_0(r)$. Note that

$$Z_0(r) = E_{f_0}[\exp(r \ln\big(\frac{g_1^{\mathrm{L}}}{f_0}\big)(\mathbf{Y}))]$$

is the generating function for the log-likelihood ratio $\ln(g_1^{\mathrm{L}}/f_0)(\mathbf{Y})$ evaluated with respect to the nominal density f_0. Accordingly, as shown in Section 3.2, $\ln(Z_0(r))$ is a convex function of r. The expression (6.68) indicates that $g_0(r)$ belongs to the exponential family with parameter r connecting the densities f_0 and g_1^{L}. From a differential geometric viewpoint, this family represents the geodesic linking f_0 and g_1^{L} [11]. The density (6.68) is normalized to one, but the parameter r needs to be selected such that the Karush-Kuhn-Tucker (KKT) condition

$$\lambda(D(g_0|f_0) - \epsilon_0) = 0 \tag{6.70}$$

holds. When $\lambda > 0$, this means that $g_0(r)$ needs to satisfy

$$D(g_0(r)|f_0) = \epsilon_0 \,. \tag{6.71}$$

In other words, we must select the value of r corresponding to the point where the geodesic linking f_0 to g_1^L crosses the boundary of the ball $\mathcal{F}_0$. By evaluating the KL divergence $d(g_0, f_0)$ along the geodesic linking f_0 to g_1^L we obtain

$$\begin{aligned} K(r) &= D(g_0(r)|f_0) \\ &= \frac{r}{Z_0(r)} \int \ln(g_1^L/f_0)(g_1^L)^r (f_0)^{1-r} d\mathbf{y} - \ln(Z_0(r)) \\ &= r\frac{d}{dr} \ln(Z_0(r)) - \ln(Z_0(r)) \,. \end{aligned} \tag{6.72}$$

Since the two end points corresponding to $r = 0$ and $r = 1$ are f_0 and g_1^L, respectively, we have $K(0) = 0$ and $K(1) = D(g_1^L|f_0)$. Also, because $\ln Z_0(r)$ is convex,

$$\frac{dK}{dr} = r\frac{d^2}{dr^2} \ln Z_0(r) \geq 0 \,, \tag{6.73}$$

so the function $K(r)$ is monotone nondecreasing. Thus as long as $\epsilon_0 < D(g_1^L|f_0)$, there exists a value of r satisfying (6.71). When $\epsilon_0 > D(g_1^L|f_0)$, the least favorable density g_1^L is located inside the uncertainty ball $\mathcal{F}_0$. In this case we need to select $\lambda = 0$, so the KKT condition (6.70) holds. This implies $r = 1$, which yields $g_0^L = g_1^L$, so in this case hypotheses H_0 and H_1 are undistinguishable.

Thus when g_1^L is located outside the uncertainty ball $\mathcal{F}_0$, by solving the convex optimization problem (6.63) we have found that the least favorable density g_0^L takes the form

$$g_0^L = \frac{(g_1^L)^r (f_0)^{1-r}}{Z_0(r)} \,, \tag{6.74}$$

where r satisfies $K(r) = \epsilon_0$. Similarly, if g_0^L is outside the ball $\mathcal{F}_1$, the solution of the optimization problem (6.64) can be expressed as

$$g_1^L = \frac{(g_0^L)^s (f_1)^{1-s}}{Z_1(s)} \,, \tag{6.75}$$

with

$$Z_1(s) \triangleq \int (g_0^L)^s (f_1)^{1-s} d\mathbf{y} \,, \tag{6.76}$$

where s is selected such that g_1^L is located on the boundary of the ball $\mathcal{F}_1$. Like lines in Euclidean geometries, geodesics in curved space are specified by two points. Since the least-favorable densities g_0^L and g_1^L are common to the geodesics specified by (6.75) and (6.76), these geodesics must coincide. If the balls $\mathcal{F}_0$ and $\mathcal{F}_1$ do not intersect, we conclude therefore that, as depicted in

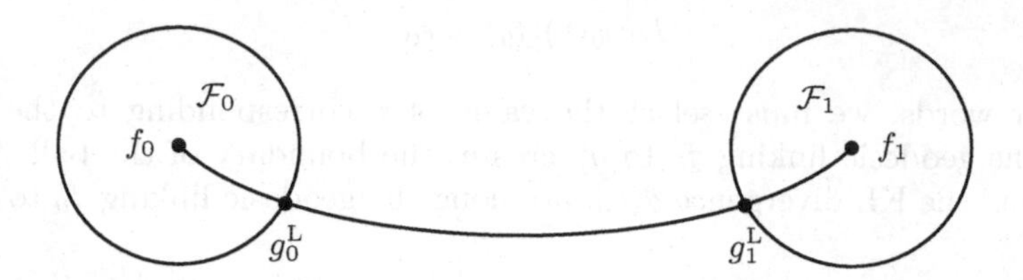

Fig. 6.3. Characterization of least favorable densities g_0^{L} and g_1^{L} as the intersections of the geodesic linking nominal densities f_0 and f_1 with uncertainty balls $\mathcal{F}_0$ and $\mathcal{F}_1$.

Fig. 6.3, g_0^{L} and g_1^{L} belong to the geodesic linking f_0 and f_1, so they can be expressed as

$$g_0^{\mathrm{L}} = \frac{(f_1)^u (f_0)^{1-u}}{G_0(u)}$$
$$g_1^{\mathrm{L}} = \frac{(f_0)^v (f_1)^{1-v}}{G_0(1-v)}\,, \tag{6.77}$$

with

$$G_0(u) = \int (f_1(\mathbf{y}))^u (f_0(\mathbf{y}))^{1-u} d\mathbf{y}\,, \tag{6.78}$$

where the parameters u and v with $0 \leq u,\ v \leq 1$ are selected such that

$$D(g_0^{\mathrm{L}}|f_0) = \epsilon_0 \quad , \quad D(g_1^{\mathrm{L}}|f_1) = \epsilon_1\,.$$

Note that expressions (6.77) can also be derived directly, without recourse to a geometric viewpoint, by eliminating g_1^{L} and g_0^{L} from identities (6.74) and (6.75). We find that the parameters u and v appearing in (6.77) can be expressed in terms of r and s as

$$u = \frac{r(1-s)}{1-rs} \quad , \quad v = \frac{s(1-r)}{1-rs}\,.$$

6.4.2 Robust Asymptotic Test

Given the least-favorable densities g_0^{L} and g_1^{L} specified by (6.77), an optimal Bayesian or NP test based on N independent observations $\mathbf{Y}_k$, $1 \leq k \leq N$ can be expressed as

$$\prod_{k=1}^{N} L_{\mathrm{L}}(\mathbf{y}_k) \underset{H_0}{\overset{H_1}{\gtrless}} \tau\,, \tag{6.79}$$

with

$$L_{\mathrm{L}}(\mathbf{y}) = \frac{g_1^{\mathrm{L}}(\mathbf{y})}{g_0^{\mathrm{L}}(\mathbf{y})} = (L(\mathbf{y}))^{1-(u+v)} \frac{G_0(u)}{G_0(1-v)}\,, \tag{6.80}$$

where

$$L(\mathbf{y}) = \frac{f_1(\mathbf{y})}{f_0(\mathbf{y})} \tag{6.81}$$

denotes the likelihood ratio function for the nominal densities. The threshold τ in (6.79) is selected to ensure that the probability of false alarm or of a miss based on the least-favorable densities does not exceed a fixed value. If we assume that $u + v < 1$, which is equivalent to requiring that the uncertainty balls $\mathcal{F}_0$ and $\mathcal{F}_1$ defined in (6.61) do not intersect, by substituting (6.80) inside (6.79), we find that the robust test (6.79) is equivalent to the nominal test

$$\prod_{k=1}^{N} L(\mathbf{y}_k) \underset{H_0}{\overset{H_1}{\gtrless}} \tau_{\mathrm{N}} \triangleq \Big(\frac{G_0(1-v)}{G_0(u)}\Big)^{N/(1-(u+v))} \tau^{1/(1-(u+v))} . \tag{6.82}$$

where the original threshold τ has been replaced by a new threshold τ_{N}.

To understand the procedure used to adjust the robust test threshold τ and replace it by the nominal threshold τ_{N}, consider the case where f_1 is known exactly, i.e. $\mathcal{F}_1 = \{f_1\}$, so that $v = 0$ and $G_0(1) = 1$ in (6.81). We seek to design an NP test of type I such that the probability of false alarm is less than or equal to α for any density $g_0 \in \mathcal{F}_0$. Since α is typically small, this implies $\tau > 1$. Then, by observing that $G_0(u) \leq 1$, we find that

$$\tau_{\mathrm{N}} = \Big(\frac{\tau}{(G_0(u))^N}\Big)^{1/(1-u)} > \tau . \tag{6.83}$$

In other words, in order to accommodate the uncertainty on the density g_0, we need to implement a nominal test with a raised threshold τ_{N}. In terms of the nominal densities, this test has a smaller probability of false alarm α_{N} than α, and a larger probability of a miss. This increased conservatism is dictated by the necessity to make the test robust against uncertainties in the location of g_0 inside $\mathcal{F}_0$.

Example 6.2: Detection of a constant in WGN

Consider the binary hypothesis testing problem

$$\begin{aligned} H_0 &: Y_k = V_k \\ H_1 &: Y_k = A + V_k , \end{aligned} \tag{6.84}$$

where V_k with $1 \leq k \leq N$ is modelled nominally as a WGN $N(0, \sigma^2)$ sequence. So $f_0 \sim N(0, \sigma^2)$ and $f_1 \sim N(A, \sigma^2)$. We assume that the uncertainty balls $\mathcal{F}_0$ and $\mathcal{F}_1$ have the same tolerance ϵ. Since

$$f_1^u(y) f_0^{1-u}(y) = \frac{G_0(u)}{(2\pi\sigma^2)^{1/2}} \exp\Big(-\frac{1}{2\sigma^2}(y - uA)^2\Big) ,$$

with

$$G_0(u) = \exp\Big(-\frac{u(1-u)A^2}{2\sigma^2}\Big)$$

we find that $g_0^{\mathrm{L}} \sim N(Au, \sigma^2)$, where by using the expression (3.45) for the KL divergence of two Gaussian distributions, we obtain

$$D(g_0^{\mathrm{L}}|f_0) = \frac{(Au)^2}{2\sigma^2} = \epsilon \, .$$

This yields

$$u = \frac{\sigma}{A}(2\epsilon)^{1/2} \, .$$

By performing a similar calculation, we also find $g_1^{\mathrm{L}} \sim N(A(1-u), \sigma^2)$. In other words, the least-favorable form of the detection problem can be expressed as

$$\begin{aligned} H_0 &: Y_k = Au + V_k \\ H_1 &: Y_k = A(1-u) + V_k \, , \end{aligned} \tag{6.85}$$

where V_k is again a $N(0, \sigma^2)$ distributed WGN sequence. So, while the nominal means of the observations were $m_0 = 0$ and $m_1 = A$, they are now $m_0^{\mathrm{L}} = Au$ and $m_1^{\mathrm{L}} = A(1-u)$. So, while the nominal distance between the two hypotheses is $d = A/\sigma$, for the least-favorable densities, it has shrunk to

$$d^{\mathrm{L}} = \frac{m_1^{\mathrm{L}} - m_0^{\mathrm{L}}}{\sigma} = \frac{A(1-2u)}{\sigma} \, .$$

In other words, the means of the two observations have been pushed towards each other to shrink the distance between the two hypotheses, and accordingly, to increase the probability of error of the test. □

The above example shows that asymptotically robust tests differ significantly from the single observation tests of Section 6.3. For a test based on a single observation, the least favorable densities are typically obtained by inserting outliers in the tails of the nominal probability distributions. Although outliers represent rare events, they can create havoc in the overall performance of tests designed to ensure that errors themselves are rare events. In contrast, for tests with repeated observations, outliers are of little consequence, since their effect is averaged out by the large number of available observations, most of which are typical. On the other hand, least-favorable densities tend to be those for which the entire mismodelling budget is aimed at disrupting the long-run detector performance. For discrimating between two Gaussian densities, the key variable affecting the long-run performance of a test is the distance d between the two hypotheses, i.e., the difference between the means of the Gaussian densities measured in units of the standard deviation. It is therefore perfectly logical that in (6.85) the entire KL tolerance budget of the least-favorable densities should go towards moving the means of the two distributions towards each other to ensure that d^{L} is as small as possible.

6.5 Robust Signal Detection

Consider the detection problem specified by

$$\begin{aligned} H_0 &: Y_k = V_k \\ H_1 &: Y_k = As_k + V_k \end{aligned} \tag{6.86}$$

for $1 \leq k \leq N$, where A and $s(k)$ are known, and the noise V_k is modelled nominally as a zero-mean WGN with variance σ^2. Thus under the two hypotheses, the observations Y_k are nominally independent with distributions

$$\begin{aligned} f_{0,k} &\sim N(0, \sigma^2) \\ f_{1,k} &\sim N(As_k, \sigma^2) \,. \end{aligned} \tag{6.87}$$

There exist several ways of formulating a robust version of this problem, depending on how many properties of the nominal probabilistic description of the noise process we seek to preserve. In essence, the fewer properties we enforce, the more robust the detection problem becomes, but also the more difficult it becomes to solve and to interpret physically. The main features of the noise V_k in (6.86) are first that sucessive samples are independent, and second, that the distribution of the noise is the same for all k and all hypotheses. In the robust version of the detection problem (6.86)–(6.87) considered by Martin and Schwartz [12], the independence of the V_k's is retained, but it is no longer assumed the V_k's are identically distributed for all k's and all hypotheses. Thus, if $g_{j,k}(y_k)$ denotes the marginal probability density of observation Y_k under hypothesis H_j with $j = 0, 1$, the joint probability density of observations Y_k under hypothesis H_j takes the form

$$g_j(\mathbf{y}) = \prod_{k=1}^{N} g_{j,k}(y_k) \,. \tag{6.88}$$

Extending slightly the setup of [12] which considered only contamination models, we assume that under H_0 and H_1, the densities $g_{0,k}$ and $g_{1,k}$ belong respectively to convex neighborhoods of the form

$$\begin{aligned} \mathcal{F}_{0,k} &= \{g_{0,k} \ : \ G_{0,k}(y) \geq (1-\epsilon)F_{0,k}(y) - \nu \text{ for all } y \in \mathbb{R}\} \\ \mathcal{F}_{1,k} &= \{g_{1,k} \ : \ G_{1,k}(y) \geq (1-\epsilon)(1 - F_{1,k}(y)) - \nu \text{ for all } y \in \mathbb{R}\} \,, \end{aligned} \tag{6.89}$$

where $F_{j,k}(y)$ and $G_{j,k}(y)$ denote the nominal and actual cumulative probability distributions corresponding respectively to the densities $f_{j,k}$ and $g_{j,k}$, with $j = 0,\ 1$. Note that since the noise V_k plays the same role in the observations under both hypotheses, the tolerances ϵ and ν are the same under both hypotheses.

The multiplicative form (6.88) for the observations density $f_j(\mathbf{y})$ and the neighborhoods $\mathcal{F}_{jk}$ for each of its components endow the overall density with the tensor neighborhood structure

$$\mathcal{F}_j = \mathop{\times}_{k=1}^{N} \mathcal{F}_{j,k} \,. \tag{6.90}$$

If $\mathcal{F} = \mathcal{F}_0 \times \mathcal{F}_1$, and if δ denotes a decision rule based on the N-dimensional observation vector $\mathbf{Y}$ with entries Y_k for $1 \leq k \leq N$, we seek again to solve the robust Bayesian hypothesis testing problem (6.14), or the robust Neyman-Pearson test (6.15). This will be accomplished again by exhibiting a robust test δ_{R} and a pair of least-favorable densities g_0^{L} and g_1^{L} with the multiplicative form

$$g_j^{\mathrm{L}}(\mathbf{y}) = \prod_{k=1}^{N} g_{j,k}^{\mathrm{L}}(y_k) \tag{6.91}$$

satisfying the saddle point identities (6.18), where the first inequality is equivalent to (6.20). The second inequality of (6.18) just indicates that the robust test δ_{R} is a likelihood ratio test, which due to the form (6.90) of the least-favorable densities, admits the multiplicative structure

$$L_{\mathrm{L}}(\mathbf{Y}) = \prod_{k=1}^{N} L_k^{\mathrm{L}}(Y_k) \underset{H_0}{\overset{H_1}{\gtrless}} \tau \,, \tag{6.92}$$

with

$$L_k^{\mathrm{L}}(Y_k) = \frac{g_{1,k}^{\mathrm{L}}(Y_k)}{g_{0,k}^{\mathrm{L}}(Y_k)} \,. \tag{6.93}$$

This reduces the solution of the minimax Bayesian detection problem (6.14) to one of exhibiting a pair of suitable least-favorable densities.

6.5.1 Least-Favorable Densities

To find each multiplicative component $g_{j,k}^{\mathrm{L}}(y_k)$ for $1 \leq k \leq N$ of the least-favorable densities $g_j^{\mathrm{L}}(\mathbf{y})$, we observe that conditioned on the knowledge of observations Y_i for all $i \neq k$, the LRT (6.92) can be rewritten as

$$L_k^{\mathrm{L}}(Y_k) \underset{H_0}{\overset{H_1}{\gtrless}} \tau(\mathbf{Y}_k^c) \tag{6.94}$$

where $\mathbf{Y}_k^c$ denotes the observation vector of dimension $N-1$ obtained by removing the k-th entry Y_k of $\mathbf{Y}$, i.e., it is the complement of Y_k inside $\mathbf{Y}$, and the threshold

$$\tau(\mathbf{Y}_k^c) \triangleq \tau / (\prod_{i \neq k} L_i^{\mathrm{L}}(Y_i)) \,.$$

The test (6.94) involves only the observation Y_k. Then denote by

$$L_k(y_k) = \frac{f_{1,k}(y_k)}{f_{0,k}(y_k)} = \exp\left(\frac{As_k}{\sigma^2}(y_k - As_k/2)\right) \tag{6.95}$$

the nominal likelihood ratio function for the k-th observation component. Depending on the sign of As_k, it is either monotone increasing or monotone decreasing. Since Y_k can always be multiplied by -1 without changing the detection problem, without loss of generality we replace As_k by $A|s_k| > 0$ in the formulation (6.86) of the detection problem. This ensures that $L_k(y_k)$ is monotone increasing and thus satisfies the condition used in Section 6.3 to construct a pair of least-favorable densities for neighborhoods of the form (6.21). Let

$$v = \frac{\epsilon + \nu}{1 - \epsilon} \; , \quad w = \frac{\nu}{1 - \epsilon} \; .$$

By following the construction procedure of Section 6.3, we can find $y_{Lk} < y_{Uk}$, which in turn specify $\ell_{Lk} = L_k(y_{Lk})$ and $\ell_{Uk} = L_k(y_{Uk})$, so that the least favorable densities

$$g^{\mathrm{L}}_{0,\,k}(y) = \begin{cases} \dfrac{1-\epsilon}{v + w\ell_{Lk}}(vf_{0,\,k}(y) + wf_{1,\,k}(y)) & y < y_{Lk} \\ (1-\epsilon_0)f_{0,\,k}(y) & y_{Lk} \leq y \leq y_{Uk} \\ \dfrac{1-\epsilon}{w + v\ell_{Uk}}(wf_{0,\,k}(y) + vf_{1,\,k}(y)) & y_{Uk} < y \end{cases} \tag{6.96}$$

and

$$g^{\mathrm{L}}_{1,\,k}(y) = \begin{cases} \dfrac{(1-\epsilon)\ell_{Lk}}{v + w\ell_{Lk}}(vf_{0,\,k}(y) + wf_{1,\,k}(y)) & y < y_{Lk} \\ (1-\epsilon)f_{1,\,k}(y) & y_{Lk} \leq y \leq y_{Uk} \\ \dfrac{(1-\epsilon)\ell_{Uk}}{w + v\ell_{Uk}}(wf_{0,\,k}(y) + vf_{1,\,k}(y)) & y_{Uk} < y \end{cases} \tag{6.97}$$

have the feature that, among all probability densities of $\mathcal{F}_{0,\,k}$, they maximize the probability of false alarm and of a miss for the conditional test (6.94).

To explain precisely what we mean by this statement, given observations Y_i for all $i \neq k$, we define the conditional probability of false alarm for the test δ_{R} and an arbitrary density $g_{0,\,k} \in \mathcal{F}_{0,\,k}$ as

$$\begin{aligned} P_F(\delta_{\mathrm{R}}, g_{0,\,k}|\mathbf{Y}^c_k) &\triangleq P_{g_{0,\,k}}[L_{\mathrm{L}}(\mathbf{Y}) \geq \tau|\mathbf{Y}^c_k] \\ &= \int u(L_{\mathrm{L}}(\mathbf{Y}) - \tau)g_{0,\,k}(Y_k)dY_k \; , \end{aligned} \tag{6.98}$$

where

$$u(z) = \begin{cases} 1 & z \geq 0 \\ 0 & z < 0 \end{cases}$$

denotes the unit step function. Similarly, the conditional probability of a miss for the test δ_{R} and an arbitrary density $g_{1,\,k} \in \mathcal{F}_{1,\,k}$ is defined as

$$\begin{aligned} P_M(\delta_{\mathrm{R}}, g_{1,\,k}|\mathbf{Y}^c_k) &\triangleq P_{g_{1,\,k}}[L_{\mathrm{L}}(\mathbf{Y}) < \tau|\mathbf{Y}^c_k] \\ &= \int \big[1 - u(L_{\mathrm{L}}(\mathbf{Y}) - \tau)\big]g_{1,\,k}(Y_k)dY_k \; . \end{aligned} \tag{6.99}$$

Based on the analysis of Section 6.3, we know that conditioned on the knowledge of all Y_i for $i \neq k$, the least-favorable densities $g^{\rm L}_{0,\,k}(y_k)$ and $g^{\rm L}_{1,\,k}(y_k)$ given by (6.96) and (6.97) satisfy

$$\begin{aligned} P_F(\delta_{\rm R}, g_{0,\,k}|\mathbf{Y}^c_k) &\leq P_F(\delta_{\rm R}, g^{\rm L}_{0,\,k}|\mathbf{Y}^c_k) \\ P_M(\delta_{\rm R}, g_{1,\,k}|\mathbf{Y}^c_k) &\leq P_M(\delta_{\rm R}, g^{\rm L}_{1,\,k}|\mathbf{Y}^c_k) \end{aligned} \tag{6.100}$$

for all densities $g_{0,\,k} \in \mathcal{F}_{0,\,k}$ and $g_{1,k} \in \mathcal{F}_{0,\,k}$.

The key feature of the least-favorable densities $g^{\rm L}_{0,\,k}$ and $g^{\rm L}_{1,\,k}$ associated to the conditional test (6.94) is that they are independent of the threshold $\tau(\mathbf{Y}^c_k)$ and thus *independent of the conditioning variables* $\mathbf{Y}^c_k$. This means they depend on y_k only. Also, the k-th component of the least-favorable likelihood ratio can be expressed as

$$L^{\rm L}_k(y_k) = C_k(L_k(y_k))\,, \tag{6.101}$$

where the clipping nonlinearity $C_k(\cdot)$ is obtained by setting $\ell_L = \ell_{Lk}$ and $\ell_U = \ell_{Uk}$ in (6.29) and Fig. 6.1.

Then the conditional property (6.100) satisfied by the least-favorable density components $g^{\rm L}_{j,\,k}$ for $j = 0,\ 1$ can be used to prove that the product densities $g^{\rm L}_0$ and $g^{\rm L}_1$ given by (6.91) are the least-favorable among all product densities

$$\begin{aligned} g_0(\mathbf{y}) &= \prod_{k=1} g_{0,\,k}(y_k) \in \mathcal{F}_0 \\ g_1(\mathbf{y}) &= \prod_{k=1} g_{1,\,k}(y_k) \in \mathcal{F}_1\,. \end{aligned} \tag{6.102}$$

To prove this result, we observe that

$$\begin{aligned} P_F(\delta_{\rm R}, g^{\rm L}_0) &= \int u(L_{\rm L}(\mathbf{y}) - \tau) \prod_{k=1}^{N} g^{\rm L}_{0,\,k}(y_k) d\mathbf{y} \\ &\geq \int u(L_{\rm L}(\mathbf{y}) - \tau) g_{0,\,1}(y_1) \prod_{k=2}^{N} g^{\rm L}_{0,\,k}(y_k) d\mathbf{y} \geq \cdots \\ \cdots &\geq \int u(L_{\rm L}(\mathbf{y}) - \tau) \prod_{k=1}^{N-1} g_{0,\,k}(y_k)\, g^{\rm L}_{0,\,N}(y_N) d\mathbf{y} \\ &\geq \int u(L_{\rm L}(\mathbf{y}) - \tau) \prod_{k=1}^{N} g_{0,\,k}(y_k) d\mathbf{y} = P_F(\delta_{\rm R}, g_0)\,, \end{aligned} \tag{6.103}$$

where we have used repeatedly the conditional property (6.100) of the least-favorable density components. Similarly, we have

$$\begin{aligned}
P_M(\delta_{\mathrm{R}}, g_1^{\mathrm{L}}) &= \int [1 - u(L_{\mathrm{L}}(\mathbf{y}) - \tau)] \prod_{k=1}^{N} g_{1,k}^{\mathrm{L}}(y_k) d\mathbf{y} \\
&\geq \int [1 - u(L_{\mathrm{L}}(\mathbf{y}) - \tau)] g_{1,1}(y_1) \prod_{k=2}^{N} g_{1,k}^{\mathrm{L}}(y_k) d\mathbf{y} \geq \dots \\
\dots &\geq \int [1 - u(L_{\mathrm{L}}(\mathbf{y}) - \tau)] \prod_{k=1}^{N-1} g_{1,k}(y_k)\, g_{1,N}^{\mathrm{L}}(y_N) d\mathbf{y} \\
&\geq \int [1 - u(L_{\mathrm{L}}(\mathbf{y}) - \tau)] \prod_{k=1}^{N} g_{1,k}(y_k) d\mathbf{y} = P_M(\delta_{\mathrm{R}}, g_1) \,. \quad (6.104)
\end{aligned}$$

This establishes that the product least-favorable densities given by (6.91) have the property (6.20), so they obey the saddle point condition (6.18), and thus solve the minimax Bayesian detection problem (6.14).

Next, consider the minimax Neyman-Pearson problem (6.15). As was already the case for the robust hypothesis testing problem of Section 6.3, the components $g_{j,k}^{\mathrm{L}}$ of the least-favorable densities g_j^{L} with $j = 0, 1$ are independent of the threshold τ of the robust test (6.92). Accordingly, if we denote by $\delta_{\mathrm{R}}(\tau)$ the test (6.92) with threshold τ, to solve the minimax NP problem (6.15), we only need to adjust the threshold τ to ensure

$$P_F(\delta_{\mathrm{R}}, g_0^{\mathrm{L}}) = \alpha \,.$$

Then, if δ_{R} denotes the resulting test, since its size is α and it satisfies (6.104), it solves the minimax NP problem (6.15).

6.5.2 Receiver Structure

Since we have established the optimality of the test (6.92), where the likelihood ratio components $L_k^{\mathrm{L}}(y_k)$ are obtained through the clipping transformation (6.101), we can now turn to its implementation. As first step, note that since the hypotheses H_0 and H_1 can always be symmetrized by subtracting $-A|s_k|/2$ from Y_k, thereby reducing the detection problem for a single observation to the form considered in Example 6.1, we conclude that the values y_{Lk} and y_{Uk} appearing in expressions (6.96)–(6.97) for the least-favorable densities satisfy

$$y_{Lk} + y_{Uk} = A|s_k| \,.$$

Taking into account expression (6.95) for the nominal LR function for the k-th observation component, this implies

$$\ell_{Lk} = (\ell_{Uk})^{-1} \,. \quad (6.105)$$

Taking the logarithm of equation (6.92) and multiplying by σ^2, the robust test can be expressed in additive form as

$$\sum_{k=1}^{N} \tilde{C}_k(Z_k) \underset{H_0}{\overset{H_1}{\gtrless}} \eta = \sigma^2 \ln(\tau) , \tag{6.106}$$

where

$$Z_k = \sigma^2 \ln(L_k(Y_k)) = A|s_k|(Y_k - A|s_k|/2) . \tag{6.107}$$

In expression (6.106),

$$\tilde{C}_k(z) = \begin{cases} -z_k & z < -z_k \\ z & -z_k \leq z \leq z_k \\ z_k & z_k \leq z , \end{cases} \tag{6.108}$$

with

$$z_k = \sigma^2 \ln(\ell_{Uk}) , \tag{6.109}$$

is a symmetric clipping function which is plotted in Fig. 6.4. Note that this function varies with k, and is thus time-varying, unless $A|s_k|$ is constant for all k, in which case the detection problem (6.86) reduces to the detection of a constant in WGN.

From expressions (6.106)–(6.109), we see that the optimum receiver can be implemented by first symmetrizing the two hypotheses by evaluating

$$\tilde{Y}_k = Y_k - A|s_k|/2 ,$$

then correlating $\tilde{Y}_k$ with the signal As_k to be detected, passing the product through the nonlinearity $C_k(\cdot)$, and then summing the output over $1 \leq k \leq N$. This receiver is depicted in Fig. 6.5. Except for the introduction of the clipping nonlinearity $C_k(\cdot)$ it takes the form of the correlator integrator/summer receiver discussed in Chapter 8 for the detection of a known signal in WGN. Accordingly, the modified structure obtained by inserting the clipper $C_k(\cdot)$ is known as the *correlator limiter* robust receiver [12].

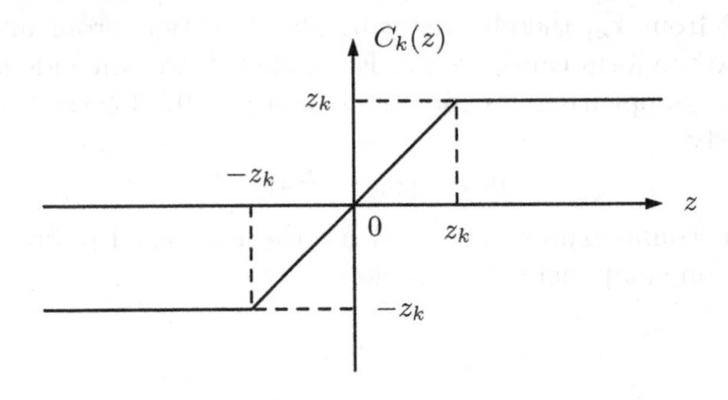

Fig. 6.4. Symmetric time-varying clipping nonlinearity $C_k(\cdot)$.

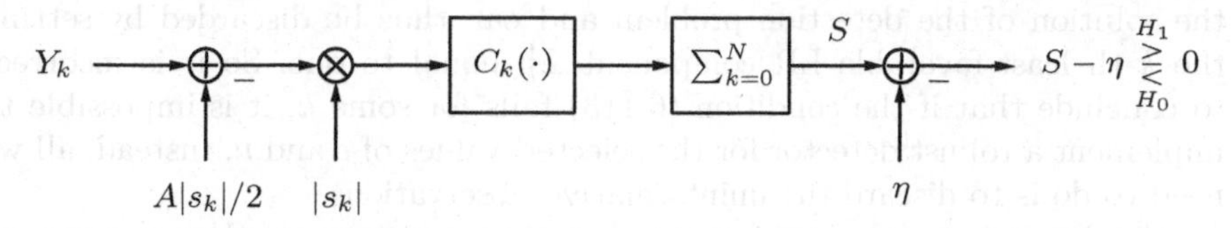

Fig. 6.5. Block diagram of the correlator limiter receiver.

In the above analysis we have assumed that the neighborhoods $\mathcal{F}_{0,\,k}$ and $\mathcal{F}_{1,\,k}$ are disjoint for all k, so that

$$y_{Lk} < A|s_k|/2 < y_{Uk} \, , \tag{6.110}$$

or equivalently,

$$\ell_{Lk} < L_k(A|s_k|/2) = 1 < \ell_{Uk} \, . \tag{6.111}$$

For the k-th observation component, y_{lk} is obtained by solving $\Gamma_k(y) = 1$ where, if $L_k(y)$ denotes the nominal LR function (6.95),

$$\Gamma_k(y) = \frac{L_k(y)\Phi(y/\sigma) - \Phi((y - A|s_k|)/\sigma)}{v + wL_k(y)} \, . \tag{6.112}$$

Since $\Gamma_k(y)$ is monotone increasing, to ensure $y_{Lk} < A|s_k|/2$, we must have

$$\Gamma_k(A|s_k|/2) > 1 \, .$$

Since $L_k(A|s_k|/2) = 1$, this implies

$$1 - 2Q(A|s_k|/(2\sigma)) > v + w = \frac{\epsilon + 2\nu}{1 - \epsilon} \, ,$$

or equivalently,

$$\frac{1/2 - \epsilon - \nu}{1 - \epsilon} > Q(A|s_k|/(2\sigma)) \, .$$

But $Q(\cdot)$ is monotone decreasing, so we must have

$$A|s_k| > 2\sigma Q^{-1}((1/2 - \epsilon - \nu)/(1 - \epsilon)) \, . \tag{6.113}$$

Consequently, to ensure that the condition (6.110) holds for all k, the values of ϵ and ν must be sufficiently small to ensure that the quantity appearing on the right-hand side of (6.113) is less than $A \min_{1 \leq k \leq N} |s_k|$.

When the condition (6.113) fails for one or more values of k, the correlator limiter receiver can still be implemented. We only need to recognize that whenever $\mathcal{F}_{0,\,k}$ and $\mathcal{F}_{1,k}$ intersect, the k-th least-favorable density components satisfy $g^{\mathrm{L}}_{0,\,k} = g^{\mathrm{L}}_{1,\,k}$, so

$$L^{\mathrm{L}}_k(y) = 1 \, .$$

Equivalently, $\ell_{Lk} = \ell_{Uk} = 1$ and thus $z_k = 0$ in (6.108). In other words, whenever $\mathcal{F}_{0,\,k}$ and $\mathcal{F}_{1,\,k}$ intersect, the k-th observation becomes useless to the solution of the detection problem and can thus be discarded by setting the k-th least-favorable LR component L_k^{L} equal to one. So it is incorrect to conclude that if the condition (6.113) fails for some k, it is impossible to implement a robust detector for the selected values of ϵ and ν. Instead, all we need to do is to discard the uninformative observations.

Finally, it is worth pointing out that there exists more than one way to robustify the detection problem (6.86). In the approach described above, following [12], we elected to preserve the independence of the noise samples V_k, but we did not require that the distributions of the noise V_k should remain the same under both hypotheses. Moustakides and Thomas [13] have presented a solution of the robust detection problem obtained when the noise V_k is required to have the same statistics under H_0 and H_1, but where successive noise samples are no longer required to be independent.

6.6 Bibliographical Notes

The foundations of robust hypothesis testing were laid by Huber who considered problems of increasing generality in [6, 7] and [10]. The asymptotic formulation of robust detection presented in Section 6.4 was first proposed by Dabak and Johnson [11], but prior to this work, asymptotic formulations of robustness had been used commonly in the robust detection literature. For example, a large deviations formulation of robust detection for repeated observations is presented in [14] for the case of constants in i.i.d. noise, where the noise distributions are required to remain the same under both hypotheses. More recently, for the case when uncertainty classes are specified by moment constraints, and when the observed i.i.d. random variables take finite values, Meyn and Veeravalli [15] obtained a complete characterization of optimal robust asymptotic tests. As one would expect, the tests are expressed in terms of the empirical probability distribution of the observations, but quite interestingly several different optimal tests exist. In our discussion of robust signal detection, we have focused on the case of a known signal in imprecisely modelled noise. However, significant effort has also been directed [12, 16, 17] at robust composite hypothesis testing problems where the amplitude A of of the signal is small but unknown in (6.86). The reader is referred to [18] for a comprehensive discussion of robust detection and estimation results. In this respect, it is worth pointing out that in parallel with robust detection, estimation problems can also be formulated in a robust manner [1]. Recently, [19] considered a minimax least-squares estimation problem for Gaussian nominal models with a Kullback-Leibler neighborhood structure and showed that the optimal estimator is in fact a risk-sensitive estimator of the type considered in [20]. Also, a minimax estimation regret approach was proposed recently in [21] for reducing the excessive conservatism of standard minimax estimation methods.

6.7 Problems

6.1. Consider two probability measures P and Q over the Borel field $\mathcal{B}$ of $\mathbb{R}^n$ specified by probability densities $f(\mathbf{y})$ and $g(\mathbf{y})$, with $\mathbf{y} \in \mathbb{R}^n$. We seek to prove the equality of expression

$$\begin{aligned} d_{\text{TV}}(P,Q) &= \max_{||h||_\infty \leq 1} |E_P[h(\mathbf{Y})] - E_Q[h(\mathbf{Y})]| \\ &= \max_{||h||_\infty \leq 1} |\int h(\mathbf{y})[f(\mathbf{y}) - g(\mathbf{y})]d\mathbf{y}| \end{aligned} \tag{6.114}$$

for the total variation with identities (6.4) and (6.5).

(a) To prove the equality of (6.114) with (6.5), use inequalities

$$\begin{aligned} |\int h(\mathbf{y})[f(\mathbf{y}) - g(\mathbf{y})]d\mathbf{y}| &\leq \int |h(\mathbf{y})||f(\mathbf{y}) - g(\mathbf{y})|d\mathbf{y} \\ &\leq ||h||_\infty \int |f(\mathbf{y}) - g(\mathbf{y})|d\mathbf{y} \,, \end{aligned}$$

and verify that the maximum in (6.114) is achieved by selecting $h(\mathbf{y}) = \text{sign}((f - g)(\mathbf{y}))$.

(b) To prove the equality of (6.114) with (6.4), as explained in the sentence leading to (6.4), observe that for $A \in \mathcal{B}$ and

$$h(\mathbf{Y}) = 1_A(\mathbf{y}) - 1_{A^c}(\mathbf{y}) \,,$$

we have

$$2|P(A) - Q(A)| = |\int h(\mathbf{y})(f(\mathbf{y}) - g(\mathbf{y}))d\mathbf{y}| \,. \tag{6.115}$$

Then evaluate the right-hand side of (6.115) by selecting $A = \{\mathbf{y} : f(\mathbf{y}) > g(\mathbf{y})\}$,

6.2. Consider the binary detection problem specified by

$$\begin{aligned} H_0 &: \mathbf{Y} = \mathbf{V} \\ H_1 &: \mathbf{Y} = \mathbf{s}_j + \mathbf{V} \,, \end{aligned}$$

where $\mathbf{V}$ is a $N(\mathbf{0}, \sigma^2 \mathbf{I}_N)$ distributed noise vector, and the received N-dimensional signal vector $\mathbf{s}_j$ is an arbitrary element of a collection $\{\mathbf{s}_j \,, 1 \leq j \leq M\}$ of M possible signals. This problem can be formulated as a robust detection problem in the following manner [22]. Let

$$\mathbf{p} = \begin{bmatrix} p_1 \dots p_j \dots p_M \end{bmatrix}$$

denote an M-dimensional probability mass distribution, where p_j denotes the probability that under H_1, the received signal is $\mathbf{s}_j$. So, if

$$f(\mathbf{y}|\mathbf{s}_j) \sim N(\mathbf{s}_j, \sigma^2 \mathbf{I}_N)$$

represents the probability density of observation vector $\mathbf{Y}$ when signal $\mathbf{s}_j$ is received, the probability density of $\mathbf{Y}$ under H_1 can be expressed as

$$f_1(\mathbf{y}) = \sum_{j=1}^{M} f(\mathbf{y}|\mathbf{s}_j) p_j \,. \tag{6.116}$$

So, under H_1, f_1 belongs to the convex set $\mathcal{F}_1$ of densities spanned by convex combinations of the form (6.116). The set $\mathcal{F}_1$ forms a polytope with vertices $f(\cdot|\mathbf{s}_j)$. Then let δ denote an arbitrary pointwise randomized test, i. e., $0 \leq \delta(\mathbf{y}) \leq 1$. We seek to construct a pair $(\delta_{\mathrm{R}}, f_1^{\mathrm{L}})$ solving the minimax Neyman-Pearson problem

$$\min_{\delta \in \mathcal{D}_\alpha} \max_{f_1 \in \mathcal{F}_1} P_M(\delta, f_1) \,, \tag{6.117}$$

where

$$P_M(\delta, f_1) = \int (1 - \delta(\mathbf{y})) f_1(\mathbf{y}) d\mathbf{y}$$

denotes the probability of a miss, and $\mathcal{D}_\alpha$ is the convex set of decision rules δ such that the probability of false alarm

$$P_F(\delta) = \int \delta(\mathbf{y}) f_0(\mathbf{y}) d\mathbf{y} \leq \alpha \,.$$

Here $f_0(\mathbf{y})$ denotes the $N(\mathbf{0}, \sigma^2 \mathbf{I}_N)$ distribution of the observations $\mathbf{Y}$ under H_0. Note that in (6.117), the density f_1 is completely specified by the probability mass distribution $\mathbf{p}$, so the maximization over f_1 can be replaced by a maximization over $\mathbf{p}$, and $P_M(\delta, f_1)$ can be rewritten equivalently as $P_M(\delta, \mathbf{p})$.

The minimax problem (6.117) is of the type examined in Section 6.3, so a saddle point $(\delta_{\mathrm{R}}, f_1^{\mathrm{L}})$ exists, which can be characterized as follows.

(a) Let $\mathbf{p}_{\mathrm{L}} = (p_j^{\mathrm{L}} \,,\, 1 \leq j \leq M\}$ be the least favorable probability mass distribution specifying f_1^{L}. Verify that the robust test δ_{R} can be expressed as the LR test

$$L_{\mathrm{L}}(\mathbf{y}) = \frac{f_1^{\mathrm{L}}(\mathbf{y})}{f_0(\mathbf{y})} = \sum_{j=1}^{M} p_j^{\mathrm{L}} L_j(\mathbf{y}) \underset{H_0}{\overset{H_1}{\gtrless}} \eta$$

where

$$L_j(\mathbf{y}) = \frac{f(\mathbf{y}|\mathbf{s}_j)}{f_0(\mathbf{y})}$$

is the LR when the received signal under H_1 is $\mathbf{s}_j$.

(b) Since a saddle point exists, the least-favorable probability mass distribution must satisfy

$$P_M(\delta_{\mathrm{R}}, \mathbf{p}_{\mathrm{L}}) \geq P_M(\delta_{\mathrm{R}}, \mathbf{p})$$

for all probability mass distributions $\mathbf{p}$. In other words, assuming that δ_{R} is known, the least favorable distribution $\mathbf{p}_{\mathrm{L}}$ is obtained by maximizing

$$P_M(\delta_R, \mathbf{p}) = \sum_{j=1}^{M} p_j P_{Mj}(\delta_R) , \tag{6.118}$$

where

$$P_{Mj}(\delta_R) \triangleq \int (1 - \delta_R(\mathbf{y})) f(\mathbf{y}|\mathbf{s}_j) d\mathbf{y} , \; 1 \leq j \leq M$$

is fixed. Of course, the probabilities p_j must satisfy

$$p_j \geq 0 \quad \text{and} \quad \sum_{j=1}^{M} p_j = 1 . \tag{6.119}$$

The objective function (6.118) and constraints (6.119) form a *linear program*. Use the method of Lagrange multipliers to verify that the solution will have the following features: Let

$$\lambda_* \triangleq \max_{1 \leq j \leq M} P_{Mj}(\delta_R) ,$$

and let J denote the set of indices j for which λ_* is attained, i.e.,

$$J = \{j \,:\, P_{Mj}(\delta_R) = \lambda_*\} .$$

Then the least favorable probability mass distribution is such that

$$p_j^L > 0 \quad \text{for } j \in J$$
$$p_j^L = 0 \text{ when } j \notin J .$$

and $P_M(\delta_R, \mathbf{p}_L) = \lambda_*$.
For readers familiar with linear programming theory [23], the above result can be interpreted by introducting the *dual program*

$$\min \lambda \quad , \quad \lambda \geq P_{Mj}(\delta_R) , \, 1 \leq j \leq M$$

corresponding to primal program (6.118)–(6.119). Then if $\mathbf{p}$ and λ are feasible solutions of the primal and dual programs, they correspond to optimal solutions of these programs if and only if they obey the *complementary slackness conditions*

$$p_j(\lambda - P_{Mj}(\delta_R)) = 0$$

for all j.

6.3. Consider the robust detection problem of Problem 6.2. The characterization of the optimal pair $(\delta_R, \mathbf{p}_L)$ obtained in parts (a) and (b) provides some guidelines concerning the structure of the solution, but guessing is still required to fully identify the solution. For example, consider the case where $M = 2$, and where under H_1, the received signal is either $\mathbf{s}_1 = \mathbf{s}$ or $\mathbf{s}_2 = -\mathbf{s}$, where $\mathbf{S}$ is a known vector. In other words, only the polarity of the signal is unknown.

(a) In this particular case, by symmetry, it is reasonable to guess that

$$p_1^{\rm L} = p_2^{\rm L} = 1/2 \, .$$

With this guess, use the result of part (a) of Problem 6.2 to verify that the robust test $\delta_{\rm R}$ is given by

$$|\mathbf{s}^T \mathbf{Y}| \underset{H_0}{\overset{H_1}{\gtrless}} \eta \, .$$

Indicate how the threshold η needs to be selected to ensure that the probability of false alarm is α. Note that, quite interestingly, the robust test coincides with the UMPI test (5.37).

(b) For the robust test $\delta_{\rm R}$, verify that

$$P_{M1}(\delta_{\rm R}) = P_{M2}(\delta_{\rm R})$$

and show that the complementary slackness conditions identified in part (b) of Problem 6.2 are satisfied.

6.4. Assume that in Problem 6.2, the signals $\mathbf{s}_i$ are orthogonal and of equal energy, i.e.,

$$||\mathbf{s}_i||_2^2 = E$$

for $1 \leq i \leq M$.

(a) By symmetry, the least favorable probability distribution can be guessed as

$$p_i^{\rm L} = \frac{1}{M}$$

for $1 \leq i \leq M$. Show that the robust test $\delta_{\rm R}$ takes the form

$$\frac{1}{M} \sum_{i=1}^{M} \exp\left(\frac{\mathbf{s}_i^T \mathbf{Y}}{\sigma^2}\right) \underset{H_0}{\overset{H_1}{\gtrless}} \eta \, . \tag{6.120}$$

Note that when $M = N$, and $\mathbf{s_i} = E^{1/2}\mathbf{e}_i$, where $\mathbf{e}_i$ is the i-th unit vector of $\mathbb{R}^N$, the test (6.120) reduces to

$$\frac{1}{N} \sum_{i=1}^{N} \exp\left(\frac{E^{1/2} Y_i}{\sigma^2}\right) \underset{H_0}{\overset{H_1}{\gtrless}} \eta \, ,$$

where Y_i denotes the i-th coordinate of $\mathbf{Y}$. This test coincides with the UMPI test obtained in Example 5.10 for the detection of an impulse in WGN.

(b) For the robust test $\delta_{\rm R}$, verify that the probability of a miss $P_{Mi}(\delta_{\rm R})$ is independent of i and check that the complementary slackness conditions identified in part (b) of Problem 6.2 are satisfied.

6.5. The robust formulation of detection problems with incompletely known received signals considered in Problem 6.2 is not restricted to situations where the received signal belongs to a discrete set. Consider the detection problem

$$H_0 : \mathbf{Y} = \mathbf{V}$$
$$H_1 : \mathbf{Y} = \mathbf{s}(\boldsymbol{\theta}) + \mathbf{V}$$

where the noise vector $\mathbf{V}$ is $N(\mathbf{0}, \sigma^2 \mathbf{I}_N)$ distributed and $\boldsymbol{\theta}$ is an unknown parameter vector of $\mathbb{R}^q$. Let

$$f(\mathbf{y}|\boldsymbol{\theta}) \sim N(\mathbf{s}(\boldsymbol{\theta}), \sigma^2 \mathbf{I}_N)$$

denote the probability density of the N-dimensional observation vector $\mathbf{Y}$ when signal $\mathbf{s}(\boldsymbol{\theta})$ is received, and assume that $\boldsymbol{\theta}$ is random with unknown probability density $f_{\boldsymbol{\Theta}}(\boldsymbol{\theta})$. Then the probability density of $\mathbf{Y}$ under H_1 can be expressed as

$$f_1(\mathbf{y}) = \int f(\mathbf{y}|\boldsymbol{\theta}) f_{\boldsymbol{\Theta}}(\boldsymbol{\theta}) d\boldsymbol{\theta} .$$

In this expression, $f(\mathbf{y}|\boldsymbol{\theta})$ is known, but $f_{\boldsymbol{\Theta}}(\boldsymbol{\theta})$ is unknown, and as it varies, $f_1(\mathbf{y})$ spans a convex set $\mathcal{F}_1$. Then, consider the minimax Neyman-Pearson test (6.115), where $P_M(\delta, f_1)$ and $\mathcal{D}(\alpha)$ are defined as in Problem 6.2. Note that since f_1 is completely specified by $f_{\boldsymbol{\Theta}}$, $P_M(\delta, f_1)$ can be rewritten equivalently as $P_M(\delta, f_{\boldsymbol{\Theta}})$.

Because $P_M(\delta, f_1)$ has a convex-concave structure, a saddle point $(\delta_{\mathrm{R}}, f_1^{\mathrm{L}})$ exists, which can be characterized as follows:

(a) Let $f_{\boldsymbol{\Theta}}^{\mathrm{L}}$ be the least-favorable probability density specifying f_1^{L}. Verify that the robust test δ_{R} can be expressed as the LR test

$$L_{\mathrm{L}}(\mathbf{y}) = \frac{f_1^{\mathrm{L}}(\mathbf{y})}{f_0(\mathbf{y})} = \int L(\mathbf{y}|\boldsymbol{\theta}) f_{\boldsymbol{\Theta}}^{L}(\boldsymbol{\theta}) d\boldsymbol{\theta} \underset{H_0}{\overset{H_1}{\gtrless}} \eta$$

where

$$L(\mathbf{y}|\boldsymbol{\theta}) = \frac{f(\mathbf{y}|\boldsymbol{\theta})}{f_0(\mathbf{y})}$$

is the LR when the received signal under H_1 is $\mathbf{s}(\boldsymbol{\theta})$.

(b) Since a saddle point exists, the least-favorable probability density function must satisfy

$$P_M(\delta_{\mathrm{R}}, f_{\boldsymbol{\Theta}}^{\mathrm{L}}) \geq P_M(\delta_{\mathrm{R}}, \mathbf{f}_{\boldsymbol{\Theta}})$$

for all density functions $f_{\boldsymbol{\Theta}}$. Thus the least favorable density $f_{\boldsymbol{\Theta}}^{\mathrm{L}}$ is obtained by maximizing

$$P_M(\delta_{\mathrm{R}}, f_{\boldsymbol{\Theta}}) = \int P_M(\delta_{\mathrm{R}}|\boldsymbol{\theta}) f_{\boldsymbol{\Theta}}(\boldsymbol{\theta}) d\boldsymbol{\theta} , \qquad (6.121)$$

with
$$P_M(\delta_{\mathrm{R}}|\boldsymbol{\theta}) \stackrel{\triangle}{=} \int (1-\delta_{\mathrm{R}}(\mathbf{y}))f(\mathbf{y}|\boldsymbol{\theta})d\mathbf{y}$$
fixed. Of course, the density f_{Θ} must satisfy
$$f_{\Theta}(\boldsymbol{\theta}) \geq 0 \tag{6.122}$$
for all $\boldsymbol{\theta}$ and
$$\int f_{\Theta}(\boldsymbol{\theta})d\boldsymbol{\theta} = 1 \,. \tag{6.123}$$
The objective function (6.121) and constraints (6.122)–(6.123) specify a linear optimization problem. Use the method of Lagrange multipliers to verify that its solution has the following features: Let
$$\lambda_* \stackrel{\triangle}{=} \max_{\boldsymbol{\theta}} P_M(\delta_{\mathrm{R}}|\boldsymbol{\theta}) \,,$$
and denote by Θ_0 the set of values of $\boldsymbol{\theta}$ for which λ_* is attained, i.e.,
$$\Theta_0 = \{\boldsymbol{\theta} \,:\, P_M(\delta_{\mathrm{R}}|\boldsymbol{\theta}) = \lambda_*\} \,.$$
Then the least-favorable probability density function f_{Θ}^{L} is such that
$$f_{\Theta}^{\mathrm{L}}(\boldsymbol{\theta}) > 0 \text{ for } \boldsymbol{\theta} \in \Theta_0$$
$$f_{\Theta}^{\mathrm{L}}(\boldsymbol{\theta}) = 0 \text{ otherwise} \,,$$
and $P_M(\delta_{\mathrm{R}}, f_{\Theta}^{\mathrm{L}}) = \lambda_*$.

6.6. Consider a detection problem such that under H_1
$$\mathbf{Y} = \begin{bmatrix} Y_c \\ Y_s \end{bmatrix} = A \begin{bmatrix} \cos(\theta) \\ \sin(\theta) \end{bmatrix} + \mathbf{V} \,,$$
where that amplitude A is known but the phase θ is unknown. The noise vector $\mathbf{V}$ is $N(\mathbf{0}, \sigma^2 \mathbf{I}_2)$ distributed, and under H_0, $\mathbf{Y} = \mathbf{V}$. We seek to design a robust test δ_{R} and identify the least-favorable phase distribution $f_{\Theta}^{L}(\theta)$ for this detection problem.

(a) Since the problem is symmetric with respect to θ, it is reasonable to guess that
$$f_{\Theta}^{\mathrm{L}}(\theta) = \frac{1}{2\pi}$$
for $0 \leq \theta < 2\pi$. By using this guess as starting point, verify that the robust test δ_{R} described in part (a) of Problem 6.5 takes the form
$$R = (Y_c^2 + Y_s^2)^{1/2} \underset{H_0}{\overset{H_1}{\gtrless}} \eta \,,$$
which coincides with the UMPI test of Example 5.6. Indicate how the threshold η needs to be selected to ensure the probability of false alarm is α.

(b) Use expression (5.58) for the probability of detection of test $\delta_{\rm R}$ to verify that the probability of a miss $P_M(\delta_{\rm R}|\theta)$ is independent of θ, so that the set Θ_0 of values of θ for which $P_M(\delta_{\rm R}|\theta)$ is maximized is the entire domain $[0, 2\pi]$. Then verify that $f_\Theta^{\rm L}$ admits the structure described in part (b) of Problem 6.5.

6.7. Consider a robust binary hypothesis testing problem with nominal Laplace densities of the form

$$\begin{aligned} f_0(y) &= \frac{a}{2} \exp(-a|y+m|) \\ f_1(y) &= \frac{a}{2} \exp(-a|y-m|) , \end{aligned}$$

where $a > 0$ and $m > 0$. The neighborhoods where the actual densities are located have the form (6.21) with $\epsilon_0 = \epsilon_1 = 0$ and $\nu_0 = \nu_1 = \nu > 0$.

(a) Evaluate and sketch the likelihood ratio

$$L(y) = \frac{f_1(y)}{f_0(y)} .$$

Verify that $L(y)$ is monotone nondecreasing but not monotone increasing, so it does not meet the condition employed to derive the clipped LR test of Section 6.3.2.

(b) In spite of this last comment, use trial least-favorable densities of the form (6.25)–(6.27) where we assume $-m < y_L < 0$, and where the problem symmetry implies $y_U = -y_L$, so that $\ell_U = \ell_L^{-1} > 1$. Consider also the test $\delta_{\rm R}$ given by (6.30). Verify that test $\delta_{\rm R}$ and the least-favorable pair $(g_0^{\rm L}, g_1^{\rm L})$ obey the saddle point condition (6.18) as long as we can find $y_L \in (-m, 0)$ such that $\Gamma(y_L) = 1$, where the function $\Gamma(y)$ is given by (6.46).

(c) Verify that $\Gamma(y) = 0$ for $y \leq -m$ and that (6.48) holds over interval $(-m, 0)$. Conclude therefore that there exists a unique $y_L \in (-m, 0)$ such that $\Gamma(y_L) = 1$ as long as $\Gamma(0) > 1$. Express the condition $\Gamma(0) > 1$ explicitly in terms of ν, a and m, and explain the role of this condition.

6.8. Consider a binary hypothesis testing problem where under hypothesis H_j with $j = 0$, 1, observation $Y \in \mathbb{R}$ admits $f_j(y)$ as nominal probability density. The actual density $g_j(y)$ belongs to neighborhood

$$\mathcal{F}_j = \{g_j : D(g_j|f_j) \leq \epsilon_j\}$$

where $D(g|f)$ denotes the KL divergence (6.11). Let $\mathcal{D}$ denote the class of pointwise randomized tests $\delta(y)$ such that if $Y = y$ we select H_1 with probability $\delta(y)$ and H_0 with probability $1 - \delta(y)$, where $0 \leq \delta(y) \leq 1$. Assume that the two hypotheses are equally likely, and denote by

$$P_E(\delta, g_0, g_1) = \frac{1}{2}[P_F(\delta, g_0) + P_M(\delta, g_1)] \qquad (6.124)$$

the probability of error of test δ when the actual densities under H_0 and H_1 are g_0 and g_1, respectively. In this expression, the probability of false alarm $P_F(\delta, g_0)$ and the probability of a miss $P_M(\delta, g_1)$ are given by (6.12). We seek to solve the minimax hypothesis testing problem [24]

$$\min_{\delta \in \mathcal{D}} \max_{(g_0,g_1) \in \mathcal{F}_0 \times \mathcal{F}_1} P_E(\delta, g_0, g_1) . \tag{6.125}$$

(a) By observing that $P_E(\delta, g_0, g_1)$ is convex in δ, concave in g_0 and g_1, and noting that sets $\mathcal{D}$, and $\mathcal{F}_j$ are convex and compact, conclude that a saddle point exists. This saddle point is formed by robust test δ_{R} and least-favorable densities $(g_0^{\mathrm{L}}, g_1^{\mathrm{L}})$, and it has the property

$$P_E(\delta, g_0^{\mathrm{L}}, g_1^{\mathrm{L}}) \geq P_E(\delta_{\mathrm{R}}, g_0^{\mathrm{L}}, g_1^{\mathrm{L}}) \geq P_E(\delta_{\mathrm{R}}, g_0, g_1) . \tag{6.126}$$

(b) Show that the first inequality of (6.126) implies that δ_{R} must be an optimum Bayesian test for the least favorable pair $(g_0^{\mathrm{L}}, g_1^{\mathrm{L}})$, so if L_{L} denotes the least-favorable likelihood ratio introduced in (6.19), we must have

$$\delta_{\mathrm{R}}(y) = \begin{cases} 1 & \text{for } L_{\mathrm{L}}(y) > 1 \\ \text{arbitrary} & \text{for } L_{\mathrm{L}}(y) = 1 \\ 0 & \text{for } L_{\mathrm{L}}(y) < 1 . \end{cases}$$

(c) Consider now the second inequality of (6.126). By taking into account the form (6.124) for $P_E(\delta, g_0, g_1)$, verify it is equivalent to the two inequalities

$$P_F(\delta_{\mathrm{R}}, g_0^{\mathrm{L}}) \geq P_F(\delta_{\mathrm{R}}, g_0)$$
$$P_M(\delta_{\mathrm{R}}, g_1^{\mathrm{L}}) \geq P_M(\delta_{\mathrm{R}}, g_1)$$

for all densities $g_0 \in \mathcal{F}_0$ and $g_1 \in \mathcal{F}_1$, respectively. Thus, given δ_{R}, g_0^{L} maximizes $P_F(\delta_{\mathrm{R}}, g_0)$ under the constraints $D(g_0|f_0) \leq \epsilon_0$ and the normalization constraint

$$I(g_0) = \int_{-\infty}^{\infty} g_0(y)dy = 1 . \tag{6.127}$$

By using the method of Lagrange multipliers, verify that the least-favorable density can be expressed as

$$g_0^{\mathrm{L}}(y) = \frac{1}{Z_0} \exp(\alpha_0 \delta_{\mathrm{R}}(y)) f_0(y) ,$$

where $Z_0 > 0$ and $\alpha_0 > 0$ are selected such that $I(g_0) = 1$ and $D(g_0^{\mathrm{L}}|f_0) = 1$. Similarly, verify that $g_1^{\mathrm{L}}(y)$ can be expressed as

$$g_1^{\mathrm{L}}(y) = \frac{1}{Z_1} \exp(\alpha_1(1 - \delta_{\mathrm{R}}(y))) f_1(y) ,$$

with $Z_1 > 0$ and $\alpha_1 > 0$.

6.9. Consider the minimax hypothesis testing problem described in Problem 6.8. Assume that the nominal likelihood ratio function

$$L(y) = \frac{f_1(y)}{f_0(y)}$$

is monotone increasing, and assume that the two nominal densities admit the symmetry

$$f_1(y) = f_0(-y) \tag{6.128}$$

Note that this property implies $L(-y) = 1/L(y)$ and thus $L(0) = 1$. For $0 \leq u \leq 1$, consider the geodesic

$$f_u(y) = \frac{f_1^u(y) f_0^{1-u}(y)}{Z(u)}$$

linking nominal densities f_0 and f_1, where the normalization constant

$$Z(u) = \int_{-\infty}^{\infty} f_1^u(y) f_0^{1-u}(y) dy \, .$$

(a) Verify that symmetry condition (6.128) ensures that the density $f_{1/2}(y)$ obtained by setting $u = 1/2$ is located midway between f_0 and f_1 in the sense of the KL divergence, i.e.,

$$D(f_{1/2}|f_0) = D(f_{1/2}|f_1) \, .$$

(b) Assume that the tolerances ϵ_j used to specify neighborhoods $\mathcal{F}_j$ with $j = 0, 1$ in Problem 6.8 are such that $\epsilon_0 = \epsilon_1 = \epsilon$ with $0 < \epsilon < D(f_{1/2}|f_0)$. This ensures that $\mathcal{F}_0$ and $\mathcal{F}_1$ do not intersect. Then given $y_U > 0$ and $l_U = L(y_U)$, verify that the decision rule δ_{R} and densities $(g_0^{\mathrm{L}}, g_1^{\mathrm{L}})$ specified by

$$\delta_{\mathrm{R}}(y) = \begin{cases} 1 & y > y_U \\ \frac{1}{2}[1 + \frac{\ln L(y)}{\ln \ell_U}] & -y_U \leq y \leq y_U \\ 0 & y < -y_U \, , \end{cases}$$

and

$$g_0^{\mathrm{L}}(y) = \begin{cases} \ell_U f_0(y)/Z(y_U) & y > y_U \\ \ell_U^{1/2} f_1^{1/2}(y) f_0^{1/2}(y)/Z(y_U) & -y_U \leq y \leq y_U \\ f_0(y)/Z(y_U) & y < -y_U \end{cases} \tag{6.129}$$

$$g_1^{\mathrm{L}}(y) = \begin{cases} f_1(y)/Z(y_U) & y > y_U \\ \ell_U^{1/2} f_1^{1/2}(y) f_0^{1/2}(y)/Z(y_U) & -y_U \leq y \leq y_U \\ \ell_U f_1(y)/Z(y_U) & y < -y_U \end{cases}$$

have the structure described in parts (b) and (c) of Problem 6.8. In these expressions the normalization constant $Z(y_U)$ is selected such that

$$I(g_0^{\rm L}) = I(g_1^{\rm L}) = 1 ,$$

where integral $I(\cdot)$ is defined by (6.127). This indicates that decision rule $\delta_{\rm R}$ and least-favorable pair $(g_0^{\rm L}, g_1^{\rm L})$ form a saddle point of minimax problem (6.125) provided $y_U > 0$ can be selected such that

$$D(g_0^{\rm L}|f_0) = \epsilon \, . \tag{6.130}$$

Note that since $g_1^{\rm L}(y) = g_0^{\rm L}(-y)$, in light of symmetry condition (6.128), the constraint

$$D(g_1^{\rm L}|f_1) = \epsilon$$

is satisfied automatically, as soon as (6.130) holds.

(c) Let $g_0^{\rm L}(\cdot|y_U)$ denote the function (6.129) where the y_U parametrization is denoted explicitly. Verify that for $y_U > 0$, the function

$$D(y_U) \triangleq D(g_0^{\rm L}(\cdot|y_U)|f_0)$$

is monotone increasing. Observing that $g_0^{\rm L}(\cdot|0) = f_0$ for $y_U = 0$ and $g_0^{\rm L}(\cdot|+\infty) = f_{1/2}$, this implies that for $0 < \epsilon < D(f_{1/2}|f_0)$ there exists a unique y_U such that $D(y_U) = \epsilon$, which yields the desired saddle point solution.

(d) Show that the least-favorable likelihood ratio $L_{\rm L}$ can be expressed in terms of the nominal likelihood ratio L as

$$L_{\rm L} = q(L)$$

where $q(\cdot)$ denotes a nonlinear transformation. Find $q(\cdot)$ and compare it to the clipping transformation $C(\cdot)$ of Fig. 6.1 for the case when the neighborhoods $\mathcal{F}_j$, $j = 0, 1$ are given by (6.21).

6.10. To illustrate the saddle point solution of the minimax problem studied in Problems 6.8 and 6.9, consider hypotheses

$$H_0 : Y = -1 + V$$
$$H_1 : Y = 1 + V$$

where noise V admits a $N(0, \sigma^2)$ nominal density. Thus in this case the nominal densities f_0 and f_1 of Y are $N(-1, \sigma^2)$ and $N(1, \sigma^2)$ distributed.

(a) Verify that the nominal likelihood ratio $L(y)$ is monotone increasing and that the symmetry condition (6.128) holds, so that the results of Problem 6.9 are applicable.
(b) Check that the half-way density $f_{1/2}$ is $N(0, \sigma^2)$ distributed.
(c) Give expressions for the robust test $\delta_{\rm R}$, and the least-favorable densities $g_j^{\rm L}$, $j = 0, 1$. In particular, verify that $g_0^{\rm L}$ includes 3 segments. For $y < -y_U$ it is an attenuated $N(-1, \sigma^2)$ density, for $-y_U \le y \le y_U$ it is a scaled $N(0, \sigma^2)$ density, and for $y > y_U$ it is an amplified $N(-1, \sigma^2)$ density. Further details can be found in [24].

6.11. For the asymptotic robustness discussed in Section 6.4, assume that f_0 and f_1 correspond to $N(0, r_0)$ and $N(0, r_1)$ densities with $r_0 < r_1$.

(a) For $0 \leq u \leq 1$, if

$$f_u(y) = \frac{f_1^u(y) f_0^{1-u}(y)}{G_0(u)} \tag{6.131}$$

is an arbitrary density of the geodesic linking f_0 and f_1 where the generating function $G_0(u)$ is given by (6.78), verify that f_u is $N(0, r(u))$ distributed with

$$r(u) = \frac{r_0 r_1}{u r_0 + (1-u) r_1} .$$

(b) By noting from (3.45) that the KL divergence of two $N(0, r')$ and $N(0, r)$ densities f' and f can be expressed as

$$D(f'|f) = \frac{1}{2}\Big[(\frac{r'}{r} - 1) - \ln(r'/r)\Big] ,$$

select least favorable densities g_0^{L} and g_1^{L} on the geodesic (6.131) such that

$$D(g_0^{\mathrm{L}} | f_0) = D(g_1^{\mathrm{L}} | f_1) .$$

Let u and v denote the geodesic parameters corresponding to g_0^{L} and g_1^{L}.

(c) Given N independent observations Y_k, $1 \leq k \leq N$, consider the robust test (6.79) and nominal test (6.82). Express nominal test threshold τ_{N} in terms of robust test threshold τ, u, v, r_0 and r_1 to ensure that the two tests coincide.

References

1. P. J. Huber, *Robust Statistics.* New York: J. Wiley & Sons, 1981.
2. A. L. Gibbs and F. E. Su, "On choosing and bounding probability metrics," *International Statistical Review*, vol. 70, no. 3, pp. 419–435, 2002.
3. S. Rachev, *Probability Metrics and the Stability of Stochastic Models.* Chichester, England: J. Wiley & Sons, 1991.
4. R. H. Schumway and D. S. Stoffer, *Time Series Analysis and its Applications.* New York: Springer-Verlag, 2000.
5. G. J. McLachlan and T. Krishnan, *The EM Algorithm and Extensions.* New York: J. Wiley & Sons, 1997.
6. P. J. Huber, "A robust version of the probability ratio test," *Annals Math. Statistics*, vol. 36, pp. 1753–1758, Dec. 1965.
7. P. J. Huber, "Robust confidence limits," *Z. Wahrcheinlichkeitstheorie verw. Gebiete*, vol. 10, pp. 269–278, 1968.
8. D. Bertsekas, A. Nedic, and A. E. Ozdaglar, *Convex Analsis and Optimization.* Belmont, MA: Athena Scientific, 2003.
9. J.-P. Aubin and I. Ekland, *Applied Nonlinear Analysis.* New York: J. Wiley, 1984.

10. P. J. Huber and V. Strassen, "Minimax tests and the Neyman-Pearson lemma for capacities," *Annals Statistics*, vol. 1, pp. 251–263, Mar. 1973.
11. A. G. Dabak and D. H. Johnson, "Geometrically based robust detection," in *Proc. Conf. Information Sciences and Systems*, (Baltimore, MD), pp. 73–77, The Johns Hopkins Univ., Mar. 1993.
12. R. D. Martin and S. C. Schwartz, "Robust detection of a known signal in nearly Gaussian noise," *IEEE Trans. Informat. Theory*, vol. 17, pp. 50–56, 1971.
13. G. V. Moustakides and J. B. Thomas, "Robust detection of signals in dependent noise," *IEEE Trans. Informat. Theory*, vol. 33, pp. 11–15, Jan. 1987.
14. G. V. Moustakides, "Robust detection of signals: a large deviations approach," *IEEE Trans. Informat. Theory*, vol. 31, pp. 822–825, Nov. 1985.
15. S. Meyn and V. V. Veeravalli, "Asymptotic robust hypothesis testing based on moment classes," in *Proc. UCSD Information Theory and its Applications (ITA) Inaugural Workshop*, (San Diego, CA), Feb. 2006.
16. A. H. El-Sawy and D. Vandelinde, "Robust detection of known signals," *IEEE Trans. Informat. Theory*, vol. 23, pp. 722–727, Nov. 1977.
17. S. A. Kassam and J. B. Thomas, "Asymptotically robust detection of a known signal in contaminated non-Gaussian noise," *IEEE Trans. Informat. Theory*, vol. 22, pp. 22–26, Jan. 1976.
18. S. A. Kassam and H. V. Poor, "Robust techniques for signal processing: a survey," *Proc. IEEE*, vol. 73, pp. 433–480, Mar. 1985.
19. B. C. Levy and R. Nikoukhah, "Robust least-squares estimation with a relative entropy constraint," *IEEE Trans. Informat. Theory*, vol. 50, pp. 89–104, Jan. 2004.
20. B. Hassibi, A. H. Sayed, and T. Kailath, *Indefinite-Quadratic Estimation and Control– A Unified Approach to H^2 and H^∞ Theories*. Philadelphia: Soc. Indust. Applied Math., 1999.
21. Y. Eldar and N. Merhav, "A competitive minimax approach to robust estimation of random parameters," *IEEE Trans. Signal Proc.*, vol. 52, pp. 1931–1946, July 2004.
22. C. W. Helstrom, "Minimax detection of signals with unknown parameters," *Signal Processing*, vol. 27, pp. 145–159, 1992.
23. D. Bertsimas and J. Tsitsiklis, *Introduction to Linear Optimization*. Belmont, MA: Athena Scientific, 1997.
24. B. C. Levy, "Robust hypothesis testing with a relative entropy tolerance." preprint, available at http://arxiv.org/abs/cs/0707.2926, July 2007.

Part II

Gaussian Detection

7

Karhunen-Loève Expansion of Gaussian Processes

7.1 Introduction

All detection problems considered up to this point have involved finite dimensional observation vectors. Unfortunately, many detection problems involve continous-time (CT) waveforms. To tackle problems of this type, we introduce the Karhunen-Loève decomposition (KLD) of Gaussian processes. This decomposition extends the familiar notion of orthogonal expansion of a deterministic signal to the class of Gaussian processes. But whereas a deterministic signal can be expanded with respect to an arbitrary orthonormal basis (Fourier, wavelet, etc...), the basis used to express a Gaussian process in orthogonal components is usually *not arbitrary* and is actually formed by the eigenfunctions of its covariance matrix/kernel. So a necessary first step in the construction of a KLD is the expansion of covariance matrices or functions in terms of their eigenvectors/functions. The KLD of a Gaussian process possesses the feature of representing the process in terms of *independent* Gaussian coefficients, so if the process has zero-mean, the coefficients form a time-varying WGN sequence. In other words the KLD is a whitening or decorrelating transformation. This feature greatly simplifies detection problems, and by combining the KLD with the observation that the signals to be detected belong to a finite dimensional subspace of the space of all functions, it is often possible to transform CT detection problems into finite dimensional problems of the type examined in the first part of this book. The KLD-based transformation technique used to convert infinite dimensional detection problems into finite-dimensional ones is usually called the "signal space" formulation of detection. After being introduced by Davenport and Root [1] and Middleton [2] in their studies of random processes and statistical detection and communications, it was popularized in the late 1960s by the influential books of Van Trees [3] and Wozencraft and Jacobs [4], and it remains to this day the most elegant device available for studying CT Gaussian detection.

Given that we live in the age of digital signal processing, it is reasonable to ask why we should even consider CT detection problems. The answer to this

B.C. Levy, *Principles of Signal Detection and Parameter Estimation*,
DOI: 10.1007/978-0-387-76544-0_7, © Springer Science+Business Media, LLC 2008

question is that information-carrying signals used in communications, radar, and sonar, are still analog and, at the very least, one needs to be concerned with the design of front end receivers, such as matched filters followed by samplers, which are often used in digital communication systems. Also, even if the front end receiver just consists of an analog to digital converter, the resulting discrete-time (DT) detection problem may have a high dimensionality, and in this case the KLD is still a useful device for reducing the dimensionality of the detection problem.

In this chapter, we present in a unified manner the KLD theory of CT and DT processes. To do so, we will try as much as possible to motivate infinite-dimensional decompositions of operators and processes in terms of the corresponding finite-dimensional matrix concepts for DT Gaussian processes defined over a finite interval. This will make it clear that there is a complete conceptual continuity in going from the finite dimensional decomposition of a DT Gaussian process over a finite interval to its infinite dimensional CT equivalent.

7.2 Orthonormal Expansions of Deterministic Signals

We consider the space $L^2[a,b]$ of DT or CT complex functions that are either square summable or square integrable over a finite interval $[a,b]$. In the DT case, a complex function $f(t)$ defined for $a \leq t \leq b$ can be represented as a vector

$$\mathbf{f} = \begin{bmatrix} f(a) \dots f(t) \dots f(b) \end{bmatrix}^T \tag{7.1}$$

of $\mathbb{C}^N$ with $N = b - a + 1$. If f and g are two such functions with vector representations $\mathbf{f}$ and $\mathbf{g}$, the inner product can be represented either in function form or in vector form as

$$< f, g >= \sum_{t=a}^{b} f(t) g^*(t) = \mathbf{g}^H \mathbf{f} \tag{7.2}$$

where $\mathbf{g}^H$ denotes the Hermitian transpose (the complex conjugate transpose) of $\mathbf{g}$. The corresponding norm is given by

$$||f|| = (< f, f >)^{1/2} = \big(\sum_{t=a}^{b} |f(t)|^2\big)^{1/2} = (\mathbf{f}^H \mathbf{f})^{1/2} \,. \tag{7.3}$$

The inner product and its associated norm statisfy the following axioms:

(i) The norm is strictly positive for nonzero vectors: $||f|| > 0$ for $f \neq 0$.
(ii) For arbitrary complex numbers a, b, the inner product is distributive with respect to the linear combination of vectors:

$$< af_1 + bf_2, g >= a < f_1, g > +b < f_2, g > \,.$$

(iii) The inner product has the Hermitian symmetry $< g, f >=< f, g >^*$.

If we consider now CT complex functions $f(t)$ and $g(t)$ of $L^2[a, b]$, their inner product is given by

$$< f, g >= \int_a^b f(t)g^*(t)dt \tag{7.4}$$

and the induced norm takes the form

$$||f|| = (< f, f >)^{1/2} = \Big(\int_a^b |f(t)|^2 dt \Big)^{1/2} . \tag{7.5}$$

The inner product and norm satisfy axioms (i)–(iii). However, an important difference between the DT and CT cases is that in the DT case $L^2[a, b]$ is finite dimensional with dimension $N = b - a + 1$, whereas in the CT case, it is an infinite dimensional Hilbert space. A Hilbert space is a complete inner product space, which means that (a) it has an inner product satisfying axioms (i)–(iii) and (b) every Cauchy sequence admits a limit. We refer the reader to [5] for an elementary exposition of Hilbert space theory. For our purposes it suffices to say that every finite-dimensional vector space is a Hilbert space, but the more interesting Hilbert spaces of CT square-integrable functions, or of random variables with a finite variance, are infinite-dimensional.

It is usually of interest to expand functions with respect to arbitrarily chosen orthonormal bases. In the DT case, we say that $\{\phi_k(t), 1 \leq k \leq N\}$ forms an orthonormal basis if

$$< \phi_k, \phi_\ell >= \sum_{t=a}^{b} \phi_k(t)\phi_\ell^*(t) = \delta_{k\ell} , \tag{7.6}$$

where

$$\delta_{k\ell} = \begin{cases} 1 & \text{for } k = \ell \\ 0 & \text{otherwise} . \end{cases}$$

Then, an arbitrary function $f(t)$ can be decomposed as

$$f(t) = \sum_{k=1}^{N} F_k \phi_k(t) , \tag{7.7}$$

where the coefficients F_k are computed by projecting f on ϕ_k, i.e.,

$$F_k =< f, \phi_k >= \sum_{t=a}^{b} f(t)\phi_k^*(t) . \tag{7.8}$$

By backsubstituting (7.8) inside (7.7), we obtain

$$f(t) = \sum_{s=a}^{b} R(t, s)f(s) , \tag{7.9}$$

where

$$R(t,s) \triangleq \sum_{k=1}^{N} \phi_k(t)\phi_k^*(s) , \tag{7.10}$$

so that both sides of (7.9) will be equal provided

$$R(t,s) = \delta(t-s) \tag{7.11}$$

for $a \leq t,\ s \leq b$. Here $\delta(t-s)$ is the Kronecker delta function in the sense that

$$\delta(t-s) = \begin{cases} 1 & t = s \\ 0 & \text{otherwise} . \end{cases}$$

The identity (7.11) means that $R(t,s)$ must be a *reproducing kernel* for $L^2[a,b]$, i.e., when this kernel operates on a function $f(t)$ of $L^2[a,b]$, it reproduces it exactly as indicated by (7.9). In the DT case, this property is just an elaborate way of saying that the basis $\{\phi_k,\ 1 \leq k \leq N\}$ is orthonormal. Let $\boldsymbol{\phi}_k$ be the vector of $\mathbb{C}^N$ corresponding to $\phi_k(t)$, and consider the matrix

$$\boldsymbol{\Phi} = \begin{bmatrix} \boldsymbol{\phi}_1 \dots \boldsymbol{\phi}_k \dots \boldsymbol{\phi}_N \end{bmatrix} .$$

Then the identity (7.10)–(7.11) can be rewritten as

$$\mathbf{I}_N = \sum_{k=1}^{N} \boldsymbol{\phi}_k \boldsymbol{\phi}_k^H = \boldsymbol{\Phi}\boldsymbol{\Phi}^H , \tag{7.12}$$

which is just a restatement of the fact that the matrix $\boldsymbol{\Phi}$ is unitary, or equivalently, that the basis $\{\phi_k,\ 1 \leq k \leq N\}$ is orthonormal.

Example 7.1: DT impulse basis

In the DT case, an arbitrary function admits a trivial decomposition in terms of the impulse functions

$$\phi_k(t) = \begin{cases} 1 \text{ for } t = a + k - 1 \\ 0 \text{ for } \qquad 0 , \end{cases}$$

with $1 \leq k \leq N$, where we have

$$f(t) = \sum_{k=1}^{N} f(a+k-1)\phi_k(t) .$$

Since the vector representation of ϕ_k is the k-th unit vector $\mathbf{e}_k$ of $\mathbb{C}^N$ which has all zero entries, except for the k-th entry which is one, this decomposition is equivalent to the usual expression of a vector in terms of its coordinates. □

Let us turn now to CT functions in $L^2[a,b]$. We say that $\{\phi_k(t),\ k \geq 1\}$ constitutes a *complete orthonormal basis* (COB) if the basis functions ϕ_k

satisfy (7.6) where the inner product is defined by (7.4), and any function $f \in L^2[a, b]$ can be decomposed as

$$f(t) = \sum_{k=1}^{\infty} F_k \phi_k(t) , \tag{7.13}$$

where the coefficients F_k are again given by

$$F_k =< f, \phi_k >= \int_a^b f(t) \phi_k^*(t) dt . \tag{7.14}$$

Note that since the summation is infinite in (7.13), this expansion should be interpreted as

$$\lim_{K \to \infty} ||f - \sum_{k=1}^{K} F_k \phi_k|| = 0 . \tag{7.15}$$

Also, since

$$||f - \sum_{k=1}^{K} F_k \phi_k||^2 = ||f||^2 - \sum_{k=1}^{K} |F_k|^2 , \tag{7.16}$$

the limit (7.15) implies Parseval's identity

$$||f||^2 = \sum_{k=1}^{\infty} |F_k|^2 . \tag{7.17}$$

To understand why the basis needs to be complete, note that in the DT case, because the space is finite dimensional, as soon as we have N orthonormal basis functions, we know that we can represent any DT function in terms of this basis. But for the CT case, it is not enough to have an infinite number of orthonormal basis functions. For example over $L^2[a, b]$, if $T = b - a$ and $\omega_0 = 2\pi/T$, the cosine functions $\{\phi_k(t) = (2/T)^{1/2} \cos(k\omega_0 t),\ k \geq 0\}$ are orthonormal, but they do not form a COB of $L^2[a, b]$, since in order to obtain a COB we also need to include the sine functions $\{\psi_k(t) = (2/T)^{1/2} \sin(k\omega_0 t),\ k \geq 1\}$. By backsubstituting (7.14) inside (7.13), we see that if

$$R(t, s) \triangleq \sum_{k=1}^{\infty} \phi_k(t) \phi_k^*(s) , \tag{7.18}$$

then $\{\phi_k,\ k \geq 1\}$ is a COB if and only if for an arbitrary function f of $L^2[a, b]$ we have

$$f(t) = \int_a^b R(t, s) f(s) ds , \tag{7.19}$$

or equivalently, if $R(t, s)$ satisfies (7.11), where $\delta(t - s)$ denotes now the Dirac impulse function, instead of the Kronecker impulse. So a set of orthonormal functions forms a COB if and only if the kernel $R(t, s)$ defined by (7.18) is a reproducing kernel for $L^2[a,b]$. Equivalently, the identity (7.11) shows that the functions ϕ_k form what is called a *resolution of the identity.*

Example 7.2: Complex exponential Fourier series

For $L^2[a,b]$, if T denotes the interval length and if $\omega_0 = 2\pi/T$ is the corresponding fundamental frequency, the complex exponential functions $\{\phi_k(t) = (1/T)^{1/2} \exp(jk\omega_0 t),\ k \in \mathbb{Z}\}$ form a COB, where the expansion

$$f(t) = (1/T)^{1/2} \sum_{k=-\infty}^{\infty} F_k \exp(jk\omega_0 t)$$

is the usual complex Fourier series expansion where the Fourier coefficients F_k are expressed as

$$F_k = (1/T)^{1/2} \int_a^b f(t) \exp(-jk\omega_0 t) dt \,.$$

For this basis, the identity (7.11) is the usual Fourier series representation

$$s(t) \triangleq \sum_{n=-\infty}^{\infty} \delta(t - nT) = \frac{1}{T} \sum_{k=-\infty}^{\infty} \exp(jk\omega_0 t)$$

for the periodic impulsive sampling function $s(t)$. Note that since (7.11) holds for $-T < t - s < T$, the infinite train of impulses appearing in $s(t)$ reduces to a single impulse. □

7.3 Eigenfunction Expansion of Covariance Kernels

Consider a zero-mean DT or CT Gaussian process $X(t)$ defined over a finite interval $[a,b]$ with covariance kernel

$$K(t,s) = E[X(t)X(s)] \,. \tag{7.20}$$

To this kernel, we can associate an operator K on functions in $L^2[a,b]$ which takes the form

$$(Kf)(t) = \sum_{s=a}^{b} K(t,s) f(s) \tag{7.21}$$

in the DT case, and

$$(Kf)(t) = \int_a^b K(t,s) f(s) ds \tag{7.22}$$

in the CT case. Then (λ, ϕ) with $\phi \in L^2[a,b]$ forms an eigenvalue/eigenfunction pair for the kernel $K(t,s)$ or the operator K, if the function $\phi \neq 0$ satisfies

$$(K\phi)(t) = \lambda \phi(t) \tag{7.23}$$

for $a \leq t \leq b$. In the CT case we shall assume that $K(t, s)$ is continuous over $[a, b]^2$. This ensures that $K(t, s)$ is square integrable over $[a, b]^2$, i.e.,

$$\int_a^b \int_a^b |K(t, s)|^2 dt ds < \infty .$$

7.3.1 Properties of Covariance Kernels

Because $K(t, s)$ is a covariance kernel, it has two important properties.

Symmetry: We have

$$K(t, s) = K(s, t) . \tag{7.24}$$

This means that the operator K defined by (7.21) or (7.22) is *self-adjoint.* We recall that the adjoint K^* of an operator K of $L^2[a, b]$ is the unique operator that satisfies

$$< Kf, g > = < f, K^* g > \tag{7.25}$$

for arbitrary functions f, g of $L^2[a, b]$. Here for real CT functions f and g, by interchanging the order of integration and using (7.24), we find

$$\begin{aligned} < Kf, g > &= \int_a^b (\int_a^b K(t, s) f(s) ds) g(t) dt \\ &= \int_a^b (\int_a^b K(s, t) g(t) dt) f(s) ds = < f, Kg > , \end{aligned} \tag{7.26}$$

which shows $K^* = K$. The DT case is obtained by replacing integrations by summations. Equivalently, in the DT case, if

$$\mathbf{X} = \begin{bmatrix} X(a) \ldots X(t) \ldots X(b) \end{bmatrix}^T$$

denotes the N-dimensional Gaussian random vector with $N = b - a + 1$ obtained by representing the process $X(t)$ for $0 \leq t \leq b$ in vector form, its $N \times N$ covariance matrix

$$\mathbf{K} = E[\mathbf{X}\mathbf{X}^T] \tag{7.27}$$

is symmetric. $\mathbf{K}$ is the matrix representation of the DT operator K given by (7.21), since whenever $g = Kf$, if

$$\begin{aligned} \mathbf{f} &= \begin{bmatrix} f(a) \ldots f(t) \ldots f(b) \end{bmatrix}^T \\ \mathbf{g} &= \begin{bmatrix} g(a) \ldots g(t) \ldots g(b) \end{bmatrix}^T \end{aligned}$$

denote the vectors representing the DT functions f and g, the operator identity (7.21) can be rewritten in matrix form as

$$\mathbf{g} = \mathbf{K}\mathbf{f} . \tag{7.28}$$

Non negativeness: The kernel $K(t,s)$ is also non negative. This means that if f is an arbitrary function of $L^2[a,b]$, we have $< Kf, f > \geq 0$. To see why this is the case, consider the CT case and define the zero-mean Gaussian random variable

$$Z = \int f(t)X(t)dt \, .$$

Its variance satisfies

$$\begin{aligned} 0 \leq E[Z^2] &= \int_a^b \int_a^b f(t)E[X(t)X(s)]f(s)dtds \\ &= \int_a^b \int_a^b f(t)K(t,s)f(s)dtds =< Kf, f > \, . \end{aligned} \quad (7.29)$$

In the DT case, the matrix equivalent of this property is that the covariance matrix $\mathbf{K}$ is nonnegative definite, i.e., for all vectors $\mathbf{f}$ of $\mathbb{R}^N$, it satisfies

$$\mathbf{f}^T \mathbf{K} \mathbf{f} \geq 0 \, . \quad (7.30)$$

When the covariance kernel $K(t,s)$ is such that $< Kf, f >> 0$ for all nonzero $f \in L^2[a,b]$, it will be said to be positive definite. An important class of random processes with this property are those that contain a WGN component. Specifically, if

$$X(t) = S(t) + V(t) \quad (7.31)$$

where $S(t)$ is a zero-mean Gaussian random process with covariance function $K_S(t,s)$, and $V(t)$ is a zero-mean WGN independent of $S(t)$ with intensity $q > 0$, the covariance of $X(t)$ is given by

$$K(t,s) = K_S(t,s) + q\delta(t-s) \, . \quad (7.32)$$

Then for $f \neq 0$, the operator K satisfies

$$< Kf, f >=< K_S f, f > + q||f||^2 > 0 \, .$$

So, although a covariance operator can be positive definite without containing a WGN component, as soon as such a component is present, the covariance operator is necessarily positive definite. Since thermal noise is an intrinsic limitation of measurement devices, most problems considered in the next few chapters will assume that a WGN component is present.

In the CT case, when the kernel has the structure (7.32) with $q > 0$, the operator Kf can be expressed as

$$(Kf)(t) = qf(t) + \int_a^b K_S(t,s)f(s)ds \quad (7.33)$$

and in this case K is called a Fredholm integral operator of the second kind, by opposition with the form (7.22), which corresponds to a Fredholm operator of the first kind. Then, if $g(t)$ denotes an arbitrary function of $L^2[a,b]$, an integral equation of the form

$$g(t) = qf(t) + \int_a^b K_S(t,s)f(s)ds$$

for $a \leq t \leq b$ is called a Fredhom equation of the second kind, by opposition with

$$g(t) = \int_a^b K(t,s)f(s)ds \,,$$

which corresponds to a Fredholm equation of the first kind. Typically, Fredholm equations of the second kind are rather easy to solve since they admit a solution with the same degree of smoothness as the function $g(t)$ appearing on the left-hand side. On the other hand, the solution f of a Fredholm equation of the first kind is less smooth than g and it may contain impulses, in which case, it needs to be defined in a weaker sense.

Properties of eigenvalues/eigenfunctions: The self-adjointness and non negativeness of the operator K implies that: (a) its eigenvalues are real and nonnegative, and (b) the eigenfunctions corresponding to distinct eigenvalues are orthogonal. To prove (a), note that if ϕ is a nonzero function of $L^2[a,b]$ such that $K\phi = \lambda\phi$, we have

$$\lambda = \frac{< K\phi, \phi >}{||\phi||^2} \tag{7.34}$$

where the right-hand side is both real and non-negative. To prove (b) let ϕ_1 and ϕ_2 be two eigenfunctions corresponding to two distinct eigenvalues λ_1 and λ_2. Then

$$\begin{aligned} \lambda_1 < \phi_1, \phi_2 > &= < K\phi_1, \phi_2 > \\ &= < \phi_1, K\phi_2 >= \lambda_2 < \phi_1, \phi_2 > \,, \end{aligned} \tag{7.35}$$

so that

$$(\lambda_1 - \lambda_2) < \phi_1, \phi_2 >= 0 \,, \tag{7.36}$$

which implies $< \phi_1, \phi_2 >= 0$ since $\lambda_1 \neq \lambda_2$.

The orthogonality property described above concerns only eigenfunctions corresponding to distinct eigenvalues. When K has several independent eigenfunctions $\{f_j,\ 1 \leq j \leq p\}$ corresponding to the same eigenvalue λ, they can always be orthogonalized by *Gram-Schmidt orthonormalization.* The Gram-Schmidt orthonormalization process can be described as follows: Let $\mathcal{F}_k$ with $1 \leq k \leq p$ be the linear space of functions spanned by $\{f_j,\ 1 \leq j \leq k\}$, so if g belongs to $\mathcal{F}_k$ it can be expressed as

$$g = \sum_{j=1}^{k} c_j f_j \,, \tag{7.37}$$

for some coefficients c_j. Note that since the eigenfunctions $\{f_j, 1 \leq j \leq p\}$ are assumed independent, the space $\mathcal{F}_k$ has dimension k for all $1 \leq k \leq p$. To orthonormalize the f_j's, we start with f_1 and select the function

$$\phi_1 = f_1/||f_1||$$

as our first basis function. Then, let

$$f_2^{||} = P_1 f_2 =< f_2, \phi_1 > \phi_1$$

be the projection of f_2 on $\mathcal{F}_1$ and denoted by

$$f_2^{\perp} = f_2 - P_1 f_2$$

be the component of f_2 perpendicular to $\mathcal{F}_1$. Note that $f_2^{\perp}$ is nonzero, since otherwise f_2 would be proportional to f_1, which contradicts the independence assumption for the f_j's. So we select

$$\phi_2 = f_2^{\perp}/||f_2^{\perp}||$$

as the second basis function. Proceeding by induction, suppose the orthonormal functions $\{\phi_j,\ 1 \leq j \leq k\}$ have been constructed by orthonormalizing the first k eigenfunctions f_j with $1 \leq kj \leq k$, and denote by P_k the orthonormal projection on the subspace $\mathcal{F}_k$. Then the projection of f_{k+1} on $\mathcal{F}_k$ is given by

$$f_{k+1}^{||} = P_k f_{k+1} = \sum_{j=1}^{k} < f_{k+1}, \phi_j > \phi_j \,,$$

and the residual which is orthogonal to $\mathcal{F}_k$ is given by

$$f_{k+1}^{\perp} = f_{k+1} - P_k f_{k+1} \,.$$

It is nonzero, since otherwise f_{k+1} would be expressible as a linear combination of the first f_j's, so that we can select as our $k+1$-th basis function

$$\phi_{k+1} = f_{k+1}^{\perp}/||f_{k+1}^{\perp}|| \,.$$

The constructed of ϕ_{k+1} is illustrated geometrically in Fig. 7.1. Continuing in this manner, we obtain an orthonormal basis $\{\phi_k,\ 1 \leq k \leq p\}$ of $\mathcal{F}_p$. Since each ϕ_k is expressible as a linear combination of eigenfunctions f_j's of K corresponding to the eigenvalue λ, the ϕ_k's are also eigenfunctions of K with eigenvalue λ. This ensures therefore that all eigenfunctions of K can be orthonormalized. When these eigenfunctions correspond to distinct eigenvalues, the orthogonality property is automatically satisfied, whereas when the eigenfunctions correspond to identical eigenvalues, a Gram-Schmidt orthonormalization is needed.

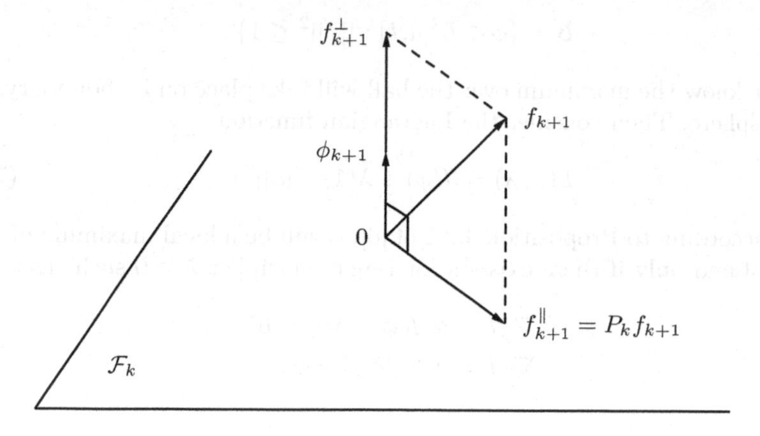

Fig. 7.1. Construction of ϕ_{k+1} by Gram-Schmidt orthonormalization.

7.3.2 Decomposition of Covariance Matrices/Kernels

To decompose a DT or CT covariance kernel in terms of its eigenfunctions, we can employ the following variational characterization of the largest eigenvalue of K and its corresponding eigenfunction. Note that in the CT case we assume that $K(t, s)$ is continuous over $[a, b]^2$, which means in particular that $X(t)$ does not have a WGN component.

Rayleigh-Ritz variational characterization of the largest eigenvalue: Consider the objective function

$$J(\phi) =< K\phi, \phi >$$

defined for $\phi \in L^2[a, b]$. Then the largest eigenvalue λ_1 and the associated eigenfunction ϕ_1 solve the maximization problem

$$\begin{aligned} \lambda_1 &= \max_{||\phi||=1} J(\phi) \\ \phi_1 &= \arg \max_{||\phi||=1} J(\phi) \,, \end{aligned} \tag{7.38}$$

and in the CT case $\phi_1(t)$ is a continuous function of t.

Proof: This result is more difficult to prove than it looks. We start by observing that since K is nonnegative definite, $J(\phi)$ is a convex function of ϕ. Because the problem (7.38) involves a *maximization* of J over the unit sphere

$$\mathcal{S} = \{\phi \in L^2[a, b] : ||\phi||^2 = 1\}$$

this problem is not convex. However, because $J(\phi)$ is convex, we can replace the maximization over the unit sphere by a maximization over the unit ball

$$\mathcal{B} = \{\phi \in L^2[a,b] : ||\phi||^2 \leq 1\},$$

since we know the maximum over the ball will take place on its boundary, i.e., on the sphere. Then consider the Lagrangian function

$$L(\phi, \lambda) = J(\phi) + \lambda(1 - ||\phi||^2). \tag{7.39}$$

Then, according to Proposition 3.2.1 of [6], ϕ will be a local maximum of $J(\phi)$ over $\mathcal{S}$ if and only if there exists a Lagrange multiplier $\lambda \geq 0$ such that

$$\begin{aligned} \nabla_\phi L &= 2(K\phi - \lambda\phi) = 0 \\ \nabla_\lambda L &= 1 - ||\phi||^2 = 0, \end{aligned} \tag{7.40}$$

and

$$< (K - \lambda I)f, f > \leq 0 \tag{7.41}$$

for all functions f of $L^2[a,b]$ orthogonal to the function ϕ satisfying (7.40). The orthogonality condition ensures the functions f are in the space tangent to the unit sphere $\mathcal{S}$ at ϕ. The conditions (7.40) indicate λ must be an eigenvalue of K and ϕ must be the corresponding normalized eigenfunction. Furthermore, the condition (7.41) implies λ must be the largest eigenvalue of K. Otherwise, if there exists an eigenvalue $\lambda' > \lambda$ corresponding to an eigenfunction ϕ', we could pick $f = \phi'$, which is orthogonal to ϕ since the eigenvalues are distinct, and in this case

$$< (K - \lambda I)f, f >= (\lambda' - \lambda) > 0, \tag{7.42}$$

so that condition (7.41) would not be satisfied. So condition (7.40)–(7.41) are satisfied only if λ is the largest eigenvalue of K, and ϕ a corresponding eigenfunction. In this case, the maximum of J is not only local but global, since even if there exists several eigenfunctions corresponding to the largest eigenvalue λ_1, the maximum of $J(\phi)$ at all these local maxima is always λ_1.

Note that the above result holds in both the DT and CT and we have $\lambda_1 > 0$ unless K is identically zero. Then, in the CT case, we have

$$\phi_1(t) = \frac{1}{\lambda_1} \int_a^b K(t,s)\phi_1(s)ds, \tag{7.43}$$

and since $K(t,s)$ is continuous in t, $\phi_1(t)$ must also be continuous.

In both the DT and CT cases, the Rayleigh-Ritz characterization of the largest eigenvalue/eigenfunction pair can be exploited to evaluate this pair numerically. We refer the reader to [7, Lecture 27] for a description of the Rayleigh quotient iteration for computing the eigenvalues of a symmetric matrix. Then, once the pair (λ_1, ϕ_1) has been evaluated, where we assume ϕ_1 has been normalized, consider the residual kernel

$$K_2(t,s) = K(t,s) - \lambda_1\phi_1(t)\phi_1(s).$$

This kernel is both nonnegative definite and, in the CT case, continuous over $[a,b]^2$. The continuity is just a consequence of the continuity of $K(t,s)$ and $\phi_1(t)$. To establish nonnegative definiteness, consider the process

$$Z(t) = X(t) - F_1\phi_1(t) \,,$$

where

$$F_1 = < X, \phi_1 > \,.$$

We have

$$E[|F_1|^2] = < K\phi_1, \phi_1 > = \lambda_1$$

and

$$E[Z(t)Z(s)] = K(t,s) - \lambda_1\phi(t)\phi(s) = K_2(t,s) \tag{7.44}$$

so $K_2(t,s)$ is the covariance kernel of $Z(t)$, and thus must be nonnegative. We also have

$$K_2\phi_1 = K\phi_1 - \lambda_1\phi_1 < \phi_1, \phi_1 > = 0 \,,$$

so ϕ_1 is a function in the null space of K_2, and thus an eigenfunction corresponding to the zero eigenvalue. Assuming that K_2 is not identically zero, let $\lambda_2 > 0$ be the largest eigenvalue of K_2 and ϕ_2 the corresponding eigenfunction. Note that ϕ_2 is necessarily orthogonal to ϕ_1, since the eigenfunctions corresponding to different eigenvalues are orthogonal. Then we can conclude that ϕ_2 is also an eigenfunction of the original kernel K corresponding to the eigenvalue λ_2, since

$$\begin{aligned} K\phi_2 &= K_2\phi_2 + \lambda_1 < \phi_1, \phi_2 > \\ &= K_2\phi_2 = \lambda_2\phi_2 \,. \end{aligned} \tag{7.45}$$

Since λ_1 was the largest eigenvalue of K, this implies $\lambda_1 \geq \lambda_2$.

So, if we proceed by induction, the kernel

$$K_k(t,s) \stackrel{\triangle}{=} K(t,s) - \sum_{j=1}^{k-1} \lambda_j\phi_j(t)\phi_j(s) \,, \tag{7.46}$$

is nonnegative and continuous in the CT case. Then if (λ_k, ϕ_k) denotes its largest eigenvalue/eigenfunction pair, the eigenvalue sequence λ_k, $k \geq 1$ is monotone decreasing, and the eigenfunctions $\{\phi_k,\ k \geq 1\}$ are orthonormal. The process either stops when K_k is identically zero for some k, or continues indefinitely.

DT covariance matrix decomposition: In the DT case, let $N = b - a + 1$ be the length of the observation interval. In this case $L^2[a,b]$ has dimension N, so the Rayleigh sequence stops for $k \leq N$. This gives a decomposition of the form

$$K(t,s) = \sum_{j=1}^{N} \lambda_j\phi_j(t)\phi_j(s) \tag{7.47}$$

where the eigenvalues are monotone decreasing and the functions $\{\phi_j, 1 \leq j \leq N\}$ are othonormal. When the operator K is positive definite, all the eigenvalues are positive, but when the kernel has a null space of dimension d, the Rayleigh sequence stops at $k = N - d + 1$, so that $\lambda_j = 0$ for $N - d + 1 \leq j \leq N$, and the orthonormal functions $\phi_j(t)$ for $N - d + 1 \leq j \leq N$ can be constructed by Gram-Schmidt orthonormalization.

The decomposition (7.47) admits a simple matrix representation. Let $\mathbf{K}$ be the $N \times N$ covariance matrix defined in (7.27), and let

$$\boldsymbol{\phi}_j = \begin{bmatrix} \phi_j(a) \dots \phi_j(t) \dots \phi_j(b) \end{bmatrix}^T$$

be the vector representing the function $\phi_j(t)$. Then if

$$\boldsymbol{\Phi} = \begin{bmatrix} \boldsymbol{\phi}_1 \dots \boldsymbol{\phi}_j \dots \boldsymbol{\phi}_N \end{bmatrix}$$

and

$$\boldsymbol{\Lambda} = \text{diag}\{\lambda_j, \ 1 \leq j \leq N\}$$

denote respectively the matrix regrouping the eigenfunctions and the eigenvalues of K, the decomposition (7.47) can be represented in matrix form as

$$\mathbf{K} = \boldsymbol{\Phi}\boldsymbol{\Lambda}\boldsymbol{\Phi}^T \tag{7.48}$$

where the orthonormality of the eigenfunctions implies $\boldsymbol{\Phi}$ is an orthonormal matrix, i.e.,

$$\boldsymbol{\Phi}\boldsymbol{\Phi}^T = \mathbf{I}_N \, . \tag{7.49}$$

The decomposition (7.48) is the familiar eigenvalue decomposition for a symmetric matrix. Note that as a consequence of (7.47) we have the identity

$$\sum_{t=a}^{b} K(t,t) = \text{tr}\,(\mathbf{K}) = \sum_{j=1}^{N} \lambda_j \, . \tag{7.50}$$

Mercer's Theorem: In the CT case, the sequence of kernels $K_k(t,s)$ given by (7.46) has the property that

$$\lim_{k \to \infty} \sup_{a \leq t, \ s \leq b} |K_k(t,s)| = 0 \, ,$$

so that

$$\lim_{k \to \infty} \sum_{j=1}^{k-1} \lambda_j \phi_j(t) \phi_j(s) = K(t,s) \tag{7.51}$$

uniformly over $[a,b]^2$. The uniform convergence of the eigenfunction expansion is a strong property which implies in particular the mean-square convergence

$$\lim_{k \to \infty} \int_a^b \int_a^b |K_k(t,s)|^2 dt ds = 0 \, .$$

Also, setting $t = s$ in (7.51) and integrating, we obtain the trace identity

$$\int_a^b K(t,t)dt = \sum_{j=1}^{\infty} \lambda_j ,$$

where the left-hand side is finite since $K(t,t)$ is a continuous function over $[a,b]$. Since the sequence of eigenvalues λ_j is summable, this indicates that these eigenvalues decay at a rate faster than j^{-1}.

Furthermore, if $f(t)$ denotes an arbitrary function of $L^2[a,b]$, it can be represented as

$$f(t) = \tilde{f}(t) + \lim_{k \to \infty} \sum_{j=1}^{k} F_j \phi_j(t) , \tag{7.52}$$

where

$$F_j =< f, \phi_j >= \int_a^b f(t)\phi_j(t)dt ,$$

and $\tilde{f}$ is a function in the null space of K, i.e.,

$$(K\tilde{f})(t) = \int_a^b K(t,s)\tilde{f}(s)ds = 0 .$$

The limit in the representation (7.52) is taken in the mean-square sense, i.e.,

$$\lim_{k \to \infty} ||f - \tilde{f} - \sum_{j=1}^{k} F_j \phi_j||^2 = 0 .$$

From the representation (7.52) we see therefore that if the operator K is positive definite, then the eigenfunctions $\{\phi_j, \ j \geq 1\}$ form a COB of $L^2[a,b]$.

Example 7.3: Quadrature amplitude modulated signal

Consider a CT random signal

$$X(t) = X_c \cos(\omega_c t) + X_s \sin(\omega_c t) ,$$

defined over $[0,T]$, where X_c and X_s are two independent zero-mean Gaussian random variables with variance

$$E[X_c^2] = E[X_s^2] = P .$$

For simplicity, we assume that T is an integer multiple of the carrier period $T_c = 2\pi/\omega_c$, i.e. $T = nT_c$ for an integer n. This ensures that the sine and cosine components of $X(t)$ are exactly orthogonal over $[0,T]$. Otherwise, when $T/T_c \gg 1$, they are only almost orthogonal. The process $X(t)$ is Gaussian and stationary with covariance

$$K(t,s) = E[X(t)X(s)] = P\cos(\omega_c|t-s|)$$
$$= P[\cos(\omega_c t)\cos(\omega_c s) + \sin(\omega_c t)\sin(\omega_c s)] \,. \tag{7.53}$$

Let

$$\phi_1(t) = (2/T)^{1/2}\cos(\omega_c t)$$
$$\phi_2(t) = (2/T)^{1/2}\sin(\omega_c t)$$

and

$$\lambda_1 = \lambda_2 = PT/2 \,.$$

Then ϕ_1 and ϕ_2 are orthonormal and (7.53) can be rewritten as

$$K(t,s) = (PT/2)[\phi_1(t)\phi_1(s) + \phi_2(t)\phi_2(s)] \,,$$

so in this case, Mercer's expansion includes only two terms, i.e., K has only two nonzero eigenvalues. The null space of K is formed by all functions that are orthogonal to $\phi_1(t)$ and $\phi_2(t)$, i.e., to the cosine and sine functions with frequency $\omega_c = n\omega_0$, where $\omega_0 = 2\pi/T$ is the fundamental frequency for periodic functions defined over $[0,T]$. To obtain a COB of $L^2[0,T]$, we need to complement ϕ_1 and ϕ_2 by basis functions obtained by Gram-Schmidt orthonormalization of the null space of K. But since ϕ_1 and ϕ_2 are just the n-th harmonic sine and cosine functions of the standard trigonometric Fourier basis of $L^2[0,T]$, a basis of the null space of K is formed by the normalized sine and cosine functions $(2/T)^{1/2}\cos(k\omega_0 t)$ and $(2/T)^{1/2}\sin(k\omega_0 t)$ with $k \neq n$. □

7.4 Differential Characterization of the Eigenfunctions

Although integral equations can be solved by employing discretization methods, there exists several classes of covariance kernels for which the integral equation satisfied by the eigenfunctions of K can be replaced by an equivalent differential system. Similarly in the DT case, the matrix equations satisfies by the eigenfunctions can be replaced by difference equations.

7.4.1 Gaussian Reciprocal Processes

For the class of Gaussian reciprocal processes, instead of computing directly the eigenvalues and eigenfunctions of a covariance kernel $K(t,s)$, we can transform the problem into one involving the computation of the eigenvalues and eigenfunctions of a positive self-adjoint Sturm-Liouville differential/difference system defined over $[a,b]$. We recall that a process $X(t)$ is Markov, if given $X(s)$, the process $X(t)$ for $t > s$ is independent of $X(r)$ for $r < s$, i.e., the past and future are conditionally independent given the present. In the Gaussian case, $X(t)$ is Markov if and only if its covariance kernel satisfies

$$K(t,r) = K(t,s)K^{-1}(s,s)K(s,r) \tag{7.54}$$

for $t \geq s \geq r$. Reciprocal processes [8] are slightly more general than Markov processes and are such that if $I = [s,u]$ denotes a finite interval contained in $[a,b]$, given the values $X(s)$ and $X(u)$ at both ends of I, the process $X(t)$ in the interior of I is independent of the process $X(r)$ in the exterior of I. In the Gaussian case, for $r \leq s \leq t \leq u$ and for $s \leq t \leq u \leq r$, the covariance kernel satisfies

$$K(t,r) = \begin{bmatrix} K(t,s) \; K(t,u) \end{bmatrix} \begin{bmatrix} K(s,s) & K(s,u) \\ K(u,s) & K(u,u) \end{bmatrix}^{-1} \begin{bmatrix} K(s,r) \\ K(u,r) \end{bmatrix} . \tag{7.55}$$

Obviously from the above definition, if a process is Markov, it is reciprocal, but not vice-versa, i.e., the reciprocal class is strictly larger than the Markov class. It is also worth noting that reciprocal processes can be viewed as the restriction to the 1-D parameter case of the class of Markov fields considered by Paul Lévy [9]. In other words, 1-D Markov fields are reciprocal processes, but not necessarily Markov processes!

For our purposes, the key property of Gaussian reciprocal processes that we shall exploit is that under various smoothness conditions for the kernel $K(t,s)$, it was shown in [10–12] that a positive definite kernel $K(t,s)$ is the covariance of a Gaussian reciprocal process if and only if it is the Green's function of a second-order positive self-adjoint Sturm-Liouville differential or difference operator with mixed Dirichlet and Neumann boundary conditions. Specifically, in the CT case, let

$$\begin{aligned} q(t) &= \frac{\partial K}{\partial t}(t^-,t) - \frac{\partial K}{\partial t}(t^+,t) \\ g(t) &= [\frac{\partial^2 K}{\partial t^2}(t^-,t) - \frac{\partial^2 K}{\partial t^2}(t^+,t)]q^{-1}(t) \\ f(t) &= [\frac{\partial^2 K}{\partial t^2}(t^+,t) - g(t)\frac{\partial K}{\partial t}(t^+,t)]K^{-1}(t,t) \, , \end{aligned} \tag{7.56}$$

and consider the second-order differential operator

$$L = q^{-1}(t)\big(- \frac{d^2}{dt^2} + g(t)\frac{d}{dt} + f(t)\big) \, . \tag{7.57}$$

Let also $\psi_1(t)$ and $\psi_2(t)$ be the transition functions satisfying the differential equation

$$L\psi_1(t) = L\psi_2(t) = 0 \tag{7.58}$$

with boundary conditions

$$\begin{bmatrix} \psi_1(a) & \psi_2(a) \\ \psi_1(b) & \psi_2(b) \end{bmatrix} = \begin{bmatrix} 1 & 0 \\ 0 & 1 \end{bmatrix} = \mathbf{I}_2 \, . \tag{7.59}$$

Consider also the matrices

$$\mathbf{P} = \begin{bmatrix} K(a,a) & K(a,b) \\ K(b,a) & K(b,b) \end{bmatrix}$$
$$\mathbf{Q} = \begin{bmatrix} q(a) & 0 \\ 0 & -q(b) \end{bmatrix}$$
$$\boldsymbol{\Psi} = \begin{bmatrix} \dot{\psi}_1(a) & \dot{\psi}_2(a) \\ \dot{\psi}_1(b) & \dot{\psi}_2(b) \end{bmatrix} , \tag{7.60}$$

and define the boundary condition (BC)

$$Bu \triangleq (\mathbf{I}_2 + \mathbf{PQ}^{-1}\boldsymbol{\Psi}) \begin{bmatrix} u(a) \\ u(b) \end{bmatrix} - \mathbf{PQ}^{-1} \begin{bmatrix} \dot{u}(a) \\ \dot{u}(b) \end{bmatrix} = \begin{bmatrix} 0 \\ 0 \end{bmatrix} . \tag{7.61}$$

Then it is shown in [10, 12] that the second-order Sturm-Liouville differential operator L with BC (7.61) is self-adjoint and $K(t,s)$ is its Green's function, i.e., it satisfies

$$LK(t,s) = \delta(t-s) \tag{7.62}$$

with $(BK)(\cdot, s) = 0$.

To see what this property implies for the eigenfunctions of K, we can apply the differential operator L and boundary operator B to the eigenfunction equation (7.23). We obtain

$$\lambda L\phi(t) = \int_a^b (LK)(t,s)\phi(s)ds = \phi(t) \tag{7.63}$$

with the BC

$$\lambda B\phi = 0 . \tag{7.64}$$

Since the operator K kas been assumed positive definite, λ is positive, so by dividing (7.63) by λ, we find that $\phi(t)$ is an eigenfunction of L with eigenvalue λ^{-1}. So instead of computing the eigenvalues and eigenfunctions of the integral operator K, we can compute those of the Sturm-Liouville operator L with boundary condition (7.61). The advantage of this approach is that it is usually much easier to solve differential or difference equations than integral equations, or matrix equations in the DT case. To illustrate this approach, we consider several examples.

Example 7.4: Wiener process over a finite interval

Consider a Wiener/Brownian motion process $X(t)$ defined over $[0, T]$. This process, which is studied in detail in [13, Chap. 7], is nonstationary zero-mean Gaussian with covariance function

$$E[X(t)X(s)] = K(t,s) = q \min(t,s) , \tag{7.65}$$

where $q > 0$ is sometimes called the "intensity" of the process. Since the kernel (7.65) satisfies the condition (7.54), $X(t)$ is Markov. We find that

$$L = -q^{-1} \frac{d^2}{dt^2} \,,$$

and the transition functions are given by

$$\psi_1(t) = 1 - t/T \quad , \quad \psi_2(t) = t/T \,.$$

We obtain

$$\mathbf{P} = \begin{bmatrix} 0 & 0 \\ 0 & qT \end{bmatrix}$$

$$\mathbf{Q} = \begin{bmatrix} q & 0 \\ 0 & -q \end{bmatrix}$$

$$\boldsymbol{\Psi} = T^{-1} \begin{bmatrix} -1 & 1 \\ -1 & 1 \end{bmatrix} ,$$

and the BC (7.61) takes the form

$$u(0) = 0 \quad , \quad u(0) + T\dot{u}(T) = 0 \,,$$

where, after elimination, the second condition reduces to $\dot{u}(T) = 0$. Note in this respect that a key feature of the subclass of Markov processes within the larger class of reciprocal processes is that the boundary condition specified by (7.61) decouples into separate boundary conditions at $t = 0$ and $t = T$. Since the Wiener process is Markov, it is therefore not surprising that its BCs take a decoupled form. Then the eigenfunction equation (7.63) can be rewritten as

$$\ddot{\phi}(t) + \frac{q}{\lambda} \phi(t) = 0 \,,$$

which, since $\lambda > 0$, admits the solution

$$\phi(t) = A \cos((q/\lambda)^{1/2} t) + B \sin((q/\lambda)^{1/2} t) \,.$$

The BC $\phi(0) = 0$ implies $A = 0$ and $\dot{\phi}(T) = 0$ implies

$$\cos((q/\lambda)^{1/2} T) = 0 \,,$$

so that

$$(q/\lambda)^{1/2} T = (2k - 1) \frac{\pi}{2} \,,$$

where k denotes a positive integer. The eigenvalues and eigenfunctions of the covariance kernel are therefore given by

$$\lambda_k = q \Big(\frac{2T}{(2k-1)\pi} \Big)^2$$

and

$$\phi_k(t) = B \sin \Big(\frac{(2k-1)\pi t}{2T} \Big)$$

for $k = 1, 2, \ldots$, where in order to ensure that the normalization

$$1 = ||\phi_k||^2 = \int_0^T \phi_k^2(t)dt$$

holds, we need to select $B = (2/T)^{1/2}$. Although the eigenfunctions $\phi_k(t)$ are sinusoids, they do not form a Fourier basis, since the frequencies are not integer multiples of the fundamental frequency $\omega_0 = 2\pi/T$. For $T = 1$, the first four eigenfunctions are plotted in Fig. 7.2. □

Example 7.5: Ornstein-Uhlenbeck process over a finite interval

Consider an Ornstein-Uhlenbeck process $X(t)$ defined over a finite interval $[0, T]$. This process is zero-mean Gaussian with the covariance function

$$E[X(t)X(s)] = K(t, s) = P \exp(-a|t - s|) , \tag{7.66}$$

where $a > 0$. Since $K(t, s)$ depends only on $t - s$, $X(t)$ is stationary, and in fact when it is defined for all times, it has for spectral density

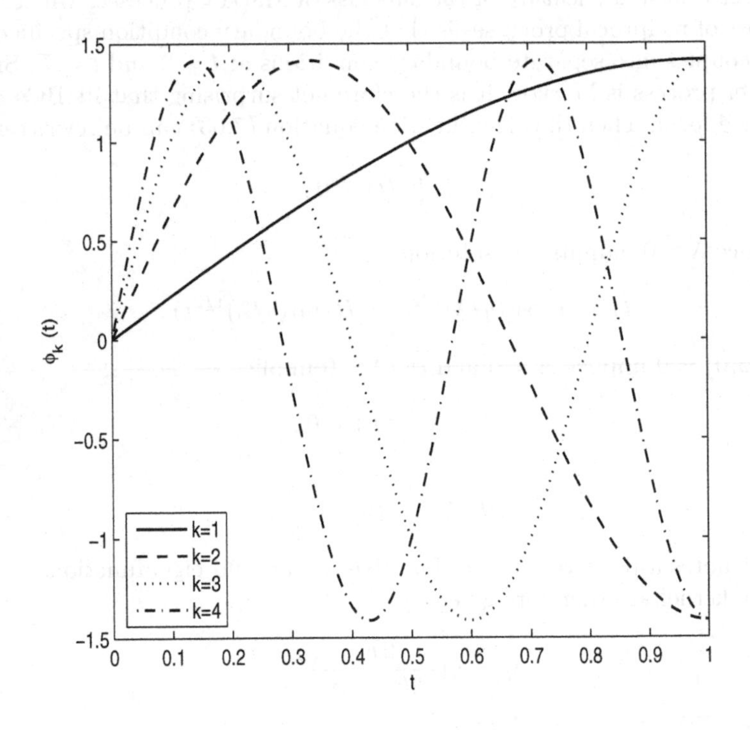

Fig. 7.2. First four eigenfunctions of the Wiener process over an interval of length $T = 1$.

$$S_X(\omega) = \frac{2aP}{\omega^2 + a^2} . \tag{7.67}$$

The covariance $K(t, s)$ satisfies the condition (7.54), so $X(t)$ is Markov. In fact as the form of the spectral density indicates, it admits a first-order dynamic model of the form

$$\frac{dX}{dt} = -aX(t) + V(t)$$

where $V(t)$ is a zero-mean WGN with intensity $q = 2aP$. Since a capacitor connected to a resistor in thermal equilibrium admits a model of this type, the Ornstein-Uhlenbeck process can be viewed as a simplified model of thermal noise in electrical circuits.

From (7.56)–(7.57), its matching Sturm-Liouville operator is given by

$$L = q^{-1}(-\frac{d^2}{dt^2} + a^2) ,$$

and the transition functions take the form

$$\begin{aligned} \psi_1(t) &= -\frac{\sinh(a(t-T))}{\sinh(aT)} \\ \psi_2(t) &= \frac{\sinh(at)}{\sinh(aT)} , \end{aligned}$$

where $\sinh(x) = (\exp(x) - \exp(-x))/2$ denotes the hyperbolic sine function. We have

$$\begin{aligned} \mathbf{P} &= P \begin{bmatrix} 1 & \exp(-aT) \\ \exp(-aT) & 1 \end{bmatrix} \\ \mathbf{Q} &= \begin{bmatrix} q & 0 \\ 0 & -q \end{bmatrix} \\ \boldsymbol{\Psi} &= \begin{bmatrix} -a\coth(aT) & a(\sinh(aT))^{-1} \\ -a(\sinh(aT))^{-1} & a\coth(aT) \end{bmatrix} , \end{aligned}$$

and after simplifications we find that the BCs (7.61) take the form

$$\begin{aligned} \dot{u}(0) - au(0) &= 0 \\ \dot{u}(T) + au(T) &= 0 . \end{aligned} \tag{7.68}$$

Then consider the eigenfunction equation

$$-\ddot{\phi}(t) + a^2\phi(t) = \frac{q}{\lambda}\phi(t) . \tag{7.69}$$

It is not difficult to verify that we must have

$$0 < \lambda < \frac{q}{a^2} = \frac{2P}{a} = S_X(0) ,$$

since otherwise it is not possible to construct a solution of the differential equation (7.69) satisfying the homogeneous BCs (7.68). Then if

$$b = (q/\lambda - a^2)^{1/2} ,$$

or equivalently

$$\lambda = \frac{q}{a^2 + b^2} = S_X(b) ,$$

the eigenfunction equation (7.69) admits the solution

$$\phi(t) = A\cos(bt) + B\sin(bt) .$$

If

$$\mathbf{M} = \begin{bmatrix} a & -b \\ a\cos(bT) - b\sin(bT) & a\sin(bT) + b\cos(bT) \end{bmatrix} , \tag{7.70}$$

the boundary conditions (7.68) can be expressed as

$$\mathbf{M} \begin{bmatrix} A \\ B \end{bmatrix} = \begin{bmatrix} 0 \\ 0 \end{bmatrix} ,$$

which has a nonzero solution if and only if

$$\det(\mathbf{M}) = 2ab\cos(bT) + (a^2 - b^2)\sin(bT) = 0 . \tag{7.71}$$

The equation (7.71) is equivalent to

$$\tan(bT) = \frac{2ab}{b^2 - a^2} = \frac{a}{b-a} + \frac{a}{b+a} ,$$

which admits an infinite number of roots $0 < b_1 < \ldots < b_k < \ldots$, as can be seen from Fig. 7.3 for the case $T = 1$ and $a = 1$.

For this choice of parameters, the first four roots are given by $b_1 = 1.3065$, $b_2 = 3.6732$, $b_3 = 6.6846$, and $b_4 = 9.6317$. By observing that the function $2ab/(b^2 - a^2)$ tends to zero as $b \to \infty$, we can also deduce that for large k

$$b_k \approx (k-1)\pi/T .$$

The eigenfunctions are then given by

$$\phi_k(t) = A[\cos(b_k t) - \frac{a}{b_k}\sin(b_k t)] ,$$

where A is selected such that $||\phi_k||^2 = 1$, and the corresponding eigenvalues

$$\lambda_k = \frac{2aP}{a^2 + b_k^2}$$

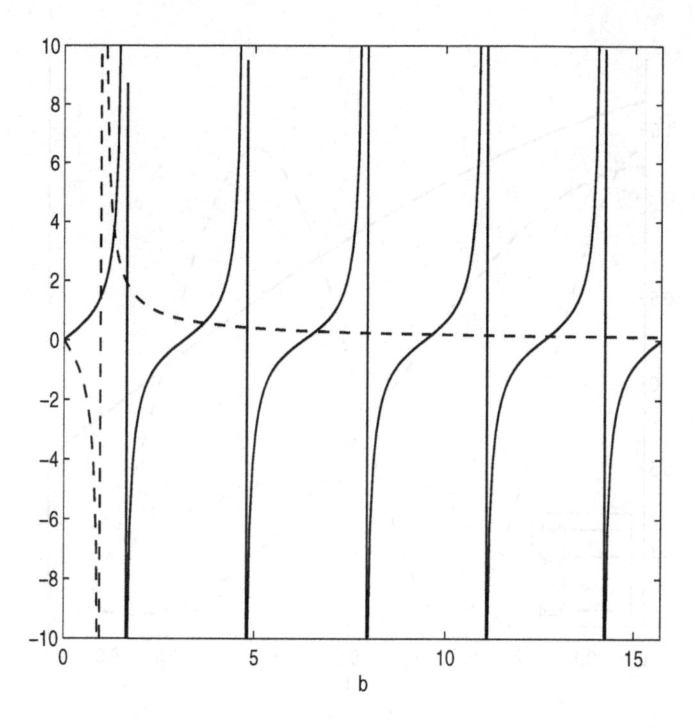

Fig. 7.3. Plot of functions $\tan(bT)$ (solid line) and $2ab/(b^2 - a^2)$ (dashed line) for $a = 1$ and $T = 1$.

are monotone decreasing. The first four eigenfunctions are plotted in Fig. 7.4 for the case $T = 1$ and $a = 1$. □

Example 7.6: Slepian process

Consider a zero-mean Gaussian process $X(t)$ defined over $[0, T]$ with the triangular covariance kernel

$$K(t, s) = E[X(t)X(s)] = P(1 - 2|t - s|/T) .$$

This process, which was studied by Slepian [14], is not Markov, but reciprocal since its covariance satisfies condition (7.55). Its matching differential operator is given by

$$L = -q^{-1} \frac{d^2}{dt^2}$$

with $q = 4P/T$, and the transition functions are given by

$$\psi_1(t) = 1 - t/T \quad , \quad \psi_2(t) = t/T .$$

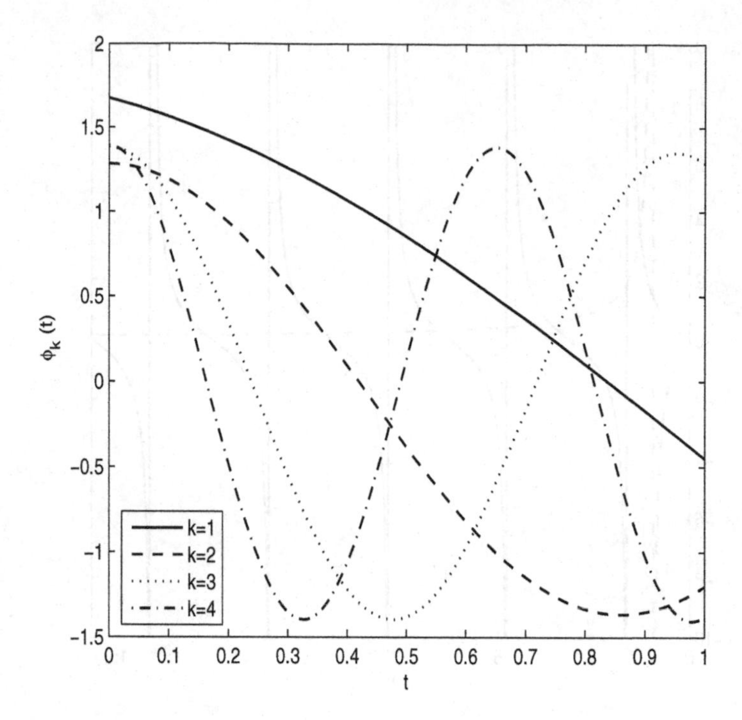

Fig. 7.4. First four eigenfunctions of the Ornstein-Uhlenbeck process with parameter $a = 1$ over an interval of length $T = 1$.

We have

$$\mathbf{P} = P \begin{bmatrix} 1 & -1 \\ -1 & 1 \end{bmatrix}$$

$$\mathbf{Q} = \begin{bmatrix} q & 0 \\ 0 & -q \end{bmatrix}$$

$$\boldsymbol{\Psi} = \frac{1}{T} \begin{bmatrix} -1 & 1 \\ -1 & 1 \end{bmatrix} .$$

Then, after simplifications, the BCs (7.61) take the form

$$\begin{aligned} u(0) + u(T) &= 0 \\ \dot{u}(0) + \dot{u}(T) &= 0 . \end{aligned} \tag{7.72}$$

Next, consider the eigenfunction equation

$$\ddot{\phi}(t) + \frac{q}{\lambda}\phi(t) = 0 .$$

Its solution takes the form

$$\phi(t) = A\cos((q/\lambda)^{1/2}t) + B\sin((q/\lambda)^{1/2}t) ,$$

and if

$$\mathbf{M} = \begin{bmatrix} 1+\cos((q/\lambda)^{1/2}T) & \sin((q/\lambda)^{1/2}T) \\ -\sin((q/\lambda)^{1/2}T) & 1+\cos((q/\lambda)^{1/2}T) \end{bmatrix} ,$$

the BC (7.72) can be expressed as

$$\mathbf{M}\begin{bmatrix} A \\ B \end{bmatrix} = \begin{bmatrix} 0 \\ 0 \end{bmatrix} .$$

This equation admits a nonzero solution if and only if

$$\det(\mathbf{M}) = 2(1+\cos((q/\lambda)^{1/2}T)) = 0 , \tag{7.73}$$

or equivalently

$$(q/\lambda)^{1/2}T = (2k-1)\pi ,$$

where k denotes a positive integer. The eigenvalues of K are therefore

$$\lambda_k = \frac{qT^2}{(2k-1)^2\pi^2} ,$$

with $k \geq 1$. The condition (7.73) implies $B = 0$, so that the k-th eigenfunction takes the form

$$\phi_k(t) = A\cos((2k-1)\pi t/T)$$

where we select $A = (2/T)^{1/2}$ to ensure ϕ_k has unit norm. For $T = 1$, the first four eigenfunctions are plotted in Fig. 7.5. □

DT Gaussian reciprocal processes admit a characterization similar to the one used above in the CT case. Specifically, let $X(t)$ be a DT zero-mean Gaussian processes over $[a, b]$, whose $N \times N$ covariance matrix $\mathbf{K}$ is defined by (7.27), with $N = b - a + 1$. It is shown in [11] that the process $X(t)$ is reciprocal if and only if the inverse $\mathbf{M} = \mathbf{K}^{-1}$ is a *cyclic tridiagonal matrix.* This means that the entries M_{ij} of $\mathbf{M}$ with $1 \leq i,\ j \leq N$ are zero unless

$$(i-j) \mod N = 0,\ \pm 1 .$$

In other words, only the diagonal and the first sub- and super-diagonals of $\mathbf{M}$ are nonzero, where the sub- and super-diagonals are extened cyclically to include the elements in the top right-hand and bottom left-hand corners of $\mathbf{M}$, i.e., the entries M_{1N} and M_{N1}. Equivalently, if $Z_{\mathrm{C}}f(t) = f((t+1) \mod N)$ denotes the cyclic forward shift operator in $L^2[a, b]$, consider the second-order difference operator let

$$M = m_0(t) + m_{-1}(t)Z_{\mathrm{C}}^{-1} + m_1(t)Z_{\mathrm{C}} , \tag{7.74}$$

where $m_0(t)$, $m_{-1}(t)$ and $m_1(t)$ are the coefficients appearing on the diagonal, subdiagonal, and superdiagonal of row $i = t-a+1$ of $\mathbf{M}$. Then the covariance function $K(t, s)$ satisfies the second-order recursion

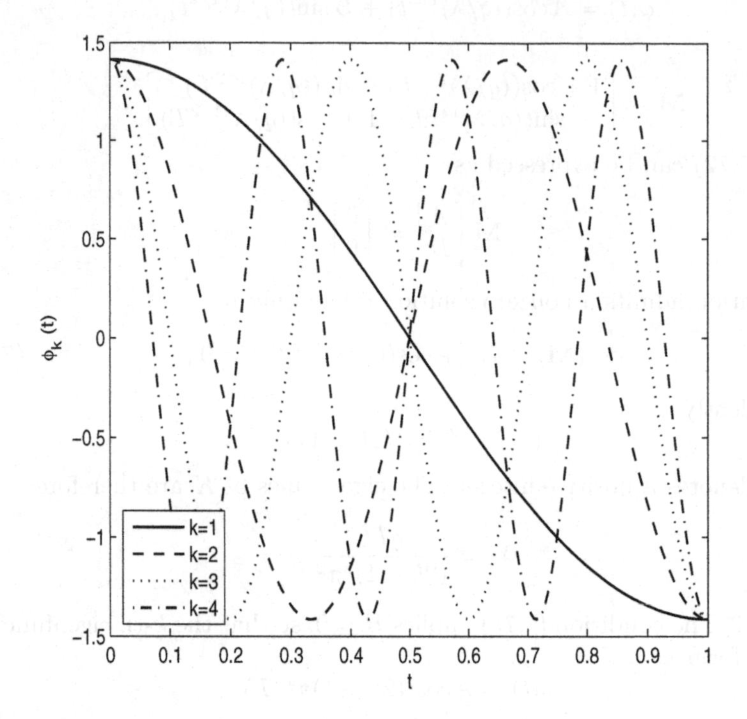

Fig. 7.5. First four eigenfunctions of the Slepian process over an interval of length $T = 1$.

$$MK(t,s) = \delta(t-s) \tag{7.75}$$

for $a \leq t \leq b$, where $\delta(t-s)$ denotes the Kronecker delta function. If we consider the vector

$$\boldsymbol{\xi}(t) \triangleq \begin{bmatrix} X(t-1) \\ X(t) \\ X(t+1) \end{bmatrix}$$

and

$$\mathbf{C}(t) = E[\boldsymbol{\xi}(t)\boldsymbol{\xi}^T(t)] = \begin{bmatrix} K(t-1,t-1) & K(t-1,t) & K(t-1,t+1) \\ K(t,t-1) & K(t,t) & K(t,t+1) \\ K(t+1,t-1) & K(t+,t) & K(t+1,t+1) \end{bmatrix} \tag{7.76}$$

represents its covariance, the coefficients $m_{-1}(t)$, $m_0(t)$ and $m_1(t)$ of the difference operator M are given by

$$\begin{bmatrix} m_{-1}(t) \; m_0(t) \; m_1(t) \end{bmatrix} = \begin{bmatrix} 0 \; 1 \; 0 \end{bmatrix} \mathbf{C}^{-1}(t)\,. \tag{7.77}$$

Note that identity (7.77) holds for $a \leq t \leq b$ provided that in (7.76) and (7.77) $t-1$ and $t+1$ are defined modulo N, which means for example

$$(a-1) \mod N = a - 1 + N = b\,.$$

Since the covariance kernel $K(t,s)$ is positive self-adjoint, the second-order difference operator M, which is its inverse, is also positive-definite self-adjoint, where the self-adjointness property reduces to

$$m_{-1}(t+1) = m_1(t)\,. \tag{7.78}$$

Then, by applying the second-order difference operator M to the eigenfunction equation (7.23) we obtain

$$M\phi(t) = \lambda^{-1}\phi(t)\,, \tag{7.79}$$

so that as in the CT case, if $\phi(t)$ is an eigenfunction of K corresponding to the eigenvalue λ, it is also an eigenfunction of the difference operator M corresponding to the eigenvalue λ^{-1}. Note that, since in the definition (7.74) of M, the shift operator Z_C is defined cyclically, the equation (7.79) holds at $t = a$ and $t = b$, where at these points it corresponds to cyclic boundary conditions which form the DT counterpart of the CT boundary condition (7.61).

In passing, it is worth noting that it is also shown in [11] that within the larger class of zero-mean DT Gaussian reciprocal processes, the subclass of Markov processes is characterized by the feature that their inverse covariance matrix $\mathbf{M} = \mathbf{K}^{-1}$ is strictly tridiagonal, instead of cyclic tridagonal, which means that the entries M_{1N} and M_{N1} in the top right and bottom left corners of $\mathbf{M}$ must be zero. This implies that the DT boundary conditions specifyng Markov processes are *separable*, in the sense that are expressible as two independent conditions involving separately $X(a)$ and $X(a+1)$ on one hand, and $X(b-1)$ and $X(b)$ on the other hand. Note that this separability property of the boundary conditions satisfied by Markov processes was already observed in the CT case for the Wiener process of Example 7.4 and the Ornstein-Uhlenbeck process of Example 7.5.

To illustrate the evaluation of the eigenfunctions of DT Gaussian reciprocal processes by solving a difference equation, we consider an example.

Example 7.7: Hyperbolic cosine process

Consider a zero-mean DT Gaussian process $X(t)$ defined over $[0,T]$ with covariance kernel

$$K(t,s) = P\frac{\cosh(\alpha(T/2 - |t-s|))}{\cosh(\alpha T/2)}\,,$$

where $\cosh(x) = (\exp(x) + \exp(-x))/2$ denotes the hyperbolic cosine function and $\alpha > 0$. It turns out this process satisfies condition (7.55) so it is reciprocal, and stationary since $K(t,s)$ depends only on $t-s$. With a slight abuse of notation, let us denote the covariance as $K(t-s)$. Since $K(0) = K(T) = P$, the process has the feature $X(0) = X(T)$. This implies that $X(t)$ is periodic and can be viewed as defined over the discretized circle obtained by taking the interval $[0, T]$ and wrapping it around in such a way that the two ends 0 and T coincide. From (7.77), if $a = \exp(-\alpha)$, we obtain

$$M = \frac{1}{P(1-a^2)}[(1+a^2) - a(Z_C + Z_C^{-1})] ,$$

so the eigenfunctions and eigenvalues satisfy the second-order difference equation

$$(1+a^2)\phi(t) - a\phi(t+1) - a\phi(t-1) = \frac{P(1-a^2)}{\lambda}\phi(t) \tag{7.80}$$

for $0 \leq t \leq T$, with the periodicity condition

$$\phi(0) = \phi(T) . \tag{7.81}$$

To solve this equation, we try a solution of the form

$$\phi(t) = A_+ \exp(jbt) + A_- \exp(-jbt) . \tag{7.82}$$

The periodicity condition (7.81) implies

$$b_k T = k2\pi$$

so that $b_k = k\omega_0$ where $\omega_0 = 2\pi/T$ denotes the fundamental frequency for the Fourier basis of $L^2[0, T]$. Then, substituting (7.82) inside (7.80), we find that the k-th eigenfunction is given by

$$\phi_k(t) = (1/T)^{1/2} \exp(jk\omega_0 t)$$

for $0 \leq k \leq T-1$ and the corresponding eigenvalue is

$$\lambda_k = \frac{P(1-a^2)}{(1+a^2) - 2a\cos(k\omega_0)} .$$

Note that since $\lambda_{T-k} = \lambda_k$, all eigenvalues with $k \neq 0$ have multiplicity two, and the second eigenfunction corresponding to λ_k is $\phi_{T-k}(t) = \phi_k^*(t)$. Thus the hyperbolic cosine process has the remarkable property that its covariance matrix is diagonalized by the discrete Fourier transform (DFT) and its eigenvalues are the values of the PSD

$$S_X(\omega) = \frac{P(1-a^2)}{(1+a^2) - 2a\cos(\omega)}$$

sampled at the discrete Fourier frequencies $b_k = k\omega_0$. This result can be explained easily if we consider the covariance matrix $\mathbf{K}$ of the process $X(t)$ defined over $[0, T-1]$, where we recall that $X(T)$ is identified with $X(0)$. If we consider the cyclic shift matrix

$$\mathbf{Z}_{\mathrm{C}} = \begin{bmatrix} 0\,0 \ldots 0\,1 \\ 1\,0 \ldots 0\,0 \\ 0\,1 \ldots 0\,0 \\ \vdots \qquad \vdots \\ 0\,0 \ldots 1\,0 \end{bmatrix},$$

of dimension $T \times T$, $\mathbf{K}$ can be written as

$$\mathbf{K} = \sum_{t=0}^{T-1} K(t)\mathbf{Z}_{\mathrm{C}}^{t}$$

so that $\mathbf{K}$ is a *circulant matrix*, i.e., it is constant along its cyclically extended diagonals. This class of matrices has the important property that it is diagonalized by the DFT [15, pp. 206–207]. □

The eigenvalues and eigenfunctions of the DT Ornstein-Uhlenbeck process, which is a zero mean Gauss-Markov process with covariance $K(t-s) = Pa^{|t-s|}$, are evaluated in [16] (see also Problem 7.6). Although the eigenfunctions are not completely identical with the cosine functions used for the discrete cosine transform (DCT), when a is larger than than 0.9, the DCT represents a close approximation to the eigenfunctions of the Ornstein-Uhlenbeck process [17]. This result represents in fact the underlying justification for the use of the DCT for coding of still images.

7.4.2 Partially Observed Gaussian Reciprocal/Markov Processes

It is worth noting that there is no reason to restrict our attention to the case where the process $X(t)$ itself is reciprocal or Markov. The technique described in the previous section can be extended to the case where $X(t)$ can be expressed as

$$X(t) = \mathbf{c}(t)\boldsymbol{\xi}(t) \tag{7.83}$$

where $\boldsymbol{\xi}(t) \in \mathbb{R}^n$ is a zero-mean Gaussian recriprocal or Markov process, and $\mathbf{c}(t)$ is a row vector of dimension n. In other words, $X(t)$ is just a linear combination of the components of a reciprocal or Markov process $\boldsymbol{\xi}(t)$, so it corresponds to a partially observed Gaussian reciprocal or Markov process. In the Markov case, it was shown by Baggeroer [18] that the eigenfunctions of the covariance operator $K(t,s)$ can be evaluated by solving a first-order two-point boundary value differential/difference Hamiltonian system of dimension $2n$.

Since the technique of [18] can be modified in a straighforward manner to deal with the case where $\boldsymbol{\xi}(t)$ is reciprocal, we consider this more general case first. If $\mathbf{K}_{\boldsymbol{\xi}}(t,s) = E[\boldsymbol{\xi}(t)\boldsymbol{\xi}^T(s)]$ denotes the covariance of $\boldsymbol{\xi}(t)$, the covariance of $X(t)$ can be expressed as

$$K(t,s) = \mathbf{c}(t)\mathbf{K}_{\boldsymbol{\xi}}(t,s)\mathbf{c}^T(s)\,. \tag{7.84}$$

Then, as in the scalar case considered in the previous section, the reciprocity property of $\boldsymbol{\xi}(t)$ manifests itself by the fact that its covariance can be viewed as the Green's function of a matrix self-adjoint differential or difference equation with two-point boundary value conditions. Specifically, in the CT case we have

$$L\mathbf{K}_{\boldsymbol{\xi}}(t,s) = \mathbf{I}_n \delta(t-s) \tag{7.85}$$

with $BK(\cdot, s) = 0$. When the matrix

$$\mathbf{Q}(t) = \frac{\partial \mathbf{K}_{\boldsymbol{\xi}}}{\partial t}(t^-, t) - \frac{\partial \mathbf{K}_{\boldsymbol{\xi}}}{\partial t}(t^+, t) \tag{7.86}$$

is invertible, which corresponds to the case when a full rank WGN component affects the dynamics of $\boldsymbol{\xi}(t)$, L is a matrix second-order differential operator of the form

$$L = \mathbf{Q}^{-1}(t)[-\mathbf{I}_n \frac{d^2}{dt^2} + \mathbf{G}(t)\frac{d}{dt} + \mathbf{F}(t)] \,,$$

where the $n \times n$ matrices $\mathbf{F}(t)$ and $\mathbf{G}(t)$ obey matrix versions of the scalar expressions (7.56). See [10, 12] for a detailed discussion of L and its properties. Similarly, the boundary operator B takes a matrix form of (7.61) which is described in [12]. When the matrix $\mathbf{Q}(t)$ appearing in (7.86) is not invertible, the operator L satisfying (7.86) has a more complicated structure and corresponds to a mixed-order model of the type discussed in [19, Chap. 6].

Let us return now to the eigenfunction equation

$$\int_0^T K(t,s)\phi(s)ds = \lambda\phi(t) \tag{7.87}$$

for $0 \leq t \leq T$, where $K(t,s)$ admits the structure (7.84). To solve this equation, we postulate the form

$$\phi(t) = \mathbf{c}(t)\boldsymbol{\theta}(t) \tag{7.88}$$

for its solution, where the vector function $\boldsymbol{\theta} \in \mathbb{R}^n$ satisfies

$$\int_0^T \mathbf{K}_{\boldsymbol{\xi}}(t,s)\mathbf{c}^T(s)\mathbf{c}(s)\boldsymbol{\theta}(s)ds = \lambda\boldsymbol{\theta}(t) \,. \tag{7.89}$$

Note that multiplying (7.89) on the left by $\mathbf{c}(t)$ and defining $\phi(t)$ through (7.88), we see that (7.87) holds whenever (7.89) is satisfied. Then by applying the differential operator L and the boundary operator B on the left to (7.89) and taking into account the Green's function property (7.85), we find that $\boldsymbol{\theta}(t)$ obeys the n-dimensional second-order differential equation

$$L\boldsymbol{\theta}(t) = \lambda^{-1}\mathbf{c}^T(t)\mathbf{c}(t)\boldsymbol{\theta}(t) \tag{7.90}$$

with boundary condition $B\boldsymbol{\theta} = 0$. In other words, if $\mu = \lambda^{-1}$ is a generalized eigenvalue of the $n \times n$ operator pencil formed by the pair $(L, \mathbf{c}^T\mathbf{c})$, then λ is an eigenvalue of K. Furthermore, the eigenfunction $\phi(t)$ corresponding to λ is expressed in terms of the generalized eigenfunction $\boldsymbol{\theta}(t)$ of $(L, \mathbf{c}^T\mathbf{c})$ as (7.88).

The extension of the above eigenfunction charaterization to the DT case is straightforward and is left as an exercise for the reader. Since the class of Markov processes is contained in the reciprocal class, the above discussion encompasses also the case where $\boldsymbol{\xi}(t)$ is Markov. However, one important feature of the technique described in [18] for the Markov case is that it does not require the invertibility of $\mathbf{Q}(t)$. As such, when $\mathbf{Q}(t)$ is not invertible, it is much easier to use than the characterization (7.90) which involves the mixed-order operator L. To describe it, note that when $\boldsymbol{\xi}(t)$ is Markov, it admits a first-order state-space model of the form

$$\dot{\boldsymbol{\xi}}(t) = \mathbf{A}(t)\boldsymbol{\xi}(t) + \mathbf{W}(t) \tag{7.91}$$

where $\mathbf{W}(t)$ is a zero-mean WGN with intensity $\mathbf{Q}(t)$, i.e.,

$$E[\mathbf{W}(t)\mathbf{W}^T(s)] = \mathbf{Q}(t)\delta(t-s)\,.$$

Then the matrix covariance function of $\boldsymbol{\xi}(t)$ is given by

$$\mathbf{K}_{\boldsymbol{\xi}}(t,s) = \begin{cases} \boldsymbol{\Psi}(t,s)\boldsymbol{\Pi}(s) & t \geq s \\ \boldsymbol{\Pi}(t)\boldsymbol{\Psi}^T(s,t) & s \geq t \end{cases} \tag{7.92}$$

where $\boldsymbol{\Psi}(t,s)$ denotes the state transition function corresponding to the dynamics (7.91) and $\boldsymbol{\Pi}(t) = E[\boldsymbol{\xi}(t)\boldsymbol{\xi}^T(t)]$ denotes the variance matrix of $\boldsymbol{\xi}(t)$. The transition function $\boldsymbol{\Psi}(t,s)$ satisfies the differential equation

$$\frac{d\boldsymbol{\Psi}}{dt}(t,s) = \mathbf{A}(t)\boldsymbol{\Psi}(t,s)$$

for $t \geq s$ with initial condition $\boldsymbol{\Psi}(s,s) = \mathbf{I}_n$, and $\boldsymbol{\Pi}(t)$ satisfies the matrix Lyapunov equation

$$\dot{\boldsymbol{\Pi}}(t) = \mathbf{A}(t)\boldsymbol{\Pi}(t) + \boldsymbol{\Pi}(t)\mathbf{A}^T(t) + \mathbf{Q}(t) \tag{7.93}$$

with initial condition $\boldsymbol{\Pi}(0) = \boldsymbol{\Pi}_0$, where $\boldsymbol{\Pi}_0$ denotes the variance matrix of the initial state $\boldsymbol{\xi}(0)$. As an aside, note that when $\mathbf{Q}(t)$ is invertible, the second-order matrix differential operator L is expressed in terms of the dynamics of the state-space model (7.91) as

$$L = -(\mathbf{I}_n\frac{d}{dt} + \mathbf{A}^T(t))\mathbf{Q}^{-1}(t)(\mathbf{I}_n\frac{d}{dt} - \mathbf{A}(t)) \tag{7.94}$$

and if $\mathbf{u}(t)$ denotes an n-dimensional vector function over $[0,T]$, the boundary operator $B\mathbf{u} = 0$ is expressed as the separable boundary conditions

$$\begin{aligned} \dot{\mathbf{u}}(0) - (\mathbf{A} + \mathbf{Q}\boldsymbol{\Pi}^{-1})(0)\mathbf{u}(0) &= \mathbf{0} \\ \dot{\mathbf{u}}(T) - \mathbf{A}(T)\mathbf{u}(T) &= \mathbf{0}\,. \end{aligned}$$

Then, by substituting expression (7.92) for the covariance $\mathbf{K}_{\boldsymbol{\xi}}(t,s)$ inside (7.89), denoting

$$\boldsymbol{\eta}(t) \triangleq \lambda^{-1} \int_t^T \boldsymbol{\Psi}^T(s,t)\mathbf{c}^T(s)\mathbf{c}(s)\boldsymbol{\theta}(s)ds \tag{7.95}$$

and differentiating, it is not difficult to verify that the pair $(\boldsymbol{\theta}(t), \boldsymbol{\eta}(t))$ obeys the $2n$-dimensional Hamiltonian system

$$\begin{bmatrix} \dot{\boldsymbol{\theta}}(t) \\ \dot{\boldsymbol{\eta}}(t) \end{bmatrix} = \begin{bmatrix} \mathbf{A}(t) & \mathbf{Q}(t) \\ -\lambda^{-1}\mathbf{c}^T(t)\mathbf{c}(t) & -\mathbf{A}^T(t) \end{bmatrix} \begin{bmatrix} \boldsymbol{\theta}(t) \\ \boldsymbol{\eta}(t) \end{bmatrix} \tag{7.96}$$

with boundary conditions

$$\boldsymbol{\eta}(0) = \boldsymbol{\Pi}_0^{-1}\boldsymbol{\theta}(0) \quad , \quad \boldsymbol{\eta}(T) = \mathbf{0} \,. \tag{7.97}$$

See Problem 7.9 for further details. Note that the system (7.96) does not require the invertibility of $\mathbf{Q}(t)$. But when $\mathbf{Q}(t)$ is invertible, eliminating $\boldsymbol{\eta}(t)$ from the above system and taking into account expression (7.94) for L gives the second-order system (7.90) for $\boldsymbol{\theta}(t)$, so that, as expected, the generalized eigenfunction characterizations (7.90) and (7.96) are completely consistent.

To summarize, we have shown in this section that when $X(t)$ is a partially observed Gaussian reciprocal or Markov process, the eigenvalues and eigenfunctions of its covariance function $K(t,s)$ can be evaluated by solving a two-point boundary value differential or difference system. If the process $\boldsymbol{\xi}(t)$ used to model $X(t)$ has dimension n, the system obtained is a second-order system of dimension n, and in the Markov case, a first-order system of dimension $2n$. It is worth noting, however, that the boundary conditions for the the system (7.90) involve the generalized eigenfunction $\boldsymbol{\theta}(t)$, not the eigenfunction $\phi(t)$ itself. Thus, although it is sometimes possible to use variable elimination to derive a differential equation of order $2n$ for $\phi(t)$, finding the boundary conditions it satisfies is essentially an impossible task. This is just an expression of the fact that for partially observed processes, the natural coordinate system for evaluating eigenfunctions is the one associated with $\boldsymbol{\theta}(t)$, not $\psi(t)$.

7.4.3 Rational Stationary Gaussian Processes

The class of stationary Gaussian processes with a rational power spectral density (PSD) has also the feature that its eigenfunctions admit a differential/difference characterization. Specifically, let $X(t)$ be a CT zero-mean process with covariance $K(t-s)$ and such that its PSD

$$\begin{aligned} S_X(\omega) &= \int_{-\infty}^{\infty} K(\tau)\exp(-j\omega\tau)d\tau \\ &= \frac{N(\omega^2)}{D(\omega^2)} \end{aligned} \tag{7.98}$$

is rational. This means that

$$D(x) = \sum_{j=0}^{n} d_j x^{n-j}$$

$$N(x) = \sum_{j=0}^{n} n_j x^{n-j} ,$$

with $d_0 = 1$, are two polynomials of degree n and of degree less than or equal to n, respectively. Then by substituting the spectral representation

$$K(t-s) = (2\pi)^{-1} \int_{-\infty}^{\infty} \frac{N(\omega^2)}{D(\omega^2)} \exp(j\omega(t-s))d\omega \tag{7.99}$$

in the eigenfunction equation (7.87), and applying the differential operator $D(-d^2/dt^2)$ to both sides of this equation, we obtain

$$\lambda D(-\frac{d^2}{dt^2})\phi(t) = \int_a^b [(2\pi)^{-1} \int_{-\infty}^{\infty} N(\omega^2) \exp(j\omega(t-s))d\omega]\phi(s)ds \, . \tag{7.100}$$

But

$$(2\pi)^{-1} \int_{-\infty}^{\infty} N(\omega^2) \exp(j\omega(t-s))d\omega = N(-\frac{d^2}{dt^2})\delta(t-s) \, , \tag{7.101}$$

so that by substituting (7.101) inside (7.100), we obtain the homogeneous differential equation

$$\lambda D(-\frac{d^2}{dt^2})\phi(t) = N(-\frac{d^2}{dt^2})\phi(t) \tag{7.102}$$

for the eigenfunction $\phi(t)$. For each $\lambda > 0$, the characteristic equation

$$\lambda D(-s^2) - N(-s^2) = 0$$

admits $2n$ roots s_i with $1 \leq i \leq 2n$, so if the roots are distinct an eigenfunction satisfying (7.102) will have the form

$$\phi(t) = \sum_{i=1}^{2n} A_i \exp(s_i t) \, . \tag{7.103}$$

Unfortunately, unlike the case of reciprocal processes, the BCs satisfied by $\phi(t)$ are not known, so that to find the $2n$ constants A_i with $1 \leq i \leq 2n$, we need to backsubstitute (7.103) inside the integral equation (7.87), which gives

$$\lambda \sum_{i=1}^{2n} A_i \exp(s_i t) = \sum_{i=1}^{2n} A_i \chi_i(t) \tag{7.104}$$

with

$$\chi_i(t) \triangleq \int_a^b K(t,u)\exp(s_i u)du\,.$$

Then the coefficients A_i are obtained by matching functions of t on both sides of (7.89). Unfortunately, the evaluation of the functions $\chi_i(t)$ can be rather tedious, so this approach should be avoided whenever the method of the previous section is applicable.

Example 7.5, continued

For the Ornstein-Uhlenbeck process of Example 7.5, the PSD (7.67) is rational with $D(x) = x + a^2$ and $N(x) = 2aP = q$ so that the eigenfunctions satisfy the differential equation

$$\lambda(-\frac{d^2}{dt^2} + a^2)\phi(t) - q\phi(t) = 0\,,$$

which matches the differential equation (7.69). In this case, if we express the solution of the differential equation (7.69) as

$$\phi(t) = A\cos(bt) + B\sin(bt)$$

for $0 \leq t \leq T$, after integration we find that the functions

$$\chi_1(t) \triangleq P\int_0^T \exp(-a|t-s|)\cos(bs)ds$$
$$\chi_2(t) \triangleq P\int_0^T \exp(-a|t-s|)\sin(bs)ds$$

can be expressed as

$$\chi_1(t) = \frac{P}{a^2+b^2}\big[2a\cos(bt) - a\exp(-at) + \exp(-a(T-t))(-a\cos(bT) + b\sin(bT))\big]$$

$$\chi_2(t) = \frac{P}{a^2+b^2}\big[2a\sin(bt) + b\exp(-at) + \exp(-a(T-t))(-a\sin(bT) - b\cos(bT))\big]\,.$$

Then the eigenfunction equation (7.22) yields

$$\lambda\big(A\cos(bt) + A\sin(bt)\big) = A\chi_1(t) + B\chi_2(t)\,,$$

and by matching the coefficients of $\cos(bt)$, $\sin(bt)$, $\exp(-at)$ and $\exp(-a(T-t))$ on both sides of this equation, we obtain

$$\lambda = \frac{q}{a^2+b^2}$$

and the matrix equation (7.70) for A and B, so, as expected, the solution obtained by this technique matches the one obtained in the last section by exploiting the Markov property of $X(t)$. □

Finally, it is worth noting that the class of rational stationary Gaussian processes discussed in this section represents the stationary form of the partially observed Gauss-Markov processes of the previous section. Specifically, consider a process $X(t)$ with PSD (7.98). Since the PSD is rational, it admits a spectral factorization [20, Sec. 6.4] of the form

$$S_X(\omega) = H(j\omega)H(-j\omega)$$

where the scalar transfer function $H(s)$ is rational, stable and of degree n. It can therefore be realized in state-space form as

$$H(s) = \mathbf{c}(s\mathbf{I}_n - \mathbf{A})^{-1}\mathbf{b}$$

with a stable matrix $\mathbf{A}$. This indicates that $X(t)$ is expressible as

$$X(t) = \mathbf{c}(t)\boldsymbol{\xi}(t)\,,$$

where $\boldsymbol{\xi}(t)$ is a stationary Gauss-Markov process with state-space realization

$$\dot{\boldsymbol{\xi}}(t) = \mathbf{A}\boldsymbol{\xi}(t) + \mathbf{W}(t)$$

where $\mathbf{W}(t)$ is a zero-mean WGN with intensity $\mathbf{Q} = \mathbf{b}\mathbf{b}^T$. Note that in this case the matrix $\mathbf{Q}$ has rank one, so it is not invertible. In this context, while we have been able to show that the eigenfunctions satisfy a differential equation of order $2n$, it is not surprising that the boundary conditions satisfied by the eigenfunctions are unknown, since as observed at the end of last section. the proper coordinate system for evaluating the eigenfunctions is really provided by the Hamiltonian system (7.96).

7.5 Karhunen-Loève Decomposition

We are now ready to prove the main result of this chapter.

Karhunen-Loève decomposition: Consider a zero-mean Gaussian process $X(t)$ defined over $[a, b]$ with covariance kernel $K(t, s)$. In the CT case we assume that $K(t, s)$ is continuous over $[a, b]^2$, so that $X(t)$ is mean-square continuous over $[a, b]$.

(a) If the eigenvalues and eigenfunctions of the kernel $K(t, s)$ are $(\lambda_k, \phi_k(t))$ with $k \geq 1$ and if

$$\begin{aligned} X_k = \langle X(t), \phi_k(t) \rangle &= \int_a^b X(t)\phi_k(t)dt \quad \text{(CT case)} \\ &= \sum_a^b X(t)\phi_k(t) \quad \text{(DT case)}\,, \end{aligned} \tag{7.105}$$

the coefficients X_k are zero-mean Gaussian with covariance

$$E[X_k X_\ell] = \lambda_k \delta_{k\ell} , \tag{7.106}$$

so that $\{X_k,\ k \geq 1\}$ is a zero-mean time-varying WGN sequence. Furthermore, we have

$$X(t) = \lim_{k \to \infty} \sum_{j=1}^{k} X_j \phi_j(t) \tag{7.107}$$

uniformly for $a \leq t \leq b$, where the limit on the right-hand side of (7.107) is taken in the mean-square sense.

(b) Conversely, if $X(t)$ admits an expansion of the form (7.107) where

$$\begin{aligned} < \phi_k, \phi_\ell > &= \delta_{k\ell} \\ E[X_k X_\ell] &= \lambda_k \delta_{k\ell} , \end{aligned} \tag{7.108}$$

then the pairs (λ_k, ϕ_k) are the eigenvalues and eigenfunctions of $K(t, s)$.

Proof: (a) By direct evaluation, we find

$$E[(X(t) - \sum_{j=1}^{k} X_j \phi_j(t))^2] = K(t,t) - \sum_{j=1}^{k} \lambda_j \phi_j^2(t) ,$$

which tends to zero as $k \to \infty$ uniformly in t according to Mercer's theorem. To prove (b), suppose the expansion (7.75) holds. Then since the coefficients X_k are independent, we have

$$K(t,s) = E[X(t)X(s)] = \sum_{j=1}^{\infty} \lambda_j \phi_j(t) \phi_j(s) .$$

This implies

$$K\phi_k = \sum_{j=1}^{\infty} \lambda_j \phi_j < \phi_j, \phi_k >= \lambda_k \phi_k ,$$

so that ϕ_k is an eigenfunction of K corresponding to the eigenvalue λ_k.

As Examples 7.4–7.6 indicate, one constraint of the KLD, when compared to the expansion of a deterministic function with respect to an orthonormal basis, is that here the basis $\{\phi_k,\ k \geq 1\}$ is not arbitrary and is formed by the eigenfunctions of the kernel $K(t, s)$. This is somewhat constraining since given a Gaussian detection problem, it often means that there is only one basis in which the problem can be represented. However, one major exception is the case of WGN, since as we shall see below, like deterministic functions, it remains white and Gaussian with respect to *any orthonormal basis.* This property is the extension to CT processes of the property of $N(\mathbf{0}, q\mathbf{I}_N)$ white Gaussian noise vectors of $\mathbb{R}^N$ that we have used repeatedly in earlier chapters, namely that these processes remain WGN with respect to any orthonormal basis of $\mathbb{R}^N$.

Example 7.8: White Gaussian noise

Consider a CT zero-mean WGN process $X(t)$ over a finite interval $[0, T]$. It can be viewed as the formal derivative of the Wiener process and its covariance kernel is given by

$$E[X(t)X(s)] = K(t, s) = \sigma^2 \delta(t - s)$$

where $\sigma^2 > 0$ is the noise intensity. Note that the Dirac impulse $\delta(t - s)$ is not a standard function, and thus not continuous, so that strictly-speaking the KLD is not applicable to this process. However the kernel $q\delta(t - s)$ has the feature that

$$\sigma^2 \int_0^T \delta(t - s) f(s) ds = \sigma^2 f(t)$$

for any function $f(t)$, so that any square integrable function is an eigenfunction of this kernel, and all eigenvalues are equal to σ^2. So in this particular case, we can perform a "KLD" of $X(t)$ with respect to *any* complete orthonormal basis $\{\phi_k,\ k \geq 1\}$ of $L^2[0, T]$ and in the Karhunen-Loève expansion

$$X(t) = \sum_{k=1}^{\infty} X_k \phi_k(t) ,$$

the coefficients $\{X_k,\ k \geq 1\}$ form a DT WGN sequence with variance σ^2. □

7.6 Asymptotic Expansion of Stationary Gaussian Processes

The eigenvalues and eigenfunctions of a stationary CT Gaussian process take a particularly simple form as the interval of observation becomes large. Let $X(t)$ be a zero-mean stationary Gaussian process observed over $[-T/2, T/2]$ with covariance function $E[X(t)X(s)] = K(t - s)$ and PSD

$$S_X(\omega) = \int_{-\infty}^{\infty} K(\tau) \exp(-j\omega\tau) d\tau . \qquad (7.109)$$

We assume that the length T of the interval of observation is much longer than the coherence time T_c of $X(t)$. Let $\omega_0 = 2\pi/T$ denote the fundamental frequency. Then if $\omega_k = k\omega_0$ with k integer, we use

$$\phi_k(t) = T^{-1/2} \exp(-j\omega_k t) \qquad (7.110)$$

as trial solution in the eigenfunction equation (7.22). This gives

$$\begin{aligned}
\lambda_k \phi_k(t) &= \int_{-T/2}^{T/2} K(t-s)\phi_k(s)ds \\
&= \int_{-T/2}^{T/2} [(2\pi)^{-1} \int_{-\infty}^{\infty} S_X(\omega) \exp(j\omega(t-s))d\omega] \phi_k(s)ds \\
&= \int_{-\infty}^{\infty} S_X(\omega) \frac{\sin((\omega-\omega_k)T/2)}{(\omega-\omega_k)\pi} T^{-1/2} \exp(j\omega t) d\omega \,. \qquad (7.111)
\end{aligned}$$

An interesting feature of the function

$$\frac{2\sin((\omega-\omega_k)T/2)}{(\omega-\omega_k)\pi}$$

appearing in the integral is that its spectral width about the center frequency ω_k is approximately $\omega_0 = 2\pi/T$ and its integral is 1, so as $T \to \infty$ it converges to a unit impulse located at ω_k. Since $T \gg T_c$, the PSD $S_X(\omega)$ is approximately constant over the width ω_0 of the impulse, so the right-hand side of identity (7.111) is approximately equal to $S_X(\omega_k)\phi_k(t)$. This implies that for a stationary Gaussian process observed over an interval much longer than the coherence time of the process, the eigenfunctions coincide with the usual complex exponential Fourier series functions, and the eigenvalues

$$\lambda_k = S_X(\omega_k) = S_X(k\omega_0) \qquad (7.112)$$

are just the values of the power spectral density function sampled at the Fourier harmonics $k\omega_0$.

7.7 Bibliographical Notes

The KLD was discovered independently by Karhunen [21] and Loève [22, 23]. The presentation we have adopted here is based loosely on Section 3.4 of [24]. The differential equations method for computing the eigenfunctions of a covariance kernels is described in detail in [3] and [25]. However, the characterization of the boundary conditions of the eigenfunctions of reciprocal covariance kernels is more recent and relies on [10–12]. In addition to its applications to detection, the KLD plays also a major role in source coding [26], particularly image compression, due to the fact that it decomposes a process in uncorrelated components that can be compressed and transmitted separately, so that transmission redundancy is eliminated. Theoretically, the KLD is restricted to Gaussian processes, and when applied to nonGaussian processes, it decorrelates the coefficients, but higher order moments may be nonzero. In such cases, there is no guarantee that the KLD minimizes the statistical dependence of the transform coefficients, and an example where the KLD is not useful is presented in [27].

7.8 Problems

7.1. The Brownian bridge process $X(t)$ is a zero-mean CT Gaussian process defined over a finite interval $[0, T]$. Its covariance function is given by

$$K(t, s) = q[\min(t, s) - \frac{ts}{T}]$$

where $q > 0$. It derives its name from the fact that it can be constructed from a Wiener process $W(t)$ with intensity q as

$$X(t) = W(t) - \frac{t}{T} W(T) .$$

From this expression we see that $X(0) = X(T) = 0$, so the Brownian bridge can be viewed as a Wiener process whose sample paths are constrained to pass through the origin at time $t = T$, hence the trajectory between 0 and T forms a "bridge" between zero values at both ends.

(a) Verify that $X(t)$ is a Markov process.
(b) Verify that the second-order Sturm-Liouville operator corresponding to $K(t, s)$ is

$$L = -q^{-1} \frac{d^2}{dt^2}$$

with the boundary conditions

$$u(0) = u(T) = 0 .$$

(c) Evaluate the eigenvalues and eigenfunctions of the covariance function $K(t, s)$.

7.2. The CT hyperbolic cosine process $X(t)$ is a zero-mean stationary Gaussian process defined over a finite interval $[0, T]$ with covariance function

$$K(t - s) = P \frac{\cosh(a(T/2 - |t - s|))}{\cosh(aT/2)} .$$

This process is reciprocal and since $K(0) = K(T)$, it is periodic, i.e.

$$X(0) = X(T) .$$

(a) Use (7.56) and (7.61) to find the second-order Sturm-Liouville operator L and boundary conditions corresponding to $K(t, s)$.
(b) Find the eigenvalues and eigenfunctions of $K(t, s)$.

7.3. The CT hyperbolic sine process $X(t)$ is a zero-mean stationary Gaussian process defined over a finite interval $[0, T]$ with covariance function

$$K(t - s) = P \frac{\sinh(a(T/2 - |t - s|))}{\sinh(aT/2)} .$$

This process is reciprocal and since $K(0) = -K(T)$, it is skew-periodic, i.e.

$$X(0) = -X(T) .$$

(a) Use (7.56) and (7.61) to find the second-order Sturm-Liouville operator L and boundary conditions corresponding to $K(t,s)$.
(b) Find the eigenvalues and eigenfunctions of $K(t,s)$.

7.4. The CT shifted sine process $X(t)$ is a zero-mean stationary Gaussian process defined over a finite interval $[0,T]$ with covariance function

$$K(t-s) = P\frac{\sin(a(T/2 - |t-s|))}{\sin(aT/2)} .$$

This process is reciprocal and since $X(0) = -X(T)$, it is skew-periodic, so $X(T) = -X(0)$.

(a) Use (7.56) and (7.61) to find the second-order Sturm-Liouville operator L and boundary conditions corresponding to $K(t,s)$.
(b) Find the eigenvalues and eigenfunctions of $K(t,s)$.

7.5. The DT Wiener process $X(t)$ is a zero-mean Gaussian process obtained by passing a zero-mean WGN $V(t)$ with intensity q through a discrete-time accumulator, i.e.,

$$X(t) = \sum_{s=0}^{t-1} V(s)$$

for $t \geq 1$

(a) Verify that $X(t)$ obeys the first-order recursion

$$X(t+1) = X(t) + V(t)$$

with initial condition $X(0) = 0$ so that $X(t)$ is a DT Gauss-Markov process.
(b) Show that the covariance kernel of $X(t)$ is

$$K(t,s) = q \min(t,s) .$$

(c) Let $\mathbf{K}$ denote the $T \times T$ covariance matrix of $X(t)$ over interval $[1,T]$. Verify that its inverse has a tridiagonal structure and use this result to show that $K(t,s)$ satisfies the second-order recursion

$$MK(t,s) = \delta(t-s)$$

with

$$M = q^{-1}[2 - Z - Z^{-1}]$$

where $Zf(t) = f(t+1)$ denotes the forward shift operator, and the boundary conditions

$$K(0, s) = 0$$

and

$$\frac{1}{q}[K(T, s) - K(T-1, s)] = \delta(T - s) .$$

(d) Use the second-order difference operator obtained in c) to evaluate the eigenvalues and eigenfunctions of $K(t, s)$.

7.6. The DT Slepian process is a zero-mean stationary Gaussian process with covariance function

$$K(t - s) = P[1 - 2\frac{|t-s|}{T}] .$$

This process is reciprocal and since $K(T) = -K(0)$, it is skew-periodic, i.e.,

$$X(0) = -X(T) .$$

(a) Verify that $K(t - s)$ satisfies the second-order recursion

$$MK(t, s) = \delta(t - s)$$

with

$$M = q^{-1}[2 - Z - Z^{-1}]$$

where $q = 4P/T$ and $Zf(t) = f(t+1)$ denotes the forward shift operator. Show that the boundary conditions

$$K(0, s) + K(T, s) = 0$$

and

$$q^{-1}[2K(0, s) - K(1, s) + K(T-1, s)] = \delta(s)$$

are satisfied.

(b) Use the second-order difference operator obtained in a) to evaluate the eigenvalues and eigenfunctions of $K(t, s)$.

7.7. The DT Ornstein-Uhlenbeck process $X(t)$ is a zero-mean stationary Gaussian process with covariance kernel

$$K(t - s) = Pa^{|t-s|} ,$$

where $|a| < 1$. This process is Markov and can be synthesized by the first order recursion

$$X(t + 1) = aX(t) + V(t)$$

where $V(t)$ is a zero-mean WGN with intensity $q = P(1 - a^2)$. The initial condition $X(0)$ is $N(0, P)$ distributed and is independent of the driving noise $V(t)$.

(a) Consider an interval $[0, T]$ and let $N = T + 1$. Denote by $\mathbf{K}$ the $N \times N$ covariance matrix of the process $X(t)$ for $0 \leq t \leq T$. Verify that $\mathbf{K}^{-1}$ has a tridiagonal structure and by reading off the coefficients of the three central diagonals of this metrix, show that $K(t, s)$ admits the second-order recursion

$$MK(t, s) = \delta(t - s)$$

with

$$M = q^{-1}[(1 + a^2) - aZ - aZ^{-1}],$$

where Z denotes the forward shift operator defined by $Zf(t) = f(t + 1)$. What are the boundary conditions for this recursion? *Hint:* Look at the first and last row of $\mathbf{K}^{-1}$.

(b) By using a trial solution of the form

$$\phi(t) = A\cos(b(t - T/2)) + B\sin(b(t - T/2))$$

find the eigenvalues and eigenfunctions of $K(t, s)$.

7.8. The DT shifted sine process is a zero-mean stationary Gaussian process defined over a finite interval $[0, T]$ with covariance function

$$K(t - s) = P\frac{\sin(\alpha(T/2 - |t - s|))}{\sin(\alpha T/2)},$$

where $\alpha < \pi/T$. This process is reciprocal and skew periodic since $X(T) = -X(0)$.

(a) Verify that $K(t - s)$ satisfies the second-order recursion

$$MK(t, s) = \delta(t - s)$$

with

$$M = q^{-1}[2\cos(\alpha) - (Z + Z^{-1})],$$

where

$$q = 2P\sin(\alpha)\cot(\alpha T/2)$$

and Z denotes the forward shift operator. Check that the boundary conditions

$$K(0, s) + K(T, s) = 0$$

and

$$q^{-1}[2\cos(\alpha)K(0, s) - K(1, s) + K(T - 1, s)] = \delta(s)$$

hold.

(b) Use the second-order difference operator obtained in a) to evaluate the eigenvalues and eigenfunctions of $K(t, s)$.

7.9. Consider a zero-mean stationary Gauss-Markov process $X(t)$ with covariance

$$K(t,s) = P\exp(-a|t-s|)\cos(b(t-s))$$

with $a > 0$.

(a) Consider the Markov process

$$\boldsymbol{\xi}(t) = \begin{bmatrix} \xi_1(t) \\ \xi_2(t) \end{bmatrix}$$

specified by the state-space model

$$\dot{\boldsymbol{\xi}}(t) = \begin{bmatrix} -a & -b \\ b & -a \end{bmatrix} \boldsymbol{\xi}(t) + \mathbf{W}(t)$$

with

$$\mathbf{W} = \begin{bmatrix} W_1(t) \\ W_2(t) \end{bmatrix}$$

where $W_1(t)$ and $W_2(t)$ are two independent zero-mean WGNs with intensity $q = 2aP$. The initial state $\boldsymbol{\xi}(0)$ is $N(\mathbf{0}, P\mathbf{I}_2)$ distributed and independent of $\mathbf{W}(t)$. Verify that the transition matrix $\boldsymbol{\Psi}(t,s)$ satisfying

$$\frac{d\boldsymbol{\Psi}}{dt}(t,s) = \begin{bmatrix} -a & -b \\ b & -a \end{bmatrix} \boldsymbol{\Psi}(t,s)$$

for $t \geq s$, with initial condition $\boldsymbol{\Psi}(s,s) = \mathbf{I}_2$ is given by

$$\boldsymbol{\Psi}(t,s) = \exp(-a(t-s)) \begin{bmatrix} \cos(b(t-s)) & -\sin(b(t-s)) \\ \sin(b(t-s)) & \cos(b(t-s)) \end{bmatrix} .$$

Observe also that the process $\boldsymbol{\xi}(t)$ is stationary with variance $\boldsymbol{\Pi}(t) = P\mathbf{I}_2$. Find the covariance function

$$K_{\boldsymbol{\xi}}(t,s) = E[\boldsymbol{\xi}(t)\boldsymbol{\xi}^T(s)]$$

of $\boldsymbol{\xi}(t)$ and conclude that the process $X(t)$ can be expressed in terms of $\boldsymbol{\xi}(t)$ as

$$X(t) = \begin{bmatrix} 1 & 0 \end{bmatrix} \boldsymbol{\xi}(t) ,$$

so it coincides with the first element of $\boldsymbol{\xi}(t)$.

(b) Obtain a first-order Hamiltonian system of the form (7.96)–(7.97) for the eigenfunctions $\boldsymbol{\theta}(t)$ of the matrix covariance kernel $K_{\boldsymbol{\xi}}(t,s)$.

(c) By eliminating the auxiliary vector $\boldsymbol{\eta}(t)$ defined by (7.95) from the Hamiltonian system of part (b), obtain a second-order differential system of the form (7.90) for the eigenfunctions $\boldsymbol{\theta}(t)$. Specify the boundary conditions for this system.

(d) Use the differential system of part (c) to obtain a fourth-order differential system for the eigenfunctions

$$\phi(t) = \begin{bmatrix} 1 \ 0 \end{bmatrix} \boldsymbol{\theta}(t)$$

of covariance function $K(t,s)$. Find the corresponding boundary conditions.

7.10. Consider again the zero-mean stationary Gauss-Markov process $X(t)$ with covariance

$$K(t,s) = P \exp(-a|t-s|) \cos(b(t-s)) \,, \ a > 0$$

of Problem 7.9.

(a) By using the Fourier transform pair

$$f(t) = \exp(-a|t|) \quad \longleftrightarrow \quad F(j\omega) = \frac{2a}{a^2 + \omega^2}$$

and the modulation property of the Fourier transform, evaluate the power spectral density $S_X(\omega)$ of $X(t)$.

(b) Write $S_X(\omega)$ in the form

$$S_X(\omega) = \frac{N(\omega^2)}{D(\omega^2)}$$

where $N(x)$ and $D(x)$ are two scalar polynomials, and obtain a differential equation of the form (7.102) for the eigenfunctions $\phi(t)$ of $K(t,s)$.

(c) Verify that the differential equation obtained in (b) coincides with the one obtained in Problem 7.9.

7.11. Consider a zero-mean bandlimited white Gaussian noise process $X(t)$ with intensity σ^2 and bandwidth B measured in Hertz. Its power spectral density can therefore be expressed as

$$S_X(\omega) = \begin{cases} \sigma^2 & |\omega| \leq 2\pi B \\ 0 & \text{otherwise} . \end{cases}$$

Its autocovariance is given by

$$K_X(\tau) = E[X(t+\tau)X(t)] = P \frac{\sin(2\pi B \tau)}{2\pi B \tau}$$

where $P = 2\sigma^2 B$ is the total signal power. We seek to construct a Karhunen-Loève expansion for this process defined over a finite interval $[-T/2, T/2]$. The eigenvalue/eigenfunction equation for its covariance function takes the form

$$P \int_{-T/2}^{T/2} \frac{\sin(2\pi B(t-s))}{2\pi B(t-s)} \phi(s) ds = \lambda \phi(t)$$

where $-T/2 \leq t \leq T/2$.

(a) By performing the change of variable $t = xT/2$ and $s = yT/2$, verify that this equation can be rewritten as

$$\int_{-1}^{1} \frac{\sin(c(x-y))}{\pi(x-y)} \psi(y) dy = \mu \psi(x)$$

where $-1 \leq x \leq 1$. Express (i) $\psi(x)$ in terms of $\phi(\cdot)$ and T, (ii) c in terms of B and T, and (iii) μ in terms of λ and σ^2.

(b) The integral operator K defined by

$$(K\psi)(x) = \int_{-1}^{1} \frac{\sin(c(x-y))}{\pi(x-y)} \psi(y) dy$$

has the interesting property [28] that it commutes with the differential operator L_x given by

$$L_x \psi = \frac{d}{dx}((1-x^2)\frac{d\psi}{dx}) - c^2 x^2 \psi .$$

Specifically, we have

$$L_x \int_{-1}^{1} \frac{\sin(c(x-y))}{\pi(x-y)} \psi(y) dy = \int_{-1}^{1} \frac{\sin(c(x-y))}{\pi(x-y)} L_y \psi(y) dy .$$

Verify that this implies that the eigenfunctions $\psi_n(x)$ of K, with $n = 1, 2, 3, \cdots$ are the *prolate spheroidal wave functions* which obey the differential eigenvalue equation

$$L_x \psi_n(x) = \kappa_n \psi_n(x) .$$

To prove this result, you may assume that all the eigenvalues of K are distinct. Are the eigenvalues κ_n of L_x related to the eigenvalues μ_n of K? By studying the prolate spheroidal functions, it can be shown that if we order the eigenvalues μ_n of K in decreasing order, i.e., $\mu_1 > \mu_2 > \mu_3 > \cdots$, then the eigenvalues μ_n are close to 1 for $n \ll 2c/\pi$, close to zero for $n \gg 2c/\pi$, and approxmately equal to $1/2$ with n in the vicinity of $2c/\pi$. Consequently $\lceil 2c/\pi \rceil$, the smallest integer greater or equal to $2c/\pi$, can be viewed as an estimate of the number of nonzero random variables X_n that need to be retained in the Karhunen-Loève expansion

$$X(t) = \sum_{n=1}^{\infty} X_n \phi_n(t) \tag{7.113}$$

of $X(t)$ over $-T/2, T/2$. Recall that assuming the eigenfunctions have been normalized to have unit energy, the random variables

$$X_n = \int_{-T/2}^{T/2} X(t) \phi(t) dt$$

are independent zero mean Gaussian with variance λ_n, i.e.,

$$E[X_n X_m] = \lambda_n \delta_{nm} .$$

Consequently, random variables with a small variance λ_n can be dropped from the sum (7.113). Compare the number of random variables needed to represent $X(t)$ with the number of samples that would be needed to reconstruct a deterministic signal of bandwidth B (in Hertz) and duration T, assuming it is sampled at the Nyquist rate. Are these results consistent?

References

1. W. B. Davenport, Jr. and W. L. Root, *An Introduction to the Theory of Random Signals and Noise.* New York: McGraw-Hill, 1958. Reprinted by IEEE Press, New York, 1987.
2. D. Middleton, *An Introduction to Satistical Communication Theory.* New York: McGraw-Hill, 1960. Reprinted by IEEE Press, New York, 1996.
3. H. L. Van Trees, *Detection, Estimation and Modulation Theory, Part I: Detection, Estimation and Linear Modulation Theory.* New York: J. Wiley & Sons, 1968. Paperback reprint edition in 2001.
4. J. M. Wozencraft and I. M. Jacobs, *Principles of Communication Engineering.* New York: J. Wiley & Sons, 1965. Reprinted by Waveland Press, Prospect Heights, IL, 1990.
5. N. Young, *An Introduction to Hilbert Space.* Cambridge, UK: Cambridge Univ. Press, 1988.
6. D. Bertsekas, *Nonlinear Programming, Second Edition.* Belmont, MA: Athena Scientific, 1999.
7. L. N. Trefethen and D. Bau, III, *Numerical Linear Algebra.* Philadelphia, PA: Soc. for Industrial and Applied Math., 1997.
8. B. Jamison, "Reciprocal processes," *Z. Wahrscheinlichkeitstheorie verw. Gebiete*, vol. 30, pp. 65–86, 1974.
9. P. Lévy, "A special problem of Brownian motion and a general theory of Gaussian random functions," in *Proc. 3rd Berkeley Symposium on Math. Statistics and Probability*, vol. 2, (Berkeley, CA), pp. 133–175, Univ. California Press, 1956.
10. A. J. Krener, R. Frezza, and B. C. Levy, "Gaussian reciprocal processes and self-adjoint stochastic differential equations of second order," *Stochastics and Stochastics Reports*, vol. 34, pp. 29–56, 1991.
11. B. C. Levy, R. Frezza, and A. J. Krener, "Modeling and estimation of discrete-time Gaussian reciprocal processes," *IEEE Trans. Automatic Control*, vol. 35, pp. 1013–1023, Sept. 1990.
12. J. M. Coleman, B. C. Levy, and A. J. Krener, "Gaussian reciprocal diffusions and positive definite Sturm-Liouville operators," *Stochastics and Stochastics Reports*, vol. 55, pp. 279–313, 1995.
13. S. Karlin and H. M. Taylor, *A First Course in Stochastic Processes.* New York: Academic Press, 1975.
14. D. Slepian, "First passage times for a particular Gaussian processes," *Annals Math. Statistics*, vol. 32, pp. 610–612, 1961.

15. C. Van Loan, *Computational Frameworks for the Fast Fourier Transform.* Philadelphia, PA: Soc. for Industrial and Applied Math., 1992.
16. W. D. Ray and R. M. Driver, "Further decomposition of the Karhunen-Loève series representation of a stationary random process," *IEEE Trans. Informat. Theory*, vol. 16, Nov. 1970.
17. N. Ahmed, T. Natarajan, and K. R. Rao, "Discrete cosine transform," *IEEE Trans. Computers*, vol. 23, pp. 90–93, 1994.
18. A. B. Baggeroer, "A state-variable approach to the solution of Fredholm integral equations," *IEEE Trans. Informat. Theory*, vol. 15, pp. 557–570, Sept. 1969.
19. R. Frezza, *Models of higher-order and mixed-order Gaussian reciprocal processes with application to the smoothing problem.* PhD thesis, Univ. California, Davis, Grad. Program in Applied Math., 1990.
20. T. Kailath, A. H. Sayed, and B. Hassibi, *Linear Estimation.* Upper Saddle River, NJ: Prentice Hall, 2000.
21. K. Karhunen, "Uber lineare methoden in der wahrcheinlichkeitsrechnung," *Ann. Acad. Sci. Sci. Fenn., Ser. A, I, Math. Phys.*, vol. 37, pp. 3–79, 1947.
22. M. Loève, "Sur les fonctions aléatoires stationnaires du second ordre," *Revue Scientifique*, vol. 83, pp. 297–310, 1945.
23. M. Loève, *Probability Theory, 3rd edition.* Princeton, NJ: Van Nostrand, 1963.
24. E. Wong and B. Hajek, *Stochastic Processes in Engineering Systems.* New York: Springer-Verlag, 1985.
25. C. W. Helstrom, *Elements of Signal Detection & Estimation.* Upper Saddle River, NJ: Prentice-Hall, 1995.
26. N. S. Jayant and P. Noll, *Digital Coding of Waveforms – Principles and Applications to Speech and Video.* Englewood Cliffs, NJ: Prentice-Hall, 1984.
27. N. Saito, "The generalized spike process, sparsity, and statistical independence," in *Modern Signal Processing* (D. Rockmore and D. Healy, Jr., eds.), Cambridge, UK: Cambridge Univ. Press, 2004.
28. D. Slepian, "Some comments on Fourier analysis, uncertainty and modeling," *SIAM Review*, vol. 25, pp. 379–393, July 1983.

8

Detection of Known Signals in Gaussian Noise

8.1 Introduction

In this chapter, we consider the detection of known signals in white or colored Gaussian noise. This problem is of course central to radar, sonar, and digital communication systems analysis. For the case of binary detection in WGN, the optimum receiver is the matched filter which was discovered independently by North, and by Van Vleck and Middleton [1, 2] during World War II. An equivalent implementation is provided by the integrator/correlator receiver. Both of these structures are described in Section 8.2, where we examine the binary detection problem for known signals in WGN. Section 8.3 considers the case of M-ary detection in WGN. The receivers obtained for this problem are applicable to higher-order digital modulation schemes such as M-ary pulse-amplitude modulation, M-ary phase shift keying, or M-QAM modulation. Finally, in Section 8.4, we consider the binary detection of known signals in colored Gaussian noise. Two receiver structures are discussed: one based on the use of a generalized integrator/correlator or matched filter structure that replaces the transmitted signal by a distorted signal which compensates for the effect of noise coloring. An alternate receiver structure passes the observations through a noise whitening filter, yielding an output to which the standard integrator/correlator receiver is applicable. Of course, the assumption that the signals to be detected are known exactly is not completely realistic. For the case of digital communication systems, this corresponds to the case of coherent detection, where all the signal parameters, such as timing, phase, and amplitude have been acquired through the use of timing recovery, phase-lock loop (PLL), or automatic gain controller blocks. Since the inclusion of such blocks increases the complexity and thus the cost of communications receivers, incoherent detection methods, where some of the above mentioned parameters are not known, will be considered in the next chapter.

B.C. Levy, *Principles of Signal Detection and Parameter Estimation*,
DOI: 10.1007/978-0-387-76544-0_8, © Springer Science+Business Media, LLC 2008

8.2 Binary Detection of Known Signals in WGN

8.2.1 Detection of a Single Signal

Consider the prototype binary detection problem where we are trying to detect the presence or absence of a known signal $s(t)$ in WGN over an interval $0 \leq t \leq T$, so that the two hypotheses are given by

$$\begin{aligned} H_0 &: Y(t) = V(t) \\ H_1 &: Y(t) = s(t) + V(t) \, . \end{aligned} \tag{8.1}$$

Here $V(t)$ is a zero-mean WGN with covariance

$$E[V(t)V(s)] = \sigma^2 \delta(t-s) \, , \tag{8.2}$$

where σ^2 denotes the power density of intensity of the noise. For the above problem, the time t can be either continuous or discrete, so that in the DT case, the impulse $\delta(t-s)$ is to be interpreted as a Kronecker delta function, whereas in the CT case, it is a Dirac impulse. To formulate this detection problem in signal space, we construct a complete orthonormal basis $\{\phi_j(t), \ j \in \mathcal{J}\}$ of the space $L^2[0,T]$ of square summable or integrable functions over $[0,T]$. In the DT case, the index set $\mathcal{J} = \{1, \ 2, \ \ldots, \ T+1\}$ is finite, and in the CT case $\mathcal{J} = \{j \geq 1\}$ is the set of positive integers, which is infinite.

As first element of the basis, we select the function

$$\phi_1(t) = s(t)/E^{1/2} \, ,$$

where

$$\begin{aligned} E = ||s||^2 &= \int_0^T s^2(t)dt \quad \text{(CT case)} \\ &= \sum_{t=0}^{T} s^2(t) \quad \text{(DT case)} \, . \end{aligned} \tag{8.3}$$

The subsequent basis functions $\phi_j(t)$ with $j \geq 2$ are constructed by Gram-Schmidt orthonormalization of the functions in $L^2[0,T]$. Because the only requirement is that these functions should be orthogonal to ϕ_1, they are not uniquely defined, but as we shall see below, this does not matter as they will ultimately play no role in the detection problem.

Once a COB $\{\phi_j, \ j \in \mathcal{J}\}$ has been identified, since $V(t)$ is WGN, it admits a KLD

$$V(t) = \sum_{j \in \mathcal{J}} V_j \phi_j(t) \tag{8.4}$$

with respect to this basis, where the coefficients

$$V_j = < V, \phi_j > = \int_0^T V(t)\phi_j(t)dt \text{ (CT case)}$$
$$= \sum_{t=0}^{T} V(t)\phi_j(t) \text{ (DT case)} \tag{8.5}$$

are independent $N(0, \sigma^2)$ distributed, so the coefficient process $\{V_j, j \in \mathcal{J}\}$ is a DT WGN with variance σ^2. Similarly, we can expand the observations as

$$Y(t) = \sum_{j \in \mathcal{J}} Y_j \phi_j(t) \tag{8.6}$$

with

$$Y_j = < Y, \phi_j > = \int_0^T Y(t)\phi_j(t)dt \text{ (CT case)}$$
$$= \sum_{t=0}^{T} Y(t)\phi_j(t) \text{ (DT case)}. \tag{8.7}$$

Then, in the Karhunen-Loève/signal space domain, the detection problem (8.1) becomes

$$\begin{aligned} H_0 &: Y_j = V_j, \ j \in \mathcal{J} \\ H_1 &: \begin{cases} Y_1 = E^{1/2} + V_1 \\ Y_j = V_j, \ j \geq 2. \end{cases} \end{aligned} \tag{8.8}$$

The transformed problem has the feature that $Y_j = V_j$ for $j \geq 2$ under both hypotheses, and since $\{V_j, \ j \in \mathcal{J}\}$ is WGN, the noises V_j for $j \geq 2$ contain no information about V_1. This means that in (8.8) we can discard all observations Y_j with $j \geq 2$, since they are noninformative, and thus in signal space, the detection problem reduces to the following *one-dimensional* problem:

$$\begin{aligned} H_0 &: Y_1 = V_1 \\ H_1 &: Y_1 = E^{1/2} + V_1 \end{aligned} \tag{8.9}$$

where $V_1 \sim N(0, \sigma^2)$.

This problem has been examined in Example 2.1. The LRT can be written as

$$L(y_1) = \frac{f(y_1|H_1)}{f(y_1|H_0)}$$
$$= \frac{\exp(-\frac{1}{2\sigma^2}(y_1 - E^{1/2})^2)}{\exp(-\frac{1}{2\sigma^2}y_1^2)} \underset{H_0}{\overset{H_1}{\gtrless}} \tau, \tag{8.10}$$

where in the Bayesian case, τ denotes the threshold (2.19). Taking logarithms and reorganizing the resulting identity gives

$$Y_1 \underset{H_0}{\overset{H_1}{\gtrless}} \eta \triangleq \frac{E^{1/2}}{2} + \frac{\sigma^2}{E^{1/2}} \ln(\tau) . \tag{8.11}$$

So the optimum detector is expressed in terms of the sufficient statistic Y_1, which in the CT case is expressed as

$$Y_1 = \int_0^T Y(t)\phi_1(t)dt = \frac{1}{E^{1/2}} \int_0^T Y(t)s(t)dt , \tag{8.12}$$

and in the DT case as

$$Y_1 = \sum_{t=0}^{T} Y(t)\phi_1(t) = \frac{1}{E^{1/2}} \sum_{t=0}^{T} Y(t)s(t)dt . \tag{8.13}$$

Observing that $Y_1 \sim N(0, \sigma^2)$ under H_0 and $Y_1 \sim N(E^{1/2}, \sigma^2)$ under H_1, if

$$d = \frac{E[Y_1|H_1] - E[Y_1|H_0]}{\sigma} = \frac{E^{1/2}}{\sigma} \tag{8.14}$$

denotes the distance between the two hypotheses measured with respect to the noise standard deviation, it was shown in (2.70) that the probabilities of detection and of false alarm for the test (8.11) are given respectively by

$$\begin{aligned} P_D &= 1 - Q(\frac{d}{2} - \frac{\ln(\tau)}{d}) \\ P_F &= Q(\frac{d}{2} + \frac{\ln(\tau)}{d}) . \end{aligned} \tag{8.15}$$

Consequently, for a Neyman-Pearson test, by eliminating the threshold τ from the above identities, we obtain

$$P_D = 1 - Q(d - Q^{-1}(P_F)) . \tag{8.16}$$

In practice, the test is implemented in terms of the unnormalized statistic

$$\begin{aligned} S = E^{1/2} Y_1 &= \int_0^T Y(t)s(t)dt \quad \text{(CT case)} \\ &= \sum_{t=0}^{T} Y(t)s(t) \quad \text{(DT case)} \end{aligned} \tag{8.17}$$

so that (8.11) becomes

$$S \underset{H_0}{\overset{H_1}{\gtrless}} \gamma \triangleq \frac{E}{2} + \sigma^2 \ln(\tau) . \tag{8.18}$$

There exists two classical implementations of this test. The *correlator receiver* is shown in Fig. 8.1. The received signal is correlated with the signal $s(t)$

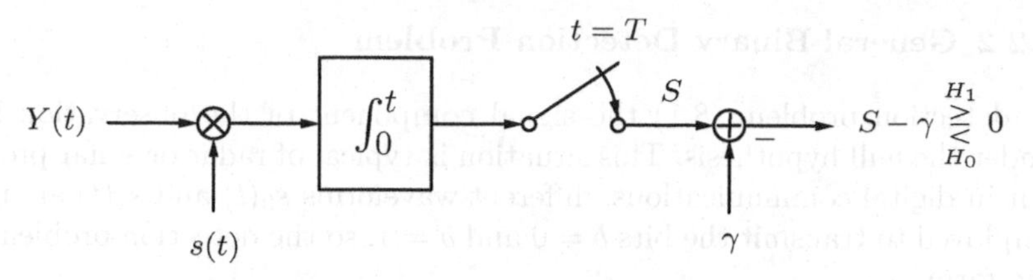

Fig. 8.1. Correlator implementation of the optimum detector.

by forming $Y(t)s(t)$ and passing it through an integrator which is sampled at $t = T$, yielding the sufficient statistic S which is then compared to the threshold γ. In the DT case, the integrator is replaced by a summer $\sum_{s=0}^{t}$. The combination of an integrator/summer with a sampler is sometimes called an integrate and dump receiver.

The second classical implementation of the optimum detector is the *matched filter* shown in Fig. 8.2. This structure, which was discovered independently by North [1] and Van Vleck and Middleton [2] during World War II, consists of passing the observed signal $Y(t)$ through a linear time-invariant filter with impulse response $h(t) = s(T - t)$. Because the impulse response is specified by the signal $s(t)$ to be detected, which is time-reversed and then shifted by T, it is deemed to be "matched" to the signal. The output of the matched filter is

$$Z(t) = \int_0^t h(t-u)Y(u)du = \int_0^t s(T+u-t)Y(u)du \,, \tag{8.19}$$

so, when it is sampled at $t = T$, it yields

$$Z(T) = \int_0^T s(u)Y(u)du = S \,, \tag{8.20}$$

which is the desired sufficient statistic. In the DT case, the detector works exactly in the same way, except that the CT convolution (8.17) is replaced by the DT convolution

$$Z(t) = \sum_{u=0}^{t} h(t-u)Y(u) = \sum_{u=0}^{t} s(T+u-t)Y(u) \tag{8.21}$$

which yields $Z(T) = S$ at $t = T$.

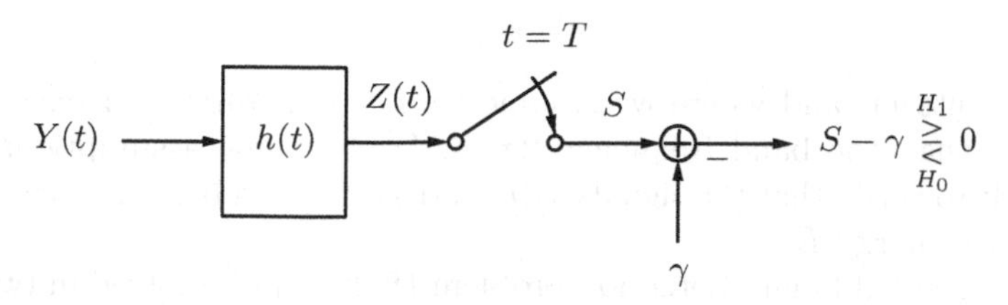

Fig. 8.2. Matched filter implementation of the optimum detector.

8.2.2 General Binary Detection Problem

In detection problem (8.1) the signal component of the observation is zero under the null hypothesis. This situation is typical of radar or sonar problems, but in digital communications, different waveforms $s_0(t)$ and $s_1(t)$ are usually employed to transmit the bits $b = 0$ and $b = 1$, so the detection problem takes the form

$$\begin{aligned} H_0 &: Y(t) = s_0(t) + V(t) \\ H_1 &: Y(t) = s_1(t) + V(t) \end{aligned} \tag{8.22}$$

for $0 \leq t \leq T$. For the case of *polar signaling*, the waveforms $s_j(t)$ are of opposite sign:

$$s_0(t) = -s_1(t) = p(t) \,, \tag{8.23}$$

so that the transmitted information is encoded in the phase of the transmitted signal. In pulse-amplitude modulation (PAM), $p(t)$ is typically a square root raised-cosine pulse $q(t)$ [3, Sec. 9.2], whereas for binary phase-shift keying (BPSK)

$$p(t) = (2E/T)^{1/2} \cos(\omega_c t) \,,$$

for $0 \leq t \leq T$, where ω_c denotes the carrier carrier frequency. For the case of *orthogonal signaling*, the signals $s_0(t)$ and $s_1(t)$ are orthogonal, i.e.,

$$\int_0^T s_0(t)s_1(t)dt = 0 \,, \tag{8.24}$$

with the same energy

$$E = \int_0^T s_j^2(t)dt \tag{8.25}$$

for $j = 0, 1$. In particular, for frequency shift-keying (FSK), the waveforms are selected as

$$\begin{aligned} s_0(t) &= (2E/T)^{1/2} \cos((\omega_c - \Delta\omega)t) \\ s_1(t) &= (2E/T)^{1/2} \cos((\omega_c + \Delta\omega)t) \end{aligned} \tag{8.26}$$

for $0 \leq t \leq T$, where

$$2\Delta\omega = n\frac{\pi}{T}$$

with n integer, and where we assume that the carrier frequency ω_c is much greater than the baud frequency $2\pi/T$. Under these assumptions, it is not difficult to verify that the signals $s_0(t)$ and $s_1(t)$ given by (8.26) are orthogonal and have energy E.

The general binary detection problem (8.22) can be solved in two different yet equivalent ways.

Single correlator/matched filter implementation: In the first approach, we consider the modified observation signal

$$\tilde{Y}(t) = Y(t) - s_0(t) \tag{8.27}$$

obtained by subtracting $s_0(t)$ from the observation $Y(t)$. Then, in terms of $\tilde{Y}(t)$, the detection problem (8.22) can be rewritten as

$$\begin{aligned} H_0 &: \tilde{Y}(t) = V(t) \\ H_1 &: \tilde{Y}(t) = s(t) + V(t) , \end{aligned} \tag{8.28}$$

with

$$s(t) \triangleq s_1(t) - s_0(t) , \tag{8.29}$$

which is in the form (8.1). Thus the optimum test is given by (8.18), where

$$\begin{aligned} S &= \int_0^T (Y(t) - s_0(t))(s_1(t) - s_0(t))dt \quad \text{(CT case)} \\ &= \sum_{t=0}^{T} (Y(t) - s_0(t))(s_1(t) - s_0(t)) \quad \text{(DT case)} , \end{aligned} \tag{8.30}$$

and

$$\begin{aligned} E &= \int_0^T (s_1(t) - s_0(t))^2 dt \quad \text{(CT case)} \\ &= \sum_{t=0}^{T} (s_1(t) - s_0(t))^2 \quad \text{(DT case)} . \end{aligned} \tag{8.31}$$

However, instead of actually forming the modified observation $\tilde{Y}(t)$, it is simpler to rewrite the test (8.18) as

$$\tilde{S} \underset{H_0}{\overset{H_1}{\gtrless}} \tilde{\gamma} \triangleq \frac{\tilde{E}}{2} + \sigma^2 \ln(\tau) , \tag{8.32}$$

where in the CT case

$$\tilde{S} \triangleq \int_0^T Y(t)(s_1(t) - s_0(t))dt \tag{8.33}$$

and

$$\tilde{E} \triangleq E + 2 \int s_0(t)(s_1(t) - s_0(t))dt = E_1 - E_0 , \tag{8.34}$$

with

$$E_j \triangleq \int_0^T s_j^2(t)dt \tag{8.35}$$

for $j = 0,\ 1$. Similar expressions can be obtained in the DT case by replacing integrals by summations. The test (8.32) can be implemented by the correlator and matched filter receivers of Fig. 8.1 and Fig. 8.2, where in Fig. 8.1 $s(t) = s_1(t) - s_0(t)$ and in Fig. 8.2 the matched filter impulse response is given by

$$h(t) = s_1(T-t) - s_0(T-t) \tag{8.36}$$

for $0 \leq t \leq T$. Also, in these figures γ needs to be replaced by $\tilde{\gamma}$.

The above approach requires only one correlator or matched filter to implement the optimum detector and it is therefore optimum from a complexity point of view. However, it is really specific to the binary detection case, and we can implement the optimum detector by employing a second approach that can be extended more easily to M-ary detection.

General method: We assume that $s_0(t)$ and $s_1(t)$ are not colinear, which rules out the polar signaling case. Let

$$\rho = \frac{1}{(E_0 E_1)^{1/2}} \int_0^T s_0(t) s_1(t) dt = < s_0, s_1 > / ||s_0|| \, ||s_1|| \tag{8.37}$$

denote the correlation coefficient of waveforms $s_0(t)$ and $s_1(t)$ in the CT case, where the corresponding DT expression is obtained by replacing the integral over $[0, T]$ by a summation. The assumption that $s_0(t)$ is not colinear with $s_1(t)$ is equivalent to $\rho \neq \pm 1$. Then, we can construct a COB of $L^2[0, T]$ where the first two basis functions are given by

$$\begin{aligned} \phi_1(t) &= \frac{s_0(t)}{E_0^{1/2}} \\ \phi_2(t) &= \frac{s_1(t) - \rho E_1^{1/2} \phi_1(t)}{E_1^{1/2}(1-\rho^2)^{1/2}} , \end{aligned} \tag{8.38}$$

and the basis functions $\phi_j(t)$ for $j \geq 3$ are arbitrary, but orthogonal to ϕ_1 and ϕ_2. Then, if we expand the noise and observations processes $V(t)$ and $Y(t)$ in the form (8.4)–(8.7), the Karhunen-Loève coefficients V_j, $j \geq 1$ still form an $N(0, \sigma^2)$ distributed DT WGN sequence. Also, under both H_0 and H_1, we have

$$Y_j = V_j$$

for $j \geq 3$, where the noise coefficients V_j, $j \geq 3$ are independent of V_1 and V_2. This means that all observations Y_j, $j \geq 3$ can be discarded, since they do not contain any information that can be used for testing purposes.
Let

$$\mathbf{V} = \begin{bmatrix} V_1 \\ V_2 \end{bmatrix} , \quad \mathbf{Y} = \begin{bmatrix} Y_1 \\ Y_2 \end{bmatrix} ,$$

and

$$\mathbf{S}_0 = E_0^{1/2} \begin{bmatrix} 1 \\ 0 \end{bmatrix} \quad , \quad \mathbf{S}_1 = E_1^{1/2} \begin{bmatrix} \rho \\ (1-\rho^2)^{1/2} \end{bmatrix} \quad .$$

Then the detection problem reduces to the two-dimensional problem

$$\begin{aligned} H_0 &: \mathbf{Y} = \mathbf{S}_0 + \mathbf{V} \\ H_1 &: \mathbf{Y} = \mathbf{S}_1 + \mathbf{V} \, , \end{aligned} \tag{8.39}$$

where $\mathbf{V} \sim N(\mathbf{0}, \sigma^2 \mathbf{I}_2)$, and $||\mathbf{S}_j||^2 = E_j$. This problem is illustrated geometrically in Fig. 8.3, where $\theta = \cos^{-1}(\rho)$ is the angle between vectors $\mathbf{S}_1$ and $\mathbf{S}_0$, or more abstractly, the angle between waveforms $s_1(t)$ and $s_0(t)$, as indicated by the definition (8.37) of ρ.

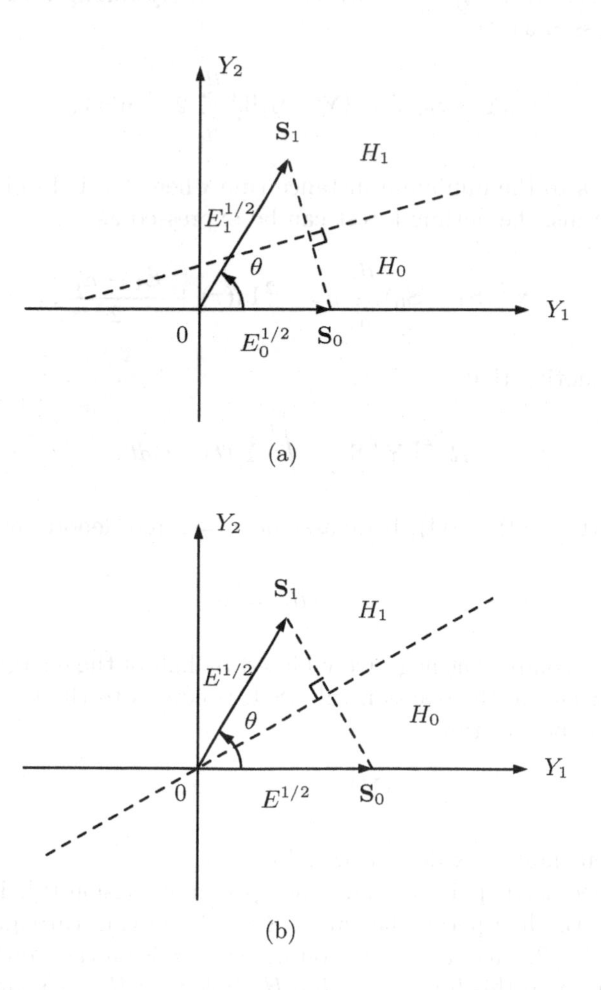

Fig. 8.3. Signal space representation of the binary detection problem for (a) signals with different energy, (b) signals with equal energy.

The optimum test for the two-dimensional detection problem (8.39) is the LRT

$$L(\mathbf{Y}) = \frac{f(\mathbf{Y}|H_1)}{f(\mathbf{Y}|H_0)}$$
$$= \exp\left[\frac{1}{2\sigma^2}(||\mathbf{Y} - \mathbf{S}_0||_2^2 - ||\mathbf{Y} - \mathbf{S}_1||_2^2)\right] \underset{H_0}{\overset{H_1}{\gtrless}} \tau\,, \tag{8.40}$$

where τ denotes the Bayesian threshold (2.19). Note that in the typical digital communications situation where the two hypotheses are equally likely, i.e., $\pi_1 = \pi_0 = 1/2$, and where we minimize the probability of error, which corresponds to selecting $C_{ij} = 1 - \delta_{ij}$, then $\tau = 1$. By taking logarithms, (8.40) can be expressed as

$$||\mathbf{Y} - \mathbf{S}_0||_2^2 - ||\mathbf{Y} - \mathbf{S}_1||_2^2 \underset{H_0}{\overset{H_1}{\gtrless}} 2\sigma^2 \ln(\tau)\,, \tag{8.41}$$

which reduces to the minimum distance rule when $\tau = 1$. Equivalently, after cancelling terms, the optimum test can be expressed as

$$\mathbf{Y}^T(\mathbf{S}_1 - \mathbf{S}_0) \underset{H_0}{\overset{H_1}{\gtrless}} \eta = \sigma^2 \ln(\tau) + \frac{E_1 - E_0}{2}\,, \tag{8.42}$$

which, after noting that

$$R_i \triangleq \mathbf{Y}^T\mathbf{S}_i = \int_0^T Y(t)s_i(t)dt\,,$$

coincides with (8.32)–(8.34). If we assume $\tau = 1$ and denote by

$$Z_i = R_i - \frac{E_i}{2} \tag{8.43}$$

the correlator component negatively biased by half of the energy of the corresponding waveform, the decision rule (8.42) reduces to the maximum biased correlator component rule

$$\delta(\mathbf{Y}) = \arg \max_{i=0,1} Z_i \tag{8.44}$$

whose implementation is described in Fig. 8.4.

From a geometric point of view, the optimum decision rule is obtained by constructing the line perpendicular to $\mathbf{S}_1 - \mathbf{S}_0$ passing through the midway point $(\mathbf{S}_1 + \mathbf{S}_0)/2$. Then depending on whether $\mathbf{Y}$ is on the same side as $\mathbf{S}_0$ or $\mathbf{S}_1$ with respect to this line, we decide H_0 holds, or H_1, as shown in part (a) of Fig. 8.3. For the special case where the two signals have the same energy, so $E_0 = E_1 = E$, it is not necessary to include the negative bias $-E/2$ in the

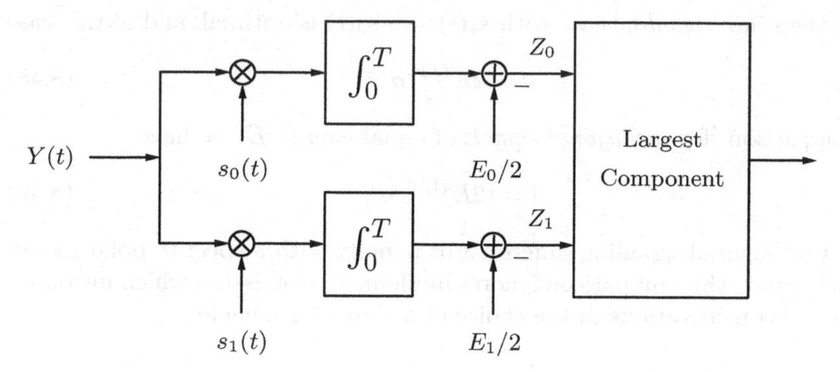

Fig. 8.4. Largest biased correlator component receiver.

receiver of Fig. 8.4, since the bias is the same for both branches of the receiver. In this case, the line specifying the decision rule in the $Y_1 - Y_2$ plane passes through the origin and bisects the angle θ separating the vectors $\mathbf{S}_1$ and $\mathbf{S}_0$ representing the two transmitted signals, as shown in part (b) of Fig. 8.3.

Receiver performance: The probablities of detection and of false alarm for the general binary detection problem are still given by (8.15), but the squared distance betwen the two hypotheses is now given by

$$d^2 = \frac{||\mathbf{S}_1 - \mathbf{S}_0||_2^2}{\sigma^2} = \frac{E_0 + E_1 - 2\rho(E_0 E_1)^{1/2}}{\sigma^2} \tag{8.45}$$

where the second equality is obtained by applying the law of cosines for calculating one side of a triangle when the angle θ opposite and the other two sides $E_0^{1/2} = ||\mathbf{S}_0||_2$ and $E_1^{1/2} = ||\mathbf{S}_1||_2$ are known, as shown in part (a) of Fig. 8.3. For the case of a minimum probability of error receiver with equally likely signals, the threshold $\tau = 1$, and the probability of error is given by

$$P[E] = Q\left(\frac{d}{2}\right). \tag{8.46}$$

The probability of error is minimized by maximizing the distance d, which for fixed E_0 and E_1 occurs when $\rho = -1$ or equivalently $\theta = \pi$.

Assuming that $\rho = -1$, it is also of interest to select the signal energies E_i with $i = 0,\ 1$ such that the average energy

$$E = \frac{E_0 + E_1}{2}$$

is fixed and d is maximized. This requires maximizing the product $E_0 E_1$ with the sum fixed, which is achieved by selecting

$$E_0 = E_1 = E. \tag{8.47}$$

Thus, the *polar signaling* case with $s_1(t) = -s_0(t)$ is optimal, and in this case

$$d = 2E^{1/2}/\sigma \,. \tag{8.48}$$

By comparison, for *orthogonal signals* of equal energy E, we have

$$d = (2E)^{1/2}/\sigma \,, \tag{8.49}$$

so that orthogonal signaling suffers a 3dB penalty with respect to polar signaling. Of course, this comparison ignores implementation issues which introduce additional considerations in the choice of a signaling scheme.

8.3 M-ary Detection of Known Signals in WGN

The general method that we have used above to tackle the binary hypothesis testing problem extends in a straightforward manner to M-ary hypothesis testing. Consider the M-ary detection problem where under hypothesis H_i with $0 \leq i \leq M-1$, the observed signal is given by

$$Y(t) = s_i(t) + V(t) \,. \tag{8.50}$$

for $0 \leq t \leq T$, where t can again be either continuous or discrete. The signals $s_i(t)$, $0 \leq i \leq M-1$ are known, and $V(t)$ is a zero-mean WGN with covariance (8.2). We assume that the signals are equally likely, i.e.,

$$\pi_i = P[H_i] = \frac{1}{M} \,,$$

and we seek to minimize the probability of error, so the cost function $C_{ij} = 1 - \delta_{ij}$.

By Gram-Schmidt orthonormalization of the signals $s_i(t)$, we can construct a set of N orthonormal signals $\{\phi_j(t),\ 1 \leq j \leq N\}$ with $N \leq M$ such that

$$s_i(t) = \sum_{j=1}^{N} s_{ij}\phi_j(t) \,. \tag{8.51}$$

Furthermore, given the N orthonormal signals $\{\phi_1(t),\ \ldots,\ \phi_N(t)\}$, we can construct a COB $\{\phi_j,\ j \in \mathcal{J}\}$ of $L^2[0,T]$ where the functions ϕ_j for $j > N$ are obtained by taking arbitrary functions of $L^2[0,T]$ and orthonormalizing them with respect to previously constructed basis functions. Then the noise $V(t)$ admits the KLD

$$V(t) = \sum_{j \in \mathcal{J}} V_j \phi_j(t)$$

where the Karhunen-Loève coefficients V_j form a zero-mean $N(0, \sigma^2)$ distributed WGN sequence. Similarly, the observed signal $Y(t)$ admits the KLD

$$Y(t) = \sum_{j \in \mathcal{J}} Y_j \phi_j(t) ,$$

where under hypothesis H_i

$$Y_j = \begin{cases} s_{ij} + V_j & 1 \leq j \leq N \\ V_j & j > N . \end{cases} \tag{8.52}$$

Since the Karhunen-Loève coefficients $Y_j = V_j$ are the same under all hypotheses for $j > N$, where the V_j's are independent, these coefficients can all be discarded. The detection problem reduces therefore to the N-dimensional problem

$$H_i \, : \, \mathbf{Y} = \mathbf{S}_i + \mathbf{V} , \tag{8.53}$$

where

$$\mathbf{Y} = \begin{bmatrix} Y_1 \\ \vdots \\ Y_j \\ \vdots \\ Y_N \end{bmatrix} , \quad \mathbf{V} = \begin{bmatrix} V_1 \\ \vdots \\ V_j \\ \vdots \\ V_N \end{bmatrix} \sim N(\mathbf{0}, \sigma^2 \mathbf{I}_N) ,$$

and

$$\mathbf{S}_i = \begin{bmatrix} s_{i1} \\ \vdots \\ s_{ij} \\ \vdots \\ s_{iN} \end{bmatrix}$$

for $1 \leq i \leq M - 1$. This problem is a standard M-ary hypothesis testing problem of the type examined in Section 2.7. Since all hypotheses are equally likely and we minimize the probability of error, the MAP decision rule reduces to the *minimum distance decision rule*:

$$\delta(\mathbf{Y}) = \arg \min_{0 \leq i \leq M-1} ||\mathbf{Y} - \mathbf{S}_i||_2^2 . \tag{8.54}$$

The squared distance can be expanded as

$$||\mathbf{Y} - \mathbf{S}_i||_2^2 = ||\mathbf{Y}||_2^2 - 2R_i + E_i , \tag{8.55}$$

with

$$R_i = \mathbf{Y}^T \mathbf{S}_i = \int_0^T Y(t) s_i(t) dt$$

$$E_i = ||\mathbf{S}_i||_2 = \int_0^T s_i^2(t) dt , \tag{8.56}$$

where the term $||\mathbf{Y}||_2^2$ can be removed since it is common to all expressions. Consequently, if we denote the i-th biased correlator component as

$$Z_i = R_i - \frac{E_i}{2}, \tag{8.57}$$

the optimum decision rule can be expressed as the maximum biased correlator component rule

$$\delta(\mathbf{Y}) = \arg \max_{0 \leq i \leq M-1} Z_i \tag{8.58}$$

which takes the same form as the binary receiver (8.44). The structure of this receiver is shown in Fig. 8.5. Note that since this receiver employs M correlators to evaluate the components R_i with $0 \leq i \leq M-1$, it is not optimal from a hardware point of view when $N \ll M$. In this case it is preferable to use N correlators to evaluate the components

$$Y_j = < Y, \phi_j >$$

with $1 \leq j \leq N$ and then evaluate the R_i's by forming the vector inner products $\mathbf{Y}^T \mathbf{S}_i$ for $0 \leq i \leq M-1$.

Example 8.1: M-ary Phase-shift keying

For M-ary phase-shift keying (PSK), the transmitted waveform under hypothesis H_i is given by

$$s_i(t) = (\frac{2E}{T})^{1/2} \cos(\omega_c t + \theta_i)$$

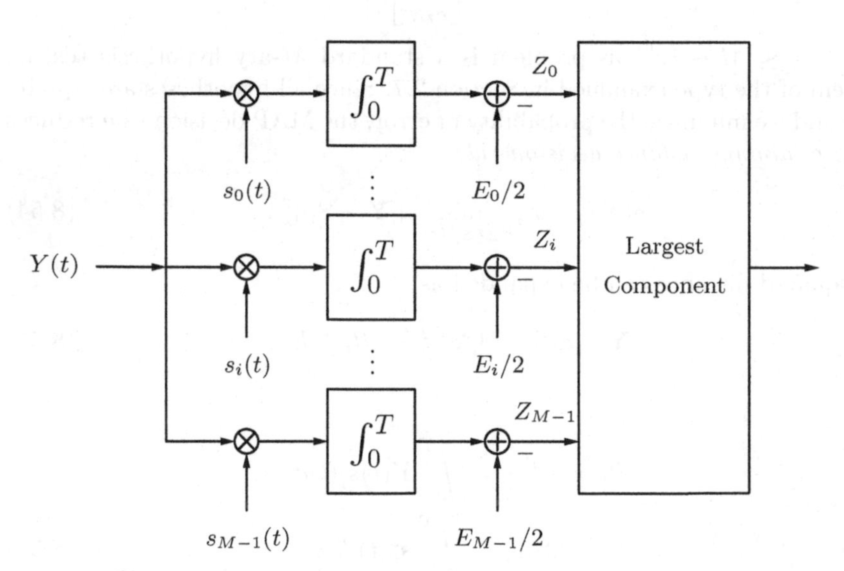

Fig. 8.5. Largest M-ary biased correlator component receiver.

for $0 \leq t \leq T$, where $\theta_i = 2\pi i/M$ with $0 \leq i \leq M-1$. Thus the information contained in $s_i(t)$ is encoded entirely in its phase ϕ_i, and during each signaling/symbol period T, $\log_2(M)$ bits are transmitted. If $\omega_s = 2\pi/T$ denotes the baud/symbol frequency, we assume that the carrier frequency ω_c is much larger than ω_s. Then consider the basis functions

$$\phi_1(t) = \left(\frac{2}{T}\right)^{1/2} \cos(\omega_c t)$$
$$\phi_2(t) = \left(\frac{2}{T}\right)^{1/2} \sin(\omega_c t) .$$

Under the assumption $\omega_c \gg \omega_s$, or equivalently $\omega_c T \gg 2\pi$, these two functions are approximately orthogonal, since we have

$$\begin{aligned} <\phi_1, \phi_2> &= \frac{2}{T} \int_0^T \sin(\omega_c t) \cos(\omega_c t) dt = \frac{1}{T} \int_0^T \sin(2\omega_c t) dt \\ &= \frac{1}{2\omega_c T} [1 - \cos(2\omega_c T)] \approx 0 . \end{aligned}$$

Furthermore, it is also easy to verify that $||\phi_1||_2^2 = ||\phi_2||_2^2 = 1$. Since we can express each signal $s_i(t)$ as

$$s_i(t) = E^{1/2} (\cos(\theta_i)\phi_1(t) - \sin(\theta_i)\phi_2(t)) ,$$

we see that $N = 2$, so that the signal space is two-dimensional. Specifically, let

$$Y_1 = \left(\frac{2}{T}\right)^{1/2} \int_0^T Y(t) \cos(\omega_c t) dt$$
$$Y_2 = \left(\frac{2}{T}\right)^{1/2} \int_0^T Y(t) \sin(\omega_c t) dt .$$

Then if

$$\mathbf{Y} = \begin{bmatrix} Y_1 \\ Y_2 \end{bmatrix} \quad , \quad \mathbf{S}_i = E^{1/2} \begin{bmatrix} \cos(\theta_i) \\ -\sin(\theta_i) \end{bmatrix} ,$$

under hypothesis H_i we have

$$\mathbf{Y} = \mathbf{S}_i + \mathbf{V} ,$$

with $\mathbf{V} \sim N(\mathbf{0}, \sigma^2 \mathbf{I}_2)$, where the points $\mathbf{S}_i$ are all located on a circle of radius $E^{1/2}$ with an angular separation of $2\pi/M$, as shown in Fig. 8.6 for $M = 8$.

By symmetry, the probability of a correct decision is the probability of a correct decision under H_0, which is the probability that under H_0, the vector $\mathbf{Y}$ falls within the cone $\mathcal{Y}_0$ corresponding to the angular range $[-\pi/M, \pi/M]$ depicted in Fig. 8.6 for the case $M = 8$. The probability of a correct decision under H_0 is therefore given by

$$P[C] = \frac{1}{2\pi\sigma^2} \int_0^\infty \exp\big(-\frac{(y_1 - E^{1/2})^2}{2\sigma^2}\big)\Big[\int_{-y_1 \tan(\pi/M)}^{y_1 \tan(\pi/M)} \exp\big(-\frac{y_2^2}{2\sigma^2}\big) dy_2\Big] dy_1$$
$$= \frac{1}{\pi} \int_0^\infty \exp\big(-\frac{(u - (E^{1/2}/\sigma))^2}{2}\big)\Big[\int_0^{u \tan(\pi/M)} \exp\big(-\frac{v^2}{2}\big) dv\Big] du \,. \quad (8.59)$$

However, this integral is difficult to evaluate. Instead of evaluating this expression, an approximation of the probability of error can be obtained by decomposing the error region in terms of pairwise error events. For the case $M = 8$, this gives

$$\mathcal{Y}_0^c = \mathcal{E}_{10} \cup \mathcal{E}_{70}$$

where $\mathcal{E}_{10}$ and $\mathcal{E}_{70}$ correspond respectively to the pairwise error events where we prefer H_1 over H_0, or H_7 over H_0. The distances d_{10} and d_{70} between $\mathbf{S}_1$ and $\mathbf{S}_0$, or between $\mathbf{S}_7$ and $\mathbf{S}_0$, measured relative to the noise standard deviation are given by

$$d_{10} = d_{70} = 2E^{1/2} \sin(\pi/M)/\sigma$$

with $M = 8$, so that

$$P[\mathcal{E}_{10}|H_0] = P[\mathcal{E}_{70}|H_0] = Q\big(\frac{E^{1/2}}{\sigma} \sin(\pi/M)\big)$$

and the improved form (2.166) of the union bound yields

$$P[E] = P[E|H_0] \leq 2Q\big(\frac{E^{1/2}}{\sigma} \sin(\pi/M)\big) \,, \quad (8.60)$$

where $P[E]$ represents the symbol error probability. Although the inequality (8.60) was derived for $M = 8$, it is in fact easy to verify that the argument consisting of decomposing errors into pairwise error events involving the two immediate neighbors of $\mathbf{S}_0$ holds for all M, so (8.60) is true in general. The inequality (8.60) is tight at high SNR since the region of overlap between the two pairwise error events ($\mathcal{E}_{10}$ and $\mathcal{E}_{70}$ for $M = 8$) has a probability mass that tends to zero as the SNR increases. □

Example 8.2: M-ary orthogonal signals

Consider the case where in (8.50), the signals $s_i(t)$ with $0 \leq i \leq M-1$ are orthogonal and of equal energy, i.e.,

$$E = ||s_i||^2 = \int_0^T s_i^2(t) dt \quad \text{(CT case)}$$
$$= \sum_{t=0}^{T} s_i^2(t) \quad \text{(DT case)} \,.$$

Signaling schemes with this property include M-ary frequency shift keying (FSK) with

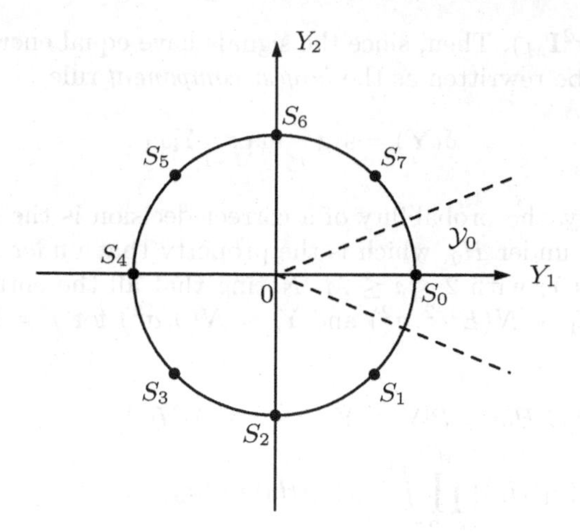

Fig. 8.6. Signal space representation of 8-PSK signals and decision region $\mathcal{Y}_0$ for H_0.

$$s_i(t) = (2E/T)^{1/2} \cos((\omega_c + \frac{i\pi}{T})t)$$

for $0 \leq t \leq T$, where the frequency separation $\Delta\omega = \pi/T$ ensures the orthogonality of the signal set. Another choice is pulse position modulation where, given a pulse $p(t)$ of duration T/M and energy E, the signals $s_i(t)$ take the form

$$s_i(t) = \begin{cases} p(t - iT/M) & \frac{iT}{M} \leq t \leq \frac{(i+1)T}{M} \\ 0 & \text{otherwise} . \end{cases}$$

Then if we select

$$\phi_i(t) = s_{i-1}(t)/E^{1/2}$$

with $1 \leq i \leq M$ as the first M basis functions of a COB, the detection problem can be expressed as an M-dimensional problem of the form (8.53), with

$$\mathbf{S}_i = E^{1/2}\mathbf{e}_{i+1}$$

for $0 \leq i \leq M-1$, where $\mathbf{e}_k$ denotes the kth unit basis vector of $\mathbb{R}^M$, i.e., its entries are given by

$$e_{kj} = \begin{cases} 1 & k = j \\ 0 & \text{otherwise} . \end{cases}$$

In (8.53), we have

$$\mathbf{Y} = \begin{bmatrix} Y_1 \dots Y_i \dots Y_M \end{bmatrix}^T$$

where

$$\begin{aligned} Y_{i+1} &= \frac{1}{E^{1/2}} \int_0^T Y(t)s_i(t)dt \quad \text{(CT case)} \\ &= \frac{1}{E^{1/2}} \sum_{t=0}^{T} Y(t)s_i(t) \quad \text{(DT case)} \end{aligned}$$

and $\mathbf{V} \sim N(\mathbf{0}, \sigma^2 \mathbf{I}_M)$. Then, since the signals have equal energy, the decision rule (8.58) can be rewritten as the *largest component* rule

$$\delta(\mathbf{Y}) = \arg \max_{0 \leq i \leq M-1} Y_{i+1} \,. \tag{8.61}$$

By symmetry, the probability of a correct decision is the probability of a correct decision under H_0, which is the property that under H_0, Y_1 is larger than the entries Y_i with $2 \leq i \leq M$. Noting that all the entries Y_i are independent with $Y_1 \sim N(E^{1/2}, \sigma^2)$ and $Y_i \sim N(0, \sigma^2)$ for $i \neq 1$ under H_0, we have

$$\begin{aligned}
P[C] &= P[C|H_0] = P[Y_1 > Y_i,\, 2 \leq i \leq M|H_0] \\
&= \int_{-\infty}^{\infty} f(y_1|H_0)\Big[\prod_{i=2}^{M} \int_{-\infty}^{y_1} f(y_i|H_0) dy_i\Big] dy_1 \\
&= \int_{-\infty}^{\infty} \frac{1}{(2\pi\sigma^2)^{1/2}} \exp\Big(-\frac{(y_1 - E^{1/2})^2}{2\sigma^2}\Big)\Big[1 - Q\Big(\frac{y_1}{\sigma}\Big)\Big]^{M-1} dy_1 \\
&= \int_{-\infty}^{\infty} \frac{1}{(2\pi)^{1/2}} \exp\Big(-\frac{(u - (E^{1/2}/\sigma))^2}{2}\Big)[1 - Q(u)]^{M-1} du \,.
\end{aligned} \tag{8.62}$$

where the function $Q(x)$ is defined in (2.69). Unfortunately, this expression is rather hard to evaluate, and instead we can decompose the error region as

$$\mathcal{Y}_0^c = \cup_{i=1}^{M-1} \mathcal{E}_{i0} \,,$$

where the pairwise error event $\mathcal{E}_{i0}$ corresponds to the case where $\mathbf{Y}$ is closer to $\mathbf{S}_i$ than $\mathbf{S}_0$. Since the distance between these two vectors is

$$d_{i0} = (2E)^{1/2}/\sigma$$

by using the union bound, we obtain

$$P[E] = P[E|H_0] \leq (M-1) Q\Big(\Big(\frac{E}{2\sigma^2}\Big)^{1/2}\Big) \,. \tag{8.63}$$

□

8.4 Detection of Known Signals in Colored Gaussian Noise

Consider the binary detection problem

$$\begin{aligned}
H_0 &: Y(t) = V(t) \\
H_1 &: Y(t) = s(t) + V(t)
\end{aligned} \tag{8.64}$$

consisting of detecting a known signal $s(t)$ in the presence of zero-mean Gaussian colored noise $V(t)$ for $0 \leq t \leq T$. The time index t can be either discrete or continuous and $V(t)$ admits a positive-definite covariance function

$$K(t,s) = E[V(t)V(s)] \tag{8.65}$$

for $0 \leq t,\ s \leq T$. In the CT case we assume either that $K(t,s)$ is continuous, or that $V(t)$ can be decomposed as

$$V(t) = V_C(t) + N(t)$$

where $V_C(t)$ is a zero-mean Gaussian component with continuous covariance function $K_C(t,s)$, and $N(t)$ is a zero-mean WGN component independent of $V_C(t)$ and with intensity $\sigma^2 > 0$. In this case the eigenfunctions of the covariance kernel

$$K(t,s) = K_C(t,s) + \sigma^2 \delta(t-s)$$

are continuous and form a COB of $L^2[0,T]$. In addition, we assume that the signal $s(t)$ is square-integrable over $[0,T]$. In both the DT and CT case, $V(t)$ admits a KLD

$$V(t) = \sum_{j \in \mathcal{J}} V_j \phi_j(t)\,, \tag{8.66}$$

where the basis functions $\{\phi_j,\ j \in \mathcal{J}\}$ are the eigenfunctions of the kernel $K(t,s)$, i.e., they satisfy

$$\begin{aligned} \lambda_j \phi_j(t) &= \int_0^T K(t,s)\phi_j(s)ds \quad \text{(CT case)} \\ &= \sum_{s=0}^{T} K(t,s)\phi_j(s) \quad \text{(DT case)} \end{aligned} \tag{8.67}$$

for $0 \leq t \leq T$. The Karhunen-Loève coefficients

$$\begin{aligned} V_j = < V, \phi_j > &= \int_0^T V(t)\phi_j(t)dt \quad \text{(CT case)} \\ &= \sum_{t=0}^{T} V(t)\phi_j(t) \quad \text{(DT case))} \end{aligned} \tag{8.68}$$

with $V_j \sim N(0, \lambda_j)$ form a zero-mean time-varying WGN sequence. Since we assume that $K(t,s)$ is positive-definite, $\{\phi_j,\ j \in \mathcal{J}\}$ forms a COB of $L^2[0,T]$, so $s(t)$ admits the generalized Fourier expansion

$$s(t) = \sum_{j \in \mathcal{J}} s_j \phi_j(t) \tag{8.69}$$

with $s_j =< s, \phi_j >$. Note that in the CT case we are assuming $s \in L^2[0,T]$. Performing a KLD

$$Y(t) = \sum_{j \in \mathcal{J}} Y_j \phi_j(t) \tag{8.70}$$

of the observations under both H_0 and H_1, we find that in the Karhunen-Loève coefficient domain, the detection problem (8.64) can be expressed as

$$\begin{aligned} H_0 &: Y_j = V_j \\ H_1 &: Y_j = s_j + V_j \end{aligned} \tag{8.71}$$

for $j \in \mathcal{J}$. Since the kernel $K(t,s)$ is positive-definite, we have $\lambda_j > 0$ for all j, so that we can introduce the normalized variables

$$\bar{Y}_j \triangleq \frac{Y_j}{\lambda_j^{1/2}} \quad \bar{s}_j \triangleq \frac{s_j}{\lambda_j^{1/2}} \quad \bar{V}_j \triangleq \frac{V_j}{\lambda_j^{1/2}}, \tag{8.72}$$

so that $\bar{V}_j \sim N(0,1)$ is now a WGN sequence with unit variance, and the detection problem (8.72) becomes

$$\begin{aligned} H_0 &: \bar{Y}_j = \bar{V}_j \\ H_1 &: \bar{Y}_j = \bar{s}_j + \bar{V}_j \end{aligned} \tag{8.73}$$

for $j \in \mathcal{J}$. This problem is a DT binary detection problem for a known signal in WGN of the type examined in Section 8.2. The only nonstandard aspect of this problem is that in the CT case the index set $\mathcal{J} = \{j \geq 1\}$ is infinite, whereas in Section 8.2 it was assumed that T is finite. Going from a finite interval to an infinite one is a nontrivial extension that needs to be analyzed carefully.

8.4.1 Singular and Nonsingular CT Detection

In the CT case, to analyze the detection problem (8.69) for all $j \geq 1$, we consider a truncated version of it with $1 \leq j \leq N$ and then let N tend to infinity. The optimum test for the truncated problem is given by (8.17)–(8.18), which takes here the form

$$S(N) \underset{H_0}{\overset{H_1}{\gtrless}} \frac{E(N)}{2} + \ln(\tau) \tag{8.74}$$

with

$$\begin{aligned} S(N) &= \sum_{j=1}^{N} \bar{Y}_j \bar{s}_j = \sum_{j=1}^{N} \frac{Y_j s_j}{\lambda_j} \\ E(N) &= \sum_{j=1}^{N} \bar{s}_j^2 = \sum_{j=1}^{N} \frac{s_j^2}{\lambda_j} . \end{aligned} \tag{8.75}$$

Under H_0 and H_1, we have respectively

$$\begin{aligned} H_0 &: \frac{S(N)}{E(N)} \sim N(0, E^{-1}(N)) \\ H_1 &: \frac{S(N)}{E(N)} \sim N(1, E^{-1}(N)) , \end{aligned} \tag{8.76}$$

so the distance between the two hypotheses is

$$d(N) = E^{1/2}(N) , \tag{8.77}$$

and the probabilities of detection and false alarm are obtained by setting $d = d(N)$ in (8.15). Then, when we let $N \to \infty$ two cases may occur:

Case 1: Suppose the total signal "energy" is infinite, i.e.,

$$\lim_{N \to \infty} E(N) = \sum_{j=1}^{\infty} \frac{s_j^2}{\lambda_j} = \infty .$$

Then the problem is said to be *singular*, in the sense that we can decide between H_0 and H_1 with zero probability of error as $\lim_{N \to \infty} d(N) = \infty$. To see why this is the case, note that from (8.76) we can conclude that under H_0

$$\frac{S(N)}{E(N)} \xrightarrow{\text{m.s.}} 0$$

and under H_1

$$\frac{S(N)}{E(N)} \xrightarrow{\text{m.s.}} 1 .$$

In other words, $S(N)/E(N)$ converges in the mean square to zero or to one depending on whether H_0 or H_1 holds, which means that we can discriminate perfectly between the two hypotheses.

Case 2: On the other hand, if the signal energy is finite, i.e.,

$$E \triangleq \lim_{N \to \infty} E(N) = \sum_{j=1}^{\infty} \frac{s_j^2}{\lambda_j} < \infty , \tag{8.78}$$

the detection problem is said to be *nonsingular*. The statistic $S(N)$ converges in the mean square to a random variable

$$S = \sum_{j=1}^{\infty} \bar{Y}_j \bar{s}_j = \sum_{j=1}^{\infty} \frac{Y_j s_j}{\lambda_j} \tag{8.79}$$

such that

$$\begin{aligned} H_0 &: S \sim N(0, E) \\ H_1 &: S \sim N(E, E) , \end{aligned} \tag{8.80}$$

and as $N \to \infty$, the test (8.74) becomes

$$S \underset{H_0}{\overset{H_1}{\gtrless}} \frac{E}{2} + \ln(\tau) \,. \tag{8.81}$$

From (8.80), the distance for this test is given by $d = E^{1/2}$.

From the above analysis we conclude that while perfect detection is possible in the singular detection case, for the nonsingular case, the CT detection problem viewed in the Karhunen-Loève domain has exactly the same form as the DT detection problem (which is always nonsingular), except that the finite summations employed for the matched filter and signal energy have to be replaced by infinite summations.

8.4.2 Generalized Matched Filter Implementation

While the transformation to the Karhunen-Loève coefficient domain proved convenient to obtain the optimum detector structure, it does not necessarily provide the best way to implement the detector, particularly if we consider that in the CT case, infinite summations are needed to evaluate the sufficient statistic S and energy E appearing in the test (8.81).

In this section we restrict our attention to the DT case and to the nonsingular CT case. Then the optimum test is given by (8.81) with

$$\begin{aligned} S &= \sum_{j \in \mathcal{J}} \frac{Y_j s_j}{\lambda_j} \\ E &= \sum_{j \in \mathcal{J}} \frac{s_j^2}{\lambda_j} \end{aligned} \tag{8.82}$$

where $\mathcal{J} = \{1, \; \ldots, \; T{+}1\}$ in the DT case, and $\mathcal{J} = \{j \geq 1\}$ in the nonsingular CT case. By interchanging the summations/integrations employed to evaluate Y_j and s_j, and the summation over j in (8.82), we find that S and E can be expressed as

$$\begin{aligned} S &= \int_0^T \int_0^T Y(t) Q(t,u) s(u) du \quad \text{(CT case)} \\ &= \sum_{t=0}^{T} \sum_{u=0}^{T} Y(t) Q(t,u) s(u) \quad \text{(DT case)} \end{aligned} \tag{8.83}$$

and

$$\begin{aligned} E &= \int_0^T \int_0^T s(t) Q(t,u) s(u) du \quad \text{(CT case)} \\ &= \sum_{t=0}^{T} \sum_{u=0}^{T} s(t) Q(t,u) s(u) \quad \text{(DT case)} \,, \end{aligned} \tag{8.84}$$

where

$$Q(t,s) \triangleq \sum_{j \in \mathcal{J}} \frac{1}{\lambda_j} \phi_j(t)\phi_j(s) \tag{8.85}$$

with $0 \leq t,\ s \leq T$ is called the *inverse kernel* of $K(t,s)$. Comparing the expansion (8.85) of $Q(t,s)$ with Mercer's expansion

$$K(t,s) = \sum_{j \in \mathcal{J}} \lambda_j \phi_j(t)\phi_j(s) \tag{8.86}$$

of the covariance kernel $K(t,s)$, we see immediately that $Q(t,s)$ has the same eigenfunctions $\{\phi_j,\ j \in \mathcal{J}\}$ as $K(t,s)$, but its eigenvalues are the inverses of the eigenvalues of $K(t,s)$. This observation implies the following property:

Inverse kernel property: The kernel $Q(t,s)$ with $0 \leq t,\ s \leq T$ satisfies the identity

$$\int_0^T Q(t,u)K(u,s)du = \delta(t-s) \tag{8.87}$$

in the CT case, and

$$\sum_{u=0}^{T} Q(t,u)K(u,s) = \delta(t-s) \tag{8.88}$$

in the DT case, where in (8.87) and (8.88), $\delta(t-s)$ represents respectively the Dirac and Kronecker delta function. We can establish the CT and DT identities simultaneously by noting that the left-hand side of (8.87) and (8.88) can be expressed as $< Q(t,\cdot), K(\cdot,s) >$ where $< f,g >$ denotes the inner product of DT or CT functions over $L^2[0,T]$. Then, taking into account (8.85) and (8.86) and exploiting the orthonormality of eigenfunctions we find

$$\begin{aligned} < Q(t,\cdot), K(\cdot,s) > &= \sum_{i \in \mathcal{J}} \sum_{j \in \mathcal{J}} \frac{\lambda_j}{\lambda_i} \phi_i(t) < \phi_i, \phi_j > \phi_j(s) \\ &= \sum_{i \in \mathcal{J}} \sum_{j \in \mathcal{J}} \frac{\lambda_j}{\lambda_i} \delta_{ij} \phi_i(t)\phi_j(s) \\ &= \sum_{j \in \mathcal{J}} \phi_j(t)\phi_j(s) = \delta(t-s)\,, \end{aligned} \tag{8.89}$$

where the last equality is due to the fact that since the functions $\{\phi_j,\ j \in \mathcal{J}\}$ form a COB of $L^2[0,T]$, they form a resolution of the identity.

In the DT case, if we consider the $(T+1) \times (T+1)$ matrices

$$\mathbf{K} = \big(K_{ij}, 1 \leq i,\ j \leq T+1\big) \quad , \quad \mathbf{Q} = \big(Q_{ij}, 1 \leq i,\ j \leq T+1\big)\,,$$

where

$$K_{ij} = K(i-1, j-1) \ \text{ and } \ Q_{ij} = Q(i-1, j-1)\,,$$

the identity (8.84) can be rewritten in matrix form as

$$\mathbf{QK} = \mathbf{I}_{T+1} , \tag{8.90}$$

so that the matrix $\mathbf{Q} = \mathbf{K}^{-1}$ is the *inverse* of the covariance matrix $\mathbf{K}$ of the noise process $V(t)$ over $[0, T]$. From this perspective, let

$$\boldsymbol{\phi}_j = \begin{bmatrix} \phi_j(0) \dots \phi_j(t) \dots \phi_j(T) \end{bmatrix}^T ,$$

be the vector representing the eigenfunction $\phi_j(t)$, and let

$$\boldsymbol{\Phi} = \begin{bmatrix} \boldsymbol{\phi}_1 \dots \boldsymbol{\phi}_j \dots \boldsymbol{\phi}_{T+1} \end{bmatrix}$$

and

$$\boldsymbol{\Lambda} = \text{diag}\,\{\lambda_j,\ 1 \leq j \leq T+1\}$$

be the matrices regrouping respectively the eigenfunctions and eigenvalues of $\mathbf{K}$. Then from the eigenvalue decomposition

$$\mathbf{K} = \boldsymbol{\Phi}\boldsymbol{\Lambda}\boldsymbol{\Phi}^T \tag{8.91}$$

where the orthonormality of the eigenfunctions implies that $\boldsymbol{\Phi}$ is an orthonormal matrix, we can deduce that the matric inverse $\mathbf{Q} = \mathbf{K}^{-1}$ satisfies

$$\mathbf{Q} = \boldsymbol{\Phi}\boldsymbol{\Lambda}^{-1}\boldsymbol{\Phi}^T , \tag{8.92}$$

which is the matrix representation of the eigenfunction expansion (8.85) of the kernel $Q(t, s)$.

Thus in the DT case, the eigenfunction expansion (8.85) and inverse kernel identity (8.88) admit straighforward matrix interpretations. However, in the CT case, the solution $Q(t, s)$ of the inverse kernel identity (8.87) needs to be defined in the sense of distributions, since the Dirac delta function appearing on the right-hand side of (8.87) is a distribution, and in the case where $K(t, s)$ is continuous, the corresponding operator K of $L^2[0, T]$ has a smoothing effect, i.e., the function Kf is smoother than f. Since the right-hand side of (8.87) is an impulse, this suggests that $Q(t, s)$ may contain higher order impulses, and thus this kernel cannot be defined as an ordinary function. A case that is relatively easy to analyze corresponds to the situation where $V(t)$ is the sum of a zero-mean mean-square continuous Gaussian component with covariance $K_{\text{C}}(t, s)$ and an independent zero-mean WGN component with intensity σ^2, so that

$$K(t, s) = K_{\text{C}}(t, s) + \sigma^2 \delta(t - s) . \tag{8.93}$$

In this case, $Q(t, s)$ admits a decomposition of the form

$$Q(t, s) = \sigma^{-2}[-H(t, s) + \delta(t - s)] , \tag{8.94}$$

where the kernel $H(t, s)$ is called the Fredholm resolvent. Then, by plugging in the operator decompositions $K = K_{\text{C}} + \sigma^2 I$ and $Q = \sigma^{-2}[-H + I]$ in the operator identity $QK = I$, we obtain

$$H + \sigma^{-2} H K_C = \sigma^{-2} K_C$$

or equivalently, in kernel form

$$H(t,s) + \sigma^{-2} \int H(t,u) K_C(u,s) du = \sigma^{-2} K_C(t,s) \tag{8.95}$$

for $0 \leq t,\ s \leq T$, which specifies a Fredholm integral equation of the second kind for $H(t,s)$. From this equation it is easy to verify that $H(t,s)$ has the same degree of smoothness as $K_C(t,s)$. We refer the reader to [4] for a discussion of the properties of Fredholm integral equations of the second kind. Note however that, even in this case, as indicated by (8.94) $Q(t,s)$ contains an impulse component.

Generalized matched filter implementation: Assuming that the kernel $Q(t,s)$ can be evaluated in an appropriate sense, the optimal test is given by (8.81), where the sufficient statistic S and energy E are specified by (8.82). However, it is possible to simplify these expressions so that the test takes a form similar to the correlator/matched filter implementation (8.17)–(8.18) used in the WGN case. To do so, we introduce the *distorted signal*

$$\begin{aligned} g(t) &= \int_0^T Q(t,u) s(u) du \quad \text{(CT case)} \\ &= \sum_{t=0}^{T} Q(t,u) s(u) \quad \text{(DT case)} . \end{aligned} \tag{8.96}$$

Then, the optimum test can be rewritten as

$$S \underset{H_0}{\overset{H_1}{\gtrless}} \gamma = \frac{E}{2} + \ln(\tau) \tag{8.97}$$

where

$$\begin{aligned} S &= \int_0^T Y(t) g(t) dt \quad \text{(CT case)} \\ &= \sum_{t=0}^{T} Y(t) g(t) \quad \text{(DT case)} , \end{aligned} \tag{8.98}$$

and

$$\begin{aligned} E &= \int_0^T s(t) g(t) dt \quad \text{(CT case)} \\ &= \sum_{t=0}^{T} s(t) g(t) \quad \text{(DT case)} . \end{aligned} \tag{8.99}$$

The test (8.97)–(8.99) is in the form (8.17)–(8.18), except that instead of employing the actual signal $s(t)$ in the correlator receiver of Fig. 8.1, we employ the distorted signal $g(t)$. Similarly for the matched filter implementation of Fig. 8.2, the filter impulse response is selected as

$$h(t) = g(T - t) \tag{8.100}$$

for $0 \leq t \leq T$, instead of $s(T-t)$, yielding what is usually called a *generalized matched filter* receiver.

8.4.3 Computation of the Distorted Signal $g(t)$

The distorted signal $g(t)$ is thus the key quantity needed to implement the generalized correlator/integrator or matched filter receiver. In the DT case, if we let

$$\mathbf{s} = \begin{bmatrix} s(0) \\ \vdots \\ s(t) \\ \vdots \\ s(T) \end{bmatrix} , \quad \mathbf{g} = \begin{bmatrix} g(0) \\ \vdots \\ g(t) \\ \vdots \\ g(T) \end{bmatrix}$$

be the $T+1$ dimensional vectors representing the functions $s(t)$ and $g(t)$ for $0 \leq t \leq T$, and if we observe that $Q(t,s)$ is the kernel representing the inverse of the covariance matrix, the expression (8.96) can be rewritten as

$$\mathbf{g} = \mathbf{K}^{-1}\mathbf{s} \tag{8.101}$$

The distorted signal $\mathbf{g}$ is obtained therefore by applying the inverse of the noise covariance $\mathbf{K}$ to the actual signal $\mathbf{s}$, so in the WGN case where $\mathbf{K} = \sigma^2 \mathbf{I}_{T+1}$, $\mathbf{g} = \mathbf{s}/\sigma^2$, and in this case the generalized matched filter reduces to the standard matched filter, as expected.

The computation of the distorted signal $\mathbf{g}$ is rather simple when the noise $V(t)$ is a DT Gauss-Markov (or Gaussian reciprocal) process, since in this case the inverse covariance matrix $\mathbf{K}^{-1}$ has a tridiagonal (or cyclic tridiagonal) structure.

Example 8.3: DT Ornstein-Uhlenbeck noise process

Assume that the noise $V(t)$ is DT zero-mean Gaussian with covariance function

$$K(t-s) = Pa^{|t-s|}$$

where $|a| < 1$. This process, which is called the DT Ornstein-Ulenbeck process, is a Gauss-Markov process since it admits a first-order state-space model

$$V(t) = aV(t-1) + N(t) \tag{8.102}$$

where $N(t)$ is a zero-mean WGN sequence with variance $q = P(1-a^2)$. Because of its Markov property, the inverse covariance matrix $\mathbf{Q} = \mathbf{K}^{-1}$ of the proces $V(t)$ defined over $[0, T]$ has a tridiagonal structure [5], and the entries Q_{ij} of $\mathbf{Q}$ are given by

$$Q_{ii} = \begin{cases} (1+a^2)q^{-1} & 1 < i < T+1 \\ q^{-1} & i = 1, T+1 \end{cases}$$

$$Q_{ii+1} = -aq^{-1} \text{ for } 1 \leq i < T$$
$$Q_{ii-1} = -aq^{-1} \text{ for } 1 < i \leq T$$

and $Q_{ij} = 0$ for $|i-j| > 1$. In this case, by taking the tridiagonal structure of $\mathbf{Q}$ into account, we find from (8.101) that the distorted function $g(t)$ can be expressed in terms of $s(t)$ as

$$g(t) = q^{-1}[(1+a^2)s(t) - a(s(t-1) + s(t+1))]$$

for $1 \leq t \leq T-1$ and

$$g(0) = q^{-1}[s(0) - as(1)]$$
$$g(T) = q^{-1}[s(T) - as(T-1)] \,.$$

□

In the case when $V(t)$ is not Markov or reciprocal, instead of evaluating $\mathbf{g}$ by first computing the matrix inverse $\mathbf{K}^{-1}$ and then multiplying it by $\mathbf{s}$, it is usually preferable to solve directly the matrix equation

$$\mathbf{K}\mathbf{g} = \mathbf{s} \,. \tag{8.103}$$

This can be accomplished in a variety of ways, depending on whether the covariance matrix $\mathbf{K}$ has any structure. For example, when $V(t)$ is stationary, its covariance function $K(t, s) = K(t-s)$, so the matrix $\mathbf{K}$ has a Toeplitz structure, and equation (8.101) can be solved by using the Levinson algorithm [6, Chap. 5]. When $\mathbf{K}$ has no structure, a standard method of solving (8.103) consists first of performing the lower times upper Cholesky factorization [7, p. 101]

$$\mathbf{K} = \mathbf{L}\mathbf{L}^T \tag{8.104}$$

where $\mathbf{L}$ is a lower triangular matrix, and then solving the systems

$$\mathbf{L}\check{\mathbf{s}} = \mathbf{s}$$

and

$$\mathbf{L}^T\mathbf{g} = \check{\mathbf{s}}$$

by backsubstitution.

Similarly, in the CT case, instead of computing the kernel $Q(t, s)$, which, as we saw earlier, is not defined as an ordinary function, it is preferable to solve the integral equation

$$\int_0^T K(t, u)g(u)du = s(t) \tag{8.105}$$

for $0 \leq t \leq T$, which is obtained by applying the integral operator K to both sides of (8.96) and taking into account identity (8.87). The solution of this equation is rather delicate in the case where $K(t, s)$ is continuous, since it corresponds to a Fredholm equation of the first kind. Then, as noted earlier, K has a smoothing effect, so $g(t)$ will be generally less smooth than $s(t)$, and in particular it may contain impulses, as we shall see below. On the other hand, if the noise $V(t)$ contains both a continuous component and a white noise component, so that $K(t, s)$ admits a decomposition of the form (8.93), the equation (8.105) can be rewritten as a Fredholm equation of the second kind

$$\sigma^2 g(t) + \int_0^T K_C(t, u)g(u)du = s(t) \tag{8.106}$$

for $0 \leq t \leq T$. Then, $g(t)$ has the same degree of smoothness as $s(t)$, and is thus impulse-free. So, in general for all CT detection problems, it is a good idea to assume that a small amount of white noise is present in the observed signal, since this component has the effect of regularizing all subsequent computations. To illustrate the lack of smoothness of solutions of Fredholm integral equations of the first kind, we coinsider the CT version of Example 8.3.

Example 8.4: CT Ornstein-Uhlenbeck noise process

Let $V(t)$ be a zero-mean Gaussian process with covariance function

$$E[V(t)V(s)] = K(t, s) = P\exp(-a|t - s|)$$

with $a > 0$. Then, as noted in Example 7.5, $V(t)$ is Markov, and if we consider the Sturm-Liouville operator

$$L = q^{-1}(-\frac{d^2}{dt^2} + a^2) \tag{8.107}$$

with $q = 2aP$, $K(t, s)$ is the Green's function of L, i.e.,

$$LK(t, s) = \delta(t - s) \tag{8.108}$$

with the separable boundary conditions

$$\begin{aligned} \frac{dK}{dt}(0, s) - aK(0, s) &= 0 \\ \frac{dK}{dt}(T, s) + aK(T, s) &= 0\,. \end{aligned} \tag{8.109}$$

If we apply the operator L to both sides of (8.105) and take into account the Greens's function identity (8.108), we obtain

$$g(t) = Ls(t) = q^{-1}(-\ddot{s}(t) + a^2 s(t)) \tag{8.110}$$

for $0 < t < T$. Unfortunately, this solution is incomplete. To see why this must be so, note that by applying the boundary operator (8.109) to both sides of (8.105) we obtain

$$\begin{aligned} 0 &= \dot{s}(0) - as(0) \\ 0 &= \dot{s}(T) + as(T) \,, \end{aligned} \tag{8.111}$$

which is obviously not satisfied by all signals $s(t)$. To fix this problem, we need to add some impulsive components to the signal $g(t)$ given by (8.110), so that

$$g(t) = Ls(t) + c_0\delta(t) + c_T\delta(t-T) \,.$$

In this case (8.105) becomes

$$\frac{c_0}{2}K(t) + \frac{c_T}{2}K(t-T) + \int_0^T K(t,u)Ls(u)du = s(t) \,, \tag{8.112}$$

where we assume that the integral from 0 to T captures only half of the impulses located at 0 and T. Then, applying the boundary condition (8.109) to both sides of (8.112), we find after some simplifications

$$\begin{aligned} c_0 &= \frac{1}{aP}(-\dot{s}(0) + as(0)) \\ c_T &= \frac{1}{aP}(\dot{s}(T) + as(T)) \,. \end{aligned}$$

In this case, if we adopt again the convention that integrals from 0 to T capture only half of the mass of impulses located at 0 and T, the sufficient statistic S of (8.98) can be expressed as

$$S = \frac{c_0}{2}Y(0) + \frac{c_T}{2}Y(T) + \int_0^T Y(t)Ls(t)dt \,,$$

which illustrates the rather delicate and somewhat heuristic computations needed to solve Fredholm equations of the first kind. □

To demonstrate the regularizing effect of the inclusion of a WGN component in the noise $V(t)$, we consider now the case where a WGN component is added to the Ornstein-Uhlenbeck process of Example 8.4.

Example 8.5: CT Ornstein-Uhlenbeck process in WGN

Consider the case where the covariance of the noise process $V(t)$ takes the form (8.93) with

$$K_C(t,s) = P\exp(-a|t-s|)$$

with $a > 0$. Then by applying the operator L given by (8.106) to both sides of the Fredholm equation the of second kind (8.105) we obtain

$$(\sigma^2 L + 1) g(t) = L s(t) \tag{8.113}$$

for $0 \leq t \leq T$, which specifies a second-order differential equation for $g(t)$. By applying the boundary operators appearing on the left-hand side of (8.108), we find that if $f(t) \triangleq \sigma^2 g(t) - s(t)$, $g(t)$ satisfies the boundary conditions

$$\begin{aligned} \dot{f}(0) - a f(0) &= 0 \\ \dot{f}(T) + a f(T) &= 0 \,. \end{aligned} \tag{8.114}$$

Then let $g_p(t)$ denote a particular solution of the differential equation (8.113), and let $\pm b$ with

$$b = \left(a^2 + \frac{q}{\sigma^2}\right)^{1/2}$$

denote the two roots of the characteristic equation

$$\frac{\sigma^2}{q}(-s^2 + a^2) + 1 = 0$$

associated to the differential equation (8.113), the distorted signal $g(t)$ is given by

$$g(t) = C_+ \exp(bt) + C_- \exp(-bt) + g_p(t) \,,$$

where the constants $C_\pm$ need to be selected so that the boundary conditions (8.114) are satisfied. □

8.4.4 Noise Whitening Receiver

The generalized integrator/correlator or matched filter receiver (8.96)–(8.98) handles the coloring of the noise $V(t)$ by replacing the signal $s(t)$ by an appropriately distorted signal $g(t)$ in the standard integrator/correlator or matched filter receiver. The signal distortion compensates for the effect of the noise coloring. An alternative approach consists of applying a noise whitening filter to the observations $Y(t)$ with $0 \leq t \leq T$ to transform the colored noise problem (8.64) to a WGN detection problem of the type examined in Section 8.2. This method is always applicable in the DT case, provided the covariance matrix $\mathbf{K}$ of the process $V(t)$ defined over $[0, T]$ is positive-definite. But in the CT case, it requires that $V(t)$ should contain a WGN component, so that $K(t, s)$ admits a decomposition of the form (8.93).

DT case: To fix ideas, we consider first the DT case, and let

$$\mathbf{Y} = \begin{bmatrix} Y(0) \\ \vdots \\ Y(t) \\ \vdots \\ Y(T) \end{bmatrix} \qquad \mathbf{s} = \begin{bmatrix} s(0) \\ \vdots \\ s(t) \\ \vdots \\ s(T) \end{bmatrix} \qquad \mathbf{V} = \begin{bmatrix} V(0) \\ \vdots \\ V(t) \\ \vdots \\ V(T) \end{bmatrix}$$

be the $T+1$ dimensional vectors representing the observations, signal and noise, so the binary-detection problem can be rewritten as

$$\begin{aligned} H_0 &: \mathbf{Y} = \mathbf{V} \\ H_1 &: \mathbf{Y} = \mathbf{s} + \mathbf{V} , \end{aligned} \tag{8.115}$$

where $\mathbf{V} \sim N(\mathbf{0}, \mathbf{K})$ under both hypotheses. Then, consider the Cholesky matrix factorization (8.104) of $\mathbf{K}$, and denote by $\ell(t,u)$ the element of row $t+1$ and column $u+1$ of $\mathbf{L}$, where, since $\mathbf{L}$ is lower triangular, we have $\ell(t,u) = 0$ for $u > t$. Note that since $\mathbf{K}$ is positive-definite, $\mathbf{L}$ is necessarily invertible and its inverse

$$\mathbf{F} = \mathbf{L}^{-1} \tag{8.116}$$

is also lower triangular. We denote by $f(t,s)$ the element of row $t+1$ and column $s+1$ of $\mathbf{F}$, where since $\mathbf{F}$ is lower triangular $f(t,u) = 0$ for $u > t$. Consequently $f(t,u)$ can be viewed as the impulse response of a causal time-varying filter.

Consider now the *noise whitening transformation*

$$\check{\mathbf{Y}} = \mathbf{L}^{-1}\mathbf{Y} \quad \check{\mathbf{s}} = \mathbf{L}^{-1}\mathbf{s} \quad \check{\mathbf{V}} = \mathbf{L}^{-1}\mathbf{V} . \tag{8.117}$$

Under this transformation, the covariance of $\check{\mathbf{V}}$ is

$$\begin{aligned} E[\check{\mathbf{V}}\check{\mathbf{V}}^T] &= \mathbf{L}^{-1}E[\mathbf{V}\mathbf{V}^T]\mathbf{L}^{-T} \\ &= \mathbf{L}^{-1}\mathbf{K}\mathbf{L}^{-T} = \mathbf{I}_{T+1}, \end{aligned} \tag{8.118}$$

so the noise vector is now white with unit variance, and accordingly the transformed detection problem

$$\begin{aligned} H_0 &: \check{\mathbf{Y}} = \check{\mathbf{V}} \\ H_1 &: \check{\mathbf{Y}} = \check{\mathbf{s}} + \check{\mathbf{V}} , \end{aligned} \tag{8.119}$$

with $\check{\mathbf{V}} \sim N(\mathbf{0}, \mathbf{I}_{T+1})$ is a conventional WGN detection problem of the type examined in Section 8.2. If $\check{Y}(t)$, $\check{s}(t)$ and $\check{V}(t)$ denote the $t+1$th entries of the vectors $\check{\mathbf{Y}}$, $\check{\mathbf{s}}$ and $\mathbf{V}$, respectively, we have

$$\begin{aligned} \check{Y}(t) &= \sum_{u=0}^{t} f(t,u)Y(u) \\ \check{s}(t) &= \sum_{u=0}^{t} f(t,u)s(u) \\ \check{V}(t) &= \sum_{u=0}^{t} f(t,u)V(u) , \end{aligned} \tag{8.120}$$

so that the transformed observation, signal and noise are obtained by applying the causal time-varying filter $f(t,u)$ to the observation process $Y(t)$, signal $s(t)$

and noise $V(t)$. Because this filter takes the colored noise $V(t)$ and transforms it causally into a white noise process $\check{V}(t)$, $\check{V}(t)$ can be interpreted as the normalized innovations process [8, Section 4.2] associated to the colored noise $V(t)$. Specifically, let

$$\begin{aligned}\hat{V}_p(t) &= E[V(t)|V(s),\, 0 \leq s \leq t-1] \\ &= -\sum_{s=0}^{t-1} a(t,s)V(s)\end{aligned} \tag{8.121}$$

denote the predicted one step ahead MSE estimate of $V(t)$ given observations $\{V(s),\ 0 \leq s \leq t-1\}$, and let

$$\nu(t) \triangleq V(t) - \hat{V}_p(t) \tag{8.122}$$

be the corresponding estimation error. This process is called the innovations process, since it represents the new information contained in $V(t)$ not predictable from past values. By observing that $\nu(s)$ is expressible in terms of the values $\{V(u), 0 \leq u \leq s\}$, and noting that $\nu(t)$ is orthogonal to the observations $V(s)$ for $0 \leq s < t$, we conclude that $\nu(t)$ is a time-varying WGN process with variance $P_\nu(t) = E[\nu^2(t)]$, i.e.,

$$K_\nu(t,s) = P_\nu(t)\delta(t-s). \tag{8.123}$$

The normalized inovations process

$$\check{V}(t) = P_\nu^{-1/2}(t)\nu(t) \tag{8.124}$$

is a WGN with unit variance. From (8.121)–(8.122) we find therefore that the filter $f(t,s)$ can be expressed as

$$f(t,s) = \begin{cases} P_\nu^{-1/2}(t) & s = t \\ P_\nu^{-1/2}(t)a(t,s) & 0 \leq s < t\,, \end{cases} \tag{8.125}$$

which shows that it can be obtained by standard optimal Wiener or Kalman filtering techniques [6, 8].

The resulting noise whitening receiver is shown in Fig. 8.7. The observed signal $Y(t)$ and transmitted signal $s(t)$ are passed through the whitening filter $f(t,\cdot)$, yielding the signals $\check{Y}(t)$ and $\check{s}(t)$, to which a standard correlator/integrator receiver is applied. The optimum test takes the form

$$S \underset{H_0}{\overset{H_1}{\gtrless}} \gamma = \frac{E}{2} + \ln(\tau) \tag{8.126}$$

where the sufficient statistic

$$S = \sum_{t=0}^{T} \check{Y}(t)\check{s}(t) \tag{8.127}$$

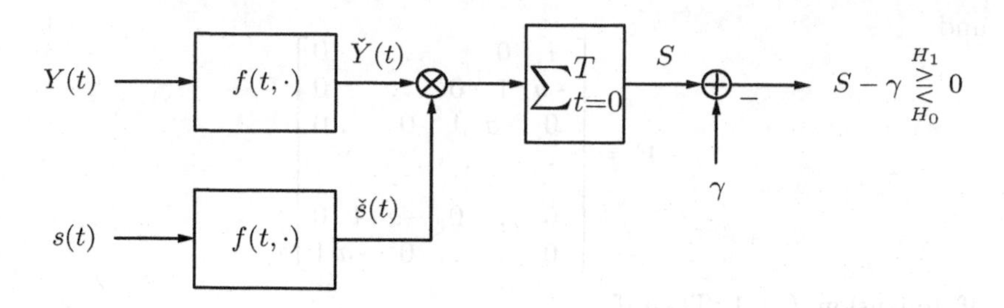

Fig. 8.7. Noise whitening receiver for colored noise detection.

and signal energy

$$E = \sum_{t=0}^{T} \check{s}^2(t) \tag{8.128}$$

are actually identical to the expressions (8.98) and (8.99) obtained for the generalized matched filter implementation of the colored noise receiver. To see why this is the case, note that

$$\begin{aligned} S &= \check{\mathbf{Y}}^T \check{\mathbf{s}} = \mathbf{Y}^T \mathbf{L}^{-T} \mathbf{L}^{-1} \mathbf{s} \\ &= \mathbf{Y}^T \mathbf{K}^{-1} \mathbf{s} = \mathbf{Y}^T \mathbf{g} = \sum_{t=0}^{T} Y(t) g(t) \end{aligned} \tag{8.129}$$

and

$$\begin{aligned} E &= \check{\mathbf{s}}^T \check{\mathbf{s}} = \mathbf{s}^T \mathbf{L}^{-T} \mathbf{L}^{-1} \mathbf{s} \\ &= \mathbf{s}^T \mathbf{K}^{-1} \mathbf{s} = \mathbf{s}^T \mathbf{g} = \sum_{t=0}^{T} s(t) g(t) \,. \end{aligned} \tag{8.130}$$

Example 8.6: Whitened DT Ornstein-Uhlenbeck noise

Consider the DT Ornstein-Uhlenbeck process of Example 8.3. Because it admits the state-space model (8.102) we conclude that the noise-whitening transformation takes the form

$$\check{V}(0) = P^{-1/2} V(0)$$

and

$$\check{V}(t) = q^{-1/2} N(t) = q^{-1/2} (V(t) - aV(t-1))$$

for $t \geq 1$. Equivalently, consider the matrices

$$\mathbf{P}_\nu = \mathrm{diag}\{P, q, \ldots, q\}$$

and

$$\bar{\mathbf{F}} = \begin{bmatrix} 1 & 0 & & \dots & & 0 \\ -a & 1 & 0 & \dots & & 0 \\ 0 & -a & 1 & 0 & \dots & 0 \\ \vdots & \ddots & \ddots & \ddots & & \vdots \\ 0 & \dots & 0 & -a & 1 & 0 \\ 0 & & \dots & 0 & -a & 1 \end{bmatrix}$$

of dimension $T+1$. Then if

$$\mathbf{F} = \mathbf{P}_{\nu}^{-1/2}\bar{\mathbf{F}} ,$$

by direct multiplication it is easy to verify that the tridiagonal matrix $\mathbf{Q}$ of Example 8.3 satisfies

$$\mathbf{Q} = \mathbf{F}^T\mathbf{F} .$$

In this case, the transformed observations and signal are given by

$$\check{Y}(t) = \begin{cases} P^{-1/2}Y(0) & t = 0 \\ q^{-1/2}(Y(t) - aY(t-1)) & t > 0 , \end{cases}$$

and

$$\check{s}(t) = \begin{cases} P^{-1/2}s(0) & t = 0 \\ q^{-1/2}(s(t) - as(t-1)) & t > 0 . \end{cases}$$

□

CT case: The noise-whitening technique is also applicable to CT detection problems where the noise $V(t)$ admits a decomposition of the form

$$V(t) = V_C(t) + N(t) \quad (8.131)$$

where $V_C(t)$ is a zero-mean mean-square continuous Gaussian process with covariance kernel $K_C(t,s)$ and $N(t)$ is a zero-mean WGN independent of $V_C(t)$ with intensity σ^2, so that the covariance $K(t,s)$ of $V(t)$ admits the decomposition (8.93). Then the counterpart of the lower times upper Cholesky factorization (8.104) is given by the operator factorization [9]

$$K_C + \sigma^2 I = \sigma^2(k + I)(k^* + I) \quad (8.132)$$

where k is a causal Volterra operator, i.e., its kernel $k(t,s)$ satisfies

$$k(t,s) = 0 \text{ for } s > t . \quad (8.133)$$

Here k^* denotes the adjoint operator of k, so that its kernel is given by

$$k^*(t,s) = k(s,t) , \quad (8.134)$$

which indicates it corresponds to an anticausal Volterra operator. Since $k + \sigma I$ is causal, its inverse can be expressed as

$$(k+I)^{-1} = -h + I \tag{8.135}$$

where the operator h is a causal Volterra operator, i.e., $h(t,s) = 0$ for $s > t$. We refer to [10] for a detailed discussion of Volterra integral operators. Accordingly, the Fredholm resolvent operator $-H + I$ admits the factorization

$$-H + I = (-h^* + I)(-h + I)\,. \tag{8.136}$$

By taking the causal part of the operator indentity

$$\sigma^{-2}(-h+I)(K_{\mathrm{C}} + \sigma^2 I) = k^* + I \tag{8.137}$$

we find that h satisfies the Wiener-Hopf equation

$$h + \sigma^{-2}\pi_+(hK_{\mathrm{C}}) = \sigma^{-2}\pi_+(K_{\mathrm{C}})\,, \tag{8.138}$$

where $\pi_+(F)$ denotes the causal part of an arbitrary integral operator F. Equivalently, in kernel form, we have

$$h(t,s) + \sigma^{-2}\int_0^t h(t,u)K_{\mathrm{C}}(u,s)du = \sigma^{-2}K_{\mathrm{C}}(t,s) \tag{8.139}$$

for $t \geq s$.

Then, assuming that $h(t,s)$ has been computed by solving the integral equation (8.139), consider the noise-whitening transformation

$$\begin{aligned}
\check{Y}(t) &= \sigma^{-1}[Y(t) - \int_0^t h(t,u)Y(u)du] \\
\check{s}(t) &= \sigma^{-1}[s(t) - \int_0^t h(t,u)s(u)du] \\
\check{V}(t) &= \sigma^{-1}[V(t) - \int_0^t h(t,u)V(u)du]\,.
\end{aligned} \tag{8.140}$$

Since

$$\sigma^{-2}(-h+I)(K_{\mathrm{C}} + \sigma^2 I)(-h^* + I) = I\,, \tag{8.141}$$

we conclude that $\check{V}(t)$ is a zero-mean WGN with unit intensity, so that the transformed problem

$$\begin{aligned}
H_0 &: \check{Y}(t) = \check{V}(t) \\
H_1 &: \check{Y}(t) = \check{s}(t) + \check{V}(t)
\end{aligned} \tag{8.142}$$

reduces to the detection of a known signal in WGN, to which the estimator/correlator or matched filter receivers of Section 8.2 can be applied. Again $h(t,s)$ admits a stochastic interpretation as an optimal Wiener filter for estimating the smooth component $V_{\mathrm{C}}(t)$ from the observations $\{V(s),\ 0 \leq s \leq t\}$. However, the implementation of a CT time-varying filter of this type is rather impractical, so it is usually preferable to implement the noise-whitening receiver in the DT domain, after applying an A/D converter to the received observations.

8.5 Bibliographical Notes

The detection of known DT or CT signals in Gaussian noise represents of course the central topic of signal detection, and it is treated at length in virtually all detection texts, such as [11–14]. While the case of DT or CT detection in WGN is relatively straightforward, CT detection in colored noise is a far more technical subject and a rigorous treatment requires either the use of reproducing kernel Hilbert spaces [15] or [14, Chap. 6] the use of Radon-Nikodym derivatives of measures and Grenander's theorem for establishing the convergence of the likelihood ratio as the number of Karhunen-Loève coefficients tends to infinity. Since the above mentioned concepts are beyond the background assumed for our readers, the approach we have followed is somewhat heuristic, in the style of [11–13]. The survey paper [16] provides a valuable historical perspective on Gaussian detection and its ramifications.

8.6 Problems

8.1. Consider a four-phase (QPSK) communication system with signals

$$\begin{aligned} s_0(t) &= A\sin(\omega_0 t) \\ s_1(t) &= A\sin(\omega_0 t + \pi/2) \\ s_2(t) &= A\sin(\omega_0 t + \pi) \\ s_3(t) &= A\sin(\omega_0 t + 3\pi/2) \end{aligned}$$

for $0 \le t \le T$. Assume the additive noise is white Gaussian with intensity σ^2. Assume equal a-priori probabilities and equal costs for errors.

(a) Obtain a signal space representation for the signals $s_i(t)$, $0 \le i \le 3$. What is the optimum receiver? Draw the decision regions in signal space. How many correlators or matched filters are required?
(b) Show that the probability of a correct decision can be expressed as

$$P[C] = (1 - Q(d))^2 ,$$

with $d = E^{1/2}/\sigma$, where the waveform energy is $E = A^2T/2$.
(c) How does this compare with the probability of a correct decision for the binary-phase (BPSK) case?

8.2. Consider an idealized single user ultrawideband (UWB) communication system where during each baud period T, a user transmits a monocycle pulse

$$s(t) = \begin{cases} A\sin(2\pi t/T_c) & 0 \le t < T_c \\ 0 & \text{otherwise} \end{cases}$$

with period $T_c = T/M$ where M is an integer. The user transmits $m = \log_2 M$ bits per time slot by employing a pulse position modulation (PPM scheme). Specifically, the monocycle is transmitted in any one of the time slots $[kT_c, (k+1)T_c]$ with $0 \leq k \leq M-1$, so that when the k-th time slot is selected, the received signal is

$$Y(t) = s(t - kT_c) + V(t) ,$$

where $V(t)$ denotes a zero-mean WGN with intensity σ^2.

(a) Formulate the design of an optimum receiver as an M-ary hypothesis testing problem. Construct a signal space representation of the detection problem and find the decision rule minimizing the probability of error.
(b) Show that the optimum receiver can be implemented with a single matched filter which is sampled at the cycle period T_c instead of the baud period T.
(c) Evaluate the probability of error of the optimum receiver. To do so, you may want to examine Example 8.2.

8.3. Consider a pulse amplitude modulated (PAM) system in which messages are coded into an odd number $M = 2n+1$ of symbols corresponding to the received signals

$$s_i(t) = A_i p(t)$$

where the amplitude levels

$$A_i = \frac{A}{n}(i - (n+1))$$

with $1 \leq i \leq M = 2n+1$ are evenly spaced about zero, and the pulse $p(t)$ has unit energy, i.e.,

$$\int_0^T p^2(t)dt = 1 .$$

A denotes here the maximum amplitude value. The signals are equally likely, so that each signal occurs with probability $1/M$. During transmission, the signal is corrupted by additive white Gaussian noise with intensity σ^2.

(a) Draw a signal space representation and obtain the optimal receiver for deciding among the M possible transmitted PAM signals.
(b) Show that the average probability of symbol error for the optimum receiver can be expressed as

$$P[E] = 2(1 - \frac{1}{M})Q\Big(\frac{h}{2\sigma}\Big)$$

where $h = A/n = 2A/(M-1)$ is the distance between two adjacent signal points. To obtain the above expression for the average probability of error, it is convenient to first evaluate the average probability P_C of a correct decision, and use the fact that $P_E = 1 - P_C$.

(c) Use the identity

$$\sum_{i=0}^{n} i^2 = \frac{1}{3} n(n+1/2)(n+1)$$

to verify that average energy of the PAM signal can be expressed as

$$E_{ave} = \frac{A^2}{3} \frac{M+1}{M-1} ,$$

and use this expression to show that the average probability of error can be expressed in terms of E_{ave} as

$$P[E] = 2(1-\frac{1}{M}) Q\Big(\sqrt{\frac{3E_{ave}}{(M^2-1)\sigma^2}}\Big) .$$

8.4. Minimum shift keying (MSK) is a form of binary frequency shift keying, where the phase of the modulated signal remains continuous. Specifically, the transmitted signal takes the form

$$s(t) = (2E/T)^{1/2} \cos(\omega_c t + \Theta(t)) ,$$

where, depending on whether the transmitted symbol over the interval $kT \leq t \leq (k+1)T$ is $I_k = 1$ or $I_k = -1$, we have

$$H_0 : \frac{d\Theta}{dt}(t) = -\frac{\pi}{2T}$$
$$H_1 : \frac{d\Theta}{dt}(t) = \frac{\pi}{2T} .$$

Integrating the above identities gives

$$\Theta(t) = \Theta(kT) + I_k \frac{\pi}{2T}(t - kT)$$

with

$$\Theta(kT) = \Theta(0) + \frac{\pi}{2} \sum_{\ell=0}^{k-1} I_\ell ,$$

so that over a signaling interval $[kT, (k+1)T]$, depending on whether $I_k = 1$ or $I_k = -1$ is transmitted, the phase Θ increases or decreases by $\pi/2$. Assuming that $\Theta(0) = 0$, this implies that the phase $\Theta(kT)$ is 0 or π for k even, and equals $\pi/2$ or $-\pi/2$ for k odd. In the following, we assume that the carrier frequency ω_c is an integer multiple of the baud frequency $\omega_b = 2\pi/T$. This implies that the frequencies $\omega_1 = \omega_c + \pi/2T$ and $\omega_0 = \omega_c - \pi/2T$ are such that $\cos(\omega_1 t + \phi_1)$ and $\cos(\omega_0 t + \phi_0)$ are orthogonal over any interval of length T, independently of the phases ϕ_0 and ϕ_1.

During transmission, the MSK signal $s(t)$ is corrupted by some additive zero-mean WGN $V(t)$ with intensity σ^2. Over the interval $kT \leq t \leq (k+1)T$, the received signal $Y(t)$ can therefore be expressed as

$$Y(t) = (2E/T)^{1/2} \cos((\omega_c + I_k \pi/2T)t + \Phi_k) + V(t)$$

with

$$\Phi_k = (\sum_{\ell=0}^{k-1} I_\ell - k I_k)\pi/2 .$$

Thus the continuity condition placed on the phase $\Theta(t)$ implies that the phase during the k-th signalling interval depends not only on the current symbol, but all the prior symbols. As a consequence of this coupling, the optimum detector is actually a maximum a posteriori sequence detector of the type discussed in Chapter 12.

However, the MAP sequence detector has a high complexity. A suboptimal receiver can be obtained by observing that, through the use of trigonometric identies, the MSK signal $s(t)$ admits the linear in-phase/quadrature (I-Q) representation

$$s(t) = s_c(t)\cos(\omega_c t) - s_s(t)\sin(\omega_c t)$$

for $t \geq 0$, where

$$s_c(t) = \sum_{k=0}^{\infty} c(k) p(t - 2kT)$$

$$s_s(t) = \sum_{k=1}^{\infty} s(k) p(t - (2k-1)T)$$

are two digitally modulated baseband signals. In this expression the signaling pulse $p(t)$ takes the form

$$p(t) = \begin{cases} (2E/T)^{1/2}\cos(\pi t/2T) & -T \leq t \leq T \\ 0 & \text{otherwise} \end{cases},$$

and

$$c(k) = \cos(\frac{\pi}{2}\sum_{\ell=0}^{2k-1} I_\ell) \qquad s(k) = \sin(\frac{\pi}{2}\sum_{\ell=0}^{2(k-1)} I_\ell)$$

for $k \geq 1$, with $c(0) = 1$. The above I-Q decomposition shows that the MSK signal can be viewed as an offset QPSK signal with baud rate $1/2T$, where the in-phase and quadrature baseband signals $s_c(t)$ and $s_s(t)$ employ the signaling pulse $p(t)$.

Consider the receiver shown in Fig. 8.8, where $h(t)$ denotes the impulse response of a linear time-invariant filter.

(a) Verify that $c(k)$ and $s(k)$ take on the values $+1$ or -1, so that the problem of estimating these coefficients can be formulated as a binary hypothesis testing problem. Find the optimum filter $h(t)$ and the optimum decision rule for deciding whether $c(k)$ (or $s(k)$) is 1 or -1.

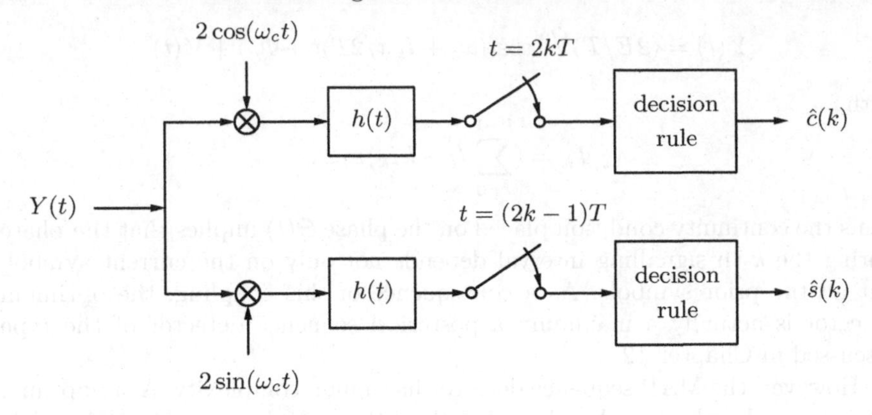

Fig. 8.8. Block diagram of a suboptimal receiver for MSK signals.

(b) Evaluate the probability of error for deciding whether $c(k)$ (or $s(k)$) is 1 or -1.

(c) Given estimates of $c(k)$ and $s(k)$, describe a procedure for recovering the binary symbols I_k. What is the probability of error of this decision rule?

8.5. A CT signal

$$s(t) = A\cos(2\pi t/T) + B\cos(4\pi t/T)$$

is to be detected in zero-mean Gaussian noise $V(t)$ with autocovariance function

$$K_V(t,s) = C\cos(2\pi t/T)\cos(2\pi s/T) + D\cos(4\pi t/T)\cos(4\pi s/T)\,,$$

where C and D are two positive constants. The received signal $Y(t)$ is observed over the interval $[0,T]$. Under hypothesis H_0, we have

$$Y(t) = V(t)$$

and under H_1,

$$Y(t) = s(t) + V(t)\,.$$

Describe in detail the optimal Neymann-Pearson detector, and calculate the probability P_D of detection, showing how it depends on the false alarm probability P_F.

8.6. Consider the CT detection problem

$$H_0 : Y(t) = A\sin(\omega_0 t) + Z(t) + V(t)$$
$$H_1 : Y(t) = A\cos(\omega_0 t) + Z(t) + V(t)$$

for $0 \le t \le T$, where the amplitude A and frequency ω_0 are known, $V(t)$ is a zero-mean WGN with variance σ^2, and $Z(t)$ is a zero-mean colored Gaussian process, independent of $V(t)$ with covariance function

$$K_Z(t,s) = E[Z(t)Z(s)] = \cos(\omega_0|t-s|)$$

for $0 \leq t, s \leq T$. Assume that H_0 and H_1 are equally likely. Derive the minimum probability of error detector, and find the corresponding probability of error.

8.7. Consider the DT detection problem

$$\begin{aligned} H_0 &: Y(t) = V(t) \\ H_1 &: Y(t) = A + V(t) \end{aligned} \tag{8.143}$$

for $0 \leq t \leq T$, where $V(t)$ is a zero-mean Gaussian process with autocorrelation function

$$K_V(t-s) = \sigma^2 a^{|t-s|} ,$$

where $|a| < 1$. Thus $V(t)$ is a DT Ornstein-Uhlenbeck process.

(a) Use the approach described in Example 8.3 to evaluate the distorted signal $g(t)$ used to implement the optimum detector (8.97)–(8.98).
(b) Evaluate the probability of error of the optimum detector in terms of A, σ^2 and T.

8.8. Consider Problem 8.7.

(a) Use the approach of Example 8.6 to construct a noise whitening transformation that transforms the DT detection problem (8.143) into

$$\begin{aligned} H_0 &: \check{Y}(t) = \check{V}(t) \\ H_1 &: \check{Y}(t) = \check{s}(t) + \check{V}(t) , \end{aligned} \tag{8.144}$$

for $0 \leq t \leq T$, where $\check{V}(t)$ is a zero-mean WGN. Express $\check{V}(t)$, $\check{s}(t)$ and $\check{Y}(t)$ in terms of the signals $V(\cdot)$, $s(\cdot)$ and $Y(\cdot)$ for $t = 0$ and $t \geq 1$.
(b) Construct an optimum detector for the transformed detection problem (8.144). Draw a complete block diagram of the dectector.
(c) Evaluate the probability of error for the optimum detector of part (b).

8.9. Consider the DT detection problem

$$\begin{aligned} H_0 &: Y(t) = V(t) \\ H_1 &: Y(t) = s(t) + V(t) \end{aligned}$$

for $0 \leq t \leq T$, where

$$s(t) = A\cos(\omega_0 t)$$

and $V(t)$ is a DT Ornstein-Uhlenbeck process with autocorrelation function

$$K_V(t-s) = \sigma^2 a^{|t-s|} ,$$

where $|a| < 1$. Assume that $T = k\pi/\omega_0$ with k integer.

(a) Construct an optimum detector for this problem. Note that you can either use the distorted signal approach of receiver (8.97)–(8.98), which requires solving equation (8.103), or you may use a noise whitening transformation.
(b) Evaluate the probability of error for this receiver in terms of A, ω_0, a, σ^2 and N.

8.10. Construct an optimum receiver and evaluate the corresponding probability of error for the case where

$$s(t) = A \sin(\omega_0 t)$$

in Problem 8.9. Assume again that $T = k\pi/\omega_0$ with k integer.

References

1. D. O. North, "An analysis of the factors which determine signal/noise discrimination in pulse-carrier systems," Tech. Rep. PTR-6C, RCA Laboratory, June 1943. Reprinted in *Proc. IEEE*, vol. 51, pp. 1016–1027, July 1963.
2. J. H. Van Vleck and D. Middleton, "A theoretical comparison of the visual, aural and meter reception of pulsed signals in the presence of noise," *J. Applied Phys.*, vol. 17, pp. 940–971, 1946. Originally published in May 1944 as a classified Harvard Radio Research Lab. technical report.
3. J. G. Proakis, *Digital Communications, Fourth Edition.* New York: McGraw-Hill, 2000.
4. H. Hochstadt, *Integral Equations.* New York: Wiley-Interscience, 1989. Reprint edition.
5. B. C. Levy, R. Frezza, and A. J. Krener, "Modeling and estimation of discrete-time Gaussian reciprocal processes," *IEEE Trans. Automatic Control*, vol. 35, pp. 1013–1023, Sept. 1990.
6. M. H. Hayes, *Statistical Digital Signal Processing and Modeling.* New York: J. Wiley & Sons, 1996.
7. A. J. Laub, *Matrix Analysis for Scientists and Engineers.* Philadelphia, PA: Soc. for Industrial and Applied Math., 2005.
8. T. Kailath, A. H. Sayed, and B. Hassibi, *Linear Estimation.* Upper Saddle River, NJ: Prentice Hall, 2000.
9. J. R. Gabriel and S. M. Kay, "On the relationship between the GLRT and UMPI tests for the detection of signals with unknown parameters," *IEEE Trans. Signal Proc.*, vol. 53, pp. 4194–4203, Nov. 2005.
10. I. C. Gohberg and M. G. Krein, *Theory and Applications of Volterra Operators in Hilbert Space*, vol. 24 of *Trans. of Mathematical Monographs.* Providence, RI: Amer. Math. Society, 1970.
11. H. L. Van Trees, *Detection, Estimation and Modulation Theory, Part I: Detection, Estimation and Linear Modulation Theory.* New York: J. Wiley & Sons, 1968. Paperback reprint edition in 2001.
12. R. N. McDonough and A. D. Whalen, *Detection of Signals in Noise, Second Edition.* San Diego, CA: Academic Press, 1995.
13. C. W. Helstrom, *Elements of Signal Detection & Estimation.* Upper Saddle River, NJ: Prentice-Hall, 1995.

14. H. V. Poor, *An Introduction to Signal Detection and Estimation, Second Edition.* New York: Springer Verlag, 1994.
15. T. Kailath, "RKHS approach to detection and estimation problems–part i: Deterministic signals in Gaussian noise," *IEEE Trans. Informat. Theory*, vol. 17, pp. 530–549, 1971.
16. T. Kailath and H. V. Poor, "Detection of stochastic processes," *IEEE Trans. Informat. Theory*, vol. 44, pp. 2230–2259, Oct. 1998.

9

Detection of Signals with Unknown Parameters

9.1 Introduction

In this chapter we investigate the detection of signals with unknown parameters in WGN. Two approaches can be employed to formulate problems of this type. If the parameters are viewed as unknown but random, one needs only to specify probability densities for the unknown parameters, and then standard Bayesian detection results are applicable. On the other hand, when the parameters are unknown but deterministic, we must either apply a UMPI test, if one exists, or design a GLR detector. In the GLR approach, the unknown parameters are first estimated, and then a likelihood ratio test is applied with the unknown parameters replaced by their ML estimates.

For CT detection problems, another issue that arises when signals have unknown parameters is whether they stay within a finite dimensional subspace as the vector $\mathbf{x}$ of unknown parameter varies. When the signals stay within a finite dimensional subspace, the signal space technique can be employed to convert the CT detection problem into an equivalent finite-dimensional composite hypothesis testing problem of the type examined in Chapter 5. On the other hand, when the signal of interest $s(t, \mathbf{x})$ spans an infinite-dimensional space, finite-dimensional techniques of the type used to identify UMPI tests are no longer applicable. Yet, as we shall see below, Bayesian and GLR detection techniques can be appropriately extended to this class of problems.

In Section 9.2 we examine the incoherent detection of bandpass signals with an unknown phase, whereas Section 9.4 considers the case where the signals have both an unknown amplitude and phase. Problems of this last type arise naturally in digital communications over fading channels. In both instances, the signal of interest stays in a two-dimensional subspace as the phase or amplitude parameters vary, thus allowing the use of signal space techniques. Both Bayesian and GLR detectors are designed, and it is shown that the GLR detectors are UMPI. Depending on the type of random channel model employed for the phase or amplitude, the Bayesian or GLR tests may coincide in some instances, and differ in other cases.

B.C. Levy, *Principles of Signal Detection and Parameter Estimation*,
DOI: 10.1007/978-0-387-76544-0_9, © Springer Science+Business Media, LLC 2008

Then in Section 9.5, we examine the detection of CT signals with arbitrary parameters, such as time delay or Doppler frequency shift, where the signal now spans an infinite dimensional space as the vector $\mathbf{x}$ of unknown parameters varies. A general likelihood ratio formula is derived, which can be used to design either Bayesian or GLR detectors. This is illustrated by the design of a GLR detector for a signal with unknown amplitude and time delay. The computation of ML estimates and associated CRLB for general CT signals is described in Section 9.6. Finally, the methods developed in this chapter are illustrated by considering the detection of radar signals in Section 9.7. Such signals include unknown phase, amplitude, time delay, and Doppler frequency shift parameters which must be estimated. A GLR detector is derived, and the ambiguity function is introduced to discuss signal pulse design.

9.2 Detection of Signals with Unknown Phase

In this section, we examine the detection of a signal $s(t, \theta)$ with an unknown phase θ in the presence of WGN. So we consider the binary detection problem

$$\begin{aligned} H_0 &: Y(t) = V(t) \\ H_1 &: Y(t) = s(t, \theta) + V(t) \end{aligned} \tag{9.1}$$

over $0 \leq t \leq T$, where $V(t)$ is a zero-mean WGN with covariance

$$E[V(t)V(s)] = \sigma^2 \delta(t - s) \, . \tag{9.2}$$

As in the previous chapter, we consider the CT and DT cases simultaneously. To model the signal $s(t, \theta)$, we assume it admits a complex envelope representation [1, Appendix A2.4] of the form

$$s(t, \theta) = \Re\{\tilde{s}(t) \exp(j\omega_c t + \theta)\} \, , \tag{9.3}$$

where

$$\tilde{s}(t) = c(t) \exp(j\psi(t)) \tag{9.4}$$

is a known complex baseband signal with amplitude $c(t)$ and phase $\psi(t)$. The bandwidth B of $\tilde{s}(t)$ is assumed to be much smaller than ω_c. This ensures that $\tilde{s}(t)$, $c(t)$, and $\psi(t)$ are approximately constant over one period of the complex carrier signal $\exp(j\omega_c t)$. Similarly, it is assumed that the baud frequency $\omega_b = 2\pi/T$ is much smaller than ω_c. The signal $s(t, \theta)$ can of course be rewritten in real form as

$$s(t, \theta) = c(t) \cos(\omega_c t + \psi(t) + \theta) \, . \tag{9.5}$$

Thus the only unknown parameter is the phase θ. In the Bayesian formulation of the above problem, it will be treated as random with the uniform probability density

$$f_\Theta(\theta) = \begin{cases} \frac{1}{2\pi} & 0 \leq \theta < 2\pi \\ 0 & \text{otherwise} . \end{cases} \tag{9.6}$$

In contrast, for the composite hypothesis testing formulation of this problem, θ will be viewed as nonrandom but unknown. In this case, as we shall see, the GLRT is UMPI and the detector obtained by this approach has the same structure as the Bayesian detector.

9.2.1 Signal Space Representation

Note that the model (9.3) we are employing for the signal $s(t,\theta)$ is quite general since it allows the use of both phase and amplitude modulation. The signal $s(t,\theta)$ can be expressed as

$$s(t,\theta) = s_c(t)\cos(\theta) - s_s(t)\sin(\theta) \tag{9.7}$$

where

$$\begin{aligned} s_c(t) &= c(t)\cos(\omega_c t + \psi(t)) \\ s_s(t) &= c(t)\sin(\omega_c t + \psi(t)) . \end{aligned} \tag{9.8}$$

denote the in-phase and quadrature signal components. By taking into account the fact that $c(t)$ and $\psi(t)$ vary slowly compared to the carrier waveform, we find that the waveforms $s_c(t)$ and $s_s(t)$ have for energy

$$\begin{aligned} E_c &= \int_0^T s_c^2(t)dt \\ &= \frac{1}{2}\int_0^T c^2(t)[1+\cos(2(\omega_c t dt + \psi(t)))]dt \approx E \end{aligned} \tag{9.9}$$

and

$$\begin{aligned} E_s &= \int_0^T s_s^2(t)dt \\ &= \frac{1}{2}\int_0^T c^2(t)[1-\cos(2(\omega_c t dt + \psi(t)))]dt \approx E , \end{aligned} \tag{9.10}$$

with

$$E \triangleq \frac{1}{2}\int_0^T c^2(t)dt . \tag{9.11}$$

The waveforms $s_c(t)$ and $s_s(t)$ are approximately orthogonal, since we have

$$\int_0^T s_c(t)s_s(t)dt = \frac{1}{2}\int_0^T c^2(t)\sin(2(\omega_c t + \psi(t))dt \approx 0 . \tag{9.12}$$

In the DT case, the integrals in (9.9)–(9.12) need to be replaced by summations. Consequently, we can construct a COB of $L^2[0,T]$ such that the first two functions are

$$\phi_1(t) = E^{-1/2} s_c(t) \ , \ \phi_2(t) = E^{-1/2} s_s(t) \ , \tag{9.13}$$

and the remaining functions are obtained by Gram-Schmidt orthonormalization. By performing a Karhunen-Loève expansion

$$\begin{aligned} V(t) &= \sum_{j \in \mathcal{J}} V_j \phi_j(t) \\ Y(t) &= \sum_{j \in \mathcal{J}} Y_j \phi_j(t) \end{aligned} \tag{9.14}$$

of $V(t)$ and $Y(t)$ with respect to this basis, where the noise coefficients

$$V_j =< V, \phi_j >$$

form an $N(0, \sigma^2)$ WGN sequence, we find that under both hypotheses $Y_j = V_j$ for $j \geq 3$, so that the coefficients Y_j with $j \geq 3$ can be discarded. Then let

$$\mathbf{Y} = \begin{bmatrix} Y_1 \\ Y_2 \end{bmatrix} \ , \ \mathbf{V} = \begin{bmatrix} V_1 \\ V_2 \end{bmatrix} \ , \tag{9.15}$$

where

$$\begin{aligned} Y_1 &= E^{-1/2} \int_0^T Y(t) c(t) \cos(\omega_c t + \psi(t)) dt \\ Y_2 &= E^{-1/2} \int_0^T Y(t) c(t) \sin(\omega_c t + \psi(t)) dt \ . \end{aligned} \tag{9.16}$$

If we denote by

$$\mathbf{S}(\theta) = E^{1/2} \begin{bmatrix} \cos(\theta) \\ -\sin(\theta) \end{bmatrix} \ , \tag{9.17}$$

the signal space representation of $s(t, \theta)$, we find that the detection problem (9.1) reduces to the two dimensional composite hypothesis testing problem

$$\begin{aligned} H_0 &: \mathbf{Y} = \mathbf{V} \\ H_1 &: \mathbf{Y} = \mathbf{S}(\theta) + \mathbf{V} \end{aligned} \tag{9.18}$$

which is of the type considered in Examples 5.6 and 5.9. Here $\mathbf{V}$ admits an $N(\mathbf{0}, \sigma^2 \mathbf{I}_2)$ distribution. Geometrically, the vector $\mathbf{S}(\theta)$ is located on a circle of radius $E^{1/2}$, with an unknown polar angle $-\theta$, as shown in Fig. 9.1.

9.2.2 Bayesian Formulation

When Θ is random with the uniform probability density (9.6), the probability density of $\mathbf{Y}$ under H_1 is given by

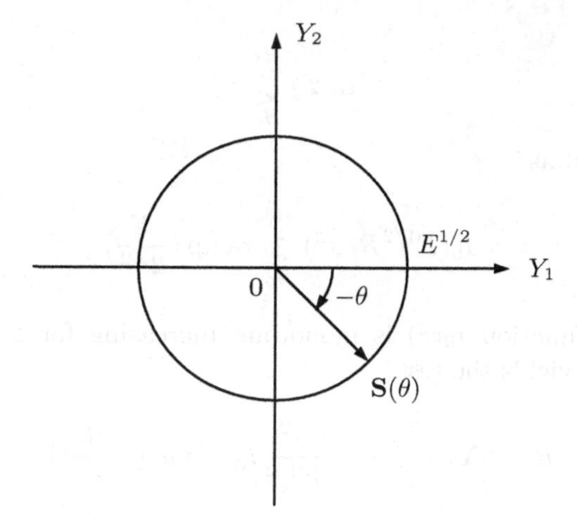

Fig. 9.1. Two-dimensional signal space representation of the signal $s(t, \theta)$.

$$f(\mathbf{y}|H_1) = \int f(\mathbf{y}|H_1, \theta) f_\Theta(\theta) d\theta$$
$$= \frac{1}{2\pi\sigma^2} \exp\big(-\frac{||\mathbf{y}||_2^2 + E}{2\sigma^2}\big) \frac{1}{2\pi} \int_0^{2\pi} \exp(\mathbf{y}^T \mathbf{S}(\theta)/\sigma^2) d\theta \,. \quad (9.19)$$

To evaluate the integral (9.19), we express the observation components Y_1 and Y_2 in polar coordinates as

$$Y_1 = R\cos(\Phi) \quad , \quad Y_2 = R\sin(\Phi)$$

with

$$R = ||\mathbf{Y}||_2 = (Y_1^2 + Y_2^2)^{1/2} \quad , \quad \Phi = \tan^{-1}(Y_2/Y_1) \,,$$

so that

$$\mathbf{Y}^T \mathbf{S}(\theta) = RE^{1/2} \cos(\Phi + \theta) \,. \quad (9.20)$$

Then, by using the definition

$$I_0(z) = \frac{1}{2\pi} \int_0^{2\pi} \exp(z\cos(\theta - \phi) d\theta$$

of the modified Bessel function of zero-th order, where ϕ denotes an arbitrary angle, we find

$$f(\mathbf{y}|H_1) = \frac{1}{2\pi\sigma^2} \exp\big(-\frac{||\mathbf{y}||_2^2 + E}{2\sigma^2}\big) I_0(E^{1/2}||\mathbf{y}||_2/\sigma^2) \,. \quad (9.21)$$

The likelihood-ratio function is therefore given by

$$L(\mathbf{y}) = \frac{f(\mathbf{y}|H_1)}{f(\mathbf{y}|H_0)} = \exp\big(-\frac{E}{2\sigma^2}\big) I_0(E^{1/2}||\mathbf{y}||_2/\sigma^2) \,. \quad (9.22)$$

Then the LRT

$$L(\mathbf{Y}) \underset{H_0}{\overset{H_1}{\gtrless}} \tau$$

can be expressed as

$$I_0(E^{1/2}R/\sigma^2) \underset{H_0}{\overset{H_1}{\gtrless}} \tau \exp\big(\frac{E}{2\sigma^2}\big) \tag{9.23}$$

and since the function $I_0(z)$ is monotone increasing for $z \geq 0$, it can be inverted, which yields the test

$$R = ||\mathbf{Y}||_2 \underset{H_0}{\overset{H_1}{\gtrless}} \eta = \frac{\sigma^2}{E^{1/2}} I_0^{-1}\big(\tau \exp\big(\frac{E}{2\sigma^2}\big)\big) . \tag{9.24}$$

So, just as we might expect, the optimum test consists of selecting H_1 when the magnitude of $||\mathbf{Y}||_2 = (Y_1^2 + Y_2^2)^{1/2}$ is large, and H_0 when it is small.

Test performance: By observing that under H_0, the amplitude R admits the Rayleigh distribution

$$f_R(r|H_0) = \frac{r}{\sigma^2} \exp\big(-\frac{r^2}{2\sigma^2}\big) ,$$

we find that the probability of false alarm can be expressed as

$$\begin{aligned} P_F = P[R > \eta|H_0] &= \int_\eta^\infty f_R(r|H_0) dr \\ &= \int_{\frac{\eta^2}{2\sigma^2}}^\infty \exp(-u) du = \exp\big(-\frac{\eta^2}{2\sigma^2}\big) . \end{aligned} \tag{9.25}$$

Consequently, to implement a Neyman-Pearson test with $P_F \leq \alpha$, the threshold η needs to be selected such that

$$\alpha = P_F = \exp(-\frac{\eta^2}{2\sigma^2}) ,$$

which yields

$$\eta = \sigma(2\ln(1/\alpha))^{1/2} . \tag{9.26}$$

Converting the density (9.21) to polar coordinates, we find that under H_1, the joint density of R and Φ can be expressed as

$$f(r, \phi|H_1) = \frac{r}{2\pi\sigma^2} \exp\big(-\frac{r^2 + E}{2\sigma^2}\big) I_0(E^{1/2} r/\sigma^2) . \tag{9.27}$$

Since it is independent of ϕ, the marginal density of R under H_1 is given by

$$\begin{aligned} f(r|H_1) &= \int_0^{2\pi} f(r, \phi|H_1) d\phi \\ &= \frac{r}{\sigma^2} \exp\big(-\frac{r^2 + E}{2\sigma^2}\big) I_0(E^{1/2} r/\sigma^2) , \end{aligned} \tag{9.28}$$

and thus the probability of detection takes the form

$$\begin{aligned} P_D &= P[R > \eta|H_1] \\ &= \int_\eta^\infty \frac{r}{\sigma^2} \exp\big(-\frac{r^2 + E}{2\sigma^2}\big) I_0(rE^{1/2}/\sigma^2) dr \\ &= Q(\frac{E^{1/2}}{\sigma}, \frac{\eta}{\sigma}) , \end{aligned} \tag{9.29}$$

where Marcum's Q function $Q(\alpha, \beta)$ is defined by (5.57).

9.2.3 GLR Test

Suppose now that we view the phase θ as unknown but nonrandom. From the signal space representation (9.17)–(9.18) of the detection problem we see that the observed signal is independent of θ under H_0, but under H_1, the probability density of $\mathbf{Y}$ is given by

$$\begin{aligned} f(\mathbf{y}|H_1, \theta) &= \frac{1}{2\pi\sigma^2} \exp\big(-\frac{||\mathbf{y} - \mathbf{S}(\theta)||_2^2}{2\sigma^2}\big) \\ &= \frac{1}{2\pi\sigma^2} \exp\big(-\frac{||\mathbf{y}||_2^2 + E}{2\sigma^2}\big) \exp(\mathbf{y}^T \mathbf{S}(\theta)/\sigma^2) . \end{aligned} \tag{9.30}$$

The ML estimate $\hat{\theta}_1$ of θ under H_1 is obtained by maximizing $f(\mathbf{Y}|H_1, \theta)$ with respect to θ, or equivalently, by maximizing

$$\mathbf{Y}^T \mathbf{S}(\theta) = RE^{1/2} \cos(\Phi + \theta) . \tag{9.31}$$

The maximum is obtained by selecting θ so that $\mathbf{S}(\theta)$ is oriented in the same direction as $\mathbf{Y}$, i.e.

$$\hat{\theta}_1(\mathbf{Y}) = -\Phi = -\tan^{-1}(Y_2/Y_1) , \tag{9.32}$$

and in this case

$$f(\mathbf{Y}|H_1, \hat{\theta}_1) = \frac{1}{2\pi\sigma^2} \exp\big(-\frac{(||\mathbf{Y}||_2 - E^{1/2})^2}{2\sigma^2}\big) . \tag{9.33}$$

Accordingly, the GLR statistic can be expressed as

$$\begin{aligned} L_G(\mathbf{Y}) &= \frac{f(\mathbf{Y}|H_1, \hat{\theta}_1)}{f(\mathbf{Y}|H_0)} \\ &= \exp\big(-\frac{E}{2\sigma^2}\big) \exp(||\mathbf{Y}||_2 E^{1/2}/\sigma^2) . \end{aligned} \tag{9.34}$$

Taking logarithms, we find that the GLRT can be expressed as

$$\frac{RE^{1/2}}{\sigma^2} - \frac{E}{2\sigma^2} \underset{H_0}{\overset{H_1}{\gtrless}} \gamma = \ln(\tau) \tag{9.35}$$

or equivalently

$$R = ||\mathbf{Y}||_2 \underset{H_0}{\overset{H_1}{\gtrless}} \eta = \frac{\sigma^2}{E^{1/2}}\gamma + \frac{E^{1/2}}{2}, \tag{9.36}$$

which coincides with the Bayesian test (9.24).

The hypothesis testing problem (9.17)–(9.18) has the feature that it is invariant under the group of rotations about the origin described in Example 5.6, since under a clockwise rotation with an angle α, the parameter θ becomes $\theta - \alpha$. The magnitude $R = ||\mathbf{Y}||_2$ is a maximal invariant statistic under this group of transformations since any two observation $\mathbf{Y}_1$ and $\mathbf{Y}_2$ with the same magnitude are located on the same circle and can thus be transformed into each other by a rotation. By expressing the probability density (9.30) in polar coordinates, we find that the joint density of R and Φ under H_1 is given by

$$f(r, \phi|H_1, \theta) = \frac{r}{2\pi\sigma^2} \exp\big(-\frac{r^2 + E}{2\sigma^2}\big) \exp(rE^{1/2}\cos(\phi + \theta))\,. \tag{9.37}$$

Proceeding as in Example 5.6, the marginal density of R under H_1 is given by

$$f(r|H_1) = \int_0^{2\pi} f(r, \phi|H_1, \theta) = \frac{r}{\sigma^2} \exp\big(-\frac{r^2 + E}{2\sigma^2}\big) I_0(rE^{1/2}/\sigma^2)\,, \tag{9.38}$$

which is independent of θ. Accordingly the LRT

$$L(R) = \frac{f(R|H_1)}{f(R|H_0)} = \exp\big(-\frac{E}{2\sigma^2}\big) I_0(RE^{1/2}/\sigma^2) \underset{H_0}{\overset{H_1}{\gtrless}} \tau \tag{9.39}$$

is UMPI. But this test coincides with the Bayesian test (9.24). Consequently for a signal with an unknown phase, we have shown that the GLRT is UMPI and coincides with the Bayesian test designed for a random phase uniformly distributed over $[0, 2\pi]$.

9.2.4 Detector Implementation

The test (9.24) relies on the projections of $Y(t)$ on the normalized basis functions $\phi_1(t)$ and $\phi_2(t)$, but it is more convenient to implement it in terms of the unnormalized projections

$$\begin{aligned} Z_c &= <Y, s_c> = \int_0^T Y(t)c(t)\cos(\omega_c t + \psi(t))dt \\ Z_s &= <Y, s_s> = \int_0^T Y(t)c(t)\sin(\omega_c t + \psi(t))dt \end{aligned} \tag{9.40}$$

of $Y(t)$ on the in-phase and quadrature components of the signal $s(t)$. Then if

$$Q \triangleq (Z_c^2 + Z_s^2)^{1/2} = E^{1/2}||\mathbf{Y}||_2 \,, \tag{9.41}$$

the test (9.24) can be rewritten as

$$Q \underset{H_0}{\overset{H_1}{\gtrless}} \kappa = E^{1/2}\eta \,. \tag{9.42}$$

A block diagram of the corresponding detector, which is the extension of the integrator/correlator detector to incoherent signals, is shown in Fig. 9.2. Because the phase θ is unknown, the received signal needs to be correlated with both the in-phase and quadrature components $s_c(t)$ and $s_s(t)$ of the transmitted signal. After integration, the outputs of the two correlators/integrators are squared and combined prior to applying a square root operation to generate the statistic Q.

From the definition of Z_c and Z_s in (9.40) we conclude that Q can be expressed as

$$\begin{aligned} Q &= \mid \int_0^T Y(t)c(t)\exp(-j(\omega_c t + \psi(t))dt \mid \\ &= \mid \int_0^T Y(t)\tilde{s}^*(t)\exp(-j\omega_c t)dt \mid \,. \end{aligned} \tag{9.43}$$

Consider the complex envelope representation

$$Y(t) = \Re\{\tilde{Y}(t)\exp(j\omega_c t)\} = \frac{1}{2}[\tilde{Y}(t)\exp(j\omega_c t) + \tilde{Y}^*(t)\exp(-j\omega_c t)] \tag{9.44}$$

of the received signal $Y(t)$. After substitution in (9.43) we find

$$Q = \frac{1}{2} \mid \int_0^T \tilde{Y}(t)\tilde{s}^*(t)dt + \int_0^T \tilde{Y}^*(t)\tilde{s}^*(t)\exp(-j2\omega_c t)dt \mid \,, \tag{9.45}$$

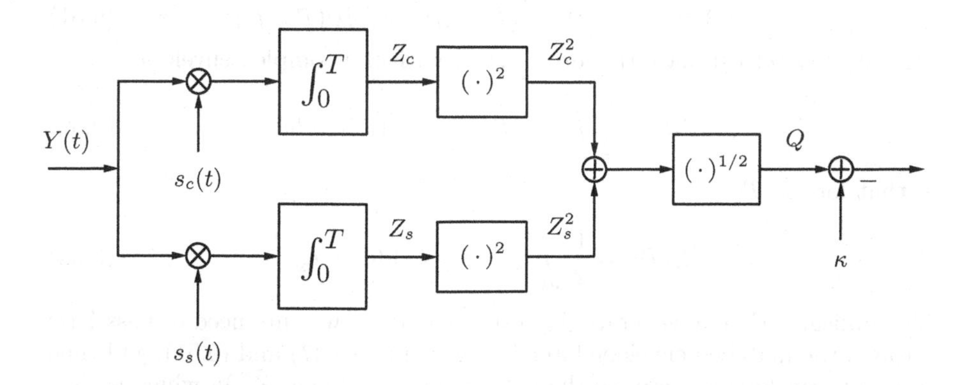

Fig. 9.2. Quadrature receiver for detection of signals with unknown phase.

where the second integral is approximately zero since the complex waveform $\tilde{Y}^*(t)\tilde{s}^*(t)$ varies very slowly compared to periodic function $\exp(-j2\omega_c t)$. This implies

$$Q \approx \frac{1}{2} \mid \int_0^T \tilde{Y}(t)\tilde{s}^*(t)dt \mid , \quad (9.46)$$

so that except for the 1/2 multiplication factor, Q is formed by correlating the complex envelope $\tilde{Y}(t)$ of the received signal $Y(t)$ with the complex conjugate of the complex envelope $\tilde{s}(t)$ of the transmitted signal and then integrating and sampling at $t = T$. So, it is the exact analog of the correlator/integrator receiver, but in terms of the complex envelopes of $Y(t)$ and $s(t, \theta)$. The expression (9.46) immediately suggests an alternate implementation of the detector incoherent detector of Fig. 9.2. Recall that if a narrowband signal $Y(t)$ with complex envelope representation (9.44) is passed through a linear time-invariant narrowband filter whose impulse response $h(t)$ admits the complex envelope representation

$$h(t) = \Re\{\tilde{h}(t) \exp(j\omega_c t)\} , \quad (9.47)$$

then the output

$$Z(t) = h(t) * Y(t) , \quad (9.48)$$

where $*$ denotes the convolution operation, is a narrowband signal with the complex envelope representation

$$Z(t) = \Re\{\tilde{Z}(t) \exp(j\omega_c t)\} , \quad (9.49)$$

where [1, p. 733]

$$\tilde{Z}(t) = \frac{1}{2}\tilde{h}(t) * \tilde{Y}(t) . \quad (9.50)$$

In other words, except for the 1/2 multiplication factor, the complex envelope of the output is the convolution of the complex envelope of the input with the complex envelope of the filter's impulse response. This suggests that if we pass the observed signal $Y(t)$ through a "matched envelope" filter of the form (9.47) with

$$\tilde{h}(t) = \tilde{s}^*(T - t) = c(T - t) \exp(-j\psi(T - t)) \quad (9.51)$$

the output $Z(t)$ will have the form (9.49), with the complex envelope

$$\tilde{Z}(t) = \frac{1}{2} \int_0^T \tilde{s}^*(T - (t - u))\tilde{Y}(u)du \quad (9.52)$$

so that for $t = T$,

$$|\tilde{Z}(T)| = \frac{1}{2} | \int_0^T \tilde{s}^*(u)\tilde{Y}(u)du| = Q . \quad (9.53)$$

This indicates that to generate the test statistic Q, we only need to pass $Y(t)$ through the matched envelope filter $h(t)$ given by (9.47) and (9.51), and then apply an envelope detector to the output $Z(t)$, yielding $|\tilde{Z}(t)|$, which is then sampled at $t = T$, as shown in Fig. 9.3.

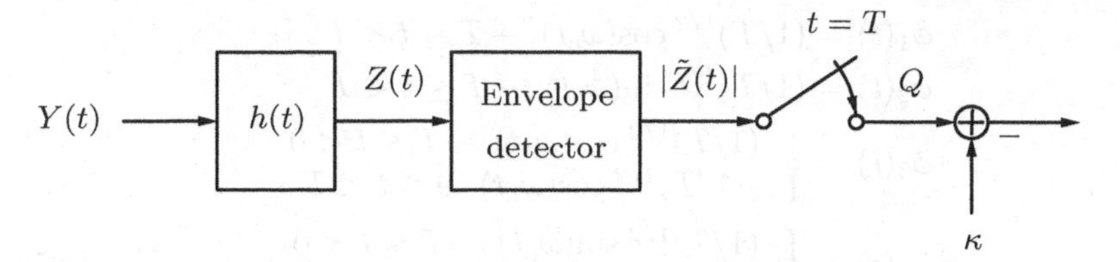

Fig. 9.3. Matched envelope detector for signals with unknown phase.

9.3 Detection of DPSK Signals

In a digital communications context, the detection problem examined in the previous section models an on-off keying (OOK) digital modulation scheme, where a modulated signal is transmitted under H_1, representing the bit $b = 1$, whereas no signal is sent under H_0, corresponding to the bit $b = 0$. The methodology employed above for the OOK case can be extended to the detection of differential phase shift keying (DPSK) or frequency-shift keying (FSK) signals with an unknown phase, but in each instance, the signal space representation of the detection problem needs to be adapted to the modulation format we consider.

To illustrate how this can be achieved, we examine the the detection of DPSK signals. DPSK modulation avoids the need to acquire the phase of a modulated signal by using the *phase changes* between signaling intervals to encode the transmitted bits. Thus a bit $b = 0$ is transmitted by keeping the phase unchanged between signaling interval $[-T, 0)$ and interval $[0, T)$, whereas a bit $b = 1$ is transmitted by a phase shift of π (a sign reversal) between these two intervals. To decide whether the bit $b = 0$ or $b = 1$ has been transmitted, we must therefore solve the detection problem

$$\begin{aligned} H_0 &: Y(t) = s_0(t, \theta_0) + V(t) \\ H_1 &: Y(t) = s_1(t, \theta_1) + V(t) \end{aligned} \tag{9.54}$$

for $-T \leq t < T$, where

$$\begin{aligned} s_0(t, \theta_0) &= \left(\frac{2E}{T}\right)^{1/2} \cos(\omega_c t + \theta_0) \quad -T \leq t < T \\ s_1(t, \theta_1) &= \begin{cases} \left(\frac{2E}{T}\right)^{1/2} \cos(\omega_c t + \theta_1) & -T \leq t < 0 \\ -\left(\frac{2E}{T}\right)^{1/2} \cos(\omega_c t + \theta_1) & 0 \leq t < T \, . \end{cases} \end{aligned} \tag{9.55}$$

The phase θ_j under hypothesis H_j with $j = 0,\ 1$ is assumed to be unknown, $V(t)$ is a WGN with intensity σ^2, and E represents the bit energy, i.e., the waveform $s_j(t, \theta_j)$ has energy E over each signaling interval of length T. To convert this problem to signal space, we consider a COB of $L^2[-T, T)$ such that the first four basis functions are given by

$$\begin{aligned}
\phi_1(t) &= (1/T)^{1/2}\cos(\omega_c t) \ \ -T \le t < T \\
\phi_2(t) &= (1/T)^{1/2}\sin(\omega_c t) \ \ -T \le t < T \\
\phi_3(t) &= \begin{cases} (1/T)^{1/2}\cos(\omega_c t) & -T \le t < 0 \\ -(1/T)^{1/2}\cos(\omega_c t) & 0 \le t < T \end{cases} \\
\phi_4(t) &= \begin{cases} (1/T)^{1/2}\sin(\omega_c t) & -T \le t < 0 \\ -(1/T)^{1/2}\sin(\omega_c t) & 0 \le t < T\,, \end{cases}
\end{aligned} \tag{9.56}$$

and the remaining basis functions ϕ_j with $j \geq 5$ are obtained by Gram-Schmit orthonormalization. With respect to this basis, the transmitted signals can be expressed as

$$\begin{aligned}
s_0(t,\theta_0) &= (2E)^{1/2}[\cos(\theta_0)\phi_1(t) - \sin(\theta_0)\phi_2(t)] \\
s_1(t,\theta_0) &= (2E)^{1/2}[\cos(\theta_1)\phi_3(t) - \sin(\theta_1)\phi_2(4)]\,.
\end{aligned} \tag{9.57}$$

Then, if we perform a KLD of the form (9.14) for both the noise $V(t)$ and observed signal $Y(t)$, we find that under both hypotheses $Y_j = V_j$ for $j \geq 5$, so that all Karhunen-Loève coefficients Y_j with $j \geq 5$ can be discarded. Then, if

$$\begin{aligned}
\mathbf{Y} &= \begin{bmatrix} Y_1 \, Y_2 \, Y_3 \, Y_4 \end{bmatrix}^T \\
\mathbf{V} &= \begin{bmatrix} V_1 \, V_2 \, V_3 \, V_4 \end{bmatrix}^T
\end{aligned}$$

and

$$\begin{aligned}
\mathbf{S}_0(\theta_0) &= (2E)^{1/2}\begin{bmatrix} \cos(\theta_0) \ -\sin(\theta_0) \ 0 \ 0 \end{bmatrix}^T \\
\mathbf{S}_1(\theta_1) &= (2E)^{1/2}\begin{bmatrix} 0 \ 0 \ \cos(\theta_1) \ -\sin(\theta_1) \end{bmatrix}^T\,,
\end{aligned} \tag{9.58}$$

the detection problem (9.54) can be rewritten in signal space as

$$\begin{aligned}
H_0 &: \mathbf{Y} = \mathbf{S}_0(\theta_0) + \mathbf{V} \\
H_1 &: \mathbf{Y} = \mathbf{S}_1(\theta_1) + \mathbf{V}
\end{aligned} \tag{9.59}$$

where $\mathbf{V} \sim N(\mathbf{0}, \sigma^2\mathbf{I}_4)$. From the form (9.59) of the observations, we see for an unknown phase angle θ_j under hypothesis H_j, the probability density of the observations is given by

$$f(\mathbf{y}|H_j,\theta_j) = \frac{1}{(2\pi\sigma^2)^2}\exp\Big(-\frac{||\mathbf{y}||_2^2 + 2E}{2\sigma^2}\Big)\exp\big(\mathbf{Y}^T\mathbf{S}_j(\theta_j)/\sigma^2\big)\,. \tag{9.60}$$

The detection problem (9.59) is invariant under the group $\mathcal{G}$ of transformations formed by the product of rotations in the $Y_1 - Y_2$ and $Y_3 - Y_4$ planes. For a rotation pair with angle α in the $Y_1 - Y_2$ plane and β in the $Y_3 - Y_4$ plane, the unknown phase angles become $\theta_0' = \theta_0 - \alpha$ and $\theta_1' = \theta_1 - \beta$. To identify a maximal invariant statistic, we express the pairs (Y_1, Y_2) and (Y_3, Y_4) in polar coordinates as

$$U = (Y_1^2 + Y_2^2)^{1/2} \quad , \quad V = (Y_3^2 + Y_4^2)^{1/2}$$
$$\gamma = \tan^{-1}(Y_2/Y_1) \quad , \quad \delta = \tan^{-1}(Y_4/Y_3) \, , \tag{9.61}$$

so that $||\mathbf{Y}||_2 = (U^2 + V^2)^{1/2}$ and

$$\begin{aligned} \mathbf{Y}^T\mathbf{S}_0(\theta_0) &= (2E)^{1/2} U \cos(\gamma + \theta_0) \\ \mathbf{Y}^T\mathbf{S}_1(\theta_1) &= (2E)^{1/2} V \cos(\delta + \theta_1) \, . \end{aligned} \tag{9.62}$$

Then (U, V) is a maximal invariant statistic under $\mathcal{G}$, and by integrating out the angular variables γ and δ to evaluate the marginal density of (U, V) under H_j we obtain

$$\begin{aligned} f(u, v|H_0, \theta_0) &= \frac{uv}{\sigma^4} \exp\big(-\frac{u^2 + v^2 + 2E}{2\sigma^2}\big) I_0((2E)^{1/2} u/\sigma^2) \\ f(u, v|H_1, \theta_1) &= \frac{uv}{\sigma^4} \exp\big(-\frac{u^2 + v^2 + 2E}{2\sigma^2}\big) I_0((2E)^{1/2} v/\sigma^2) \, , \end{aligned} \tag{9.63}$$

which do not depend on the unknown angles θ_j with $j = 0, 1$.

As in Section 9.2, we can use either Bayesian or composite hypothesis testing approaches to solve the detection problem (9.59).

Bayesian formulation: If we assume that the phase angles θ_j are uniformly distributed over $[0, 2\pi)$, i.e.,

$$f(\theta_j|H_j) = \begin{cases} \frac{1}{2\pi} & 0 \leq \theta_j < 2\pi \\ 0 & \text{otherwise} \, , \end{cases}$$

by evaluating the marginal densities

$$f(\mathbf{y}|H_j) = \int f(\mathbf{y}|H_j, \theta_j) f(\theta_j|H_j) d\theta_j$$

we obtain

$$\begin{aligned} f(\mathbf{y}|H_0) &= \frac{1}{(2\pi\sigma^2)^2} \exp\big(-\frac{||\mathbf{y}||_2^2 + 2E}{2\sigma^2}\big) I_0((2E)^{1/2} u/\sigma^2) \\ f(\mathbf{y}|H_1) &= \frac{1}{(2\pi\sigma^2)^2} \exp\big(-\frac{||\mathbf{y}||_2^2 + 2E}{2\sigma^2}\big) I_0((2E)^{1/2} v/\sigma^2) \, . \end{aligned} \tag{9.64}$$

Then, if the two hypotheses are equally likely, and if we minimize the probability of error, so that the test threshold $\tau = 1$, the LRT takes the form

$$L(\mathbf{Y}) = \frac{f(\mathbf{Y}|H_1)}{f(\mathbf{Y}|H_0)} = \frac{I_0((2E)^{1/2} V/\sigma^2)}{I_0((2E)^{1/2} U/\sigma^2)} \underset{H_0}{\overset{H_1}{\gtrless}} 1 \tag{9.65}$$

or equivalently,

$$V \underset{H_0}{\overset{H_1}{\gtrless}} U \,. \tag{9.66}$$

To evaluate the probability of error, we note that since hypotheses H_0 and H_1 are symmetric under the exchange of coordinates (Y_1, Y_2) and (Y_3, Y_4), the probabilities of errors under the two hypotheses are equal, so that

$$\begin{aligned} P[E] &= P[E|H_0] = P[V > U|H_0] \\ &= \int_0^\infty f_U(u|H_0)[\int_u^\infty f_V(v|H_0)dv]du \,. \end{aligned} \tag{9.67}$$

But

$$\begin{aligned} \int_u^\infty f_V(v|H_0)dv &= \int_u^\infty \frac{v}{\sigma^2} \exp\big(-\frac{v^2}{2\sigma^2}\big)dv \\ &= \exp\big(-\frac{u^2}{2\sigma^2}\big) \,, \end{aligned} \tag{9.68}$$

so (9.67) yields

$$P[E] = \exp\big(-\frac{E}{\sigma^2}\big) \int_0^\infty \frac{u}{\sigma^2} \exp\big(-\frac{u^2}{\sigma^2}\big) I_0((2E)^{1/2}u/\sigma^2)du \,. \tag{9.69}$$

Using the integral

$$\int_0^\infty z \exp(-a^2 z^2) I_0(bz)dz = \frac{1}{2a^2} \exp\big(\frac{b^2}{4a^2}\big) \,, \tag{9.70}$$

we find

$$P[E] = \frac{1}{2} \exp\big(-\frac{E}{2\sigma^2}\big) \,. \tag{9.71}$$

GLR Test: Instead of viewing the angles θ_j as random, we can treat them as unknown but nonrandom. From expression (9.60) for the density $f(\mathbf{y}|H_j, \theta_j)$, we find that under H_0 and H_1 the ML estimates of θ_0 and θ_1 are given respectively by

$$\begin{aligned} \hat{\theta}_0 &= -\gamma = -\tan^{-1}(Y_2/Y_1) \\ \hat{\theta}_1 &= -\delta = -\tan^{-1}(Y_4/Y_3) \,. \end{aligned} \tag{9.72}$$

The maximized densities can be expressed as

$$\begin{aligned} f(\mathbf{y}|H_0, \hat{\theta}_0) &= \frac{1}{(2\pi\sigma^2)^2} \exp\big(-\frac{||\mathbf{y}||_2^2 + 2E}{2\sigma^2}\big) \exp((2E)^{1/2}u/\sigma^2) \\ f(\mathbf{y}|H_1, \hat{\theta}_1) &= \frac{1}{(2\pi\sigma^2)^2} \exp\big(-\frac{||\mathbf{y}||_2^2 + 2E}{2\sigma^2}\big) \exp((2E)^{1/2}v/\sigma^2) \,, \end{aligned} \tag{9.73}$$

so the GLR test takes the form

$$L_G(\mathbf{Y}) = \frac{f(\mathbf{Y}|H_1, \hat{\theta}_1)}{f(\mathbf{Y}|H_0, \hat{\theta}_0)} = \exp((2E)^{1/2}(V - U)/\sigma^2) \underset{H_0}{\overset{H_1}{\gtrless}} 1 \tag{9.74}$$

After taking logarithms, it reduces to (9.66), so the GLRT coincides with the Bayesian test with uniformly distributed densities for the unknown angles. We also found earlier that (U, V) is a maximal invariant statistic under the group $\mathcal{G}$ formed by the product of the rotation groups in the Y_1-Y_2 and Y_3-Y_4 planes. Since the densities (9.65) for (U, V) under H_j with $j = 0,\ 1$ do not depend on the unknown angles, the test

$$L(U, V) = \frac{f(U, V|H_1)}{f(U, V|H_0} = \frac{I_0((2E)^{1/2}V/\sigma^2)}{I_0((2E)^{1/2}U/\sigma^2)} \underset{H_0}{\overset{H_1}{\gtrless}} 1 \tag{9.75}$$

is UMPI. Since it coincides with the LRT (9.65), we can conclude the GLRT is UMPI.

Detector implementation: Because the signal space coordinates Y_j with $1 \leq j \leq 4$ require the integration of the received signal $Y(t)$ over two signaling intervals, it is not convenient to implement the optimum detector in the form (9.66). A simpler implementation can be obtained by using the integrated quadrature components

$$\begin{aligned} Z_c &= \int_0^T Y(t)\cos(\omega_c t)dt \\ Z_s &= \int_0^T Y(t)\sin(\omega_c t)dt \end{aligned} \tag{9.76}$$

and their one-symbol delayed versions

$$\begin{aligned} Z_c^- &= \int_{-T}^0 Y(t)\cos(\omega_c t)dt \\ Z_s^- &= \int_{-T}^0 Y(t)\sin(\omega_c t)dt\,. \end{aligned} \tag{9.77}$$

Then, since

$$\begin{aligned} Y_1 &= (1/T)^{1/2}(Z_c + Z_c^-)\ , \quad Y_3 = (1/T)^{1/2}(-Z_c + Z_c^-) \\ Y_2 &= (1/T)^{1/2}(Z_s + Z_s^-)\ , \quad Y_4 = (1/T)^{1/2}(-Z_s + Z_s^-)\,, \end{aligned}$$

after squaring both sides of (9.66), the test can be expressed as

$$S = Z_c Z_c^- + Z_s Z_s^- \underset{H_1}{\overset{H_0}{\gtrless}} 0\,, \tag{9.78}$$

which involves only the quadrature components of the received signal integrated over a single baud period and their delayed versions. A block diagram of the resulting detector is shown in Fig. 9.4, where D denotes a unit delay.

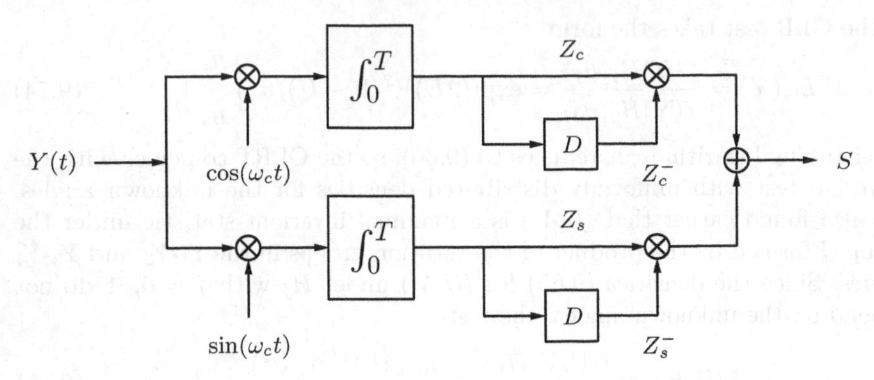

Fig. 9.4. Detector structure for QPSK signals.

9.4 Detection of Signals with Unknown Amplitude and Phase

In addition to phase uncertainty, most communications channels introduce an unknown fading gain, so that both the amplitude and phase of the received signal are unknown. In this case, the detection problem (9.1) becomes

$$\begin{aligned} H_0 &: Y(t) = V(t) \\ H_1 &: Y(t) = As(t,\theta) + V(t) \end{aligned} \tag{9.79}$$

for $0 \leq t \leq T$, where the amplitude $A \geq 0$ is unknown, the signal $s(t,\theta)$ takes the form (9.3)–(9.4), and $V(t)$ is a zero-mean WGN with intensity σ^2. Under H_1, the signal component of the observations can be expressed in terms of the in-phase and quadrature signals $s_c(t)$ and $s_s(t)$ defined in (9.8) as

$$As(t,\theta) = X_c s_c(t) + X_s s_s(t) \tag{9.80}$$

with

$$X_c = A\cos(\theta) \quad , \quad X_s = -A\sin(\theta) \,. \tag{9.81}$$

Then we can select a COB $\{\phi_j, j \in \mathcal{J}\}$ of $L^2[0,T]$ where the first two basis functions are given by (9.13). By performing a KLD of the noise and observation signals $V(t)$ and $Y(t)$ with respect to this basis, and discarding all coefficients beyond the first two, since these coefficients contain noise only, we find that the detection problem admits the two-dimensional signal space representation

$$\begin{aligned} H_0 &: \mathbf{Y} = \mathbf{V} \\ H_1 &: \mathbf{Y} = E^{1/2}\mathbf{X} + \mathbf{V} \,, \end{aligned} \tag{9.82}$$

where the vectors $\mathbf{Y}$ and $\mathbf{V}$ are defined as in (9.15) and

$$\mathbf{X} = \begin{bmatrix} X_c \\ X_s \end{bmatrix} = A \begin{bmatrix} \cos(\theta) \\ -\sin(\theta) \end{bmatrix} . \tag{9.83}$$

Note that, unlike the unknown phase case where the signal vector $\mathbf{S}(\theta)$ is constrained to belong to a circle of radius $E^{1/2}$, here the signal $\mathbf{S} = E^{1/2}\mathbf{X}$ is arbitrary. As in the examples considered earlier in this chapter, the detection problem (9.82) can be formulated by using either Bayesian or composite hypothesis testing techniques.

9.4.1 Bayesian Formulation

The key element of a Bayesian approach is the selection of an appropriate probability density for the unknown parameters. For the phase only case, the selection of a uniform probability density for θ is natural, based on symmetry considerations. The selection of a probability density for the amplitude A is more difficult as it needs to reflect the underlying physical nature of the channel.

Rayleigh channel: For a scattering-rich wireless channel with no direct path between the transmitter and receiver, it is customary to assume that A and θ are independent, where θ is uniformly distributed over $[0, 2\pi)$ and A admits the Rayleigh density

$$f_A(a) = \frac{a}{q} \exp\left(-\frac{a^2}{2q}\right) u(a) . \tag{9.84}$$

Equivalently, the vector $\mathbf{X}$ is $N(\mathbf{0}, q\mathbf{I}_2)$ distributed, so the detection problem (9.82) takes the form

$$\begin{aligned} H_0 &: \mathbf{Y} \sim N(\mathbf{0}, \sigma^2 \mathbf{I}_2) \\ H_1 &: \mathbf{Y} \sim N(\mathbf{0}, (qE + \sigma^2)\mathbf{I}_2) . \end{aligned} \tag{9.85}$$

This test has exactly the same form as Example 2.2, for which the optimum test was shown to be

$$||\mathbf{Y}||_2 \underset{H_0}{\overset{H_1}{\gtrless}} \eta = \kappa^{1/2} , \tag{9.86}$$

with

$$\kappa = \frac{2\sigma^2(qE + \sigma^2)}{qE} \ln((1 + qE/\sigma^2)\tau) . \tag{9.87}$$

Except for the choice of threshold κ, it is worth noting that this test is identical to the test (9.24) obtained for the incoherent phase channel. Since the probability density of the vector $\mathbf{Y}$ under the null hypothesis is the same for the case of an unknown phase only, and the case where both the phase and amplitude are unknown, we conclude from (9.25) that the probability of false alarm is given by

$$P_F = P[||\mathbf{Y}||_2 > \eta | H_0] = \exp\big(-\frac{\eta^2}{2\sigma^2}\big) . \tag{9.88}$$

Also, since the only difference between hypotheses H_0 and H_1 is that under H_1 the entries of $\mathbf{Y}$ have variance $\sigma^2 + qE$, instead of σ^2 under H_0, the probability of detection is given by

$$P_D = P[||\mathbf{Y}||_2 > \eta | H_1] = \exp\big(-\frac{\eta^2}{2(\sigma^2 + qE)}\big) . \tag{9.89}$$

In fact, as was observed in Example 2.2, the ROC can be expressed in closed form as

$$P_D = (P_F)^{\frac{1}{1+\mathrm{SNR}}} , \tag{9.90}$$

where

$$\mathrm{SNR} = \frac{E[||E^{1/2}\mathbf{X}||_2^2]}{E[||\mathbf{V}||_2^2]} = \frac{qE}{\sigma^2}$$

represents the signal to noise ratio under H_1.

Ricean channel: For a wireless channel with either a line of sight path or fixed scatterers, under H_1, the received signal takes the form

$$Y(t) = Bs(t, \chi) + As(t, \theta) + V(t) \tag{9.91}$$

where the specular component $Bs(t, \chi)$ of the signal is fixed, with B and χ known, whereas the fading component $As(t, \theta)$ represents the effect of diffuse scatterers, and is random. Without loss of generality, we can assume $\chi = 0$. In this case, the detection problem still takes the form (9.82) in signal space, but

$$\mathbf{X} = \begin{bmatrix} B \\ 0 \end{bmatrix} + A \begin{bmatrix} \cos(\theta) \\ -\sin(\theta) \end{bmatrix} . \tag{9.92}$$

Consequently, if we assume again that θ is uniformly distributed over $[0, 2\pi)$ and A admits the Rayleigh distribution (9.84), we find that

$$\mathbf{X} \sim N\Big(\begin{bmatrix} B \\ 0 \end{bmatrix}, q\mathbf{I}_2 \Big) \tag{9.93}$$

has a nonzero mean B for its first component. Accordingly, the LRT for the detection problem can now be expressed as

$$\begin{aligned} L(\mathbf{Y}) &= \frac{f(\mathbf{Y}|H_1)}{f(\mathbf{Y}|H_0)} \\ &= \frac{\sigma^2}{(\sigma^2 + qE)} \exp\Big(\frac{||\mathbf{Y}||_2^2}{2\sigma^2} - \frac{(Y_1 - B)^2 + Y_2^2}{2(\sigma^2 + qE)} \Big) \underset{H_0}{\overset{H_1}{\gtrless}} \tau , \end{aligned} \tag{9.94}$$

so, taking logarithms and reorganizing the resulting expression we obtain the test

$$(1 + \frac{\sigma^2}{qE})Y_1^2 - \frac{\sigma^2}{qE}(Y_1 - B)^2 + Y_2^2 \underset{H_0}{\overset{H_1}{\gtrless}} \kappa \tag{9.95}$$

where κ is given by (9.87). Unfortunately, the probabilities of detection and of false alarm for this test are difficult to evaluate.

9.4.2 GLR Test

It is also possible to formulate the detection problem (9.82) from a composite hypothesis testing viewpoint. Note that $\mathbf{X}$ is an arbitrary vector of $\mathbb{R}^2$ under both the parametrization (9.83) of $\mathbf{X}$ in terms of A and θ, or the parametrization (9.92) in terms of B, A and θ. So the composite hypothesis testing approach is incapable of making a distinction between Rayleigh and Ricean channel models, since it views all model parameters as unknown but nonrandom. Under H_1, the probability density of the observation vector $\mathbf{Y}$ is given by

$$f(\mathbf{Y}|H_1, \mathbf{X}) = \frac{1}{2\pi\sigma^2} \exp\big(- \frac{||\mathbf{Y} - E^{1/2}\mathbf{X}||_2^2}{2\sigma^2}\big) , \tag{9.96}$$

so that the ML estimate of the parameter vector $\mathbf{X}$ is given by

$$\hat{\mathbf{X}}_1 = \mathbf{Y}/E^{1/2} . \tag{9.97}$$

Accordingly, the GLRT can be expressed as

$$L_G(\mathbf{Y}) = \frac{f(\mathbf{Y}|H_1, \hat{\mathbf{X}}_1)}{f(\mathbf{Y}|H_0)} = \exp\big(\frac{||\mathbf{Y}||_2^2}{2\sigma^2}\big) \underset{H_0}{\overset{H_1}{\gtrless}} \tau . \tag{9.98}$$

Taking logarithms and then square roots, we obtain the test (9.86) with

$$\eta = 2\sigma^2 \ln(\tau) . \tag{9.99}$$

So the GLRT coincides with the Bayesian test for a Rayleigh channel model, but is different from the test (9.95) obtained for the Ricean channel case. It is also worth noting that the detection problem (9.82) is invariant under the group of two-dimensional rotations. For this problem, it was shown in Example 5.6 that $||\mathbf{Y}||_2$ is a maximal invariant statistic, so the GLR test (9.86) is UMPI.

9.5 Detection with Arbitrary Unknown Parameters

All the detection problems we have discussed up to this point have the feature that the signal $s(t, \mathbf{x})$ stays in a *finite dimensional* subspace of $L^2[0, T]$ as the unknown parameter vector $\mathbf{x} \in \mathbb{R}^m$ varies. So by appropriate choice of a COB, as the vector $\mathbf{x}$ varies, the signal space vector $\mathbf{S}(\mathbf{x})$ representing $s(t,\mathbf{x})$ stays in a finite-dimensional space, allowing the analysis of the detection problem by finite-dimensional techniques. This feature always holds, of course, in the DT case, but it is not guaranteed in the CT case. So in this section, we describe a general methodology for handling detection problems where the range of $s(\cdot,\mathbf{x})$ is infinite dimensional as $\mathbf{x}$ varies. One limitation of this methodology is, however, that the observed signal under the null hypothesis should be just WGN, i.e., we consider the detection problem

$$\begin{aligned} H_0 &: Y(t) = V(t) \\ H_1 &: Y(t) = s(t,\mathbf{x}) + V(t) \end{aligned} \tag{9.100}$$

for $0 \leq t \leq T$, where $V(t)$ is a WGN with intensity σ^2.

Let

$$E(\mathbf{x}) = \int_0^T s^2(t,\mathbf{x})dt \tag{9.101}$$

denote the energy of signal $s(t,\mathbf{x})$. To represent the detection problem (9.100), we employ a *parameter dependent* COB of $L^2[0,T]$ with

$$\phi_1(t,\mathbf{x}) = s(t,\mathbf{x})/E^{1/2}(\mathbf{x}) , \tag{9.102}$$

and where the basis functions $\phi_j(t,\mathbf{x})$ with $j \geq 2$ are obtained by Gram-Schmidt orthonormalization. Note that since ϕ_1 is x dependent, these functions also depend on $\mathbf{x}$. We can then perform KLDs

$$\begin{aligned} V(t) &= \sum_{j\in\mathcal{J}} V_j(\mathbf{x})\phi_j(t,\mathbf{x}) \\ Y(t) &= \sum_{j\in\mathcal{J}} Y_j(\mathbf{x})\phi_j(t,\mathbf{x}) \end{aligned} \tag{9.103}$$

of the noise and received signal with respect to this basis, where

$$V_j(\mathbf{x}) = < V, \phi_j(\mathbf{x}) > \quad , \quad Y_j(\mathbf{x}) = < Y, \phi_j(\mathbf{x}) > \; .$$

The coefficients $V_j(\mathbf{x})$ form a zero-mean WGN sequence with variance σ^2, so even though the coefficients $V_j(\mathbf{x})$ are parameter dependent, their statistics are fixed. Under both hypotheses, the KL coefficients satisfy $Y_j(\mathbf{x}) = V_j(\mathbf{x})$ for $j \geq 2$, so these coefficients can be discarded, as they do not influence the detection problem. The signal space representation of the detection problem is therefore given by

$$\begin{aligned} H_0 &: Y_1(\mathbf{x}) = V_1(\mathbf{x}) \\ H_0 &: Y_1(\mathbf{x}) = E^{1/2}(\mathbf{x}) + V_1(\mathbf{x}) \, . \end{aligned} \tag{9.104}$$

Hence, conditioned on $\mathbf{x}$, the LR is given by

$$\begin{aligned}
L(Y(\cdot)|\mathbf{x}) &= \exp\left(\frac{1}{\sigma^2}[E^{1/2}(\mathbf{x})Y_1(\mathbf{x}) - \frac{E(\mathbf{x})}{2}]\right) \\
&= \exp\left(\frac{1}{\sigma^2}[\int_0^T Y(t)s(t,\mathbf{x})dt - \frac{1}{2}\int_0^T s^2(t,\mathbf{x})dt]\right) \quad \text{(CT case)} \\
&= \exp\left(\frac{1}{\sigma^2}[\sum_{t=0}^{T} Y(t)s(t,\mathbf{x}) - \frac{1}{2}\sum_{t=0}^{T} s^2(t,\mathbf{x})]\right) \quad \text{(DT case)}
\end{aligned} \tag{9.105}$$

Then, if we adopt a Bayesian approach and assume that $\mathbf{X}$ admits the density $f_{\mathbf{X}}(\mathbf{x})$, the likelihood ratio is given by

$$L(Y(\cdot)) = \int L(Y(\cdot)|\mathbf{x}) f_{\mathbf{X}}(\mathbf{x}) d\mathbf{x} \,. \tag{9.106}$$

On the other hand, if $\mathbf{x}$ is viewed as unknown but nonrandom, the GLR statistic is given by

$$L_{\mathrm{G}}(Y(\cdot)) = \max_{\mathbf{x} \in \mathcal{X}} L(Y(\cdot)|\mathbf{x}) \,, \tag{9.107}$$

where $\mathcal{X}$ denotes the range of values of $\mathbf{x}$ under H_1. Note that the derivation of formulas (9.106) and (9.107) relies on the fact that the observed signal $Y(t)$ is independent of $\mathbf{x}$ under H_0. Also, the above formulation assumes implicitly that the correlation of $Y(t)$ with $s(t,\mathbf{x})$ can be evaluated for all $\mathbf{x}$. The computational complexity of this task can be formidable, but there exists problems for which this computation can be accomplished in a simple manner. This is the case for the example considered below, which illustrates the design of a GLR test based on identity (9.107).

Example 9.1: Signal with unknown amplitude and time delay

Assume that the received signal

$$s(t,\mathbf{x}) = As(t-d) \tag{9.108}$$

has an unknown amplitude A and time delay $d \in [0, D]$, so

$$\mathbf{x} = \begin{bmatrix} A \ d \end{bmatrix}^T$$

is a two-dimensional vector. We assume that the pulse $s(\cdot)$ is completely contained in the open interval $(0, T-D)$ so that the delayed pulse $s(t-d)$ remains in the interior of observation interval $(0,T)$. This ensures that

$$E = \int_0^T s^2(t-d)dt$$

is independent of d. Then to design a GLR test, instead of maximizing the likelihood ratio $L(Y(\cdot)|A,d)$, we can maximize its logarithm

$$\ln L(Y(\cdot)|A,d) = \frac{1}{\sigma^2}[A\int_0^T Y(t)s(t-d)dt - \frac{A^2 E}{2}] \,. \tag{9.109}$$

This gives

$$\hat{d} = \arg \max_{0 \leq d \leq D} | \int_0^T Y(t)s(t-d)dt | \tag{9.110}$$

$$\hat{A} = \frac{1}{E} \int_0^T Y(t)s(t-\hat{d})dt \,, \tag{9.111}$$

and

$$\ln L_G(Y(\cdot)) = \frac{\hat{A}^2 E}{2\sigma^2} \,. \tag{9.112}$$

The maximization (9.110) can be implemented easily with a matched filter. Specifically, consider the matched filter with impulse response

$$h(t) = \begin{cases} s(T-t) & 0 \leq t \leq T \\ 0 & \text{otherwise} \,, \end{cases} \tag{9.113}$$

and let $Z(t)$ be the output signal obtained by passing the observed signal $Y(\cdot)$ through $h(\cdot)$, as shown in Fig. 8.2. Note that as long as we set $Y(t) = 0$ for $t > T$, the output $Z(t)$ is defined for all $t \geq 0$. Then, according to (8.19), the matched filter output at $t = T + d$ is given by

$$Z(T+d) = \int_0^T s(T+u-(T+d))Y(u)du = \int_0^T s(u-d)Y(u)du \tag{9.114}$$

so that

$$\hat{d} = \arg \max_{0 \leq d \leq D} |Z(T+d)| \,. \tag{9.115}$$

In other words, the ML estimate of the delay d is obtained by finding the peak of the magnitude of the output signal $Z(t)$ over the interval $T \leq t \leq T + D$. So, as shown in Fig. 9.5, $\hat{d}$ is obtained by a matched filter, rectifier, and peak detector and

$$\hat{A} = Z(T+\hat{d})/E$$

is obtained by scaling the peak value by E.

From (9.112), we see that the GLRT can be expressed as

$$|\hat{A}| \underset{H_0}{\overset{H_1}{\gtrless}} \eta \,. \tag{9.116}$$

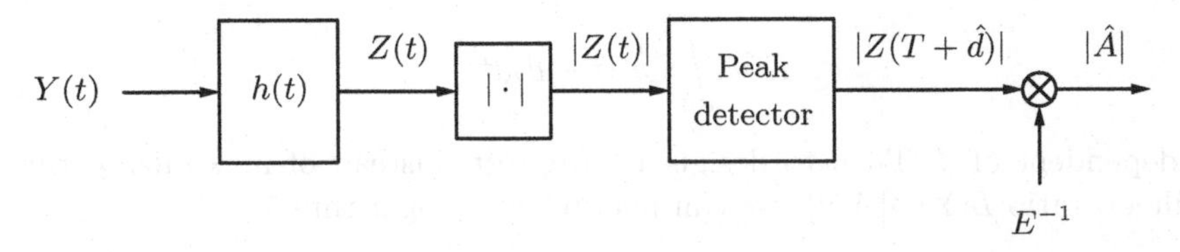

Fig. 9.5. Matched filter and peak detector receiver for a signal with unknown amplitude and time delay.

Note that this test has a form similar to the one obtained in Example 5.10 for a DT impulse with unknown amplitude and location in noise. It turns out that the probability of false alarm of this detector is difficult to evaluate in closed form. To see why this is the case, observe that under H_0, the matched filter output $Z(T+d)$ given by (9.114) reduces to

$$Z(T+d) = \int_0^T s(u-d)V(u)du\,.$$

Viewed as a function of the delay d, it is a zero-mean Gaussian process with autocorrelation

$$E[Z(T+d)Z(T+d')] = \sigma^2 E\rho(d-d')$$

where

$$\begin{aligned}\rho(d-d') &\triangleq \frac{1}{E}\int_0^T s(u-d)s(u-d')du \\ &= \frac{1}{E}\int_0^T s(v)s(v+d-d')dv \end{aligned} \tag{9.117}$$

denotes the correlation coefficient of waveforms $s(\cdot - d)$ and $s(\cdot - d')$. Since $|\hat{A}|$ is obtained by dividing by E the peak value of $|Z(T+d)|$ for $0 \leq d \leq D$, we see that a false alarm will occur whenever the magnitude process $|Z(T+d)|$ exceeds the level ηE for some d in the interval $[0, D]$. So P_F is a level-crossing probability for the random process $|Z(T+d)|$ over the interval $[0, D]$. Level-crossing problems for random processes have been studied extensively [2], [3, Chaps. 10–13] but most estimates of level-crossing probabilities require coarse approximations.

A rough approximation for P_F that does not rely explicitly on the theory of level crossings can be obtained as follows. Let $S(\omega)$ denote the Fourier transform of the pulse $s(t)$. By applying Parseval's identity, the energy of $s(\cdot)$ can be expressed as

$$E = \int_0^T s^2(t)dt = \frac{1}{2\pi}\int_{-\infty}^{\infty} |S(\omega)|^2 d\omega\,. \tag{9.118}$$

Then the root mean-square (RMS) bandwidth B of $s(\cdot)$ is defined by

$$B^2 = \frac{1}{E}\int_0^T \dot{s}^2(t)dt = \frac{1}{2\pi E}\int_{-\infty}^{\infty} \omega^2 |S(\omega)|^2 d\omega\,. \tag{9.119}$$

If we sample the process $Z(T+d)/E$ at the Nyquist period π/B, the samples are approximately uncorrelated and thus independent, since the process is Gaussian. So we need to find the probability that the maximum absolute value of $N = \lceil DB/\pi \rceil$ independent $N(0, \sigma^2/E)$ random variables exceeds the test threshold η. But this probability can be evaluated easily by using the theory of order statistics [4, p. 246]. Since the events that the maximum exceeds η, or the minimum exceeds $-\eta$ are approximately independent, if $Z_{\max}/E$ denotes the maximum value of the samples, we find

$$\begin{aligned} P_F &\approx 2P[Z_{\max}/E > \eta | H_0] \\ &= \frac{2N}{(2\pi)^{1/2})} \int_{\eta E^{1/2}/\sigma}^{\infty} (1 - Q(z))^N \exp\big(-\frac{z^2}{2}\big) dz \,. \end{aligned} \quad (9.120)$$

Except for a factor 2, this identity is the same as expression (5.90) we obtained for the pulse detection problem of Example 5.10. The factor 2 difference is due to the fact that here the sign of the amplitude A is unknown, so we must test $\hat{A}$ in both directions. This multiplies by 2 the probability that the threshold will be exceeded. Yet, the estimate (9.120) is only a rough approximation. In practice, for a ranging system with known pulse $s(\cdot)$, the best approach to set the threshold η would be to perform a Monte Carlo simulation to obtain the curve $P_F(\eta)$. Then, given a desired false alarm probability α, we need only to select η such that $\alpha = P_F(\eta)$.

Under H_1, if A and d denote the true parameter values, and $\hat{d}$ is the delay estimate, we have

$$\hat{A} = \rho(d - \hat{d})A + \frac{1}{E} \int_0^T s(u - \hat{d})V(u) du \,. \quad (9.121)$$

If we assume that the estimate $\hat{d}$ is approximately correct, and the signal to noise ratio (SNR) is large, we have

$$\hat{A} \approx A + \frac{1}{E} \int_0^T s(u - \hat{d})V(u) du \sim N(A, \sigma^2/E) \,. \quad (9.122)$$

Note that the noise term remains Gaussian distributed for high SNR, since it is dominated by the constant term when the maximization (9.115) is performed. Consequently, by using the distribution (9.122), the power of the test can be expressed as

$$\begin{aligned} P_D(A) &= P[|\hat{A}| > \eta | H_1] \\ &= Q((\eta - A)\frac{E^{1/2}}{\sigma}) + Q((\eta + A)\frac{E^{1/2}}{\sigma}) \\ &= Q(Q^{-1}(\frac{\alpha}{2}) - \frac{AE^{1/2}}{\sigma}) + Q(Q^{-1}(\frac{\alpha}{2}) + \frac{AE^{1/2}}{\sigma}) \,. \end{aligned} \quad (9.123)$$

Note that under the assumption that the delay is estimated correcty, the problem reduces to the detection of a signal with unknown amplitude in WGN, and accordingly (9.123) coincides with expression (5.40) we obtained earlier in Example 5.7 for detecting a DT signal of unknown amplitude in WGN. □

9.6 Waveform Parameter Estimation

Since the design of GLR detectors for signals with arbitrary parameters in WGN relies on the evaluation of ML parameter estimates, we discuss waveform parameter estimation in further detail. Consider first the DT case and assume we observe

$$Y(t) = s(t,\mathbf{x}) + V(t) \tag{9.124}$$

for $0 \leq t \leq T$, where $V(t)$ is a zero-mean WGN with variance σ^2. By writing the observation and signal as the $T+1$ dimensional vectors

$$\begin{aligned}\mathbf{Y} &= \left[\, Y(0) \dots Y(t) \dots Y(T) \,\right]^T \\ \mathbf{S}(\mathbf{x}) &= \left[\, s(0,\mathbf{x}) \dots s(t,\mathbf{x}) \dots s(T,\mathbf{x}) \,\right]^T ,\end{aligned}$$

the logarithm of the joint probability density of the observations can be expressed as

$$\begin{aligned}\ln(f(\mathbf{Y}|\mathbf{x})) &= -\frac{1}{2\sigma^2}||\mathbf{Y} - \mathbf{S}(\mathbf{x})||^2 - \frac{T+1}{2}\ln(2\pi\sigma^2) \\ &= \ln(L(\mathbf{Y}|\mathbf{x})) + C(\mathbf{Y}) ,\end{aligned} \tag{9.125}$$

with

$$\begin{aligned}\ln(L(\mathbf{Y}|\mathbf{x})) &= \frac{1}{\sigma^2}[\mathbf{Y}^T\mathbf{S}(\mathbf{x}) - \frac{1}{2}||\mathbf{S}(\mathbf{x})||_2^2] \\ &= \frac{1}{\sigma^2}[\sum_{t=0}^{T} Y(t)s(t,\mathbf{x}) - \frac{1}{2}\sum_{t=0}^{T} s^2(t,\mathbf{x})] ,\end{aligned} \tag{9.126}$$

where the function

$$C(\mathbf{Y}) = -\frac{1}{2\sigma^2}||\mathbf{Y}||_2^2 - \frac{T+1}{2}\ln(2\pi\sigma^2) = \ln f(\mathbf{Y}|H_0) \tag{9.127}$$

which regroups all terms independent of $\mathbf{x}$ can be interpreted as the log-density of $\mathbf{Y}$ under hypothesis

$$H_0 \; : \; Y(t) = V(t) . \tag{9.128}$$

So, removing all terms independent of $\mathbf{x}$ from $\ln(f(\mathbf{Y}|\mathbf{x}))$ is equivalent to forming the log-likelihood ratio for testing hypothesis H_1 corresponding to (9.124) against hypothesis H_0 represented by (9.128). Maximizing the log likelihood ratio $\ln L(\mathbf{Y}|\mathbf{x})$ is clearly equivalent to maximizing the log density $\ln f(\mathbf{Y}|\mathbf{x})$ with respect to $\mathbf{x}$. Also, since

$$\nabla_{\mathbf{x}} \ln(L(\mathbf{Y}|\mathbf{x})) = \nabla_{\mathbf{x}} \ln(f(\mathbf{Y}|\mathbf{x})) , \tag{9.129}$$

we can use the gradient of the log-likelihood function to evaluate the Fisher information matrix $\mathbf{J}(\mathbf{x})$ given by (4.102).

In the CT case, if we convert the observation model (9.124) to the Karhunen-Loève domain by selecting an arbitrary *fixed* COB of $L^2[0,T]$, we obtain an infinite sequence

$$Y_j = s_j(\mathbf{x}) + V_j \tag{9.130}$$

with $j \geq 1$. We cannot form the joint probability density for the full sequence, since it is infinite, but by truncating the sequence to its first N elements, we obtain a DT waveform estimation problem of the type considered above, for which by eliminating the terms independent of $\mathbf{x}$ in the log-density, we obtain the log-likelihood ratio

$$\ln(L_N(Y(\cdot)|\mathbf{x})) = \frac{1}{\sigma^2}[\sum_{j=1}^{N} Y_j s_j(\mathbf{x}) - \frac{1}{2}\sum_{j=1}^{N} s_j^2(\mathbf{x})] \tag{9.131}$$

which as $N \to \infty$ converges in the mean-square sense to

$$\begin{aligned} \ln(L(Y(\cdot)|\mathbf{x})) &= \frac{1}{\sigma^2}[\sum_{j=1}^{\infty} Y_j s_j - \frac{1}{2}\sum_{j=1}^{\infty} s_j^2(\mathbf{x})] \\ &= \frac{1}{\sigma^2}\Big[\int_0^T Y(t)s(t,\mathbf{x})dt - \frac{1}{2}\int_0^T s^2(t,\mathbf{x})dt\Big] . \end{aligned} \tag{9.132}$$

So, even though the joint density of the truncated observation sequence does not have a limit as the number of observations $N \to \infty$, its $\mathbf{x}$ dependent component does and converges to the log-likelihood ratio (9.132).

Accordingly, the ML estimate of $\mathbf{x}$ is obtained by maximizing this function or its DT counterpart, and the Fisher information matrix can be expressed as

$$\mathbf{J}(\mathbf{x}) = E[\nabla_{\mathbf{x}} \ln L(Y(\cdot)|\mathbf{x})(\nabla_{\mathbf{x}} \ln L(Y(\cdot)|\mathbf{x}))^T] . \tag{9.133}$$

We have

$$\begin{aligned} \nabla_{\mathbf{x}} L(Y(\cdot)|\mathbf{x}) &= \frac{1}{\sigma^2}\int_0^T (Y(t) - s(t,\mathbf{x}))\nabla_{\mathbf{x}} s(t,\mathbf{x})dt \\ &= \frac{1}{\sigma^2}\int_0^T V(t)\nabla_{\mathbf{x}} s(t,\mathbf{x})dt , \end{aligned} \tag{9.134}$$

so that

$$\mathbf{J}(\mathbf{x}) = \frac{1}{\sigma^2}\int_0^T \nabla_{\mathbf{x}} s(t,\mathbf{x})(\nabla_{\mathbf{x}} s(t,\mathbf{x}))^T dt . \tag{9.135}$$

Furthermore, $\hat{\mathbf{x}}(Y(\cdot))$ is an efficient estimator if and only if it satisfies

$$\hat{\mathbf{x}}(Y(\cdot)) = \mathbf{x} + \mathbf{J}^{-1}(\mathbf{x})\nabla_{\mathbf{x}} \ln L(Y(\cdot)|\mathbf{x}) . \tag{9.136}$$

In the case when $\mathbf{X}$ is viewed as random with a-priori probability density $f_{\mathbf{X}}(\mathbf{x})$, the MAP estimate of $\mathbf{X}$ is obtained by maximizing

$$\ln L(Y(\cdot)|\mathbf{x}) + \ln f_{\mathbf{X}}(\mathbf{x}) .$$

Example 9.2: Amplitude and phase estimation

Let

$$s(t, \mathbf{x}) = Ac(t) \cos(\omega_c t + \psi(t) + \theta)$$

be a signal with unknown amplitude A and phase θ of the type considered in Section 9.4, so the parameter vector

$$\mathbf{x} = \begin{bmatrix} A \ \theta \end{bmatrix}^T .$$

The signal energy can be expressed as

$$E(\mathbf{x}) = \int_0^T s^2(t, \mathbf{x}) dt = A^2 E ,$$

where E is given by (9.11). We have

$$\nabla_{\mathbf{x}} s(t, \mathbf{x}) = c(t) \begin{bmatrix} \cos(\omega_c t + \psi(t) + \theta) \\ -A \sin(\omega_c t + \psi(t) + \theta) \end{bmatrix} , \tag{9.137}$$

so by taking into account the orthogonality of the sine and cosine waveforms over $[0, T]$, the expression (9.135) yields

$$\mathbf{J}(\mathbf{x}) = \frac{1}{\sigma^2} \begin{bmatrix} E & 0 \\ 0 & A^2 E \end{bmatrix} . \tag{9.138}$$

Consequently the CRLB for any unbiased estimate

$$\hat{\mathbf{x}} = \begin{bmatrix} \hat{A} \ \hat{\theta} \end{bmatrix}^T$$

is given by

$$E[(\mathbf{x} - \hat{\mathbf{x}})(\mathbf{x} - \hat{\mathbf{x}})^T] \geq \mathbf{J}^{-1}(\mathbf{x}) = \begin{bmatrix} \frac{\sigma^2}{E} & 0 \\ 0 & \frac{\sigma^2}{A^2 E} \end{bmatrix} .$$

To find the ML estimate, we need to solve the ML equation

$$\nabla_{\mathbf{x}} \ln L(Y(\cdot)|\hat{x}) = \frac{1}{\sigma^2} [\int_0^T Y(t) \nabla_{\mathbf{x}} s(t, \hat{\mathbf{x}}) dt - \frac{1}{2} \nabla_{\mathbf{x}} E(\hat{\mathbf{x}})] = 0 , \tag{9.139}$$

where $\nabla_{\mathbf{x}} s(t, \mathbf{x})$ is given by (9.137) and

$$\frac{1}{2} \nabla_{\mathbf{x}} E(\mathbf{x}) = \begin{bmatrix} AE \\ 0 \end{bmatrix} . \tag{9.140}$$

Note that this equation is only necessary, but not sufficient, and accordingly, it may admit several solutions, only one of which is the ML estimate. The second entry of the vector equation (9.139) gives

$$\begin{aligned} 0 &= \int_0^T Y(t)c(t)\sin(\omega_c t + \psi(t) + \hat{\theta})dt \\ &= Z_s \cos(\hat{\theta}) + Z_c \sin(\hat{\theta}) \end{aligned}$$

where Z_c and Z_s are defined in (9.40). Consequently, if we employ the polar coordinates

$$Q = (Z_c^2 + Z_s^2)^{1/2} \quad , \quad \Phi = \tan^{-1}(Z_s/Z_c) ,$$

we obtain

$$\hat{\theta} = -\Phi . \tag{9.141}$$

Similarly, the first entry of the ML equation (9.139) yields

$$\begin{aligned} \hat{A} &= \frac{1}{E}\int_0^T Y(t)c(t)\cos(\omega_c t + \phi(t) + \hat{\theta})dt \\ &= \frac{1}{E}(Z_c \cos(\hat{\theta}) - Z_s \sin(\hat{\theta})) , \end{aligned}$$

where by substituting

$$\cos(\hat{\theta}) = Z_c/Q \quad , \quad \sin(\hat{\theta}) = -Z_s/Q ,$$

we find

$$\hat{A} = \frac{Q}{E} . \tag{9.142}$$

Note that the ML estimates (9.141)–(9.142 match the Cartesian coordinate estimate (9.97) obtained by a signal space approach in Section 9.4. To determine whether the ML estimate is efficient, we express the right-hand side of (9.136) as

$$\begin{aligned} &\begin{bmatrix} A \\ \theta \end{bmatrix} + \begin{bmatrix} \dfrac{\sigma^2}{E} & 0 \\ 0 & \dfrac{\sigma^2}{A^2 E} \end{bmatrix} \frac{1}{\sigma^2}\left(\begin{bmatrix} Q\cos(\Phi+\theta) \\ -AQ\sin(\Phi+\theta) \end{bmatrix} - \begin{bmatrix} AE \\ 0 \end{bmatrix} \right) \\ &\qquad = \begin{bmatrix} \dfrac{Q}{E}\cos(\Phi+\theta) \\ \theta - \dfrac{Q}{AE}\sin(\Phi+\theta) \end{bmatrix} . \end{aligned} \tag{9.143}$$

When $\Phi + \theta$ is small and $Q/AE \approx 1$, this last expression can be approximated by

$$\begin{bmatrix} Q/E \\ -\Phi \end{bmatrix} = \begin{bmatrix} \hat{A} \\ \hat{\theta} \end{bmatrix} , \tag{9.144}$$

so, under these conditions, the identity (9.136) holds and the ML estimate is efficient. But from (9.82)–(9.83) we observe that

$$\mathbf{Z} \triangleq \begin{bmatrix} Z_c \\ Z_s \end{bmatrix} = AE \begin{bmatrix} \cos(\theta) \\ -\sin(\theta) \end{bmatrix} + E^{1/2}\mathbf{V}$$

with $\mathbf{V} \sim N(0, \sigma^2 \mathbf{I}_2)$. So the vector $\mathbf{Z}$ with magnitude Q and polar angle Φ is obtained by adding to a vector of magnitude AE and polar angle $-\theta$ a noise vector whose components are independent and $N(0, \sigma^2 E)$ distributed. Obviously, in order for Q to approximate AE and Φ to be close to $-\theta$, the noise vector must be small compared to the signal vector, i.e., the signal to noise ratio

$$\text{SNR} = \frac{(AE)^2}{E[||E^{1/2}\mathbf{V}||_2^2]} = \frac{A^2 E}{2\sigma^2}$$

must be large. So provided the SNR is large, we find that the the ML estimate of the signal amplitude and phase is efficient. This conclusion is similar to the one obtained for the phase estimation problem of Example 4.9. □

Example 9.3: Amplitude and time delay estimation

Let

$$s(t, \mathbf{x}) = As(t - d)$$

with $\mathbf{x} = \begin{bmatrix} A \; d \end{bmatrix}^T$ be a signal with unknown amplitude A and delay $d \in [0, D]$ where, as in Example 9.1, we assume that the signal $s(\cdot)$ is contained in $(0, T - D)$, so that the delayed signal $s(t - d)$ remains in the interior of observation interval $[0, T]$. This type of estimation problem arises when an autonomous vehicle is equipped with an acoustic ranging system whose function is to detect obstacles ahead of the vehicle. When a pulse $s(t)$ is transmitted, if an obstacle located at distance h is present, a reflected pulse $As(t - d)$ is received by the acoustic transducer in the presence of noise. The amplitude A measures the reflection coefficient of the obstacle and the round-trip wave propagation attenuation, and if v denotes the wave propagation velocity, the delay $d = 2h/v$ represents the two-way travel time of the acoustic pulse to the obstacle and back. So the distance h can be evaluated directly from the delay d. Then the energy of the signal $s(t, \mathbf{x})$ takes the form

$$E(\mathbf{x}) = \int_0^T s^2(t, \mathbf{x}) dt = A^2 E \,, \tag{9.145}$$

where E denotes the energy of the transmitted pulse given by (9.120). The gradient of $s(t, \mathbf{x})$ with respect to $\mathbf{x}$ can be expressed as

$$\nabla_{\mathbf{x}} s(t, \mathbf{x}) = \begin{bmatrix} s(t - d) \\ -A\dot{s}(t - d) \end{bmatrix} . \tag{9.146}$$

But

$$-A \int_0^T s(t-d)\dot{s}(t-d)dt = -\frac{A}{2}\int_0^T \frac{d}{du}s^2(u)du$$
$$= -\frac{A}{2}(s^2(T) - s^2(0)) = 0\,, \qquad (9.147)$$

since the pulse $s(\cdot - d)$ remains in the interior of $[0, T]$, and

$$A^2 \int_0^T \dot{s}^2(t-d)dt = A^2 \int_0^T \dot{s}^2(u)du = (AB)^2 E\,,$$

where B denotes the RMS bandwidth of $s(\cdot)$ defined in (9.119). Consequently, the expression (9.135) for the Fisher information matrix yields

$$\mathbf{J}(\mathbf{x}) = \frac{1}{\sigma^2}\begin{bmatrix} E & 0 \\ 0 & (AB)^2 E \end{bmatrix} . \qquad (9.148)$$

Thus, the CRLB for any unbiased estimate

$$\hat{\mathbf{x}} = \begin{bmatrix} A \; d \end{bmatrix}^T$$

is given by

$$E[(\mathbf{x}-\hat{\mathbf{x}})(\mathbf{x}-\hat{\mathbf{x}})^T] \geq \mathbf{J}^{-1}(\mathbf{x}) = \begin{bmatrix} \frac{\sigma^2}{E} & 0 \\ 0 & \frac{\sigma^2}{(AB)^2 E} \end{bmatrix} .$$

Note that the CRLB for the variance $E[(d-\hat{d})^2]$ of the delay estimate is inversely proportional to the product of the signal to noise ratio

$$\text{SNR} = A^2 E/\sigma^2$$

and the squared RMS bandwith B^2, so assuming the CRLB can be reached, more accurate delay estimates require a probing pulse $s(\cdot)$ with a higher bandwidth.

To find the ML estimate, we solve the ML equation (9.139) where $\nabla_{\mathbf{x}} s(t, \mathbf{x})$ and $\nabla_{\mathbf{x}} E(\mathbf{x})$ are given by (9.146) and (9.145), respectively. The first entry of this equation yields the amplitude estimate (9.111), and the second entry gives

$$\int_0^T Y(t)\dot{s}(t-\hat{d})dt = 0\,. \qquad (9.149)$$

Taking into account expression (9.114) for the output $Z(t)$ of the matched filter with impulse response (9.113), we see that (9.149) is equivalent to

$$\dot{Z}(T+\hat{d}) = 0\,. \qquad (9.150)$$

Note that this equation does not specify the ML estimate uniquely, since it only indicates that $\hat{d}$ must be a local maximum or minimum of $Z(T+d)$. But from (9.115) it turns out that the ML estimate is the maximum of the absolute value $|Z(T+d)|$ for $0 \leq d \leq D$. In fact, the structure of the ML delay estimator can be used to explain intuitively why the RMS bandwidth of the signal controls the accuracy of the delay estimate. According to (9.114) the output of the matched filter for an estimated delay $\hat{d}$ can be expressed as

$$Z(T+\hat{d}) = AE\rho(e) + \int_0^T s(t-\hat{d})V(t)dt \tag{9.151}$$

where $e = d - \hat{d}$ denotes the delay estimation error and the noise term is $N(0, \sigma^2 E)$ distributed. The peak of the first term is achieved for $e = 0$, i.e., when $\hat{d} = d$, but because of the presence of the noise term, our ability to find this peak depends on two factors. First, on how high the peak stands compared to the noise, which is fixed by the SNR A^2E/σ^2, but also on the narrowness of the correlation function $\rho(e)$ in the vicinity of $e = 0$. If the function $\rho(e)$ has a broad maximum around $e = 0$, this increases the chances that due to noise fluctuations, we will pick a maximum $Z(T+\hat{d})$ where $\hat{d}$ will be significantly different from d. On the other hand, if $\rho(e)$ is very narrow, only small errors can occur. But since

$$\rho(e) = s(e) * s(-e)/E$$

where $*$ denotes the convolution operation, its width is inversely proportional to the width of its Fourier transform $|S(\omega)|^2/E$, which is measured by the RMS bandwidth.

To examine the performance of the ML estimator, we assume that the error $e = d - \hat{d}$ is small and we substitute the first-order Taylor series approximation

$$\dot{s}(t-\hat{d}) = \dot{s}(t-d) + e\ddot{s}(t-d)$$

inside (9.149). This gives

$$-e\int_0^T Y(t)\ddot{s}(t-d)dt = \int_0^T Y(t)\dot{s}(t-d)dt \tag{9.152}$$

In this expression, at high SNR, the integral multiplying e on the left-hand side can be approximated by its mean value

$$-A\int_0^T s(t-d)\ddot{s}(t-d)dt = \frac{A}{2\pi}\int_{-\infty}^{\infty} \omega^2 |S(\omega)|^2 d\omega = AB^2E\,.$$

The right-hand side of (9.152) can be decomposed as

$$\begin{aligned}\int_0^T Y(t)\dot{s}(t-d)dt &= A\int_0^T s(t-d)\dot{s}(t-d)dt \\ &+ \int_0^T V(t)\dot{s}(t-d)dt\,,\end{aligned}$$

where the first term is zero according to (9.147) and the second term is $N(0, \sigma^2 B^2 E)$ distributed. So we have found that

$$e \approx \frac{1}{AB^2E} \int_0^T V(t)\dot{s}(t-d)dt \sim N\big(0, \frac{\sigma^2}{(AB)^2 E}\big)\,, \tag{9.153}$$

which shows that the ML estimate $\hat{d}$ is efficient at high SNR. Similarly, the distribution (9.122) of the amplitude estimate $\hat{A}$ indicates it is efficient. □

9.7 Detection of Radar Signals

The detection of radar signals presents an interesting challenge, since when a moving target is present, in addition to requiring the estimation of amplitude, phase, and time delay parameters, it involves the the estimation of a Doppler frequency shift. Specifically, assume that a radar transmits a probing signal

$$s(t) = \Re\{\tilde{s}(t)\exp(j\omega_c t)\} \tag{9.154}$$

whose complex envelope

$$\tilde{s}(t) = c(t)\exp(j\psi(t)))$$

has bandwidth $B \ll \omega_c$. This ensures that $\tilde{s}(t)$ is approximately constant over one period of the carrier signal $\exp(j\omega_c t)$. When the radar illuminates a point target located at distance Z and moving away from the radar with the radial velocity V at the instant of illumination, the return signal produced by the target can be expressed as

$$s(t,\mathbf{x}) = \Re\{\Gamma\tilde{s}(t-d)\exp(j(\omega_c + \omega_0)t)\}\,. \tag{9.155}$$

If c denotes the speed of light,

$$d = 2Z/c \quad \text{and} \quad \omega_0 = -2V\omega_c/c \tag{9.156}$$

represent respectively the round-trip signal delay and the Doppler frequency shift due to the radial velocity of the target. We assume that ω_0 is small compared to the bandwidth B of the complex envelope signal. $\Gamma = A\exp(j\theta)$ is an unknown complex coefficient specifying the amplitude $A \geq 0$ of the return signal and its phase shift θ. Accordingly, the signal $s(t,\mathbf{x})$ depends on the 4-dimensional real parameter vector

$$\mathbf{x} = \begin{bmatrix} A\;\theta\;d\;\omega_0 \end{bmatrix}^T\,.$$

Then the radar detection problem can be expressed as

$$\begin{aligned} H_0 &: Y(t) = V(t) \\ H_0 &: Y(t) = s(t,\mathbf{x}) + V(t) \end{aligned} \tag{9.157}$$

for $0 \leq t \leq T$, where $V(t)$ is a zero-mean bandpass white Gaussian noise with power spectral density (PSD)

$$S_V(\omega) = \sigma^2 \tag{9.158}$$

for $|\omega \pm \omega_c| < B$. As in Example 9.1, we assume that the pulse $\tilde{s}(t)$ is such that $\tilde{s}(t-d)$ remains contained in the interval $[0, T]$ for all posslble values of the delay d.

9.7.1 Equivalent Baseband Detection Problem

Instead of considering the radar detection problem in its bandpass form (9.157), it is convenient to convert it into an equivalent baseband form expressed in terms of the complex envelopes of the signals $Y(t)$, $s(t, \mathbf{x})$, and $V(t)$. To do so we employ Rice's representation

$$V(t) = \Re\{\tilde{V}(t)\exp(j\omega_c t)\} \tag{9.159}$$

with

$$\tilde{V}(t) = V_c(t) + jV_s(t)$$

of the bandpass WGN $V(t)$ in terms of its in phase and quadrature components $V_c(t)$ and $V_s(t)$. This representation is discussed in [1, Sec. 1.11] and [4, pp. 465–469]. The processes $V_c(t)$ and $V_s(t)$ are independent zero-mean baseband WGNs with PSD

$$S_{V_c}(\omega) = S_{V_s}(\omega) = 2\sigma^2 \tag{9.160}$$

for $|\omega| \leq B$. This means that as long as these processes are passed through filters with bandwidth less than or equal to B, they can be treated as WGNs with intensity $2\sigma^2$, i.e. with autocorrelation

$$E[V_c(t)V_c(s)] = E[V_s(t)V_s(s)] = 2\sigma^2\delta(t-s)\,.$$

Equivalently, the complex envelope process $\tilde{V}(t)$ is a zero-mean circular WGN with intensity $4\sigma^2$, i.e.,

$$\begin{aligned} E[\tilde{V}(t)\tilde{V}^*(s)] &= 4\sigma^2\delta(t-s) \\ E[\tilde{V}(t)\tilde{V}(s)] &= 0\,, \end{aligned} \tag{9.161}$$

where the second identity expresses the circularity property of the process.

Then, if we consider Rice's representation

$$Y(t) = \Re\{\tilde{Y}(t)\exp(j\omega_c t)\} \tag{9.162}$$

of the observed signal, with

$$\tilde{Y}(t) = Y_c(t) + jY_s(t)\,,$$

by combining the bandpass representations (9.155), (9.159) and (9.162), the detection problem (9.157) can be converted to the equivalent complex baseband detection problem

$$\begin{aligned} H_0 &: \tilde{Y}(t) = \tilde{V}(t) \\ H_1 &: \tilde{Y}(t) = \tilde{s}(t, \mathbf{x}) + \tilde{V}(t) \end{aligned} \tag{9.163}$$

for $0 \leq t \leq T$, with

$$\tilde{s}(t, \mathbf{x}) = \Gamma \tilde{s}(t - d) \exp(j\omega_0 t) . \tag{9.164}$$

Note that, although the problem (9.163) is expressed in terms of complex random signals, by separating the real and imaginary parts of these signals, it can be converted into an equivalent real detection problem expressed in terms of the two-dimensional observation waveform

$$\bar{Y}(t) = \left[Y_c(t) \; Y_s(t) \right]^T$$

for $0 \leq t \leq T$. So although the analysis below is performed in the complex domain, readers who wish to verify the correctness of the results presented may want to carry out an equivalent analysis of the real detection problem expressed in terms of the vector observation waveform $\bar{Y}(t)$.

Let

$$E \triangleq \frac{1}{2} \int_0^T |\tilde{s}(t)|^2 dt = \frac{1}{2} \int_0^T c^2(t) dt \tag{9.165}$$

denote the energy of the transmitted pulse $s(t)$. The factor 2 appearing in this definition just reflects the fact that the rms power of a sinusoidal waveform is half of its peak power. To formulate the baseband detection problem (9.163), we perform a KLD of the processes $\tilde{V}(t)$ and $\tilde{Y}(t)$ with respect to a COB of the space of *complex* square-integrable functions over $[0, T]$. As the first basis function, we select

$$\phi_1(t, \boldsymbol{\xi}) = (2E)^{-1/2} \tilde{s}(t - d) \exp(j\omega_0 t) , \tag{9.166}$$

where

$$\boldsymbol{\xi} = \left[d \; \omega_0 \right]^T \tag{9.167}$$

denotes the two-dimensional vector formed by the last two entries of $\mathbf{x}$. This function is normalized since

$$\begin{aligned} \int_0^T |\phi_1(t, \boldsymbol{\xi})|^2 dt &= (2E)^{-1} \int_0^T |\tilde{s}(t - d)|^2 dt \\ &= (2E)^{-1} \int_0^T |\tilde{s}(t)|^2 dt = 1 \end{aligned}$$

where we have used the fact that the pulse $\tilde{s}(t-d)$ remains contained in the interval $[0, T]$ for all values of the delay. The function $\phi_1(t, \boldsymbol{\xi})$ depends on the unknown parameter vector $\boldsymbol{\xi}$, so the functions $\phi_j(t, \boldsymbol{\xi})$ with $j \geq 2$ obtained by completing the basis will also depend on $\boldsymbol{\xi}$. We can then perform KLDs

$$\begin{aligned} \tilde{V}(t) &= \sum_{j=1}^{\infty} \tilde{V}_j \phi_j(t, \boldsymbol{\xi}) \\ \tilde{Y}(t) &= \sum_{j=1}^{\infty} \tilde{Y}_j \phi_j(t, \boldsymbol{\xi}) , \end{aligned} \tag{9.168}$$

of the complex baseband signals $\tilde{V}(t)$ and $\tilde{Y}(t)$ with respect to this basis, where

$$\begin{aligned} \tilde{V}_j &= \int_0^T \tilde{V}(t) \phi_j^*(t, \boldsymbol{\xi}) dt \\ \tilde{Y}_j &= \int_0^T \tilde{Y}(t) \phi_j^*(t, \boldsymbol{\xi}) dt . \end{aligned}$$

The coefficients $\tilde{V}_j(\boldsymbol{\xi})$ form a zero-mean complex circular WGN sequence with variance $4\sigma^2$, i.e., $\tilde{V}_j(\boldsymbol{\xi})$ admits the $CN(0, 4\sigma^2)$ probability density

$$f_{\tilde{V}_j}(v) = \frac{1}{4\pi\sigma^2} \exp\big(-\frac{|v|^2}{4\sigma^2}\big) \tag{9.169}$$

where $v = v_c + jv_s$ is a complex number. Under both hypotheses, the KL coefficients satisfy $\tilde{Y}_j(\boldsymbol{\xi}) = \tilde{V}_j(\boldsymbol{\xi})$ for $j \geq 2$, so these coefficients can be discarded. Accordingly, the complex baseband detection problem admits the signal space representation

$$\begin{aligned} H_0 &: \tilde{Y}_1(\boldsymbol{\xi}) = \tilde{V}_1(\boldsymbol{\xi}) \\ H_1 &: \tilde{Y}_1(\boldsymbol{\xi}) = (2E)^{1/2} \Gamma + \tilde{V}_1(\boldsymbol{\xi}) . \end{aligned} \tag{9.170}$$

Hence the likelihood ratio function of the detection problem takes the form

$$\begin{aligned} L(\tilde{Y}(\cdot)|\mathbf{x}) &= \exp\left(\frac{1}{4\sigma^2}\big[-|\tilde{Y}_1(\boldsymbol{\xi}) - (2E)^{1/2}\Gamma|^2 + |\tilde{Y}_1(\boldsymbol{\xi})|^2\big]\right) \\ &= \exp\left(\frac{1}{2\sigma^2}\Re\{(2E)^{1/2}\tilde{Y}_1(\boldsymbol{\xi})\Gamma^*\} - \frac{E|\Gamma|^2}{2\sigma^2}\right) \\ &= \exp\left(\frac{1}{2\sigma^2}\big[\Re\{\int_0^T \tilde{Y}(t)\tilde{s}^*(t, \mathbf{x}) dt\} - \frac{1}{2}\int_0^T |\tilde{s}(t, \mathbf{x})|^2 dt\big]\right) . \end{aligned} \tag{9.171}$$

9.7.2 Cramér-Rao Bound

To evaluate the Cramér-Rao bound for $\mathbf{x}$, by taking the gradient of the log-likelihood function, we obtain

$$\begin{aligned}\nabla_{\mathbf{x}} \ln L(\tilde{Y}(\cdot)|\mathbf{x}) &= \frac{1}{2\sigma^2}\Re\{\int_0^T \nabla_{\mathbf{x}}\tilde{s}(t,\mathbf{x})(\tilde{Y}(t)-\tilde{s}(t,\mathbf{x}))^* dt\} \\ &= \frac{1}{2\sigma^2}\Re\{\int_0^T \nabla_{\mathbf{x}}\tilde{s}(t,\mathbf{x})\tilde{V}^*(t)dt\}\,. \end{aligned} \tag{9.172}$$

Taking into account the correlation properties (9.161) of the circular WGN $\tilde{V}(t)$, we find that the Fisher information matrix can be expressed as

$$\begin{aligned}\mathbf{J}(\mathbf{x}) &= E[\nabla_{\mathbf{x}} \ln L(\tilde{Y}(\cdot)|\mathbf{x})\big(\nabla_{\mathbf{x}} \ln L(Y(\cdot)|\mathbf{x})\big)^T] \\ &= \frac{1}{2\sigma^2}\Re\{\int_0^T \nabla_{\mathbf{x}}\tilde{s}(t,\mathbf{x})(\nabla_{\mathbf{x}}\tilde{\mathbf{s}}(t,\mathbf{x}))^H dt\}\,. \end{aligned} \tag{9.173}$$

But

$$\nabla_{\mathbf{x}}\tilde{s}(t,\mathbf{x}) = \begin{bmatrix} \tilde{s}(t-d) \\ jA\tilde{s}(t-d) \\ -A\dot{\tilde{s}}(t-d) \\ jAt\tilde{s}(t-d) \end{bmatrix} \exp(j(\psi(t-d)+\omega_0 t+\theta))\,. \tag{9.174}$$

Consequently,

$$\begin{aligned} J_{11}(\mathbf{x}) &= \frac{1}{2\sigma^2}\int_0^T |\tilde{s}(t-d)|^2 dt = \frac{E}{\sigma^2} \\ J_{12}(\mathbf{x}) &= J_{14}(\mathbf{x}) = 0 \\ J_{13}(\mathbf{x}) &= -\frac{A}{4\sigma^2}\int_0^T \frac{d}{dt}|\tilde{s}(t-d)|^2 dt = 0\,, \end{aligned} \tag{9.175}$$

where the last identity uses the fact that the pulse $\tilde{s}(t-d)$ is completely contained in the interior of the interval $[0,T]$. Similarly, we have

$$\begin{aligned} J_{22}(\mathbf{x}) &= \frac{A^2}{2\sigma^2}\int_0^T |\tilde{s}(t-d)|^2 dt = \frac{A^2 E}{\sigma^2} \\ J_{23}(\mathbf{x}) &= -\frac{A^2}{2\sigma^2}\Im\{\int_0^T \dot{\tilde{s}}(t-d)\tilde{s}^*(t-d)dt\} \\ &= -\frac{A^2 E}{\sigma^2}\bar{\omega} \\ J_{24}(\mathbf{x}) &= \frac{A^2}{2\sigma^2}\int_0^T t|\tilde{s}(t-d)|^2 dt \\ &= \frac{A^2 E}{\sigma^2}(\bar{t}+d) \end{aligned} \tag{9.176}$$

where, if $\tilde{S}(\omega)$ denotes the Fourier transform of $\tilde{s}(t)$,

$$\bar{\omega} \triangleq \frac{1}{2E}\Im\{\int_0^T \dot{\tilde{s}}(t)\tilde{s}^*(t)dt\} = \frac{1}{4\pi E}\int_{-\infty}^{\infty} \omega|\tilde{S}(\omega)|^2 d\omega \tag{9.177}$$

and

$$\bar{t} \triangleq \frac{1}{2E}\int_0^T t|\tilde{s}(t)|^2 dt \tag{9.178}$$

are the centers of mass of the pulse $\tilde{s}(\cdot)$ in the frequency domain and the time domain, respectively. Note that when $\tilde{s}(t)$ is real, the magnitude of its Fourier transform is an even function, so $\bar{\omega} = 0$. It is also worth noting that the time origin and the carrier frequency ω_c can always be selected so that $\bar{t}$ and $\bar{\omega}$ are both zero. In this case, the observation interval becomes $[a, a+T]$ for some a. Finally, we have

$$\begin{aligned} J_{33}(\mathbf{x}) &= \frac{A^2}{2\sigma^2}\int_0^T |\dot{\tilde{s}}(t-d)|^2 dt = \frac{A^2 E}{\sigma^2}\overline{\omega^2} \\ J_{34}(\mathbf{x}) &= -\frac{A^2}{2\sigma^2}\Im\{\int_0^T t\dot{\tilde{s}}(t-d)\tilde{s}^*(t-d)dt\} = -\frac{A^2 E}{\sigma^2}\overline{\omega(t+d)} \\ J_{44}(\mathbf{x}) &= \frac{A^2}{2\sigma^2}\int_0^T t^2|\tilde{s}(t-d)|^2 dt = \frac{A^2 E}{\sigma^2}\overline{(t+d)^2}, \end{aligned} \tag{9.179}$$

where

$$\begin{aligned} \overline{\omega^2} &\triangleq \frac{1}{2E}\int_0^T |\dot{\tilde{s}}(t)|^2 dt = \frac{1}{4\pi E}\int_{-\infty}^{\infty} \omega^2|\tilde{S}(\omega)|^2 d\omega \\ \overline{\omega t} &\triangleq \frac{1}{2E}\Im\{\int_0^T t\dot{\tilde{s}}(t)\tilde{s}^*(t)dt\} = \frac{1}{2E}\int_0^T c^2(t)\dot{\psi}(t)dt \\ \overline{t^2} &\triangleq \frac{1}{2E}\int_0^T t^2|\tilde{s}(t)|^2 dt\,. \end{aligned} \tag{9.180}$$

To check these results, note that the submatrix obtained by retaining only the first and second rows and columns of $\mathbf{J}(\mathbf{x})$ coincides with the Fisher information matrix (9.138) obtained in Example 9.2 for the joint amplitude and phase estimation of a bandpass signal. Similarly, the matrix obtained by retaining only the first and third rows and columns of $\mathbf{J}(\mathbf{x})$ coincides with the Fisher information matrix (9.148) obtained for the joint amplitude and time delay estimation of a real baseband signal.

Next, if we let

$$\gamma = [A\ \theta]^T \tag{9.181}$$

be the vector formed by the first two entries of $\mathbf{x}$ and if, as indicated in (9.167), the vector $\boldsymbol{\xi}$ corresponds to the last two entries of $\mathbf{x}$, we consider the partition

$$\mathbf{J}(\mathbf{x}) = \begin{bmatrix} \mathbf{J}_{\gamma}(\mathbf{x}) & \mathbf{J}_{\gamma\xi}(\mathbf{x}) \\ \mathbf{J}_{\xi\gamma}(\mathbf{x}) & \mathbf{J}_{\xi}(\mathbf{x}) \end{bmatrix} \tag{9.182}$$

of the Fisher information matrix, where the 2×2 blocks are given by

$$\mathbf{J}_{\gamma}(\mathbf{x}) = \frac{E}{\sigma^2} \begin{bmatrix} 1 & 0 \\ 0 & A^2 \end{bmatrix}$$

$$\mathbf{J}_{\boldsymbol{\xi}\gamma}(\mathbf{x}) = \mathbf{J}_{\gamma\boldsymbol{\xi}}^T(\mathbf{x}) = \frac{A^2 E}{\sigma^2} \begin{bmatrix} 0 & -\bar{\omega} \\ 0 & \overline{t+d} \end{bmatrix}$$

$$\mathbf{J}_{\boldsymbol{\xi}}(\mathbf{x}) = \frac{A^2 E}{\sigma^2} \begin{bmatrix} \overline{\omega^2} & -\overline{\omega(t+d)} \\ -\overline{\omega(t+d)} & \overline{(t+d)^2} \end{bmatrix} . \tag{9.183}$$

Then the CRLB for an unbiased estimate $\hat{\mathbf{x}}$ of the vector $\mathbf{x}$ is given by

$$E[(\mathbf{x} - \hat{\mathbf{x}})(\mathbf{x} - \hat{\mathbf{x}})^T] \geq \mathbf{J}^{-1}(\mathbf{x}) . \tag{9.184}$$

We are particularly interested in the lower bound for the error covariance of the estimate

$$\hat{\boldsymbol{\xi}} = \begin{bmatrix} \hat{d} \ \hat{\omega}_0 \end{bmatrix}^T$$

of the time delay and Doppler frequency. We observe [5, p. 48] that the $(2,2)$ block of the matrix inverse $\mathbf{J}^{-1}(\mathbf{x})$ is the inverse of the Schur complement

$$\begin{aligned} \mathbf{C}(\mathbf{x}) &= (\mathbf{J}_{\boldsymbol{\xi}} - \mathbf{J}_{\boldsymbol{\xi}\gamma}\mathbf{J}_{\gamma}^{-1}\mathbf{J}_{\gamma\boldsymbol{\xi}})(\mathbf{x}) \\ &= \frac{A^2 E}{\sigma^2} \left(\begin{bmatrix} \overline{\omega^2} & -\overline{\omega(t+d)} \\ -\overline{\omega(t+d)} & \overline{(t+d)^2} \end{bmatrix} - \begin{bmatrix} -\bar{\omega} \\ \overline{t+d} \end{bmatrix} \begin{bmatrix} -\bar{\omega} & \overline{t+d} \end{bmatrix} \right. \\ &= \begin{bmatrix} \overline{\omega^2} - (\bar{\omega})^2 & -(\overline{\omega t} - \bar{\omega}\bar{t}) \\ -(\overline{\omega t} - \bar{\omega}\bar{t}) & \overline{t^2} - (\bar{t})^2 \end{bmatrix} \end{aligned} \tag{9.185}$$

of the block $\mathbf{J}_{\gamma}$ inside $\mathbf{J}$. So the CRLB for the time delay and Doppler frequency estimates takes the form

$$E[(\boldsymbol{\xi} - \hat{\boldsymbol{\xi}})(\boldsymbol{\xi} - \hat{\boldsymbol{\xi}})^T] \geq \mathbf{C}^{-1}(\mathbf{x}) , \tag{9.186}$$

where $\mathbf{C}(\mathbf{x})$ is given by (9.185).

Assume that the carrier frequency and time origin are selected such that $\bar{\omega}$ and $\bar{t}$ are zero. Then

$$B^2 = \overline{\omega^2} \quad \text{and} \quad W^2 = \overline{t^2}$$

represent respectively the squared rms bandwidth and the squared width of the complex pulse $\tilde{s}(t)$, and in (9.185) we have

$$\mathbf{C}^{-1}(\mathbf{x}) = \frac{\sigma^2}{A^2 E(B^2 W^2 - (\overline{\omega t})^2)} \begin{bmatrix} W^2 & \overline{\omega t} \\ \overline{\omega t} & B^2 \end{bmatrix} . \tag{9.187}$$

This expression can be simplified further when $\tilde{s}(t)$ is real, i.e., $\psi(t) = 0$, since according to (9.180), this implies $\overline{\omega t} = 0$, and thus

$$\mathbf{C}^{-1}(\mathbf{x}) = \frac{\sigma^2}{A^2 E} \begin{bmatrix} B^{-2} & 0 \\ 0 & W^{-2} \end{bmatrix} . \tag{9.188}$$

In this case the time delay CRLB

$$E[(d - \hat{d})^2] \geq \frac{\sigma^2}{(AB)^2 E}$$

coincides with the bound obtained in Example 9.3, and the Doppler frequency CRLB is given by

$$E[(\omega_0 - \widehat{\omega}_0)^2] \geq \frac{\sigma^2}{(AW)^2 E} .$$

From these expressions, we see that, as a general rule of thumb, to obtain accurate estimates of the time delay and Doppler frequency, we need a high SNR, as well as a signal with large bandwidth and a large time spread.

9.7.3 ML Estimates and GLR Detector

Since

$$\ln(L(\tilde{Y}(\cdot)|\mathbf{x})) = \frac{1}{4\sigma^2}\big[- |\tilde{Y}_1(\boldsymbol{\xi}) - (2E)^{1/2}\Gamma|^2 + |\tilde{Y}_1(\boldsymbol{\xi})|^2\big] \tag{9.189}$$

we conclude that the ML estimates of the signal parameters are given by

$$\begin{aligned} \hat{\boldsymbol{\xi}} &= \arg\max_{\boldsymbol{\xi}} |\tilde{Y}_1(\boldsymbol{\xi})| \\ \hat{\Gamma} &= (2E)^{-1/2}\tilde{Y}_1(\hat{\boldsymbol{\xi}}) \end{aligned} \tag{9.190}$$

so the logarithm of the GLR function can be expressed as

$$\ln(L_{\mathrm{G}}(\tilde{Y}(\cdot))) = \frac{|\tilde{Y}_1(\hat{\boldsymbol{\xi}})|^2}{4\sigma^2} . \tag{9.191}$$

Let

$$\begin{aligned} \tilde{Z}(d, \omega_0) &= (E/2)^{1/2}\tilde{Y}_1(\boldsymbol{\xi}) \\ &= \frac{1}{2}\int_0^T \tilde{Y}(t)\tilde{s}^*(t-d)\exp(-j\omega_0 t)dt . \end{aligned} \tag{9.192}$$

Then the estimates (9.190) can be rewritten as

$$(\hat{d}, \widehat{\omega}_0) = \arg\max_{d,\, \omega_0} |\tilde{Z}(d, \omega_0)| \tag{9.193}$$

$$\hat{\Gamma} = \tilde{Z}(\hat{d}, \widehat{\omega}_0)/E . \tag{9.194}$$

Equivalently, if we express $\tilde{Z}(\hat{d}, \widehat{\omega}_0)$ in polar coordinates as

$$\tilde{Z}(\hat{d}, \widehat{\omega}_0) = Q \exp(j\Phi) ,$$

from (9.194), the ML estimates of the amplitude and phase are given by

$$\hat{A} = \frac{Q}{E} \quad \text{and} \quad \hat{\theta} = \Phi . \tag{9.195}$$

Furthermore,

$$\ln L_{\mathrm{G}}(\tilde{Y}(\cdot)) = \frac{\hat{A}^2 E}{2\sigma^2} , \tag{9.196}$$

which is identical to the expression (9.112) obtained in Example 9.1 for a signal with unknown amplitude and time delay. Accordingly, the GLRT can be expressed as

$$\hat{A} \underset{H_0}{\overset{H_1}{\gtrless}} \eta . \tag{9.197}$$

As for the GLRT for a signal of unknown amplitude and time delay considered in Example 9.1, the probability of false alarm of the GLR detector (9.197) cannot be evaluated in closed form. This is due to the fact that P_F is a level crossing probability for the magnitude of the two-dimensional (2D) complex random field $\tilde{Z}(d, \omega_0)$. Under hypothesis H_0

$$\tilde{Z}(d, \omega_0) = \frac{1}{2} \int_0^T \tilde{V}(t)\tilde{s}^*(t-d)\exp(-j\omega_0 t)dt , \tag{9.198}$$

so $\tilde{Z}(d, \omega_0)$ is circular Gaussian with zero-mean and autocorrelation

$$\begin{aligned} E[\tilde{Z}(d,\omega_0)\tilde{Z}^*(d',\omega_0')] &= \sigma^2 \int_0^T \tilde{s}(t-d')\tilde{s}^*(t-d)\exp(j(\omega_0' - \omega_0)t)dt \\ &= 2\sigma^2 E\chi(d-d', \omega_0 - \omega_0')\exp(-j(\omega_0 - \omega_0')(d+d')/2) , \end{aligned} \tag{9.199}$$

where

$$\chi(e, \nu) \triangleq \frac{1}{2E} \int_0^T \tilde{s}(u+e/2)\tilde{s}^*(u-e/2)\exp(-j\nu u)du \tag{9.200}$$

denotes the *ambiguity function* of the pulse $\tilde{s}(t)$. In this context, P_F is the probability that $|\tilde{Z}(d,\omega_0)|/E$ exceeds η at some point of the domain $[0, D] \times [-\Omega_0/2, \Omega_0/2]$, where $[-\Omega_0/2, \Omega_0/2]$ denotes the range of Doppler frequencies of interest. To approximate P_F, it is possible to employ an approach similar to the one used to derive the approximation (9.120). Details are omitted, as this method relies on a 2-D extension of the Nyquist sampling theorem, which is beyond the scope of this text.

Under H_1, if d, ω_0 are the correct time delay and Doppler shift, and $\hat{d}$, $\widehat{\omega}_0$ represent their estimates, we have

$$\hat{\Gamma} = \Gamma\chi(e,\nu)\exp(-j\nu\bar{d}t) + \tilde{N} \qquad (9.201)$$

with $e = \hat{d} - d$, $\bar{d} = (\hat{d} + d)/2$, and $\nu = \widehat{\omega}_0 - \omega_0$, where

$$\tilde{N} = \frac{1}{2E}\int_0^T \tilde{V}(t)\tilde{s}^*(t-\hat{d})\exp(-j\widehat{\omega}_0 t)dt\,.$$

When the SNR $= A^2E/\sigma^2$ is large, the first term of (9.201) dominates, so the maximization used to obtain $\hat{\Gamma}$ does not affect the statistics of the noise term $\tilde{N}$, which can therefore be modelled as a $CN(0, 2\sigma^2/E)$ distributed complex random variable. Next, if we assume that the time delay and Doppler shift estimates are correct, i.e., $\hat{d} = d$ and $\hat{\omega}_0 = \omega_0$, (9.201) reduces to

$$\hat{\Gamma} = \Gamma + \tilde{N}\,. \qquad (9.202)$$

Since $|\Gamma| = A$, this implies $\hat{A} = |\hat{\Gamma}|$ admits the Ricean probability density

$$f(\hat{a}|H_1) = \frac{\hat{a}E}{\sigma^2}\exp\big(-\frac{(\hat{a}^2 + A^2)E}{2\sigma^2}\big)I_0(A\hat{a}E/\sigma^2)$$

with $\hat{a} \geq 0$, so the probability of detection

$$P_D(A) = P[\hat{A} \geq \eta|H_1] = Q(\frac{AE^{1/2}}{\sigma}, \frac{\eta E^{1/2}}{\sigma})\,. \qquad (9.203)$$

Detector implementation: The joint time delay and Doppler frequency estimation scheme specified by (9.193) can be implemented by using a bank of matched envelope filters followed by envelope detectors and then a peak detector that finds the largest peak among all the filter bank outputs, as shown in Fig. 9.6. Specifically, compare the expression (9.192) for $\tilde{Z}(d,\omega_0)$ with the output (9.52) for a matched envelope detector. By noting that $\tilde{Y}(t)\exp(-j\omega_0 t)$ is the envelope of the observed signal when the carrier frequency is shifted to $\omega_c + \omega_0$, we employ the matched envelope filter with impulse response

$$\begin{aligned} h(t,\omega_0) &= \Re\{\tilde{s}^*(T-t)\exp(j(\omega_c+\omega_0)t) \\ &= c(t-t)\cos((\omega_c+\omega_0)t - \psi(T-t))\,. \end{aligned} \qquad (9.204)$$

Then, if we express the output signal

$$Z(t,\omega_0) = h(t,\omega_0) * Y(t)\,,$$

in its bandpass form

$$Z(t,\omega_0) = \Re\{\tilde{Z}(t,\omega_0)\exp(j(\omega_c+\omega_0)t)\}$$

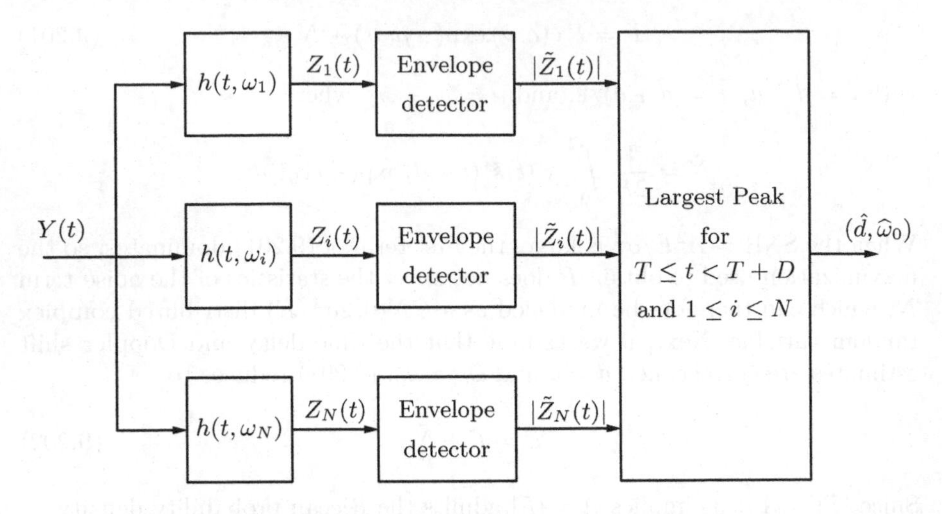

Fig. 9.6. Matched envelope filter bank for the joint estimation of time delay and Doppler frequency shift.

with respect to the center frequency $\omega_c + \omega_0$, we have

$$\begin{aligned}
\tilde{Z}(t,\omega_0) &= \frac{1}{2}\int_0^t \tilde{h}(t-u)\tilde{Y}(u)\exp(-j\omega_0 u)du \\
&= \frac{1}{2}\int_0^t \tilde{Y}(u)\tilde{s}^*(u+T-t)\exp(-j\omega_0 u)du\,. \qquad (9.205)
\end{aligned}$$

Consequently, when the envelope $\tilde{Z}(t,\omega_0)$ is sampled at $t = T+d$, it yields the statistic (9.192). To implement the maximization (9.193), we need therefore to pass the received signal $Y(t)$ through a bank of matched envelope filters $h(t,\omega_i)$ with $1 \leq i \leq N$, where the frequencies ω_i discretize the range of Doppler frequencies. The output $Z_i(t)$ of each of these filters is then passed through an envelope detector, yielding

$$|\tilde{Z}_i(T+d)| = |\tilde{Z}(d,\omega_i)|$$

for each possible Doppler frequency ω_i. Then the delay d and frequency ω_i that yield the largest peak value are the desired estimates.

9.7.4 Ambiguity Function Properties

From expression (9.201), we see that our ability to estimate the time delay d and Doppler shift ω_0 accurately depends on the shape of the ambiguity function $\chi(e,\nu)$ in the vicinity of $e = 0$ and $\nu = 0$. Since our discussion will not be restricted to pulses of finite duration, we rewrite the ambiguity function as

$$\chi(e,\nu) \triangleq \frac{1}{2E} \int_{-\infty}^{\infty} \tilde{s}(t+e/2)\tilde{s}^*(t-e/2)\exp(-j\nu t)dt\,, \tag{9.206}$$

where the integration is now over the entire real line. Ideally, we would like this function to have a sharp and narrow peak at $e=0$ and $\nu=0$. To fix ideas, we start by observing that for two function

$$\begin{aligned}\tilde{s}_1(t) &= \tilde{s}(t-d_1)\exp(j\omega_1 t)\\ \tilde{s}_2(t) &= \tilde{s}(t-d_2)\exp(j\omega_2 t)\end{aligned} \tag{9.207}$$

with the same pulse, but different delays and Doppler shifts, we have

$$\begin{aligned}<\tilde{s}_1,\tilde{s}_2> &= \int_{-\infty}^{\infty} \tilde{s}(t-d_1)\tilde{s}^*(t-d_2)\exp(j(\omega_1-\omega_2)t)\\ &= 2E\chi(d_2-d_1,\omega_2-\omega_1)\exp(-j(\omega_2-\omega_1)(d_1+d_2)/2)\end{aligned} \tag{9.208}$$

with

$$||\tilde{s}_1||^2 = ||\tilde{s}_2||^2 = 2E\,.$$

Consequently, by applying the Cauchy-Schartz inequality, we find

$$0 \leq \frac{|<\tilde{s}_1,\tilde{s}_2>|}{||\tilde{s}_1||\,||\tilde{s}_2||} = |\chi(d_2-d_1,\omega_2-\omega_1)| \leq 1\,. \tag{9.209}$$

Thus the magnitude of the ambiguity function lies between 0 and 1 and for zero error we have $\chi(0,0)=1$. It is also of interest to examine the slices of the ambiguity function obtained by setting $\nu=0$ and $e=0$, respectively. We have

$$\chi(e,0) = \frac{1}{2E}\int_{-\infty}^{\infty} \tilde{s}(t+e/2)\tilde{s}^*(t-e/2)dt = \rho(e)\,,$$

so with $\nu=0$, the ambiguity function just reduces to the correlation function of the pulse $\tilde{s}(\cdot)$. Setting $e=0$ gives

$$\chi(0,\nu) = \frac{1}{2E}\int_{-\infty}^{\infty} |\tilde{s}(t)|^2 \exp(-j\nu t)dt\,,$$

which is just the Fourier transform of the time distribution of the energy. To illustrate our discussion we consider two simple examples.

Example 9.4: Ambiguity function of a rectangular pulse

Consider the pulse

$$\tilde{s}_{\mathrm{R}}(t) = \frac{1}{W^{1/2}}\mathrm{rect}(t/W)\,,$$

where the unit rectangular function is defined as

$$\mathrm{rect}(t) = \begin{cases} 1 & |t| \leq 1/2 \\ 0 & \text{otherwise}. \end{cases}$$

In this case

$$\chi_{\mathrm{R}}(e, \nu) = (1 - \frac{|e|}{W})\mathrm{Sa}(\nu\frac{W}{2}(1 - \frac{|e|}{W}))$$

for $0 \leq |e| \leq W$, where

$$\mathrm{Sa}(x) \triangleq \frac{\sin(x)}{x}$$

denotes the sine over argument function, and $\chi_{\mathrm{R}}(e, \nu) = 0$ otherwise. So the rectangular pulse admits a triangular autocorrelation function

$$\rho_{\mathrm{R}}(e) = \chi_{\mathrm{R}}(e, 0) = 1 - \frac{|e|}{W}$$

of width W, and the $e = 0$ slice of the ambiguity function is given by

$$\chi_{\mathrm{R}}(0, \nu) = Sa(\nu\frac{W}{2}).$$

If we approximate its width by its first zero at $2\pi/W$, we see that it is inversely proportional to the width W of the ambiguity function along the horizontal axis. So, by selecting W, we can estimate accurately either the delay or the Doppler frequency, but not both. The function $\chi_{\mathrm{R}}(e, \nu)$ is plotted in Fig. 9.7 for $W = 2$. □

Example 9.5: Ambiguity function of a Gaussian pulse

For a Gaussian pulse

$$\tilde{s}_{\mathrm{G}}(t) = \frac{1}{(\pi W^2)^{1/4}} \exp\big(-\frac{t^2}{2W^2}\big),$$

of width W, the ambiguity function is given by

$$\chi_{\mathrm{G}}(e, \nu) = \exp\big(-(\frac{e^2}{4W^2} + \frac{\nu^2 W^2}{4})\big).$$

Its width is W along the e axis and W^{-1} along the the ν axis, so again, by adjusting W we can ensure that either d or ω_0 are estimated accurately, but not both. The function χ_{G} is plotted in Fig. 9.8 for $W = 2$. □

The above examples suggest that the ambiguity function cannot be shaped arbitrarily and must satisfy a constraint which, for the case of the pulses considered above, manifests itself by the fact that if we attempt to squeeze the ambiguity function in one direction, it necessarily becomes broader in the

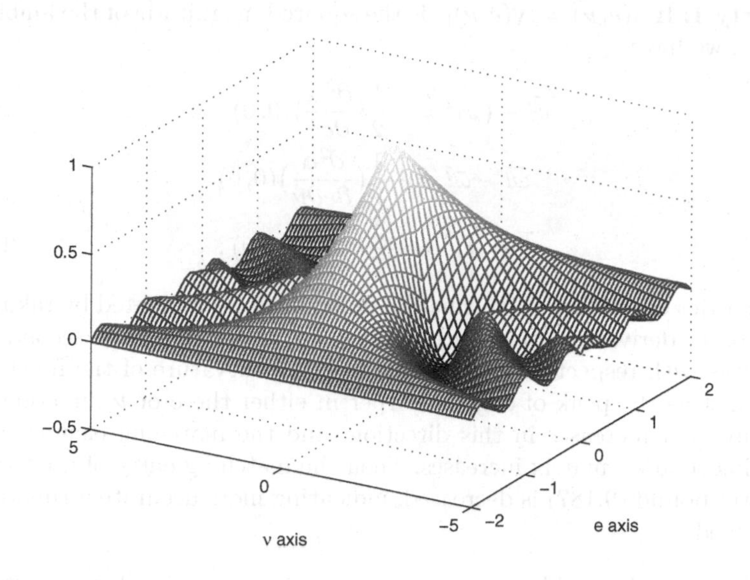

Fig. 9.7. Ambiguity function $\chi_{\mathrm{R}}(e, \nu)$ for $W = 2$.

other direction. Before describing this contraint, we first present a property of the ambiguity function that quantifies our intuition that accurate estimates of the delay and Doppler frequency depend on the sharpness of its peak at the point $(e, \nu) = (0, 0)$.

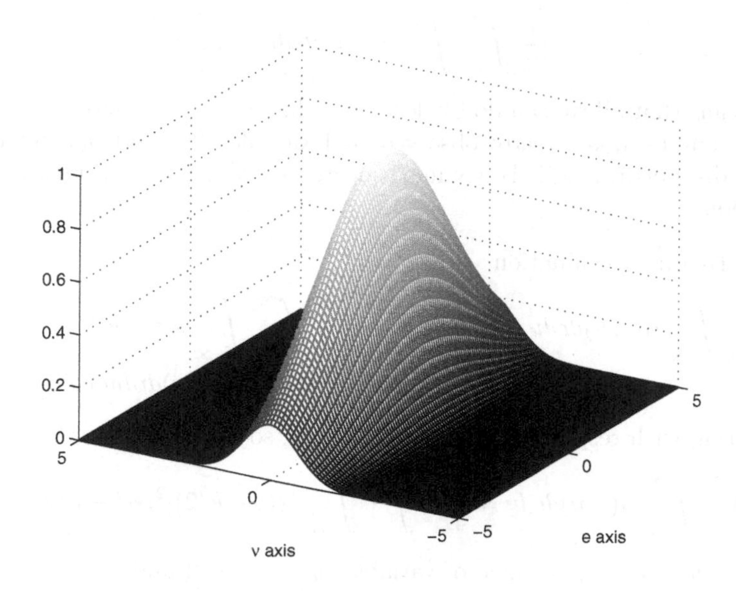

Fig. 9.8. Ambiguity function $\chi_{\mathrm{G}}(e, \nu)$ for $W = 2$.

Property 1: If $\alpha(e, \nu) = |\chi(e, \nu)|^2$ is the squared magnitude of the ambiguity function, we have

$$\begin{aligned}
\overline{\omega^2} - (\bar{\omega})^2 &= -\frac{1}{2}(\frac{\partial^2 \alpha}{\partial e^2})(0,0) \\
\overline{\omega t} - \bar{\omega}\bar{t} &= -\frac{1}{2}(\frac{\partial^2 \alpha}{\partial e \partial \nu})(0,0) \\
\overline{t^2} - (\bar{t})^2 &= -\frac{1}{2}(\frac{\partial^2 \alpha}{\partial \nu^2})(0,0) \,.
\end{aligned} \tag{9.210}$$

So the entries of the Fisher information matrix can be evaluated by taking the second order derivatives of $\alpha(e, \nu)$ at the point $(0, 0)$. Since the second-order derivatives with respect to e and ν measure the curvature of the function α, we see that as the peak of α gets sharper in either the e or ν directions, the curvature of α increases in this direction, and the matching element of the Fisher information matrix increases. Then the matching entry of the Cramér-Rao lower bound (9.187) is decreased, indicating more accurate estimates can be obtained.

So Property 1 provides a precise analytical justification for our intuition that to obtain accurate estimates of e and/or ν, the ambiguity function must have a sharp peak in the corresponding direction. The identities (9.210) can be obtained by direct differentiation as indicated in Problem 9.10. Next, the key constraint affecting the design of ambiguity functions is as follows:

Property 2: The total volume under the squared magnitude function $\alpha(e, \nu)$ is constant and is not affected by the choice of signal, i.e.,

$$\frac{1}{2\pi} \int_{-\infty}^{\infty} \int_{-\infty}^{\infty} \alpha(e, \nu) de d\nu = 1 \,. \tag{9.211}$$

This means that all we can do by designing the pulse $\tilde{s}$ is to move the volume around, but as was already observed in Examples 9.4 and 9.5, we cannot make it disappear, i.e., if the volume is decreased somewhere it must reappear elsewhere.

Proof: By direct evaluation

$$\begin{aligned}
\frac{1}{2\pi} \int_{-\infty}^{\infty} \int_{-\infty}^{\infty} \alpha(e, \nu) de d\nu = \frac{1}{8\pi E^2} \int_{-\infty}^{\infty} \int_{-\infty}^{\infty} \int_{-\infty}^{\infty} \int_{-\infty}^{\infty} \tilde{s}(t + e/2) \tilde{s}^*(t - e/2) \\
\tilde{s}^*(u + e/2) \tilde{s}(u - e/2) \exp(-j(t - u)\nu) dt du de d\nu \,.
\end{aligned} \tag{9.212}$$

Integrating with respect to ν gives $2\pi\delta(t - u)$, so (9.212) becomes

$$\frac{1}{2\pi} \int_{-\infty}^{\infty} \int_{-\infty}^{\infty} \alpha(e, \nu) de d\nu = \frac{1}{4E^2} \int_{-\infty}^{\infty} \int_{-\infty}^{\infty} |\tilde{s}(t - e/2)|^2 |\tilde{s}(t + e/2)|^2 dt de \,.$$

Then, performing the change of variables $y = t - e/2$ and $z = t + e/2$, and noting that the Jacobian of the transformation is 1, we obtain

$$\frac{1}{2\pi}\int_{-\infty}^{\infty}\int_{-\infty}^{\infty}\alpha(e,\nu)ded\nu = \frac{1}{4E^2}\Big(\int_{-\infty}^{\infty}|\tilde{s}(y)|^2dy\Big)^2 = 1\,.$$

The Property 2 is often called the *radar uncertainty principle.* Since it is now clear that we can never get rid of the volume of the squared ambiguity function, this leads us to a strategy consisting of moving the volume to places that will not affect significantly the peak at $(0,0)$. Two schemes that achieve this goal, but in different ways, are pulse compression and pulse repetition/sequencing. The pulse compression idea involves linear frequency modulation and relies on the following property of the ambiguity function:

Property 3: Consider a pulse $\tilde{s}(t)$ with ambiguity function $\chi(e,\nu)$. Then if we multiply $\tilde{s}(t)$ by the chirp signal $\exp(jbt^2/2)$, the ambiguity function of the modulated signal

$$\tilde{s}_{\mathrm{C}}(t) = \tilde{s}(t)\exp(jbt^2/2)$$

is

$$\chi_{\mathrm{C}}(e,\nu) = \chi(e,\nu - be)\,. \tag{9.213}$$

The identity (9.213) can be verified by inspection. Note that since the derivative of the phase $bt^2/2$ is bt, the signal $\chi_{\mathrm{C}}(t)$ is obtained from $\tilde{s}(t)$ by linear frequency modulation. If the width of $\tilde{s}(t)$ is W, the total frequency deviation over the duration of the pulse is $\Delta\omega = bW$. From (9.213), we see that the volume of the ambiguity function χ located along the e axis has been moved to the $\nu = be$ ridge. This can be used to improve the resolution of delay estimates, as can be seen from the following Gaussian pulse example.

Example 9.5, continued

According to (9.213), the ambiguity function of the pulse

$$\tilde{s}_{\mathrm{C}}(t) = \frac{1}{(\pi W^2)^{1/4}}\exp\Big(-\big(\frac{1}{W^2}+jb\big)\frac{t^2}{2}\Big)$$

takes the form

$$\chi_{\mathrm{C}}(e,\nu) = \exp\Big(-\big(\frac{e^2}{4W^2}+\frac{(\nu-be)^2W^2}{4}\big)\Big)\,.$$

This function is plotted below in Fig. 9.9 for $W=2$ and $b=1$. It is then easy to verify that

$$\begin{aligned}\overline{\omega^2} &= \frac{1}{2}(W^{-2}+(bW)^2)\\ \overline{\omega t} &= \frac{bW^2}{2}\\ \overline{t^2} &= \frac{W^2}{2}\,.\end{aligned} \tag{9.214}$$

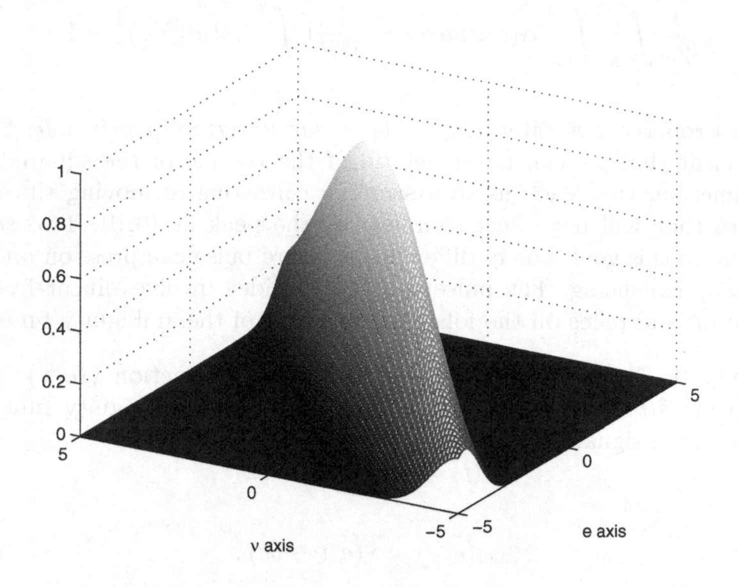

Fig. 9.9. Ambiguity function $\chi_C(e,\nu)$ for $W=2$ and $e=1$.

From this expression, we see that while the width W^{-1} of the pulse along the ν axis has not been affected by frequency modulation, for large b the width W along the e axis has become $(\Delta\omega)^{-1} = (bW)^{-1}$. The ratio $\Delta\omega W$ is called the pulse compression ratio. Accordingly, by selecting both b and W large, it is possible to ensure that the width of the ambiguity function is small along both the e and ν axes. This means that the corresponding range and Doppler radar will be able to estimate accurately the range of a target with known velocity, or the velocity of a target with known range. However, there is a catch. Because we have moved a large portion of the volume of the ambiguity function along the $\nu = be$ line, we already can guess that estimating d and ω_0 simultaneously will be difficult. In fact, if we consider the CRLB (9.189), by observing that

$$\overline{\omega^2 t^2} - (\overline{\omega t})^2 = \frac{1}{4} \, ,$$

we find that

$$\mathbf{C}^{-1}(\mathbf{x}) = \frac{4\sigma^2}{A^2 E} \begin{bmatrix} \overline{\omega^2} & -\overline{\omega t} \\ -\overline{\omega t} & \overline{t^2} \end{bmatrix} \, .$$

The expressions (9.214) then imply that the MSE errors of the joint delay and Doppler frequency estimates are quite large! So the pulse we have designed works very well if we know either the range or target velocity, but its performance is awful if we know neither. □

Pulse repetition: The second approach that can be used to displace the volume of the ambiguity function away from the point $(e,\nu) = (0,0)$ is pulse repetition, or even better, the design of coded pulse sequences where the phase of each successive pulse is selected appropriately [6]. We focus here on pulse repetition. Let $\tilde{p}(t)$ be a pulse of temporal width W, and consider the signal

$$\tilde{s}(t) = \frac{1}{M^{1/2}} \sum_{0}^{M-1} \tilde{p}(t - kT_r) \tag{9.215}$$

obtained by repeating it M times with repetition period $T_r \gg W$, where the scaling factor $M^{-1/2}$ ensures the pulse energy is unchanged. The ambiguity function of $\tilde{s}(\cdot)$ can be expressed in terms of the ambiguity function of $\tilde{p}(\cdot)$ as

$$\begin{aligned} \chi_{\tilde{s}}(e, \nu) &= \frac{1}{M} \sum_{k=0}^{M-1} \sum_{\ell=0}^{M-1} \int_{-\infty}^{\infty} \tilde{p}(t - kT_r + e/2)\tilde{p}^*(t - \ell T_r - e/2) \exp(-j\nu t) dt \\ &= \frac{1}{M} \sum_{k=0}^{M-1} \sum_{\ell=0}^{M-1} \exp(-j\nu(k+\ell)T_r/2) \chi_{\tilde{p}}(e + (\ell - k)T_r, \nu) \,. \end{aligned} \tag{9.216}$$

Then, setting $m = \ell - k$, we obtain

$$\chi_{\tilde{s}}(e, \nu) = \sum_{m=-(M-1)}^{M-1} C(m, \nu) \chi_{\tilde{p}}(e + mT_r, \nu) \tag{9.217}$$

with

$$\begin{aligned} C(m, \nu) &= \frac{\exp(-j|m|T_r\nu/2))}{M} \sum_{n=0}^{M-1-|m|} \exp(-jnT_r\nu) \\ &= \exp(-j(M-1)T_r\nu/2) \frac{\sin((M - |m|)T_r\nu/2)}{M \sin(T_r\nu/2)} \,. \end{aligned} \tag{9.218}$$

But since $\chi_{\tilde{p}}(e, \nu) \approx 0$ for $|e| > W$, we conclude from (9.217) that for $|e| < W$ we have

$$|\chi_{\tilde{s}}(e, \nu)| \approx |C(0, \nu)||\chi_{\tilde{p}}(e, \nu)| \,, \tag{9.219}$$

so that in the vicinity of $(0, 0)$, $|\chi_{\tilde{s}}(e, \nu)|$ varies with e in the same manner as $|\chi_{\tilde{p}}(e, \nu)|$ but its ν dependence is changed by the multiplicative factor

$$|C(0, \nu)| = \left| \frac{\sin(MT_r\nu/2)}{M \sin(T_r\nu/2)} \right|$$

which is periodic with period $2\pi/T_r$, and has a main lobe of width $2\pi/MT_r$. Consequently, by selecting M large enough, we can make the ambiguity function quite narrow in the ν direction. Of course, the price we are paying is that secondary maxima have been created. In the e direction, these secondary maxima occur with period T_r and correspond to the $2M+1$ terms in the sum (9.217), where we observe that the maxima decrease according to $1 - |m|/M$

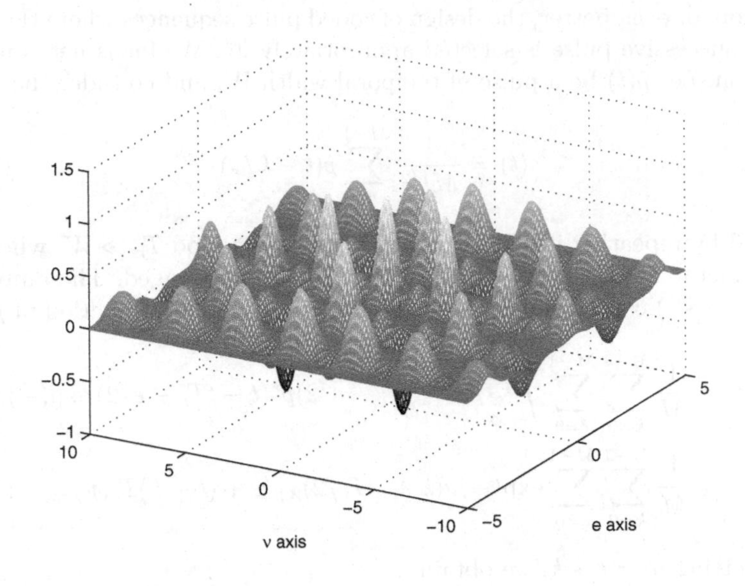

Fig. 9.10. Ambiguity function $\chi(e, \nu)$ for a Gaussian pulse sequence with $W = 0.2$, $T_r = 2$, and $M = 5$.

for $|e + mT_r| < W$. In the ν direction, the secondary maxima occur every $2\pi/T_r$. So, in essence, pulse repetition clears the area in the immediate vicinity of $(0, 0)$ but creates a large number of secondary peaks that may create difficulties in case radar echoes are located in these positions. To illustrate this phenomenon, the ambiguity function for a sequence of $M = 5$ Gaussian pulses of width $W = 0.2$ and repetition period $T_r = 2$ is plotted in Fig. 9.10. As can be seen from the figure, the central peak at $(0, 0)$ is quite sharp in both the e and ν directions, but it is surrounded by a multitude of secondary peaks in both directions.

9.8 Bibliographical Notes

Early works on the detection of signals with unknown parameters include the papers by Middleton and Van Meter [7, 8] (see also [9]), where a Bayesian perspective is adopted, and those of Kelly, Reed and Root [10, 11] which rely on a composite hypothesis testing formulation. The estimation of radar signal parameters is discussed in [12]. Among signal detection textbooks, some of the earlier texts [13] tend to favor Bayesian methods, whereas more recent presentations [14] favor the GLRT. Like [15, 16], we have attempted to steer our presentation midway between both approaches since, depending on the target application, both styles of detector design have a role to play. In wireless communications, detailed statistical channel models are often available, and data records tend to be short, as channels change rapidly, so Bayesian techniques tend to be appealing. On the other hand, for surveillance applications, long records are often available, making possible the use of a frequentist prespective and the design of GLR detectors. So what is the "right" detection formulation depends really on the specific problem at hand and on the physical or statistical modelling information available. The importance of the ambiguity function in radar detection theory was recognized early by Woodward [17] and Siebert [18]. Constraints affecting the volume and height of the ambiguity function are discussed in [19]. Readers wishing to study radar signal processing problems in further detail are encouraged to consult [20–22], among other works.

9.9 Problems

9.1. For binary noncoherent frequency shift keying (FSK), the received signals under the two hypotheses are

$$\begin{aligned} H_0 &: Y(t) = A\cos(\omega_0 t + \Theta_0) + V(t) \\ H_1 &: Y(t) = A\cos(\omega_1 t + \Theta_1) + V(t) \end{aligned}$$

for $0 \leq t \leq T$. The amplitude A is known, and $V(t)$ is a WGN with zero mean and intensity σ^2. The frequencies ω_1 and ω_0 are chosen such that $\omega_1 = m\omega_b$ and $\omega_0 = n\omega_b$ with m and n integer, where $\omega_b = 2\pi/T$ is the baud frequency. This ensures that the waveforms $\cos(\omega_1 t + \Theta_1)$ and $\cos(\omega_0 t + \Theta_0)$ are orthogonal over $[0, T]$ independently of the values of the phases Θ_0 and Θ_1. The phases Θ_1 and Θ_0 are independent and uniformly distributed over $[0, 2\pi]$, as well as independent of the noise $V(t)$. The two hypotheses are equally likely, and we seek to contruct a receiver that minimizes the probability of error.

(a) Obtain a signal space representation of this detection problem, and find the optimum receiver. The optimum receiver will depend on the envelopes R_1 and R_0 of the signals obtained by passing $Y(t)$ through bandpass filters centered at ω_1 and ω_0, respectively, and sampling at $t = T$.
(b) Evaluate the probability of error, i.e., the probability of false alarm, under H_i with $i = 0,\ 1$. Note that by symmetry this probability will be the same for the two hypotheses. This yields the probability of error for noncoherent FSK detectors.

9.2. Transmission diversity in a Rayleigh fading channel can be achieved in a variety of ways. For example consider an on-off keyed direct spread spectrum system. In this system, for each 0 bit, no signal is transmitted, whereas for a bit equal to 1, two different spreading codes $c(t)$ and $d(t)$ are used to transmit the same information. So if T denotes the baud period, and if we assume a chip period $T_c = T/N$, after demodulation and sampling with period T_c, when a 1 is transmitted, the observations admit the *complex* DT baseband representation

$$H_1 : Y(t) = Bb(t) + Cc(t) + V(t)$$

for $1 \leq t \leq N$. The spreading waveforms $b(t)$ and $c(t)$ are mutually orthogonal and have the same energy, i.e.,

$$\sum_{t=1}^{N} b^2(t) = \sum_{t=1}^{N} c^2(t) = E$$

and

$$\sum_{t=1}^{N} b(t)c(t) = 0 .$$

Since the waveforms $b(t)$ and $c(t)$ have different spectral characteristics they undergo different fading, so the complex coefficients B and C are independent identically distributed circularly symmetric Gaussian random variables with $N(0, \sigma^2/2)$ distributed real and imaginary parts. The additive noise $V(t)$ is a zero-mean complex circularly symmetric WGN, where the real and imaginary parts are $N(0, N_0/4)$ distributed. When a zero is transmitted, the demodulated DT baseband signal reduces to a zero-mean circularly symmetric complex WGN $V(t)$, i.e.,

$$H_0 : Y(t) = V(t)$$

for $1 \leq t \leq T$.

(a) Construct a signal space representation for the observation signal $Y(t)$ under H_0 and H_1.
(b) Construct a minimum probability of error receiver for deciding whether a 0 or a 1 was transmitted.
(c) Evaluate the probability of error. To do so, you may want to use the fact that if X_k, $1 \leq k \leq N$ is a sequence of independent $N(0, \sigma_X^2)$ random variables, then

$$Z = \sum_{k=1}^{N} X_k^2$$

admits a χ-squared distribution with N degrees of freedom, i.e.,

$$f_Z(z) = \frac{1}{(2\sigma_X^2)^{N/2}\Gamma(N/2)} z^{N/2-1} \exp\left(\frac{-z}{2\sigma_X^2}\right) u(z)$$

where $\Gamma(x)$ is the Gamma function and $u(\cdot)$ denotes the unit step function.
(d) Compare the probability of error of part (c) to the one obtained for a system where only one spreading code $b(t)$ is used to a transmit a 1, so that under H_1, the received signal is

$$Y(t) = Bb(t) + V(t) .$$

9.3. Consider an FSK communication system with space diversity over a Rayleigh channel, where two antenna elements are used to measure the transmitted signal. Under hypothesis H_0, the signals received by the two antenna elements are:

$$\mathbf{Y}(t) = \begin{bmatrix} Y_1(t) \\ Y_2(t) \end{bmatrix} = \begin{bmatrix} A_1 \cos(\omega_0 t + \Theta_1) \\ A_2 \cos(\omega_0 t + \Theta_2) \end{bmatrix} + \begin{bmatrix} V_1(t) \\ V_2(t) \end{bmatrix},$$

with $0 \leq t \leq T$, and under hypothesis H_1, the two signals can be expressed as

$$\mathbf{Y}(t) = \begin{bmatrix} Y_1(t) \\ Y_2(t) \end{bmatrix} = \begin{bmatrix} A_1 \cos(\omega_1 t + \Theta_1) \\ A_2 \cos(\omega_1 t + \Theta_2) \end{bmatrix} + \begin{bmatrix} V_1(t) \\ V_2(t) \end{bmatrix}.$$

Over a signaling interval $[0, T]$, the amplitudes A_i and phases Θ_i with $i = 1, 2$ are approximately constant. Typically, the amplitudes and phases vary slowly with time (slow fading), so that it is reasonable to assume that they do not change significantly within a pulse, but vary from pulse to pulse. We assume that the two antenna elements are sufficiently separated to ensure that the received signals undergo different fading. Thus, amplitudes A_1 and A_2 are independent and Rayleigh distributed with variance σ^2, i.e., they admit a probability density of the form

$$f_A(a) = \frac{a}{\sigma^2} \exp(-\frac{a^2}{2\sigma^2}).$$

The phases Θ_1 and Θ_2 are independent and uniformly distributed over the interval $[0, 2\pi]$, and are independent of the amplitudes. $V_1(t)$ and $V_2(t)$ are two independent white Gaussian noises with intensity $N_0/2$. which are independent of the amplitudes and phases of the two received signals. Assume that the frequencies ω_0 and ω_1 are selected such that $\cos(\omega_0 t + \theta)$ and $\cos(\omega_1 t + \phi)$ are orthogonal for all choices of θ and ϕ. The two hypotheses are equally likely, and we seek to minimize the probability of error.

(a) Construct a signal space representation for the binary hypothesis testing problem described above. Note that the received signal has a vector form, so that if $\{\phi_k(t); k \geq 0\}$ represents a complete orthonormal basis for square-integrable scalar functions over the interval $[0, T]$, the decomposition of $\mathbf{Y}(t)$ requires the orthonormal vector functions

$$\left\{ \begin{bmatrix} 1 \\ 0 \end{bmatrix} \phi_k(t) \, , \, \begin{bmatrix} 0 \\ 1 \end{bmatrix} \phi_k(t) \, ; \, k \geq 0 \right\}.$$

(b) Obtain the likelihood ratio test and describe the structure of the optimum receiver.

(c) Show that the probability of error is

$$P[E] = \frac{4 + 3\rho}{(2 + \rho)^3},$$

where $\rho = 2\bar{E}/N_0$ denotes the signal to noise ratio, with $\bar{E} = \sigma^2 T/2$.

9.4. Consider the CT detection problem

$$H_0 : Y(t) = V(t)$$
$$H_1 : Y(t) = A + Bt + V(t)$$

consisting of detecting a straight line in WGN for $0 \leq t \leq T$. The variance σ^2 of the zero-mean WGN $V(t)$ is known.

(a) By reparametrizing the line as $A' + B(t - T/2)$ with $A' = A + BT/2$, where the functions

$$f_1(t) = 1 \quad , \quad f_2(t) = t - T/2$$

are orthogonal, obtain a two-dimensional signal space representation of the detection problem.
(b) Assume first that A and B are known. Use the representation obtained in part (a) to design a Neyman-Pearson detector. Explain how the threshold η of the detector needs to be selected to ensure that the false alarm probability P_F is less than or equal to α.
(c) Unfortunately A and B are unknown. Explain why no UMP test exists and design a GLRT test. This will require finding the ML estimates $\hat{A}_{\mathrm{ML}}$ and $\hat{B}_{\mathrm{ML}}$ of A and B. To do so, you may wish to evaluate first ML estimates of A' and B. Set the threshold η of the GLR test to ensure that the false alarm probability P_F does not exceed α.

9.5. Consider the CT version of detection Problem 5.6. We wish to decide between

$$H_0 : Y(t) = B\cos(\omega_i t + \Theta) + V(t)$$
$$H_1 : Y(t) = As(t) + B\cos(\omega_i t + \Theta) + V(t)$$

for $0 \leq t \leq T$, where $s(t)$ is a known square-integrable signal with unknown amplitude $A > 0$. The observations are contaminated by an interfering sinusoidal signal with known frequency ω_i, but unknown amplitude B and phase Θ, as well as by a zero-mean WGN with intensity σ^2. For simplicity, we assume that the length T of the interval of observation is an integer multiple of the period $T_i = 2\pi/\omega_i$ of the intefering signal. This assumption ensures that the sine and cosine functions with frequency ω_i are exactly orthogonal, instead of being just approximately so.

(a) By using an orthonormal basis where the first three basis functions span the subspace of linear combinations of $\{\cos(\omega_i t), \sin(\omega_i t), s(t)\}$, find a finite-dimensional signal space representation of the detection problem.
(b) Obtain a maximal invariant statistic for the test under the group $\mathcal{G}$ of transformations corresponding to the contamination of the observations by a sinusoidal signal with frequency ω_i and arbitrary amplitude and phase.

(c) Show that a UMPI test exists. Set its threshold to ensure a probability of false alarm below α.
(d) Evaluate the power $P_D(\delta_{\text{UMPI}}, A)$ of the test of part (c).

9.6. Recall that a square integrable function f of $L^2[0,T]$ admits a trigonometric Fourier series expansion of the form

$$\begin{aligned} f(t) &= a_0 + \sum_{k=1}^{\infty} (a_k \cos(k\omega_0 t) + b_k \sin(k\omega_0 t)) \\ &= a_0 + \sum_{k=1}^{\infty} c_k \cos(k\omega_0 t + \theta_k) \end{aligned}$$

where $\omega_0 = 2\pi/T$ denotes the fundamental frequency,

$$\begin{aligned} a_0 &= \frac{1}{T}\int_0^T f(t)dt \\ a_k &= \frac{2}{T}\int_0^T f(t)\cos(k\omega_0 t)dt \ , \ k \geq 1 \\ b_k &= \frac{2}{T}\int_0^T f(t)\sin(k\omega_0 t)dt \ , \ k \geq 1 \end{aligned}$$

and $c_k = (a_k^2 + b_k^2)^{1/2}$, $\theta_k = -\tan(b_k/a_k)$. Then consider the CT detection problem

$$\begin{aligned} H_0 &: Y(t) = i(t) + V(t) \\ H_1 &: Y(t) = s(t) + i(t) + V(t) \end{aligned}$$

for $0 \leq t \leq T$, where $V(t)$ is a zero-mean WGN with intensity σ^2, and the signal $s(t)$ to be detected and interference signal $i(t)$ are unknown. However, it is known that the spectral support of $s(t)$ is restricted to $[k_L\omega_0, k_U\omega_0]$ and the spectral support of $i(t)$ to $[j_L\omega_0, j_U\omega_0]$, with $0 < k_L < j_L < k_U < j_U$. Thus, the only nonzero trigonometric Fourier coefficients a_k and b_k of $s(t)$ are for $k_L \leq k \leq k_U$. Similarly, the only nonzero Fourier coefficients of $i(t)$ are for $j_L \leq k \leq j_U$, where intervals $K = [k_L, k_U]$ and $J = [j_L, j_U]$ overlap but do not coincide.

(a) Obtain a finite dimensional signal space representation of the detection problem.
(b) By evaluating the ML estimates of all unknown parameters under H_0 and H_1, obtain a GLRT test.
(c) Verify that the GLRT can be written as

$$Z \underset{H_0}{\overset{H_1}{\gtrless}} \eta ,$$

where

$$Z = \int_0^T \int_0^T Y(t)Q(t-s)Y(s)dtds\,. \tag{9.220}$$

Evaluate the impulse response $Q(\cdot)$ appearing in this expression. Describe the operation performed by this filter.

(d) Use expression (9.220) to obtain an estimator-correlator implementation of the detector.

9.7. For Problem 9.6, assume that the interference signal $i(t)$ is still unknown with spectral support over $[j_L\omega_0, j_U\omega_0]$, and that the signal $s(t)$ to be detected has spectral support $[k_L\omega_0, k_U\omega_0]$ with $1 \leq k_L < j_L < k_U < j_U$. However assume now that the magnitude spectrum

$$c_k = (a_k^2 + b_k^2)^{1/2}$$

of $s(t)$ is known but its phase spectrum is unknown for $k_L \leq k \leq K_U$.

(a) Obtain a finite dimensional signal space representation of the detection problem.
(b) By evaluating the ML estimates of all unknown parameters under H_0 and H_1, obtain a GLRT test.
(c) Draw a detailed block diagram showing the implementation of the GLRT test.

9.8. Let $s(t)$ be the CT triangular wave with unit amplitude and period T_0 shown in Fig. 9.11.

We observe a scaled and delayed version $As(t-d)$ of $s(t)$ in the presence of a zero-mean WGN $V(t)$ of intensity σ^2, which yields

$$Y(t) = As(t-d) + V(t)$$

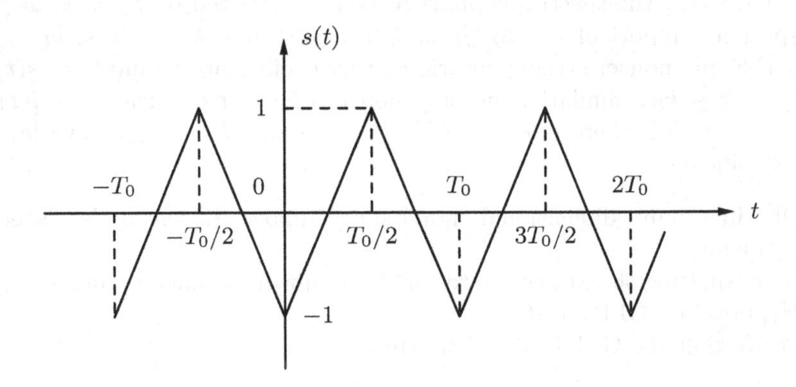

Fig. 9.11. Triangular wave with unit amplitude and period T_0.

for $0 \leq t \leq T$, where the interval of observation T is an integer multiple of the period T_0 of $s(t)$, i.e., $T = NT_0$ with N integer. The delay d is assumed to be such that $0 \leq d < T_0$. Note that $s(t)$ is defined for all t, so if $s(t)$ is shifted to the right by d to form $s(t-d)$, the part of the signal $s(t)$ which is pushed out of the interval $[0, T]$ is replaced on the left by an identical waveform segment (due to the periodicity of $s(t)$). This assumption is different from the one used in Section 9.6 of the textbook.

(a) Evaluate the Fisher information matrix of the vector parameter

$$\mathbf{x} = \begin{bmatrix} A \\ d \end{bmatrix} ,$$

and obtain a Cramér-Rao lower bound (CRLB) for unbiased estimates

$$\hat{\mathbf{x}} = \begin{bmatrix} \hat{A} \\ \hat{d} \end{bmatrix}$$

of $\mathbf{x}$ based on observations $\{Y(t),\ 0 \leq t \leq T\}$.

(b) Use the approach of Examples 9.1 and 9.3 to obtain the ML estimates $\hat{A}_{\text{ML}}$ and $\hat{d}_{\text{ML}}$ of A and d. Describe precisely the receiver used to produce these estimates.

(c) To find the ML estimate of d, instead of using a matched filter with impulse response $h(t) = s(T-t)$ and then examining the maximum of the output $Z(t)$ for $T \leq t \leq T + T_0$, show that the ML estimate of d can also be evaluated by passing $Y(t)$ through a linear time-invariant filter with impulse response $k(t) = \dot{s}(T-t)$ and then detecting the zero crossings of the output $R(t)$ for $T \leq t \leq T + T_0$. In other words, let $R(t) = k(t) * Y(t)$. Then $\hat{d}$ with $0 \leq \hat{d} < T_0$ is a ML estimate of d if it satisfies $R(T + \hat{d}) = 0$, i.e., if it is a zero crossing of $R(t)$. The reason why this implementation is attractive is that it is much easier to detect zero crossings than to find the maximum of a waveform. Specifically, to find $T+\hat{d}$ such that $R(T+\hat{d}) = 0$, we only need to apply a one-bit quantizer

$$q(r) = \begin{cases} 1 & r > 0 \\ -1 & r < 0 \end{cases}$$

to the output $R(t)$ of filter $k(t)$ and then detect the transition times at which signal $Q(t) = q(R(t))$ changes sign, as shown in Fig. 9.12.
Note that the filter $k(t)$ is easy to implement since the derivative $\dot{s}(t)$ of the triangular waveform is the square wave shown in Fig. 9.13.

(d) Verify that the output of filter $k(\cdot)$ can be decomposed into signal and noise components as

$$R(T + \hat{d}) = Az(e) + (k * V)(T + \hat{d})$$

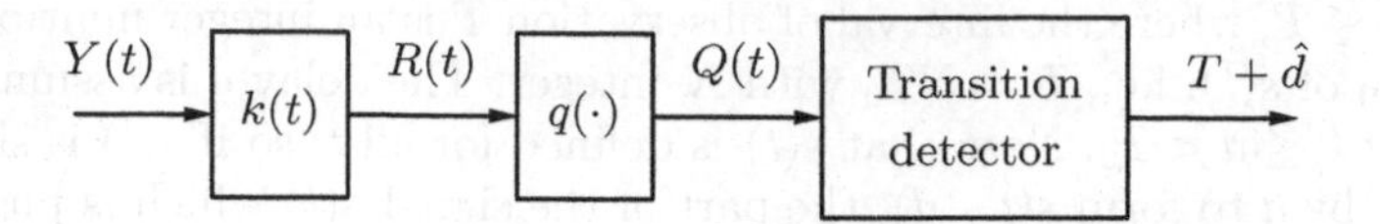

Fig. 9.12. Zero-crossing time-delay estimator.

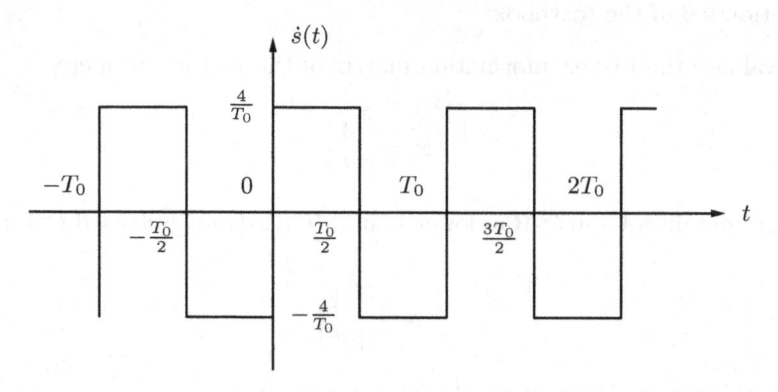

Fig. 9.13. Square wave obtained by differentiating the triangular wave $s(t)$.

where the zero-crossing function

$$z(e) \triangleq k(t) * s(t-d) \mid_{t=T+\hat{d}}$$

is a function of the error $e = d - \hat{d}$ only. Evaluate and sketch $z(e)$ and verify that it is zero for $e = mT_0/2$ with m integer. Indicate whether the zeros at $T_0/2 + nT_0$ with n integer represent a valid estimate of the delay (note that the above formulation allows A to be negative).

(e) For the case when e is small, use the expression you have obtained for $z(e)$ to obtain the probability distribution of the error and determine if the ML estimate $\hat{d}_{\text{ML}}$ produced by the zero-crossing detector is efficient. Similarly, is the amplitude estimate $\hat{A}_{\text{ML}}$ efficient?

9.9. An M-element radar antenna array is employed to detect a moving target in the presence of noise and interference. After demodulation, the received M-dimensional observations admit the complex baseband model [23]

$$\mathbf{Y}(t) = \Gamma \mathbf{M}(\phi) e^{j\omega_0 t} + \mathbf{V}(t)$$

for $0 \leq t \leq T$, where $\Gamma = Ae^{j\theta}$ is a complex reflection coefficient measuring the strength of the return signal, ϕ denotes the angle of arrival of the signal, and ω_0 is a Doppler frequency shift due to the relative motion of the array platform and target. The complex M-dimensional vector $\mathbf{M}(\phi)$ represents the array manifold. It depends on the array geometry and is assumed to be known.

It is scaled such that its first element equals one for all values of ϕ. The noise $\mathbf{V}(t) \in \mathbb{C}^M$ is temporally white, with each $\mathbf{V}(t) \sim CN(\mathbf{0}, \mathbf{K})$, so the density of $\mathbf{V}(t)$ can be expressed as

$$f_{\mathbf{V}}(\mathbf{v}) = \frac{1}{\pi^M |\mathbf{K}|} \exp\left(- \mathbf{v}^H \mathbf{K}^{-1} \mathbf{v} \right) .$$

The Hermitian covariance matrix $\mathbf{K}$ characterizes the statistical dependence of the noises affecting the M array elements. It depends on electromagnetic coupling between elements. It is evaluated by calibration and is assumed known. The reflection coefficient Γ, angle ϕ and Doppler frequency ω_0 are all unknown, so that for this detection problem the received signal is parameterized by vector

$$\mathbf{x} = \begin{bmatrix} A \, \theta \, \phi \, \omega_0 \end{bmatrix}^T ,$$

and we seek to decide between $H_1 \; : \; A > 0$ and $H_0 \; : \; A = 0$.

(a) Verify that the log-likelihood function for the observation sequence $\mathbf{Y}(t)$, $0 \leq t \leq T$ can be expressed as

$$\ln f(\mathbf{y}|\mathbf{x}) = - \sum_{t=0}^{T} ||\mathbf{y}(t) - \Gamma \mathbf{M}(\phi) e^{j\omega_0 t}||^2_{\mathbf{K}^{-1}} + c$$

where

$$||\mathbf{z}||^2_{\mathbf{K}^{-1}} \stackrel{\triangle}{=} \mathbf{z}^H \mathbf{K}^{-1} \mathbf{z}$$

and c is a constant independent of $\mathbf{x}$. Thus, finding the ML estimate of $\mathbf{x}$ reduces to the minimization of

$$J(\mathbf{x}) = \sum_{t=0}^{T} ||(\mathbf{y}(t) - \Gamma \mathbf{M}(\phi) e^{j\omega_0 t}||^2_{\mathbf{K}^{-1}} .$$

Let $N = T + 1$. By minimizing $J(\mathbf{x})$, verify that the ML estimate of $\mathbf{x}$ is given by

$$\hat{\Gamma} = \frac{\mathbf{M}^H(\hat{\phi}) \mathbf{K}^{-1} \bar{\mathbf{Y}}(e^{j\hat{\omega}_0})}{N \mathbf{M}^H(\hat{\phi}) \mathbf{K}^{-1} \mathbf{M}(\hat{\phi})}$$

and

$$\hat{\phi}, \, \hat{\omega}_0 = \arg \max_{\phi, \omega} \frac{|\mathbf{M}^H(\phi) \mathbf{K}^{-1} \bar{\mathbf{Y}}(e^{j\omega})|}{(\mathbf{M}^H(\phi) \mathbf{K}^{-1} \mathbf{M}(\phi))^{1/2}} ,$$

where

$$\bar{\mathbf{Y}}(e^{j\omega}) = \sum_{t=0}^{T} \mathbf{y}(t) e^{-j\omega t}$$

denotes the DTFT of the sequence $\{\mathbf{y}(t), 0 \leq t \leq T\}$.

(b) Obtain a matrix CRLB for the vector $\mathbf{x}$. To do so, you may want to parametrize the Fisher information matrix in terms of

$$\mathbf{d}(\phi) \triangleq \frac{d}{d\phi}\mathbf{M}(\phi)\,.$$

(c) Show that the GLR test can be expressed as

$$N^{1/2}||\mathbf{M}(\hat{\phi})||_{\mathbf{K}^{-1}}|\hat{\Gamma}| \underset{H_0}{\overset{H_1}{\gtrless}} \eta\,.$$

Explain how the threshold η should be selected.

9.10. Consider the time domain expression (9.206) for the ambiguity function $\chi(e,\nu)$. Let $\tilde{S}(\omega)$ denote the CT Fourier transform of pulse $\tilde{s}(t)$. By using Parseval's identity and the time and frequency shift properties of the Fourier transform, verify that the ambiguity function admits the equivalent frequency-domain expression

$$\chi(e,\nu) = \frac{1}{4\pi E}\int_{-\infty}^{\infty} \tilde{S}(\omega+\nu/2)\tilde{S}^*(\omega-\nu/2)\exp(j\omega e)d\omega\,. \tag{9.221}$$

Use this identity together with (9.206) to derive relations (9.210) forming Property 1 of the ambiguity function.

9.11. In addition to the properties of the ambiguity function described in Section 9.7.4, $\chi(e,\nu)$ admits several additional properties which are sometimes useful for pulse synthesis.

(a) Show that the ambiguity function $\chi(e,\nu)$ of an arbitrary pulse $\tilde{s}(t)$ satisfies

$$\chi(e,\nu) = \chi^*(-e,-\nu)\,,$$

i.e., it is invariant under complex conjugation and symmetry with respect to the origin.

(b) If an arbitrary complex pulse $\tilde{s}(t)$ has for ambiguity function $\chi(e,\nu)$, the scaled pulse $\tilde{s}(at)$ with $a > 0$ has for ambiguity function

$$\chi(ae,\frac{\nu}{a})\,.$$

To derive this result, do not forget to take into account the effect of the scaling parameter a in the energy scale factor E appearing in definition (9.206).

(c) Consider a pulse $\tilde{s}(t)$ with CT Fourier transform $\tilde{S}(\omega)$. Use identity (9.221) to verify that if $\tilde{s}(t)$ has for ambiguity function $\chi(e,\nu)$, then the dual pulse $\tilde{s}_D(t) = \tilde{S}(t)$ has for ambiguity function

$$\chi_D(e,\nu) = \chi(-\nu,e)\,.$$

Thus $\chi_D(e,\nu)$ is obtained by rotating the ambiguity function $\chi(e,\nu)$ by 90 degrees in the clockwise direction. To evaluate $\chi_D(e,\nu)$, do not forget to include the energy E_D of the dual pulse as a scale factor.

(d) Let $\tilde{s}_1(t)$ and $\tilde{s}_2(t)$ denote two complex pulses with ambiguity functions $\chi_1(e,\nu)$ and $\chi_2(e,\nu)$, respectively. Then the ambiguity function of the product

$$\tilde{s}(t) = \tilde{s}_1(t)\tilde{s}_2(t)$$

is the scaled convolution

$$\chi(e,\nu) = C \int_{-\infty}^{\infty} \chi_1(e,\mu)\chi_2(e,\nu-\mu)d\mu$$

of the two ambiguity functions with respect to the frequency variable, where the scaling constant C depends on the energies of pulses $\tilde{s}_1(t)$, $\tilde{s}_2(t)$ and $\tilde{s}(t)$.

9.12. Consider the decaying exponential pulse

$$\tilde{s}(t) = \frac{1}{W^{1/2}} \exp(-\frac{t}{2W})u(t) \,.$$

(a) Evaluate the ambiguity function $\chi(e,\nu)$ and its squared magnitude $\alpha(e,\nu) = |\chi(e,\nu)|^2$.

(b) Sketch the $\nu = 0$ slice $\alpha(e,0)$ of the squared magnitude and estimate its width in function of W. Similarly, sketch and estimate the width of the $e = 0$ slice $\alpha(0,\nu)$.

(c) Given a baseband signal model of the form (9.164), is it possible to select the pulse parameter W so that both the delay d and Doppler shift ω_0 are estimated accurately?

References

1. S. Haykin, *Communication Systems, Fouth Edition.* New York: J. Wiley & Sons, 2001.
2. I. F. Blake and W. C. Lindsey, "Level-crossing problems for random processes," *IEEE Trans. Informat. Theory,* vol. 19, pp. 295–315, May 1973.
3. H. Cramér and M. R. Leadbetter, *Stationary and Related Stochastic Processes–Sample Function Properties and Their Applications.* New York: J. Wiley & Sons, 1967. Reprinted by Dover Publ., Mineola, NY, 2004.
4. A. Papoulis and S. U. Pillai, *Probability, Random Variables and Stochastic Processes, Fourth Edition.* New York: McGraw Hill, 2002.
5. A. J. Laub, *Matrix Analysis for Scientists and Engineers.* Philadelphia, PA: Soc. for Industrial and Applied Math., 2005.
6. N. Levanon and E. Mozeson, *Radar Signals.* New York: J. Wiley/IEEE Press, 2004.
7. D. Middleton and D. Van Meter, "Detection and extraction of signals in noise from the point of view of statistical decision theory. I," *J. Soc. Indust. Applied Math.*, vol. 3, pp. 192–253, Dec. 1955.
8. D. Middleton and D. Van Meter, "Detection and extraction of signals in noise from the point of view of statistical decision theory. II," *J. Soc. Indust. Applied Math.*, vol. 4, pp. 86–119, June 1956.

9. D. Middleton, *An Introduction to Satistical Communication Theory.* New York: McGraw-Hill, 1960. Reprinted by IEEE Press, New York, 1996.
10. E. J. Kelly, I. S. Reed, and W. L. Root, "The detection of radar echoes in noise. I," *J. Soc. for Indust. Applied Math.*, vol. 8, pp. 309–341, June 1960.
11. E. J. Kelly, I. S. Reed, and W. L. Root, "The detection of radar echoes in noise. II," *J. Soc. for Indust. Applied Math.*, vol. 8, pp. 481–507, Sept. 1960.
12. E. J. Kelly, "The radar measurement of range, velocity and acceleration," *IRE Trans. on Military Electron.*, vol. 5, pp. 51–57, 1961.
13. H. L. Van Trees, *Detection, Estimation and Modulation Theory, Part I: Detection, Estimation and Linear Modulation Theory.* New York: J. Wiley & Sons, 1968. Paperback reprint edition in 2001.
14. S. M. Kay, *Fundamentals of Statistical Signal Processing: Detection Theory.* Prentice-Hall, 1998.
15. C. W. Helstrom, *Elements of Signal Detection & Estimation.* Upper Saddle River, NJ: Prentice-Hall, 1995.
16. R. N. McDonough and A. D. Whalen, *Detection of Signals in Noise, Second Edition.* San Diego, CA: Academic Press, 1995.
17. P. M. Woodward, *Probability and Information Theory, with Applications to Radar.* New York: Pergamon Press, 1953.
18. W. Siebert, "A radar detection philosophy," *IEEE Trans. Informat. Theory,* vol. 2, pp. 204–221, Sept. 1956.
19. R. Price and E. M. Hofstetter, "Bounds on the volume and heights distributions of the ambiguity function," *IEEE Trans. Informat. Theory,* vol. 11, pp. 207–214, Apr. 1965.
20. H. L. Van Trees, *Detection, Estimation and Modulation Theory, Part III: Radar-Sonar Signal Processing and Gaussian Signals in Noise.* New York: J. Wiley & Sons, 1971. Paperback reprint edition in 2001.
21. A. Rihaczek, *Principles of High-Resolution Radar.* New York: McGraw-Hill, 1969.
22. P. Z. Peebles, Jr., *Radar Principles.* New York: J. Wiley & Sons, 1998.
23. A. L. Swindlehurst and P. Stoica, "Maximum likelihood methods in radar array signal processing," *Proeedings of the IEEE*, vol. 86, pp. 421–441, Feb. 1998.

10

Detection of Gaussian Signals in WGN

10.1 Introduction

Up to this point we have considered the detection of known or partially known signals in Gaussian noise. This type of situation is common in digital communications or active radar/sonar applications. However there exists other situations where the signal to be detected has a physical origin and is random. For example, in industrial applications, the onset of a malfunction in heavy machinery is often preceded by vibrations, which typically appear random, but nevertheless have a characteristic spectral signature. Similarly, in naval surveillance applications, the propeller noise of surface ships or submarines is typically random, but its spectral characteristics can often be used to identify not only the ship type, but even its model. In covert communications applications, to ensure that transmitted signals cannot be intercepted and demodulated by potential eavesdroppers, users often use spread spectrum modulation and encryption techniques. This produces waveforms that appear random to any user other than the intended one. From this perspective, the goal of a surveillance system becomes to detect the presence of such signals, and perhaps to characterize some of their features that might provide clues about the identity of covert users.

This chapter tackles therefore the detection of Gaussian signals in the presence of WGN. We consider both CT and DT signals, but since optimum receiver structures rely on waveform estimators, which are significantly easier to implement in the DT domain, we ultimately focus our attention on the DT case. The CT and DT detection problems can be treated in a unified manner by expressing them in the Karhunen-Loève coefficient domain. In Section 10.2, we show that the optimum detector can be implemented as a noncausal estimator-correlator receiver. This detector has a form reminiscent of the correlator/integrator receiver for detecting a known signal in WGN, except that the known signal needs now to be replaced by the estimate of the Gaussian signal to be detected based on observations in the interval $[0, T]$ over which measureaments are available. Unfortunately, this estimate is noncausal, i.e.,

B.C. Levy, *Principles of Signal Detection and Parameter Estimation*,
DOI: 10.1007/978-0-387-76544-0_10, © Springer Science+Business Media, LLC 2008

it is a smoothed estimate, since to obtain the estimate of the signal to be detected at a time t, we need to wait until observations have been collected over the entire interval. Fortunately, a rather ingeniuous receiver was developed independently by Schweppe [1] and Stratonovich and Sosulin [2] in the mid-1960s, that relies entirely on a causal estimate of the signal. This receiver architecture is described in Section 10.3. It has a structure remarkably similar to the correlator/integrator detector used for known signals in WGN and can therefore be called the "true" extension of this detector to the Gaussian signal case. Unfortunately, the performance of detectors for Gaussian signals in WGN is rather difficult to characterize, since the probability distribution of the sufficient statistic used by these detectors cannot be evaluated easily. However, for the class of stationary Gaussian processes, it turns out that the detector performance can be characterized asymptotically. Like the case of i.i.d. measurements considered in Section 3.2, the asymptotic performance analysis relies on the theory of large deviations [3,4]. However, instead of employing Cramer's theorem, which is valid only for i.i.d. observations, it relies on a powerful extension of this result, called the Gärtner-Ellis theorem, which is applicable to dependent observations. In this context, the analog of the Kullback-Leibler divergence for characterizing the rate of decay of the probability of a miss and of false alarm of NP tests is the Itakura-Saito spectral distortion measure [5] between power spectral densities (PSDs) of stationary Gaussian processes.

10.2 Noncausal Receiver

In Chapter 8 we considered the detection of known signals in Gaussian noise. We start this chapter by examining a binary detection problem of the form

$$\begin{aligned} H_0 &: Y(t) = V(t) \\ H_1 &: Y(t) = Z(t) + V(t) \end{aligned} \tag{10.1}$$

for $0 \leq t \leq T$, where the signal $Z(t)$ to be detected is zero-mean and Gaussian. The signal $Z(t)$ has covariance

$$E[Z(t)Z(s)] = K(t,s) , \tag{10.2}$$

and the noise $V(t)$ is a zero-mean WGN uncorrelated with $Z(t)$ and with covariance

$$E[V(t)V(s)] = \sigma^2 \delta(t-s) . \tag{10.3}$$

As in the last two chapters, we consider simultaneously the DT and CT cases, and thus the impulse $\delta(t-s)$ appearing in (10.3) denotes either a Kronecker or a Dirac delta function, depending on whether we consider the DT or CT case. In the CT case we assume that the covariance kernel $K(t,s)$ is continuous over $[0,T]^2$. However, since optimal receivers rely on least-squares estimation filters, we will ultimately focus on the DT case, since DT Wiener and Kalman filters are much easier to implement than their CT counterparts.

10.2.1 Receiver Structure

By evaluating the eigenfunctions of the kernel $K(t,s)$ and, in the case when $K(t,s)$ is not positive-definite, by Gram-Schmidt orthonormalization of square-integrable/summable functions in its null space, we can construct a COB $\{\phi_j, j \in \mathcal{J}\}$ of $L^2[0,T]$ such that the eigenfunctions ϕ_j satisfy

$$\begin{aligned} \lambda_j \phi_j(t) &= \int_0^T K(t,s)\phi_j(s)ds \quad \text{(CT case)} \\ &= \sum_{t=0}^{T} K(t,s)\phi_j(s) \quad \text{(DT case)} \end{aligned} \tag{10.4}$$

with $\lambda_j \geq 0$. We can then perform Karhunen-Loève (KL) decompositions

$$Z(t) = \sum_{j\in\mathcal{J}} Z_j \phi_j(t) \tag{10.5}$$

$$V(t) = \sum_{j\in\mathcal{J}} V_j \phi_j(t) \tag{10.6}$$

of both the signal and noise process with respect to this basis, where the KL coefficients

$$\begin{aligned} Z_j =< X, \phi_j > &= \int_0^T Z(t)\phi_j(t)dt \quad \text{(CT case)} \\ &= \sum_{t=0}^{T} Z(t)\phi_j(t) \quad \text{(DT case)} \end{aligned} \tag{10.7}$$

and

$$\begin{aligned} V_j =< V, \phi_j > &= \int_0^T V(t)\phi_j(t)dt \quad \text{(CT case)} \\ &= \sum_{t=0}^{T} V(t)\phi_j(t) \quad \text{(DT case)} \end{aligned} \tag{10.8}$$

form two independent WGN sequences with $Z_j \sim N(0, \lambda_j)$ and $V_j \sim N(0, \sigma^2)$. Note that while the variance σ^2 of V_j is constant, the variance λ_j of Z_j is time-varying. Then, if we perform a Karhunen-Loève decomposition (KLD)

$$Y(t) = \sum_{j\in\mathcal{J}} Y_j \phi_j(t) \tag{10.9}$$

of the observations under both hypotheses, the detection problem (10.1) can be expressed in the KL domain as

$$\begin{aligned} H_0 &: Y_j = V_j \sim N(0, \sigma^2) \\ H_1 &: Y_j = Z_j + V_j \sim N(0, \lambda_j + \sigma^2) \,. \end{aligned} \tag{10.10}$$

Thus, the transformed detection problem reduces to a test between WGN sequences with different intensities. Under H_0, the WGN sequence Y_j has variance σ^2, but under H_1, it has the time-varying variance $\lambda_j + \sigma^2$.

In the DT case, we have $\mathcal{J} = \{1, \dots, N\}$ with $N = T + 1$, but in the CT case, $\mathcal{J} = \mathbb{N}_+ = \{1, 2, \dots\}$ is infinite. So, to tackle the detection problem, we first consider a truncated version where only the first N coefficients of (10.10) are retained. This, of course, solves the DT problem, but in the CT case, we will then need to let the truncation length N tend to infinity. If $\mathbf{Y}_1^N$ denotes the N-dimensional vector whose j-th entry is Y_j, with $1 \leq j \leq N$, the likelihood ratio function for the test based on the first N KL coefficients of $Y(t)$ is given by

$$L(\mathbf{Y}_1^N) = \frac{f(\mathbf{Y}_1^N|H_1)}{f(\mathbf{Y}_1^N|H_0)} = \prod_{j=1}^{N} \frac{\sigma}{(\lambda_j + \sigma^2)^{1/2}} \exp\left(\frac{Y_j^2}{2}\left[\frac{1}{\sigma^2} - \frac{1}{\lambda_j + \sigma^2}\right]\right). \tag{10.11}$$

Taking logarithms, we find

$$\Lambda(\mathbf{Y}_1^N) \triangleq \ln L(\mathbf{Y}_1^N) = -\frac{1}{2}\sum_{j=1}^{N} \ln(1 + \lambda_j/\sigma^2) + \frac{1}{2\sigma^2}\sum_{j=1}^{N} \frac{\lambda_j Y_j^2}{\lambda_j + \sigma^2} \,, \tag{10.12}$$

and the LRT based on the first N KLD coefficients can be expressed as

$$\Lambda(\mathbf{Y}_1^N) \underset{H_0}{\overset{H_1}{\gtrless}} \eta = \ln(\tau) \,. \tag{10.13}$$

In the Bayesian case, the threshold τ is given by (2.19), whereas for a Neyman-Pearson test, it needs to be adjusted to ensure a fixed false alarm rate. This solves completely the DT detection problem, but we need now to examine the behavior of the above test in the CT case as $N \to \infty$. In the decomposition (10.12), the sum

$$\sum_{j=1}^{N} \ln(1 + \lambda_j/\sigma^2)$$

is monotone increasing as a function of N, and since $0 \leq \ln(1 + x) \leq x$ for $x \geq 0$, we have

$$\sum_{j=1}^{\infty} \ln(1 + \lambda_j/\sigma^2) \leq \sum_{j=1}^{\infty} \lambda_j/\sigma^2 = \int_0^T K(t,t)dt/\sigma^2 \,,$$

where because the kernel $K(t, s)$ is continuous over $[0, T]^2$, it is integrable over the diagonal $t = s$ for $0 \leq t \leq T$. This proves that the first sum on the right-hand side of (10.12) admits a finite limit as N tends to infinity. Let us turn now our attention to the second term, which after scaling by a factor of σ^2, can be expressed as

$$S(N) = \frac{1}{2} \sum_{j=1}^{N} \frac{\lambda_j Y_j^2}{\lambda_j + \sigma^2} . \tag{10.14}$$

We shall show that this term converges in the mean-square sense under H_1 and H_0. Since the proof is the same for both hypotheses, we consider H_1 only. To prove the mean-square convergence of $S(N)$, we show that it is a Cauchy sequence, i.e.,

$$\lim_{N,M \to \infty} E[(S(N) - S(M))^2 | H_1] = 0 . \tag{10.15}$$

But for $N \geq M$, we have

$$S(N) - S(M) = \frac{1}{2} \sum_{j=M}^{N} \frac{\lambda_j Y_j^2}{\lambda_j + \sigma^2} \tag{10.16}$$

where the random variables Y_j^2 are independent. This implies

$$E[(S(N) - S(M))^2 | H_1] = \frac{1}{4} \Big(\sum_{j=N}^{M} \frac{\lambda_j E[Y_j^2 | H_1]}{\lambda_j + \sigma^2} \Big)^2 + \frac{1}{4} \sum_{j=N}^{M} \frac{\lambda_j^2 (E[Y_j^4 | H_1] - E[Y_j^2 | H_1]^2)}{(\lambda_j + \sigma^2)^2} , \tag{10.17}$$

where since Y_j is zero-mean Gaussian with variance $\lambda_j + \sigma^2$, we have

$$E[Y_j^4 | H_1] = 3(E[Y_j^2 | H_1])^2 = 3(\lambda_j + \sigma^2)^2 .$$

This yields

$$E[(S(N) - S(M))^2 | H_1] = \frac{1}{4} \big(\sum_{j=N}^{M} \lambda_j \big)^2 + \frac{1}{2} \sum_{j=N}^{M} \lambda_j^2 , \tag{10.18}$$

But, since the series

$$\sum_{j=1}^{\infty} \lambda_j = \int_0^T K(t,t) dt$$

converges, the series $\sum_{j=1}^{\infty} \lambda_j^2$ converges as well, and thus the two terms on the right-hand side of (10.18) tend to zero as $N, M \to \infty$. This proves that $S(N)$ converges in the mean-square sense to a random variable that we denote as

$$S = \frac{1}{2} \sum_{j=1}^{\infty} \frac{\lambda_j Y_j^2}{\lambda_j + \sigma^2} . \tag{10.19}$$

This establishes that the log-likelihood function $\Lambda(\mathbf{Y}_1^N)$ converges in the mean-square as N tends to infinity, so it makes sense to consider the limit of the LR test (10.12) as $N \to \infty$.

Optimal Test: Combining the DT and CT cases, we have found that the optimal test can be expressed as

$$S = \frac{1}{2} \sum_{j \in \mathcal{J}} \frac{\lambda_j Y_j^2}{\lambda_j + \sigma^2} \underset{H_0}{\overset{H_1}{\gtrless}} \gamma \tag{10.20}$$

with

$$\gamma = \sigma^2 \ln(\tau) + \frac{\sigma^2}{2} \sum_{j \in \mathcal{J}} \ln(1 + \lambda_j / \sigma^2) \,. \tag{10.21}$$

Even though this test is well defined, the above expressions do not necessarily provide the best test implementation, since they involve infinite summations in the CT case. To obtain a simpler implementation, we observe that by substituting expression

$$\begin{aligned} Y_j &= \int_0^T \phi_j(t) Y(t) dt \quad \text{(CT case)} \\ &= \sum_{t=0}^{T} \phi_j(t) Y(t) \quad \text{(DT case)} \end{aligned}$$

for the KL coefficients inside (10.20), and exchanging the summation over j with the integrations/summations over the time variables, we find that the sufficient statistic S can be expressed as

$$\begin{aligned} S &= \frac{1}{2} \int_0^T \int_0^T Y(t) H(t,s) Y(s) dt ds \quad \text{(CT case)} \\ &= \frac{1}{2} \sum_{t=0}^{T} \sum_{s=0}^{T} Y(t) H(t,s) Y(s) \quad \text{(DT case)} \end{aligned} \tag{10.22}$$

where

$$H(t,s) = \sum_{j \in \mathcal{J}} \frac{\lambda_j}{\lambda_j + \sigma^2} \phi_j(t) \phi_j(s) \tag{10.23}$$

denotes the resolvent kernel corresponding to the covariance kernel $K(t,s) + \sigma^2 \delta(t-s)$.

Note that in the CT case, expression (10.23) for the kernel $H(t,s)$ is just formal. In order to justify this expression one needs to show that the sum

$$H_N(t,s) = \sum_{j=1}^{N} \frac{\lambda_j}{\lambda_j + \sigma^2} \phi_j(t) \phi_j(s)$$

converges in $L^2([0,T]^2)$ as $N \to \infty$. For $N \geq M$, we have

$$
\begin{aligned}
||H_N - H_M||^2 &\triangleq \int_0^T \int_0^T |H_N(t,s) - H_M(t,s)|^2 dt ds \\
&= \sum_{j=M}^{N} \frac{\lambda_j^2}{(\lambda_j + \sigma^2)^2}
\end{aligned}
\tag{10.24}
$$

where, to obtain the equality in (10.24), we have used the orthonormality of the eigenfunctions ϕ_j. Since the sum of the squared eigenvalues converges, the last expression in (10.24) converges to zero as $M, N \to \infty$. This shows that the kernels $H_N(t,s)$ form a Cauchy sequence in $L^2([0,T]^2)$, so they converge to a limit $H(t,s)$ with norm

$$
||H||^2 = \int_0^T \int_0^T |H(t,s)|^2 dt ds = \sum_{j=1}^{\infty} \frac{\lambda_j^2}{(\lambda_j + \sigma^2)^2} \, . \tag{10.25}
$$

Next we proceed to interpret expression (10.22) for the sufficient statistic S. We consider separately the DT and CT cases.

DT case: Let

$$
\mathbf{Y} = \begin{bmatrix} Y(0) \\ \vdots \\ Y(t) \\ \vdots \\ Y(T) \end{bmatrix} \quad \mathbf{Z} = \begin{bmatrix} Z(0) \\ \vdots \\ Z(t) \\ \vdots \\ Z(T) \end{bmatrix} \quad \mathbf{V} = \begin{bmatrix} V(0) \\ \vdots \\ V(t) \\ \vdots \\ V(T) \end{bmatrix}
$$

denote the $N = T+1$ dimensional vectors representing the observations, signal and noise over $0 \leq t \leq T$. The detection problem (10.1) can be expressed in vector form as

$$
\begin{aligned}
H_0 &: \mathbf{Y} = \mathbf{V} \\
H_1 &: \mathbf{Y} = \mathbf{Z} + \mathbf{V}
\end{aligned}
\tag{10.26}
$$

which is in the form considered in Section 2.6.2. The vectors $\mathbf{V}$ and $\mathbf{Z}$ are independent with $\mathbf{V} \sim N(0, \sigma^2 \mathbf{I}_N)$ and $\mathbf{Z} \sim N(0, \mathbf{K})$, where the element of row $t+1$ and column $s+1$ of $\mathbf{K}$ is $K(t,s)$. Then if

$$
\hat{\mathbf{Z}} = \mathbf{H}\mathbf{Y} \tag{10.27}
$$

denotes the MSE estimate of $\mathbf{Z}$ given $\mathbf{Y}$, we have

$$
\mathbf{H} = \mathbf{K}_{\mathbf{ZY}} \mathbf{K}_{\mathbf{Y}}^{-1} = \mathbf{K}(\mathbf{K} + \sigma^2 \mathbf{I}_N)^{-1} . \tag{10.28}
$$

Note that if $H(t,s)$ denotes the element or row $t+1$ and column $s+1$ of $\mathbf{H}$, and if $\hat{Z}(t)$ is the MSE estimate of $Z(t)$ given observations $\{Y(s), 0 \leq s \leq T\}$, the vector estimate (10.27) can be rewritten componentwise as

$$\hat{Z}(t) = \sum_{s=0}^{T} H(t,s)Y(s) . \tag{10.29}$$

To relate the DT estimation filter $H(t,s)$ to expression (10.23), we note that since (ϕ_j, λ_j) with $1 \leq j \leq N$ correspond to the eigenfunctions and eigenvalues of the covariance kernel $K(t,s)$, the covariance matrix $\mathbf{K}$ admits the eigendecomposition

$$\mathbf{K} = \boldsymbol{\Phi}\boldsymbol{\Lambda}\boldsymbol{\Phi}^T , \tag{10.30}$$

where if

$$\boldsymbol{\phi}_j = \begin{bmatrix} \phi_j(0) \dots \phi_j(t) \dots \phi_j(T) \end{bmatrix}^T$$

denotes the vector representing the j-th eigenfunction, we have

$$\boldsymbol{\Phi} = \begin{bmatrix} \boldsymbol{\phi}_1 \dots \boldsymbol{\phi}_j \dots \boldsymbol{\phi}_N \end{bmatrix}$$

and

$$\boldsymbol{\Lambda} = \text{diag}\,\{\lambda_j,\ 1 \leq j \leq N\} .$$

Since the eigenfunctions $\{\phi_j, 1 \leq j \leq N\}$ are orthonormal, $\boldsymbol{\Phi}$ is an orthonormal matrix, i.e.,

$$\boldsymbol{\Phi}\boldsymbol{\Phi}^T = \mathbf{I}_N , \tag{10.31}$$

and consequently, by substituting (10.30) inside (10.28) and taking into account (10.31), we can express the estimation gain matrix as

$$\mathbf{H} = \boldsymbol{\Phi}\boldsymbol{\Lambda}(\boldsymbol{\Lambda} + \sigma^2\mathbf{I}_N)^{-1}\boldsymbol{\Phi}^T , \tag{10.32}$$

which is the matrix form of expression (10.23) for the estimation kernel $H(t,s)$. Accordingly, in the DT case, the sufficient statistic S which is used in the optimal receiver (10.20) can be expressed as

$$S = \frac{1}{2}\sum_{t=0}^{T}\sum_{s=0}^{T} Y(t)H(t,s)Y(s) = \frac{1}{2}\sum_{t=0}^{T} Y(t)\hat{Z}(t) . \tag{10.33}$$

Thus, it can be implemented as an estimator/correlator receiver as depicted in Fig. 10.1. It has a structure similar to the correlator/integrator receiver

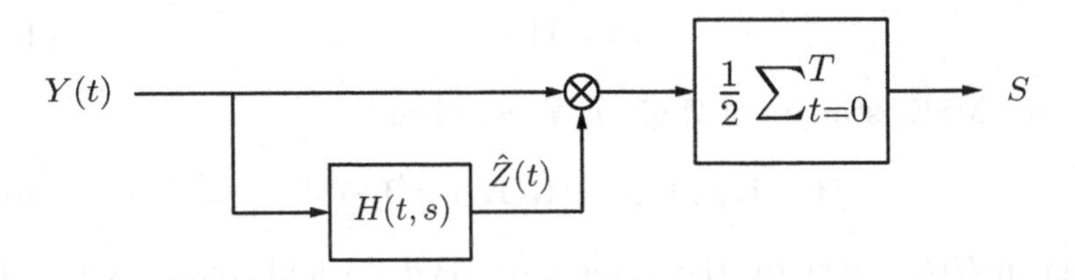

Fig. 10.1. Noncausal estimator/correlator receiver.

used for detecting a known signal, except that the transmitted signal $Z(t)$ is no longer known, but random. So, the optimum receiver just substitutes the estimate $\hat{Z}(t)$ of the signal for its actual value. However, a significant limitation of this receiver is that the estimate $\hat{Z}(t)$ depends on the values of the observations $Y(\cdot)$ over the entire observation interval $[0, T]$. In other words, $\hat{Z}(t)$ is a *smoothed estimate*, since it depends on observations $Y(s)$ both in the past and the future of time t. Accordingly, to evaluate $\hat{Z}(t)$, it is necessary to wait until all observations have been received, so that a delay is introduced between the collection of observations $\{Y(s),\ 0 \leq s \leq T\}$ and the time when a decision can be made for the detection problem (10.1). We shall show in the next section that it is actually possible to implement the optimum receiver by using only causal operations, which removes the above limitation.

CT Case: To interpret the kernel $H(t, s)$ in the CT case, let

$$\hat{Z}(t) = \int_0^T H(t, u)Y(u)du \tag{10.34}$$

denote the MSE estimate of $Z(t)$ based on observations $\{Y(s),\ 0 \leq s \leq T\}$. By using the orthogonality property

$$E[(Z(t) - \hat{Z}(t))Y(s)] = 0$$

of MSE estimates for $0 \leq s \leq T$, and noting that

$$E[Z(t)Y(s)] = K(t, s) \quad , \quad E[Y(t)Y(s)] = K(t, s) + \sigma^2\delta(t - s)$$

we find that $H(t, s)$ satisfies the Fredholm integral equation of the second kind

$$\sigma^2 H(t, s) + \int_0^T H(t, u)K(u, s)du = K(t, s) \tag{10.35}$$

for $0 \leq t, s \leq T$. This identity can be rewritten in operator form as

$$H = K(\sigma^2 I + K)^{-1}$$

or equivalently, as

$$\sigma^{-2}(I - H) = (\sigma^2 I + K)^{-1} ,$$

so that the smoothing filter $H(t, s)$ is the Fredholm resolvent of the Fredholm operator of the second kind $\sigma^2 I + K$.

To show that the solution $H(t, s)$ of the Fredholm equation (10.35) can be expressed in the form (10.23), note that operators K, $\sigma^2 I + K$ and H have all the same eigenfunctions. Hence $H(t, s)$ admits an eigenfunction expansion of the form

$$H(t, s) = \sum_{j=0}^{\infty} \mu_j \phi_j(t)\phi_j(s) \tag{10.36}$$

and, by substituting (10.36) inside (10.35) and identifying eigenvalues on both sides, we find

$$\mu_j = \frac{\lambda_j}{\lambda_j + \sigma^2}\,, \tag{10.37}$$

which verifies that the Fredholm resolvent $H(t, s)$ admits expression (10.23). Then, the sufficient statistic S given by (10.22) can be expressed as

$$S = \frac{1}{2}\int_0^T \int_0^T Y(t)H(t,s)Y(s)dtds = \frac{1}{2}\int_0^T Y(t)\hat{Z}(t)\,, \tag{10.38}$$

which is similar to the estimator/correlator expression obtained in the DT case, except that the summation is replaced by an integration.

10.2.2 Smoother Implementation

In Section 7.4 it was observed that the covariance function of a partially observed Gaussian reciprocal or Markov process has the feature that its eigenfunctions can be obtained by solving a difference/differential system with two-point boundary value conditions. It turns out that covariance kernels with this structure have also the property that the smoothing kernel $H(t, s)$ or the smoothed estimates $\hat{Z}(t)$ can be obtained by solving a difference/differential system with boundary value conditions. In practice, it is easier to compute the estimates $\hat{Z}(t)$ directly, but it is still of interest to know that the smoothing kernel $H(t, s)$ can be expressed in terms of the Green's function of a self-adjoint difference/differential operator.

To fix ideas, let $Z(t)$ be a DT process that can be expressed as the partial observation

$$Z(t) = \mathbf{c}\mathbf{X}(t) \tag{10.39}$$

of a zero-mean Gaussian reciprocal process $\mathbf{X}(t) \in \mathbb{R}^n$ whose matrix covariance kernel $\mathbf{K}_\mathbf{X}(t, s)$ satisfies the matrix second-order difference equation

$$M\mathbf{K}_\mathbf{X}(t,s) = \mathbf{I}_n\delta(t-s) \tag{10.40}$$

with

$$M = \mathbf{M}_0(t) + \mathbf{M}_{-1}(t)Z_C^{-1} + \mathbf{M}_1(t)Z_C\,, \tag{10.41}$$

where Z_C represents the forward cyclic shift operator defined by $Z_C f(t) = f((t+1) \mod (T+1))$, Note that since Z_C is defined cyclically, the boundary conditions for the difference equation (10.40) are obtained by setting $t = 0$ and $t = T$ in (10.40). Thus the relation obtained by setting $t = 0$ couples the values of $\mathbf{K}_\mathbf{X}(t, s)$ at $t = T$, $t = 0$, and $t = 1$. Similarly, $t = T$ in (10.40) couples the values of $\mathbf{K}_\mathbf{X}(t, s)$ at $t = T - 1$, $t = T$ and $t = 0$. It is also worth noting that the matrix operator M is positive self-adjoint. This implies that the $n \times n$ matrix functions $\mathbf{M}_0(t)$, $\mathbf{M}_{-1}(t)$ and $\mathbf{M}_1(t)$ satisfy

$$\mathbf{M}_0(t) = \mathbf{M}_0^T(t) \quad , \quad \mathbf{M}_1(t) = \mathbf{M}_{-1}^T(t+1)\,.$$

Then, it was shown in [6] that the smoothed estimate

$$\hat{\mathbf{X}}(t) = E[\mathbf{X}(t)|Y(s), 0 \leq s \leq T]$$

of $\mathbf{X}(t)$, given the observations over $[0, T]$, obeys the second-order equation

$$M\hat{\mathbf{X}}(t) = \mathbf{c}^T(t)\sigma^{-2}[Y(t) - \hat{Z}(t)] , \tag{10.42}$$

where

$$\hat{Z}(t) = \mathbf{c}\hat{\mathbf{X}}(t) \tag{10.43}$$

is the smoothed estimate of the partially observed process. In the light of (10.29), the identities (10.42) and (10.43) imply that the smoothing kernel $H(t, s)$ can be expressed as

$$H(t, s) = \mathbf{c}\mathbf{G}(t, s)\mathbf{c}^T \tag{10.44}$$

where the $n \times n$ matrix function $\mathbf{G}(t, s)$ is the Green's function of the self-adjoint second-order smoothing operator

$$M_s \triangleq \sigma^2 M + \mathbf{c}^T(t)\mathbf{c}(t) , \tag{10.45}$$

so that we have

$$M_s\mathbf{G}(t, s) = \mathbf{I}_n\delta(t - s) . \tag{10.46}$$

At this point it is worth noting that only DT Gaussian reciprocal processes with full rank noise admit a second-order covariance model of the form (10.40)–(10.41) and the full class of reciprocal processes actually requires the use of mixed-order difference operators, in the sense that some components of $\mathbf{X}(t)$ may require models of order higher than two. Since Markov models do not have this restriction, let us consider briefly the case where in (10.39), $\mathbf{X}(t)$ is a DT Gauss-Markov process with state-space model

$$\mathbf{X}(t + 1) = \mathbf{A}(t)\mathbf{X}(t) + \mathbf{W}(t) \tag{10.47}$$

such that $\mathbf{X}(0) \sim N(\mathbf{0}, \boldsymbol{\Pi}_0)$. Here $\mathbf{W}(t)$ is a zero-mean WGN independent of $\mathbf{X}(0)$ whose covariance matrix $\mathbf{Q}(t) = E[\mathbf{W}(t)\mathbf{W}^T(t)]$ is not necessarily invertible. Then, it is shown in [7, Sec. 10.5], [8] that the smoothed estimate $\hat{\mathbf{X}}(t)$ obeys the Hamiltonian system

$$\begin{bmatrix} \hat{\mathbf{X}}(t+1) \\ \boldsymbol{\eta}(t) \end{bmatrix} = \begin{bmatrix} \mathbf{A}(t) & \mathbf{Q}(t) \\ -\mathbf{c}^T(t)\sigma^{-2}\mathbf{c}(t) & \mathbf{A}^T(t) \end{bmatrix} \begin{bmatrix} \hat{\mathbf{X}}(t) \\ \boldsymbol{\eta}(t+1) \end{bmatrix} + \begin{bmatrix} \mathbf{0} \\ \mathbf{c}^T(t)\sigma^{-2} \end{bmatrix} Y(t) \tag{10.48}$$

with boundary conditions

$$\hat{\mathbf{X}}(0) = \boldsymbol{\Pi}_0\boldsymbol{\eta}(0) \quad , \quad \boldsymbol{\eta}(T + 1) = \mathbf{0} . \tag{10.49}$$

This system represents the counterpart in the DT case and for the smoothing problem of the CT Hamiltonian system (7.96)–(7.97) for the eigenfunctions of a partially observed Markov process. Since equations (10.42)–(10.43) and (10.48)–(10.49) involve two-point boundary value conditions, they do not by themselves specify a recursive procedure for computing the smoothed estimates $\hat{\mathbf{X}}(t)$. However a variety of methods have been developed over the years to solve this noncausal system by performing causal processing operations. Roughly speaking, in the Markov case these techniques involve the block triangularization or block diagonalization of the Hamiltonian system (10.48)–(10.49). Triangularization methods evaluate the smoothed estimates by performing a double-sweep over the interval $[0, T]$. In a first pass, a forward Kalman filter is employed to compute the filtered estimate $\hat{\mathbf{X}}_f(t)$ of $\mathbf{X}(t)$ based on observations up to time t, and after the end T of the interval has been reached, a backward recursion is implemented which uses the filtered estimates $\hat{\mathbf{X}}_f(t)$ obtained in the forward pass in order to evaluate the smoothed estimates $\hat{\mathbf{X}}(t)$. In contrast, the diagonalization approach implements two Kalman filters running in the forward and backward directions and obtain the smoothed estimate as a matrix-weighted sum of the estimates provided by the two Kalman filters. We refer the reader to [7, 8] for further details. Similar techniques are also available for the case of reciprocal processes [6, 9], except that the counterpart of the triangularization of the Hamiltonian is a multiplicative factorization of the second-order difference operator M as a product of causal and anticausal first-order operators.

To illustrate the previous discussion, we consider two examples.

Example 10.1: DT Ornstein-Uhlenbeck process in WGN

Let $Z(t)$ be a CT Ornstein-Uhlenbeck process. This process is zero-mean Gaussian and Markov, with covariance function

$$K(t, s) = Pa^{|t-s|} ,$$

where $|a| < 1$. If $q = P(1 - a^2)$, $K(t, s)$ satisfies the difference equation

$$MK(t, s) = \delta(t - s) \tag{10.50}$$

where for $1 \leq t \leq T - 1$, the difference operator M is given by

$$M = q^{-1}[(1 + a^2) - a(Z + Z^{-1})] . \tag{10.51}$$

Here Z denotes the ordinary forward shift operator $Zf(t) = f(t + 1)$. In (10.50), the operator M at $t = 0,\ T$ is specified by the separable boundary conditions

$$\begin{aligned} q^{-1}[K(0, s) - aK(1, s)] &= \delta(s) \\ q^{-1}[K(T, s) - aK(T - 1, s)] &= \delta(s - T) . \end{aligned} \tag{10.52}$$

Then, if

$$M_s = \sigma^2 M + 1$$

the smoothed estimate $\hat{Z}(t)$ satisfies

$$M_s \hat{Z}(t) = Y(t) \tag{10.53}$$

for $0 \leq t \leq T$. To evaluate $\hat{Z}(t)$, instead of implementing the usual Rauch-Tung-Striebel (RTS) double-sweep smoother, which is time-varying and requires the solution of a Riccati equation [7, Sec. 10.3], one can apply the following version of the double-sweep technique, which is time-invariant, but requires the computation of two initial conditions.

Let $b < 1$ be the root inside the unit circle of the characteristic equation $M_s(z) = 0$, where

$$M_s(z) \triangleq \frac{\sigma^2}{q}[(1+a^2) - a(z+z^{-1})] + 1 ,$$

and factor $M_s(z)$ as

$$M_s(z) = \frac{1}{\Gamma(1-b^2)}(1-bz^{-1})(1-bz) . \tag{10.54}$$

Then, since the smoothing kernel $H(t,s)$ is symmetric and satisfies the difference equation

$$M_s H(t,s) = \delta(t-s) \tag{10.55}$$

its solution admits the form

$$H(t,s) = C_- b^{|t-s|} + C_+ b^{-|t-s|} + D_- b^{t+s} + D_+ b^{-(t+s)} . \tag{10.56}$$

Let

$$\rho \triangleq -\frac{(1+\sigma^2/q) - a\sigma^2/(bq)}{(1+\sigma^2/q) - ab\sigma^2/q} .$$

By applying the boundary conditions obtained by setting $t = 0$ and $t = T$ in (10.55) and evaluating (10.55) along the diagonal $t = s$, we find

$$C_- = \frac{\Gamma}{1-\rho^2 b^{2T}} \quad , \quad D_- = \rho C_-$$

and

$$C_+ = \rho^2 b^{2T} C_- \quad , \quad D_+ = b^{2T} D_- .$$

Then if

$$\xi(t) \triangleq \frac{1}{1-b^2}(\hat{Z}(t) - b\hat{Z}(t+1)) \tag{10.57}$$

for $0 \leq t \leq T-1$, we have

$$\xi(t) - b\xi(t-1) = \Gamma Y(t) \tag{10.58}$$

for $1 \leq t \leq T$. Together (10.57) and (10.58) define a double-sweep recursion. Specifically, starting with the initial condition $\xi(0)$, (10.58) can be propagated in the forward direction to evaluate $\xi(t)$ for $1 \leq t \leq T-1$. Then, given $\xi(t)$ and given the initial condition $\hat{Z}(T)$, the backward recursion (10.57) can be used to compute the smoothed estimates $\hat{Z}(t)$ for $0 \leq t \leq T-1$. So to implement the double-sweep smoother we only need to evaluate the initial conditions $\xi(0)$ and $\hat{Z}(T)$. This can be accomplished by using expression (10.29) for the smoothed estimate, where the smoothing kernel $H(t,s)$ admits expression (10.56). The evaluation of $\xi(0)$ and $\hat{Z}(T)$ requires of the order of T operations each, which increases slightly the complexity of the usual RTS double-sweep smoother, but on the other hand it is not necessary to solve a Riccati equation and only time-invariant digital filtering operations are required. □

Example 10.2: DT hyperbolic cosine process in WGN

Let $Z(t)$ be the DT hyperbolic process over $[0,T]$ considered in Example 7.7. It has zero-mean and covariance

$$K(t,s) = P\frac{\cosh(\alpha(T/2 - |t-s|))}{\cosh(\alpha T/2)} \,,$$

with $\alpha > 0$. $Z(t)$ is stationary and obeys the periodicity condition $Z(0) = Z(T)$, so it can be viewed as defined over the discretized circle obtained by wrapping around the interval $[0,T]$ and identifying its extremities 0 and T. Note that after this identification, $Z(t)$ is effectively defined over $[0,T-1]$ only. If $a = \exp(-\alpha)$, the covariance kernel $K(t,s)$ satisfies (10.50), where the operator M is obtained by replacing the ordinary forward shift Z by its cyclic version Z_C in (10.51). Thus, the hyperbolic cosine process and the Ornstein-Uhlenbeck process of Example 10.1 differ only by the boundary conditions which are statisfied by the covariance kernel $K(t,s)$. This difference in boundary conditions accounts for the fact that the hyperbolic cosine process is not Markov, but reciprocal. In this case, the smoothing kernel $H(t,s)$ still satisfies the difference equation (10.55), but with the periodic boundary condition $H(0,s) = H(T,s)$. To solve this equation, condider the factorization (10.54) and denote $b = \exp(-\beta)$ with $\beta > 0$. By matching coefficients of z on both sides of (10.54) and recalling that $q = P(1-a^2)$, we find

$$\frac{\sigma^2}{P(a^{-1}-a)} = \frac{1}{\Gamma(b^{-1}-b)} \,,$$

which implies

$$\Gamma = \frac{P\sinh(\alpha)}{\sigma^2 \sinh(\beta)} \,.$$

Next, observing that the difference equation (10.54) with the periodic condition satisfied by $H(t,s)$ takes exactly the same form as the equation satisfied by $K(t,s)$, but with a replaced by b and P by Γ, we deduce that its solution can be expressed as

$$H(t, s) = \Gamma \frac{\cosh(\beta(T/2 - |t - s|))}{\cosh(\beta T/2)} . \tag{10.59}$$

The matrix $\mathbf{H}$ of dimension T corresponding to this kernel has a circulant structure. This is expected, since it was observed in Example 7.7 that the covariance matrix $\mathbf{K}$ of the process $Z(t)$ over $[0, T-1]$ is circulant, and since the algebra of circulant matrices is closed under addition and inversion, so

$$\mathbf{H} = \mathbf{K}(\mathbf{K} + \sigma^2 \mathbf{I}_T)^{-1}$$

must be circulant. Consequently, $\mathbf{H}$ is diagonalized by the DFT matrix of dimension T and its eigenvalues μ_j can be expressed in terms of the eigenvalues

$$\lambda_j = \frac{P(1 - a^2)}{(1 + a^2) - 2a \cos(j2\pi/T)}$$

of $\mathbf{K}$ as

$$\mu_j = \frac{\lambda_j}{\lambda_j + \sigma^2} ,$$

which in the light of the definitions of b and Γ, yields

$$\mu_j = \frac{\Gamma(1 - b^2)}{(1 + b^2) - 2b \cos(j2\pi/T)} .$$

To implement the smoother (10.53), we then have several options. We can of course implement the double-sweep recursions (10.57), (10.58), where the evaluation of the initial conditions $\xi(0)$ and $\hat{Z}(T)$ needs now to be performed with the kernel (10.59). On the other hand, we can also use an FFT and an inverse FFT to implement the matrix multiplication by $\mathbf{H}$. □

Finally, It is worth noting that in the DT case, if the signal $Z(t)$ to be detected is stationary, FFT-based algorithms can be employed to implement the smoother (10.29). To see why this is the case, note that the smoothing matrix $\mathbf{H}$ in (10.28) can be expressed as

$$\mathbf{H} = \mathbf{I}_N - \sigma^2 \mathbf{K}_{\mathbf{Y}}^{-1} \tag{10.60}$$

where, since $Z(t)$ is stationary, the covariance matrix $\mathbf{K}_{\mathbf{Y}} = \mathbf{K} + \sigma^2 \mathbf{I}_N$ has a Toeplitz structure, i.e., it is constant along diagonals. But the inverse of a Toeplitz matrix has the important property [10, 11] that it is a close to Toeplitz matrix, in the sense that it is a perturbation of a Toeplitz matrix by a matrix of rank less than or equal to two. For example, in Example 10.1, the first two terms of the smoothing kernel $H(t, s)$ have a Toeplitz structure, while the last two terms represent a perturbation of rank two. On the other hand, in Example 10.2, the smoothing kernel $H(t, s)$ given by (10.9) is Toeplitz. Thus, from expression (10.60) we conclude that when $Z(t)$ is stationary, the smoother $\mathbf{H}$ has a close to Toeplitz structure. This structure can be exploited by observing that a Toeplitz matrix can be embedded in a circulant matrix of twice its size. Since a circulant matrix is diagonalized by the FFT, this means that multiplication by a Toeplitz matrix can be achieved by performing an FFT, followed by a diagonal scaling, and then an inverse FFT, where the FFT and its inverse have twice the size of the Toeplitz matrix.

10.3 Causal Receiver

In spite of the existence of efficient smoothing techniques, it turns out that ultimately the best way to implement the optimal receiver (10.20)–(10.21) consists of employing only causal filtering operations. We consider the DT case first.

DT case: If

$$\hat{Z}_p(t) = E[Z(t)|Y(s),\ 0 \leq s \leq t-1] = -\sum_{s=0}^{t-1} a(t,s)Y(s) \tag{10.61}$$

denotes the predicted estimate of the signal $Z(t)$ based on the observations up to time $t-1$, consider the innovations process

$$\nu(t) \triangleq Y(t) - \hat{Z}_p(t) = \sum_{s=0}^{t} a(t,s)Y(s) \tag{10.62}$$

with $a(t,t) = 1$. Since $V(t)$ is WGN, the predicted estimate $\hat{Z}_p(t)$ of the signal coincides with the predicted estimate

$$\hat{Y}_p(t) = E[Y(t)|Y(s),\ 0 \leq s \leq t-1]$$

of the observations, and consequently from the orthogonality property of mean-square estimates, $\nu(t)$ is a time-varying WGN process with variance $P_\nu(t) = E[\nu^2(t)]$, i.e.,

$$E[\nu(t)\nu(s)] = P_\nu(t)\delta(t-s)\,. \tag{10.63}$$

Then, introduce the normalized innovations

$$\bar{\nu}(t) = P_\nu^{-1/2}(t)\nu(t) = \sum_{s=0}^{t} f(t,s)Y(s)\,, \tag{10.64}$$

where the causal filter $f(t,s)$ is defined as

$$f(t,s) = P_\nu^{-1/2}(t)a(t,s) \tag{10.65}$$

for $0 \leq s \leq t$. The process $\bar{\nu}(t)$ is now zero-mean WGN with unit intensity. Consider the $T+1$ dimensional vectors

$$\bar{\boldsymbol{\nu}} = \begin{bmatrix} \bar{\nu}(0) \\ \vdots \\ \bar{\nu}(t) \\ \vdots \\ \bar{\nu}(T) \end{bmatrix} \qquad \mathbf{Y} = \begin{bmatrix} Y(0) \\ \vdots \\ Y(t) \\ \vdots \\ Y(T) \end{bmatrix}$$

and the lower triangular matrix $\mathbf{F}$ such that the entry of row $t+1$ and column $s+1$ is $f(t,s)$. The identity (10.64) can be written in matrix form as

$$\bar{\boldsymbol{\nu}} = \mathbf{F}\mathbf{Y} \tag{10.66}$$

and since $\bar{\nu}(t)$ is a WGN with unit intensity, we have

$$\mathbf{I}_{T+1} = E[\bar{\boldsymbol{\nu}}\bar{\boldsymbol{\nu}}^T] = \mathbf{F}\mathbf{K}_{\mathbf{Y}}\mathbf{F}^T .$$

This implies that

$$\mathbf{K}_{\mathbf{Y}}^{-1} = \mathbf{F}^T\mathbf{F} \tag{10.67}$$

represents an upper times lower Cholesky factorization of the inverse of the covariance matrix $\mathbf{K}_{\mathbf{Y}}$ of the observations.

Then, by substituting the factorization (10.67) inside expression (10.60) for the smoothing matrix, we find that the sufficient statistic S given by (10.22) can be written as

$$\begin{aligned} S &= \frac{1}{2}\mathbf{Y}^T\mathbf{H}\mathbf{Y} = \mathbf{Y}^T\mathbf{Y} - \sigma^2\bar{\boldsymbol{\nu}}^T\bar{\boldsymbol{\nu}} \\ &= \frac{1}{2}\sum_{t=0}^{T}[Y^2(t) - \frac{\sigma^2}{P_\nu(t)}\nu^2(t)] . \end{aligned} \tag{10.68}$$

This expression depends only on the observations and the innovations process $\nu(t)$, which is generated by causal filtering operations. An even simpler expression can be obtained if we introduce the filtered estimate

$$\hat{Z}_f(t) = E[Z(t)|Y(s),\ 0 \leq s \leq t]$$

based on the observations up to time t. Consider the a-posteriori estimation residual

$$\begin{aligned} \delta(t) &= Y(t) - \hat{Z}_f(t) \\ &= \tilde{Z}_f(t) + V(t) \end{aligned} \tag{10.69}$$

where $\tilde{Z}_f(t) = Z(t) - \hat{Z}_f(t)$ denotes the filtering error. From the first line of (10.69) we know that the residual $\delta(t)$ depends linearly on the observations $\{Y(s),\ 0 \leq s \leq t\}$, but from the second line we also know it is orthogonal to $Y(s)$ for $0 \leq s \leq t-1$. This implies that $\delta(t)$ must be proportional to the innovations $\nu(t)$, i.e.,

$$\delta(t) = G\nu(t) \tag{10.70}$$

where the coefficient

$$G = E[\delta(t)\nu(t)]/P_\nu(t) .$$

But, from the second line of (10.69), the orthogonality property of the error $\tilde{Z}(t)$, and the whiteness of $V(t)$, we find

$$E[\delta(t)\nu(t)] = E[\delta(t)Y(t)] = E[V^2(t)] = \sigma^2 ,$$

which proves

$$G = \sigma^2/P_\nu(t) . \tag{10.71}$$

Then, substituting (10.70)–(10.71) inside expression (10.68) for the sufficient statistic S and simplifying, we obtain

$$\begin{aligned} S &= \frac{1}{2}\sum_{t=0}^{T}[Y^2(t) - \nu(t)\delta(t)] \\ &= \sum_{t-0}^{T} Y(t)(\hat{Z}_f(t) + \hat{Z}_p(t))/2 - \frac{1}{2}\sum_{t=0}^{T} \hat{Z}_f(t)\hat{Z}_p(t) . \end{aligned} \tag{10.72}$$

A block diagram of the receiver is shown in Fig. 10.2. The receiver uses both the predicted and filtered estimates $\hat{Z}_p(t)$ and the $\hat{Z}_f(t)$, and unlike the estimator/correlator receiver of the previous section, which relied on the noncausal signal estimate $\hat{Z}(t)$, it can be viewed as the true analog of the matched filter detector for known signals in WGN. Specifically, it was found in Chapter 8 that

$$S = \sum_{t=0}^{T} Y(t)s(t) - \frac{1}{2}\sum_{t=0}^{T} s^2(t) \tag{10.73}$$

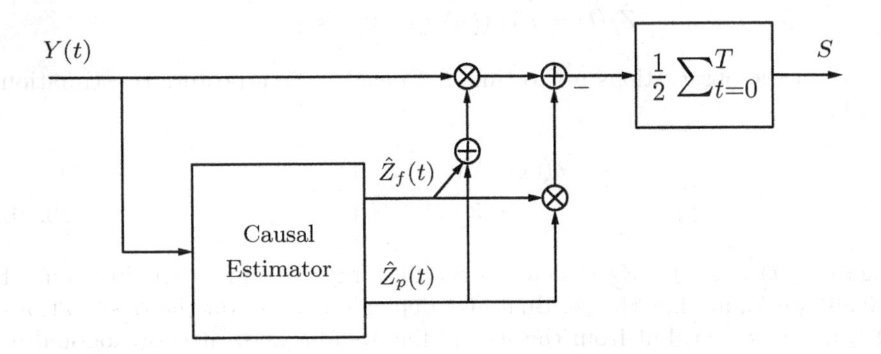

Fig. 10.2. Causal DT estimator-correlator receiver for Gaussian signal detection.

is a sufficient statistic for detecting a known signal $s(t)$ in WGN from a set of observations $Y(t)$ with $0 \leq t \leq T$. Comparing (10.73) and (10.72), we see that the causal Gaussian signal detector replaces $s(t)$ by the arithmetic mean

$$\hat{Z}_{\text{ari}}(t) \triangleq \frac{1}{2}(\hat{Z}_f(t) + \hat{Z}_p(t))$$

of the filtered and predicted estimates of $Z(t)$ in evaluating the correlation of the signal with the observations $Y(t)$. Similarly, to evaluate the squared energy of the signal over $[0, T]$, the causal detector replaces the signal $s(t)$ by the geometric mean

$$\hat{Z}_{\text{geo}}(t) \triangleq (\hat{Z}_f(t)\hat{Z}_p(t))^{1/2}$$

of the filtered and predicted estimates. Note that this interpretation assumes that $\hat{Z}_f(t)$ and $\hat{Z}_p(t)$ have the same sign, which is usually the case. As will be shown below, the corresponding CT expression takes a simpler form, since in the CT case the filtered and predicted estimates coincide.

Typically, the estimates $\hat{Z}_p(t)$ and $\hat{Z}_f(t)$ are evaluated by using causal Wiener or Kalman filters. For example, suppose that $Z(t)$ is a partially observed Markov process, so it can be expressed as

$$Z(t) = \mathbf{c}(t)\mathbf{X}(t) \tag{10.74}$$

where $\mathbf{X}(t)$ is an n-dimensional zero-mean Gauss-Markov process with state-space model

$$\mathbf{X}(t+1) = \mathbf{A}(t)\mathbf{X}(t) + \mathbf{W}(t)\,. \tag{10.75}$$

The initial state $\mathbf{X}(0)$ is $N(\mathbf{0}, \boldsymbol{\Pi}_0)$ distributed, and the input $\mathbf{W}(t)$ is a zero-mean WGN independent of $\mathbf{X}(0)$ and of the measurement noise $\mathbf{V}(t)$, with variance matrix $\mathbf{Q}(t) = E[\mathbf{W}(t)\mathbf{W}^T(t)]$. Then if

$$\begin{aligned}\hat{\mathbf{X}}_p(t) &= E[\mathbf{X}(t)|Y(s),\ 0 \leq s \leq t-1]\\ \hat{\mathbf{X}}_f(t) &= E[\mathbf{X}(t)|Y(s),\ 0 \leq s \leq t]\end{aligned}$$

denote respectively the predicted and filtered estimates of the state $\mathbf{X}(t)$, the predicted and filtered signals

$$\begin{aligned}\hat{Z}_p(t) &= \mathbf{c}(t)\hat{\mathbf{X}}_p(t)\\ \hat{Z}_f(t) &= \mathbf{c}(t)\hat{\mathbf{X}}_f(t)\end{aligned} \tag{10.76}$$

can be evaluated recursively by implementing the time-update and measurement-update form of the Kalman filter [7, Sec. 9.3]. Specifically, if

$$\begin{aligned}\tilde{\mathbf{X}}_p(t) &\triangleq \mathbf{X}(t) - \hat{\mathbf{X}}_p(t)\\ \tilde{\mathbf{X}}_f(t) &\triangleq \mathbf{X}(t) - \hat{\mathbf{X}}_f(t)\end{aligned}$$

denote respectively the predicted and filtered state estimation errors and $\mathbf{P}_p(t) = E[\tilde{\mathbf{X}}_p(t)\tilde{\mathbf{X}}_p^T(t)]$, $\mathbf{P}_f(t) = E[\tilde{\mathbf{X}}_f(t)\tilde{\mathbf{X}}_f^T(t)]$ are the corresponding error covariance matrices, the *time-update recursions* are given by

$$\begin{aligned}\hat{\mathbf{X}}_p(t+1) &= \mathbf{A}(t)\hat{\mathbf{X}}_f(t) \\ \mathbf{P}_p(t+1) &= \mathbf{A}(t)\mathbf{P}_f(t)\mathbf{A}^T(t) + \mathbf{Q}(t)\end{aligned} \tag{10.77}$$

with initial conditions

$$\hat{\mathbf{X}}_p(0) = \mathbf{0} \ , \ \mathbf{P}_p(0) = \mathbf{\Pi}_0 \, .$$

Similarly, the information form (see identities (4.73)–(4.74)) of the *measurement-update recursions* is given by

$$\begin{aligned}\mathbf{P}_f^{-1}(t)\hat{\mathbf{X}}_f(t) &= \mathbf{P}_p^{-1}(t)\hat{\mathbf{X}}_p(t) + \sigma^{-2}\mathbf{c}^T(t)\mathbf{Y}(t) \\ \mathbf{P}_f^{-1}(t) &= \mathbf{P}_p^{-1}(t) + \sigma^{-2}\mathbf{c}^T(t)\mathbf{c}(t)\, .\end{aligned} \tag{10.78}$$

Clearly, starting with the predicted estimate and error variance at time $t = 0$, the filtered and predicted estimates can be evaluated by applying the measurement- and time-update recursions in alternance. Furthermore, by observing that the inovations can be expressed as

$$\nu(t) = \mathbf{c}\tilde{\mathbf{X}}_p(t) + V(t)$$

where the predicted state error $\tilde{X}_p(t)$ and measurement noise $V(t)$ are uncorrelated, the innovations variance can be expressed as

$$P_\nu(t) = \mathbf{c}(t)\mathbf{P}_p(t)\mathbf{c}^T(t) + \sigma^2 \, , \tag{10.79}$$

and hence it can be evaluated in the course of the Kalman filter implementation. Together, the recursions (10.76)–(10.78) and expression (10.72) for the sufficient statistic S represent a fully recursive and causal implementation of the Gaussian detector

$$S \underset{H_0}{\overset{H_1}{\gtrless}} \gamma \, . \tag{10.80}$$

In fact, this implementation represents the DT counterpart of the original Kalman filter-based causal receiver proposed by Schweppe [1] in 1965.

Of course, a causal implementation of the receiver (10.80) based on expression (10.72) for the sufficient statistic S is possible even if $Z(t)$ is not a partially observed Markov process. For example, if $Z(t)$ is stationary, the predicted and filtered estimates $\hat{Z}_p(t)$ and $\hat{Z}_f(t)$ can be computed by using transversal or lattice prediction filters of the type discussed in [12].

At this point, it is worth saying a few words about the evaluation of the term

$$\ln(D) = \ln(|\mathbf{K_Y}|/\sigma^{2(T+1)}) = \sum_{j=1}^{T+1} \ln(1 + \lambda_j/\sigma^2) \tag{10.81}$$

appearing in expression (10.21) for the threshold γ of the test (10.80), where $|\mathbf{K_Y}|$ denotes the determinant of $\mathbf{K_Y}$. Instead of employing expression (10.80) which requires the evaluation of the eigenvalues $\{\lambda_j,\ 1 \leq j \leq T+1\}$ of the covariance matrix $\mathbf{K}$ of $\mathbf{Z}$, we observe from the factorization (10.67) of $\mathbf{K_Y^{-1}}$ and the fact that $\mathbf{F}$ is lower triangular with diagonal terms $\{P_\nu^{-1/2}(t),\ 0 \leq t \leq T\}$ that

$$|\mathbf{K_Y^{-1}}| = |\mathbf{F}|^2 = \prod_{t=0}^{T} P_\nu^{-1}(t).$$

Consequently, since the determinant of a matrix inverse is the inverse of the determinant, we find

$$|\mathbf{K_Y}| = \prod_{t=0}^{T} P_\nu(t) \tag{10.82}$$

so that

$$\ln(D) = \sum_{t=0}^{T} \ln(P_\nu(t)/\sigma^2)\,. \tag{10.83}$$

This shows that the the term $\ln(D)$ can be expressed entirely in terms of the innovations variance $P_\nu(t)$ for $0 \leq t \leq T$. But this variance is computed automatically as part of either the Levinson recursions of linear prediction [12, Chap. 5], or as indicated by (10.79), as part of the Kalman filter recursions.

CT case: Let us turn now to the CT case. We denote by

$$\hat{Z}_f(t) = E[Z(t)|Y(s),\ 0 \leq s \leq t] = \int_0^t h(t,u)Y(u)du \tag{10.84}$$

the filtered estimate of $Z(t)$ based on the observations up to time t. The orthogonality property of the least-squares estimates implies

$$E[Z(t)Y(s)] = E[\hat{Z}_f(t)Y(s)]$$

for $0 \leq s \leq t$, so $h(t,\cdot)$ satisfies the Volterra integral equation of the second kind

$$K(t,s) = \sigma^2 h(t,s) + \int_0^t h(t,u)K(u,s)du \tag{10.85}$$

for $0 \leq s \leq t$. In this equation, the continuity of $K(t,s)$ implies that the causal kernel $h(t,s)$ is continuous in both t and s [13, Theorem 4.19]. Then let $\tilde{Z}_f(t) = Z(t) - \hat{Z}_f(t)$ and consider the innovations process

$$\nu(t) = Y(t) - \hat{Z}_f(t) = \tilde{Z}_f(t) + V(t)\,. \tag{10.86}$$

By using the orthognality property of the filtered error $\tilde{Z}_f(t)$ and the whiteness of $V(t)$, together with the fact that $\nu(s)$ depends only on the observations $\{Y(u),\ 0 \leq u \leq s\}$, we find

$$E[\nu(t)\nu(s)] = 0$$

for $t > s$. By symmetry, this is also true for $t < s$, so that $\nu(t)$ is a WGN process. Quite remarkably, the intensity of the innovations process $\nu(t)$ is the same as that of the white noise component $V(t)$ of the observations, i.e.,

$$E[\nu(t)\nu(s)] = \sigma^2\delta(t-s)\,. \tag{10.87}$$

Although this property can be derived heuristically [7, p. 630], it is really best explained if we consider the integrated forms of the two decompositions

$$\begin{aligned} Y(t) &= Z(t) + V(t) \\ &= \hat{Z}_f(t) + \nu(t)\,. \end{aligned} \tag{10.88}$$

The integrated process $\int_0^t Y(s)$ is a continuous semimartingale [14, p. 234] and given a sequence $\{\mathcal{F}_t, t \geq 0\}$ of sigma fields, it can be decomposed into a component which is predictable from $\mathcal{F}_t$ and a martingale with respect to $\mathcal{F}_t$. The martingale part has the property that it has the same quadratic variation independently of the choice of sigma-field sequence $\{\mathcal{F}_t,\ t \geq 0\}$. Then, if we consider the integrated form of the first line of (10.87) and denote

$$N(t) = \int_0^t V(s)ds\,,$$

we observe that $N(t)$ is a Wiener process and thus a martingale with respect to the sigma-field sequence

$$\mathcal{G}_t \triangleq \sigma\{Z(s), V(s),\ 0 \leq s \leq t\}\,,$$

and its quadratic variation is $<N, N>_t = \sigma^2 t$. Similarly, the integrated form of the second line of (10.87) also yields a semimartingale decomposition, but with respect to the sigma-field sequence

$$\mathcal{Y}_t \triangleq \sigma\{Y(s),\ 0 \leq s \leq t\}\,.$$

Then $\int_0^t \nu(s)ds$ is a martingale with respect to $\mathcal{Y}_t$, and accordingly, its quadratic variation must be $\sigma^2 t$, which proves (10.86). Since the above discussion requires advanced stochastic calculus concepts, we refer the reader to [14, 15] for further information.

Then, since the innovations process can be expressed in operator notation as

$$\nu = (I - h)Y$$

where Y has covariance $\sigma^2 I + K$, the whiteness property (10.87) of the innovations implies the operator identity

$$\sigma^2 I = (I - h)(\sigma^2 I + K)(I - h^*)\,, \tag{10.89}$$

where the adjoint h^* of h has kernel $h^*(t,s) = h(s,t)$, so it is anticausal, since $h^*(t,s) = 0$ for $t > s$. Recalling that $(\sigma^2 I + K)^{-1} = \sigma^{-2}(I - H)$, this implies the operator factorization

$$I - H = (I - h^*)(I - h)\,, \tag{10.90}$$

which is the CT counterpart of the inverse covariance matrix factorization (10.67). Equivalently, we have

$$H = h + h^* - h^* h\,,$$

and, noting that expression (10.22) can be expressed in operator form as $S =< Y, HY >$, we obtain

$$\begin{aligned} S &= < Y, hY > - \frac{1}{2}||hY||^2 \\ &= \int_0^T Y(t)\hat{Z}_f(t)dt - \frac{1}{2}\int_0^T \hat{Z}_f^2(t)dt \end{aligned} \tag{10.91}$$

where we have used the fact that $\hat{Z}_f = hY$. Identity (10.91) is the CT counterpart of the DT expression (10.72). It shows that S can be evaluated directly from the filtered estimate $\hat{Z}_f(t)$, without recourse to the smoothed estimate $\hat{Z}(t)$ appearing in the correlator/integrator expression (10.38). In this respect, it is worth noting that the first integral in (10.91) is really a stochastic integral, and since all operations used to derive (10.91) have relied on the ordinary rules of integral calculus, this integral should be interpreted in the sense of Stratonovich. See the discussion of the causal receiver (10.91) given in [16,17]. For a comprehensive presentation of Ito and Stratonovich stochastic integrals, the reader is referred to [15,18]. The CT estimator/correlator receiver (10.91) is depicted in Fig. 10.3. Note again that it is similar to the matched filter structure used for the detection of a known signal $s(t)$, except that $s(t)$ needs to be replaced by the causal estimate $\hat{Z}_f(t)$ of the Gaussian signal $Z(t)$.

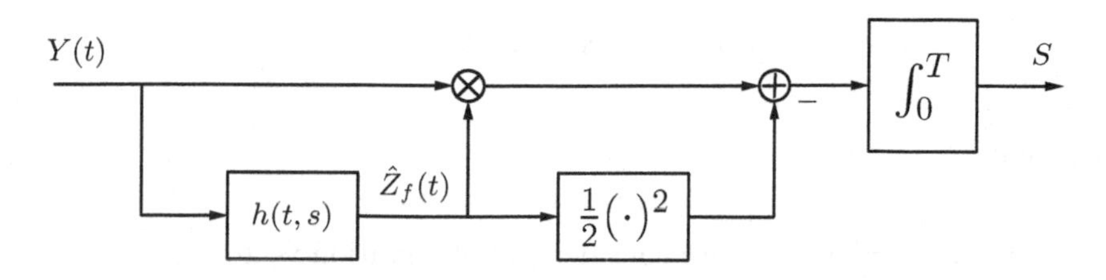

Fig. 10.3. Causal CT estimator-correlator receiver for Gaussian signal detection.

10.4 Asymptotic Stationary Gaussian Test Performance

It is rather difficult to characterize analytically the performance of the optimal detector given by (10.20)–(10.21). To see why this is the case, note that in the DT case the sufficient statistic takes the form

$$S = \frac{1}{2} \sum_{j=1}^{N} \frac{\lambda_j Y_j^2}{\lambda_j + \sigma^2} \tag{10.92}$$

with $N = T + 1$, where Y_j is $N(0, \sigma^2)$ distributed under H_0 and $N(0, \lambda_j + \sigma^2)$ distributed under H_1. Thus, S is a weighted sum of N χ_1^2 distributed random variables and its probability distribution does not admit a closed-form expression under either H_0 or H_1. Consequently, the probabilities of a miss or of false alarm cannot be evaluated analytically, except in rare cases. For the case of stationary Gaussian processes, it is however possible to characterize the test performance asymptotically, by using the theory of large deviations [3,4].

The results presented in this section are, in fact, similar to those obtained in Section 3.2 for the asymptotic performance of LR tests based on independent observations, except that here the observations are not independent, but form a zero-mean stationary Gaussian process. In spite of this difference, it is worth noting that since the observed Gaussian process has typically a finite coherence time, by subsampling the observations sufficiently slowly, we can extract an uncorrelated and thus independent subsequence to which the results of Section 3.2 are applicable. Thus, we can already guess that the probabilities of a miss and of false alarm for a Bayesian test will decrease exponentially with the length of the observation period.

The analysis presented is restricted to the DT case, but we examine a more general version of the detection problem (10.1), where under H_i the observed process $Y(t)$ with $0 \leq t \leq T$ is zero-mean, stationary, and Gaussian, with covariance

$$K_i(k) = E[Y(t+k)Y(t)|H_i] .$$

So the observed process need not be white under the null hypothesis, and it need not contain a white noise component under the alternate hypothesis. Under H_i, we assume that the covariance sequence $\{K_i(k),\ k \in \mathbb{Z}\}$ admits a power spectral density

$$S_i(e^{j\omega}) = \sum_{k=-\infty}^{\infty} K_i(k) e^{-jk\omega} \tag{10.93}$$

such that

$$0 < m_i < S_i(e^{j\omega}) < M_i$$

for all $0 \leq \omega \leq 2\pi$. Then, if we consider the observation vector

$$\mathbf{Y}_T = \begin{bmatrix} Y(0) \dots Y(t) \dots Y(T) \end{bmatrix}^T ,$$

the detection problem takes the form

$$\begin{aligned} H_0 &: \mathbf{Y}_T \sim N(\mathbf{0}, \mathbf{K}_T^0) \\ H_1 &: \mathbf{Y}_T \sim N(\mathbf{0}, \mathbf{K}_T^1) , \end{aligned} \tag{10.94}$$

where for $i = 0,\ 1$

$$\mathbf{K}_T^i = \begin{bmatrix} K_i(0) & K_i(1) & \dots & & K_i(T) \\ K_i(1) & K_i(0) & K_i(1) & \dots & K_i(T-1) \\ & \ddots & \ddots & \ddots & \\ K_i(T-1) & \dots & K_i(1) & K_j(0) & K_i(1) \\ K_i(T) & & \dots & K_i(1) & K_i(0) \end{bmatrix}$$

is a Toeplitz covariance matrix of dimension $T+1$. Then, if $f(\mathbf{y}_T|H_i)$ denotes the conditional density of the observation vector $\mathbf{Y}_T$ under hypothesis H_i, the log-likelihood ratio scaled by the number of observations can be expressed as

$$\begin{aligned} S_T &= \frac{1}{T+1} \ln \big(\frac{f(\mathbf{Y}_T|H_1)}{f(\mathbf{Y}_T|H_0)} \big) \\ &= \frac{1}{2(T+1)} \big[\mathbf{Y}_T^T [(\mathbf{K}_T^0)^{-1} - (\mathbf{K}_T^1)^{-1}] \mathbf{Y}_T - \ln(|K_T^1|/|K_T^0|) \big] , \end{aligned} \tag{10.95}$$

so the LR test takes the form

$$S_T \underset{H_0}{\overset{H_1}{\gtrless}} \gamma . \tag{10.96}$$

For a Bayesian test, γ is actually T dependent and equals $\ln(\tau(T))/(T+1)$ where, as indicated at the beginning of Section 3.2, $\tau(T)$ is approximately constant as T becomes large. So, for Bayesian tests, as T becomes large, the threshold γ can be selected equal to zero. However, our asymptotic study of the LR test will not be limited to Bayesian tests, so we will also examine nonzero values of the threshold γ.

10.4.1 Asymptotic Equivalence of Toeplitz and Circulant Matrices

The asymptotic analysis of the test (10.96) will rely extensively on the concept of asymptotic equivalence of matrix sequences of increasing dimensions, and specifically on equivalence results between Toeplitz and circulant matrices. All the results that we employ are described in Gray's elegant tutorial report [19] on Toeplitz and circulant matrices, which has been updated regularly over the last 35 years, and can be downloaded from the author's web site.

To start, note that two sequences $\{A_N,\ N \geq 1\}$, $\{B_N, N \geq 1\}$ of matrices, where A_N and B_N have size $N \times N$, are said to be asymptotically equivalent if the spectral norm (the largest singular value) of A_N and B_N is uniformly bounded from above by a common constant M for all N, and if

$$\lim_{N\to\infty} \frac{1}{N^{1/2}}||A_N - B_N||_F = 0 \tag{10.97}$$

where the Frobenius or Euclidean norm of a matrix P of dimension $n \times n$ is defined

$$||P||_F = (\sum_{i=1}^{n}\sum_{j=1}^{n} |p_{ij}|^2)^{1/2} .$$

Asymptotic equivalence is denoted as $A_N \sim B_N$. This relation is transitive, and has the following properties [19, Th. 2.1]:

(i) If $A_N \sim B_N$, then
$$\lim_{N\to\infty} |A_N| = \lim_{N\to\infty} |B_N| .$$
(ii) If $A_N \sim B_N$ and $C_N \sim D_N$, then $A_N B_N \sim C_N D_N$.
(iii) If $A_N \sim B_N$ and the spectral norm of A_N^{-1} and B_N^{-1} is uniformly bounded from above by a constant K for all N, then $A_N^{-1} \sim B_N^{-1}$.
(iv) If $A_N \sim B_N$, there exists finite constants m and M such that the eigenvalues $\lambda_{k,N}$ and $\mu_{k,N}$ with $1 \leq k \leq N$ of A_N and B_N, respectively, satisfy
$$m \leq \lambda_{k,N},\ \mu_{k,N} \leq M . \tag{10.98}$$
(v) If $A_N \sim B_N$ are two asymptotically equivalent sequences of Hermitian matrices, and $F(x)$ is a continuous function over the interval $[m, M]$ specified by the contants m and M of (10.98), then

$$\lim_{N\to\infty} \frac{1}{N}\sum_{k=1}^{N} F(\lambda_{k,N}) = \lim_{N\to\infty} \frac{1}{N}\sum_{k=1}^{N} F(\mu_{k,N}) . \tag{10.99}$$

Note that by selecting $F(x) = x$ and $F(x) = \ln(x)$, property (10.99) implies that if A_N and B_N are two asymptotically equivalent Hermitian matrix sequences, then

$$\lim_{N\to\infty} \frac{1}{N}\mathrm{tr}(A_N) = \lim_{N\to\infty} \frac{1}{N}\mathrm{tr}(B_N)$$
$$\lim_{N\to\infty} \frac{1}{N} \ln(|A_N|) = \lim_{N\to\infty} \frac{1}{N} \ln(|B_N|) .$$

Then, consider a covariance function $K(k)$ with $k \in \mathbb{Z}$ whose power spectral density (PSD)

$$S(e^{j\omega}) = \sum_{k=-\infty}^{\infty} K(k) e^{-jk\omega}$$

satisfies $0 < m < S(e^{j\omega}) < M$ over $[0, 2\pi]$. It is shown in [19] that the sequence of Toeplitz matrices

$$\mathbf{K}_T = \begin{bmatrix} K(0) & K(1) & \dots & & K(T) \\ K(1) & K(0) & K(1) & \dots & K(T-1) \\ & \ddots & \ddots & \ddots & \\ K(T-1) & \dots & K(1) & K(0) & K(1) \\ K(T) & & \dots & K(1) & K(0) \end{bmatrix}$$

is asymptotically equivalent to the sequence of circulant matrices

$$\mathbf{C}_T = \mathbf{W}_T^H \mathbf{D}_T \mathbf{W}_T \tag{10.100}$$

where $\mathbf{W}_T$ is the DFT matrix of dimension $(T+1) \times (T+1)$, whose entry in row k and column ℓ is $w_T^{(k-1)(\ell-1)}/(T+1)^{1/2}$ with $w_T = e^{-j2\pi/(T+1)}$, and

$$\mathbf{D}_T = \text{diag}\,\{S(e^{j\omega_{k-1}}),\ 1 \leq k \leq T+1\}\,, \tag{10.101}$$

with $\omega_k = k2\pi/(T+1)$.

10.4.2 Mean-square Convergence of S_T

The first step in our asymptotic analysis of the LRT (10.96) is to show that as $T \to \infty$, S_T converges in the mean-square sense under both H_0 and H_1. To do so, we introduce the Itakura-Saito spectral distortion measure between two PSDs $S(e^{j\omega})$ and $T(e^{j\omega})$ defined over $[0, 2\pi]$ as

$$I(S,T) = \frac{1}{2\pi} \int_0^{2\pi} \Big[\frac{S(e^{j\omega})}{T(e^{j\omega})} - 1 - \ln\Big(\frac{S(e^{j\omega})}{T(e^{j\omega})}\Big)\Big]\,d\omega\,. \tag{10.102}$$

Note that as long as $S(e^{j\omega})$ and $T(e^{j\omega})$ admit strictly positive lower bounds and finite upper bounds, both $I(S,T)$ and $I(T,S)$ are finite. The Itakura-Saito (IS) spectral distortion measure represents the equivalent for stationary Gaussian processes of the Kullback-Leibler (KL) divergence of probability distributions (see [20, p. 123]). In particular, it is a standard distortion measure used in speech compression [5]. The inequality $x - 1 - \ln(x) \geq 0$ for $x > 0$ implies that $I(S,T) \geq 0$ with equality if and only if $S = T$, but like the KL divergence, the IS spectral distortion measure is not symmetric and does not satisfy the triangle inequality.

Our first observation concerning the test (10.96) is that as $T \to \infty$

$$S_T \xrightarrow{\text{m.s.}} e_1 = \frac{1}{2} I(S_1, S_0) \tag{10.103}$$

under H_1 and

$$S_T \xrightarrow{\text{m.s.}} e_0 = -\frac{1}{2} I(S_0, S_1) \tag{10.104}$$

under H_0. These limits represent the analog for Gaussian stationary processes of expressions (3.4) and (3.5) for the LR function of i.i.d. observations. To establish this result, consider the complex zero-mean Gaussian vector

$$\hat{\mathbf{Y}}_T = \mathbf{W}_T \mathbf{Y}_T = \left[\hat{Y}(e^{j\omega_0}) \dots \hat{Y}(e^{j\omega_k}) \dots \hat{Y}(e^{j\omega_T})\right]^T , \tag{10.105}$$

with $\omega_k = k2\pi/(T+1)$ and

$$\hat{Y}(e^{j\omega}) = (T+1)^{-1/2} \sum_{t=0}^{T} Y(t) e^{-jt\omega} ,$$

which is obtained by evaluating the DFT of length $T+1$ of observations $Y(t)$ for $0 \le t \le T$. Under hypothesis H_i we have

$$E[\hat{\mathbf{Y}}_T \hat{\mathbf{Y}}_T^H | H_i] = \mathbf{W}_T \mathbf{K}_T^i \mathbf{W}_T^H \sim \mathbf{D}_T^i , \tag{10.106}$$

with

$$\mathbf{D}_T^i = \operatorname{diag}\{S_i(e^{j\omega_k}),\ 0 \le k \le T\}$$

for large T. Since the DFT components satisfy $\hat{Y}(e^{j\omega_{N-k}}) = \hat{Y}^*(e^{j\omega_k})$ with $N = T+1$, we deduce from (10.106) that as T becomes large, components $\hat{Y}(e^{j\omega_k})$ are independent for $0 \le k \le N/2$ with N even, and for $0 \le k \le (N-1)/2$ with N odd. For $k = 0$ and $k = N/2$ with N even, $\hat{Y}(e^{j\omega_k}) \sim N(0, S_i(e^{j\omega_k}))$; otherwise $\hat{Y}(e^{j\omega_k}) \sim CN(0, S_i(e^{j\omega_k}))$. Then consider the large T approximation

$$\begin{aligned} S_T \approx \frac{1}{2(T+1)} \sum_{k=0}^{T} &\left[(S_0^{-1}(e^{j\omega_k}) - S_1^{-1}(e^{j\omega_k})) |\hat{Y}(e^{j\omega_k})|^2 \right. \\ &\left. - \ln(S_1(e^{j\omega_k})/S_0(e^{j\omega_k})\right] \end{aligned} \tag{10.107}$$

of test statistic S_T. This approximation is obtained by replacing the Toeplitz matrices $\mathbf{K}_T^i$ by their circulant approximations $\mathbf{C}_T^i$ with $i = 0, 1$ and using the definition (10.105) of the DFT $\hat{\mathbf{Y}}_T$ of the observations. Then taking into account (10.106), we find

$$E[S_T|H_1] \approx \frac{1}{4\pi} \sum_{k=0}^{T} \Big[\frac{S_1(e^{j\omega_k})}{S_0(e^{j\omega_k})} - 1 - \ln(S_1(e^{j\omega_k})/S_0(e^{j\omega_k})) \Big] \Delta\omega_k$$

with $\Delta\omega_k = \omega_{k+1} - \omega_k = 2\pi/(T+1)$. So, as $T \to \infty$, we have

$$\lim_{T\to\infty} E[S_T|H_1] = \frac{1}{2} I(S_1, S_0) . \tag{10.108}$$

Similarly, we find that under H_0

$$\lim_{T\to\infty} E[S_T|H_0] = -\frac{1}{2} I(S_0, S_1) . \tag{10.109}$$

Next, we need to evaluate the variance of S_T for large T. By observing that

$$E[|\hat{Y}(e^{j\omega_k})|^2 |\hat{Y}(e^{j\omega_\ell})|^2 | H_i] = \begin{cases} 3(S_i(e^{j\omega_k}))^2 & k = \ell \\ S_i(e^{j\omega_k}) S_i(e^{j\omega_\ell}) & k \ne \ell , \end{cases}$$

for $k = 0$ and $k = N/2$ with N even, and

$$E[|\hat{Y}(e^{j\omega_k})|^2|\hat{Y}(e^{j\omega_\ell})|^2|H_i] = \begin{cases} 2(S_i(e^{j\omega_k}))^2 & k = \ell \\ S_i(e^{j\omega_k})S_i(e^{j\omega_\ell}) & k \neq \ell \,, \end{cases}$$

for $1 \leq k, \ell \leq N/2 - 1$ with N even, and for $1 \leq k, \ell \leq (N-1)/2$ with N odd, we find

$$E[(S_T - E[S_T|H_1])^2|H_1] \approx \frac{1}{T+1}\frac{1}{4\pi}\sum_{k=0}^{T}\Big(\frac{S_1(e^{j\omega_k})}{S_0(e^{j\omega_k})} - 1\Big)^2 \Delta\omega_k$$

$$\approx \frac{1}{T+1}\frac{1}{4\pi}\int_0^{2\pi}\Big(\frac{S_1(e^{j\omega})}{S_0(e^{j\omega})} - 1\Big)^2 d\omega \rightarrow 0 \qquad (10.110)$$

as $T \rightarrow \infty$. A similar result holds under H_0. This establishes the mean-square convergence of S_T expressed by (10.103) and (10.104) under H_1 and H_0. These expressions immediately indicate that, as long as the threshold γ of the test (10.95) satisfies

$$e_0 = -I(S_0, S_1)/2 < \gamma < e_1 = I(S_1, S_0)/2,$$

the test becomes error free as $T \rightarrow \infty$.

10.4.3 Large Deviations Analysis of the LRT

To characterize the rate of decay of the probability of a miss, and of the probability of false alarm as $T \rightarrow \infty$ we employ a generalization of Cramer's theorem called the Gärtner-Ellis theorem [3,4] which is applicable to sequences of random variables with an asymptotic log-generating function.

Specifically, consider a sequence $\{Z_N,\ N \geq 1\}$ of random variables with generating function

$$G_N(u) = E[\exp(uZ_N)]$$

Then, assume that the asymptotic log-generating function

$$\Lambda(u) \stackrel{\triangle}{=} \lim_{N\rightarrow\infty} \frac{1}{N}\ln(G_N(Nu)) \qquad (10.111)$$

exists for all $u \in \mathbb{R}$, where $\Lambda(u)$ is defined over the extended real line $(-\infty, \infty]$. If $\mathcal{D} = \{u : \ \Lambda(u) < \infty$, we assume also that $0 \in \mathcal{D}$, that $\Lambda(u)$ is differentiable over $\mathcal{D}$, and that $\Lambda(\cdot)$ is steep [3, p. 44], i.e., that

$$\lim_{k\rightarrow\infty} |\frac{d\Lambda}{du}(u_k)| = \infty$$

whenever $\{u_k,\ k \geq 1\}$ is a sequence converging to a point on the boundary of $\mathcal{D}$. The function $\Lambda(u)$ is convex, and its Legendre transform

$$I(z) \stackrel{\triangle}{=} \sup_{u\in\mathbb{R}}(zu - \Lambda(u)) \qquad (10.112)$$

is also convex. Then Cramer's theorem can be extended as follows:

Gärtner-Ellis Theorem: if C denotes an arbitrary closed set of $\mathbb{R}$,

$$\lim_{N\to\infty} \frac{1}{N} \ln P[Z_N \in C] \leq - \inf_{z \in C} I(z) \tag{10.113}$$

and if O an open set of $\mathbb{R}$

$$\lim_{N\to\infty} \frac{1}{N} \ln P[Z_N \in O] \geq - \inf_{z \in O} I(z) \,. \tag{10.114}$$

The reader is referred to [3, Sec. 2.3] or [4, Chap. 5] for a proof. For the problem at hand, the sequence of random variables we consider is S_T with $T \geq 0$. The log-generating function of $(T+1)S_T$ under H_0 was evaluated in Example 3.2, and is given by

$$\begin{aligned} \ln(G_T^0((T+1)u)) = -\frac{1}{2}\Big[& \ln(|u(\mathbf{K}_T^1)^{-1} + (1-u)(\mathbf{K}_T^0)^{-1}|) \\ & + u\ln(|\mathbf{K}_T^1|) + (1-u)\ln(|\mathbf{K}_T^0|)\Big] \,. \end{aligned} \tag{10.115}$$

But the Toeplitz matrices $\mathbf{K}_T^i$ with $i = 0, 1$ can again be replaced by their asymptotically equivalent circulant approximations $\mathbf{C}_T^i$, and since these two matrices are jointly diagonalized by the DFT matrix $\mathbf{W}_T$, we have

$$\begin{aligned} & \ln(G_T^0((T+1)u)) \\ & \approx -\frac{1}{2}\Big[\ln(|u(\mathbf{D}_T^1)^{-1} + (1-u)(\mathbf{D}_T^0)^{-1}|) + u\ln(|\mathbf{D}_T^1|) + (1-u)\ln(|\mathbf{D}_T^0|)\Big] \\ & = -\frac{1}{2}\sum_{k=0}^{T} \big[\ln\left(uS_1^{-1}(e^{j\omega_k}) + (1-u)S_0^{-1}(e^{j\omega_k})\right) \\ & \qquad + u\ln(S_1(e^{j\omega_k})) + (1-u)\ln(S_0(e^{j\omega_k}))\big] \end{aligned} \tag{10.116}$$

with $\omega_k = k2\pi/(T+1)$. This implies that under H_0, S_T admits the asymptotic log-generating function

$$\begin{aligned} \Lambda_0(u) &= \lim_{T\to\infty} \frac{1}{T+1} \ln(G_T^0((T+1)u) \\ &= -\frac{1}{4\pi}\int_0^{2\pi} \Big[\ln\Big(\frac{u}{S_1(e^{j\omega})} + \frac{(1-u)}{S_0(e^{j\omega})}\Big) \\ & \qquad + u\ln(S_1(e^{j\omega})) + (1-u)\ln(S_0(e^{j\omega}))\Big] d\omega \\ &= \frac{1}{4\pi}\int_0^{2\pi} \Big[u\ln(S_0(e^{j\omega})) + (1-u)\ln(S_1(e^{j\omega})) \\ & \qquad - \ln(uS_0(e^{j\omega}) + (1-u)S_1(e^{j\omega}))\Big] d\omega \,. \end{aligned} \tag{10.117}$$

Note that

$$\Lambda_0(0) = \Lambda_0(1) = 0$$

and

$$\begin{aligned} \frac{d}{du}\Lambda_0(0) &= e_0 = -\frac{1}{2}I(S_0, S_1) \\ \frac{d}{du}\Lambda_0(1) &= e_1 = \frac{1}{2}I(S_1, S_0)\,. \end{aligned} \tag{10.118}$$

Accordingly, the asymptotic log-generating function plotted in Fig. 10.4 has a shape similar to the case of i.i.d. observations shown in Fig. 3.2, but with $m_0 = -D(f_0|f_1)$ replaced by e_0 and $m_1 = D(f_1|f_0)$ replaced by e_1.

Similarly, the asymptotic log-generating function of S_T under H_1 is given by

$$\Lambda_1(u) = \Lambda_0(u+1)\,, \tag{10.119}$$

i.e., it is obtained by shifting $\Lambda_0(u)$ to the left by -1. This implies that its Legendre transform $I_1(z)$ is related to the Legendre transform $I_0(z)$ of $\Lambda_0(u)$ by

$$I_1(z) = I_0(z) - z\,. \tag{10.120}$$

Then, if $P_F(T) = P[S_T \geq \gamma|H_0]$ and $P_M(T) = P[S_T < \gamma|H_1]$ denote respectively the probability of false alarm, and the probability of a miss for the test (10.105) based on observations over $[0, T]$, by applying the Gärtner-Ellis theorem we find

$$\begin{aligned} \lim_{T\to\infty} \frac{1}{T}\ln P_F(T) &= -I_0(\gamma) \\ \lim_{T\to\infty} \frac{1}{T}\ln P_M(T) &= -I_1(\gamma) = -[I_0(\gamma) - \gamma]\,. \end{aligned} \tag{10.121}$$

Hence, as was the case for i.i.d. observations, the probabilities of false alarm and of a miss decay exponentially with the number of observations. The decay rates are specified by the Legendre transforms of the log-generating functions $\Lambda_i(u)$ evaluated at the threshold γ. Unfortunately, even though the log-generating function $\Lambda_0(u)$ admits the simple expression (10.117) in terms of

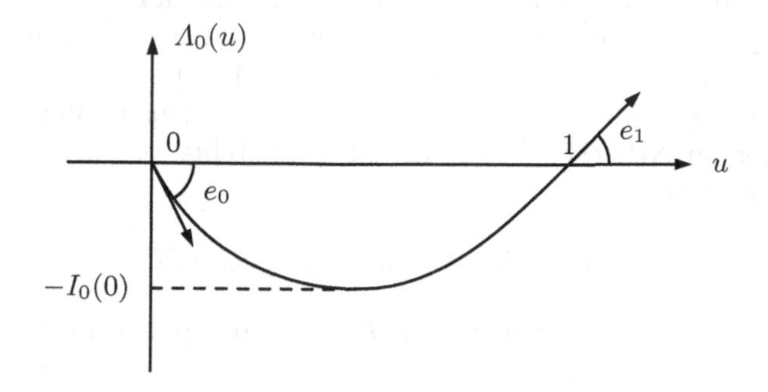

Fig. 10.4. Asymptotic log-generating function under H_0.

the PSDs $S_0(\omega)$ and $S_1(\omega)$ of the observations under H_0 and H_1, it is usually impossible to evaluate its Legendre transform in closed form, except in a few special cases.

Let us consider now several special cases for the threshold γ. For an NP test of type I, the probability P_M of a miss is minimized while keeping the probability P_F of false alarm less or equal to a fixed bound α. Since we know that as T goes to infinity, the test statistic S_T converges in the mean-square sense to $e_0 = -I(S_0, S_1)/2$ under H_0, we can deduce that to ensure that the probability of false alarm falls below α, we can select the threshold as

$$\gamma_{\text{NP}}^I = e_0 + \epsilon \qquad (10.122)$$

with $\epsilon > 0$ arbitrarily small. With this choice, we have $I_0(e_0) = 0$ since, as seen from Fig. 10.4, the line with slope e_0 which is tangent to $\Lambda_0(u)$ has $u = 0$ as its tangency point, and has a zero intercept with the vertical axis. Accordingly, we have $I_1(e_0) = -e_0$, so for an NP test of type I, the probability of a miss decays exponentially according to

$$P_M(T) \approx K_M(T) \exp(-I(S_0, S_1)T/2) \qquad (10.123)$$

where the function $K_M(T)$ depends on the spectral densities $S_i(\omega)$ with $i = 0,\ 1$, but it varies with T at a sub-exponential rate, in the sense that

$$\lim_{T\to\infty} \frac{1}{T} \ln(K_M(T)) = 0 \,.$$

The exponential rate of decay (10.123) for the probability of a miss represents the fastest rate of decay that can be achieved by any statistical test based only on the observations over $[0, T]$. Similarly, for an NP test of type II, the probability of false alarm P_F is minimized while keeping the probability P_M of a miss below a fixed constant β. Since, as T goes to infinity, S_T converges in the mean-square to $e_1 = I(S_1, S_0)/2$ under H_1, we can select the threshold as

$$\gamma_{\text{NP}}^{II} = e_1 - \epsilon \qquad (10.124)$$

with $\epsilon > 0$ arbitrarily small. By examining the plot for $\Lambda_1(u)$ obtained by shifting the plot $\Lambda_0(u)$ of Fig. 10.4 to the left by -1, we see that $I_1(e_1) = 0$, since the line with slope e_1 which is tangent to $\Lambda_1(u)$ has $u = 0$ as point of tangency, and a zero intercept with the vertical axis. This implies $I_0(e_1) = e_1$, and hence for an NP test of type II, the probability of false alarm decays exponentially according to

$$P_F(T) \approx K_F(T) \exp(-I(S_1, S_0)T/2) \qquad (10.125)$$

where the functiom $K_F(T)$ varies with T at a sub-exponential rate, i.e.,

$$\lim_{T\to\infty} \frac{1}{T} \ln(K_F(T)) = 0 \,.$$

Again, the exponential rate of decay (10.125) is the fastest that can be achieved by any statistical test depending only on the observations over $[0, T]$.

If we rewrite (10.121) as

$$\begin{aligned} P_F(T) &\approx K_F(T) \exp(-I_0(\gamma)T) \\ P_M(T) &\approx K_M(T) \exp(-I_1(\gamma)T) , \end{aligned} \tag{10.126}$$

where the functions $K_F(T)$ and $K_M(T)$ are subexponential, the probability of error for the test (10.96) can be written as

$$\begin{aligned} P_E(T) &= P_F(T)\pi_0 + P_M(T)(1-\pi_0) \\ &= K_F(T)\pi_0 \exp(-I_0(\gamma)T) + K_M(T)(1-\pi_0) \exp(-I_1(\gamma)T)) \, . \end{aligned} \tag{10.127}$$

Then, as in Section 3.2 we may ask what value of γ equalizes the rates of decay of the two types of errors, and thus maximizes the overall rate of decay of the probability of error. The optimum threshold is again given by $\gamma = 0$, which is expected, since we saw earlier that this is the threshold obtained for Bayesian tests, which are explicitly designed to minimize the probability of error. For $\gamma = 0$ the rate of decay of the probability of error is given by

$$\begin{aligned} I_0(0) &= -\min_{0 \leq u \leq 1} \Lambda_0(u) \\ &= \min_{0 \leq u \leq 1} \frac{1}{4\pi} \int_0^{2\pi} \Big[\ln(uS_0(e^{j\omega}) + (1-u)S_1(e^{j\omega})) \\ &\qquad -u \ln(S_0(e^{j\omega})) - (1-u)\ln(S_1(e^{j\omega})) \Big] d\omega \, . \end{aligned} \tag{10.128}$$

This expression is the analog for stationary Gaussian processes of the Chernoff distance between two probability densities which was used in (3.36) to characterize the maximum rate of decay of the probability of error for the case of i.i.d. observations.

Although the Legendre transforms $I_0(z)$ and $I_1(z)$ cannot be evaluated in closed form, they admit a simple geometric interpretation in terms of the Itakura-Saito spectral distortion measure. Specifically, consider the family of zero-mean stationary Gaussian processes specified by the power spectral density

$$S_u(e^{j\omega}) = \Big[\frac{u}{S_1(e^{j\omega})} + \frac{1-u}{S_0(e^{j\omega})} \Big]^{-1} \tag{10.129}$$

for $0 \leq u \leq 1$. This family represnts the geodesic linking the process with PSD $S_0(e^{j\omega})$ to the one with PSD $S_1(e^{j\omega})$ as u varies from 0 to 1. Note that $S_u(e^{j\omega})$ is obtained by performing the harmonic mean of PSDs $S_0(e^{j\omega})$ and $S_1(e^{j\omega})$ with weights u and $1-u$. Then consider the tangent with slope z to the curve $\Lambda_0(u)$ and denote the tangency point by $u(z)$, as shown in Fig. 10.5. The relation between z and $u(z)$ is obtained by setting the derivative of

$$F(z, u) = zu - \Lambda_0(u)$$

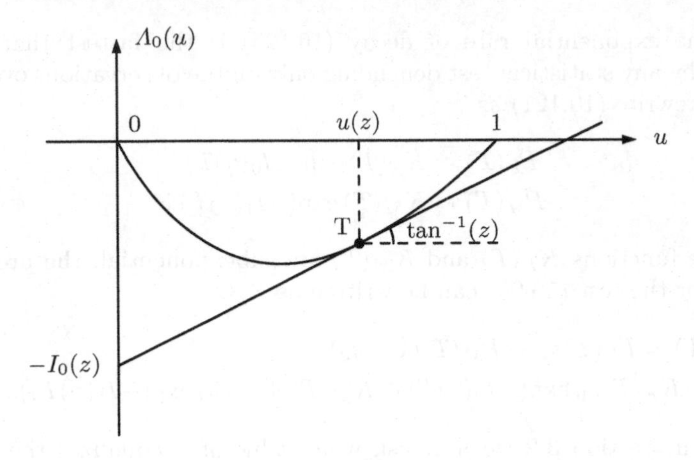

Fig. 10.5. Point of tangency with coordinate $u(z)$ corresponding to a tangent with slope z to $\Lambda_0(u)$.

with respect to u equal to zero, which gives

$$\frac{dF}{du} = 0 = z + \frac{1}{4\pi} \int_0^{2\pi} \Big[\ln \Big(\frac{S_1(e^{j\omega})}{S_0(e^{j\omega})} \Big) - \frac{(S_1 - S_0)(e^{j\omega})}{uS_0(e^{j\omega}) + (1-u)S_1(e^{j\omega})} \Big] d\omega \,. \quad (10.130)$$

Then, substituting (10.130) inside $F(z, u)$ gives after some algebra

$$\begin{aligned} I_0(z) &= \frac{1}{4\pi} \int_0^{2\pi} \Big[\frac{S_{u(z)}(e^{j\omega})}{S_0(e^{j\omega})} - 1 - \ln \Big(\frac{S_{u(z)}(e^{j\omega})}{S_0(e^{j\omega})} \Big) \Big] d\omega \\ &= \frac{1}{2} I(S_{u(z)}, S_0) \,. \end{aligned} \quad (10.131)$$

Similarly, we obtain

$$I_1(z) = \frac{1}{2} I(S_{u(z)}, S_1) \,. \quad (10.132)$$

This shows that $I_0(z)$ and $I_1(z)$ represent the IS distortions between $S_{u(z)}$ and S_0, and between $S_{u(z)}$ and S_1, respectively. Furthermore, when $z = 0$, we have $I_0(0) = I_1(0)$, so the corresponding PSD $S_{u(0)}$ has the feature that

$$I(S_{u(0)}, S_0) = I(S_{u(0)}, S_1) \,,$$

i.e., it is located midway between S_0 and S_1 with respect to the IS metric between PSDs.

10.4.4 Detection in WGN

Consider now the special case where the detection problem takes the form (10.1). Under H_0, the observed signal $Y(t)$ is a zero-mean WGN with PSD

$$S_0(e^{j\omega}) = \sigma^2 , \tag{10.133}$$

and under H_1, $Y(t)$ is the sum of a zero-mean statinary Gaussian process $Z(t)$ with PSD $S_Z(e^{j\omega})$ and an independent zero-mean WGN with intensity σ^2. Thus

$$S_1(e^{j\omega}) = S_Z(e^{j\omega}) + \sigma^2 . \tag{10.134}$$

Then, the expressions obtained in the previous subsection for $\Lambda_0(u)$, $I(S_1, S_0)$ and $I(S_0, S_1)$ can be simplified. Specifically, by substituting (10.133)–(10.134) inside (10.117) and subtracting $\ln(\sigma^2)$ from the first term of the integrand and adding it to the second term, we obtain

$$\Lambda_0(u) = \frac{1}{4\pi} \int_0^{2\pi} \Big[(1-u) \ln \Big(1 + \frac{S_Z(e^{j\omega})}{\sigma^2} \Big) - \ln \Big(1 + (1-u) \frac{S_Z(e^{j\omega})}{\sigma^2} \Big) \Big] d\omega . \tag{10.135}$$

Since $\Lambda_1(u) = \Lambda_0(u+1)$, $\Lambda_1(u)$ admits an expression of the same form as (10.135), but with $1-u$ replaced by $-u$.

To obtain expressions for $I(S_1, S_0)$ and $I(S_0, S_1)$, we denote by

$$P_Y = E[Y^2(t)|H_1] = \frac{1}{2\pi} \int_0^{2\pi} S_1(e^{j\omega}) d\omega \tag{10.136}$$

the power of the observed signal under H_1, and we consider two estimation problems under H_1. First, we denote by

$$\hat{Y}_p(t) = E[Y(t)|Y(t-k),\ k = 1, 2, \dots]$$

the one-step ahead predicted estimate of $Y(t)$ based on the observations up to time $t-1$. Then

$$\nu(t) = Y(t) - \hat{Y}_p(t)$$

is the innovations process, and its variance is given by the Kolmogorov-Szegö formula [12, p. 366]

$$P_\nu = E[\nu^2(t)|H_1] = \exp \Big(\frac{1}{2\pi} \int_0^{2\pi} \ln(S_1(e^{j\omega})) d\omega \Big) . \tag{10.137}$$

Similarly, consider the symmetric interpolation estimate

$$\hat{Y}_{\mathrm{int}}(t) = E[Y(t)|Y(t-k),\ k = \pm 1, \pm 2, \dots]$$

obtained by estimating $Y(t)$ based on all observations on both sides of $Y(t)$. It is shown in Problem 10.4 that the variance of the error

$$\tilde{Y}_{\mathrm{int}}(t) = Y(t) - \hat{Y}_{\mathrm{int}}(t)$$

can be expressed as

$$P_{\rm int} = E[\tilde{Y}_{\rm int}^2(t)|H_1] = \Big[\frac{1}{2\pi}\int_0^{2\pi} S_1^{-1}(e^{j\omega})d\omega\Big]^{-1} . \tag{10.138}$$

Then, since the error variances P_Y, P_ν and $P_{\rm int}$ are based on progressively more information: no observations for P_Y, past observations for P_ν, and two-sided observations for $P_{\rm int}$, we necessarily have

$$P_Y \geq P_\nu \geq P_{\rm int} . \tag{10.139}$$

These inequalities can also be verified directly by applying Jensen's inequality

$$\phi\Big(\frac{1}{2\pi}\int_0^{2\pi} g(\omega)d\omega\Big) \geq \frac{1}{2\pi}\int_0^{2\pi} \phi(g(\omega))d\omega$$

which holds for any function $\phi(\cdot)$ concave over the range of g. Then if we consider the concave function $\phi(x) = \ln(x)$, the first inequality in (10.139) is obtained by setting $g(\omega) = S_1(e^{j\omega})$, and the second by choosing $g(\omega) = S_1^{-1}(e^{j\omega})$.

Then, by applying the definitions of P_Y, P_ν and $P_{\rm int}$, we find

$$\begin{aligned} I(S_1, S_0) &= \frac{1}{4\pi}\int_0^{2\pi} \Big[\frac{S_1(e^{j\omega})}{\sigma^2} - 1 - \ln\Big(\frac{S_1(e^{j\omega})}{\sigma^2}\Big)\Big] d\omega \\ &= \frac{1}{2}\Big[\frac{P_Y}{\sigma^2} - 1 - \ln\Big(\frac{P_\nu}{\sigma^2}\Big)\Big] , \end{aligned} \tag{10.140}$$

and

$$\begin{aligned} I(S_0, S_1) &= \frac{1}{4\pi}\int_0^{2\pi} \Big[\frac{\sigma^2}{S_1(\omega)} - 1 - \ln\Big(\frac{\sigma^2}{S_1(e^{j\omega})}\Big)\Big] d\omega \\ &= \frac{1}{2}\Big[\frac{\sigma^2}{P_{\rm int}} - 1 - \ln\Big(\frac{\sigma^2}{P_\nu}\Big)\Big] , \end{aligned} \tag{10.141}$$

So, rather interestingly, the IS metrics $I(S_1, S_0)$ and $I(S_0, S_1)$ are parametrized entirely by the variances P_Y, P_ν and $P_{\rm int}$.

Next, observe that for the detection of a stationary Gaussian signal in WGN, the parametrization (10.129) for PSDs on the geodesic linking S_0 and S_1 can be expressed as

$$\begin{aligned} S_u(e^{j\omega}) &= \frac{\sigma^2(S_Z(e^{j\omega}) + \sigma^2)}{(1-u)S_Z(e^{j\omega}) + \sigma^2} \\ &= S_{Z_u}(e^{j\omega}) + \sigma^2 \end{aligned} \tag{10.142}$$

with

$$S_{Z_u}(e^{j\omega}) = \frac{u\sigma^2 S_Z(e^{j\omega})}{\sigma^2 + (1-u)S_Z(e^{j\omega})} , \tag{10.143}$$

so $S_u(e^{j\omega})$ is the PSD of the observations $Y(t)$ for a hypothesis

$$H_u \; : \; Y(t) = Z_u(t) + V(t)$$

where the zero-mean Gaussian stationary signal $Z_u(t)$ is independent of $V(t)$ and its PSD S_{Z_u} is given by (10.143). Thus we can associate to the PSD S_u a hypothesis H_u in Gaussian signal plus WGN form, which connects continuously H_0 to H_1. To illustrate the previous discussion, we consider two examples:

Example 10.3: Bandlimited WGN detection

Consider first the case where the signal $Z(t)$ to be detected is a bandlimited WGN with PSD

$$S_Z(e^{j\omega}) = \begin{cases} P\pi/W & |\omega| \leq W \\ 0 & W < |\omega| \leq \pi \end{cases}$$

where $P = E[Z^2(t)]$ denotes the power of $Z(t)$. This model can sometimes be used as a rough approximation for a bandlimited process with power P and bandwidth W. The PSD $S_1(e^{j\omega}) = S_Z(e^{j\omega}) + \sigma^2$ of the observations under H_1 is sketched in Fig. 10.6.

Then, by substituting S_Z inside expression (10.143) for S_{Z_u}, we find that the process $Z_u(t)$ is also a bandlimited WGN of bandwidth W, but its power is now

$$P_u = \frac{uP}{1 + (1-u)P\pi/(W\sigma^2)} \, .$$

The signal variance, innovations variance, and inperpolation error variance satisfy respectively

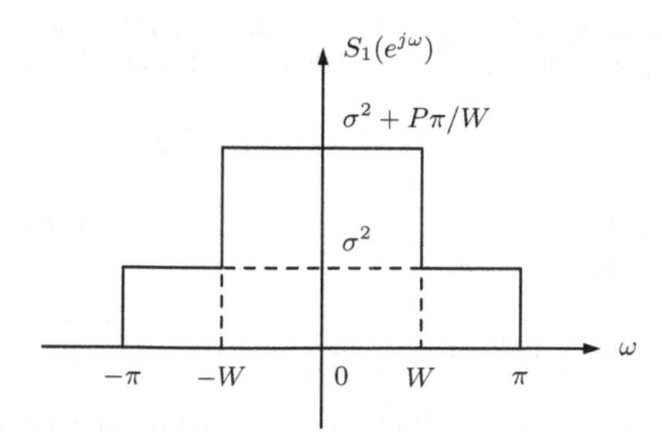

Fig. 10.6. Power spectral density of a bandlimited noise with bandwidth W observed in WGN.

$$P_Y = P + \sigma^2$$
$$\ln\left(\frac{P_\nu}{\sigma^2}\right) = \frac{W}{\pi}\ln\left(1 + \frac{P\pi}{W\sigma^2}\right)$$
$$P_{\text{int}}^{-1} = \frac{1}{\sigma^2}\left(1 - \frac{W}{\pi}\right) + \frac{W}{\pi}\frac{1}{\sigma^2 + P\pi/W}$$

which then yield

$$I(S_1, S_0) = \frac{W}{2\pi}\left[\frac{P\pi}{W\sigma^2} - \ln\left(1 + \frac{P\pi}{W\sigma^2}\right)\right]$$
$$I(S_0, S_1) = \frac{W}{2\pi}\left[-1 + \frac{1}{1 + P\pi/(W\sigma^2)} + \ln\left(1 + \frac{P\pi}{W\sigma^2}\right)\right]. \quad (10.144)$$

Similarly, $I_0(z) = I(S_{u(z)}, S_0)$ can be obtained by replacing P by P_u in the expression for $I(S_1, S_0)$.

It is instructive to examine the behavior of LRTs for the low SNR case with $P\pi/W\sigma^2 \ll 1$. In this case we have $P_u \approx uP$, and by using the Taylor series expansion

$$\ln(1 + x) \approx x - \frac{x^2}{2}$$

with $x = P\pi/W\sigma^2$ in (10.144) we find

$$I(S_1, S_0) \approx I(S_0, S_1) \approx \frac{P^2\pi}{4\sigma^4 W}. \quad (10.145)$$

So for low SNR, the probability of false alarm and of a miss for NP tests of type I and II decay like the *square* of the SNR P/σ^2. In the light of the observation that $P_u \approx uP$, a similar conclusion holds for all choices of the threshold γ. In contrast, for a known signal in WGN, the probability of a miss and false alarm are specified by (8.15). Setting $E = PT$, where P denotes the average signal power, letting T go to infinity, and using the approximation

$$Q(x) \approx \frac{1}{(2\pi)^{1/2}x}\exp(-x^2/2)$$

this implies that, for a known signal in WGN, as the length T of the detection interval becomes large, we have

$$\lim_{T\to\infty}\frac{1}{T}\ln P_F(T) = \lim_{T\to\infty}\frac{1}{T}\ln P_M(T) = -\frac{PT}{8\sigma^2},$$

and thus, for a known signal in WGN, the probabilities of false alarm and of a miss decay like the SNR. This indicates that at low SNR, the performance of a zero-mean Gaussian signal detector is much worse than the matched filter performance for a known signal in WGN. □

Example 10.4: Ornstein-Uhlenbeck process in WGN

Consider the case where the signal $Z(t)$ is an Ornstein-Uhlenbeck process with covariance function $E[Z(t)Z(s)] = Pa^{|t-s|}$, and PSD

$$S_Z(e^{j\omega}) = \frac{q}{(1+a^2) - 2a\cos(\omega)}$$

with $q = P(1-a^2)$. By substituting $S_Z(e^{j\omega})$ inside (10.143), we find that the PSD of $Z_u(t)$ takes the form

$$S_{Z_u}(e^{j\omega}) = \frac{u\sigma^2 q}{\sigma^2[(1+a^2) - 2a\cos(\omega)] + (1-u)q} ,$$

so $Z_u(t)$ is also an Ornstein-Uhlenbeck process. The variance of $Y(t)$, is given by

$$P_Y = P + \sigma^2 .$$

Since $Z(t)$ is Markov, the predicted estimate $\hat{Z}_p(t)$ can be evaluated by employing the Kalman filtering recursion

$$\hat{Z}_p(t+1) = a\hat{Z}_p(t) + K_p\nu(t) , \tag{10.146}$$

where

$$\nu(t) = Y(t) - \hat{Z}_p(t) \tag{10.147}$$

denotes the innovations process and

$$K_P = aP_p(P_p + \sigma^2)^{-1}$$

is the Kalman gain. In this expression P_p represents the variance of the prediction error

$$\tilde{Z}_p(t) = Z(t) - \hat{Z}_p(t) .$$

It is the unique positive solution of the algebraic Riccati equation

$$P_p = a^2[P_p^{-1} + \sigma^{-2}]^{-1} + q ,$$

which is given by

$$P_p = \frac{1}{2}\Big[q - (1-a^2)\sigma^2 + \big([(1-a^2)\sigma^2 - q]^2 + 4q\sigma^2\big)^{1/2}\Big] .$$

Noting that the innovations can be expressed as

$$\nu(t) = \tilde{Z}_p(t) + V(t) ,$$

where $\tilde{Z}_p(t)$ and $V(t)$ are independent, we obtain

$$P_\nu = P_p + \sigma^2 .$$

Rewriting the recursion (10.146) as

$$\hat{Z}_p(t+1) = b\hat{Z}_p(t) + K_p Y(t) \tag{10.148}$$

where

$$b = a - K_p = a\sigma^2(P_p + \sigma^2)^{-1}$$

we find from (10.147) and (10.148) that the whitening filter $H(z)$, which under H_1 produces the innovations process $\nu(t)$ at its output from the input $Y(t)$, is given by

$$H(z) = -\frac{K_p}{(z-b)} + 1 = \frac{1 - az^{-1}}{1 - bz^{-1}}$$

Since $\nu(t)$ is a WGN process with intensity P_ν, we have therefore

$$S_1^{-1}(z) = H(z)P_\nu^{-1}H(z^{-1}) = P_\nu^{-1}\frac{(1-az^{-1})(1-az)}{(1-bz^{-1})(1-bz)} . \tag{10.149}$$

Then, by noting that

$$S_1^{-1}(z) = P_\nu^{-1}((1+a^2) - a(z+z^{-1}))F(z) ,$$

where the inverse z transform of

$$F(z) = \frac{1}{(1-bz^{-1})(1-bz)}$$

is

$$f(t) = \frac{1}{1-b^2}b^{|t|} ,$$

by using the shift property of the z transform, we find

$$P_{\text{int}}^{-1} = \frac{(1+a^2) - 2ab}{(1-b^2)P_\nu} , \tag{10.150}$$

so P_{int} can be evaluated as a byproduct of the Kalman filter implementation. An interesting interpretation for P_{int} was proposed in [21]. Let

$$\bar{H}(z) = P_\nu^{-1/2}H(z) = P_\nu^{-1/2}\frac{1 - az^{-1}}{1 - bz^{-1}}$$

denote the normalized innovations filter. Under H_1, if the input to this filter is $Y(t)$, the ouput is the normalized innovations process $\bar{\nu}(t) = P_\nu^{-1/2}\nu(t)$, which is a WGN with unit intensity. But under H_0, if the input to this filter is $Y(t)$, since $Y(t) = V(t)$ is a WGN with intensity σ^2, the PSD of the output $M(t)$ of $\bar{H}(z)$ is

$$S_M(z) = \bar{H}(z)\sigma^2\bar{H}(z^{-1}) = \frac{\sigma^2}{S_1(z)} .$$

This implies that the power of the output $M(t)$ is

$$E[M^2(t)] = \frac{\sigma^2}{P_{\text{int}}} \,. \tag{10.151}$$

Consequently, P_{int} can be interpreted as the inverse power (scaled by σ^2) of the ouput of the normalized innovations filter excited by the mismatched input $V(t)$, where $V(t)$ corresponds to the observations under H_0. □

10.5 Bibliographical Notes

The optimal noncausal estimator/correlator, receiver of Section 10.2 was first proposed by Price [22] in the CT case and Middleton [23] in the DT case. This structure was extended by Shepp [24] to detection problems where in (10.1) the signal $Z(t)$ is correlated with the noise $V(t)$. The causal detector of Section 10.3 was discovered independently by Schweppe [1] and Stratonovich and Soluslin [2] for partially observed Gauss-Markov processes, and then extended to the case of nonGaussian signals in WGN by Duncan and Kailath [16, 25]. The insight that the causal detector can be best derived by adopting an inovations viewpoint is due to Kailath [26]. A detailed historical account of random processes detection is given in [17]. The asymptotic analysis of Gaussian LRTs that we have presented finds its origin in an extension of Chernoff's theorem to stationary Gaussian processes proposed by Coursol and Dacunha-Castelle in [27]. Bahr [28] subsequently recast this work in terms of the Gärtner-Ellis theorem, and extended it to Gaussian procesees with nonzero-mean. Sung et al. [21] considered the class of partially observed Gauss-Markov processes and showed that in this case, the Itakura-Saito spectral distortion measure governing the asymptotic performance of NP tests can be evaluated by Kalman filtering techniques.

10.6 Problems

10.1. Consider the DT detection problem

$$\begin{aligned} H_0 &: Y(t) = V(t) \\ H_1 &: Y(t) = X(t) + V(t) \end{aligned}$$

for $0 \leq t \leq T$, where $V(t)$ is a zero-mean WGN process with variance σ^2, and $X(t)$ is a DT Wiener process, so it satisfies the state-space model

$$X(t+1) = X(t) + W(t)$$

with $X(0) = 0$, where the driving noise $W(t)$ is a zero-mean WGN independent of $V(t)$, with variance q.

(a) Find an optimal noncausal detector of the form described in Section 10.2. Specify the Hamiltonian system (10.48) and the corresponding boundary conditions satisfied by the smoothed estimate $\hat{X}(t)$.

(b) Obtain a double-sweep implementation of the smoother of part (a). To do so, you should use a transformation of the form

$$\begin{bmatrix} \hat{X}(t) \\ \xi(t) \end{bmatrix} = \begin{bmatrix} \hat{X}(t) \\ \hat{X}(t) - S(t)\eta(t) \end{bmatrix}$$

to triangularize the Hamiltonian system (10.48). The transformation matrix $S(t)$ is selected such that the transformed Hamiltonian dynamics admit the triangular form

$$\begin{bmatrix} \xi(t+1) \\ \hat{X}(t) \end{bmatrix} = \begin{bmatrix} * & 0 \\ * & * \end{bmatrix} \begin{bmatrix} \xi(t) \\ \hat{X}(t+1) \end{bmatrix} + \begin{bmatrix} * \\ * \end{bmatrix} Y(t) ,$$

where the $*$'s indicate unspecified entries. Compare the equation satisfied by $S(t)$ with the equation for the predicted error variance $P_p(t)$ obtained by combining the Kalman filtering time and measurement updates (10.77) and (10.78).

10.2. Consider Problem 10.1. Instead of using the Hamiltonian system (10.48) to specify the smoothed estimate $\hat{X}(t)$, note (see Problem 7.5) that the covariance

$$E[X(t)X(s)] = K(t,s) = q \min(t,s)$$

satisfies the second-order difference equation

$$MK(t,s) = \delta(t-s)$$

with

$$M = q^{-1}[2 - (Z + Z^{-1})]$$

as well as the boundary conditions

$$K(0,s) = 0$$

and

$$\frac{1}{q}[K(T,s) - K(T-1,s)] = \delta(T-s)$$

(a) Use this result to obtain a second-order difference equation of the form (10.42)–(10.43) for the smoothed estimate $\hat{X}(t)$.
(b) By employing a factorization of the form (10.54) derive a double-sweep implementation of the smoother similar to the one described in Example 10.1 for the case of an Ornstein-Uhlenbeck process.

10.3. Consider the DT detection problem

$$\begin{aligned} H_0 &: Y(t) = V(t) \\ H_1 &: Y(t) = X(t) + V(t) \end{aligned}$$

for $0 \leq t \leq T$, where $V(t)$ is a zero-mean WGN process with variance σ^2, and $X(t)$ is a first-order Ornstein-Uhlenbeck process with state-space model

$$X(t+1) = aX(t) + W(t)$$

where $|a| < 1$. The driving noise $W(t)$ is a zero-mean WGN process independent of $X(0)$ and $V(t)$ with variance q, and the initial state $X(0)$ is $N(0, P)$ distributed, with

$$P = \frac{q}{1-a^2} .$$

This choice of initial state variance ensures that $X(t)$ is stationary with covariance function

$$E[X(t)X(s)] = K(t-s) = Pa^{|t-s|} .$$

If we consider the optimal detector

$$S \underset{H_0}{\overset{H_1}{\gtrless}} \gamma ,$$

it was shown in Section 10.3 that the sufficient statistic S and threshold γ can be evaluated causally by using identities (10.72) and

$$\gamma = \sigma^2 \ln(\tau) + \frac{\sigma^2}{2} \ln(D)$$

where $\ln(D)$ is given by (10.83).

For this problem, construct a DT Kalman filter of the form (10.76)–(10.79) to evaluate the optimal predicted and filtered estimates $\hat{X}_p(t)$ and $\hat{X}_f(t)$ of $X(t)$, and to compute the innovations variance $P_\nu(t)$ for $0 \leq t \leq T$. Use these quantities to obtain a causal implementation of the optimum detector.

10.4. In this problem, we derive identity (10.138) for the symmetric interpolation error variance. Let $Y(t)$ be a zero-mean stationary Gaussian process with covariance

$$K(k) = E[Y(t+k)Y(t)]$$

and power spectral density

$$S(e^{j\omega}) = \sum_{k=-\infty}^{\infty} K(k)e^{-jk\omega} .$$

Let

$$\begin{aligned} \hat{Y}_{\text{int}}(t) &= E[Y(t)|Y(t-k),\ k = \pm 1, \pm 2, \ldots] \\ &= \sum_{k \neq 0} h(k)Y(t-k) \end{aligned} \tag{10.152}$$

denote the mean-squares error interpolation estimate of $Y(t)$, and write the corresponding interpolation filter as

$$H(z) = \sum_{k \neq 0} h(k) z^{-k} .$$

If

$$\tilde{Y}_{\text{int}}(t) = Y(t) - \hat{Y}_{\text{int}}(t)$$

denotes the corresponding interpolation error, its variance is written as

$$P_{\text{int}} = E[\tilde{Y}_{\text{int}}^2(t)] .$$

(a) Use the orthogonality property $\tilde{Y}(t) \perp Y(t-k)$ for $k \neq 0$ of the least-squares estimate to show that the filter $h(k)$ satisfies equation

$$P_{\text{int}} \delta(k) = K(k) - \sum_{\ell \neq 0} h(\ell) K(k-\ell) , \tag{10.153}$$

where

$$\delta(k) = \begin{cases} 1 & k = 0 \\ 0 & k \neq 0 \end{cases}$$

denotes the Kronecker delta function.

(b) By z-transforming (10.153), verify that the interpolation filter $H(z)$ can be expressed as

$$H(z) = 1 - \frac{P_{\text{int}}}{S(z)} .$$

In this expression, P_{int} is still unknown, but by noting that the coefficient $h(0)$ of filter $H(z)$ corresponding to $k = 0$ must be zero, verify that

$$P_{\text{int}} = \left[\frac{1}{2\pi} \int_0^{2\pi} S^{-1}(e^{j\omega}) d\omega \right]^{-1} .$$

(c) By observing that the interpolation error process $\tilde{Y}_{\text{int}}(t)$ is obtained by passing $Y(t)$ through the filter

$$1 - H(z) = \frac{P_{\text{int}}}{S(z)} ,$$

conclude that $\tilde{Y}_{\text{int}}(t)$ has for power spectrum

$$S_{\tilde{Y}_{\text{int}}}(z) = \frac{P_{\text{int}}^2}{S(z)} .$$

Thus $S_{\tilde{Y}_{\text{int}}}(z)$ is proportional to the inverse of $S(z)$, and for this reason, $\tilde{Y}_{\text{int}}(t)$ is sometimes called the *conjugate process* of $Y(t)$.

10.5. Consider a DT detection problem where under H_0 the observed process $Y(t)$ with $0 \leq t \leq T$ is a zero-mean WGN $V(t)$ with variance σ^2, and under H_1 it is a first-order moving average process given by

$$Y(t) = V(t) - bV(t-1)$$

with $|b| < 1$, where $V(t)$ is again a zero-mean WGN process with variance σ^2. Thus under H_1, $Y(t)$ is a zero-mean stationary Gaussian process with power spectrum

$$S_1(z) = (1 - bz^{-1})(1 - bz)\sigma^2$$

and its variance is given by

$$P_Y = (1 + b^2)\sigma^2 .$$

For this detection problem, we seek to evaluate the exponential decay rates (10.123) and (10.125) of NP tests of type I and II as T becomes large. This requires evaluating identities (10.140) and (10.141) for the Itakura-Saito (IS) spectral distortion measures.

(a) By observing that under H_1, the innovations process $\nu(t) = V(t)$ is obtained by passing the observations $Y(t)$ through the causal stable whitening filter

$$H(z) = \frac{1}{1 - bz^{-1}} ,$$

evaluate the innovations variance P_ν.

(b) By evaluating the inverse z-transform $f(t)$ of $F(z) = S_1^{-1}(z)$, find $f(0) = P_{\text{int}}^{-1}$.

(c) Use the results of parts (a) and (b) to evaluate $I(S_1, S_0)$ and $I(S_0, S_1)$.

10.6. Consider a DT Gaussian detection problem, where under H_0,

$$Y(t) = V(t)$$

is a zero-mean WGN with intensity σ^2, whereas under H_1, $Y(t)$ admits an autoregressive model of order p of the form

$$Y(t) + \sum_{k=1}^{p} a_k Y(t-k) = V(t) ,$$

where the polynomial

$$a(z) = 1 + \sum_{k=1}^{p} a_k z^{-k}$$

is stable, i.e., all its zeros are inside the unit circle, and $V(t)$ is a zero-mean WGN with variance σ^2. For this problem, we seek to evaluate the IS distortion measures $I(S_1, S_0)$ and $I(S_0, S_1)$ characterizing the asymptotic decay rates of the probabilities of a miss and of false alarm for NP tests of type I and II, respectively.

(a) Verify that $S_0(z) = \sigma^2$ and

$$S_1(z) = \frac{\sigma^2}{a(z)a(z^{-1})} .$$

Also, by observing that under H_1 the predicted estimate

$$\begin{aligned}\hat{Y}_p(t) &= E[Y(t)|Y(s),\, s \leq t-1] \\ &= -\sum_{k=1}^{p} a_k Y(t-k) ,\end{aligned} \tag{10.154}$$

conclude that the innovations process

$$\nu(t) = Y(t) - \hat{Y}_p(t) = V(t) ,$$

which implies $P_\nu = E[V^2(t)] = \sigma^2$.

(b) Evaluate $S_1^{-1}(z)$ and P_{int}.

(c) To evaluate $P_Y = E[Y^2(t)|H_1]$, we can employ the following procedure proposed in [29]. Write as

$$K_1(k) = E[Y(t+k)Y(t)|H_1]$$

the autocovariance function of $Y(t)$ under H_1, where $P_Y = K_1(0)$. By using the orthogonality property $E[\nu(t)Y(t-k)] = 0$ for $k \geq 1$ of the innovations process, verify that the vector

$$\mathbf{a} = \begin{bmatrix} 1 \; a_1 \; a_2 \; \cdots \; a_p \end{bmatrix}^T$$

satisfies the Yule-Walker equation

$$\mathbf{K}_1 \mathbf{a} = \sigma^2 \mathbf{e} , \tag{10.155}$$

where

$$\mathbf{K}_1 = \begin{bmatrix} K_1(0) & K_1(1) & \dots & & K_1(p) \\ K_1(1) & K_1(0) & K_1(1) & \dots & K_1(p-1) \\ & \ddots & \ddots & \ddots & \\ K_1(p-1) & \dots & K_1(1) & K_1(0) & K_1(1) \\ K_1(p) & & \dots & K_1(1) & K_1(0) \end{bmatrix}$$

is the $(p+1) \times (p+1)$ covariance matrix of $Y(t)$ under H_1, and

$$\mathbf{e} = \begin{bmatrix} 1 \; 0 \; 0 \cdots 0 \end{bmatrix}^T$$

denotes the first unit vector of $\mathbb{R}^{p+1}$. Show that equation (10.155) can be rewritten as

$$(\mathbf{A} + \mathbf{B})\mathbf{c} = \sigma^2 \mathbf{e} \tag{10.156}$$

with

$$\mathbf{A} = \begin{bmatrix} 1 & 0 & 0 & \dots & 0 \\ a_1 & 1 & 0 & \dots & 0 \\ & \ddots & \ddots & & 0 \\ a_{p-1} & \dots & a_1 & 1 & 0 \\ a_p & a_{p-1} & \dots & a_1 & 1 \end{bmatrix},$$

$$\mathbf{B} = \begin{bmatrix} 1 & a_1 & a_2 & \dots & a_p \\ a_1 & a_2 & \dots & a_p & 0 \\ a_2 & \dots & a_p & 0 & 0 \\ \vdots & & & & \vdots \\ a_p & 0 & 0 & \dots & 0 \end{bmatrix},$$

and

$$\mathbf{c} = \left[K_1(0)/2 \; K_1(1) \; \dots \; K_1(p-1) K_1(p) \right]^T .$$

Conclude from (10.156) that

$$P_Y = K_1(0) = 2\sigma^2 \mathbf{e}^T (\mathbf{A} + \mathbf{B})^{-1} \mathbf{e} .$$

(d) Use the results of parts (a) to (c) to evaluate $I(S_1, S_0)$ and $I(S_0, S_1)$.

10.7. As in Example 10.3, consider the DT Gaussian detection problem

$$\begin{aligned} H_0 &: Y(t) = V(t) \\ H_1 &: Y(t) = Z(t) + V(t) \end{aligned}$$

for $0 \leq t \leq T$, where $V(t)$ is a zero-mean WGN with variance σ^2, and $Z(t)$ is a zero-mean stationary bandlimited WGN independent of $V(t)$ with PSD

$$S_Z(e^{j\omega}) = \begin{cases} q & |\omega| \leq W \\ 0 & W < |\omega| \leq \pi . \end{cases}$$

Assume that T is large and that the bandwidth W is an integer multiple of the resolution $2\pi/(T+1)$ of the DFT of length $T+1$, i.e., $W = m2\pi/(T+1)$ with m integer. Then for large T, the optimal Bayesian test takes the form

$$S_T \underset{H_0}{\overset{H_1}{\gtrless}} 0$$

where the sufficient statistic S_T specified by (10.95) admits approximation (10.107).

(a) Verify that this test can be rewritten as

$$U_T \underset{H_0}{\overset{H_1}{\gtrless}} \eta_T$$

where the sufficient statistic

$$U_T \triangleq \frac{1}{(T+1)} \sum_{k=-m}^{m} |\hat{Y}(e^{j\omega_k})|^2 .$$

In this expression, $\hat{Y}(e^{j\omega_k})$ is the $k+1$-th entry of the DFT vector specified by (10.105). Evaluate the threshold η_T.

(b) It turns out that U_T measures the power of the signal $Y(t)$, $0 \le t \le T$ in the frequency band $[-W, W]$. To verify this fact, observe that for large T, U_T admits the approximation

$$U_T \approx \frac{1}{2\pi(T+1)} \int_{-W}^{W} |\bar{Y}(e^{j\omega})|^2 d\omega ,$$

where $\bar{Y}(e^{j\omega})$ denotes the DTFT of signal $Y(t)$ with $0 \le t \le T$. Let

$$Q(e^{j\omega}) = \begin{cases} 1 & |\omega| \le W \\ 0 & \text{otherwise} \end{cases}$$

denote the ideal lowpass filter which passes frequencies over the band $[-W, W]$. If $q(t)$ represents its impulse response, we can write as

$$R(t) = q(t) * Y(t)$$

the low-passed version of $Y(t)$, where $*$ denotes the convolution operation. Show that

$$\begin{aligned} U_T &= \frac{1}{2\pi(T+1)} \int_{-\pi}^{\pi} |\bar{R}(e^{j\omega})|^2 d\omega \\ &= \frac{1}{T+1} \sum_{t=0}^{T} |R(t)|^2 , \end{aligned}$$

where $\bar{R}(e^{j\omega})$ denotes the DTFT of $R(t)$, and conclude that U_T represents the average power of $R(t)$.

(c) Find the limit of the threshold η_T as T tends to infinity, and evaluate it in terms of W, q and σ^2.

References

1. F. C. Schweppe, "Evaluation of likelihood functions for Gaussian signals," *IEEE Trans. Informat. Theory*, vol. 11, pp. 61–70, Jan. 1965.
2. R. L. Stratonovich and Y. G. Sosulin, "Optimal detection of a diffusion process in white noise," *Radio Eng. Electron. Phys.*, vol. 10, pp. 704–713, June 1965.
3. A. Dembo and O. Zeitouni, *Large Deviations Techniques and Applications, Second Edition.* New York: Springer Verlag, 1998.
4. F. den Hollander, *Large Deviations.* Providence, RI: American Mathematical Soc., 2000.

5. R. M. Gray, A. Buzo, A. H. Gray, Jr., and Y. Matsuyama, "Distortion measures for speech processing," *IEEE Trans. Acoust., Speech, Signal Proc.*, vol. 28, pp. 367–376, Aug. 1980.
6. B. C. Levy, R. Frezza, and A. J. Krener, "Modeling and estimation of discrete-time Gaussian reciprocal processes," *IEEE Trans. Automatic Control*, vol. 35, pp. 1013–1023, Sept. 1990.
7. T. Kailath, A. H. Sayed, and B. Hassibi, *Linear Estimation.* Upper Saddle River, NJ: Prentice Hall, 2000.
8. H. L. Weinert, *Fixed Interval Smoothing for State Space Models.* Boston, MA: Kluwer Publ., 2001.
9. C. D. Greene and B. C. Levy, "Some new smoother implementations for discrete-time Gaussian reciprocal processes," *Int. J. Control*, pp. 1233–1247, 1991.
10. T. Kailath, S. Kung, and M. Morf, "Displacement ranks of matrices and linear equations," *J. Math. Anal. and Appl.*, vol. 68, pp. 395–407, 1979.
11. T. Kailath and A. H. Sayed, "Displacement structure: theory and applications," *SIAM Review*, vol. 37, pp. 297–386, Sept. 1995.
12. M. H. Hayes, *Statistical Digital Signal Processing and Modeling.* New York: J. Wiley & Sons, 1996.
13. P. A. Ruymgaart and T. T. Soong, *Mathematics of Kalman-Bucy Filtering, Second Edition.* Berlin: Springer-Verlag, 1988.
14. E. Wong and B. Hajek, *Stochastic Processes in Engineering Systems.* New York: Springer-Verlag, 1985.
15. L. C. G. Rogers and D. Williams, *Diffusions, Markov Processes, and Martingales, Vol. 2: Ito Calculus.* Chichester, England: J. Wiley & Sons, 1987.
16. T. Kailath, "A general likelihood-ratio formula for random signals in Gaussian noise," *IEEE Trans. Informat. Theory*, vol. 15, pp. 350–361, May 1969.
17. T. Kailath and H. V. Poor, "Detection of stochastic processes," *IEEE Trans. Informat. Theory*, vol. 44, pp. 2230–2259, Oct. 1998.
18. B. Oksendal, *Stochastic Differential Equations: An Introduction with Applications, Sixth Edition.* Berlin: Springer Verlag, 2003.
19. R. M. Gray, "Toeplitz and circulant matrices: A review." Technical report available on line at: http://www-ee.stanford.edu/ gray/toeplitz.html, Aug. 2005.
20. S.-I. Amari and H. Nagaoka, *Methods of Information Geometry.* Providence, RI: American Mathematical Soc., 2000.
21. Y. Sung, L. Tong, and H. V. Poor, "Neyman-Pearson detection of Gauss-Markov signals in noise: closed form error exponent and properties," *IEEE Trans. Informat. Theory*, vol. 52, pp. 1354–1365, Apr. 2006.
22. R. Price, "Optimum detection of random signals in noise, with application to scatter-multipath communications, I," *IEEE Trans. Informat. Theory*, vol. 2, pp. 125–135, Dec. 1956.
23. D. Middleton, "On the detection of stochastic signals in additive normal noise–part I," *IEEE Trans. Informat. Theory*, vol. 3, pp. 86–121, June 1957.
24. L. A. Shepp, "Radon-Nikodym derivatives of Gaussian measures," *Annals Math. Stat.*, vol. 37, pp. 321–354, Apr. 1966.
25. T. E. Duncan, "Likelihood functions for stochastic signals in white noise," *Information and Control*, vol. 16, pp. 303–310, June 1970.
26. T. Kailath, "The innovations approach to detection and estimation theory," *Proc. IEEE*, vol. 58, pp. 680–695, May 1970.

27. J. Coursol and D. Dacunha-Castelle, "Sur la formule de Chernoff pour deux processus Gaussiens stationnaires," *Comptes Rendus Académie des Sciences, Paris*, vol. 288, pp. 769–770, May 1979.
28. R. K. Bahr, "Asymptotic analysis of error probabilities for the nonzero-mean Gaussian hypothesis testing problem," *IEEE Trans. Informat. Theory*, vol. 36, pp. 597–607, May 1990.
29. B. Porat and B. Friedlander, "Parametric techniques for adaptive detection of Gaussian signals," *IEEE Trans. Acoust., Speech, Signal Proc.*, vol. 32, pp. 780–790, Aug. 1984.

11

EM Estimation and Detection of Gaussian Signals with Unknown Parameters

11.1 Introduction

In the last chapter, it was assumed that the statistics of the Gaussian signal to be detected and the variance of the additive WGN are known exactly. This assumption is rather unrealistic since in practice, even if a statistical model of the signal to be detected is available, it is quite common that some of its parameters are unknown. For such problems, if we attempt to implement a GLRT detector by evaluating the ML estimate of the signal and noise parameters under each hypothesis, we encounter a problem that has not occurred until now. Specifically, whereas for the detection of a known signal in WGN, it is usually possible to solve the ML estimation equation in closed form, this is rarely the case for Gaussian signals in WGN. Since the GLRT detector assumes the availability of ML estimates, one must therefore attempt to solve the ML equation iteratively.

While the ML estimation equation can be solved iteratively by applying Newton's method or any one of its variants [1, Chap. 1], a systematic and elegant scheme for evaluating ML estimates is provided by the *Expectation-Maximization* (EM) method of Dempster, Laird and Rubin [2]. The key idea behind this procedure is rather simple: whenever the maximization of the likelihood function becomes untractable, we can postulate that the available data is only part of a larger data set, called the complete data, where the missing data, representing the difference between the complete and available data, is selected judiciously to ensure that the maximization of the complete data likelihood function over the unknown parameters can be performed easily. Then the EM method is implemented as an iteration consisting of two phases, where in the expectation phase (E-phase), the conditional density of the missing data, given the actual observations, is estimated based on the current values of the unknown parameters and is used to evaluate the expected value of the complete log-likelihood function. Next, a maximization phase (M-phase) finds the maximum of the estimated complete log-likelihood function with respect to the unknown system parameters, yielding new estimates of

B.C. Levy, *Principles of Signal Detection and Parameter Estimation*,
DOI: 10.1007/978-0-387-76544-0_11, © Springer Science+Business Media, LLC 2008

the parameters that can be used in the E-phase of the next iteration. Since its introduction, the EM method has been studied in detail. A comprehensive discussion of its properties and variants is given in [3]. For our purposes the two most important features of the EM method are first that it converges to a local maximum of the likelihood function [4], but not necessarily to a global maximum. Accordingly, proper initialization of the EM method is important, so it is desirable to have a rough guess of the values of the system parameters as a starting point. Otherwise, application of the EM method with multiple start points is recommended. The second important aspect of the EM method is that the selection of the "augmented data" is rather flexible. The key goal of data augmentation is to ensure that the maximization of the complete log-likelihood function is easy to perform. Ideally, it shoud admit a closed form solution. However, it has been observed [5] that, as the dimension of the missing data increases, the convergence of the EM algorithm slows down. So, one should try to augment the given data just enough to allow an easy maximization of the likelihood function. In this respect, it is useful to note that there exists variants of the EM algorithm called the space-alternating generalized EM (SAGE) algorithm [6] or the alternating expectation-conditional maximization (AECM) method [5] which perform a partial maximization of the complete likelihood function over different components of the full parameter vector. These methods enjoy all the convergence properties of the EM algorithm as long as successive iterations cycle over all components of the parameter vector, and as long as the subspaces spanned by the different component-wise maximizations span the entire space. These methods are attractive when the simultaneous maximization of the complete likelihood function over all parameters is difficult to accomplish, but component-wise maximization is easy to achieve.

For the detection of Gaussian signals with unknown parameters, the selection of the augmentation data relies on the observation that the maximization of the likelihood function for the detection of *known signals* with unknown parameters is usually easy to perform. So, if under hypothesis H_1, we observe a discrete-time signal

$$Y(t) = Z(t, \boldsymbol{\theta}) + V(t)$$

for $0 \leq t \leq T$, where $Z(t, \boldsymbol{\theta})$ is a zero-mean Gaussian signal depending on the parameter vector $\boldsymbol{\theta}$ and $V(t)$ is a zero-mean WGN with variance v, we can select as missing data the entire waveform $Z(t, \boldsymbol{\theta})$ for $0 \leq t \leq T$, since this has the effect of converting the Gaussian detection problem with unknown parameters into a detection problem for a known signal with unknown parameters of the type discussed in Chapter 9.

The chapter is organized as follows: Section 11.2 examines the case where a Gaussian signal of unknown amplitude is observed in a WGN of unknown power. For this problem, it is shown that the ML equation cannot be solved in closed form. The EM parameter estimation method is described in Section 11.3 and then applied to the amplitude and noise variance estimation example of the previous section. The EM algorithm requires the evaluation of the smoothed estimate and error covariance matrix for the entire signal record, and as such it has a high computational complexity. It is shown in Section 11.4 that this complexity can be reduced drastically when the observed signal is a hidden Gauss-Markov model, i.e., a partially observed Gauss-Markov process with unknown parameters, since in this case, Kalman filtering techniques can be used to evaluate the quantities needed for the EM iteration. Finally, the EM implementation of the ML method is used to design GLRT detectors in Section 11.5.

11.2 Gaussian Signal of Unknown Amplitude in WGN of Unknown Power

To illustrate the difficulty associated with the evaluation of ML estimates for the system parameters of Gaussian signals, consider the DT observations

$$Y(t) = AX(t) + V(t) \tag{11.1}$$

for $0 \leq t \leq T$, where $X(t)$ is a zero-mean Gaussian process with covariance function

$$E[X(t)X(s)] = K(t,s)$$

for $0 \leq t,\ s \leq T$ and $V(t)$ is a zero-mean WGN with variance v. The amplitude A of the received signal is unknown, and the variance v of the noise $V(t)$ may or may not be known.

Let

$$\mathbf{Y} = \begin{bmatrix} Y(0) \\ \vdots \\ Y(t) \\ \vdots \\ Y(T) \end{bmatrix} \qquad \mathbf{X} = \begin{bmatrix} X(0) \\ \vdots \\ X(t) \\ \vdots \\ X(T) \end{bmatrix} \qquad \mathbf{V} = \begin{bmatrix} V(0) \\ \vdots \\ V(t) \\ \vdots \\ V(T) \end{bmatrix} \tag{11.2}$$

be the $N = T+1$ dimensional vectors representing the observations, signal and noise. The vectors $\mathbf{X}$ and $\mathbf{V}$ are $N(\mathbf{0}, \mathbf{K})$ and $N(\mathbf{0}, v\mathbf{I}_N)$ distributed, where $\mathbf{K}$ is the $N \times N$ covariance matrix such that the entry of row i and column j is $K(i-1, j-1)$. Let

$$\boldsymbol{\theta} \triangleq \begin{bmatrix} A \\ v \end{bmatrix}$$

denote the vector of unknown parameters. Then, since the observation equation (11.1) can be rewritten in vector form as

$$\mathbf{Y} = A\mathbf{X} + \mathbf{V}\,, \tag{11.3}$$

$\mathbf{Y}$ is $N(\mathbf{0}, A^2\mathbf{K} + v\mathbf{I}_N)$ distributed and the logarithm of its density can be expressed as

$$L(\mathbf{y}|\boldsymbol{\theta}) = \ln f(\mathbf{y}|\boldsymbol{\theta}) = -\frac{1}{2}\mathbf{y}^T(A^2\mathbf{K} + v\mathbf{I}_N)^{-1}\mathbf{y} - \frac{1}{2}\ln(|A^2\mathbf{K} + v\mathbf{I}_N|)\,. \tag{11.4}$$

Note that $f(\mathbf{y}|\boldsymbol{\theta})$ depends only on the magnitude $|A|$ of the amplitude A. This is due to the fact that $\mathbf{X}$ and $-\mathbf{X}$ have the same Gaussian distribution. Thus, only the magnitude of A can be estimated, but not its sign.

Then the gradient

$$\nabla L(\mathbf{y}|\boldsymbol{\theta}) = \begin{bmatrix} \frac{\partial L}{\partial A}(\mathbf{y}|\boldsymbol{\theta}) \\ \frac{\partial L}{\partial v}(\mathbf{y}|\boldsymbol{\theta}) \end{bmatrix}$$

of the likelihood function with respect to parameter vector $\boldsymbol{\theta}$ is specified by

$$\frac{\partial L}{\partial A}(\mathbf{y}|\boldsymbol{\theta}) = A\Big[\mathbf{y}^T(A^2\mathbf{K}+v\mathbf{I}_N)^{-1}\mathbf{K}(A^2\mathbf{K}+v\mathbf{I}_N)^{-1}\mathbf{y} - \mathrm{tr}\left((A^2\mathbf{K}+v\mathbf{I}_N)^{-1}\mathbf{K}\right)\Big] \tag{11.5}$$

$$\frac{\partial L}{\partial v}(\mathbf{y}|\boldsymbol{\theta}) = \frac{1}{2}\Big[\mathbf{y}^T(A^2\mathbf{K}+v\mathbf{I}_N)^{-2}\mathbf{y} - \mathrm{tr}\left((A^2\mathbf{K}+v\mathbf{I}_N)^{-1}\right)\Big]\,. \tag{11.6}$$

To derive expressions (11.5) and (11.6) we have employed the matrix differentiation identities [7, Chap. 8]

$$\frac{d}{dx}\mathbf{M}^{-1}(x) = -\mathbf{M}^{-1}(x)\frac{d\mathbf{M}}{dx}(x)\mathbf{M}^{-1}(x)$$

$$\frac{d}{dx}\ln\left(|\mathbf{M}(x)|\right) = \mathrm{tr}\left[\mathbf{M}^{-1}(x)\frac{d\mathbf{M}}{dx}(x)\right],$$

where $\mathbf{M}(x)$ is a square invertible matrix depending on the scalar parameter x. Then it is easy to verify that the ML equation

$$\nabla L(\mathbf{y}|\boldsymbol{\theta}) = 0 \tag{11.7}$$

cannot be solved in closed form, and, in fact, if either v or A is known, the separate equations

$$\frac{\partial L}{\partial A}(\mathbf{y}|\boldsymbol{\theta}) = 0 \quad , \quad \frac{\partial L}{\partial v}(\mathbf{y}|\boldsymbol{\theta}) = 0$$

cannot be solved either. To tackle the ML estimation of A, or v, or both, some new tools are obviously needed.

11.3 EM Parameter Estimation Method

The key idea of the EM method is to expand the available data vector $\mathbf{Y}$ into a larger complete data vector

$$\mathbf{Z} = \begin{bmatrix} \mathbf{Y} \\ \mathbf{X} \end{bmatrix} \tag{11.8}$$

where $\mathbf{X}$ is called the missing data vector. The augmentation is performed in such a way that the probability density $f_c(\mathbf{z}|\boldsymbol{\theta})$ of the complete data satisfies

$$f(\mathbf{y}|\boldsymbol{\theta}) = \int f_c(\mathbf{z}|\boldsymbol{\theta}) d\mathbf{x} , \tag{11.9}$$

i.e., the density $f(\mathbf{y}|\boldsymbol{\theta})$ of the actual data is the marginal obtained by integrating out the missing data $\mathbf{x}$ from the complete data density $f_c(\mathbf{z}|\boldsymbol{\theta})$. One requirement of the augmentation scheme is that the maximization of $f_c(\mathbf{Z}|\boldsymbol{\theta})$, or equivalently, of the complete data log-likelihood

$$L_c(\mathbf{Z}|\boldsymbol{\theta}) \triangleq \ln f_c(\mathbf{Z}|\boldsymbol{\theta}) , \tag{11.10}$$

over $\boldsymbol{\theta}$ should be easy to accomplish. Of course, the log-likelihood function $L_c(\mathbf{z}|\boldsymbol{\theta})$ cannot be evaluated, since the missing data component $\mathbf{X}$ of $\mathbf{Z}$ is unknown. So instead of evaluating $L_c(\mathbf{Z}|\boldsymbol{\theta})$ we estimate it. Given the observations $\mathbf{Y} = \mathbf{y}$ and an earlier estimate $\boldsymbol{\theta}'$ of the parameter vector $\boldsymbol{\theta}$, this is accomplished by first obtaining the conditional density

$$f_c(\mathbf{x}|\mathbf{y}, \boldsymbol{\theta}') = \frac{f_c(\mathbf{z}|\boldsymbol{\theta}')}{f(\mathbf{y}|\boldsymbol{\theta}')} \tag{11.11}$$

and then using it to evaluate the estimated complete log-likelihood function

$$\begin{aligned} Q(\boldsymbol{\theta}, \boldsymbol{\theta}') &= E[L_c(\mathbf{Z}|\boldsymbol{\theta})|\mathbf{y}, \boldsymbol{\theta}'] \\ &= \int L_c(\mathbf{z}|\boldsymbol{\theta}) f_c(\mathbf{x}|\mathbf{y}, \boldsymbol{\theta}') d\mathbf{x} . \end{aligned} \tag{11.12}$$

EM iteration: Assume that the unknown parameter vector $\boldsymbol{\theta}$ belongs to $\mathbb{R}^q$, and denote it by

$$L(\mathbf{y}|\boldsymbol{\theta}) = \ln f(\mathbf{y}|\boldsymbol{\theta}) \tag{11.13}$$

the true log-likelihood function. Let

$$\hat{\boldsymbol{\theta}}_{\mathrm{ML}}(\mathbf{y}) = \arg \max_{\boldsymbol{\theta} \in \mathbb{R}^q} L(\mathbf{y}|\boldsymbol{\theta}) \tag{11.14}$$

be the ML estimate of $\boldsymbol{\theta}$ based on the observation $\mathbf{Y} = \mathbf{y}$. Instead of attempting to compute $\hat{\boldsymbol{\theta}}_{\mathrm{ML}}$ directly, the EM method applies the following iteration:

(a) Initialization: Select an initial estimate $\boldsymbol{\theta}^0$ of $\hat{\boldsymbol{\theta}}_{\mathrm{ML}}$ and let K be the maximum number of iterations allowed. Set $k = 0$.
(b) Expectation phase: Evaluate the conditional density $f_c(\mathbf{x}|\mathbf{y}, \boldsymbol{\theta}^k)$ for the missing data based on the observations $\mathbf{Y} = \mathbf{y}$ and the current parameter vector estimate $\boldsymbol{\theta}^k$ and use it to obtain the estimated complete log-likelihood function

$$Q(\boldsymbol{\theta}, \boldsymbol{\theta}^k) = E[L_c(\mathbf{Z}|\boldsymbol{\theta})|\mathbf{y}, \boldsymbol{\theta}^k] . \tag{11.15}$$

(c) Maximization phase: Evaluate

$$\boldsymbol{\theta}^{k+1} = \arg \max_{\boldsymbol{\theta} \in \mathbb{R}^q} Q(\boldsymbol{\theta}, \boldsymbol{\theta}^k) . \tag{11.16}$$

(d) Looping decision: If $k+1 < K$ and $||\boldsymbol{\theta}^{k+1} - \boldsymbol{\theta}^k||^2 > \epsilon$, where ϵ is a preset tolerance parameter, set $k = k+1$, and return to step (b). Otherwise, terminate the iteration.

11.3.1 Motonicity Property

The EM iteration has the feature that the log-likelihood function increases motonically at successive iterates $\boldsymbol{\theta}^k$ of the parameter vector, i.e.,

$$L(\mathbf{y}|\boldsymbol{\theta}^{k+1}) \geq L(\mathbf{y}|\boldsymbol{\theta}^k) . \tag{11.17}$$

To establish this property, we introduce the function

$$\begin{aligned} H(\boldsymbol{\theta}, \boldsymbol{\theta}') &\stackrel{\triangle}{=} -E[\ln f_c(\mathbf{X}|\mathbf{y}, \boldsymbol{\theta})|\mathbf{y}, \boldsymbol{\theta}'] \\ &= -\int \ln f_c(\mathbf{x}|\mathbf{y}, \boldsymbol{\theta}) f_c(\mathbf{x}|\mathbf{y}, \boldsymbol{\theta}') d\mathbf{x} . \end{aligned} \tag{11.18}$$

Note that

$$\begin{aligned} H(\boldsymbol{\theta}, \boldsymbol{\theta}') - H(\boldsymbol{\theta}', \boldsymbol{\theta}') &= \int \ln \Big(\frac{f_c(\mathbf{x}|\mathbf{y}, \boldsymbol{\theta}')}{f_c(\mathbf{x}|\mathbf{y}, \boldsymbol{\theta})} \Big) f_c(\mathbf{x}|\mathbf{y}, \boldsymbol{\theta}') d\mathbf{x} \\ &= D(\boldsymbol{\theta}'|\boldsymbol{\theta}), \end{aligned} \tag{11.19}$$

where $D(\boldsymbol{\theta}', \boldsymbol{\theta})$ denotes the Kullback-Leibler divergence of the missing data conditional densities $f_c(\mathbf{x}|\mathbf{y}, \boldsymbol{\theta}')$ and $f_c(\mathbf{x}|\mathbf{y}, \boldsymbol{\theta})$. Since the KL divergence is always nonnegative, we conclude that

$$H(\boldsymbol{\theta}, \boldsymbol{\theta}') - H(\boldsymbol{\theta}', \boldsymbol{\theta}') \geq 0 \tag{11.20}$$

for all pairs of vectors $\boldsymbol{\theta}'$, $\boldsymbol{\theta}$ in $\mathbb{R}^q$.

By observing that

$$\begin{aligned} Q(\boldsymbol{\theta}, \boldsymbol{\theta}') + H(\boldsymbol{\theta}, \boldsymbol{\theta}') &= \int \ln f(\mathbf{y}|\boldsymbol{\theta}) f_c(\mathbf{x}|\mathbf{y}, \boldsymbol{\theta}') d\mathbf{x} \\ &= L(\mathbf{y}|\boldsymbol{\theta}) , \end{aligned} \tag{11.21}$$

we find that for two successive EM iterates

$$\begin{aligned} L(\mathbf{y}|\boldsymbol{\theta}^{k+1}) - L(\mathbf{y}|\boldsymbol{\theta}^k) &= Q(\boldsymbol{\theta}^{k+1}|\boldsymbol{\theta}^k) - Q(\boldsymbol{\theta}^k, \boldsymbol{\theta}^k) \\ &\quad + H(\boldsymbol{\theta}^{k+1}, \boldsymbol{\theta}^k) - H(\boldsymbol{\theta}^k, \boldsymbol{\theta}^k) . \end{aligned} \tag{11.22}$$

Consequently, taking into account (11.20) and observing from (11.16) that

$$Q(\boldsymbol{\theta}^{k+1}, \boldsymbol{\theta}^k) \geq Q(\boldsymbol{\theta}^k, \boldsymbol{\theta}^k) \tag{11.23}$$

we conclude that (11.17) is satisfied. If we assume that $L(\mathbf{y}|\boldsymbol{\theta}^k)$ is bounded from above, this implies that the sequence $\{L(\mathbf{y}|\boldsymbol{\theta}_k),\ k \geq 0\}$ converges as $k \to \infty$. Note, however, that for practical purposes, we truncate the number of iterations to a finite number K. Although the convergence of the likelihood values does not by itself ensure the convergence of the iterates $\boldsymbol{\theta}^k$, under relatively mild smoothness conditions for the log-likelihood function $L(\mathbf{y}|\boldsymbol{\theta})$, it was shown by Wu [4] that the sequence $\{\boldsymbol{\theta}^k, k \geq 0\}$ converges to a local maximum of $L(\mathbf{y}|\boldsymbol{\theta})$. However, it is worth noting that even if we start with an initial iterate located close to a local maximum of $L(\mathbf{y}|\boldsymbol{\theta})$, the EM convergence result does not explicitly rule out the possibility that at some point the EM iterates might jump to the vicinity of another local maximum of L. Fortunately, for signal estimation problems, it turns out that the EM iteration can often be rewritten as a quasi-Newton iteration, for which it is then possible to show that if we consider a local maximum $\boldsymbol{\theta}_*$ of the log-likelihood function, $\boldsymbol{\theta}_*$ admits a capture set such that if the initial iterate $\boldsymbol{\theta}^0$ is located within the capture set, then the EM iteration necessarily converges to $\boldsymbol{\theta}_*$.

11.3.2 Example

To illustrate the application of the EM method, we analyze in detail a special case.

Example 11.1: Signal Amplitude and/or Noise Variance Estimation

Consider the signal amplitude and noise variance estimation problem described in Section 11.2. We select as missing data the signal $X(t)$ for $t \leq t \leq T$, or equivalently the vector $\mathbf{X}$ of (11.2). This selection transforms the parameter estimation problem into one involving a known signal with unknown amplitude A in a WGN of unknown variance v. To verify that this choice of missing data meets our goal of ensuring that the complete likelihood function is easy to minimize, note that

$$f_c(\mathbf{z}|\boldsymbol{\theta}) = f_c(\mathbf{y}|\mathbf{x}, \boldsymbol{\theta}) f_c(\mathbf{x}) \tag{11.24}$$

where, based on the observation model (11.3) and the assumption that $\mathbf{X}$ is $N(\mathbf{0}, \mathbf{K})$ distributed, we have

$$\begin{aligned} f_c(\mathbf{y}|\mathbf{x}, \boldsymbol{\theta}) &= \frac{1}{(2\pi v)^{N/2}} \exp\Big(-\frac{1}{2v} ||\mathbf{y} - A\mathbf{x}||_2^2 \Big) \\ f_c(\mathbf{x}) &= \frac{1}{(2\pi)^{N/2} |\mathbf{K}|^{1/2}} \exp\Big(-\frac{1}{2} \mathbf{x}^T \mathbf{K}^{-1} \mathbf{x} \Big) . \end{aligned} \tag{11.25}$$

Then, the conditional density of $\mathbf{X}$ based on the observation of $\mathbf{Y}$ and

$$\boldsymbol{\theta}^k = \begin{bmatrix} A_k \; v_k \end{bmatrix}^T$$

is $N(\hat{\mathbf{X}}_k, \mathbf{P}_k)$ distributed with

$$\begin{aligned} \hat{\mathbf{X}}_k &= \mathbf{P}_k \frac{A_k}{v_k} \mathbf{Y} \\ \mathbf{P}_k &= \Big(\mathbf{K}^{-1} + \frac{A_k^2}{v_k} \mathbf{I}_N \Big)^{-1} . \end{aligned} \tag{11.26}$$

Then,

$$\begin{aligned} Q(\boldsymbol{\theta}, \boldsymbol{\theta}^k) &= -\frac{1}{2v} E[||\mathbf{Y} - A\mathbf{X}||_2^2 | \mathbf{Y}, \boldsymbol{\theta}^k] - \frac{N}{2} \ln(v) + C(\boldsymbol{\theta}^k) \\ &= -\frac{1}{2v} \Big[||\mathbf{Y} - A\hat{\mathbf{X}}_k||_2^2 + A^2 \mathrm{tr}(\mathbf{P}_k) \Big] - \frac{N}{2} \ln(v) + C(\boldsymbol{\theta}^k) , \end{aligned} \tag{11.27}$$

where $C(\boldsymbol{\theta}^k)$ denotes a constant independent of $\boldsymbol{\theta}$. The gradient of $Q(\boldsymbol{\theta}, \boldsymbol{\theta}^k)$ with respect to $\boldsymbol{\theta}$ is then given by

$$\nabla Q(\boldsymbol{\theta}, \boldsymbol{\theta}^k) = \begin{bmatrix} v^{-1} \Big[\mathbf{Y}^T \hat{\mathbf{X}}_k - A(||\hat{\mathbf{X}}_k||_2^2 + \mathrm{tr}\,(\mathbf{P}_k)) \Big] \\ \frac{1}{2} \Big[v^{-2} (||\mathbf{Y} - A\hat{\mathbf{X}}_k||_2^2 + A^2 \mathrm{tr}\,(\mathbf{P}_k)) - N v^{-1} \Big] \end{bmatrix} . \tag{11.28}$$

The maximization equation

$$\nabla Q(\boldsymbol{\theta}, \boldsymbol{\theta}^k) = 0$$

is easy to solve in closed form and yields

$$A_{k+1} = \frac{\mathbf{Y}^T \hat{\mathbf{X}}_k}{||\hat{\mathbf{X}}_k||_2^2 + \mathrm{tr}(\mathbf{P}_k)} \tag{11.29}$$

$$v_{k+1} = \frac{1}{N} [||\mathbf{Y} - A_{k+1} \hat{\mathbf{X}}_k||_2^2 + A_{k+1}^2 \mathrm{tr}(\mathbf{P}_k)] \tag{11.30}$$

as the next iterate of the parameter vector.

To analyze the EM iteration, we consider a modified version of the above update where (11.30) is replaced by

$$v_{k+1} = \frac{1}{N} [||\mathbf{Y} - A_k \hat{\mathbf{X}}_k||_2^2 + A_k^2 \mathrm{tr}(\mathbf{P}_k)] , \tag{11.31}$$

i.e., we freeze the value of A_k while updating v_k, which corresponds to an AECM version of the EM algorithm. Then, to analyze the iteration formed by (11.29) and (11.31), we observe that

$$\begin{aligned}\mathbf{Y}^T\hat{\mathbf{X}}_k &= A_k\mathbf{Y}^T\mathbf{K}(A_k^2\mathbf{K}+v_k\mathbf{I}_N)^{-1}\mathbf{Y} \\ &= A_k\mathbf{Y}^T\mathbf{K}(A_k^2\mathbf{K}+v_k\mathbf{I}_N)^{-1}(A_k^2\mathbf{K}+v_k\mathbf{I}_N)(A_k^2\mathbf{K}+v_k\mathbf{I}_N)^{-1}\mathbf{Y} \\ &= A_k||\hat{\mathbf{X}}_k||^2 + v_kA_k\mathbf{Y}^T(A_k^2\mathbf{K}+v_k\mathbf{I}_N)^{-1}\mathbf{K}(A_k^2\mathbf{K}+v_k\mathbf{I}_N)^{-1}\mathbf{Y} \quad (11.32)\end{aligned}$$

and

$$\mathbf{Y} - A_k\hat{\mathbf{X}}_k = v_k(A_k^2\mathbf{K}+v_k\mathbf{I}_N)^{-1}\mathbf{Y} \quad (11.33)$$

Next, by substituting (11.32) and (11.33) in (11.29) and (11.31), respectively, and taking into account identity (11.5) for the gradient of the likelihood function, we find that the modified EM iteration formed by (11.29) and (11.31) can be rewritten as

$$\boldsymbol{\theta}^{k+1} = \boldsymbol{\theta}^k + \mathbf{D}(\boldsymbol{\theta}^k)\nabla L(\mathbf{Y}|\boldsymbol{\theta}^k) \quad (11.34)$$

with

$$\mathbf{D}(\boldsymbol{\theta}^k) = \frac{1}{N}\begin{bmatrix}\mathrm{SNR}_k^{-1} & 0 \\ 0 & 2v_k^2\end{bmatrix}, \quad (11.35)$$

where

$$\begin{aligned}\mathrm{SNR}_k &= \frac{||\hat{\mathbf{X}}_k||_2^2 + \mathrm{tr}\,(\mathbf{P}_k)}{Nv_k} \\ &= \frac{E[||\mathbf{X}||_2^2|\mathbf{Y},\boldsymbol{\theta}^k]}{Nv_k} \quad (11.36)\end{aligned}$$

is the estimated signal to noise ratio at the k-th iteration. The expression (11.34) indicates that the modified EM iteration is just a quasi-Newton scheme. In fact, provided the diagonal entries of $\mathbf{D}(\boldsymbol{\theta}_k)$ admit a finite upper bound independent of k, Proposition 1.2.5 of [1] ensures the existence of a capture set for any local maximum of the likelihood function $L(\mathbf{y}|\boldsymbol{\theta})$. Specifically, if $\boldsymbol{\theta}_*$ is a local maximum of L, there exists an open neighborhood $\mathcal{S}$ of $\boldsymbol{\theta}_*$ such that if the initial iterate $\boldsymbol{\theta}^0$ is located in $\mathcal{S}$, then all subsequent iterates $\boldsymbol{\theta}^k$ are located in $\mathcal{S}$, and $\boldsymbol{\theta}^k \to \boldsymbol{\theta}_*$ as $k \to \infty$. This implies that if the initial iterate $\boldsymbol{\theta}^0$ is located close enough to the ML estimate $\hat{\boldsymbol{\theta}}_{\mathrm{ML}}$, the EM algorithm will converge to $\hat{\boldsymbol{\theta}}_{\mathrm{ML}}$.

To illustrate the EM iteration (11.29)–(11.30), let $X(t)$ in (11.1) be an Ornstein-Uhlenbeck (OU) process, i.e., a zero-mean stationary Gaussian process with covariance function

$$K(t,s) = a^{|t-s|}S$$

where we select $a = 0.5$ and the signal power $S = 1$. For our simulations, the amplitude $A = 1$ and the WGN $V(t)$ has variance $v = 1$. The length of the observed sequence is $T = 400$. Note that this length is rather short if we consider that the EM method is just an implementation of the ML estimation method, and given our earlier observation that in the large sample case, the variance of ML estimates is proportional to $T^{-1/2}$. So $T = 400$ is actually too short to yield accurate estimates of A and v. However, expression (11.26) for $\mathbf{P}_k$ requires the inversion of a matrix of size $N = T + 1$, and accordingly, the EM iteration (11.29)–(11.30) is applicable only to short data sequences. With initial parameter values $A_0 = 0.7$ and $v_0 = 1.5$, 100 EM iterations are performed, and the plots of Figs. 11.1 and 11.2 show the values of A_k and v_k as a function of iteration k. The iterates converge, but as expected, the estimates $\hat{A}_{\mathrm{ML}}$ and $\hat{v}_{\mathrm{ML}}$ have a percentage error of 5% to 10%.

Since the Ornstein-Uhlenbeck process is a Gauss-Markov process, it turns out there exists a more efficient implementation of the EM algorithm that relies on Kalman filtering techniques. This method, which is described in Section 11.4, allows the use of long data records, and consequently it is preferable to the iteration (11.29)–(11.30) we discussed in this section. □

11.3.3 Convergence Rate

The EM iteration specifies a vector mapping

$$\boldsymbol{\theta}^{k+1} = \mathbf{m}(\boldsymbol{\theta}^k) \tag{11.37}$$

expressing the $k+1$-th iterate in terms of its predecessor. The i-th component of this mapping is denoted as $m_i(\boldsymbol{\theta}^k)$ with $1 \leq i \leq q$. As $k \to \infty$, the iterates $\boldsymbol{\theta}^k$ converge to a local maximum $\boldsymbol{\theta}_*$ of the likelihood function $L(\mathbf{y}|\boldsymbol{\theta})$. The limit $\boldsymbol{\theta}_*$ is obviously a fixed point of the mapping (11.37), so it satisfies

$$\boldsymbol{\theta}_* = \mathbf{m}(\boldsymbol{\theta}_*) \,. \tag{11.38}$$

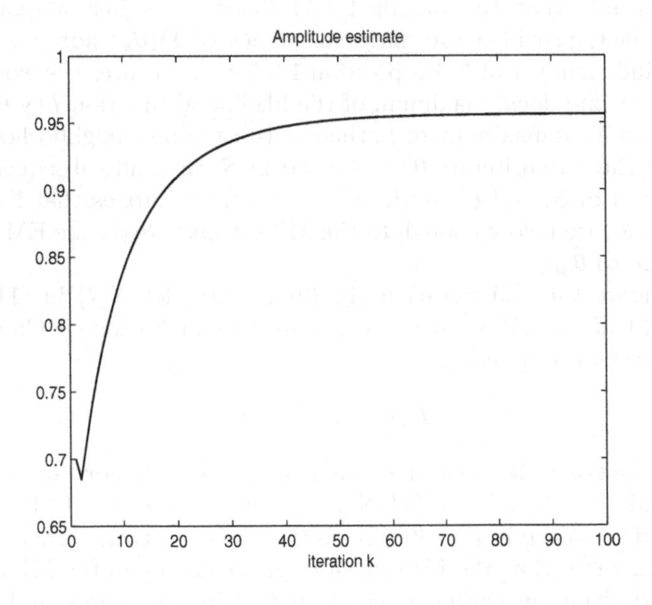

Fig. 11.1. Iterates A_k as a function of iteration k.

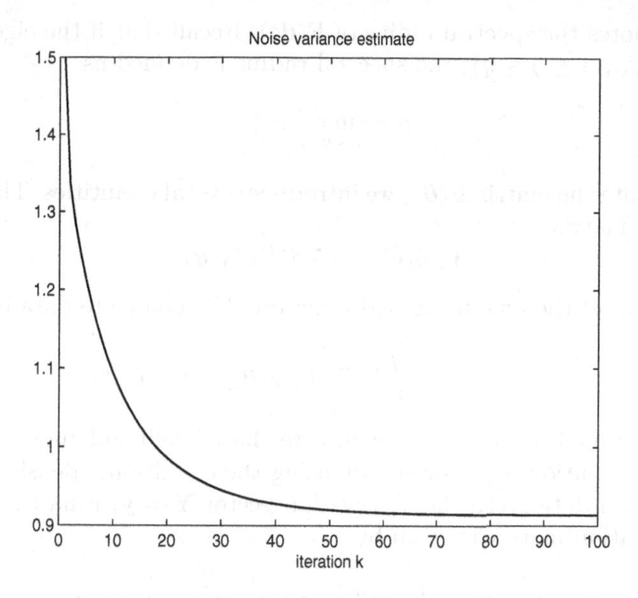

Fig. 11.2. Iterates v_k as a function of iteration k.

As k becomes large, the distance $||\boldsymbol{\theta}_k - \boldsymbol{\theta}_*||_2$ becomes small, so, by subtracting (11.38) from (11.37) and performing a Taylor series expansion of $\mathbf{m}(\boldsymbol{\theta}^k)$ in the vicinity of $\boldsymbol{\theta}_*$, we obtain

$$\begin{aligned}\boldsymbol{\theta}^{k+1} - \boldsymbol{\theta}_* &= \mathbf{m}(\boldsymbol{\theta}^k) - \mathbf{m}(\boldsymbol{\theta}_*) \\ &= \mathbf{F}(\boldsymbol{\theta}_*)(\boldsymbol{\theta}^k - \boldsymbol{\theta}_*) + O(||\boldsymbol{\theta}_k - \boldsymbol{\theta}_*||_2^2)\,,\end{aligned} \tag{11.39}$$

where the element of row i and column j of the $q \times q$ matrix $\mathbf{F}(\boldsymbol{\theta})$ is

$$F_{ij}(\boldsymbol{\theta}) = \frac{\partial m_i}{\partial \theta_j}(\boldsymbol{\theta})$$

with $1 \leq i,\, j \leq q$. $\mathbf{F}(\boldsymbol{\theta})$ can be written compactly as

$$\mathbf{F}(\boldsymbol{\theta}) = \nabla^T \mathbf{m}(\boldsymbol{\theta})\,, \tag{11.40}$$

where

$$\nabla^T = \left[\frac{\partial}{\partial \theta_1} \cdots \frac{\partial}{\partial \theta_j} \cdots \frac{\partial}{\partial \theta_q}\right].$$

The expression (11.38) implies that the rate of convergence of the EM iteration in the vicinity of a convergence point $\boldsymbol{\theta}_*$ is specified by

$$\lim_{k \to \infty} \frac{||\boldsymbol{\theta}^{k+1} - \boldsymbol{\theta}_*||}{||\boldsymbol{\theta}^k - \boldsymbol{\theta}_*||} = \rho\,, \tag{11.41}$$

where ρ denotes the spectral radius of $\mathbf{F}(\boldsymbol{\theta}_*)$. Recall that if the eigenvalues of $\mathbf{F}(\boldsymbol{\theta}_*)$ are $\{\lambda_j,\ 1 \leq j \leq q\}$, the spectral radius is defined as

$$\rho = \max_{1 \leq j \leq q} |\lambda_j| \,.$$

To evaluate the matrix $\mathbf{F}(\boldsymbol{\theta}_*)$ we introduce several quantities. The observed information matrix

$$\mathbf{J}_o(\boldsymbol{\theta}|\mathbf{y}) \triangleq -\nabla\nabla^T L(\mathbf{y}|\boldsymbol{\theta}) \tag{11.42}$$

is the Hessian of the true likelihood function. The complete data information matrix

$$\mathbf{J}_c(\boldsymbol{\theta}|\mathbf{y}) \triangleq -\int \nabla\nabla^T L_c(\mathbf{z}|\boldsymbol{\theta}) f_c(\mathbf{x}|\mathbf{y},\boldsymbol{\theta}) d\mathbf{x} \tag{11.43}$$

is the estimated Hessian of the complete data likelihood function $L_c(\mathbf{z}|\boldsymbol{\theta})$, where the estimation is performed by using the conditional density $f_c(\mathbf{x}|\mathbf{y},\boldsymbol{\theta})$ for the missing data given the observation vector $\mathbf{Y} = \mathbf{y}$. Finally, the missing data information matrix is given by

$$\mathbf{J}_m(\boldsymbol{\theta}|\mathbf{y}) \triangleq -\int \nabla\nabla^T \ln f_c(\mathbf{x}|\mathbf{y},\boldsymbol{\theta}) f_c(\mathbf{x}|\mathbf{y},\boldsymbol{\theta}) d\mathbf{x} \,. \tag{11.44}$$

From the decomposition

$$L_c(\mathbf{z}|\boldsymbol{\theta}) = \ln f_c(\mathbf{x}|\mathbf{y},\boldsymbol{\theta}) + L(\mathbf{y}|\boldsymbol{\theta}) \,,$$

it follows that

$$\mathbf{J}_o(\boldsymbol{\theta}|\mathbf{y}) = \mathbf{J}_c(\boldsymbol{\theta}|\mathbf{y}) - \mathbf{J}_m(\boldsymbol{\theta}|\mathbf{y}) \,. \tag{11.45}$$

Since the observation vector $\mathbf{y}$ is fixed, unless the $\mathbf{y}$ dependence needs to be considered, in the following discussion we will drop $\mathbf{y}$ as an argument of information matrices $\mathbf{J}_o$, $\mathbf{J}_c$ and $\mathbf{J}_m$.

Then, consider the function $Q(\boldsymbol{\theta},\boldsymbol{\theta}')$ and denote respectively by ∇ and ∇' the gradients with respect to its first and second argument, respectively. We perform a Taylor series expansion of $\nabla Q(\boldsymbol{\theta},\boldsymbol{\theta}')$ in the vicinity of the point $(\boldsymbol{\theta}_*,\boldsymbol{\theta}_*)$. This gives

$$\begin{aligned}\nabla Q(\boldsymbol{\theta},\boldsymbol{\theta}') &= \nabla Q(\boldsymbol{\theta}_*,\boldsymbol{\theta}_*) + \nabla\nabla^T Q(\boldsymbol{\theta}_*,\boldsymbol{\theta}_*)(\boldsymbol{\theta}-\boldsymbol{\theta}_*) \\ &\quad + \nabla(\nabla')^T Q(\boldsymbol{\theta}_*,\boldsymbol{\theta}_*)(\boldsymbol{\theta}'-\boldsymbol{\theta}_*) + \mathbf{e}(\boldsymbol{\theta}-\boldsymbol{\theta}_*,\boldsymbol{\theta}'-\boldsymbol{\theta}_*) \,,\end{aligned} \tag{11.46}$$

where the error $\mathbf{e}$ regroups terms of order two or higher. Then set $\boldsymbol{\theta} = \boldsymbol{\theta}^{k+1}$ and $\boldsymbol{\theta}' = \boldsymbol{\theta}^k$, and observe that since $\boldsymbol{\theta}^{k+1}$ maximizes $Q(\boldsymbol{\theta},\boldsymbol{\theta}^k)$ over $\boldsymbol{\theta}$ it satisfies

$$\nabla Q(\boldsymbol{\theta}^{k+1},\boldsymbol{\theta}^k) = 0 \,. \tag{11.47}$$

Also, letting $k \to \infty$ in this identity, we obtain

$$\nabla Q(\boldsymbol{\theta}_*,\boldsymbol{\theta}_*) = 0 \,. \tag{11.48}$$

Substituting (11.47)–(11.48) inside (11.46) yields

$$-\nabla\nabla^T Q(\boldsymbol{\theta}_*, \boldsymbol{\theta}_*)(\boldsymbol{\theta}^{k+1} - \boldsymbol{\theta}_*) \approx \nabla(\nabla')^T Q(\boldsymbol{\theta}_*, \boldsymbol{\theta}_*)(\boldsymbol{\theta}_k - \boldsymbol{\theta}_*) . \tag{11.49}$$

By direct evaluation, we find

$$-\nabla\nabla^T Q(\boldsymbol{\theta}, \boldsymbol{\theta}) = \mathbf{J}_c(\boldsymbol{\theta}|\mathbf{y}) \tag{11.50}$$

for an arbitrary parameter vector $\boldsymbol{\theta} \in \mathbb{R}^q$. The identity (11.21) implies also

$$\nabla(\nabla')^T Q(\boldsymbol{\theta}, \boldsymbol{\theta}') = -\nabla(\nabla')^T H(\boldsymbol{\theta}, \boldsymbol{\theta}') . \tag{11.51}$$

But, by differentiating

$$\int f_c(\mathbf{x}|\mathbf{y}, \boldsymbol{\theta}) d\mathbf{x} = 1$$

under the integral, we find

$$-\nabla(\nabla')^T H(\boldsymbol{\theta}, \boldsymbol{\theta}) = \nabla\nabla^T H(\boldsymbol{\theta}, \boldsymbol{\theta}) = \mathbf{J}_m(\boldsymbol{\theta}|\mathbf{y}) \tag{11.52}$$

for an arbitrary vector $\boldsymbol{\theta} \in \mathbb{R}^q$. Assume that $\mathbf{J}_c(\theta_*)$ is invertible. By substituting (11.50) and (11.52) inside (11.49) and inverting $\mathbf{J}_c(\boldsymbol{\theta}_*|\mathbf{y})$, we find that in the vicinity of $\boldsymbol{\theta}_*$

$$\boldsymbol{\theta}^{k+1} - \boldsymbol{\theta}_* \approx \mathbf{J}_c^{-1}(\boldsymbol{\theta}_*)\mathbf{J}_m(\boldsymbol{\theta}_*)(\boldsymbol{\theta}^k - \boldsymbol{\theta}_*) , \tag{11.53}$$

which implies

$$\mathbf{F}(\boldsymbol{\theta}_*) = \mathbf{J}_c^{-1}(\boldsymbol{\theta}_*)\mathbf{J}_m(\boldsymbol{\theta}_*) . \tag{11.54}$$

So, the iteration matrix $\mathbf{F}(\boldsymbol{\theta}_*)$ at a convergence point can be expressed in terms of the missing data and complete data information matrices at that point. Also, substituting (11.45) inside (11.54), we find also

$$\mathbf{F}(\boldsymbol{\theta}_*) = \mathbf{I}_q - \mathbf{J}_c^{-1}(\boldsymbol{\theta}_*)\mathbf{J}_o(\boldsymbol{\theta}_*) . \tag{11.55}$$

Example 11.1, continued

To illustrate the evaluation of information matrices $\mathbf{J}_c(\boldsymbol{\theta}_*)$ and $\mathbf{J}_m(\boldsymbol{\theta}_*)$, consider the amplitude and noise variance estimation problem of Section 11.2 and Example 11.1. Since

$$\ln(f_c(\mathbf{z}|\boldsymbol{\theta})) = -\frac{1}{2v}||\mathbf{y} - A\mathbf{x}||_2^2 - \frac{N}{2}\ln(2\pi v) + \ln f_c(\mathbf{x}) ,$$

we find

$$-\nabla\nabla^T \ln(f_c(\mathbf{z}|\boldsymbol{\theta})) = \begin{bmatrix} v^{-1}||\mathbf{x}||_2^2 & v^{-2}(\mathbf{y} - A\mathbf{x})^T\mathbf{x} \\ v^{-2}(\mathbf{y} - A\mathbf{x})^T\mathbf{x} & v^{-3}||\mathbf{y} - A\mathbf{x}||_2^2 - N/(2v^2) \end{bmatrix} . \tag{11.56}$$

Then, since the conditional density $f_c(\mathbf{x}|\mathbf{y}, \boldsymbol{\theta})$ is $N(\hat{\mathbf{x}}, \mathbf{P})$ distributed with

$$\hat{\mathbf{x}} = \mathbf{P}\frac{A}{v}\mathbf{y} \tag{11.57}$$

$$\mathbf{P} = \Big(\mathbf{K}^{-1} + \frac{A^2}{v}\mathbf{I}_N\Big)^{-1} , \tag{11.58}$$

we find

$$\mathbf{J}_c(\boldsymbol{\theta}) = \tag{11.59}$$

$$\begin{bmatrix} v^{-1}\big(||\hat{\mathbf{x}}||_2^2 + \mathrm{tr}\,(\mathbf{P})\big) & v^{-2}\big[\mathbf{y}^T\hat{\mathbf{x}} - A\big(||\hat{\mathbf{x}}||_2^2 + \mathrm{tr}\,(\mathbf{P})\big)\big] \\ v^{-2}\big[\mathbf{y}^T\hat{\mathbf{x}} - A\big(||\hat{\mathbf{x}}||_2^2 + \mathrm{tr}\,(\mathbf{P})\big)\big] & v^{-3}\big[||\mathbf{y} - A\hat{\mathbf{x}}||_2^2 + A^2\mathrm{tr}\,(\mathbf{P})\big] - N/(2v^2) \end{bmatrix} .$$

But, if

$$\boldsymbol{\theta}_* = \big[A_* \; v_*\big]^T$$

is a fixed point of the iteration (11.29), (11.30), its entries satisfy

$$A_* = \frac{\mathbf{y}^T\hat{\mathbf{x}}}{||\hat{\mathbf{x}}||^2 + \mathrm{tr}\,(\mathbf{P})}$$

$$v_* = \frac{1}{N}\big[||\mathbf{y} - A_*\hat{\mathbf{x}}||^2 + A_*^2\mathrm{tr}\,(\mathbf{P})\big] ,$$

which, after substitution inside (11.59) gives

$$\mathbf{J}_c(\boldsymbol{\theta}_*) = \begin{bmatrix} v_*^{-1}\big(||\hat{\mathbf{x}}||_2^2 + \mathrm{tr}\,(\mathbf{P})\big) & 0 \\ 0 & N/(2v_*^2) \end{bmatrix} . \tag{11.60}$$

Next, let us turn to the evaluation of $\mathbf{J}_m(\boldsymbol{\theta})$. We have

$$-\ln f_c(\mathbf{x}|\mathbf{y},\boldsymbol{\theta}) = \frac{1}{2}(\mathbf{x} - \hat{\mathbf{x}})^T\mathbf{P}^{-1}(\mathbf{x} - \hat{\mathbf{x}}) + \frac{1}{2}\ln|\mathbf{P}| + \frac{N}{2}\ln(2\pi) , \tag{11.61}$$

where $\hat{\mathbf{x}}$ and $\mathbf{P}$ depend on $\boldsymbol{\theta}$ and are given by (11.57) and (11.58). By differentiating twice (11.61) and taking the expectation with respect to $f_c(\mathbf{x}|\mathbf{y},\boldsymbol{\theta})$ we find

$$\begin{aligned} (\mathbf{J}_m)_{11}(\boldsymbol{\theta}) &= E[-\frac{\partial^2}{\partial A^2}\ln f_c(\mathbf{x}|\mathbf{y},\boldsymbol{\theta}) \mid \mathbf{y},\boldsymbol{\theta}] \\ &= \big(\frac{\partial\hat{\mathbf{x}}}{\partial A}\big)^T\mathbf{P}^{-1}\frac{\partial\hat{\mathbf{x}}}{\partial A} + \frac{1}{2}\mathrm{tr}\Big(\frac{\partial^2\mathbf{P}^{-1}}{\partial A^2}\mathbf{P} + \frac{\partial\mathbf{P}^{-1}}{\partial A}\frac{\partial\mathbf{P}}{\partial A} + \mathbf{P}^{-1}\frac{\partial^2\mathbf{P}}{\partial A^2}\Big) \\ &= \big(\frac{\partial\hat{\mathbf{x}}}{\partial A}\big)^T\mathbf{P}^{-1}\frac{\partial\hat{\mathbf{x}}}{\partial A} - \frac{1}{2}\mathrm{tr}\Big(\frac{\partial\mathbf{P}^{-1}}{\partial A}\frac{\partial\mathbf{P}}{\partial A}\Big) \\ &= \big(\frac{\partial\hat{\mathbf{x}}}{\partial A}\big)^T\mathbf{P}^{-1}\frac{\partial\hat{\mathbf{x}}}{\partial A} + \frac{1}{2}\mathrm{tr}\Big(\frac{\partial\mathbf{P}^{-1}}{\partial A}\mathbf{P}\frac{\partial\mathbf{P}^{-1}}{\partial A}\mathbf{P}\Big) , \end{aligned} \tag{11.62}$$

where, to go from the second to the third line of (11.62), we have differentiated twice the matrix inversion identity

$$\mathbf{P}^{-1}\mathbf{P} = \mathbf{I}_N .$$

Similarly we have

$$
\begin{aligned}
(\mathbf{J}_m)_{12}(\boldsymbol{\theta}) &= E[-\frac{\partial^2}{\partial A \partial v} \ln f_c(\mathbf{x}|\mathbf{y},\boldsymbol{\theta}) \mid \mathbf{y},\boldsymbol{\theta}] \\
&= (\frac{\partial \hat{\mathbf{x}}}{\partial A})^T \mathbf{P}^{-1} \frac{\partial \hat{\mathbf{x}}}{\partial v} + \frac{1}{2}\mathrm{tr}\Big(\frac{\partial^2 \mathbf{P}^{-1}}{\partial A \partial v}\mathbf{P} + \frac{\partial \mathbf{P}^{-1}}{\partial v}\frac{\partial \mathbf{P}}{\partial A} + \mathbf{P}^{-1}\frac{\partial^2 \mathbf{P}}{\partial A \partial v}\Big) \\
&= (\frac{\partial \hat{\mathbf{x}}}{\partial A})^T \mathbf{P}^{-1} \frac{\partial \hat{\mathbf{x}}}{\partial v} - \frac{1}{2}\mathrm{tr}\Big(\frac{\partial \mathbf{P}^{-1}}{\partial A}\frac{\partial \mathbf{P}}{\partial v}\Big) \\
&= (\frac{\partial \hat{\mathbf{x}}}{\partial A})^T \mathbf{P}^{-1} \frac{\partial \hat{\mathbf{x}}}{\partial v} + \frac{1}{2}\mathrm{tr}\Big(\frac{\partial \mathbf{P}^{-1}}{\partial A}\mathbf{P}\frac{\partial \mathbf{P}^{-1}}{\partial v}\mathbf{P}\Big), \qquad (11.63)
\end{aligned}
$$

and

$$
\begin{aligned}
(\mathbf{J}_m)_{22}(\boldsymbol{\theta}) &= E[-\frac{\partial^2}{\partial v^2} \ln f_c(\mathbf{x}|\mathbf{y},\boldsymbol{\theta}) \mid \mathbf{y},\boldsymbol{\theta}] \\
&= (\frac{\partial \hat{\mathbf{x}}}{\partial v})^T \mathbf{P}^{-1} \frac{\partial \hat{\mathbf{x}}}{\partial v} + \frac{1}{2}\mathrm{tr}\Big(\frac{\partial^2 \mathbf{P}^{-1}}{\partial v^2}\mathbf{P} + \frac{\partial \mathbf{P}^{-1}}{\partial v}\frac{\partial \mathbf{P}}{\partial v} + \mathbf{P}^{-1}\frac{\partial^2 \mathbf{P}}{\partial v^2}\Big) \\
&= (\frac{\partial \hat{\mathbf{x}}}{\partial v})^T \mathbf{P}^{-1} \frac{\partial \hat{\mathbf{x}}}{\partial v} - \frac{1}{2}\mathrm{tr}\Big(\frac{\partial \mathbf{P}^{-1}}{\partial v}\frac{\partial \mathbf{P}}{\partial v}\Big) \\
&= (\frac{\partial \hat{\mathbf{x}}}{\partial v})^T \mathbf{P}^{-1} \frac{\partial \hat{\mathbf{x}}}{\partial v} + \frac{1}{2}\mathrm{tr}\Big(\frac{\partial \mathbf{P}^{-1}}{\partial v}\mathbf{P}\frac{\partial \mathbf{P}^{-1}}{\partial v}\mathbf{P}\Big). \qquad (11.64)
\end{aligned}
$$

Note that since $\mathbf{J}_m(\boldsymbol{\theta})$ is an expected Hessian, it is symmetric, so

$$(\mathbf{J}_m)_{21} = (\mathbf{J}_m)_{12} .$$

Next, by premultiplying (11.57) by $\mathbf{P}^{-1}$ and differentiating with respect to A and v, respectively, we find

$$
\begin{aligned}
\mathbf{P}^{-1}\frac{\partial \hat{\mathbf{x}}}{\partial A} + 2Av^{-1}\hat{\mathbf{x}} &= v^{-1}\mathbf{y} \\
\mathbf{P}^{-1}\frac{\partial \hat{\mathbf{x}}}{\partial v} - A^2 v^{-2}\hat{\mathbf{x}} &= -Av^{-2}\mathbf{y} ,
\end{aligned}
$$

so that

$$
\begin{aligned}
\frac{\partial \hat{\mathbf{x}}}{\partial A} &= v^{-1}\mathbf{P}(\mathbf{y} - 2A\hat{\mathbf{x}}) \\
\frac{\partial \hat{\mathbf{x}}}{\partial v} &= -v^{-2}\mathbf{P}(\mathbf{y} - A\hat{\mathbf{x}}) . \qquad (11.65)
\end{aligned}
$$

Similarly, inverting (11.58) and differentiating with respect to A and v respectively yields

$$
\begin{aligned}
\frac{\partial \mathbf{P}^{-1}}{\partial A} &= 2Av^{-1}\mathbf{I}_N \\
\frac{\partial \mathbf{P}^{-1}}{\partial v} &= -A^2 v^{-2}\mathbf{I}_N . \qquad (11.66)
\end{aligned}
$$

Then, by substituting (11.65) and (11.66) inside (11.62)–(11.64) we obtain

$$\begin{aligned}(\mathbf{J}_m)_{11}(\boldsymbol{\theta}) &= \frac{1}{v^2}(\mathbf{y}-2A\hat{\mathbf{x}})^T\mathbf{P}(\mathbf{y}-2A\hat{\mathbf{x}}) + \frac{2A^2}{v^2}\mathrm{tr}\,(\mathbf{P}^2) \\ (\mathbf{J}_m)_{12}(\boldsymbol{\theta}) &= -\frac{A}{v^3}(\mathbf{y}-2A\hat{\mathbf{x}})^T\mathbf{P}(\mathbf{y}-A\hat{\mathbf{x}}) - \frac{A^3}{v^3}\mathrm{tr}\,(\mathbf{P}^2) \\ (\mathbf{J}_m)_{22}(\boldsymbol{\theta}) &= \frac{A^2}{v^4}(\mathbf{y}-A\hat{\mathbf{x}})^T\mathbf{P}(\mathbf{y}-A\hat{\mathbf{x}}) + \frac{A^4}{2v^4}\mathrm{tr}\,(\mathbf{P}^2)\,. \end{aligned} \tag{11.67}$$

This shows that all entries of the matrix $\mathbf{J}_m(\boldsymbol{\theta})$ can all be evaluated as a byproduct of the EM algorithm, since both the estimate $\hat{\mathbf{x}}$ and the covariance matrix $\mathbf{P}$ are evaluated as part of EM iteration.

The expressions (11.60) and (11.67) for $\mathbf{J}_c$ and $\mathbf{J}_m$ can be used to monitor the rate of convergence of the EM iteration as it approaches a convergence point. This is accomplished by approximating the matrix $\mathbf{F}(\boldsymbol{\theta}_*)$ by

$$\mathbf{F}(\boldsymbol{\theta}^k) = \mathbf{J}_c^{-1}(\boldsymbol{\theta}^k)\mathbf{J}_m(\boldsymbol{\theta}^k)\,,$$

and using the iterate θ^k, and the estimate $\hat{\mathbf{x}}^k$ and error covariance $\mathbf{P}^k$ to evaluate expressions (11.60) and (11.67) for $\mathbf{J}_c$ and $\mathbf{J}_m$. Then the spectral radius of $\mathbf{F}(\boldsymbol{\theta}^k)$ provides a measure of the speed of convergence of the EM iteration as it gets close to a convergence point. This information can be exploited to decide when to terminate the EM iteration.

To illustrate the convergence rate calculation for the EM iteration, consider the case discussed earlier in Example 11.1 where $X(t)$ is an Ornstein-Uhlenbeck process with $a = 0.5$ and $S = 1$, the amplitude $A = 1$, and the WGN $V(t)$ has variance $v = 1$. Then, for a time window of length $T = 400$ and $K = 100$ iterations, the spectral radius of the matrix $\mathbf{F}(\boldsymbol{\theta}_k)$ is plotted in Fig. 11.3 as a function of the iteration k. For this case, as the iterate $\boldsymbol{\theta}_k$ approaches its limit, the spectral radius equals 0.925, which suggests that fewer than 100 iterations were truly necessary to reach a fixed point of the iteration.

□

11.3.4 Large-Sample Covariance Matrix

When the size N of the observation vector $\mathbf{Y}$ is large, Meng and Rubin [8] made the interesting observation that the covariance of the ML estimate can be evaluated as a byproduct of the EM algorithm. The additional computations required by this evaluation form the *Supplemented EM* (SEM) algorithm. The computation described below relies on two assumptions:

(i) The number N of observations is large.
(ii) The EM algorithm converges to the ML estimate, i.e., the local maximum $\boldsymbol{\theta}_* = \hat{\boldsymbol{\theta}}_{\mathrm{ML}}$. In other words, the EM algorithm was initialized so that the EM iterates $\boldsymbol{\theta}^k$ converge to the global maximum $\hat{\boldsymbol{\theta}}_{\mathrm{ML}}$ of the likelihood function, instead of a suboptimal local maximum.

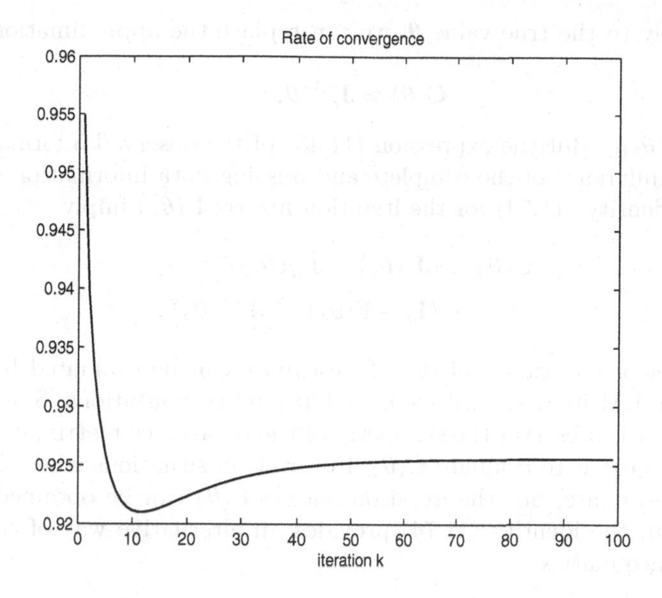

Fig. 11.3. Spectral radius of the matrix $F(\boldsymbol{\theta}_k)$ as a function of iteration k.

As starting point, recall that for the case of independent identically distributed observations, it was shown in Section 4.5 that as the number N of observations becomes large, the covariance $\mathbf{C}(\boldsymbol{\theta})$ of the ML estimate satisfies

$$\mathbf{C}(\boldsymbol{\theta}) = E[(\hat{\boldsymbol{\theta}}_{\mathrm{ML}} - \boldsymbol{\theta})(\hat{\boldsymbol{\theta}}_{\mathrm{ML}} - \boldsymbol{\theta})^T] \approx \frac{\mathbf{J}^{-1}(\boldsymbol{\theta})}{N} , \tag{11.68}$$

where the Fisher information matrix $\mathbf{J}(\boldsymbol{\theta})$ is the expectation with respect to $\mathbf{Y}$ of the observed information matrix $\mathbf{J}_o(\boldsymbol{\theta}|\mathbf{Y})$, i.e.,

$$\mathbf{J}(\boldsymbol{\theta}) = E_{\mathbf{Y}}[\mathbf{J}_o(\boldsymbol{\theta}|\mathbf{Y})] = -\int \nabla\nabla^T \ln(f(\mathbf{y}|\boldsymbol{\theta}))f(\mathbf{y}|\boldsymbol{\theta})d\mathbf{y} . \tag{11.69}$$

For the case of independent, identically distributed observations, by the strong law of large numbers we have

$$\mathbf{J}(\boldsymbol{\theta}) = \lim_{N\to\infty} \frac{1}{N}\mathbf{J}_o(\boldsymbol{\theta}) . \tag{11.70}$$

Combining (11.68) and (11.70) we conclude that as the number N of samples becomes large

$$\mathbf{C}(\boldsymbol{\theta}) \approx \mathbf{J}_o^{-1}(\boldsymbol{\theta}) . \tag{11.71}$$

Note also that the assumption that the observations are i.i.d. that we have used here to justify the approximation (11.71) can be relaxed significantly. Also, since it was shown in Section 4.5 that the ML estimate $\hat{\boldsymbol{\theta}}_{\mathrm{ML}}$ converges almost surely to the true value $\boldsymbol{\theta}$, we can replace the approximation (11.71) by

$$\mathbf{C}(\boldsymbol{\theta}) \approx \mathbf{J}_o^{-1}(\boldsymbol{\theta}_*) , \tag{11.72}$$

where $\boldsymbol{\theta}_* = \hat{\boldsymbol{\theta}}_{\mathrm{ML}}$. But the expression (11.45) of the observed information matrix as the difference of the complete and missing data information matrices, as well as identity (11.54) for the iteration matrix $\mathbf{F}(\boldsymbol{\theta}_*)$ imply

$$\mathbf{C}(\boldsymbol{\theta}) \approx (\mathbf{J}_c(\boldsymbol{\theta}_*) - \mathbf{J}_m(\boldsymbol{\theta}_*))^{-1} \tag{11.73}$$

$$= \left(\mathbf{I}_q - \mathbf{F}(\boldsymbol{\theta}_*)\right)^{-1} \mathbf{J}_c^{-1}(\boldsymbol{\theta}_*) . \tag{11.74}$$

Thus, the error covariance of the ML estimate can be evaluated by supplementing the EM iteration with some additional computations. Since the expression (11.73) preserves the symmetry of the covariance matrix, it is usually preferable to use it to evaluate $\mathbf{C}(\boldsymbol{\theta})$. However, in situations where $\mathbf{J}_m(\boldsymbol{\theta}_*)$ is difficult to evaluate, but the iteration matrix $\mathbf{F}(\boldsymbol{\theta})$ can be obtained directly from (11.40), the identity (11.74) provides an alternative way of computing the covariance matrix.

11.4 Parameter Estimation of Hidden Gauss-Markov Models

The description of the EM algorithm given in Section 11.3 is quite general, but it relies implicitly on the assumption that the conditional density $f_c(\mathbf{x}|\mathbf{y}, \boldsymbol{\theta})$ for the missing data vector $\mathbf{X}$ can be evaluated easily. This means that the conditional density must admit a finite parametrization, and that a convenient algorithm should exist for its evaluation. With respect to the parametrization issue, observe that Gaussian densities are finitely parametrized and remain Gaussian after conditioning. However, even if we restrict our attention to Gaussian observations and missing data, the resulting computations can still be prohibitive. For example, if we consider Example 11.1, when the number N of observations is large, the evaluation of the covariance $\mathbf{P}_k$ in (11.26) requires the inversion of a matrix of dimension $N \times N$, which requires of order N^3 operations. Fortunately, this computational burden can be reduced substantially if we consider the class of *Hidden Gauss-Markov Models* (HGMMs). To explain our terminology, HGMMs are partially observed Gauss-Markov processes for which some of the parameters of the internal state-space model

$$\mathbf{X}(t+1) = \mathbf{A}\mathbf{X}(t) + \mathbf{W}(t) \tag{11.75}$$

$$Y(t) = \mathbf{c}\mathbf{X}(t) + V(t) \tag{11.76}$$

need to be estimated. Since we have assumed up to this point that $Y(t)$ is a scalar observation, we retain this assumption, although extension to the vector case is straighforward. On the other hand, the state vector $\mathbf{X}(t)$ is assumed to have dimension n. Equations (11.75) and (11.76) hold for $0 \leq t \leq T-1$ and $0 \leq t \leq T$, respectively. We assume that $V(t)$ and $\mathbf{W}(t)$ are two independent zero-mean WGN sequences of intensity r and $\mathbf{Q}$ respectively, i.e.,

$$\begin{aligned} E[V(t)V(s)] &= r\delta(t-s) \\ E[\mathbf{W}(t)\mathbf{W}^T(s)] &= \mathbf{Q}\delta(t-s)\,, \end{aligned} \tag{11.77}$$

where to keep things simple, it is assumed that $\mathbf{Q}$ is invertible, i.e., we consider the full-rank noise case. The initial state $\mathbf{X}(0)$ is independent of the noise sequences $\mathbf{W}(t)$ and $V(t)$ and admits the Gaussian distribution

$$\mathbf{X}(0) \sim N(\mathbf{0}, \boldsymbol{\Pi}_0)\,.$$

To ensure that the available data exceeds the number of parameters to be estimated, the matrices $\mathbf{A}$, $\mathbf{Q}$, the row vector $\mathbf{c}$ and the scalar r are assumed constant.

Note that the terminology regarding what constitutes a hidden Markov model is not entirely settled, and in this text we have elected to call a model of the form (11.75)–(11.76) a partially observed Gauss-Markov process when the model is completely known, and to call it a HGMM whenever some of its parameters need to be identified, (in other words, it is the model that is hidden). Most authors call processes satisfying (11.75)–(11.76) an HGMM irrespective of whether the parameters are known or not. However, since there is some obvious merit to distinguishing situations where the model is unknown from those where it is known, we have chosen here to restrict the HGMM label to the unknown parameter case.

11.4.1 EM iteration

For the class of HGMMs, Schumway and Stoffer [9, 10] observed that the Markov structure of the model (11.75)–(11.76) allows the implementation of the EM iteration in terms of a recursive Kalman smoother. However, before describing these results, it is worth reviewing briefly what we are trying to achieve. Given the observations $\{Y(t),\ 0 \leq t \leq T\}$, we seek to construct a state-space model $\{\mathbf{A}, \mathbf{c}, \mathbf{Q}, r\}$ maximizing the likelihood function of the observations. In the stochastic literature, this problem is usually referred to as an ML state-space identification problem [11, Chap. 7], [12, Chap. 7]. One key feature of this problem is that, given a model of the form (11.75)-(11.76), there exists a class of equivalent models that produce exactly the same output. Specifically, if we apply the state transformation

$$\mathbf{X}'(t) = \mathbf{T}\mathbf{X}(t) \tag{11.78}$$

where $\mathbf{T}$ is an invertible $n \times n$ matrix, the original model is transformed into the equivalent model

$$\mathbf{X}'(t+1) = \mathbf{A}'\mathbf{X}'(t) + \mathbf{W}'(t)$$
$$Y(t) = \mathbf{c}'\mathbf{X}'(t) + V(t)$$

with

$$\mathbf{A}' = \mathbf{T}\mathbf{A}\mathbf{T}^{-1} \quad , \quad \mathbf{c}' = \mathbf{c}\mathbf{T}^{-1} , \tag{11.79}$$

and where $\mathbf{W}'(t)$ is a $N(\mathbf{0}, \mathbf{Q}')$ distributed WGN sequence with

$$\mathbf{Q}' = \mathbf{T}\mathbf{Q}\mathbf{T}^T . \tag{11.80}$$

The transformation (11.79)–(11.80) is called a similarity transformation. It defines an equivalence relation among state-space models since it does not affect the observation statistics.

When selecting a parameter vector $\boldsymbol{\theta}$ representing the state-space model (11.75)–(11.76), it is therefore important to ensure that all models related by a similarity transformation (11.79)–(11.80) correspond to a single vector $\boldsymbol{\theta}$. This is often accomplished by bringing the state-space model to a canonical form. However, canonical forms tend to destroy the natural structure of state-space systems, which is usually obtained by invoking basic physical laws. When such a structure exists, typically only a small number of parameters are genuinely unknown, so that the vector $\boldsymbol{\theta}$ needs to include only those entries of $\mathbf{A}$, $\mathbf{c}$, $\mathbf{Q}$, and r that truly require estimation. For the discussion here, the state-space estimation problem we consider is completely general and is not associated to a specific physical system. In this context, we observe that, by properly selecting the similarity transformation $\mathbf{T}$, we can ensure that the matrix $\mathbf{Q}$ is fixed. For example, if

$$\mathbf{Q} = \mathbf{L}\mathbf{L}^T$$

is a lower times upper Cholesky factorization of $\mathbf{Q}$, by choosing $\mathbf{T} = \mathbf{L}^{-1}$ we can ensure that after transformation $\mathbf{Q}' = \mathbf{I}_n$. This does not exhaust completely all the degrees of freedom afforded by the application of a similarity transformation, since after $\mathbf{Q}$ has been normalized by setting it equal to the identity matrix, any orthonormal transformation $\mathbf{T}$ still leaves $\mathbf{Q}$ equal to the identity, but can be used to bring $\mathbf{A}$ to Schur canonical form [13, p. 98]. However, to keep our discussion as simple as possible, we assume $\mathbf{Q}$ is fixed, so it does not need to be estimated, but we leave $\mathbf{A}$ free, even though as explained above, this allows the possibility that different state-space models might correspond to the same system. So the parameter vector $\boldsymbol{\theta}$ is formed by the coefficients of $\{\mathbf{A}, \mathbf{c}, r\}$. For simplicity, we assume that the covariance $\boldsymbol{\Pi}_0$ is known, although its estimation could be included in the EM procedure. Note, however, that since the dependence of the observations $Y(t)$ on the initial covariance $\boldsymbol{\Pi}_0$ becomes progressively weaker as t increases, the problem of estimating $\boldsymbol{\Pi}_0$ from the observations is ill-conditioned. For the HGMM (11.75)–(11.76), we select as missing data the state sequence $\{\mathbf{X}(t),\ 0 \leq t \leq T\}$, i.e., the vector $\mathbf{X}$ is given by

$$\mathbf{X} = \left[\mathbf{X}^T(0) \cdots \mathbf{X}^T(t) \cdots \mathbf{X}^T(T)\right]^T .$$

Then, the logarithm of the density of the complete data can be expressed as

$$-\ln f_c(\mathbf{y},\mathbf{x}\mid\boldsymbol{\theta}) = \frac{1}{2}\sum_{t=0}^{T-1}(\mathbf{X}(t+1)-\mathbf{A}\mathbf{X}(t))^T\mathbf{Q}^{-1}(\mathbf{X}(t+1)-\mathbf{A}\mathbf{X}(t))$$

$$+\frac{T}{2}\ln(|\mathbf{Q}|)+\frac{1}{2r}\sum_{t=0}^{T}(Y(t)-\mathbf{c}\mathbf{X}(t))^2+\frac{(T+1)}{2}\ln(r)+\gamma\,, \qquad (11.81)$$

where the constant γ does not depend on $\boldsymbol{\theta}$. Let

$$\hat{\mathbf{X}}_k(t) = E[\mathbf{X}(t)|\mathbf{Y},\boldsymbol{\theta}^k]$$
$$\tilde{\mathbf{X}}_k(t) = \mathbf{X}(t)-\hat{\mathbf{X}}_k(t) \qquad (11.82)$$

denote respectively the smoothed estimate and smoothing error of $\mathbf{X}(t)$ based on observations $\{Y(s),\, 0\leq s\leq T\}$ and on the parameter vector $\boldsymbol{\theta}^k$ formed by the model $\{\mathbf{A}_k,\mathbf{c}_k,r_k\}$ at the k-th EM iteration. Let also

$$P_k(t) = E[\tilde{\mathbf{X}}_k(t)\tilde{\mathbf{X}}_k^T(t)\mid\mathbf{Y},\boldsymbol{\theta}^k]$$
$$P_k(t,t-1) = E[\tilde{\mathbf{X}}_k(t)\tilde{\mathbf{X}}_k^T(t-1)\mid\mathbf{Y},\boldsymbol{\theta}^k] \qquad (11.83)$$

be the conditional covariance and cross-covariance matrices of the error at time t, and at times t and $t-1$, respectively. Then the estimated complete log-likelihood function based on the observations and parameter vector $\boldsymbol{\theta}^k$ can be expressed as

$$Q(\boldsymbol{\theta},\boldsymbol{\theta}^k) = -\frac{1}{2}\mathrm{tr}\left\{\mathbf{Q}^{-1}[\boldsymbol{\Phi}_k-\mathbf{A}\boldsymbol{\Psi}_k^T-\boldsymbol{\Psi}_k\mathbf{A}^T+\mathbf{A}\boldsymbol{\Omega}_k\mathbf{A}^T]\right\}$$
$$-\frac{1}{2r}\sum_{t=0}^{T}\left[(Y(t)-\mathbf{c}\hat{\boldsymbol{X}}_k(t))^2+\mathbf{c}\mathbf{P}_k(t)\mathbf{c}^T\right]$$
$$-\frac{T}{2}\ln(|\mathbf{Q}|)-\frac{T+1}{2}\ln(r) \qquad (11.84)$$

where

$$\boldsymbol{\Phi}_k \triangleq \sum_{t=1}^{T}\left[\hat{\mathbf{X}}_k(t)\hat{\mathbf{X}}_k^T(t)+\mathbf{P}_k(t)\right]$$

$$\boldsymbol{\Psi}_k \triangleq \sum_{t=1}^{T}\left[\hat{\mathbf{X}}_k(t)\hat{\mathbf{X}}_k^T(t-1)+\mathbf{P}_k(t,t-1)\right]$$

$$\boldsymbol{\Theta}_k \triangleq \sum_{t=0}^{T-1}\left[\hat{\mathbf{X}}_k(t)\hat{\mathbf{X}}_k^T(t)+\mathbf{P}_k(t)\right]$$

$$\boldsymbol{\Omega}_k \triangleq \sum_{t=0}^{T}\left[\hat{\mathbf{X}}_k(t)\hat{\mathbf{X}}_k^T(t)+\mathbf{P}_k(t)\right]$$

$$\boldsymbol{\zeta}_k \triangleq \sum_{t=0}^{T}Y(t)\hat{\mathbf{X}}_k(t)\,. \qquad (11.85)$$

Next, we maximize $Q(\boldsymbol{\theta}, \boldsymbol{\theta}_k)$ with respect to $\boldsymbol{\theta}$, i. e., with respect to $\mathbf{A}$, $\mathbf{c}$, and r. Since $Q(\boldsymbol{\theta}, \boldsymbol{\theta}^k)$ is quadratic in $\mathbf{A}$ and $\mathbf{c}$, this is accomplished by forming complete squares. The minimum is achieved for

$$\begin{aligned} \mathbf{A}_{k+1} &= \boldsymbol{\Psi}_k \boldsymbol{\Theta}_k^{-1} \\ \mathbf{c}_{k+1}^T &= \boldsymbol{\Omega}_k^{-1} \zeta_k . \end{aligned} \tag{11.86}$$

Substituting the values (11.86) inside $Q(\boldsymbol{\theta}, \boldsymbol{\theta}^k)$ and differentiating with respect to r, we then find

$$r_{k+1} = \frac{1}{T+1} \sum_{t=0}^{T} \left[(Y(t) - \mathbf{c}_{k+1} \hat{\mathbf{X}}_k(t))^2 + \mathbf{c}_{k+1} \mathbf{P}_k(t) \mathbf{c}_{k+1}^T \right] . \tag{11.87}$$

Together, expressions (11.86) and (11.87) form the EM iteration. As indicated by identities (11.85), they require the evaluation of the smoothed estimates $\hat{X}(t)$, the smoothed error covariance $\mathbf{P}(t)$ and the cross-covariance $\mathbf{P}(t, t-1)$ between the errors at times t and $t-1$. The quantities can be evaluated in a variety of ways, but as indicated in [9], the double-sweep or Rauch-Tung-Striebel smoother [14, Sec. 10.3] is particularly well adapted to this task, and, accordingly, we review its structure below.

11.4.2 Double-sweep smoother

The double-sweep smoother consists of two phases. First the causal predicted and filtered estimates $\hat{\mathbf{X}}_p(t)$ and $\hat{\mathbf{X}}_f(t)$ are propagated in the forward direction by using the time and measurement-update recursions (10.77)–(10.78). In a second phase, the smoothed estimates $\hat{\mathbf{X}}(t)$ are propagated backwards in time starting from

$$\hat{\mathbf{X}}(T) = \hat{\mathbf{X}}_f(T) . \tag{11.88}$$

The backwards recursion can be derived by employing a simple repeated conditioning argument. Let

$$\mathbf{Y}_t = \begin{bmatrix} Y(0)\ Y(1) \dots Y(t) \end{bmatrix}^T \tag{11.89}$$

denote the vector formed by the observations up to time t, and let $\mathbf{Y} = \mathbf{Y}_T$ denote the vector of observations over the entire interval $[0, T]$. Then, by repeated conditioning we clearly have

$$\hat{\mathbf{X}}(t) = E[\hat{\mathbf{X}}_c(t) | \mathbf{Y}] \tag{11.90}$$

with

$$\hat{\mathbf{X}}_c(t) \triangleq E[\mathbf{X}(t) | \mathbf{X}(t+1), \mathbf{Y}] .$$

However, since $\mathbf{X}(\cdot)$ is a Markov process, given $\mathbf{X}(t+1)$, the vector $\mathbf{X}(t)$ is conditionally independent of the observations $\{Y(s), t+1 \leq s \leq T\}$. Accordingly, the constrained estimated $\hat{\mathbf{X}}_c(t)$ can be expressed as

$$\hat{\mathbf{X}}_c(t) = E[\mathbf{X}(t)|\mathbf{X}(t+1), \mathbf{Y}_t] \,. \tag{11.91}$$

To evaluate this estimate, we note that the conditional distribution of $\mathbf{X}(t)$ based on $\mathbf{Y}_t$ is $N(\hat{\mathbf{X}}_f(t), \mathbf{P}_f(t))$. Thus $\hat{\mathbf{X}}_c(t)$ can be viewed as the new estimate based on $\mathbf{Y}_t$ and the additional "observation"

$$\mathbf{X}(t+1) = \mathbf{A}\mathbf{X}(t) + \mathbf{W}(t) \tag{11.92}$$

where $\mathbf{W}(t)$ is independent of $\mathbf{Y}_t$ and the vector $\mathbf{X}(t+1)$ on the left-hand side of (11.92) is viewed as a measurement. According to (4.73)–(4.74) the information form of this estimate can be expressed as

$$\mathbf{P}_c^{-1}(t)\hat{\mathbf{X}}_c(t) = \mathbf{P}_f^{-1}(t)\hat{\mathbf{X}}_f(t) + \mathbf{A}^T\mathbf{Q}^{-1}\mathbf{X}(t+1) \tag{11.93}$$

$$\mathbf{P}_c^{-1}(t) = \mathbf{P}_f^{-1}(t) + \mathbf{A}^T\mathbf{Q}^{-1}\mathbf{A} \,, \tag{11.94}$$

where

$$\mathbf{P}_c(t) = E[\tilde{\mathbf{X}}_c(t)\tilde{\mathbf{X}}_c^T(t)]$$

is the covariance matrix of the error

$$\tilde{\mathbf{X}}_c(t) = \mathbf{X}(t) - \hat{\mathbf{X}}_c(t) \,.$$

Then, by taking the expectation with respect to $\mathbf{Y}$ on both sides of (11.93) and using the repeated conditioning identity (11.89), we obtain the Rauch-Tung-Striebel recursion

$$\hat{\mathbf{X}}(t) = \mathbf{P}_c(t)\mathbf{A}^T\mathbf{Q}^{-1}\hat{\mathbf{X}}(t+1) + \mathbf{P}_c(t)\mathbf{P}_f^{-1}(t)\hat{\mathbf{X}}_f(t) \tag{11.95}$$

that can be used to propagate the smoothed estimate $\mathbf{X}(t)$ backwards in time starting with the initial value (11.88).

Let

$$\mathbf{A}_s(t) \triangleq \mathbf{P}_c(t)\mathbf{A}\mathbf{Q}^{-1}$$

denote the matrix specifying the smoother dynamics (11.95). By multiplying (11.93) on the left by $\mathbf{P}_c(t)$ and substracting (11.95) we obtain

$$\hat{\mathbf{X}}_c(t) - \hat{\mathbf{X}}(t) = \mathbf{A}_s\tilde{\mathbf{X}}(t+1) \tag{11.96}$$

where

$$\tilde{\mathbf{X}}(t) \triangleq \mathbf{X}(t) - \hat{\mathbf{X}}(t)$$

denotes the smoothing error. By adding and substracting $\mathbf{X}(t)$ on the left-hand side of identity (11.96), it can be rewritten as

$$\tilde{\mathbf{X}}(t) = \mathbf{A}_s\tilde{\mathbf{X}}(t+1) + \tilde{\mathbf{X}}_c(t) \,. \tag{11.97}$$

But, in this expression, since $\tilde{X}_c(t)$ represents the error for estimating $\mathbf{X}(t)$ based on $\mathbf{X}(t+1)$ and $\mathbf{Y}$ and since $\tilde{\mathbf{X}}(t+1)$ is in the space spanned by $\mathbf{X}(t+1)$ and $\mathbf{Y}$, the orthogonality property of least-squares estimates implies

$$E[\tilde{\mathbf{X}}_c(t)\tilde{\mathbf{X}}^T(t+1)] = 0\,. \tag{11.98}$$

If

$$\mathbf{P}(t) = E[\tilde{\mathbf{X}}(t)\tilde{\mathbf{X}}^T(t)]$$

denotes the smoothing error covariance matrix, the identity (11.97) and the orthogonality property (11.98) imply

$$\mathbf{P}(t) = \mathbf{A}_s(t)\mathbf{P}(t+1)\mathbf{A}_s^T(t) + \mathbf{P}_c(t) \tag{11.99}$$

which can be propagated backwards in time to evaluate the smoothing error covariance, starting with

$$\mathbf{P}(T) = \mathbf{P}_f(T)\,.$$

In addition to the smoothing estimate $\hat{\mathbf{X}}(t)$ and the error covariance matrix $\mathbf{P}(t)$, the EM iteration requires also the evaluation of the cross-covariance

$$\mathbf{P}(t+1,t) = E[\tilde{\mathbf{X}}(t+1)\tilde{\mathbf{X}}^T(t)].$$

By substituting (11.97) and taking into account the orthogonality property (11.98) we find

$$\mathbf{P}(t+1,t) = \mathbf{P}(t+1)\mathbf{A}_s^T(t)\,, \tag{11.100}$$

which completes the derivation of the Rauch-Tung-Striebel or double-sweep smoother. The double-sweep terminology arises from the fact that the smoother requires first a forward Kalman filter sweep from 0 to T, followed by a backward sweep where recursions (11.97), (11.99) and (11.100) are propagated from T to 0.

Remark: Although this is not widely known, if $\mathbf{X}(t)$ is a Gauss-Markov process, the corresponding smoothing error $\tilde{\mathbf{X}}(t)$ is also Gauss-Markov. This result was first etablished in a stochastic realization context by Badawi, Lindquist and Pavon [15], and later derived by a quasi-martingale decomposition argument in [16,17]. To quickly verify this result, following [18] observe that a zero-mean DT Gaussian process $\{X(t),\ 0 \le t \le T\}$ is Markov if and only if the inverse $\mathbf{K}^{-1}$ of the covariance matrix $\mathbf{K}$ of $\mathbf{X}$ has a tridiagonal structure. Specifically, if $\mathbf{X}(t)$ has dimension n, and if we consider a partition of $\mathbf{M} = \mathbf{K}^{-1}$ into blocks $\mathbf{M}_{ij}$ of size $n \times n$, only the blocks $\mathbf{M}_{ij}$ with $|i-j| \le 1$ can be nonzero. Assume therefore that the covariance $\mathbf{K}$ of $\mathbf{X}$ has this property. Then the covariance $\mathbf{P}$ of the smoothing error vector $\tilde{\mathbf{X}}$ based on the observation vector $\mathbf{Y}$ satisfies

$$\mathbf{P}^{-1} = \mathbf{K}^{-1} + \mathbf{I}_n \otimes \mathbf{c}^T r^{-1}\mathbf{c} \tag{11.101}$$

where $\otimes$ denotes the matrix Kronecker product [13, p. 139]. Since $\mathbf{I}_n \otimes \mathbf{c}^T r^{-1}\mathbf{c}$ is block diagonal, the identity (11.101) ensures that $\mathbf{P}^{-1}$ has a tridiagonal structure, so $\tilde{\mathbf{X}}(t)$ is Gauss-Markov. Having established this property, consider now the problem of constructing a backwards state-space model for $\tilde{\mathbf{X}}(t)$. We have

$$\begin{aligned} E[\tilde{X}(t) \mid \tilde{\mathbf{X}}(s), t+1 \leq s \leq T] &= E[\tilde{X}(t) \mid \tilde{\mathbf{X}}(t+1)] \\ &= \mathbf{A}_s(t)\tilde{\mathbf{X}}(t+1) \end{aligned} \quad (11.102)$$

where to go from the first to the second line of (11.102) we have used the orthogonality of $\tilde{\mathbf{X}}(t+1)$ and $\tilde{\mathbf{X}}_c(t)$ in (11.97). The identity (11.102) establishes that (11.97) forms a backward state-space model for $\tilde{\mathbf{X}}(t)$ and thus the error sequence $\tilde{\mathbf{X}}_c(t)$ is a WGN, i.e.,

$$E[\tilde{\mathbf{X}}_c(t)\tilde{\mathbf{X}}_c^T(s)] = \mathbf{P}_c(t)\delta(t-s) \, .$$

11.4.3 Example

To illustrate the EM estimation of HGMM models with a double-sweep smoother, we consider a simple example.

Example 11.2: Ornstein Uhlenbeck process in WGN

We are given the observations

$$Y(t) = Z(t) + V(t) \quad (11.103)$$

for $0 \leq t \leq T$, where $Z(t)$ is a stationary DT Ornstein-Uhlenbeck process, so it has zero-mean and covariance

$$K(t,s) = E[Z(t)Z(s)] = a^{|t-s|}S, \quad (11.104)$$

and $V(t)$ is a zero-mean WGN of variance r independent of $Z(t)$. We assume here that, while the covariance structure (11.104) is known, the correlation parameter a and signal power S are unknown. To estimate these parameters, observe that $Z(t)$ admits a state-space model of the form

$$\begin{aligned} X(t+1) &= aX(t) + W(t) \\ Z(t) &= cX(t) \end{aligned} \quad (11.105)$$

for $0 \leq t \leq T-1$, where $W(t)$ is a scalar WGN sequence with unit variance, i.e.,

$$E[W(t)W(s)] = \delta(t-s) \, ,$$

and

$$c = ((1-a^2)S)^{1/2} \, . \quad (11.106)$$

We assume that the initial state $X(0)$ admits the stationary distribution

$$X(0) \sim N(0, \Pi)$$

with $\Pi = (1-a^2)^{-1}$, which ensures that $X(t)$ is stationary. As explained at the beginning of subsection 11.4.1, the choice $q = 1$ for the variance of the input noise $W(t)$ has the effect of fixing the internal coordinate system of the state-space model for $Z(t)$. Accordingly, the state-space model (11.105) is just parametrized by the two parameters a and c. Let $\hat{a}_{\text{ML}}$ and $\hat{c}_{\text{ML}}$ denote the ML estimates of a and c obtained by the EM iteration. Taking into account (11.106) and the fact that ML estimates do not depend on the choice of a coordinate system, we conclude that the ML estimate of the power S will be given by

$$\hat{S}_{\text{ML}} = \frac{\hat{c}^2_{\text{ML}}}{1 - \hat{a}^2_{\text{ML}}} .$$

So, estimating a and c is equivalent to estimating a and S.

Let a_k and c_k be the estimates of a and c obtained at the k-th EM iteration. They can be used to implement a double-sweep smoother. In the forward sweep the Kalman filter recursions

$$\begin{aligned}
\hat{X}_{fk}(t) &= a_k \hat{X}_{fk}(t-1) + G_k(t)\nu_k(t) \\
\nu_k(t) &= Y(t) - c_k a_k \hat{X}_{fk}(t-1) \\
G_k(t) &= [(a_k^2 P_{fk}(t-1) + 1)c_k]/P_{\nu k}(t) \\
P_{fk}(t) &= a_k^2 P_{fk}(t-1) + 1 - G_k^2(t) P_{\nu k}(t) \\
P_{\nu k}(t) &= c_k^2(a_k^2 P_{fk}(t-1) + 1) + r
\end{aligned} \tag{11.107}$$

are propagated for $0 \leq t \leq T$, starting with

$$\hat{X}_{fk}(-1) = 0 \quad , \quad P_{fk}(-1) = \frac{1}{1 - a_k^2} .$$

Here $\nu_k(t)$ and $G_k(t)$ denote respectively the innovations process and the Kalman filter gain,

$$P_{fk}(t) = E[(X(t) - \hat{X}_{fk}(t))^2]$$

represents the filtering error variance, and

$$P_{\nu k}(t) = E[\nu_k^2(t)]$$

is the innovations variance.

Then, let

$$P_{ck}(t) = (P_{fk}^{-1}(t) + a_k^2)^{-1}$$

and denote

$$a_{sk}(t) = a_k P_{ck}(t) .$$

The smoothed estimate and error variance satisfy the backward recursions

$$\begin{aligned}
\hat{X}_k(t) &= a_{sk}(t)\hat{X}_k(t+1) + P_{fk}(t)\hat{X}_{fk}(t)/P_{ck}(t) \\
P_k(t) &= a_{sk}^2(t) P_k(t+1) + P_{ck}(t)
\end{aligned} \tag{11.108}$$

with initial conditions

$$\hat{X}_k(T) = \hat{X}_{fk}(T) \quad , \quad P_k(T) = P_{fk}(T) .$$

Also, as noted in (11.100), the autocorrelation of the smoothing error at times t and $t-1$ is given by

$$P_k(t,t-1) = P_k(t)a_{sk}(t-1) .$$

After completion of the double-sweep smoothing pass, the $k+1$-th iterates are given by

$$a_{k+1} = \psi_k/\theta_k \quad , \quad c_{k+1} = \zeta_k/\omega_k , \tag{11.109}$$

where

$$\begin{aligned}
\psi_k &= \sum_{t=1}^{T} \hat{X}_k(t)\hat{X}(t-1) + P_k(t,t-1) \\
\theta_k &= \sum_{t=0}^{T-1} \hat{X}_k^2(t) + P_k(t) \\
\omega_k &= \sum_{t=0}^{T} \hat{X}_k^2(t) + P_k(t) \\
\zeta_k &= \sum_{t=0}^{T} Y(t)\hat{X}_k(t) .
\end{aligned}$$

Given the pair (a_{k+1}, c_{k+1}), an estimate of the signal power at the $k+1$-th iteration is given by

$$S_{k+1} = \frac{\hat{c}_{k+1}^2}{1-\hat{a}_{k+1}^2} .$$

To illustrate the performance of the HGMM estimation algorithm for the Ornstein-Uhlenbeck process, consider the case $a = 0.8$, $S = 1$, $T = 10^4$. Since the variance of ML estimates is proportional to $T^{-1/2}$, this choice of T ensures that the estimates of a and S will have an accuracy of a few percents. The initial parameter estimates are selected as $a_0 = 0.6$ and $S_0 = 1.2$, and 300 EM iterations are performed for measurement noise variance values $r = 1$, $r = 0.1$, and $r = 0.01$, which correspond respectively to SNR values of 0dB, 10dB, and 20dB. The values of the parameter estimate a_k at successive iterations k are plotted in Fig. 11.4, and the values of S_k as function of k are plotted in Fig. 11.5. For both plots, SNR values of 0dB, 10dB and 20dB are considered. The simulations show that while an increased SNR is beneficial for the estimation of a, since the EM iteration converges faster and is more accurate for a large SNR, the converse holds for estimates of the signal power S. As indicated in Fig. 11.5, for SNR values of 20dB, the estimate of the signal power S overshoots significantly the true value at low iterations before finally converging to the correct value.

The estimates $\hat{a}$ and $\hat{S}$ of the parameters a and S obtained after convergence of the EM iteration are listed in Table 11.1 as a function of the SNR. Note that the estimates become more accurate as the SNR increases. □

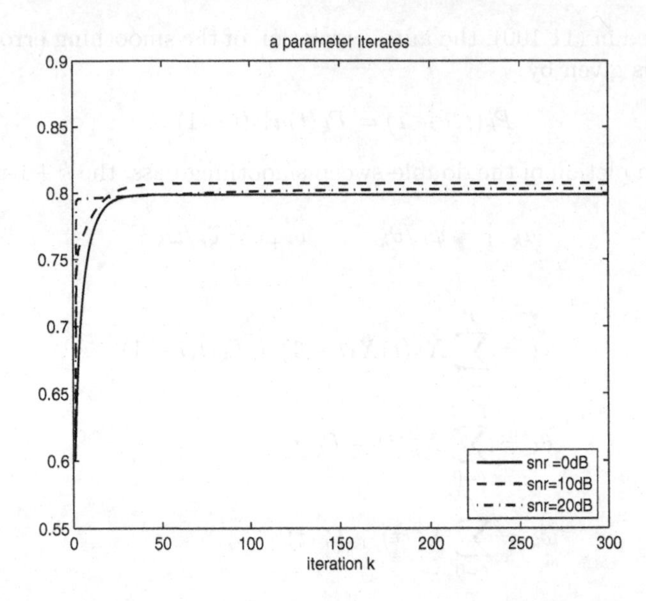

Fig. 11.4. Iterates a_k as a function of iteration k for SNR values of 0dB, 10dB and 20dB.

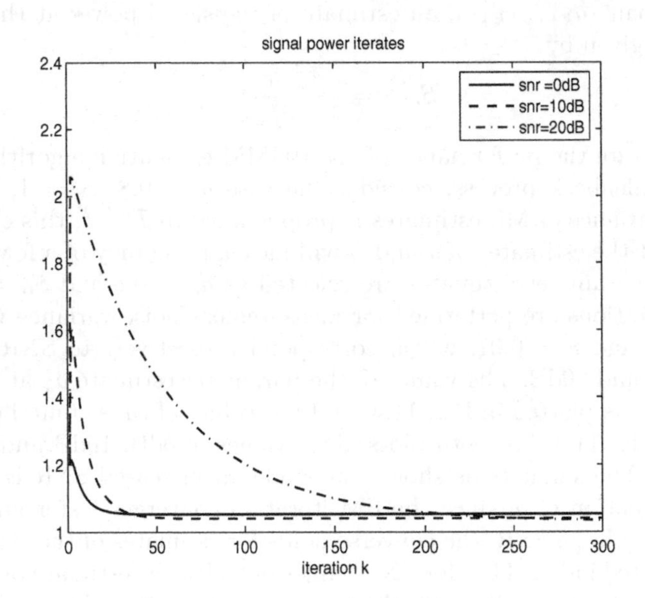

Fig. 11.5. Iterates S_k as a function of iteration k for SNR values of 0dB, 10dB and 20dB.

Table 11.1. Estimates of a and S for SNR values of 0dB, 10dB and 20dB.

SNR	$\hat{a}$	$\hat{S}$
0dB	0.7988	1.0533
10dB	0.8074	1.0402
20dB	0.8031	1.0344

11.5 GLRT Implementation

To illustrate the GLRT implementation for detecting a zero-mean Gaussian signal with unknown parameters in WGN, consider the detection problem

$$\begin{aligned} H_0 &: Y(t) = V(t) \\ H_1 &: Y(t) = AX(t) + V(t) \end{aligned} \tag{11.110}$$

for $0 \leq t \leq T$, where $X(t)$ is a zero-mean Gaussian process with covariance function

$$E[X(t)X(s)] = K(t,s)$$

for $0 \leq t,\ s \leq T$ and $V(t)$ is a zero-mean WGN with variance v. We assume that $X(t)$ and $V(t)$ are independent. The amplitude A of the signal under H_1 is unknown and the WGN variance v may or may not be known. The EM estimation of the parameters A and v under H_1 has been discussed in Example 11.1. In the analysis below, we assume the EM algorithm has been initialized properly, so the estimates $\hat{A}_1$ and $\hat{v}_1$ of A and v under H_1 correspond to the ML estimates.

Case 1: Unknown amplitude, known noise variance

Let us consider first the case where only A is unknown. Then, after convergence of the EM iteration, the amplitude estimate under H_1 can be expressed as

$$\hat{A}_1 = \frac{\mathbf{Y}^T \hat{\mathbf{X}}_1}{||\hat{\mathbf{X}}_1||_2^2 + \mathrm{tr}(\mathbf{P}_1)}, \tag{11.111}$$

where if $\mathbf{X}$ is the $N = T + 1$ dimensional vector defined in (11.2),

$$\begin{aligned} \hat{\mathbf{X}}_1 &= \mathbf{P}_1 \frac{\hat{A}_1}{v} \mathbf{Y} \\ \mathbf{P}_1 &= \Big(\mathbf{K}^{-1} + \frac{\hat{A}_1^2}{v} \mathbf{I}_N\Big)^{-1} \end{aligned} \tag{11.112}$$

denote respectively the conditional mean and covariance matrix of $\mathbf{X}$ given $\mathbf{Y}$ and H_1, evaluated with the estimated amplitude $\hat{A}_1$. Note that the pair $(\hat{\mathbf{X}}_1, \mathbf{P}_1)$ and amplitude estimate $\hat{A}_1$ form a fixed point of the EM iteration in the sense that $\hat{A}_1$ is used in the E-phase to evaluate $(\hat{\mathbf{X}}_1, \mathbf{P}_1)$, which are in turn used in the M-phase to obtain $\hat{A}_1$. Then consider the GLR statistic

$$L_G(\mathbf{Y}) = \frac{f(\mathbf{Y}|\hat{A}_1, H_1)}{f(\mathbf{Y}|H_0)} . \qquad (11.113)$$

Taking logarithms, the GLR statistic can be expressed as

$$2 \ln L_G(\mathbf{Y}) = -\mathbf{Y}^T(\hat{A}_1^2 \mathbf{K} + v\mathbf{I}_N)^{-1}\mathbf{Y} + \frac{||\mathbf{Y}||_2^2}{v}$$
$$- \ln |\hat{A}_1^2 \mathbf{K} + v\mathbf{I}_N| - N \ln(v) . \qquad (11.114)$$

By employing the matrix inversion identity (4.75), we obtain

$$(v\mathbf{I}_N + \hat{A}_1^2 \mathbf{K})^{-1} = \frac{1}{v}\mathbf{I}_N - \frac{\hat{A}_1}{v}(\mathbf{K}^{-1} + \frac{\hat{A}_1^2}{v}\mathbf{I}_N)^{-1}\frac{\hat{A}_1}{v}$$
$$= \frac{1}{v}\mathbf{I}_N - \frac{\hat{A}_1^2}{v^2}\mathbf{P}_1 .$$

Substituting this expression inside (11.114), we find that the GLRT takes the form

$$2 \ln L_G(|bfY) = \frac{\hat{A}_1}{v}\mathbf{Y}^T\hat{\mathbf{X}}_1 - \ln |\mathbf{I}_N + \frac{\hat{A}_1^2}{v}\mathbf{K}| \underset{H_0}{\overset{H_1}{\gtrless}} \gamma = 2\ln(\tau) , \qquad (11.115)$$

where the threshold γ is selected to ensure that the probability of false alarm is less than or equal to α.

Consider the eigendecomposition

$$\mathbf{K} = \boldsymbol{\Phi}\boldsymbol{\Lambda}\boldsymbol{\Phi}^T$$

of the covariance matrix $\mathbf{K}$, where the eigenvector matrix $\boldsymbol{\Phi}$ is orthonormal and

$$\boldsymbol{\Lambda} = \mathrm{diag}\,\{\lambda_j,\ 1 \le j \le N\}$$

is the diagonal matrix formed by the eigenvalues of $\mathbf{K}$. By observing that

$$\mathbf{I}_N + \frac{\hat{A}_1^2}{v}\mathbf{K} = \boldsymbol{\Phi}(\mathbf{I}_N + \frac{\hat{A}_1^2}{v}\boldsymbol{\Lambda})\boldsymbol{\Phi}^T ,$$

we conclude that

$$\ln |\mathbf{I}_N + \frac{\hat{A}_1^2}{v}\mathbf{K}| = \sum_{j=1}^{N} \ln(1 + \frac{\hat{A}_1^2}{v}\lambda_j) .$$

The GLRT (11.115) can be simplified in the *low SNR case* where

$$\frac{\hat{A}_1^2}{v}\lambda_j \ll 1$$

for all j. In this case

$$\ln(1 + \frac{\hat{A}_1^2}{v}\lambda_j) \approx \frac{\hat{A}_1^2}{v}\lambda_j$$

so that

$$\ln |\mathbf{I}_N + \frac{\hat{A}_1^2}{v}\mathbf{K}| \approx \frac{\hat{A}_1^2}{v}\sum_{j=1}^{N}\lambda_j = \frac{\hat{A}_1^2}{v}\mathrm{tr}(\mathbf{K}) . \tag{11.116}$$

Substituting approximation (11.116) inside the GLRT and taking into account expression (11.111) for $\hat{A}_1$, we find

$$2\ln L_{\mathrm{G}}(\mathbf{Y}) \approx \frac{\hat{A}_1^2}{v}\Big(E[||\mathbf{X}||_2^2|\mathbf{Y}, \hat{A}_1, H_1] - E[||\mathbf{X}||_2^2|H_1]\Big) , \tag{11.117}$$

where we have used

$$\mathrm{tr}(\mathbf{K}) = \mathrm{tr}\big(E[\mathbf{X}\mathbf{X}^T|H_1]\big) = E[||\mathbf{X}||_2^2|H_1] .$$

The term

$$E[||\mathbf{X}||_2^2|\mathbf{Y}, \hat{A}_1, H_1] - E[||\mathbf{X}||_2^2|H_1] = ||\hat{\mathbf{X}}_1||_2^2 + \mathrm{tr}\,(\mathbf{P}_1 - \mathbf{K})$$

appearing in expression (11.117) for $2\ln L_{\mathrm{G}}(\mathbf{Y})$ corresponds to the difference between the a posteriori and a priori expected energies of the signal $X(t)$ under H_1. So, in essence, the GLRT uses the *increase* in the expected energy of the observed signal based on the observations $\mathbf{Y}$ as the test statistic to decide between H_1 and H_0.

Case 2: Unknown amplitude and noise variance

Assume now that both A and v are unknown. In this case, after convergence of the EM iteration, the amplitude and noise variance estimates under H_1 are given respectively by (11.111) and

$$\begin{aligned}\hat{v}_1 &= \frac{1}{N}[||\mathbf{Y} - \hat{A}_1\hat{\mathbf{X}}_1||_2^2 + \hat{A}_1^2\mathrm{tr}\,(\mathbf{P}_1)] \\ &= \frac{1}{N}\mathbf{Y}^T(\mathbf{Y} - \hat{A}_1\hat{\mathbf{X}}_1) ,\end{aligned} \tag{11.118}$$

where the signal estimate and error covariance matrix are now expressed as

$$\begin{aligned}\hat{\mathbf{X}}_1 &= \mathbf{P}_1\frac{\hat{A}_1}{\hat{v}_1}\mathbf{Y} \\ \mathbf{P}_1 &= \Big(\mathbf{K}^{-1} + \frac{\hat{A}_1^2}{\hat{v}_1}\Big)^{-1} ,\end{aligned} \tag{11.119}$$

and where, to go from the first to second line of (11.118), we have taken into account expression (11.111) for $\hat{A}_1$. Under H_0, the ML estimate of v is given by

$$\hat{v}_0 = \frac{||\mathbf{Y}||_2^2}{N} .$$

Then the GLR statistic takes the form

$$L_G(\mathbf{Y}) = \frac{f(\mathbf{Y}|\hat{A}_1, \hat{v}_1, H_1)}{f(\mathbf{Y}|\hat{v}_0, H_0)} \tag{11.120}$$

where

$$\begin{aligned} f(\mathbf{Y}|\hat{v}_0, H_0) &= \frac{1}{(2\pi\hat{v}_0)^{N/2}} \exp\Big(-\frac{||\mathbf{Y}||_2^2}{2\hat{v}_0}\Big) \\ &= \frac{1}{(2\pi\hat{v}_0)^{N/2}} \exp(-N/2) \end{aligned} \tag{11.121}$$

and

$$\begin{aligned} f(\mathbf{Y}|\hat{A}_1, \hat{v}_1, H_1) = &\frac{1}{(2\pi)^{N/2}|\hat{A}_1^2\mathbf{K} + \hat{v}_1\mathbf{I}_N|^{1/2}} \\ &\cdot \exp\Big(-\frac{1}{2}\mathbf{Y}^T(\hat{A}_1^2\mathbf{K} + \hat{v}_1\mathbf{I}_N)^{-1}\mathbf{Y}\Big) . \end{aligned} \tag{11.122}$$

Application of the matrix identity (4.75) gives

$$\begin{aligned} (\hat{v}_1\mathbf{I}_N + \hat{A}_1^2\mathbf{K})^{-1} &= \frac{1}{\hat{v}_1}\mathbf{I}_N - \frac{\hat{A}_1}{\hat{v}_1}\Big(\mathbf{K}^{-1} + \frac{\hat{A}_1^2}{\hat{v}_1}\mathbf{I}_N\Big)^{-1}\frac{\hat{A}_1}{\hat{v}_1} \\ &= \frac{1}{\hat{v}_1}\mathbf{I}_N - \frac{\hat{A}_1}{\hat{v}_1}\mathbf{P}_1\frac{\hat{A}_1}{\hat{v}_1} . \end{aligned}$$

This implies

$$\mathbf{Y}^T(\hat{v}_1\mathbf{I}_N + \hat{A}_1^2\mathbf{K})^{-1}\mathbf{Y} = \frac{1}{\hat{v}_1}\mathbf{Y}^T(\mathbf{Y} - \hat{A}_1\hat{\mathbf{X}}_1) = N ,$$

so that (11.122) reduces to

$$f(\mathbf{Y}|\hat{A}_1, \hat{v}_1, H_1) = \frac{1}{(2\pi)^{N/2}|\hat{A}_1^2\mathbf{K} + \hat{v}_1\mathbf{I}_N|^{1/2}} \exp(-N/2) . \tag{11.123}$$

Substituting (11.121) and (11.123) inside (11.120), we then find

$$L_G(\mathbf{Y}) = \Big(\frac{\hat{v}_0}{\hat{v}_1}\Big)^{N/2} \frac{1}{|\mathbf{I}_N + (\hat{A}_1^2/\hat{v}_1)\mathbf{K}|^{1/2}} . \tag{11.124}$$

It is interesting to compare this expression with the GLR statistic (5.81) obtained in Chapter 5 for the detection of a *known signal* with an unknown amplitude in a WGN of unknown variance. We see that the first factor $(\hat{v}_0/\hat{v}_1)^{N/2}$ is common to both detectors, but the GLR expression (11.123) includes also a second factor involving the covariance $\mathbf{K}$ of the *random signal* $\mathbf{X}$.

In (11.124), observe that

$$\hat{v}_0 = \hat{v}_1 + \frac{\hat{A}_1}{N}\mathbf{Y}^T\hat{\mathbf{X}}_1$$

so that

$$\frac{\hat{v}_0}{\hat{v}_1} = 1 + \frac{\hat{A}_1}{N\hat{v}_1}\mathbf{Y}^T\hat{\mathbf{X}}_1 = 1 + \frac{\hat{A}_1^2}{N\hat{v}_1}E[||\mathbf{X}||_2^2|\mathbf{Y}, \hat{A}_1, \hat{v}_1, H_1] \tag{11.125}$$

with

$$E[||\mathbf{X}||_2^2|\mathbf{Y}, \hat{A}_1, \hat{v}_1, H_1] = ||\hat{\mathbf{X}}_1||_2^2 + \mathrm{tr}\,(\mathbf{P}_1)\,.$$

For the *low SNR* case, the second term in (11.125) is much smaller than 1, so

$$\left(\frac{\hat{v}_0}{\hat{v}_1}\right)^{N/2} \approx 1 + \frac{\hat{A}_1^2}{2\hat{v}_1}E[||\mathbf{X}||_2^2|\mathbf{Y}, \hat{A}_1, \hat{v}_1, H_1]$$

Similarly, as shown in (11.116),

$$|\mathbf{I}_N + \frac{\hat{A}_1^2}{\hat{v}_1}\mathbf{K}| \approx 1 + \frac{\hat{A}_1^2}{\hat{v}_1}\mathrm{tr}(\mathbf{K})\,.$$

Combining these two expressions, we find that in the low SNR case, the GLR statistic takes the form

$$2L_G(\mathbf{Y}) \approx \frac{\hat{A}_1^2}{\hat{v}_1}\Big(E[||\mathbf{X}||_2^2|\mathbf{Y}, \hat{A}_1, \hat{v}_1, H_1] - E[||\mathbf{X}||_2^2|H_1]\Big)\,, \tag{11.126}$$

which is identical to the statistic derived in (11.117) for the case when v is known, except that v is now replaced by its estimate $\hat{v}_1$.

11.6 Bibliographical Notes

The detection of Gaussian signals with unknown parameters is discussed in [19, Chaps. 8–9], where several examples are presented for which ML estimates cannot be expressed in closed form. After its introduction in [2], the EM algorithm has been studied extensively. Its convergence is proved in [4], and several variants have been proposed. The SAGE and AECM forms of the algorithm [5, 6] simplify the maximization stage by performing the maximization in a componentwise manner and cycling over different parameter components at successive iterations. In [20], it is shown that the EM algorithm can be interpreted as a proximal point algorithm [21] implemented with the estimated Kullback-Leibler divergence for the a-posteriori complete data density based on the observations. This interpretation allows the introduction of variants of the EM algorithm that have a faster rate of convergence. The expression (11.54) for the EM iteration matrix $\mathbf{F}(\boldsymbol{\theta}_*)$ in the vicinity of a stationary point is derived in [2], and its implication for the study of the convergence rate of the EM algorithm is examined in [22]. Finally, the supplemented EM algorithm is described in [8]. A full account of the EM algorithm, its variants, and properties is given in [3]. The EM procedure described in Section 11.4 for estimating the parameters of an HGMM was proposed by Schumway and Stoffer [9]. It relies on a recursive state-space smoother due to Rauch, Tung, and Striebel [23]. However the smoother derivation presented here is different from the one given in [23]. Another method of constructing this smoother consists of finding a triangularizing transformation [14, p. 386] for the Hamiltonian (10.48)–(10.49). Finally, it is worth noting that, although we have focused here on parameter estimation for Gaussian signals in WGN, the EM methodology is completely general, and in Chapter 13 it will be used for the estimation of hidden Markov chain models (HMCMs), i.e., partially observed Markov chains with unknown parameters.

11.7 Problems

11.1. Consider the DT Gaussian detection problem

$$\begin{aligned} H_0 &: Y(t) = V(t) \\ H_0 &: Y(t) = Z(t) + V(t) \end{aligned}$$

for $0 \leq t \leq T$. As in Problem 10.7, we assume that $V(t)$ is a zero-mean WGN with variance σ^2 and $Z(t)$ is a zero-mean stationary bandlimited WGN independent of $V(t)$ with PSD

$$S_Z(e^{j\omega}) = \begin{cases} q & |\omega| \leq W \\ 0 & W < |\omega| \leq \pi \,, \end{cases}$$

where the intensity q is unknown. We assume that T is large and that the bandwidth W is known and is an integer multiple of the resolution $2\pi/(T+1)$, i.e., $W = m2\pi/(T+1)$ with m integer.

When q is known, the LRT can be expressed as

$$S_T \underset{H_0}{\overset{H_1}{\gtrless}} 0 \,,$$

where S_T is given by (10.95) and admits approximation (10.107).

(a) Use approximation (10.107) to find the ML estimate $\hat{q}$ of q under H_1.
(b) Obtain a GLRT for the detection problem. By observing that the function $g(x) = x - 1 - \ln(x)$ is monotone increasing for $x > 1$, show that the test can be expressed as

$$\widehat{SNR} \underset{H_1}{\overset{H_1}{\gtrless}} \eta$$

where

$$\widehat{SNR} = \frac{1}{\sigma^2(2m+1)} \sum_{k=-m}^{m} |\hat{Y}(e^{j\omega_k})|^2$$

represents the estimated SNR of the received signal $Y(t)$ over the frequency band $[-W, W]$ under H_1. Here

$$\hat{Y}(e^{-j\omega_k}) = (T+1)^{-1/2} \sum_{t=0}^{T} Y(t) e^{-jt\omega_k} ,$$

with $\omega_k = k2\pi/(T+1)$ and $0 \leq k \leq T$, denotes the DFT of length $T+1$ of observations $Y(t)$, $0 \leq t \leq T$.

(c) By comparing the GLRT of part (b) with the test obtained in Problem 10.7, determine whether the GLRT is UMP.

11.2. Consider a noise cancellation problem, where the DT observations can be expressed as

$$Y_1(t) = X(t) + AN(t) + V_1(t)$$

for $0 \leq t \leq T$. The component $X(t)$ represents the desired signal, whereas $N(t)$ is an undesired noise which corrupts the observations with mixing constant A. $V_1(t)$ represents an additive zero-mean WGN with variance v_1 and independent of $X(t)$ and $N(t)$. A common method to remove the undesired noise component $AN(t)$ consists of placing a sensor close to the noise source, which measures

$$Y_2(t) = N(t) + V_2(t)$$

for $0 \leq t \leq T$, where $V_2(t)$ is a zero-mean WGN independent of $X(t)$, $N(t)$ and $V_1(t)$, with variance v_2. The resulting system is depicted in Fig. 11.6.

Then, assuming that $V_2(t)$ is small, $Y_2(t)$ can be used as a proxy for $N(t)$ in order to cancel the component $AN(t)$ of $Y_1(t)$. However, this requires an estimate of the mixing constant A. To do so, we propose to employ an EM iteration. The desired signal $X(t)$ and noise $N(t)$ can be written in vector form as

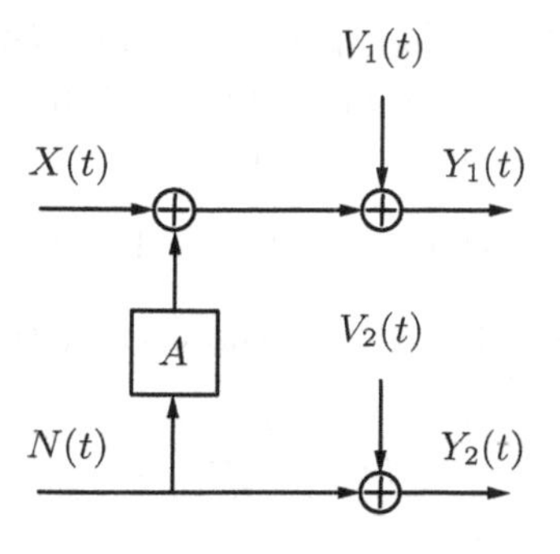

Fig. 11.6. Noise cancellation system.

$$\mathbf{X} = \begin{bmatrix} X(0) \cdots X(t) \cdots X(T) \end{bmatrix}^T$$
$$\mathbf{N} = \begin{bmatrix} N(0) \cdots N(t) \cdots N(T) \end{bmatrix}^T ,$$

and although vectors $\mathbf{X}$ and $\mathbf{N}$ are unknown, we assume they are statistically independent with the known statistics

$$\mathbf{X} \sim N(\mathbf{0}, \mathbf{K}_X) \quad , \quad \mathbf{N} \sim N(\mathbf{0}, \mathbf{K}_N) .$$

(a) The given data for this problem is formed by vectors $\mathbf{Y}_1$ and $\mathbf{Y}_2$ representing observations $Y_1(t)$ and $Y_2(t)$ for $0 \leq t \leq T$. The "missing" data can be selected as the noise vector $\mathbf{N}$. Let A_k denote the estimate of the coupling constant A obtained at the k-th EM iteration. Evaluate the conditional density
$$f(\mathbf{n}|\mathbf{y}_1, \mathbf{y}_2, A_k) \sim N(\hat{\mathbf{N}}_k, \mathbf{P}_k)$$
of vector $\mathbf{N}$ given observations $\mathbf{Y}_1$ and $\mathbf{Y}_2$ and estimate A_k. Note that, since all random vectors are Gaussian, the conditional density is Gaussian with mean $\hat{\mathbf{N}}_k$ and covariance matrix $\mathbf{P}_k$, so the only quantities that need to be evaluated are $\hat{\mathbf{N}}_k$ and $\mathbf{P}_k$.

(b) Given the conditional density obtained in part (a), evaluate the estimated complete log-likelihood function
$$Q(A, A_k) = E[\ln f(\mathbf{y}_1, \mathbf{y}_2, \mathbf{N}|A)|\mathbf{y}_1, \mathbf{y}_2, A_k] .$$

(c) Obtain the $k+1$-th EM iterate A_{k+1} by maximizing $Q(A, A_k)$.

11.3. Consider the DT signal

$$Y(t) = AX(t) + V(t)$$

for $0 \leq t \leq T$, where $X(t)$ is an Ornstein-Uhlenbeck (OU) process with known covariance function

$$E[X(t)X(s)] = K(t, s) = a^{|t-s|}P \ , \ |a| < 1 ,$$

and $V(t)$ is a zero-mean WGN independent of $X(t)$ with variance v. Both the amplitude A and noise variance v are unknown. Since $X(t)$ is an OU process, it admits a state-space model of the form

$$X(t+1) = aX(t) + W(t)$$

where $W(t)$ is a zero-mean WGN independent of $V(t)$ with variance

$$q = \frac{P}{1 - a^2} .$$

The initial state $X(0)$ is $N(0, P)$ distributed.

(a) Use the technique of Section 11.4 to obtain an EM iteration procedure for A and v. Express iterates A_k and v_k in terms of the smoothing estimates $\hat{X}_k(t)$ and error variance $P_k(t)$ obtained by the double-sweep smoother at the k-th iteration.
(b) Compare the complexity of the state-space based technique of part a) with the complexity of iteration (11.26), (11.29)–(11.30), which employs the covariance matrix $\mathbf{K}$ of the vector $\mathbf{X}$ reqrouping all values of $X(t)$ for $0 \leq t \leq T$.

11.4. We observe a signal

$$Y(t) = Z(t) + V(t)$$

for $0 \leq t \leq T$, where $V(t)$ is zero-mean WGN with known variance v, and $Z(t)$ is a first-order autoregressive moving average process modelled by the recursion

$$Z(t) + aZ(t-1) = W(t) + bW(t-1)$$

for $1 \leq t \leq T$, where the scalar parameters a and b are unknown and need to be estimated, and $W(t)$ is a zero-mean WGN independent of $V(t)$ with known variance q. Assume $|a| < 1$, so the transfer function

$$H(z) = \frac{1 + bz^{-1}}{1 + az^{-1}}$$

is stable.

(a) Verify that the process $Z(t)$ admits a first-order state-space realization of the form

$$\begin{aligned} X(t+1) &= -aX(t) + W(t) \\ Z(t) &= (b-a)X(t) + W(t) \,. \end{aligned}$$

(b) Use the state-space model obtained in (a) to develop an EM iteration procedure for estimating a and b from the given observations $Y(t)$ with $0 \leq t \leq T$.

11.5. The EM estimation technique can be applied to amplitude and time delay estimation problems. Consider the DT observations

$$\begin{bmatrix} Y_1(t) \\ Y_2(t) \end{bmatrix} = \begin{bmatrix} X(t) \\ AX(t-D) \end{bmatrix} + \begin{bmatrix} V_1(t) \\ V_2(t) \end{bmatrix} \tag{11.127}$$

obtained by two different sensors for $0 \leq t \leq T$. The received signal $X(t)$ is a zero-mean stationary Gaussian process of known spectral density $S_X(e^{j\omega})$, and $V_1(t)$ and $V_2(t)$ are two independent zero-mean WGNs with variance v, which are independent of $X(t)$. We seek to estimate the amplitude A and delay D, so for this problem, the parameter vector

$$\boldsymbol{\theta} = \begin{bmatrix} A \\ D \end{bmatrix} .$$

The above problem arises in array processing, where if a target emits a signal $X(t)$ of known spectral signature, it is desired to estimate the target location from the delays D_1 and D_2 at which the signal is received at two sensors compared to a reference sensor. Here the reference sensor measures $Y_1(t)$, and we consider the estimation of a single time delay D from the observations $Y_2(t)$ at a second sensor whose location with respect to the reference sensor is known.

The number $N = T+1$ of observations is assumed to be large. Accordingly, if we evaluate the DFTs

$$\hat{X}(e^{j\omega_\ell}) = \frac{1}{N^{1/2}} \sum_{t=0}^{T} X(t) e^{-jt\omega_\ell}$$

$$\hat{Y}_i(e^{j\omega_\ell}) = \frac{1}{N^{1/2}} \sum_{t=0}^{T} Y_i(t) e^{-jt\omega_\ell}$$

$$\hat{V}_i(e^{j\omega_\ell}) = \frac{1}{N^{1/2}} \sum_{t=0}^{T} V_i(t) e^{-jt\omega_\ell} ,$$

with $\omega_\ell = \ell 2\pi/N$, $0 \leq \ell \leq T$, observations (11.127) can be rewritten as

$$\begin{bmatrix} \hat{Y}_1(e^{j\omega_\ell}) \\ \hat{Y}_2(e^{j\omega_\ell}) \end{bmatrix} = \begin{bmatrix} 1 \\ Ae^{-j\omega_\ell D} \end{bmatrix} \hat{X}(e^{j\omega_\ell}) + \begin{bmatrix} \hat{V}_1(e^{j\omega_\ell}) \\ \hat{V}_2(e^{j\omega_\ell}) \end{bmatrix} \tag{11.128}$$

with $0 \leq \ell \leq T$. In this expression, since the amplitude A and delay D remains the same for the different frequency components $\hat{X}(e^{j\omega_\ell})$ of signal $X(t)$, $X(t)$ is assumed to be a narrowband signal. Note that since the sequences $X(t)$, $V_i(t)$ and $Y_i(t)$ with $i = 1, 2$ are real, the DFT coefficients admit the symmetry

$$\begin{aligned} \hat{X}(e^{j\omega_\ell}) &= \hat{X}^*(e^{j\omega_{N-\ell}}) \\ \hat{V}_i(e^{j\omega_\ell}) &= \hat{V}_i^*(e^{j\omega_{N-\ell}}) \\ \hat{Y}_i(e^{j\omega_\ell}) &= \hat{Y}_i^*(e^{j\omega_{N-\ell}}) , \end{aligned} \tag{11.129}$$

so that only nonredundant coefficients need to be retained. Specifically, for N even, only coefficients with $0 \leq \ell \leq N/2$ are independent, whereas for N odd, only indices $0 \leq \ell \leq (N-1)/2$ yield independent observations. So in both cases only $\lfloor N/2 \rfloor$ coefficients need to be retained, where for a real number x, $\lfloor x \rfloor$ represents the largest integer less than or equal to x. The advantage of transformation (11.128) is that since the data record is long, as explained in Section 10.4, the DFT whitens the sequence $X(t)$, so the transformed sequences $\hat{X}(e^{j\omega_\ell})$ and $\hat{V}_i(e^{j\omega_\ell})$ are circular complex white Gaussian noise sequences with

$$\hat{X}(e^{j\omega_\ell}) \sim CN(0, S_X(e^{j\omega_\ell})) \qquad \hat{V}_i(e^{j\omega_\ell}) \sim CN(0, v)$$

for $i = 1, 2$ and $1 < \ell < N/2$. For indices $\ell = 0$ and $\ell = N/2$ when N is even, the DFT coefficients are real and admit the normal distributions

$$\hat{X}(e^{j\omega_\ell}) \sim N(0, S_X(e^{j\omega_\ell})) \qquad \hat{V}_i(e^{j\omega_\ell}) \sim N(0, v)\,.$$

This implies that the estimate of each DFT coefficient $\hat{X}(e^{j\omega_\ell})$ from observations $\hat{Y}_i(e^{j\omega_m})$, with $i = 1, 2$ and $0 \leq m \leq T$, will depend only on the observation coefficients $Y_i(e^{j\omega_\ell})$, $i = 1, 2$ with the same frequency. Thus the estimation problem for the DFT coefficients $\hat{X}(e^{j\omega_\ell})$ is completely decoupled.

To obtain an EM iteration for estimating A and D, the given data

$$\mathbf{Y}(t) = \begin{bmatrix} Y_1(t) \\ Y_2(t) \end{bmatrix}$$

with $0 \leq t \leq T$ can be augmented by the "missing data" $X(t)$, $0 \leq t \leq T$. Equivalently, in the DFT domain, we augment

$$\hat{\mathbf{Y}}_\ell = \begin{bmatrix} \hat{Y}_1(e^{j\omega_\ell}) \\ \hat{Y}_2(e^{j\omega_\ell}) \end{bmatrix}$$

with $0 \leq \ell \leq \lfloor N/2 \rfloor$, by $\hat{X}_\ell = \hat{X}(e^{j\omega_\ell})$ with $0 \leq \ell \leq \lfloor N/2 \rfloor$.

(a) Let

$$\boldsymbol{\theta}^k = \begin{bmatrix} A_k \\ D_k \end{bmatrix}$$

denote the k-th iterate of parameter vector $\boldsymbol{\theta}$, and let $f(\hat{x}_\ell|\hat{\mathbf{y}}_\ell, \boldsymbol{\theta}^k)$ be the conditional density of $\hat{X}_\ell$ given observation vector $\hat{\mathbf{Y}}_\ell$ based on parameter vector $\boldsymbol{\theta}^k$. By using the model (11.128), verify that

$$f(\hat{x}_\ell|\hat{\mathbf{y}}_\ell, \boldsymbol{\theta}^k) \sim CN(m_\ell^k, P_\ell^k)$$

with

$$\begin{aligned} P_\ell^k &= \left[S(e^{j\omega_\ell})^{-1} + (1 + A_k^2)v^{-1}\right]^{-1} \\ m_\ell^k &= \frac{P_\ell^k}{v}(\hat{Y}_1(e^{j\omega_\ell}) + A_k e^{j\omega_\ell D_k} Y_2(e^{j\omega_\ell})\,. \end{aligned}$$

For further information on estimation problems involving complex random variables, the reader is referred to [24, Sec. 15.8] and [25, Chap. 5].

(b) Use the conditional density obtained in part (a) to evaluate

$$\begin{aligned} Q(\boldsymbol{\theta}, \boldsymbol{\theta}^k) &\triangleq \sum_{\ell=0}^{\lfloor N/2 \rfloor} E[\ln f(\hat{\mathbf{y}}_\ell, \hat{X}_\ell|\boldsymbol{\theta})|\hat{\mathbf{y}}_\ell, \boldsymbol{\theta}^k] \\ &= -\frac{1}{2v} \sum_{\ell=0}^{T} E[||\hat{\mathbf{y}}_\ell - \begin{bmatrix} 1 \\ Ae^{-j\omega_\ell D} \end{bmatrix} \hat{X}_\ell||_2^2 | \hat{\mathbf{y}}_\ell, \boldsymbol{\theta}^k] + c\,, \end{aligned} \qquad (11.130)$$

where the constant c is independent of A and D. Together, parts (a) and (b) form the E-phase of the EM iteration. To explain the different summation ranges in going from the first to the second line of (11.130), note first that due to the symmetry property (11.129), only the DFT coefficients $\hat{\mathbf{Y}}(e^{j\omega_\ell})$ and $\hat{X}(e^{j\omega_\ell})$ for $0 \leq \ell \leq \lfloor N/2 \rfloor$ are independent. However, while almost all coefficients are complex with a circular Gaussian distribution, the coefficients for $\ell = 0$ and $\ell = N/2$ when N is even and real. To use a common notation for the real and complex coefficients, instead of using a scaling of $(2v)^{-1}$ for real coefficients and v^{-1} for complex coefficients, we double count the complex coefficients by replacing the summation over $0 \leq \ell \leq \lfloor N/2 \rfloor$ by one for $0 \leq \ell \leq N-1$, but with a weighting of $(2v)^{-1}$ to compensate for the double counting.

(c) Maximize $Q(\boldsymbol{\theta}|\boldsymbol{\theta}^k)$ and obtain expressions for the new iterates A_{k+1} and D_{k+1}. These expressions form the M-phase of the EM iteration.

11.6. For the noise contamination model of Problem 11.2, it is useful to determine if noise contamination is taking place or not. Design a GLR test to decide between hypotheses $H_0 : A = 0$ and $H_1 : A \neq 0$.

11.7. A zero-mean Gaussian process $Y(t)$ is observed for $0 \leq t \leq T$. The process $Y(t)$ obeys a first-order autoregressive model

$$Y(t) + aY(t-1) = V(t) ,$$

with $|a| < 1$, where $V(t)$ is zero-mean WGN with variance v. The coefficient a is unknown. Design a GLR test for choosing between $H_0 : a = 0$ (in which case $Y(t)$ is just a WGN) and $H_1 : a \neq 0$.

References

1. D. Bertsekas, *Nonlinear Programming, Second Edition.* Belmont, MA: Athena Scientific, 1999.
2. A. P. Dempster, N. M. Laird, and D. B. Rubin, “Maximum likelihood from incomplete data via the EM algorithm,” *J. Royal Stat. Society, Series B*, vol. 39, no. 1, pp. 1–38, 1977.
3. G. J. McLachlan and T. Krishnan, *The EM Algorithm and Extensions.* New York: J. Wiley & Sons, 1997.
4. C. F. J. Wu, “On the convergence properties of the EM algorithm,” *Annals Statistics*, vol. 11, pp. 95–103, 1983.
5. X.-L. Meng and D. van Dyk, “The EM algorithm – an old folk-song sung to a fast new tune,” *J. Royal Stat. Soc., Series B*, vol. 59, no. 3, pp. 511–567, 1997.
6. J. A. Fessler and A. Hero, “Space-alternating generalized expectation-maximization algorithm,” *IEEE Trans. Signal Proc.*, vol. 42, pp. 2664–2677, Oct. 1994.
7. J. R. Magnus and H. Neudecker, *Matrix Differential Calculus with Applications in Statistics and Econometrics.* Chichester, England: J. Wiley & Sons, 1988.

8. X.-L. Meng and D. B. Rubin, "Using EM to obtain asymptotic variance-covariance matrices: The SEM algorithm," *J. Amer. Stat. Assoc.*, vol. 86, pp. 899–909, Dec. 1991.
9. R. H. Schumway and D. S. Stoffer, "An approach to time series smoothing and forecasting using the EM algorithm," *J. Time Series Analysis*, vol. 3, pp. 253–264, 1982.
10. R. H. Schumway and D. S. Stoffer, *Time Series Analysis and its Applications.* New York: Springer-Verlag, 2000.
11. P. E. Caines, *Linear Stochastic Systems.* New York: J. Wiley & Sons, 1988.
12. L. Ljung, *System Identification: Theory for the User, Second edition.* Upper Saddle River, NJ: Prentice-Hall, 1999.
13. A. J. Laub, *Matrix Analysis for Scientists and Engineers.* Philadelphia, PA: Soc. for Industrial and Applied Math., 2005.
14. T. Kailath, A. H. Sayed, and B. Hassibi, *Linear Estimation.* Upper Saddle River, NJ: Prentice Hall, 2000.
15. F. A. Badawi, A. Lindquist, and M. Pavon, "A stochastic realization approach to the smoothing problem," *IEEE Trans. Automat. Control*, vol. 24, pp. 878–888, 1979.
16. M. G. Bello, A. S. Willsky, B. C. Levy, and D. A. Castanon, "Smoothing error dynamics and their use in the solution of smoothing and mapping problems," *IEEE Trans. Informat. Theory*, vol. 32, pp. 483–495, 1986.
17. M. G. Bello, A. S. Willsky, and B. C. Levy, "Construction and application of discrete-time smoothing error models," *Int. J. Control*, vol. 50, pp. 203–223, 1989.
18. B. C. Levy, R. Frezza, and A. J. Krener, "Modeling and estimation of discrete-time Gaussian reciprocal processes," *IEEE Trans. Automatic Control*, vol. 35, pp. 1013–1023, Sept. 1990.
19. S. M. Kay, *Fundamentals of Statistical Signal Processing: Detection Theory.* Prentice-Hall, 1998.
20. S. Chrétien and A. O. Hero, "Kullback proximal algorithms for maximum likelihood estimation," *IEEE Trans. Informat. Theory*, vol. 46, pp. 1800–1810, Aug. 2000.
21. R. T. Rockafellar, "Monotone operators and the proximal point algorithm," *SIAM J. Control Optim.*, vol. 14, pp. 877–898, 1976.
22. X.-L. Meng and D. B. Rubin, "On the global and componentwise rates of convergence of the EM algorithm," *Linear Algebra Appl.*, vol. 199, pp. 413–425, 1994.
23. H. E. Rauch, F. Tung, and C. T. Striebel, "Maximum likelihood estimates of linear dynamic systems," *J. Amer. Inst. Aero. Astro.*, vol. 3, pp. 1445–1450, 1965.
24. S. M. Kay, *Fundamentals of Statistical Signal Processing: Estimation Theory.* Prentice-Hall, 1993.
25. K. S. Miller, *Complex Stochastic Processes.* Readings, MA: Addison-Wesley, 1974.

Part III

Markov Chain Detection

12

Detection of Markov Chains with Known Parameters

12.1 Introduction

Although the Gaussian signal model used in the second part of the book is applicable to a wide class of radar and sonar detection problems, it has become increasingly ill-adapted to the study of modern digital communications systems. Specifically, it assumes that over each observation interval, the received signal is free of intersymbol interference (ISI). Although for a fixed channel and a matched filter receiver, ISI can be theoretically avoided by proper design of a signaling pulse, many channels such as wireless channels change rapidly, so for such channels, ISI is unavoidable and an equalizer needs to be implemented. Another implicit assumption of the detection problems we have considered up to this point is that channel coding and modulation are two separate operations, so that signal demodulation and decoding can be handled separately. While this viewpoint was appropriate in the 1960s and 1970s when signal detection was first developed as a coherent theory, subsequent developments have made this picture obsolete. Specifically, the introduction of trellis-coded modulation [1] in order to transmit data reliably over bandlimited channels has made it impossible to keep the demodulation and decoding operations separate. More recently, turbo equalizers [2, 3] have also demonstrated the benefit of cooperation between the equalization and decoding blocks of communication receivers. For all the above mentioned examples, the received signal can no longer be modelled as a deterministic signal (possibly with unknown parameters) in WGN. Instead, it needs to be viewed as a function of the state of a Markov chain or of its state transitions, possibly with unknown parameters, observed in WGN. We refer the reader to [4, 5] for recent presentations of detection problems of digital communications from this perspective.

In addition to digital communications, Markov chain (MC) detection problems arise in the study of speech recognition and segmentation problems [6, 7]. So in this chapter we examine detection problems where the observed signal is a function of the states or transitions of a Markov chain. The chapter is

B.C. Levy, *Principles of Signal Detection and Parameter Estimation*,
DOI: 10.1007/978-0-387-76544-0_12, © Springer Science+Business Media, LLC 2008

organized as follows: In Section 12.2, we examine the binary hypothesis testing problem for exactly observed Markov chains. It is found that the empirical probability distribution for two successive states of the Markov chain forms a sufficient statistic, and by analogy with the minimum Euclidean distance rule for testing Gaussian signals in WGN, the optimum test can be expressed as a mimimum discrimination function test, where the discrimination function is a measure of proximity of joint probability distributions for pairs of discrete random variables. The asymptotic rate of decay of the two types of errors for the optimum test as the number of observations increases is evaluated by using the theory of large deviations. The detection of partially observed Markov chains is examined in Section 12.3. It turns out that, given noisy observations of an MC over a fixed time interval, we can consider either an MAP sequence detection problem or a pointwise MAP detection problem. The MAP sequence detection problem seeks to maximize the a-posteriori probability of the entire state trajectory, whereas pointwise MAP detection maximizes the a-posteriori probability of the state value at each instant in time. The calculation of the MAP sequence is performed by the Viterbi algorithm, which is an instance of the method of dynamic programming. The evaluation of the pointwise MAP state probabilities is accomplished recursively by the forward-backward algorithm. Although the Viterbi algorithm is slightly more efficient than the forward-backward algorithm, the latter algorithm evaluates the a-posteriori probability of each state at each time. This soft information makes it ideally suited for use in iterative/turbo decoding algorithms of the type discussed in [4, 5]. To illustrate the detection of partially observed Markov chains, Section 12.4 examines the MAP equalization of ISI signals in noise.

12.2 Detection of Completely Observed Markov Chains

12.2.1 Notation and Background

For implementation reasons, the Markov chains arising in engineering systems have usually a finite number of states. So throughout this chapter, we restrict our attention to DT Markov chains $X(t)$ with $t \geq 1$ where the state $X(t)$ takes values in a finite set $\mathcal{S} = \{1, 2, \ldots, k, \ldots, K\}$. We also assume that the Markov chain is homogeneous, i.e., time-invariant. Then the Markov chain is specified by the row vector

$$\boldsymbol{\pi}(1) = \left[\pi_1(1)\ \pi_2(1) \ldots \pi_k(1) \ldots \pi_K(1)\right]$$

representing its initial probability distribution

$$\pi_k(1) = P[X(1) = k] \,, \tag{12.1}$$

and by its one-step transition probability matrix $\mathbf{P}$. This matrix has dimension $K \times K$ and its (k, ℓ)-th element is given by

$$p_{k\ell} = P[X(t+1) = \ell | X(t) = k] \,. \tag{12.2}$$

The set of allowed state transitions is denoted as

$$\mathcal{T} = \{(k, \ell) \in \mathcal{S}^2 \, : \, p_{k\ell} > 0\} \,. \tag{12.3}$$

To this set, we can associate a directed graph $\mathcal{G}$ which is called the transition diagram of the Markov chain (MC). This graph has for vertices the states of $\mathcal{S}$ and for edges the elements of $\mathcal{T}$, i.e., an edge goes from vertex k to vertex ℓ if $p_{k\ell} > 0$. The graph $\mathcal{G}$ is used to determine several important properties of the MC. Specifically, the MC is irreducible if there exists a directed path from any vertex k to any other vertex ℓ. In other words, if an MC is irreducible, all its states are connected to each other. In this case, independently of how the MC is initialized, all the states are visited an infinite number of times, and the MC has a single recurrent class. The period d of an MC is the greatest common divisor of the lengths of the cycles of its graph. The MC is said to be aperiodic if $d = 1$. To fix ideas, we consider a simple example.

Example 12.1: Transition diagram of a Markov chain

Consider a Markov chain with one-step transition probability matrix

$$\mathbf{P} = \begin{bmatrix} 0 & 0 & 1/2 & 1/2 \\ 0 & 0 & 1/2 & 1/2 \\ 1/2 & 1/2 & 0 & 0 \\ 1/2 & 1/2 & 0 & 0 \end{bmatrix} \,.$$

This MC can be used to model the phase states of a continuous-phase frequency-shift-keyed (CPFSK) signal with modulation index $h = 1/2$ [8, p. 193]. Its state transition diagram is shown in Fig. 12.1.

By inspection it can be verified that all states are connected, so the MC is irreducible. The MC has several elementary cycles of length 2, such as the cycle 1-3-1. Recall that a cycle is called elementary if only the first and last

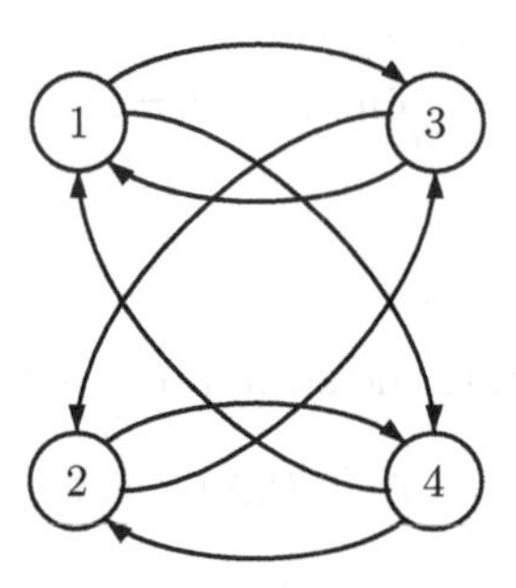

Fig. 12.1. Transition diagram $\mathcal{G}$ of the MC with transition matrix $\mathbf{P}$.

vertex are repeated. Similarly, it has several elementary cycles of length 4, such as 1-4-2-3-1. Accordingly, the period of the MC is $d = 2$. Note that this is easy to see both from the block structure of $\mathbf{P}$ and from the transition diagram $\mathcal{G}$. Specifically, the set $\mathcal{S} = \{1, 2, 3, 4\}$ of states can be partitioned into two groups: $\mathcal{S}_1 = \{1, 2\}$ and $\mathcal{S}_2 = \{3, 4\}$, and if the state $X(t)$ belongs to one of the two groups at time t, it jumps to a member of the other group at time $t+1$. In the block partition

$$\mathbf{P} = \begin{bmatrix} \mathbf{0} & \mathbf{P}_T \\ \mathbf{P}_T & \mathbf{0} \end{bmatrix}$$

of $\mathbf{P}$, with

$$\mathbf{P}_T = \frac{1}{2}\begin{bmatrix} 1 & 1 \\ 1 & 1 \end{bmatrix} ,$$

the zero block diagonal matrices clearly indicate that there is no transition internal to either $\mathcal{S}_1$ or $\mathcal{S}_2$. Similarly, the graph $\mathcal{G}$ has a bipartite structure, where all edges originating from states in $\mathcal{S}_1$ terminate in $\mathcal{S}_2$, and vice versa. □

In this section it is assumed that the MCs we consider are irreducible and aperiodic, so that all states form a single recurrent class. Then, as a consequence of the Perron-Frobenius theorem for matrices with nonnegative entries [9, Chap. 8], [10, Sec. 4.4], $\lambda = 1$ is the eigenvalue of $\mathbf{P}$ with the largest magnitude, it has multiplicity one, its corresponding right eigenvector is

$$\mathbf{e} = \begin{bmatrix} 1 \; 1 \ldots 1 \end{bmatrix}^T \tag{12.4}$$

up to a scale factor, and when the sum of its entries is normalized to one, its corresponding left eigenvector is a probability distrbution

$$\boldsymbol{\pi} = \begin{bmatrix} \pi_1 \ldots \pi_k \ldots \pi_K \end{bmatrix} \tag{12.5}$$

with $\pi_k > 0$ for all $1 \leq k \leq K$. This implies that the t-step probability transition matrix P^t satisfies

$$\lim_{t \to \infty} P^t = \mathbf{e}\boldsymbol{\pi} . \tag{12.6}$$

Let also

$$\boldsymbol{\pi}(t) = \begin{bmatrix} \pi_1(t) \ldots \pi_k(t) \ldots \pi_K(t) \end{bmatrix} \tag{12.7}$$

be the row vector representing the state probability distribution

$$\pi_k(t) = P[X(t) = k]$$

of the Markov chain at time t. Since it obeys

$$\boldsymbol{\pi}(t) = \boldsymbol{\pi}(1)\mathbf{P}^{t-1} , \tag{12.8}$$

we conclude from (12.6) that

$$\lim_{t \to \infty} \boldsymbol{\pi}(t) = \boldsymbol{\pi} . \tag{12.9}$$

In other words, $\boldsymbol{\pi}$ is the steady-state probability distribution of the Markov chain, independently of the choice of initial probability distribution $\boldsymbol{\pi}(1)$.

Next, suppose that we observe the Markov chain $X(t)$ over a time interval $1 \leq t \leq N+1$. The observed state at time t is $x_t \in \mathcal{S}$. Let $N_{k\ell}(\mathbf{x})$ denote the number of transitions from state k to state ℓ observed in the sequence

$$\mathbf{x} = (x_1, \ldots, x_t, \ldots, x_{N+1}) .$$

Then the number

$$N_k(\mathbf{x}) = \sum_{\ell=1}^{K} N_{k\ell}(\mathbf{x}) \tag{12.10}$$

of transitions out of state k corresponds also to the number of times the state k appears in the truncated state sequence

$$\mathbf{x}_T = (x_1, \ldots, x_t, \ldots, x_N)$$

obtained by removing the last observation from the sequence $\mathbf{x}$. We define the empirical successive state probability distribution of the Markov chain as the function

$$q_{k\ell}(\mathbf{x}) = \frac{N_{k\ell}(\mathbf{x})}{N} \tag{12.11}$$

defined for $1 \leq k, \ell \leq K$. By using the strong law of large numbers, it is easy to verify that as $N \to \infty$

$$q_{k\ell}(\mathbf{X}) \xrightarrow{\text{a.s.}} \pi_k p_{k\ell} \tag{12.12}$$

and

$$\sum_{\ell=1}^{K} q_{k\,\ell}(\mathbf{X}) = \frac{N_k(\mathbf{X})}{N} \xrightarrow{\text{a.s.}} \pi_k , \tag{12.13}$$

so both the one-step transition probability matrix $\mathbf{P}$ and the steady-state probability distribution $\boldsymbol{\pi}$ can be recovered asymptotically from the empirical distribution q.

The empirical probability distribution q belongs to the class $\mathcal{Q}$ of probability mass distributions over $\mathcal{S}^2$. This class is formed by functions $\alpha_{k\ell}$ such that $\alpha_{k\ell} \geq 0$ and

$$\sum_{k\ell} \alpha_{k\ell} = 1.$$

For a distribution α of $\mathcal{Q}$, we define the marginal distribution with respect to the first variable as

$$\alpha_k = \sum_{\ell} \alpha_{k\ell} , \tag{12.14}$$

and the conditional distribution of the second variable given the first as

$$\alpha_{\ell|\,k} = \alpha_{k\ell}/\alpha_k . \tag{12.15}$$

Equivalently, if we consider the $K \times K$ matrix with entries $\alpha_{k,\ell}$, α_k represents the sum of the entries of the k-th row of this matrix and $\alpha_{\cdot|\,k}$ is the row obtained after normalizing the k-th row by α_k. Thus $\alpha_{\cdot|\,k}$ is itself a probability distribution, i.e.,

$$\sum_{\ell=1}^{K} \alpha_{\ell|\,k} = 1 .$$

Then, given two distributions α and β of $\mathcal{Q}$, the information discrimination measure between α and β is defined as

$$\begin{aligned} J(\alpha, \beta) &= \sum_{k\,\ell} \alpha_{k\,\ell} \ln \left(\frac{\alpha_{\ell|\,k}}{\beta_{\ell|\,k}} \right) \\ &= \sum_{k=1}^{K} \alpha_k D(\alpha_{\cdot|\,k} | \beta_{\cdot|\,k}) . \end{aligned} \tag{12.16}$$

In other words, $J(\alpha, \beta)$ is a weighted sum of the Kullback-Leibler divergences of the normalized rows of the matrices representing α and β, where the weights are given by the marginal distribution of α. Since $\alpha_k \geq 0$ and the KL divergence is nonnegative, we deduce that $J(\alpha, \beta) \geq 0$. It is of particular interest to examine the situation where α and β are the steady-state joint probability distributions for two successive states $X(t)$ and $X(t+1)$ of Markov chains with one-step transition probability matrices $\mathbf{P}^0$ and $\mathbf{P}^1$. We assume both chains are irreducible and aperiodic with the same set $\mathcal{T}$ of allowed transitions, and thus the same transition diagram $\mathcal{G}$. Let $\boldsymbol{\pi}^0$ and $\boldsymbol{\pi}^1$ be the steady-state marginal probability distributions of state $X(t)$ corresponding to $\mathbf{P}^0$ and $\mathbf{P}^1$. Note that, since they are the left eigenvectors of $\mathbf{P}^0$ and $\mathbf{P}^1$ corresponding to $\lambda = 1$, they are uniquely specified by the one-step transition probability matrices. Then the steady state probability distributions of two successive states for the two Markov chains are given by

$$\alpha_{k\ell} = \pi^0_k p^0_{k\ell} \quad , \quad \beta_{k\ell} = \pi^1_k p^1_{k\ell} \tag{12.17}$$

and the information discrimination measure can be expressed as

$$J(\alpha, \beta) = \sum_{k=1}^{K} \pi^0_k D(p^0_{k\cdot} | p^1_{k\cdot}) . \tag{12.18}$$

Note that the requirement that both chains have the same set $\mathcal{T}$ of allowed transitions ensures that $J(\alpha, \beta)$ is finite. If the allowed transitions $\mathcal{T}^0$ and $\mathcal{T}^1$ of both chains are different, $J(\alpha, \beta) = \infty$. Then, since $\pi^0_k > 0$ for all k, we conclude that $J(\alpha, \beta) = 0$ if and only if $\mathbf{P}^0 = \mathbf{P}^1$, i.e., if and only if the Markov chains are identical. Thus the function $J(\alpha, \beta)$ is well adapted to the task of discriminating between two irreducible aperiodic Markov chains. Of course, like the Kullback-Leibler divergence and the Itakura-Saito spectral distortion measure considered earlier, it is not a true metric, since it is not symmetric, i.e, $J(\alpha, \beta) \neq J(\beta, \alpha)$, and it does not satisfy the triangle inequality. Since the one-step transition probability matrix of an irreducible aperiodic Markov chain specifies uniquely its steady-state probability distribution, whenever the probability distribution α corresponds to the steady-state distribution of two consecutive states of a Markov chain with one-step transition probability matrix $\mathbf{P}$, by some abuse of notation, we will replace α by $\mathbf{P}$ as an argument of the discrimination function J. For example, if $\alpha = q$ is the empirical probability distribution of two consecutive states, and β is the steady-state probability distribution of two consecutive states specified by $\mathbf{P}$, the discrimination function is denoted as $J(q, \mathbf{P})$. Similarly, if α and β are the steady-state probability distributions (12.17) specified by the Markov chains with transition matrices $\mathbf{P}^0$ and $\mathbf{P}^1$, their discrimination function is denoted as $J(\mathbf{P}^0, \mathbf{P}^1)$.

12.2.2 Binary Hypothesis Testing

Consider now the binary hypothesis testing problem where under hypothesis H_j with $j = 0, 1$, we observe $N+1$ consecutive states $Y(t)$ with $1 \leq t \leq N+1$ of a Markov chain with one-step transition probability matrix $\mathbf{P}^j$. We assume that the two Markov chains have K states, are irreducible and aperiodic, and that the set $\mathcal{T}$ of allowed transitions is the same for both chains. The motivation for this assumption is that otherwise, the detection problem is singular. Specifically, since the Markov chain is recurrent under both hypotheses, each possible transition occurs sooner or later. So, by waiting long enough, we are certain that a transition allowed by one chain, and not by the other, will take place, which allows a perfect decision. So we remove this trivial case by assuming that both chains have the same set of allowable transitions. To simplify our analysis, we also assume that the probability distribution $\pi(1)$ of the initial state is the same under both hypotheses.

Then, let $Y(t) = y_t$ with $1 \leq t \leq N+1$ be the observed sequence of states with $y_t \in \mathcal{S}$. The log-likelihood ratio for the binary hypothesis testing problem can be expressed as

$$\ln\big(L(\mathbf{y})\big) = \sum_{t=1}^{N} \ln\left(\frac{p^1_{y_t y_{t+1}}}{p^0_{y_t y_{t+1}}}\right) \tag{12.19}$$

$$= \sum_{(k,\ell)\in\mathcal{T}} N_{k\ell}(\mathbf{y}) \ln\left(\frac{p^1_{k\ell}}{p^0_{k\ell}}\right)$$

$$= N\big[J(q(\mathbf{y}), \mathbf{P}^0) - J(q(\mathbf{y}), \mathbf{P}^1)\big] , \tag{12.20}$$

where $q(\mathbf{y})$ denotes the empirical probability distribution for two consecutive states specified by the sequence $\mathbf{y}$. From expression (12.20) we see that

$$S_N = J(q(\mathbf{y}), \mathbf{P}^0) - J(q(\mathbf{y}), \mathbf{P}^1) \tag{12.21}$$

is a sufficient statistic for the test, and the LRT takes the form

$$S_N \underset{H_0}{\overset{H_1}{\gtrless}} \gamma_N = \frac{\ln(\tau)}{N} \tag{12.22}$$

where τ denotes the Bayesian threshold specified by (2.19). Note that if the costs C_{ij} are function of the number N of observations, τ will depend on N.

For the case when the costs are symmetric and the two hypotheses are equally likely, the threshold $\gamma_N = 0$, so the test (12.21)–(12.22) takes the geometrically appealing form

$$J(q(\mathbf{y}), \mathbf{P}^0) \underset{H_0}{\overset{H_1}{\gtrless}} J(q(\mathbf{y}), \mathbf{P}^1) \, . \tag{12.23}$$

Thus, hypothesis H_j is selected if the observed empirical distribution q is closest to $\mathbf{P}^j$ in the sense of the discrimination function J. In other words, J plays the same role for Markov chain detection as the Euclidean distance for the detection of known signals in WGN.

12.2.3 Asymptotic Performance

The empirical probability distribution $q_{k\ell}(\mathbf{y})$ has the property that under hypothesis H_j, it converges almost surely to the steady state probability distribution $\pi_k^j p_{k\ell}^j$ for two consecutive states of the Markov chain specified by $\mathbf{P}^j$. This implies that as N tends to infinity

$$S_N \xrightarrow{\text{a.s.}} J(\mathbf{P}^1, \mathbf{P}^0) \tag{12.24}$$

under H_1, and

$$S_N \xrightarrow{\text{a.s.}} -J(\mathbf{P}^0, \mathbf{P}^1) \tag{12.25}$$

under H_0. If we assume that τ is either constant or tends to a finite constant as N tends to infinity, the threshold γ_N in (12.23) tends to zero as N goes to infinity. Assuming that $\mathbf{P}^1 \neq \mathbf{P}^0$, the limits (12.24) and (12.25) indicate that it is possible to make decisions with vanishingly small probability of error as N increases.

The objective of this section is to show that for a test of the form (12.22), where the threshold γ on the right-hand side is a constant independent of N, the probabilities of false alarm and of a miss decay exponentially with the number N of observations. Note that for finite N, the probability of false alarm and of a miss do not admit simple closed form expressions. So the expressions obtained below for large N will provide some rough estimates of the test performance. The approach that we use to analyze the asymptotic behavior of the test is based again on the theory of large deviations. Specifically, observe that for $N+1$ observations $Y(t)$ with $1 \leq t \leq N+1$, the sufficient statistic S_N can be written as

$$S_N = \frac{1}{N} \sum_{t=1}^{N} Z_t \tag{12.26}$$

with

$$Z_t = \ln(p^1_{Y(t)Y(t+1)}/p^0_{Y(t)Y(t+1)}) . \tag{12.27}$$

In this expression, the random variables Z_t are not independent, so Cramér's theorem is not applicable. However, as will be shown below, the sequence S_N admits an asymptotic log-generating function under H_j for $j = 0,\ 1$, so the Gärtner-Ellis theorem can be employed to evaluate the asymptotic probabilities of false alarm and of a miss of the test as N tends to infinity.

Consider the generating functions

$$G^j_N(u) = E[\exp(uS_N)|H_j] \tag{12.28}$$

for $j = 0,\ 1$. We have

$$G^j_N(Nu) = E\big[\prod_{t=1}^{N} \Big(\frac{p^1_{Y(t)Y(t+1)}}{p^0_{Y(t)Y(t+1)}} \Big)^u \mid H_j \big] . \tag{12.29}$$

Then, at stage t, assume that the expectations with respect to $Y(t+1), \ldots, Y(N+1)$ have already been taken, conditioned on the knowledge of $Y(1), \ldots, Y(t)$. If we now take the expectation with respect to $Y(t)$ conditioned on the knowledge of $Y(1), \ldots, Y(t-1)$, and proceed recursively in this manner until $t = 1$, we obtain

$$G^0_N(Nu) = \boldsymbol{\pi}(1)\mathbf{M}(u)^N \mathbf{e} \tag{12.30}$$

where $\mathbf{M}(u)$ denotes the $K \times K$ matrix with (k, ℓ)-th element

$$m_{k\ell}(u) = (p^1_{k\ell})^u (p^0_{k\ell})^{1-u} , \tag{12.31}$$

$\boldsymbol{\pi}(1)$ is the row vector representing the probability distribution of $Y(1)$, and $\mathbf{e}$ is the vector (12.4) whose entries are all equal to one. The matrix $\mathbf{M}(u)$ can be expressed as a Hadamard product [11, Chap. 5]

$$\mathbf{M}(u) = (\mathbf{P}^1)^{(u)} \circ (\mathbf{P}^0)^{(1-u)} \tag{12.32}$$

where $\mathbf{P}^{(v)} = (p^v_{k\ell}, 1 \leq k, \ell \leq K)$ denotes the v-th Hadamard power of a matrix $\mathbf{P}$. Although $\mathbf{M}(u)$ is not a stochastic matrix, since its row sums are not necessarily equal to one for all u, it is a nonegative matrix since $m_{k\ell}(u) \geq 0$ for all $1 \leq k, \ell \leq K$. It admits a directed graph $\mathcal{G}$ constructed in the same way as for a stochastic matrix. Its vertices are the elements of $\mathcal{S}$, and a directed edge links vertices k and ℓ if $m_{k\ell}(u) > 0$. Obviously $\mathcal{G}$ does not depend on u and coincides with the transition diagram of $\mathbf{P}^0$ and $\mathbf{P}^1$, which is the same since both chains have the same set of allowed transitions. Since $\mathbf{P}^0$ and $\mathbf{P}^1$

were assumed irreducible and aperiodic, and since the irreducibility and period of a nonnegative matrix depend only on its graph, this implies that the matrix $\mathbf{M}(u)$ is irreducible and aperiodic.

Accordingly, the Perron-Frobenius theorem is applicable to $\mathbf{M}(u)$, so it has a positive real eigenvalue $\lambda(u)$ which is greater than the magnitude of all the other eigenvalues. The multiplicity of $\lambda(u)$ is one, and if we consider the left and right eigenvectors

$$\begin{aligned} \mathbf{a}^T(u) &= \big[a_1(u)\ a_2(u)\ \dots\ a_K(u)\big] \\ \mathbf{b}^T(u) &= \big[b_1(u)\ b_2(u)\ \dots\ b_K(u)\big] \end{aligned}$$

of $\mathbf{M}(u)$ corresponding to $\lambda(u)$, i.e.,

$$\begin{aligned} \mathbf{a}^T(u)\mathbf{M}(u) &= \lambda(u)\mathbf{a}^T(u) \\ \mathbf{M}(u)\mathbf{b}(u) &= \lambda(u)\mathbf{b}(u) \end{aligned} \tag{12.33}$$

all the entries $a_k(u)$ and $b_k(u)$ are real and positive for $1 \leq k \leq K$. Since the eigenvectors $\mathbf{a}(u)$ and $\mathbf{b}(u)$ can be scaled arbitrarily, we can always ensure that

$$\mathbf{a}^T(u)\mathbf{b}(u) = \sum_{k=1}^{K} a_k(u)b_k(u) = 1\,. \tag{12.34}$$

In addition, since $\mathbf{M}(0) = \mathbf{P}^0$ and $\mathbf{M}(1) = \mathbf{P}^1$, we have

$$\lambda(0) = \lambda(1) = 1\,, \tag{12.35}$$

and we can always select

$$\mathbf{a}^T(0) = \boldsymbol{\pi}^0 \quad , \mathbf{a}^T(1) = \boldsymbol{\pi}^1 \tag{12.36}$$

$$\mathbf{b}(0) = \mathbf{b}(1) = \mathbf{e} \tag{12.37}$$

where $\boldsymbol{\pi}^j$ denotes the steady-state probability distribution of the Markov chain with transition matrix $\mathbf{P}^j$ for $j = 0,\ 1$.

Then as $N \to \infty$, we have

$$\frac{G_N^0(Nu)}{\lambda^N(u)} \to \boldsymbol{\pi}(1)\mathbf{b}(u)\mathbf{a}^T(u)\mathbf{e} \tag{12.38}$$

where the inner products $\boldsymbol{\pi}(1)\mathbf{b}(u)$ and $\mathbf{a}^T(u)\mathbf{e}$ are both nonzero. Accordingly, under H_0 the sequence S_N admits the asymptotic log-generating function

$$\Lambda_0(u) \triangleq \lim_{N\to\infty} \frac{1}{N} \ln G_N^0(Nu) = \ln \lambda(u)\,, \tag{12.39}$$

i.e., the log-generating function is the logarithm of the spectral radius of $\mathbf{M}(u)$. Similarly, since the generating function of S_N under H_1 can be expressed as

$$G_N^1(u) = G_N^0(u+1)\,,$$

we find that under H_1, S_N admits the log-generating function

$$\Lambda_1(u) \triangleq \lim_{N\to\infty} \frac{1}{N} \ln G_N^1(Nu) = \ln \lambda(u+1)\,. \tag{12.40}$$

To verify that the function $\ln \lambda(u)$ is convex, we employ the following property of Hadamard products of nonegative matrices. Let $\mathbf{M}_1$ and $\mathbf{M}_2$ be two $K \times K$ matrices with nonnegative entries. It is shown in [11, p. 361–362] that for $0 \leq \alpha \leq 1$

$$\rho(\mathbf{M}_1^{(\alpha)} \circ \mathbf{M}_2^{(1-\alpha)}) \leq \rho(\mathbf{M}_1)^{\alpha} \rho(\mathbf{M}_2)^{1-\alpha} \tag{12.41}$$

where $\rho(\mathbf{M})$ denotes the spectral radius of an arbitrary matrix $\mathbf{M}$. Then, for $u_1,\ u_2 \in \mathbb{R}$ and $0 \leq \alpha \leq 1$, by observing that

$$\mathbf{M}(\alpha u_1 + (1-\alpha)u_2) = \mathbf{M}(u_1)^{(\alpha)} \circ \mathbf{M}(u_2)^{(1-\alpha)} \tag{12.42}$$

we deduce

$$\begin{aligned}\lambda(\alpha u_1 + (1-\alpha)u_2) &= \rho(\mathbf{M}(\alpha u_1 + (1-\alpha)u_2) \\ &\leq \rho(\mathbf{M}(u_1))^{\alpha} \rho(\mathbf{M}(u_2))^{1-\alpha} = \lambda(u_1)^{\alpha}\lambda(u_2)^{1-\alpha}\,,\end{aligned} \tag{12.43}$$

so $\ln \lambda(u)$ is convex.

By multiplying $\mathbf{M}(u)$ on the left by $\mathbf{a}^T(u)$ and on the right by $\mathbf{b}(u)$, and taking into account the normalization (12.34) we obtain the identity

$$\lambda(u) = \mathbf{a}^T(u)\mathbf{M}(u)\mathbf{b}(u)\,. \tag{12.44}$$

Differentiating with respect to u gives

$$\frac{d\lambda}{du} = \mathbf{a}^T(u)\frac{d\mathbf{M}}{du}(u)\mathbf{b}(u) = \sum_{k\ell} a_k(u)(p_{k\ell}^1)^u (p_{k\ell}^0)^{1-u} b_\ell(u) \ln\left(\frac{p_{k\,\ell}^1}{p_{k\,\ell}^0}\right)\,, \tag{12.45}$$

where we have used the fact that $\mathbf{a}^T(u)$ and $\mathbf{b}(u)$ are left and right eigenvectors of $\mathbf{M}(u)$, as well as

$$\frac{d}{du}\mathbf{a}^T(u)\mathbf{b}(u) = 0\,.$$

Then, setting $u = 0$ and $u = 1$ inside (12.45) and taking into account (12.35)–(12.37), we find

$$\begin{aligned}\frac{d}{du}\Lambda_0(0) &= \frac{d\lambda}{du}(0) = -J(\mathbf{P}^0, \mathbf{P}^1) \\ \frac{d}{du}\Lambda_0(1) &= \frac{d\lambda}{du}(1) = J(\mathbf{P}^1, \mathbf{P}^0)\,.\end{aligned} \tag{12.46}$$

Accordingly, the asymptotic log-generating function $\Lambda_0(u)$ has a graph of the form shown in Fig. 10.4, where the only difference is that the slopes e_0 and e_1 at $u = 0$ and $u = 1$ are given now by $e_0 = -J(\mathbf{P}^0, \mathbf{P}^1)$ and $e_1 = J(\mathbf{P}^1, \mathbf{P}^0)$.

Consider now the Legendre transform

$$I_j(z) = \sup_{u \in \mathbb{R}} (zu - \Lambda_j(u)) \tag{12.47}$$

for $j = 0,\ 1$. The identity

$$\Lambda_1(u) = \Lambda_0(u+1) \tag{12.48}$$

implies

$$I_1(z) = I_0(z) - z\,. \tag{12.49}$$

Then, let $P_F(N) = P[S_N \geq \gamma | H_0]$ and $P_M(N) = P[S_N < \gamma | H_1]$ denote respectively the probability of false alarm, and the probability of a miss, for the test (12.22) based on the observations $Y(t)$ for $1 \leq t \leq N+1$. By applying the Gärtner-Ellis theorem we obtain

$$\begin{aligned}
\lim_{N \to \infty} \frac{1}{N} \ln P_F(N) &= -I_0(\gamma) \\
\lim_{N \to \infty} \frac{1}{N} \ln P_M(N) &= -I_1(\gamma) = -[I_0(\gamma) - \gamma]\,.
\end{aligned} \tag{12.50}$$

We can again consider several special cases for the threshold γ. For an NP test of type I, the probability P_M of a miss is minimized while keeping the probability P_F of false alarm below a bound α. Since we know that under H_0 the test satistic S_N converges almost surely to e_0, to guarantee that the probability of false alarm stays below α, we can select as threshold

$$\gamma_{\mathrm{NP}}^{I} = e_0 + \epsilon$$

with $\epsilon > 0$ arbitrarily small. In this case $I_0(e_0) = 0$, since the line with slope e_0 is tangent to $\Lambda_0(u)$ at the origin. In the light of (12.50), this implies $I_1(e_0) = -e_0$, so for an NP test of type I, the probability of a miss decays exponentially according to

$$P_M(N) = K_M(N) \exp(-J(\mathbf{P}^1, \mathbf{P}^0)N) \tag{12.51}$$

where the constant $K_M(N)$ varies with N at a sub-exponential rate, i.e.,

$$\lim_{N \to \infty} \frac{1}{N} \ln(K_M(N)) = 0\,.$$

Similarly, for an NP test of type II, the probability of false alarm P_F is minimized while requiring that the probability P_M of a miss should stay below a constant β. Since S_N converges almost surely to $e_1 = J(\mathbf{P}^1, \mathbf{P}^0)$ under H_1, the test threshold can be selected as

$$\gamma_{\mathrm{NP}}^{II} = e_1 - \epsilon$$

with $\epsilon > 0$ arbitrary small. In this case, $I_1(e_1) = 0$ and $I_0(e_1) = e_1$, so for an NP test of type II, the probability of false alarm decays exponentially according to

$$P_F(N) = K_F(N) \exp(-J(\mathbf{P}^0, \mathbf{P}^1)N) \tag{12.52}$$

where the constant $K_F(N)$ varies with N at a sub-exponential rate.

For a test with threshold γ, identities (12.50) imply that for large N

$$\begin{aligned} P_F(N) &\approx K_F(N) \exp(-I_0(\gamma)N) \\ P_M(N) &\approx K_M(N) \exp(-I_1(\gamma)N) \end{aligned} \tag{12.53}$$

where the coefficients $K_F(N)$ and $K_M(N)$ vary with N at a sub-exponential rate. If π_0 denotes the a-priori probability of H_0, the probability of error for the test (12.22) can be expressed as

$$P_E(N) \approx K_F(N)\pi_0 \exp(-I_0(\gamma)N) + K_M(N)(1-\pi_0) \exp(-I_1(\gamma)N) \tag{12.54}$$

The threshold γ that maximizes the overall rate of decay of $P_E(N)$ is again achieved by equalizing the rates of decay for the two types of errors and setting $\gamma = 0$. With this choice, the rate of decay of the probability of error is given by

$$I_0(0) = - \min_{0 \leq u \leq 1} \Lambda_0(u) = - \min_{0 \leq u \leq 1} \ln \lambda(u) \,. \tag{12.55}$$

Although the Legendre transforms $I_0(z)$ and $I_1(z)$ cannot usually be evaluated in closed form, they admit an interesting geometric interpretation in terms of the information discrimination measure J. For $0 \leq u \leq 1$, consider the family of Markov chains with state space $\mathcal{S}$ and transition diagram $\mathcal{G}$ specified by the one-step transition matrix

$$\mathbf{P}(u) = \frac{1}{\lambda(u)} \mathbf{T}^{-1}(u) \mathbf{M}(u) \mathbf{T}(u) \,, \tag{12.56}$$

where

$$\mathbf{T}(u) \stackrel{\triangle}{=} \mathrm{diag}\,\{b_k(u),\, 1 \leq k \leq K\} \tag{12.57}$$

denotes the diagonal similarity transformation specified by the entries of the right eigenvector of $\mathbf{M}(u)$ corresponding to $\lambda(u)$. The (k,ℓ)-th element of $\mathbf{P}(u)$ is given by

$$p_{k\ell}(u) = \frac{1}{\lambda(u)} (p^1_{k\ell})^u (p^0_{k\ell})^{1-u} \frac{b_\ell(u)}{b_k(u)} \,. \tag{12.58}$$

Clearly $\mathbf{P}(u)$ is a stochastic matrix with the same graph $\mathcal{G}$ as $\mathbf{P}^0$, $\mathbf{P}^1$ and $\mathbf{M}(u)$. Its right and left eigenvectors corresponding to $\lambda = 1$ are given respectively by $\mathbf{e}$ and

$$\begin{aligned} \pi(u) &= \mathbf{a}^T(u) \mathbf{T}(u) \\ &= \big[a_1(u) b_1(u) \; a_2(u) b_2(u) \ldots a_K(u) b_K(u) \big] \,. \end{aligned} \tag{12.59}$$

The family $\mathbf{P}(u)$ forms the geodesic linking the Markov chain with transition matrix $\mathbf{P}^0$ to the one with transition matrix $\mathbf{P}^1$ as u varies from 0 to 1. By direct evaluation, we find

$$\begin{aligned} J(\mathbf{P}(u), \mathbf{P}^0) &= \sum_{k\ell} a_k(u) b_k(u) p_{k\ell}(u) \ln \left(\frac{p_{k\ell}(u)}{p^0_{k\ell}} \right) \\ &= \frac{1}{\lambda(u)} \sum_{k\ell} a_k(u) (p^1_{k\ell})^u (p^0_{k\ell})^{1-u} b_\ell(u) \ln \left(\left(\frac{p^1_{k\ell}}{p^0_{k\ell}} \right)^u \frac{b_k(u)}{\lambda(u) b_\ell(u)} \right) \\ &= -\ln \lambda(u) + u \frac{d}{du} \ln \lambda(u) \, . \end{aligned} \quad (12.60)$$

Then, if we refer to the definition (12.47) of the Legendre transform $I_0(z)$ of $\Lambda_0(u)$, and to the Fig. 10.5 showing its geometric construction, for a line with slope z and vertical intercept $-I_0(z)$, the horizontal coordinate $u(z)$ of the tangency point of the line to $\Lambda_0(u)$ satisfies the equation

$$z = \frac{d}{du} \Lambda_0(u) = \frac{d}{du} \ln \lambda(u) \, . \quad (12.61)$$

Substituting this identity inside (12.60) yields

$$I_0(z) = J(\mathbf{P}(u(z)), \mathbf{P}^0) \, , \quad (12.62)$$

so the Legendre transform represents the information discrimination measure existing between the Markov chain with transition matrix $\mathbf{P}(u(z))$ and the one with transition matrix $\mathbf{P}^1$. We can show in a similar manner that

$$I_1(z) = J(\mathbf{P}(u(z)), \mathbf{P}^1) \, . \quad (12.63)$$

When $z = 0$, we have $I_0(0) = I_1(0)$, and in this case the transition matrix $\mathbf{P}(u(0))$ satisfies

$$J(\mathbf{P}(u(0)), \mathbf{P}^1) = J(\mathbf{P}(u(0)), \mathbf{P}^0) \, , \quad (12.64)$$

so in terms of the information discrimination measure J, $\mathbf{P}(u(0))$ is located at the midway point on the geodesic linking $\mathbf{P}^0$ and $\mathbf{P}^1$ (see [12] for a discussion of the differential geometry of statistical models).

Example 12.2: Discrimination of symmetric 2-state MCs

Consider a Markov chain with 2 states, which under hypotheses H_0 and H_1 admits the transition matrices

$$\mathbf{P}^0 = \begin{bmatrix} 1/2 & 1/2 \\ 1/2 & 1/2 \end{bmatrix} \quad , \quad \mathbf{P}^1 = \begin{bmatrix} p & q \\ q & p \end{bmatrix}$$

with $q = 1 - p$, $0 < p < 1$, and $p \neq 1/2$. Under H_0 and H_1, the MC has the feature that it remains invariant under a permutation of states 1 and 2, so both states play a symmetrical role. Both transition matrices have for steady-state probability distribution

$$\boldsymbol{\pi} = \begin{bmatrix} 1/2 \; 1/2 \end{bmatrix} .$$

This implies that the frequency of occurrence of each state does not provide any information about the correct hypothesis. Instead, the optimum test depends on the transition rates between states. Specifically, if we consider the empirical probability distribution $q_{k\ell}$ with $k, \ell = 1, 2$ of successive states of the Markov chain, by taking into account the normalization

$$q_{11} + q_{12} + q_{21} + q_{22} = 1$$

we find that the sufficient test statistic

$$\begin{aligned} S_N &= (q_{11} + q_{22}) \ln(2p) + (q_{12} + q_{21}) \ln(2q) \\ &= (q_{11} + q_{22})(\ln(2p) - \ln(2q)) + \ln(2q) \end{aligned}$$

depends only on the fraction $q_{11}+q_{22}$ of successive states which are unchanged. By direct evaluation, we find that the information discrimination functions of $\mathbf{P}^0$ and $\mathbf{P}^1$ are given by

$$\begin{aligned} J(\mathbf{P}^0, \mathbf{P}^1) &= -\ln(2(pq)^{1/2}) \\ J(\mathbf{P}^1, \mathbf{P}^0) &= p \ln(2p) + q \ln(2q) . \end{aligned}$$

It is also easy to check that the spectral radius of

$$\mathbf{M}(u) = \frac{1}{2} \begin{bmatrix} (2p)^u & (2q)^u \\ (2q)^u & (2p)^u \end{bmatrix}$$

is given by

$$\lambda(u) = \frac{1}{2}[(2p)^u + (2q)^u]$$

and its corresponding left and right eigenvectors $\mathbf{a}^T(u)$ and $\mathbf{b}(u)$ are given by

$$\mathbf{a}(u) = \mathbf{b}(u) = \begin{bmatrix} 1 \\ 1 \end{bmatrix} .$$

To illustrate the convexity of the log-generating function $\Lambda_0(u) = \ln \lambda(u)$, it is plotted in Fig. 12.2 for $p = 3/4$, $q = 1/4$ and $-0.5 \leq u \leq 1.5$.

For this example, the Legendre transform can be evaluated in closed form. By solving

$$\frac{d}{du}(zu - \ln \lambda(u)) = z - \frac{d}{du}\lambda(u) = 0$$

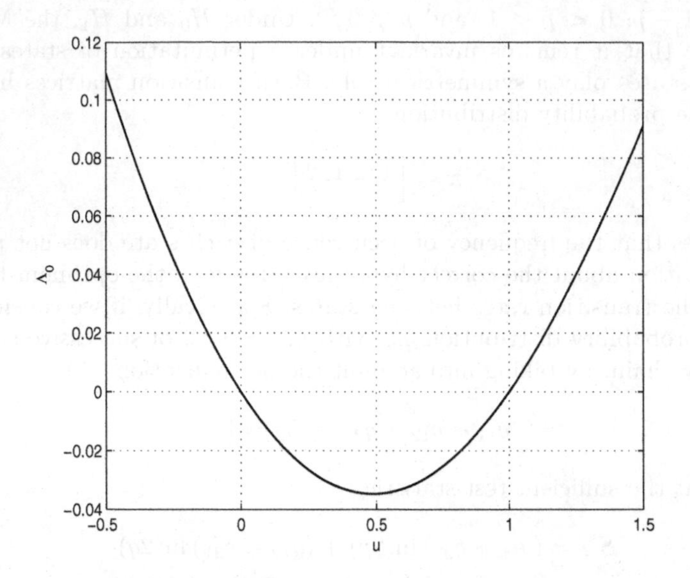

Fig. 12.2. Asymptotic log-generating function $\Lambda_0(u)$ for the LRT between $\mathbf{P}^1$ and $\mathbf{P}^0$.

with

$$\frac{d}{du}\lambda(u) = \frac{(2p)^u \ln(2p) + (2q)^u \ln(2q)}{(2p)^u + (2q)^u} \ ,$$

we obtain

$$u(z) = \frac{\ln[(\ln(2q) - z)/(z - \ln(2p))]}{\ln(p/q)} \ .$$

The corresponding rate functions

$$I_0(z) = zu(z) - \ln \lambda(u(z))$$

and

$$I_1(z) = I_0(z) - z$$

are plotted below in Fig. 12.3 for $p = 3/4$ and $q = 1/4$ and $-0.2 \leq z \leq 0.2$. Note that for the above values of p and q, we have $e_0 = -J(\mathbf{P}^0, \mathbf{P}^1) = -0.1438$ and $e_1 = J(\mathbf{P}^1, \mathbf{P}^0) = 0.1308$, so the parts of the plot that are relevant for hypothesis testing purposes extend only between e_0 and e_1, for which $I_0(e_0) = 0$ and $I_1(e_1) = 0$, as expected. □

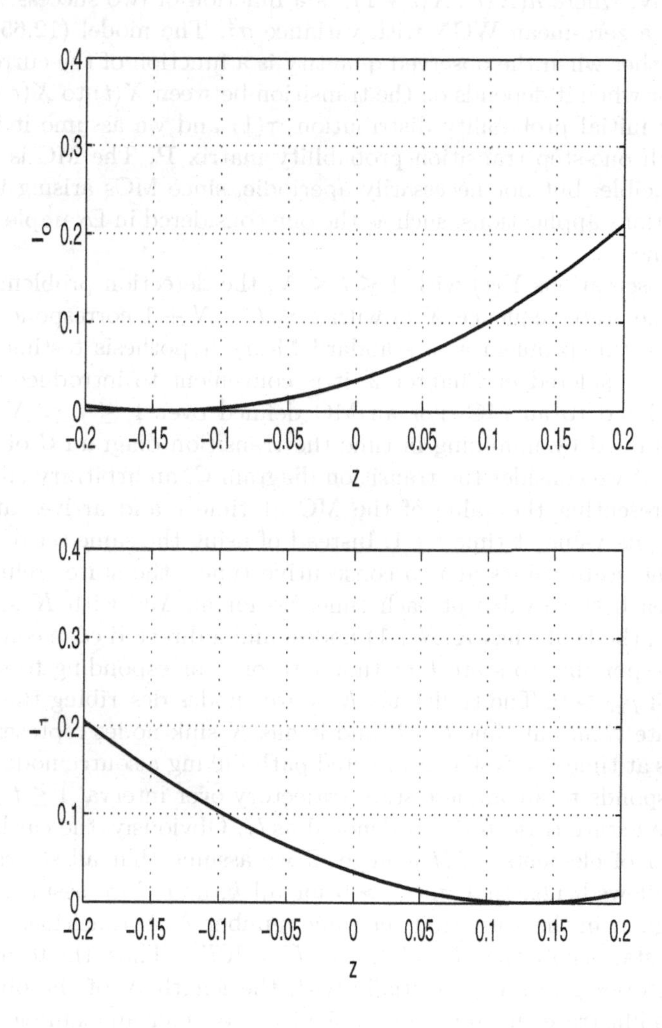

Fig. 12.3. Rate functions $I_0(z)$ and $I_1(z)$ for the asymptotic LRT between $\mathbf{P}^1$ and $\mathbf{P}^0$.

12.3 Detection of Partially Observed Markov Chains

Consider now the situation where instead of observing directly the Markov chain state $X(t)$, we observe

$$Y(t) = h(X(t), X(t+1)) + V(t) \tag{12.65}$$

for $1 \leq t \leq N$, where $h(X(t), X(t+1))$ is a function of two successive states, and $V(t)$ is a zero-mean WGN with variance σ^2. The model (12.65) can be employed either when the observed quantity is a function of the current state $X(t)$ only, or when it depends on the transition between $X(t)$ to $X(t+1)$. The MC has for initial probability distribution $\boldsymbol{\pi}(1)$ and we assume it is homogeneous with one-step transition probability matrix $\mathbf{P}$. The MC is assumed to be irreducible, but not necessarily aperiodic, since MCs arising in digital communications applications, such as the one considered in Example 12.1, are often periodic.

Given observations $Y(t)$ with $1 \leq t \leq N$, the detection problem consists of finding the state sequence $X(t)$ with $1 \leq t \leq N+1$ corresponding to it. To formulate this problem as a standard M-ary hypothesis testing problem of the type considered in Chapter 2, it is convenient to introduce the state trellis associated to an MC. For an MC defined over $1 \leq t \leq N+1$, the trellis is obtained by unfolding in time the transition diagram $\mathcal{G}$ of the MC. Specifically, if we consider the transition diagram $\mathcal{G}$, an arbitrary edge leaves a node representing the value of the MC at time t and arrives at a node representing its value at time $t+1$. Instead of using the same set of nodes to represent the state values at two consecutive times, the state trellis assigns a node to each state value at each time. So for an MC with K states and $N+1$ times, the trellis has $K(N+1)$ nodes, and a directed edge connects the vertex corresponding to state k at time t to one corresponding to state ℓ at time $t+1$ if $p_{k\ell} > 0$. The trellis has K source nodes describing the different possible state values at time $t = 1$, and it has K sink nodes representing the state values at time $t = N+1$. A directed path linking a source node to a sink node corresponds to a complete state trajectory over interval $1 \leq t \leq N+1$. The set of all state trajectories is denoted as $\mathcal{U}$. Obviously, the cardinality U (the number of elements) of $\mathcal{U}$ is huge. If we assume that all states at time $t = 0$ are allowed, i.e., that $\pi_k(1) > 0$ for all k, and if we assume that the trellis is regular in the sense that the same number T of transitions is allowed out of each state (note that $T \leq K$), then $U = KT^N$. Thus, the total number U of trajectories grows exponentially with the length N of the observation interval. To illustrate the above discussion, we consider an example.

Example 12.1, continued

Consider the MC of Example 12.1. After time-unfolding the state transition diagram of Fig. 12.1, we obtain the state trellis shown in Fig. 12.4. The trellis is regular, since $P = 2$ transitions are allowed out of each state, so the total number of trellis trajectories is $U = 4\,2^N$. Since all state transitions have probability 1/2, if the initial states are equally likely, i.e.,

$$\boldsymbol{\pi}(1) = \frac{1}{K}\left[1\,1 \cdots 1\right] ,$$

then all state trajectories are equally likely, a situation which is actually rather common in digital communications applications.

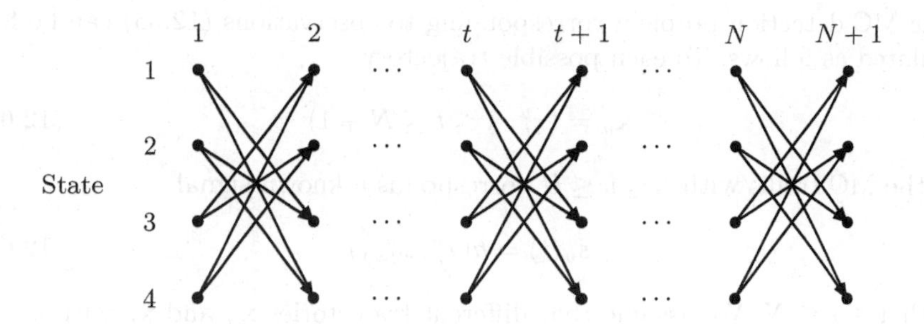

Fig. 12.4. State trellis for the Markov chain of Example 12.1.

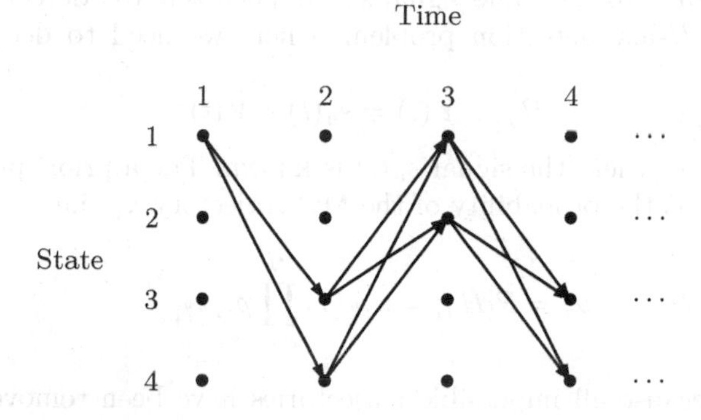

Fig. 12.5. Pruned state trellis obtained by removing impossible states/transitions.

Another situation that occurs frequently is when the initial state or the terminal state, or both, are known. For example, in wireless communications, blocks of transmitted data are divided into training data and payload sub-blocks. So if the received signal needs to be equalized, the fact that the training block is known has the effect of fixing the initial state of the MC representing the symbols stored by the channel. When the initial state is known, the trellis needs to be pruned to remove impossible states and/or transitions. For example, for the trellis of Fig. 12.4, suppose the initial state is $X(1) = 1$. Then the pruned trellis obtained by removing trajectories starting from states 2, 3 and 4 is shown in Fig. 12.5. Note that because the MC has period $d = 2$ and the initial state is known, for $t \geq 2$, $X(t)$ must belong to $\mathcal{S}_2 = \{3, 4\}$ for t even, and to $\mathcal{S}_1 = \{1, 2\}$ for t odd. It is worth observing that after all trajectories starting from initial states other than 1 have been removed, all the remaining trajectories are equally likely. When the final state $X(N+1)$ is known, a similar pruning operation can be performed to remove trajectories terminating in forbidden final states. $\square$

12.3.1 MAP Sequence Detection

The MC detection problem corresponding to observations (12.65) can be formulated as follows: To each possible trajectory

$$\mathbf{x}_u = (x_t^u,\ 1 \leq t \leq N+1) \tag{12.66}$$

of the MC trellis with $1 \leq u \leq U$ corresponds a known signal

$$s_u(t) = h(x_t^u, x_{t+1}^u) \tag{12.67}$$

with $1 \leq t \leq N$. We assume that different trajectories $\mathbf{x}_u$ and $\mathbf{x}_v$ with $u \neq v$ give rise to different signals $s_u(t)$ and $s_v(t)$. Otherwise the detection problem is ill-posed, since it is impossible to discriminate between state trajectories corresponding to the same signal $s_u(t)$. Then the MC detection problem is a standard U-ary detection problem, where we need to decide between hypotheses

$$H_u \ :\ Y(t) = s_u(t) + V(t) \tag{12.68}$$

with $1 \leq u \leq U$, where the signal $s_u(t)$ is known. The a-priori probability of hypothesis H_u is the probability of the MC trajectory $\mathbf{x}_u$, i.e.,

$$p_u = P[H_u] = \pi_{x_1^u}(1) \prod_{t=1}^{N} p_{x_t^u x_{t+1}^u} \,. \tag{12.69}$$

Recall that because all impossible trajectories have been removed from the state trellis, $p_u > 0$ for all $1 \leq u \leq U$. Then, as indicated by (2.140) to find the MAP state trajectory, we need to apply the decision rule

$$\delta_{\mathrm{MAP}}(\mathbf{y}) = \arg \max_{1 \leq u \leq U} [\ln(f(\mathbf{y}|H_u)) + \ln(p_u)] \,, \tag{12.70}$$

where according to (12.69), $\ln(p_u)$ can be expressed as

$$\ln(p_u) = \ln \pi_{x_1^u}(1) + \sum_{t=1}^{N} \ln p_{x_t^u x_{t+1}^u} \,. \tag{12.71}$$

In this expression $\mathbf{y} \in \mathbb{R}^N$ is the observed value of the random vector

$$\mathbf{Y} = \begin{bmatrix} Y(1) \cdots Y(t) \cdots Y(N) \end{bmatrix}^T \tag{12.72}$$

and if

$$\mathbf{s}_u = \begin{bmatrix} s_u(1) \cdots s_u(t) \cdots s_u(N) \end{bmatrix}^T \tag{12.73}$$

denotes the N-dimensional vector representing the signal $s_u(t)$, since $X(t)$ is a zero-mean WGN with variance σ^2, the random vector $\mathbf{Y}$ is $N(\mathbf{s}_u, \sigma^2 \mathbf{I}_N)$ distributed under H_u, i.e.,

$$f(\mathbf{y}|H_u) = \frac{1}{(2\pi\sigma^2)^N} \exp\Big(-\frac{1}{2\sigma^2}||\mathbf{y} - \mathbf{s}_u||_2^2\Big) . \tag{12.74}$$

Multiplying the quantity to be maximized by $-2\sigma^2$, replacing the maximization by a minimization, and dropping terms that do not depend on u, the MAP decision rule (12.70) can be expressed as

$$\delta(\mathbf{y}) = \arg \min_{1\leq u\leq U} [||\mathbf{y} - \mathbf{s}_u||_2^2 - 2\sigma^2 \ln(p_u)] . \tag{12.75}$$

For the special case when all trajectories are equally likely, the MAP decision rule (12.75) reduces to the minimum distance rule

$$\delta(\mathbf{y}) = \arg \min_{1\leq u\leq U} ||\mathbf{y} - \mathbf{s}_u||_2^2 . \tag{12.76}$$

Up to this point, there is nothing to distinguish the MC detection problem from any conventional U-ary detection problem, except that the total number U of trellis trajectories grows exponentially with the length N of the observation block. Since U is a huge number, it is impossible to examine the trajectories one by one to find the one that maximizes the maximum a-posteriori probability. So clever search strategies must be devised to explore the state trellis in such a way that only the most promising trajectories are considered. In 1967, Viterbi [13] proposed a search technique which is optimal in the sense that it always yields the MAP trellis trajectory. This method has since become known as the Viterbi algorithm. Its key idea is, for each possible state value at time t, to find the MAP trellis path between time $t = 1$ and this state. Thus the Viterbi algorithm can be viewed as a breadth-first trellis search since it examines as many paths as there are possible state values at time t, i.e., the search is performed over an entire cross-section of the trellis. Furthermore, all paths under consideration are extended simultaneously. By comparison, a depth-first strategy would focus on one promising trellis path and extend it until it reveals to be less promising than initially thought, at which point backtracking would be required prior to a forward thrust in another search direction, and so on. Although depth-first strategies typically have a low computational cost, they can never guarantee optimality. The Viterbi algorithm was originally proposed for decoding convolutional codes [13]. It was later extended by Forney [14] to the ML equalization of ISI signals. Then in [15], Forney observed that the method used by Viterbi to extend MAP paths from time t to time $t + 1$ was just an instance of the optimality principle employed by Bellman [16, 17] to develop the theory of of dynamic programming for solving optimization problems with an additive cost structure. It was also shown in [15] that the Viterbi algorithm is applicable to the MAP sequence detection problem for a partially observed MC, which opened the way to its use in speech recognition [6] and computational biology.

Before describing the Viterbi algorithm in detail, it is useful to highlight the specific property of the MAP trellis search problem (12.75) that it exploits. Observe that the squared Euclidean distance between $\mathbf{y}$ and $\mathbf{s}_u$ can be expressed as

$$||\mathbf{y} - \mathbf{s}_u||_2^2 = \sum_{t=1}^{N} (y_t - h(x_t^u, x_{t+1}^u))^2 . \tag{12.77}$$

Then, if we drop the index u identifying the trellis trajectory, by combining (12.71) and (12.77), we find that the MAP trellis path $\hat{\mathbf{x}}_{\mathrm{MAP}}(\mathbf{y})$ can be expressed as the solution of the minimization problem

$$\hat{\mathbf{x}}_{\mathrm{MAP}}(\mathbf{y}) = \arg \min_{\mathbf{x} \in \mathcal{U}} J(\mathbf{x}) \tag{12.78}$$

with

$$J(\mathbf{x}) = -2\sigma^2 \ln p_{x_1} + \sum_{t=1}^{N} c(x_t, x_{t+1}, t) , \tag{12.79}$$

where each incremental cost

$$c(x_t, x_{t+1}, t) = (y_t - h(x_t, x_{t+1}))^2 - 2\sigma^2 \ln p_{x_t x_{t+1}} \tag{12.80}$$

depends only on the path segment (x_t, x_{t+1}). For the MAP sequence detection problem, the cost $c(x_t, x_{t+1}, t)$ is known as the *branch metric* and $J(\mathbf{x})$ is called the *path metric*. For applications where all allowed transitions $(k, \ell) \in \mathcal{T}$ are equally likely, the term $\ln p_{x_t x_{t+1}}$ is constant, so it can be dropped from the cost since it is the same for all branches. In this case

$$c(x_t, x_{t+1}, t) = (y_t - h(x_t, x_{t+1}))^2 \tag{12.81}$$

is the Euclidean metric. Two properties allow the application of Bellman's optimality principle to the MAP sequence detection problem. First, the decomposition (12.79) of the cost function $J(\mathbf{x})$ is *additive*. Second, each incremental cost $c(x_t, x_{t+1}, t)$ is *local* since it depends only on the segment (x_t, x_{t+1}) of the trajectory. These two features characterize optimization problems that can be handled by dynamic programming.

The optimality principle can be stated as follows [17, p. 83]: "An optimal policy has the property that whatever the initial state and the initial decision, the remaining decisions must constitute an optimal policy with regard to the state resulting from the first decision." Since this is a very general prescription, let us describe how it can be applied to the problem at hand.

Remark: To keep our analysis as simple as possible, it has been assumed in observation model (12.65) that the observation $Y(t)$ and function $h(X(t), X(t+1)$ are scalar. However, for applications such as decoding of convolutional codes with rate $1/M$, or MAP fractional equalization of modulated signals sampled at rate M/T, where T denotes the baud interval, $Y(t)$ and $h(X(t), X(t+1))$ are vectors of dimension M. For such problems, the only change required consists of replacing the scalar quadratic term $(y_t - h(x_t, x_{t+1}))^2$ by the quadratic vector norm $||\mathbf{Y}_t - \mathbf{h}(x_t, x_{t+1})||_2^2$ in the definition (12.80) of the branch metric $c(x_t, x_{t+1}, t)$. This change does not affect the foregoing derivation of the Viterbi algorithm.

Viterbi algorithm: Let $\mathcal{U}(k,t)$ denote the set of all trellis trajectories starting from an arbitrary state at time $t=1$ and terminating in the state k at time t. Consider also the function

$$J_t(\mathbf{x}) = -2\sigma^2 \ln \pi_{x_1}(1) + \sum_{s=1}^{t-1} c(x_s, x_{s+1}, s) \tag{12.82}$$

representing the accumulated cost for the portion of trellis trajectories extending from time 1 to time t. Note that if there are t states on a partial trajectory, there are $t-1$ transitions between them. Then at time t, and for each state k with $1 \leq k \leq K$, let

$$\hat{\mathbf{x}}(k,t) = \arg \min_{\mathbf{x} \in \mathcal{U}(k,t)} J_t(\mathbf{x}) \tag{12.83}$$

denote the MAP trajectory terminating in state k. The minimum value of $J_t(\mathbf{x})$ achieved by this trajectory is called the value function

$$V(k,t) \triangleq J_t(\hat{\mathbf{x}}(k,t)) \,. \tag{12.84}$$

In the context of MAP sequence detection, this function is called the accumulated path metric, and the MAP trajectories $\hat{\mathbf{x}}(t,k)$ with $1 \leq k \leq K$ are called the survivor trajectories. For the case when there is no path terminating in state k at time t, the set $\mathcal{U}(k,t)$ is empty. If we consider the pruned trellis of Fig. 12.5, we see for example that no path terminates in states 1 and 2 for t even, or in states 3 and 4 for t odd. When this situation arises we set the value function

$$V(k,t) = \infty \,.$$

So suppose that at time t, the MAP trajectories $\hat{\mathbf{x}}(k,t)$ and value function $V(k,t)$ are known for $1 \leq k \leq K$, and we seek to find the MAP trajectories and value function at time $t+1$. According to the optimality principle, the MAP trajectory terminating in state ℓ at time $t+1$ must be formed by an MAP trajectory terminating in one of the K states, say k, at time t, to which we concatenate an allowed transition $(k,\ell) \in \mathcal{T}$. Then for all possible transitions from a state k to state ℓ, we select the one with the smallest accumulated path metric, i.e.,

$$V(\ell, t+1) = \min_{k:\,(k,\ell)\in\mathcal{T}} (V(k,t) + c(k,\ell,t)) \,. \tag{12.85}$$

The argument

$$b(\ell, t+1) = \arg \min_{k:\,(k,\ell)\in\mathcal{T}} (V(k,t) + c(k,\ell,t)) \tag{12.86}$$

of the minimization (12.85) forms what is called the backpointer function. Specifically, the state $b(\ell, t+1) \in \mathcal{S}$ represents the predecessor state at time

t on the MAP path terminating in state ℓ at time $t+1$, and the MAP path terminating in ℓ at time $t+1$ can be expressed as

$$\hat{\mathbf{x}}(\ell, t+1) = (\hat{\mathbf{x}}(b(\ell, t+1), t), \ell) \,. \tag{12.87}$$

In other words, it is obtained by concatenating the segment $(b(\ell, t+1), \ell)$ to the MAP path terminating in $b(\ell, t+1)$ at time t.

The recursion (12.85)–(12.87) is implemented until the end of the trellis is reached, corresponding to time $t = N+1$. At this point the survivor paths $\hat{\mathbf{x}}(k, N+1)$ and the value function $V(k, N+1)$ are available. These paths represent the final candidates for selecting the overall MAP path. The optimal terminal state

$$k_0 = \arg \min_{k \in \mathcal{S}} V(k, N+1) \tag{12.88}$$

is selected by minimizing the value function, and once this terminal state has been identified, the MAP path is the corresponding survivor path, i.e.

$$\hat{\mathbf{x}}_{\mathrm{MAP}} = \hat{\mathbf{x}}(k_0, N+1) \,. \tag{12.89}$$

The associated path metric

$$V(k_0, N+1) = J(\hat{\mathbf{x}}_{MAP}) \tag{12.90}$$

represents the minimum of the function $J(\mathbf{x})$ over the entire set $\mathcal{U}$ of trellis paths.

Viterbi algorithm implementation: From the above presentation of the Viterbi algorithm, we see that the key steps are recursions (12.85)–(12.87). Equation (12.85) is usually implemented by what is called an add, compare and select (ACS) unit. Observe for a given state ℓ, equation (12.85) adds to the value function $V(k, t)$ for each possible predecessor state the transition cost $c(k, \ell, t)$. Then the sums $V(k, t) + c(k, \ell, t)$ are compared and the smallest is selected, so the acronym ACS captures precisely all the steps required to implement (12.85).

As for equations (12.86) and (12.87), they are employed by the survivor memory unit (SMU) to keep track of the survivor paths. Two different approaches can be used to accomplish this task. The register exchange method employs a $K \times N+1$ register array to store the MAP survivor paths. At stage t, the k-th row of the array stores the survivor path $\hat{\mathbf{x}}(k, t)$. Since this path has length t, this means that at stage t, only the first t columns of the array are filled, and the remaining columns are zero. Then at stage $t+1$, as indicated by equation (12.87), the MAP survivor path $\hat{\mathbf{x}}(\ell, t+1)$ which is stored in row ℓ of the array is obtained by taking the MAP path stored in row $b(\ell, t+1)$ at stage t and appending the state ℓ in column $t+1$ as the last trajectory element. Since this scheme involves swapping the contents of different array rows as we go from stage t to stage $t+1$, it is called the register exchange algorithm. Note that after the new survivor trajectories have been stored at stage

$t+1$, all the stage t survivor paths can be discarded since they are no longer needed. Right at the end, at stage $N+1$, the array has been filled. Then, after the terminal state k_0 of the MAP path has been identified by performing the minimization (12.88), the entire MAP trajectory is obtained instantaneously by reading off the k_0-th row of the array. So the register exchange method has a high memory management overhead since it requires moving information continuously from one set of registers to another, but it has no latency.

The second approach that can be used to evaluate the MAP trajectory is based on an observation of Rader [18], and is called the trace-back method [19,20]. It employs a register array of size K by N, but in this case the contents of the array do not need to be moved around. Specifically, the element of row ℓ and column t is the backpointer $b(\ell, t+1)$, which indicates the predecessor state of ℓ on the survivor trajectory terminating in ℓ at time $t+1$. So as stage $t+1$, the t-th column of the array is evaluated, and as the Viterbi recursion (12.85)–(12.86) advances, the array is filled one column at a time. At stage $N+1$, the array is full, and after the MAP terminal state k_0 has been identified, we need to identify the MAP trajectory

$$\hat{\mathbf{x}}_{\text{MAP}} = (\hat{x}_t^{\text{MAP}}, 1 \leq t \leq N+1) .$$

This trajectory is identified by applying the trace-back recursion

$$\hat{x}_t^{\text{MAP}} = b(\hat{x}_{t+1}^{\text{MAP}}, t+1) \tag{12.91}$$

with initial condition

$$\hat{x}_{N+1}^{\text{MAP}} = k_0 . \tag{12.92}$$

So at stage $t+1$, the recursion involves reading the element in row $\hat{x}_{t+1}^{\text{MAP}}$ and column t of the array to determine the state of the MAP trajectory at time t. Proceeding sequentially, we see that the entire MAP trajectory is traced back, starting from the terminal state k_0 at time $N+1$. From the above discussion, we see that the trace-back method has no memory management overhead, but it incurs some latency due to the need to trace-back the entire trajectory. So depending on the power consumption and latency requirements of the application one considers, either the register exchange method or the trace-back technique may be preferred.

As explained in [21], a variety of additional computation tricks, such as parallelization or pipelining, can be applied to the implementation of the Viterbi algorithm.

Example 12.3: Viterbi decoding of AMI line code

To illustrate the Viterbi algorithm, consider a two-state Markov chain with transition matrix

$$\mathbf{P} = \frac{1}{2} \begin{bmatrix} 1 & 1 \\ 1 & 1 \end{bmatrix} .$$

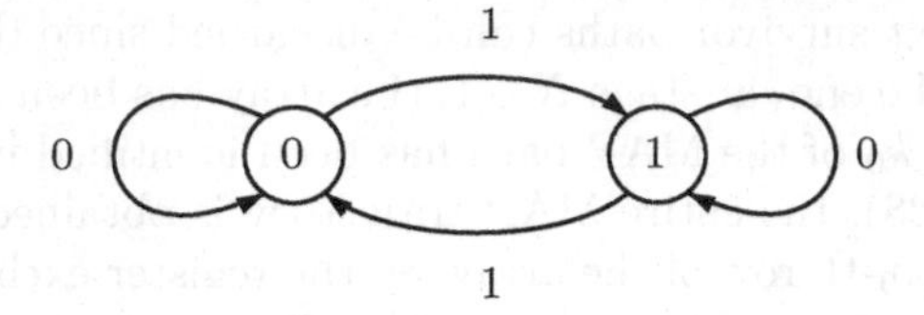

Fig. 12.6. State transition diagram for the AMI MC, with transitions labeled by the corresponding input bits.

Its state-transition diagram is shown in Fig. 12.6. Note that to simplify the following discussion, the state labels are $\{0, 1\}$. The signal

$$s(t) = X(t+1) - X(t) \tag{12.93}$$

takes on the ternary values $\{1, 0, -1\}$ and corresponds to the signal generated by an alternate mark inversion (AMI) line encoder. In this context, if $\{U(t),\, t \geq 1\}$ denotes the bit sequence at the input of the AMI encoder, the state of the MC satisfies

$$X(t+1) = X(t) \oplus U(t)\,, \tag{12.94}$$

where $\oplus$ denotes the modulo two addition, so that $X(t)$ can be interpreted as the running digital sum of all binary inputs prior to time t. The state transitions of Fig. 12.6 are labeled by the value of the input $U(t)$ giving rise to the transition. As can be seen from the diagram, a 0 input leaves the state unchanged, whereas a 1 forces a transition to the other state. In setting up the MC model, it is assumed that input bits $U(t)$ are independent and take values 0 and 1 with probability 1/2.

Given observations

$$Y(t) = s(t) + V(t)\,, \tag{12.95}$$

we seek to reconstruct the state sequence $X(t)$. Note that once the sequence $X(t)$ is known, the input $U(t)$ is given by

$$U(t) = X(t+1) \oplus X(t)\,. \tag{12.96}$$

The standard AMI decoder consists of a ternary slicer followed by a memoryless mapper [22, pp. 564–568]. As indicated by the input-output characteristic shown in Fig. 12.7, the slicer applies a minimum distance rule to observation $Y(t)$ to select the value 1, 0, or -1 of the signal estimate $\hat{s}(t)$. Then the mapper associates points $\hat{s}(t) = 1$ or -1 to the estimated input $\hat{U}(t) = 1$, and point $\hat{s}(t) = 0$ to $\hat{U}(t) = 0$. However, this detector is not optimal, and it was shown in [23] that Viterbi decoding yields a gain of up to 3dB over this decoder.

We focus here on the mechanics of the Viterbi algorithm. The initial state of the MC is assumed to be $X(1) = 0$. The observed sequence is

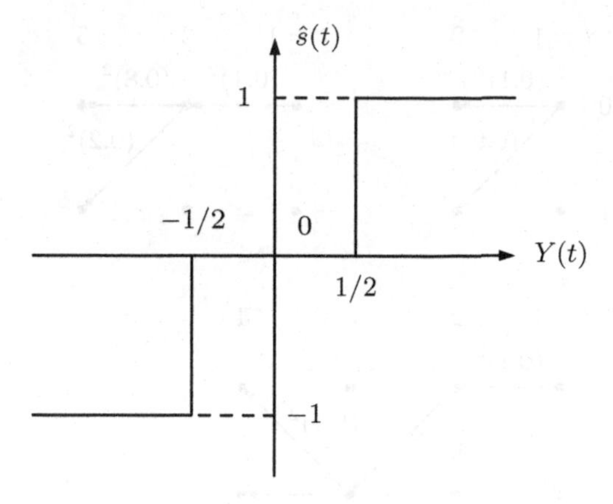

Fig. 12.7. Input-output characteristic of the ternary slicer used to map the observation $Y(t)$ into a signal estimate $\hat{s}(t)$.

$$\begin{aligned}\mathbf{y} &= \begin{bmatrix} y_1 \; y_2 \; y_3 \; y_4 \; y_5 \; y_6 \; y_7 \; y_8 \end{bmatrix}^T \\ &= \begin{bmatrix} 0.1 \; 0.8 \; -1.1 \; 0.4 \; -0.2 \; -0.9 \; 0.2 \; 0.1 \end{bmatrix}^T .\end{aligned}$$

Since all trellis paths are equally likely, we can use

$$c(x_{t+1}, x_t, t) = (y_t - x_{t+1} + x_t)^2 \tag{12.97}$$

as the metric for the trellis branches. Then the survivor paths obtained at successive stages of the Viterbi algorithm are shown in Figs. 12.8 and 12.9. In this figure, each branch of the survivor paths is labeled by its Euclidean metric (12.97).

When the end of the trellis has been reached, we see from Fig. 12.9 that the survivor terminating in state 0 has a smaller metric than the survivor terminating in 1, so the MAP state sequence is given by

$$\begin{aligned}\hat{\mathbf{x}}_{\text{MAP}} &= \begin{bmatrix} \hat{x}_1^{\text{MAP}} \; \hat{x}_2^{\text{MAP}} \; \hat{x}_3^{\text{MAP}} \; \hat{x}_4^{\text{MAP}} \; \hat{x}_5^{\text{MAP}} \; \hat{x}_6^{\text{MAP}} \; \hat{x}_7^{\text{MAP}} \; \hat{x}_8^{\text{MAP}} \; \hat{x}_9^{\text{MAP}} \end{bmatrix}^T \\ &= \begin{bmatrix} 0 \; 0 \; 1 \; 0 \; 1 \; 1 \; 0 \; 0 \; 0 \end{bmatrix}^T .\end{aligned} \tag{12.98}$$

Applying the decoding rule (12.96), we find that the corresponding input sequence is given by

$$\begin{aligned}\hat{\mathbf{u}}_{\text{MAP}} &= \begin{bmatrix} \hat{u}_1^{\text{MAP}} \; \hat{u}_2^{\text{MAP}} \; \hat{u}_3^{\text{MAP}} \; \hat{u}_4^{\text{MAP}} \; \hat{u}_5^{\text{MAP}} \; \hat{u}_6^{\text{MAP}} \; \hat{u}_7^{\text{MAP}} \; \hat{u}_8^{\text{MAP}} \end{bmatrix}^T \\ &= \begin{bmatrix} 0 \; 1 \; 1 \; 1 \; 0 \; 1 \; 0 \; 0 \end{bmatrix}^T .\end{aligned}$$

Note that that for $t = 4$, this sequence differs from the one produced by the conventional AMI line decoder, which yields

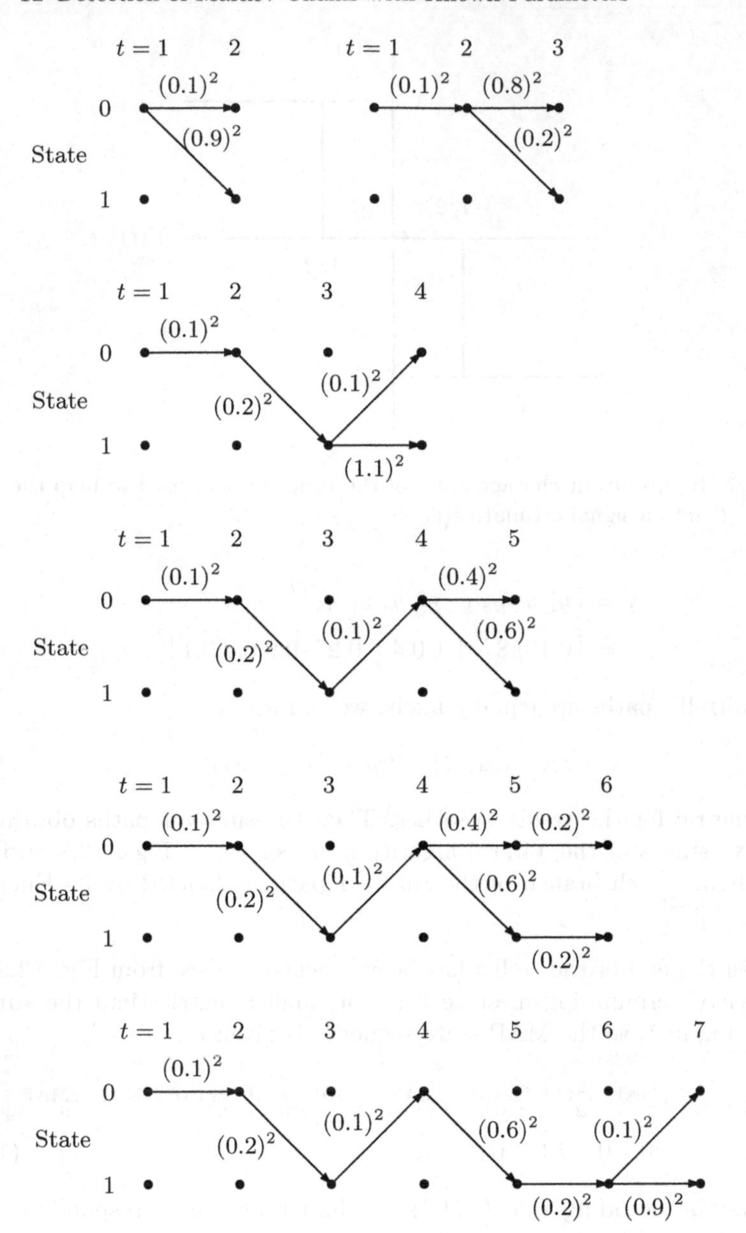

Fig. 12.8. Survivor paths at the first six stages of the Viterbi algorithm for Example 12.3.

$$\hat{u}_4^{\text{AMI}} = 0 \, .$$

An important feature of survivor paths which is in evidence in Figs. 12.8 and 12.9 is that as the Viterbi algorithm progresses, the survivors merge, i.e.,

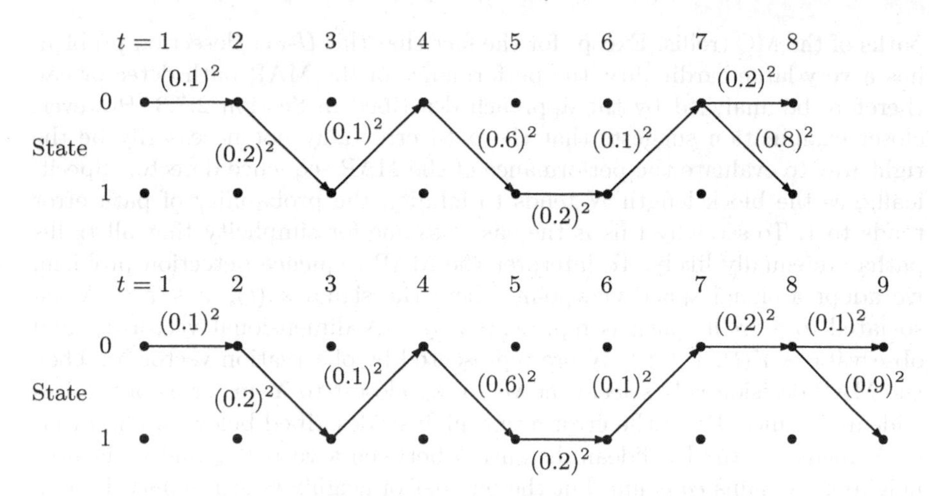

Fig. 12.9. Survivor paths at the last two stages of the Viterbi algorithm for Example 12.3.

while there may be some disagreement about recent states among different survivors, all survivors tend to have a common remote past. This allows the online implementation of the Viterbi algorithm, where, as the Viterbi algorithm processes the current observation $Y(t)$, decisions about the value of state $X(t-D)$ can be made. The value of the delay D is based on computer simulations or on an analysis of error events for the MC. What constitutes an "error event" is discussed later in this section. For the AMI example considered here, a delay $D = 2$ suffices.

Whereas Figs. 12.8 and 12.9 implement the register exchange version of the Viterbi algorithm, a trace-back implementation would store the back-pointer function $b(t, k)$ shown in Table 12.1, where the columns represent the time index t, and the rows the state index k. Recall that $b(t, k)$ specifies the state at time $t-1$ of the survivor path terminating in state k at time t.

Then, if we observe that the terminal survivor path metric is minimized by $k_0 = 0$, the MAP path (12.98) can be generated by using the entries of Table 12.1 to implement the trace-back recursion (12.91). □

Probability of error: We have seen above that the MAP sequence detection problem for Markov chains can be viewed as a U-ary hypothesis testing problem among the different signals $s_u(t)$ with $1 \leq u \leq U$ corresponding to the

Table 12.1. Backpointer $b(t, k)$ as a function of time t and terminal state k.

$b(t,k)$	2	3	4	5	6	7	8	9
0	0	0	1	0	0	1	0	0
1	0	0	1	0	1	1	0	0

paths of the MC trellis. Except for the fact that this U-ary detection problem has a very large cardinality, the performance of the MAP path detector can therefore be analyzed by the approach described in Section 2.7.3. However, closer examination suggests that the path error may not necessarily be the right way to evaluate the performance of the MAP sequence detector. Specifically, as the block length N tends to infinity, the probability of path error tends to 1. To see why this is the case, assume for simplicity that all trellis paths are equally likely. To interpret the MAP sequence detection problem, we adopt a signal space viewpoint where the signal $s_u(t)$, $1 \leq t \leq N$ associated to a trellis path is represented by a N-dimensional vector $\mathbf{s}_u$, and observations $Y(t)$, $1 \leq t \leq N$ are represented by observation vector $\mathbf{Y}$. Then the MAP decision rule selects the vector $\mathbf{s}_u$ closest to $\mathbf{Y}$ in terms of the Euclidean distance. From the error event analysis described below, we find that as N increases, the Euclidean distance h between a vector $\mathbf{s}_u$ and its nearest neighbors remains *constant*, but the number of neighbors grows linearly with N. Under H_u, the observation vector $\mathbf{Y}$ can be expressed as

$$\mathbf{Y} = \mathbf{s}_u + \mathbf{V} \tag{12.99}$$

where $\mathbf{V} \sim N(0, \sigma^2 \mathbf{I}_N)$ is the noise vector with entries $V(t)$ for $1 \leq t \leq N$. Then as N becomes large, the probability of a correct decision can be roughly approximated as

$$P[C|H_u] \approx P[\mathbf{V} \in \mathcal{B}] \tag{12.100}$$

where

$$\mathcal{B} = \{\mathbf{v} : \, |v_t| < h/2 \text{ for } 1 \leq t \leq N\}$$

denotes a square box centered at the origin with width h in each direction. The probability that the noise vector $\mathbf{V}$ falls within this box is given by

$$\begin{aligned} P[\mathbf{V} \in \mathcal{B}] &= \prod_{k=1}^{N} P[|V(t)| < h/2] \\ &= [1 - 2Q(\frac{h}{2\sigma})]^N \, , \end{aligned} \tag{12.101}$$

which tends to zero as $N \to \infty$. This shows that $P[C|H_u] \to 0$ as $N \to \infty$.

Thus the probability that the MAP sequence detector selects the correct sequence is zero as the block length N goes to infinity. While this result may appear discouraging, it is worth noting that a path may be viewed as incorrect even though it contains only a small number of times t such that $\hat{X}_t^{\mathrm{MAP}} \neq X(t)$. So instead of focusing on the probability that the entire MAP sequence is incorrect, we ought to to focus on the probability that any one of its components, say $\hat{X}_t^{\mathrm{MAP}}$ differs from the state $X(t)$. While the pointwise error probability

$$P[E] = P[\hat{X}_t^{\mathrm{MAP}} \neq X(t)]$$

cannot be evaluated in closed form, it was shown by Forney [14] that it admits an upper bound constructed by using the improved union bound method of Section 2.7.4.

First, note that by conditioning with respect to the entire state sequence

$$\mathbf{X} = (X(t),\, 1 \leq t \leq N+1)\,,$$

the probability of detecting incorrectly the state at time t can be expressed as

$$P[E] = \sum_{u=1}^{U} P[\hat{X}_t^{\mathrm{MAP}} \neq x_t^u | \mathbf{X} = \mathbf{x}_u] p_u \,, \tag{12.102}$$

where

$$p_u = P[\mathbf{X} = \mathbf{x}_u]$$

is given by (12.69). In our analysis, we assume that the time t we consider is located far from both ends of the trellis, so that edge effects do not come into play. Then for a fixed trellis path $\mathbf{x}_u$ and a fixed time t, we introduce the concept of error event. From the definition (12.78) of the MAP path, we see that given that the actual path is $\mathbf{x}_u$, a pairwise error event occurs at time t if there exists a path $\mathbf{x}_v$ such that $J(\mathbf{x}_v) < J(\mathbf{x}_u)$ and $x_t^v \neq x_t^u$. The set of paths of $\mathcal{U}$ satisfying these conditions can be denoted as $\mathcal{E}(t, \mathbf{x}_u)$. Among pairwise error events of $\mathcal{E}(t, \mathbf{x}_u)$, if we refer to the improved union bound (2.166), in order to obtain a tight bound, it is of interest to identify a sufficient set, i.e., a set of paths of $\mathcal{E}(t, \mathbf{x}_u)$ that are neighbors of $\mathbf{x}_u$ and which cover all circumstances under which the MAP path differs from $\mathbf{x}_u$ at time t. A sufficient set is provided by the *simple error events* $\mathcal{S}(t, \mathbf{x}_u)$ introduced in [24]. These are the paths $\mathbf{x}_v$ of $\mathcal{E}(t, \mathbf{x}_u)$ that differ from $\mathbf{x}_u$ in a connected pattern. For such a path, there exists times t_L snd t_R such that $t_L < t < t_R$ and

$$\begin{aligned} x_r^v &= x_r^u \;\; r \leq t_L \\ x_r^v &\neq x_r^u \;\; t_L < r < t_R \\ x_r^v &= x_r^u \;\; t_R \leq r \,. \end{aligned} \tag{12.103}$$

In other words, the paths $\mathbf{x}_v$ and $\mathbf{x}_u$ merge at a time t_L to the left of t and remain the same thereafter. Similarly, $\mathbf{x}_v$ and $\mathbf{x}_u$ merge at a time t_R to the right of t and remain the same thereafter. Accordingly, the value $x_t^v \neq x_t^u$ is part of a single excursion of the path $\mathbf{x}_v$ away from the true path $\mathbf{x}_u$. For example, if we consider a Markov chain with two states $\{0,\, 1\}$ and if we assume that the true path $\mathbf{x}_u$ is the all zero state path, a simple error event is shown in Fig. 12.10. For this example, we see that the left and right merging times are $t_L = t - 2$ and $t_R = t + 1$, respectively.

To explain why the set $\mathcal{S}(t, \mathbf{x}_u)$ of simple error events forms a sufficient set of pairwise error events covering all cases where the MAP path $\hat{\mathbf{x}}_{\mathrm{MAP}}$ differs from $\mathbf{x}_u$ at time t, we use the optimality principle. For a Markov chain with three states $\{0,\, 1,\, 2\}$ and where the true path $\mathbf{x}_u$ is the all zero state path, suppose the MAP path is shown in Fig. 12.11. It contains several excursions

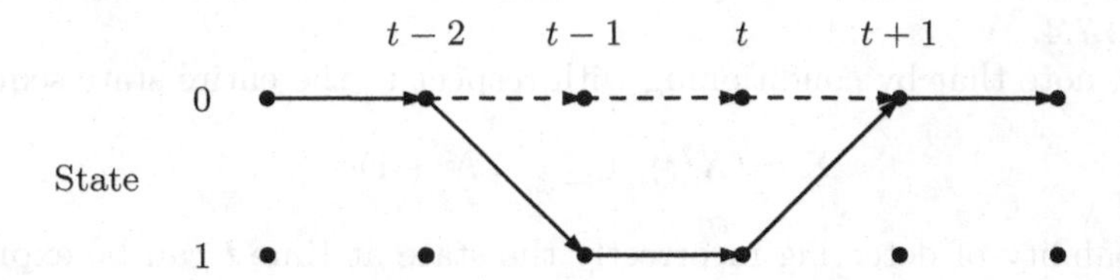

Fig. 12.10. Simple error event $\mathbf{x}_v$ with respect to the all-zero path in a two-state MC.

away from the true path, two of which are depicted in the figure. Then consider a time t at which the MAP path differs from $\mathbf{x}_u$, and let $\mathbf{x}_v$ be the path obtained by retaining only the local excursion of $\hat{\mathbf{x}}_{\text{MAP}}$ surrounding time t, and denote by t_L and t_R the left and right merge times for this excursion. Since $\hat{\mathbf{x}}_{\text{MAP}} \neq \mathbf{x}_u$, we know that $J(\hat{\mathbf{x}}_{\text{MAP}}) < J(\mathbf{x}_u)$. Then observe that between times t_L and t_R, the paths $\hat{\mathbf{x}}_{\text{MAP}}$ and $\mathbf{x}_v$ coincide. By the optimality principle of the MAP path, $\mathbf{x}_v$ must have a lower metric than $\mathbf{x}_u$, i.e. $J(\mathbf{x}_v) < J(\mathbf{x}_u)$, since otherwise the $\mathbf{x}_u$ segment would have been preferred over $\mathbf{x}_v$ between times t_L and t_R. So we have shown that if the MAP path differs from the

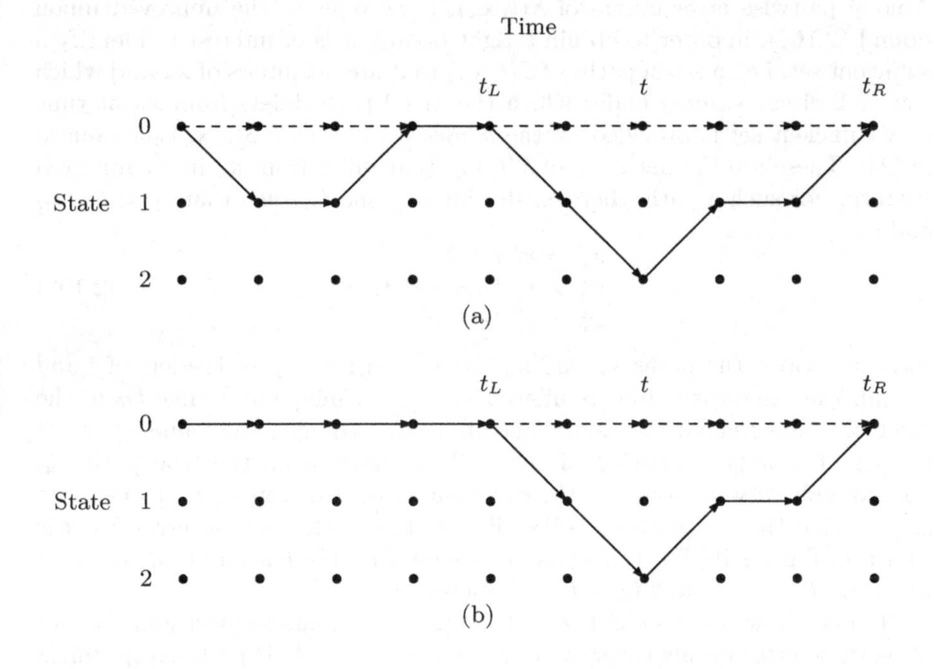

Fig. 12.11. (a) MAP path $\hat{\mathbf{x}}_{\text{MAP}}$ differing from the all zero path at time t, and (b) simple error event $\mathbf{x}_v$ obtained by retaining only the local excursion away from the all zero path.

true path at time t, the local excursion $\mathbf{x}_v$ of the MAP path surrounding t is a simple error event. Consequently, simple error events form a sufficient set of pairwise error events for the decision $\hat{x}_t^{\text{MAP}} \neq x_t^u$.

Then, according to the improved union bound (2.166), the conditional probability of error

$$P[\hat{X}_t^{\text{MAP}} \neq x_t^u | \mathbf{X} = \mathbf{x}_u] \leq \sum_{\mathbf{x}_v \in \mathcal{S}(t,\mathbf{x}_u)} P[J(\mathbf{x}_v) < J(\mathbf{x}_u) | \mathbf{X} = \mathbf{x}_u]$$
$$= \sum_{\mathbf{x}_v \in \mathcal{S}(t,\mathbf{x}_u)} Q\Big(\frac{d_{uv}}{2} - \frac{\ln(p_v/p_u)}{d_{uv}}\Big) , \quad (12.104)$$

where

$$d_{uv} = \frac{||\mathbf{s}_u - \mathbf{s}_v||_2}{\sigma} \quad (12.105)$$

is the Euclidean distance between signals $\mathbf{s}_u$ and $\mathbf{s}_v$, scaled by the standard deviation σ of the noise, i.e., d_{uv} represents the Euclidean distance of the two signals measured in units of the standard deviation. To go from the first to the second line of (12.104), consider the random variable

$$Z \triangleq \frac{J(\mathbf{x}_v) - J(\mathbf{x}_u)}{2\sigma^2 d_{uv}}$$
$$= \frac{1}{2\sigma^2 d_{uv}} (||\mathbf{Y} - \mathbf{s}_v||_2^2 - ||\mathbf{Y} - \mathbf{s}_u||_2^2) - \frac{\ln(p_v/p_u)}{d_{uv}} . \quad (12.106)$$

Under H_u, by substituting (12.99) inside (12.106), we find

$$Z = \frac{\mathbf{V}^T(\mathbf{s}_u - \mathbf{s}_v)}{\sigma^2 d_{uv}} + \frac{d_{uv}}{2} - \frac{\ln(p_v/p_u)}{d_{uv}} . \quad (12.107)$$

Since $\mathbf{V} \sim N(\mathbf{0}, \sigma^2 \mathbf{I}_N)$, this implies that under H_u

$$Z \sim N(\frac{d_{uv}}{2} - \frac{\ln(p_v/p_u)}{d_{uv}}, 1) . \quad (12.108)$$

But $\mathbf{x}_v$ is preferred to $\mathbf{x}_u$ whenever $Z < 0$, and from the distribution (12.108) we find

$$P[Z < 0 | H_u] = Q\Big(\frac{d_{uv}}{2} - \frac{\ln(p_v/p_u)}{d_{uv}}\Big) , \quad (12.109)$$

which proves the second line of (12.104).

In theory, to obtain an upper bound for the probability of error $P[E]$ given by (12.103), we need to characterize the simple error events $\mathcal{S}(t, \mathbf{x}_u)$ for *each* trellis path $\mathbf{x}_u$ and average the bound (12.104) for path $\mathbf{x}_u$ by weighting it with the path probability p_u. However, in some applications, such as decoding of convolutional codes, there exists a transformation that maps the set of simple error events for one path into the set of simple error events for any other path while, at the same time, preserving the distance d_{uv}. In such a case, if we assume also that all trellis paths are equally likely, the probability of error admits the upper bound

$$P[E] \leq \sum_{\mathbf{x}_v \in \mathcal{S}(t,\mathbf{x}_u)} Q(d_{uv}/2) , \tag{12.110}$$

where $\mathbf{x}_u$ is an arbitrary reference path, which is usually selected as the all zero path for decoding of convolutional codes.

At this point it is worth noting that the set $\mathcal{S}(t, \mathbf{x}_u)$ contains paths that are translates of each other. For example, if we consider the simple error event $\mathbf{x}_v$ of Fig. 12.10, the path $\mathbf{x}_w$ obtained in Fig. 12.12 by shifting $\mathbf{x}_v$ to the right by one time unit belongs also to $\mathcal{S}(t, \mathbf{x}_u)$. But, since the mapping $h(x_t, x_{t+1})$ used to generate the output signal from a state transition (x_t, x_{t+1}) is time-invariant, time-shifting error events do not affect their distance to the reference path $\mathbf{x}_u$, i.e., $d_{uw} = d_{uv}$. So, once we have considered a simple error event of $\mathcal{S}(t, \mathbf{x}_u)$, it is unnecessary to examine its translates. Let $\mathbf{S}_r(t, \mathbf{x}_u)$ be the reduced set of error events obtained under translation equivalence. If $\mathbf{x}_v$ is an element of $\mathbf{S}_r(t, \mathbf{x}_u)$, let

$$l_{uv} = t_R - t_L - 1 \tag{12.111}$$

denote the length of time for which $\mathbf{x}_s^v \neq \mathbf{x}_s^u$. Then the upper bound (12.110) can be written as

$$P[E] \leq \sum_{\mathbf{x}_v \in \mathcal{S}_r(t,\mathbf{x}_u)} l_{uv} Q(d_{uv}/2) . \tag{12.112}$$

For high SNR, the union bound sum is dominated by the terms corresponding to simple error events $\mathbf{x}_v$ for which the normalized distance d_{uv} between $\mathbf{s}_v$ and $\mathbf{s}_u$ is minimum. These events form the *nearest neighbors* of $\mathbf{x}_u$. Let $d_{\min}$ denote the minimal normalized distance between $\mathbf{s}_u$ and signals $\mathbf{s}_v$ corresponding to simple error events $\mathbf{x}_v \in \mathcal{S}_r(t, \mathbf{x}_u)$. Then for high SNR (high $d_{\min}$), the upper bound (12.112) reduces to

$$P[E] \leq C_U Q(d_{\min}/2) . \tag{12.113}$$

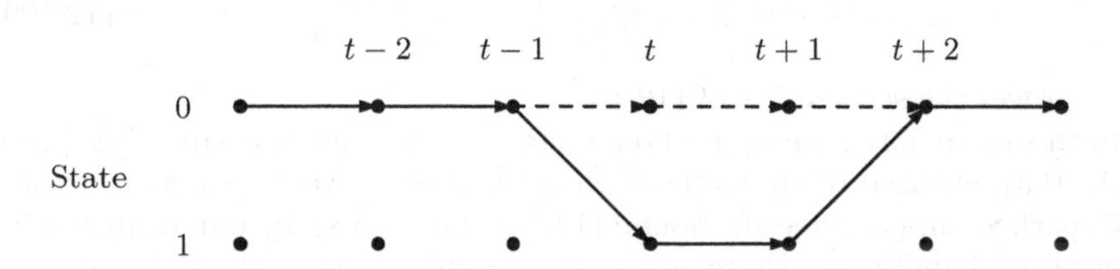

Fig. 12.12. Error event $\mathbf{x}_w$ obtained by shifting the event $\mathbf{x}_v$ of Fig. 12.10 to the right by one time unit.

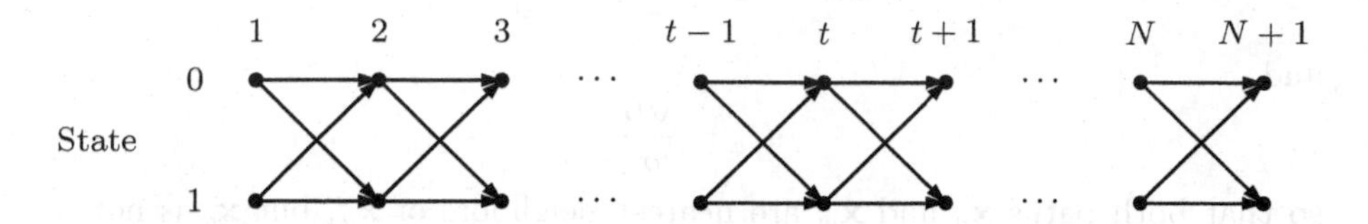

Fig. 12.13. Trellis for the MC representing the AMI line encoder.

where the constant C_U represents the sum of the lengths l_{uv} of all different nearest neighbors of $\mathbf{x}_u$ in $\mathcal{S}_r(t, \mathbf{x}_u)$.

Example 12.3, continued

Consider the trellis for the AMI line code shown in Fig. 12.13. For this example, there does not exist a transformation that maps an arbitrary trellis path $\mathbf{x}_u$ into the all zero state path in such a way that the scaled Euclidean distance d_{uv} between signal $\mathbf{s}_u$ corresponding to $\mathbf{x}_u$ and the signal $\mathbf{s}_v$ corresponding to another path $\mathbf{x}_v$ is preserved by the transformation. This is due to the fact that, unlike error-correcting codes where the observed signal $s(t)$ is generated by binary arithmetic, the AMI signal $s(t)$ given by (12.93) takes ternary values $\{1, 0, -1\}$. This means that the set $\mathcal{S}(t, \mathbf{x}_u)$ of simple error events and the subset of nearest neighbors of $\mathbf{x}_u$ depends on the reference path $\mathbf{x}_u$.

Specifically, let $\mathbf{x}_u$ be the dashed path in Fig. 12.14. For this path, the paths $\mathbf{x}_v$, $\mathbf{x}_w$ and $\mathbf{x}_q$ represented by solid trajectories in parts (a) to (c) of Fig. 12.14 correspond all to simple error events in $\mathcal{S}(t, \mathbf{x}_u)$. Over subinterval $[t, t+3]$, the signals corresponding to $\mathbf{x}_u$ and paths $\mathbf{x}_v$, $\mathbf{x}_w$ and $\mathbf{x}_q$ are given respectively by

$$\begin{bmatrix} s_u(t) \\ s_u(t+1) \\ s_u(t+2) \\ s_u(t+3) \end{bmatrix} = \begin{bmatrix} -1 \\ 0 \\ 1 \\ -1 \end{bmatrix} , \quad \begin{bmatrix} s_v(t) \\ s_v(t+1) \\ s_v(t+2) \\ s_v(t+3) \end{bmatrix} = \begin{bmatrix} 0 \\ -1 \\ 1 \\ -1 \end{bmatrix}$$

$$\begin{bmatrix} s_w(t) \\ s_w(t+1) \\ s_w(t+2) \\ s_w(t+3) \end{bmatrix} = \begin{bmatrix} 0 \\ 0 \\ 0 \\ -1 \end{bmatrix} , \quad \begin{bmatrix} s_q(t) \\ s_q(t+1) \\ s_q(t+2) \\ s_q(t+3) \end{bmatrix} = \begin{bmatrix} 0 \\ 0 \\ -1 \\ 0 \end{bmatrix} ,$$

and

$$s_u(r) = s_v(r) = s_w(r) = s_q(r)$$

for all values of r not in $[t, t+3]$. We find therefore that

$$d_{\min} = d_{uv} = d_{uw} = \frac{\sqrt{2}}{\sigma}$$

and

$$d_{uq} = \frac{\sqrt{6}}{\sigma} ,$$

so that both paths $\mathbf{x}_v$ and $\mathbf{x}_w$ are nearest neighbors of $\mathbf{x}_u$, but $\mathbf{x}_q$ is not a nearest neighbor even though it belongs to the set $\mathcal{S}(t, \mathbf{x}_u)$ of simple error events.

More generally, for an arbitrary path $\mathbf{x}_u$, the path $\mathbf{x}_v$ such that $x_r^u = x_r^v$ for $r \neq t$ and $x_t^v = 1 - x_t^u$ is a nearest neighbor at distance $d_{\min} = \sqrt{2}/\sigma$, but it is not necessarily the only nearest neighbor. For example, for the path $\mathbf{x}_u$

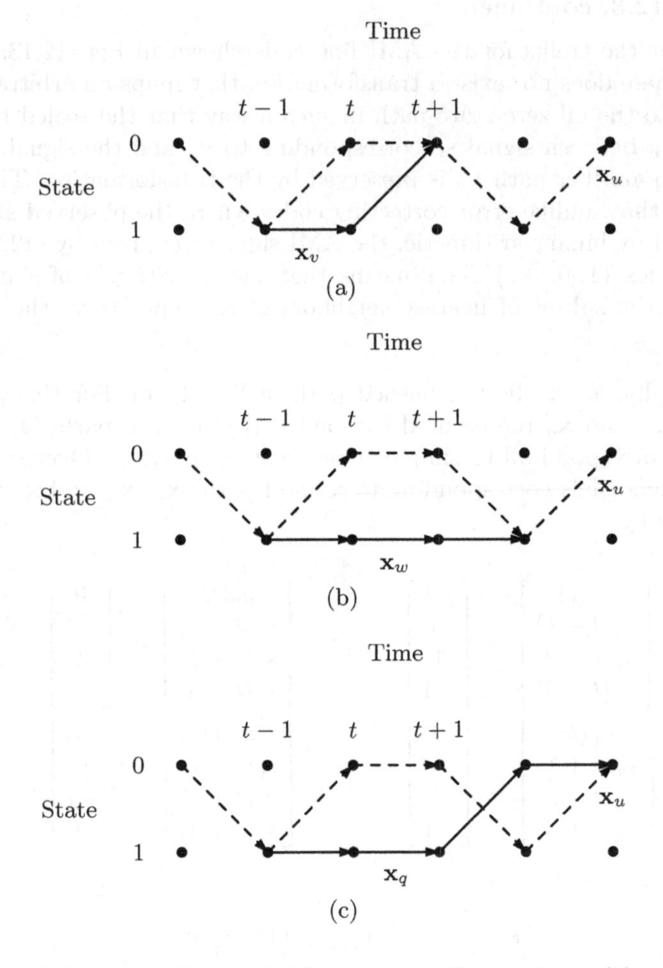

Fig. 12.14. Reference path $\mathbf{x}_u$ and (a) nearest neighbor path $\mathbf{x}_v$, (b) nearest neighbor path $\mathbf{x}_w$, and (c) simple error event $\mathbf{x}_q$.

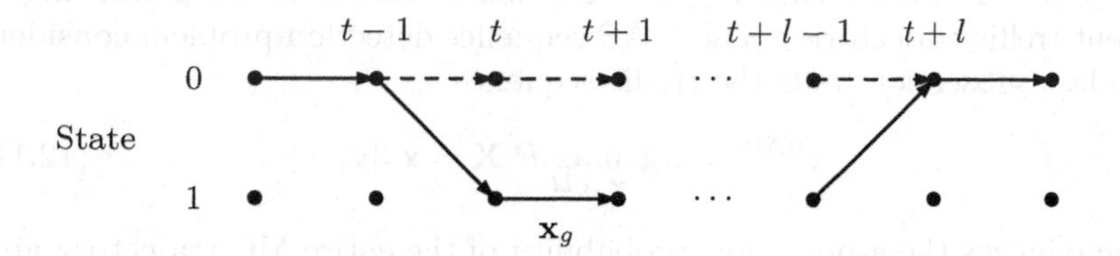

Fig. 12.15. Nearest neighbor path $\mathbf{x}_g$ of the zero path $\mathbf{0}$ specified by (12.114).

of Fig. 12.14, the only two nearest neighbors are $\mathbf{x}_v$ and $\mathbf{x}_w$. However, if the true path is the all zero path $\mathbf{0}$, then the path

$$x_r^g = \begin{cases} 0 & r \leq t_L = t-1 \\ 1 & t \leq r \leq t+l-1 \\ 0 & t+l \leq r \,, \end{cases} \tag{12.114}$$

which is depicted in Fig. 12.15 is a nearest neighbor of $\mathbf{0}$. Observe indeed that the signal corresponding to $\mathbf{x}_g$ is given by

$$s_g(r) = \begin{cases} 1 & r = t-1 \\ -1 & r = t+l-1 \\ 0 & \text{otherwise} \,, \end{cases}$$

so the normalized distance between $\mathbf{s}_g$ and the vector $\mathbf{0}$ representing the all zero path is $d_{\min} = \sqrt{2}/\sigma$. Note that in the specification (12.114), the merge time $t+l$ is such that $l > 1$ is arbitrary. Furthermore, all translates of $\mathbf{x}_g$ to the left by less than l steps are also nearest neighbors.

So for the AMI line code, the number of nearest neighbors in $\mathcal{S}(t, \mathbf{x}_u)$ depends on the reference path $\mathbf{x}_u$, but each path $\mathbf{x}_u$ admits at least one nearest neighbor at distance $d_{\min} = \sqrt{2}/\sigma$. This implies that at high SNR, the probability of error admits an upper bound of the form (12.113), where the constant C_U is obtained by averaging the number of neighbors for each path $\mathbf{x}_u$ with the path probability p_u. In this bound the normalized distance $d_{\min}$ is $\sqrt{2}$ larger than the distance $1/\sigma$ controlling the probability of error

$$P_E^{\text{AMI}} = \frac{3}{2} Q(1/(2\sigma))$$

for the usual AMI detector formed by the ternary slicer of Fig. 12.7 followed by a mapper. This $\sqrt{2}$ increase in signal space distance explains the 3dB improvement of the Viterbi decoder over the standard AMI decoder observed in [23]. □

12.3.2 Pointwise MAP Detection

Let

$$\mathbf{X} = \begin{bmatrix} X(1) \cdots X(t) \cdots X(N+1) \end{bmatrix}^T \tag{12.115}$$

be the random vector representing the trajectory of the MC over interval $1 \leq t \leq N+1$ and let $\{\mathbf{x}_u, 1 \leq u \leq U\}$ denote the vectors representing the different trellis trajectories. The MAP sequence detection problem considered in the last subsection finds the trellis sequence

$$\hat{\mathbf{x}}^{\mathrm{MAP}} = \arg \max_{\mathbf{x}_u \in \mathcal{U}} P[\mathbf{X} = \mathbf{x}_u | \mathbf{y}] \tag{12.116}$$

that maximizes the a-posteriori probability of the entire MC trajectory given observation vector $\mathbf{y}$. However, instead of maximizing the a-posteriori probability for an entire trajectory, it is also possible to perform the maximization at each time instant. This leads to a K-ary detection problem, where at time t, we seek to find the state

$$\hat{x}_t^{\mathrm{P}} = \arg \max_{1 \leq k \leq K} P[X(t) = k | \mathbf{y}] \tag{12.117}$$

maximizing the a-posteriori distribution of state $X(t)$ given $\mathbf{y}$.

At this point it is important to recognize that the two problems (12.116) and (12.117) are different. Specifically, the trajectory

$$\hat{\mathbf{x}}^{\mathrm{P}} \triangleq (\hat{x}_t^{\mathrm{P}}, 1 \leq t \leq N+1) \tag{12.118}$$

traced by performing the maximization (12.117) for each t need not coincide with solution $\hat{\mathbf{x}}^{MAP}$ of the pathwise maximization. In fact $\hat{\mathbf{x}}^{\mathrm{P}}$ need not form a trellis trajectory of $\mathcal{U}$, since it may contain some transitions that are not allowed, i.e., which are assigned a zero probability by the one-step transition probability distribution $\mathbf{P}$. This is, of course, rare, but this phenomenon may occur at low SNR.

Forward–backward algorithm: The pointwise MAP estimate can be evaluated by using the forward-backward or BCJR algorithm which was introduced independently by Baum et al. [25] and by Bahl et al. [26]. The first step in the derivation of this algorithm consists in recognizing that the maximization of the discrete a-posteriori probability distribution

$$p_{X(t)|\mathbf{Y}}(k,t|\mathbf{y}) \triangleq P[X(t) = k | \mathbf{Y} = \mathbf{y}]$$

is equivalent to the maximization of the joint hybrid probability distribution/density

$$p_{X(t),\mathbf{Y}}(k,t;\mathbf{y}) \triangleq p_{X(t)|\mathbf{Y}}(k,t|\mathbf{y}) f_{\mathbf{Y}}(\mathbf{y}) \tag{12.119}$$

where $f_{\mathbf{Y}}(\mathbf{y})$ denotes the probability density of the observation vector $\mathbf{Y}$ which, as shown in (12.72), regroups all observations over the interval $1 \leq t \leq N$. This vector can be split into the vectors

$$\begin{aligned} \mathbf{Y}_t^- &= \big[Y(1)\, Y(2) \cdots\, Y(t-1)\big]^T \\ \mathbf{Y}_t^+ &= \big[Y(t)\, Y(t+1) \cdots\, Y(N)\big]^T \end{aligned} \tag{12.120}$$

representing the observations in the past and future of the current time t. Note that because $Y(t)$ depends on the transition between states $X(t)$ and $X(t+1)$, it must be allocated to the future, not the past, of time t. Then the hybrid distribution/density (12.119) can be denoted as $p_{X(t),\mathbf{Y}_t^-,\mathbf{Y}_t^+}(k,t;\mathbf{y}_t^-,\mathbf{y}_t^+)$, and by conditioning jointly with respect to $X(t)$ and $\mathbf{Y}_t^-$, we obtain

$$p_{X(t),\mathbf{Y}_t^-,\mathbf{Y}_t^+}(k,t;\mathbf{y}_t^-,\mathbf{y}_t^+) = f_{\mathbf{Y}_t^+|X(t),\mathbf{Y}_t^-}(\mathbf{y}_t^+|k,t;\mathbf{y}_t^-) \cdot p_{X(t),\mathbf{Y}_t^-}(k,t;\mathbf{y}_t^-) \,. \tag{12.121}$$

But since $X(t)$ is a Markov process, its future and past are conditionally independent given its present value. Consequently, given $X(t)$, $\mathbf{Y}_t^+$ and $\mathbf{Y}_t^-$ are conditionally independent, so we have

$$f_{\mathbf{Y}_t^+|X(t),\mathbf{Y}_t^-}(\mathbf{y}_t^+|k,t;\mathbf{y}_t^-) = f_{\mathbf{Y}_t^+|X(t)}(\mathbf{y}_t^+|k,t) \,. \tag{12.122}$$

Let us denote

$$\alpha(k,t) \triangleq p_{X(t),\mathbf{Y}_t^-}(k,t;\mathbf{y}_t^-) \tag{12.123}$$

$$\beta(k,t) \triangleq f_{\mathbf{Y}_t^+|X(t)}(\mathbf{y}_t^+|k,t) \tag{12.124}$$

with $1 \leq k \leq K$, where for simplicity, the observation vectors $\mathbf{y}_t^-$ and $\mathbf{y}_t^+$ have been dropped from the arguments of α and β, since they are known. The factorization (12.121)–(12.122) can be rewritten as

$$\begin{aligned} P(k,t) &\triangleq P[X(t)=k|\mathbf{Y}=\mathbf{y}]f_{\mathbf{Y}}(\mathbf{y}) \\ &= \alpha(k,t)\beta(k,t) \,, \end{aligned} \tag{12.125}$$

so that the pointwise MAP estimate can be expressed as

$$\hat{x}_t^{\mathrm{P}} = \arg \max_{1\leq k\leq K} (\alpha(k,t)\beta(k,t)) \,. \tag{12.126}$$

It turns out that the discrete valued functions $\alpha(k,t)$ and $\beta(k,t)$ admit forward and backward recursions, respectively, which justify the name of the algorithm. To derive the forward recursion for $\alpha(k,t)$, observe that the joint hybrid probability distribution/density of $X(t)$, $X(t+1)$, and

$$\mathbf{Y}_{t+1}^- = \begin{bmatrix} \mathbf{Y}_t^- \\ Y(t) \end{bmatrix}$$

admits the factorization

$$p_{X(t),X(t+1),\mathbf{Y}_{t+1}^-}(k,\ell,t;\mathbf{y}_{t+1}^-) = f_{Y(t)|X(t),X(t+1)}(y_t|k,\ell) \cdot p_{X(t+1)|X(t)}(\ell|k) p_{X(t),\mathbf{Y}_t^-}(k,t;\mathbf{y}_t^-) \,, \tag{12.127}$$

where we have used again the fact that the Markov property of $X(t)$ implies that $X(t+1)$ and $\mathbf{Y}_t^-$ are conditionally independent given $X(t)$. In this factorization, we recognize $p_{k\ell} = p_{X(t+1)|X(t)}(\ell|k)$ and the observation model (12.65) implies

$$f_{Y(t)|X(t),X(t+1)}(y_t|k,\ell) = \frac{1}{(2\pi\sigma^2)^{1/2}} \exp\big(-\frac{1}{2\sigma^2}(y_t - h(k,\ell))^2\big) . \quad (12.128)$$

Taking into account the definition (12.123) of $\alpha(k,t)$, by marginalizing the distribution $p_{X(t),X(t+1),\mathbf{Y}_{t+1}^-}$ with respect to $X(t)$, we obtain

$$\alpha(\ell,t+1) = \frac{1}{(2\pi\sigma^2)^{1/2}} \sum_{k=1}^{K} \alpha(k,t) p_{k\ell} \exp\big(-\frac{1}{2\sigma^2}(y_t - h(k,\ell))^2\big) . \quad (12.129)$$

The recursion (12.129) can be used to propagate the probability distribution $\alpha(k,t)$ forward in time, starting with the initial condition

$$\alpha(k,1) = \pi_k(1) , \quad (12.130)$$

where $\pi_k(1) = P[X(1) = k]$ is the initial probability distribution of the MC.
Similarly, by observing that the vector $\mathbf{Y}_t^+$ admits the partition

$$\mathbf{Y}_t^+ = \begin{bmatrix} Y(t) \\ \mathbf{Y}_{t+1}^+ \end{bmatrix}$$

we find that the hybrid conditional probability distribution/density of $X(t+1)$ and $\mathbf{Y}_t^+$ given $X(t)$ can be factored as

$$\begin{aligned} p_{X(t+1),\mathbf{Y}_t^+|X(t)}(\ell;\mathbf{y}_t^+|k,t) &= f_{\mathbf{Y}_{t+1}^+|X(t),X(t+1),Y(t)}(\mathbf{y}_{t+1}^+|k,\ell,t;y_t) \\ &\cdot f_{Y(t)|X(t),X(t+1)}(y_t|k,\ell) p_{X(t+1)|X(t)}(\ell|k) . \end{aligned} \quad (12.131)$$

Again, the Markov property of $X(t)$ implies that given $X(t+1)$, the future observations $\mathbf{Y}_{t+1}^+$ are conditionally independent of the previous state $X(t)$ and observation $Y(t)$, so in (12.131) we have

$$f_{\mathbf{Y}_{t+1}^+|X(t),X(t+1),Y(t)}(\mathbf{y}_{t+1}^+|k,\ell,t;y_t) = f_{\mathbf{Y}_{t+1}^+|X(t+1)}(\mathbf{y}_{t+1}^+|\ell,t) . \quad (12.132)$$

Recognizing that this last function is $\beta(\ell,t+1)$, by marginalizing (12.131) with respect to $X(t+1)$, we obtain

$$\beta(k,t) = \frac{1}{(2\pi\sigma^2)^{1/2}} \sum_{\ell=1}^{K} \beta(\ell,t+1) p_{k\ell} \exp\big(-\frac{1}{2\sigma^2}(y_t - h(k,\ell))^2\big) , \quad (12.133)$$

which can be used to propagate $\beta(k,t)$ backwards in time. The initial condition for this recursion can be selected as

$$\beta(\ell, N+1) = 1 \tag{12.134}$$

for all ℓ. To justify this choice, observe that

$$\begin{aligned}
\beta(k, N) &= \sum_{\ell=1}^{K} p_{X(N+1),Y(N)|X(N)}(\ell, y_N|k, N) \\
&= \sum_{\ell=1}^{K} f_{Y(N)|X(N),X(N+1)}(y_N|k, \ell, N) p_{X(N+1)|X(N)}(\ell|k) \\
&= \frac{1}{(2\pi\sigma^2)^{1/2}} \sum_{\ell=1}^{K} p_{k\ell} \exp\big(-\frac{1}{2\sigma^2}(y_N - h(k,\ell))^2\big)
\end{aligned} \tag{12.135}$$

coincides with recursion (12.133) for $t = N$, provided $\beta(\ell, N+1)$ satisfies (12.134).

It is interesting to note that the forward and backward recursions (12.129) and (12.133) admit a compact matrix representation. Let $\mathbf{M}(t)$ be the $K \times K$ matrix whose (k,ℓ)-th entry is given by

$$m_{k\ell}(t) = \begin{cases} \dfrac{1}{(2\pi\sigma^2)^{1/2}} \exp\big(-\dfrac{c(k,\ell,t)}{2\sigma^2}\big) \text{for } (k,\ell) \in \mathcal{T} \\ 0 \qquad\qquad\qquad \text{otherwise}\,, \end{cases} \tag{12.136}$$

where $c(k,\ell,t)$ is the path metric defined in (12.80). Note that all entries of $\mathbf{M}(t)$ are nonnegative, and $\mathbf{M}(t)$ has the same graph as $\mathbf{P}$. Let also

$$\boldsymbol{\alpha}(t) = \big[\alpha(1,t) \cdots \alpha(k,t) \cdots \alpha(K,t)\big] \tag{12.137}$$

denote the $1 \times K$ row vector representing the discrete function $\alpha(\cdot, t)$. Then the forward recursion (12.129)–(12.130) can be rewritten in matrix form as

$$\boldsymbol{\alpha}(t+1) = \boldsymbol{\alpha}(t)\mathbf{M}(t) \tag{12.138}$$

with

$$\boldsymbol{\alpha}(1) = \boldsymbol{\pi}(1)\,. \tag{12.139}$$

Similarly, let

$$\boldsymbol{\beta}(t) = \big[\beta(1,t) \cdots \beta(k,t) \cdots \beta(K,t)\big]^T \tag{12.140}$$

denote the $K \times 1$ column vector representing the function $\beta(\cdot, t)$. Then the backward recursion (12.133) can be expressed as

$$\boldsymbol{\beta}(t) = \mathbf{M}(t)\boldsymbol{\beta}(t+1) \tag{12.141}$$

with initial condition

$$\boldsymbol{\beta}(N+1) = \mathbf{e}\,, \tag{12.142}$$

where $\mathbf{e}$ is the vector whose entries are all one, as shown in (12.4). In this respect, it is worth observing that the forward and backward recursions can be expressed in terms of the same matrix $\mathbf{M}(t)$, but the forward recursion involves row vectors, whereas the backward expression involves column vectors.

By observing that the a-posteriori probability distribution $P[X(t) = k|\mathbf{y}]$ must be normalized, we find from (12.125) that it can be expressed as

$$P[X(t) = k|\mathbf{y}] = \alpha_k(t)\beta_k(t)/\gamma(t) \tag{12.143}$$

where the normalizing constant $\gamma(t)$ is the row times column inner product

$$\gamma(t) = \boldsymbol{\alpha}(t)\boldsymbol{\beta}(t) = \sum_{k=1}^{K} \alpha_k(t)\beta_k(t) \,. \tag{12.144}$$

Substituting recursions (12.138) and (12.141) inside (12.144) yields

$$\gamma(t) = \boldsymbol{\alpha}(t)\mathbf{M}(t)\boldsymbol{\beta}(t+1) = \boldsymbol{\alpha}(t+1)\boldsymbol{\beta}(t+1) = \gamma(t+1) \,, \tag{12.145}$$

so γ is actually constant.

Computational complexity: Assume that the MC trellis is regular, so that the same number T of transitions is allowed out of each state. For example, for the MC of Fig. 12.4, $T = 2$. Then the matrix notation (12.138), (12.141) for the forward and backward recursions indicates that KT scalar multiplications are required to advance vectors $\boldsymbol{\alpha}(t)$ and $\boldsymbol{\beta}(t)$ by one time unit. In contrast, each stage of the Viterbi algorithm requires implementing the ACS operation (12.85) for each state, so that $O(K)$ operations are required by the Viterbi algorithm. Thus the Viterbi algorithm has a factor of T advantage in computational complexity over the forward-backward algorithm. When the transition matrix $\mathbf{P}$ is dense, i.e., when $T = K$, so that all state transitions are allowed, this difference can be significant. This explains why, even though the Viterbi and forward-backward algorithms were discovered during the same time frame, for a long time the Viterbi algorithm was preferred. In this respect it is useful to observe that, because it maximizes the a-posteriori probability $P[X(t) = k|\mathbf{Y}]$, the probability of error of the pointwise MAP estimate $\hat{X}_t^{\mathrm{P}}$ is slighly lower than that of the t-th component $\hat{X}_t^{\mathrm{MAP}}$ of the sequence estimate $\hat{\mathbf{X}}_{\mathrm{MAP}}$, i.e.,

$$P[\hat{X}_t^{\mathrm{P}} \neq X(t)] \leq P[\hat{X}_t^{\mathrm{MAP}} \neq X(t)] \,.$$

However this advantage is very small, typically a small fraction of a dB. So in spite of its slight performance advantage, until recently, the forward-backward algorithm was rarely implemented in practical systems, due to its higher computational cost. This situation changed overnight when it was discovered [2,27] that iterative equalization and/or decoding algorithms can offer an advantage of several dBs over systems that treat equalization and decoding, or decoding stages of concatenated codes, as separate noninteracting blocks. The key feature of iterative decoding algorithms is that they require the exchange of soft information, representing a relative likelihood measure of different state values, based on the given observations. Since the forward–backward algorithm computes explicitly the a-posteriori distribution of the different state values given the observation vector $\mathbf{Y}$, it is ideally suited for the implementation of iterative decoding algorithms. By comparison, the classical Viterbi algorithm does not provide an explicit reliability estimate for the values of the states at different times based on $\mathbf{Y}$. However, Hagenauer and Hoeher [28] introduced a variant of the Viterbi algorithm called the soft Viterbi algorithm (SOVA), which computes state reliability values by performing a backward recursion over a time window selected to ensure that survivors have merged. But these reliability values are only rough approximations of the a-posteriori probability distribution $P[X(t) = k|\mathbf{Y}]$, so it is often preferable to implement the forward-backward algorithm in spite of its slightly larger computational load.

Max-log-MAP algorithm: Since the a-posteriori state probability distribution is obtained by performing the normalization (12.143), the common scale factor $(2\pi\sigma^2)^{-1/2}$ can be removed from the forward and backward recursions (12.129) and (12.133). Then if we denote

$$V_f(k,t) \triangleq -2\sigma^2 \ln(\alpha(k,t)) \tag{12.146}$$

$$V_b(k,t) \triangleq -2\sigma^2 \ln(\beta(k,t)) \, , \tag{12.147}$$

the recursions (12.129) and (12.133) can be rewritten in logarithmic form as

$$V_f(\ell,t+1) = \\ -2\sigma^2 \ln\Big(\sum_{k:\,(k,\ell)\in\mathcal{T}} \exp\big(-\frac{1}{2\sigma^2}(V_f(k,t) + c(k,\ell,t))\big)\Big) \tag{12.148}$$

$$V_b(k,t) = \\ -2\sigma^2 \ln\Big(\sum_{\ell:\,(k,\ell)\in\mathcal{T}} \exp\big(-\frac{1}{2\sigma^2}(V_b(\ell,t+1) + c(k,\ell,t))\big)\Big) \, . \tag{12.149}$$

These recursions specify the log-MAP algorithm, which is strictly equivalent to the forward-backward algorithm. The a-posteriori probability distribution (12.143) can be expressed in terms of the forward and backward value functions V_f and V_b as

$$P[X(t) = k|\mathbf{y}] = \frac{\exp(-(V_f(k,t) + V_b(k,t))/(2\sigma^2))}{\sum_{\ell=1}^{K} \exp(-(V_f(\ell,t) + V_b(\ell,t))/(2\sigma^2))} \, . \tag{12.150}$$

Note that, strictly speaking, the functions $V_f(k,t)$ and $V_b(k,t)$ defined by (12.146) and (12.137) are the *negative* of the scaled logarithms of forward and backward components $\alpha(k,t)$ and $\beta(b,t)$ of the a-posteriori probability $P[X(t) = k|\mathbf{y}]$ of state k at time t. So, applying minimization operations to these functions is performing a maximization on the logarithmic components of the a-posteriori probability. At this point, it is convenient to introduce the $\min^*$ function which for n positive real numbers x_i is defined as

$$\min{}^*(x_1, \cdots, x_n) \triangleq -\ln\big(\sum_{i=1}^{n} \exp(-x_i)\big) . \tag{12.151}$$

This function satisfies the properties

$$\min{}^*(x, y) = \min(x, y) - \ln(1 + \exp(-|x - y|)) \tag{12.152}$$
$$\min{}^*(x, y, z) = \min{}^*(\min{}^*(x, y), z) . \tag{12.153}$$

With this notation, the log-MAP algorithm (12.148)–(12.149) can be written as

$$V_f(\ell, t+1) = 2\sigma^2 \min{}^*_{k:\,(k,\ell)\in\mathcal{T}} \big(\frac{1}{2\sigma^2}(V_f(k,t) + c(k,\ell,t))\big) \tag{12.154}$$
$$V_b(k, t) = 2\sigma^2 \min{}^*_{\ell:\,(k,\ell)\in\mathcal{T}} \big(\frac{1}{2\sigma^2}(V_b(\ell,t+1) + c(k,\ell,t))\big) , \tag{12.155}$$

where the forward and backward recursions are formally identical to Viterbi algorithms, except that the min operator in the usual Viterbi recursion is replaced by the operator $\min^*$. But the decomposition (12.152) indicates that whenever the arguments of the $\min^*$ operator are sufficiently far apart, the $\min^*$ operator can be approximated by the min operator. By applying this approximation to recursions (12.154)–(12.155) and noting that for an arbitrary positive scale factor s

$$s \min(\frac{x}{s}, \frac{y}{s}) = \min(x, y)$$

we obtain the max-log-MAP algorithm

$$V_f(\ell, t+1) = \min_{k:\,(k,\ell)\in\mathcal{T}} (V_f(k,t) + c(k,\ell,t)) \tag{12.156}$$
$$V_b(k, t) = \min_{\ell:\,(k,\ell)\in\mathcal{T}} (V_b(\ell,t+1) + c(k,\ell,t)) , \tag{12.157}$$

which is now expressed in the form of forward and backward Viterbi algorithms for the MC trellis with path metric (12.80). So although the forward-backward/log-MAP algorithm has a higher complexity than the Viterbi algorithm, by approximating the $\min^*$ operator by a min operator, it is possible to approximate the log-MAP algorithm by the max-log-MAP algorithm, whose complexity is only twice that of the Viterbi algorithm. It was shown in [29] that the max-log-MAP algorithm is identical to a form of the soft Viterbi algorithm proposed by Battail [30].

Remark: Note that, strictly speaking, the functions $V_f(k,t)$ and $V_b(k,t)$ defined by (12.146) and (12.137) are the *negative* of the scaled logarithms of forward and backward components $\alpha(k,t)$ and $\beta(b,t)$ of the a-posteriori probability $P[X(t) = k|\mathbf{y}]$ of state k at time t. So, even though recursions (12.156

and (12.157) are expressed in terms of minimization operations, they really represent maximimizations of logarithmic components of the a-posteriori state probability, which explains why the algorithm is called the max-log-MAP algorithm.

12.4 Example: Channel Equalization

To illustrate the detection of partially observed Markov chains, we consider in this section the MAP sequence equalization of a binary pulse amplitude modulated (PAM) signal transmitted though a linear ISI channel. The DT observation signal at the output of the receiver front-end can be expressed as

$$Y(t) = s(t) + V(t) \tag{12.158}$$

with

$$s(t) = \sum_{k=0}^{L} h_k I(t-k) , \tag{12.159}$$

where the transmitted symbols $\{I(t),\ t \geq 1\}$ are independent identically distributed and take the values ± 1 with probability 1/2. The noise $V(t)$ is a zero mean WGN sequence with variance σ^2. Note that the design of an analog receiver front-end leading to a DT model of the form (12.158)–(2.159) is not a trivial matter. It was shown by Forney [14] that an optimal structure consists of applying a CT filter matched to the received pulse (the convolution of the channel impulse response with the transmit pulse), followed by a baud rate sampler and a DT noise whitening filter. For transmission systems such that the received pulse has less than 100% excess bandwidth, i.e., with a bandwidth less than the baud frequency, another optimal front-end structure [31] consists of passing the signal through a lowpass filter with bandwidth equal to the baud frequency, followed by a Nyquist sampler with sampling period $T/2$, where T denotes the baud period. The advantage of this structure is that it remains optimal when the channel is not known exactly [32]. However, in this case the observation $Y(t)$ and noise $V(t)$, as well as the impulse response h_k are 2-dimensional vectors. Also, the variance of the additive WGN is twice as large as for the whitened matched filter front end receiver, due to the 100% excess bandwidth of the analog front end filter. For simplicity, we restrict our attention to the case when $Y(t)$ is scalar, leaving the extension to the case of vector observations as an excercise for the reader.

12.4.1 Markov Chain Model

Note that the transmitted symbol sequence $\{I(t)\}$ is in one-to-one correspondence with a bit stream $\{B(t)\}$ where

$$B(t) = \frac{1}{2}(I(t) + 1) \tag{12.160}$$

is either zero or one, so with respect to the bit sequence $\{B(t)\}$, the ISI signal $s(t)$ can be expressed as

$$s(t) = 2\sum_{k=0}^{L} h_k B(t-k) - \sum_{k=0}^{L} h_k \,. \tag{12.161}$$

In the model (12.159), it is assumed that L is finite, so that the channel transfer function

$$H(z) = \sum_{k=0}^{L} h_k z^{-k} \tag{12.162}$$

is FIR. This property plays an essential role in the constructing of an MC model of equation (12.159), and thus in formulating the equalization problem as an MAP sequence detection problem. In practice, the sampled channel h_k does not have finite length, but it decays sufficiently rapidly to allow the truncation of its tail.

The Markov chain state $X(t)$ corresponding to the observation model (12.159) represents an encoding of the symbols $I(t-k)$ (or equivalently the bits $B(t-k)$) with $1 \leq k \leq L$ stored in the channel. For example, we can use for $X(t)$ the binary expansion

$$\begin{aligned} X(t) &= 1 + \sum_{k=1}^{L} 2^{k-1} B(t-k) \\ &= \frac{1}{2}[1 + 2^L + \sum_{k=1}^{L} 2^{k-1} I(t-k)] \,. \end{aligned} \tag{12.163}$$

Under this mapping, $X(t)$ takes values in the set $\mathcal{S} = \{1, \cdots, 2^L\}$, so the total number of states $K = 2^L$ grows exponentially with the order L of the FIR filter $H(z)$. Furthermore, it can be verified from (12.163) that, if symbols $(i_{t-1}, i_{t-2}, \cdots, i_{t-L})$ are mapped into state x_t, then the opposite symbols $(-i_{t-1}, -i_{t-2}, \cdots, -i_{t-L})$ are mapped into state $1+2^L - x_t$. In other words, reversing the sign of the symbols stored in the channel maps the state x_t into its mirror image with respect to the center $(1+2^L)/2$ of the state labels. Observe also that the latest bit $B(t)$ does not appear in the encoding of $X(t)$, but in the encoding of $X(t+1)$, and the state dynamics can be expressed as

$$X(t+1) = 1 + [2(X(t) - 1) + B(t)] \bmod 2^L \,, \tag{12.164}$$

which clearly indicates that $X(t)$ is a Markov process. The recursion (12.164) just implements a shift register, so as a new bit $B(t)$ enters the register, the oldest bit $B(t-L)$ is pushed out. Note that, since $B(t)$ is binary, there are only $T = 2$ transitions out of each state, each with probability 1/2, thus all trellis paths are equally likely.

Table 12.2. State mapping for $L = 2$.

State	Bits $B(t-1)$, $B(t-2)$	Symbols $I(t-1)$, $I(t-2)$
1	0 , 0	-1, -1
2	1 , 0	1, -1
3	0, 1	-1, 1
4	1, 1	1 , 1

Example 12.4

For $L = 2$ the mapping (12.163) is illustrated in Table 12.2 where the first column lists all four possible states, the second column lists the corresponding bits $(B(t-1), B(t-2))$, and the third column the matching symbols $(I(t-1), I(t-2))$.
In this case, the state transition diagram corresponding to dynamics (12.164) is shown in Fig. 12.16. Each transition is labeled by the corresponding value of $B(t)$. The one-step transition matrix for the corresponding MC is given by

$$\mathbf{P} = \begin{bmatrix} 1/2 & 1/2 & 0 & 0 \\ 0 & 0 & 1/2 & 1/2 \\ 1/2 & 1/2 & 0 & 0 \\ 0 & 0 & 1/2 & 1/2 \end{bmatrix},$$

and it is easy to verify that the MC is irreducible and aperiodic. □

Since the state $X(t)$ encodes the transmitted symbols $I(t-k)$ with $1 \leq k \leq L$, the ISI signal

$$s(t) = \sum_{k=0}^{L} h_k I(t-k) = h(X(t), X(t+1)) \tag{12.165}$$

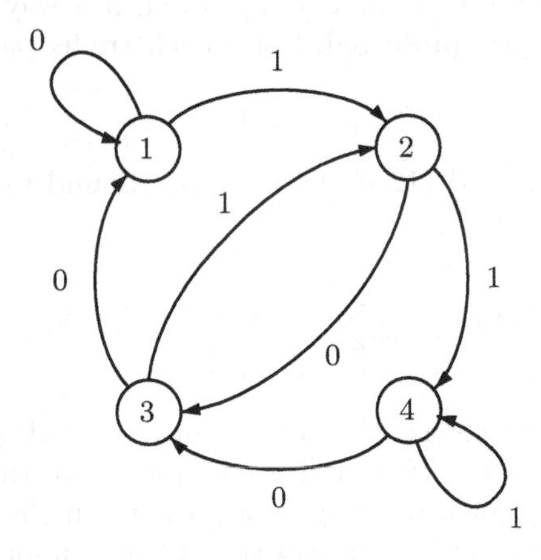

Fig. 12.16. State transition diagram of the MC modelling the channel memory.

is clearly a function of state transitions, so the equalization problem can be viewed as a partially observed MC detection problem.

Example 12.4, continued

For the case when

$$H(z) = 0.8 - 0.6z^{-1} + 0.1z^{-2} , \tag{12.166}$$

the matrix $\mathbf{H}$ with entries $h(k, \ell)$ for $1 \leq k, \ell \leq 4$ is given by

$$\mathbf{H} = \begin{bmatrix} -0.3 & 1.3 & x & x \\ x & x & -1.5 & 0.1 \\ -0.1 & 1.5 & x & x \\ x & x & -1.3 & 0.3 \end{bmatrix} ,$$

where the entries labeled with x's correspond to forbidden transitions. Note that the matrix $\mathbf{H}$ is skew-centrosymmetric, which means that entries of $\mathbf{H}$ on opposite sides of its center have an opposite value. This property is due to the mirror image symmetry of the state labeling (12.163) under symbol sign reversal, and to the fact that, if the symbols $\{i_t, i_{t-1}, i_{t-2}\}$ produce an output

$$s_t = h_0 i_t + h_1 i_{t-1} + h_2 i_{t-2},$$

the opposite symbols $\{-i_t, -i_{t-1}, -i_{t-2}\}$ will produce output $-s_t$. □

12.4.2 Performance Analysis

Consider now the error analysis of MAP sequence equalization. Unlike the decoding problem of convolutional codes, given an MC trellis path $\mathbf{x}_u$ and its set $\mathcal{S}(t, \mathbf{x}_u)$ of simple error events for an arbitrary time t, there does not exist a mapping which transforms $\mathbf{x}_u$ and $\mathcal{S}(t, \mathbf{x}_u)$ into any other trellis path $\mathbf{x}_w$ and its set of simple error events $\mathcal{S}(t, \mathbf{x}_w)$ in such a way that distances d_{uv} between trellis paths are preserved. Since each trellis path is equally likely with probability

$$p_u = 1/U ,$$

by combining (12.102) and (12.104), the upper bound for the probability of error can be expressed as

$$P[E] \leq \frac{1}{U} \sum_{u=1}^{U} \sum_{\mathbf{x}_v \in \mathcal{S}(t, \mathbf{x}_u)} Q(\frac{d_{uv}}{2}) . \tag{12.167}$$

Thus in theory, to characterize the performance of MAP sequence detection, for each reference path $\mathbf{x}_u$, we would need to evaluate the distance between $\mathbf{s}_u$ and each signal $\mathbf{s}_v$ corresponding to a path $\mathbf{x}_v$ in the simple error event set $\mathcal{S}(t, \mathbf{x}_u)$. This represents an unrealistically large amount of computation. Fortunately, for high SNR, the bound becomes

$$P[E] \leq C_U Q(\frac{d_{\min}}{2}) , \tag{12.168}$$

where $d_{\min}$ is the minimum normalized distance betwen signals $\mathbf{s}_u$ and $\mathbf{s}_v$ corresponding to paths $\mathbf{x}_u$ and $\mathbf{x}_v$ which differ from each other in a connected pattern.

Due to the linearity of expressions (12.159) and (12.161) relating the ISI signal $s(t)$ to the corresponding symbols $\{I(t)\}$ or bits $\{B(t)\}$, it turns out that the evaluation of $d_{\min}$ can be performed without examining each reference path $\mathbf{x}_u$. Specifically, let $\{B_u(t)\}$ and $\{B_v(t)\}$ denote respectively the bit sequences corresponding to paths $\mathbf{x}_u$ and $\mathbf{x}_v$. According to (12.161) the signal difference

$$s_u(t) - s_v(t) = 2 \sum_{k=0}^{L} h_k D(t-k) , \tag{12.169}$$

where

$$D(t) = B_u(t) - B_v(t) \tag{12.170}$$

denotes the difference between the two bit sequences. Since $B_u(t)$ and $B_v(t)$ take values 0 or 1, $D(t)$ takes values in the ternary set $\{1, 0, -1\}$.

Therefore, the squared minimum normalized distance $d_{\min}^2$ can be expressed as

$$d_{\min}^2 = \frac{4}{\sigma^2} \min_{D(\cdot)\in\mathcal{D}} \sum_{t=1}^{\infty} \Big(\sum_{k=0}^{L} h_k D(t-k) \Big)^2 , \tag{12.171}$$

where $\mathcal{D}$ denotes the set of nonzero sequences $\{D(t),\, t \geq 1\}$. Without loss of generality, since the ISI channel is time-invariant, we can assume $D(t) \equiv 0$ for $t \leq 0$ and $D(1) \neq 0$ in (12.171). The minimization (12.171) can be accomplished by associating to it an enlarged state trellis, and then recognizing that the resulting trellis minimization is a shortest path problem.

Since $D(t)$ takes values over $\{-1, 0, 1\}$, the number of states needed to encode the values $D(t-k)$ for $1 \leq t \leq L$ is 3^L instead of 2^L for the state mapping (12.163). For instance, we can use the mapping

$$\begin{aligned} \xi(t) &= 1 + \sum_{k=1}^{L} 3^{k-1}(D(t-k)+1) \\ &= \frac{1}{2}(1+3^L) + \sum_{k=1}^{L} 3^{k-1} D(t-k) . \end{aligned} \tag{12.172}$$

Note again that the mapping (12.172) is such that if state ξ_t encodes $\{d_{t-1}, d_{t-2}, \cdots, d_{t-L}\}$, the sign reversed sequence $\{-d_{t-1}, -d_{t-2}, \cdots, -d_{t-L}\}$ is represented by state $1 + 3^L - \xi_t$, so the mapping is symmetric with respect to the middle label under sign reversal of the difference values $D(t-k)$ for $1 \leq k \leq L$.

The difference $D(t)$ does not appear in the encoding of $\xi(t)$, but in the encoding of $\xi(t+1)$, and the state transitions are specified by the recursion

$$\xi(t+1) = 1 + [3\xi(t) + D(t) - 2] \mod 3^L , \tag{12.173}$$

so the dynamics of $\xi(t)$ can be described by a finite state machine where three different transitions can take place out of a fixed state ξ_t depending on whether $D(t)$ takes values -1, 0 or 1.

Example 12.5

For $L = 2$, the mapping (12.172) is depicted in Table 12.3. Each column shows the state $\xi(t)$, and the differences $D(t-1)$ and $D(t-2)$ it encodes. The trellis diagram for the state transition dynamics (12.173) is shown in Fig. 12.17 for $L = 2$. □

Table 12.3. State mapping (12.172) for $L = 2$.

State	1	2	3	4	5	6	7	8	9
$D(t-1)$	–1	0	1	–1	0	1	–1	0	1
$D(t-2)$	–1	–1	–1	0	0	0	1	1	1

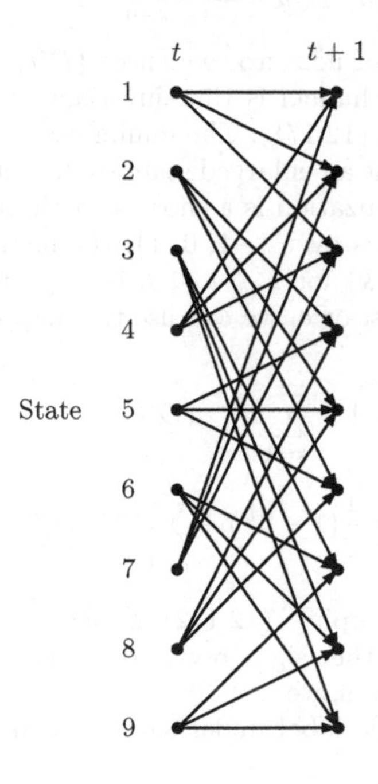

Fig. 12.17. Trellis diagram corresponding to the state dynamics (12.173) for $L = 2$.

Given the state mapping (12.172), we can express the difference signal

$$\sum_{k=1}^{L} h_k D(t-k) = g(\xi(t), \xi(t+1)) \qquad (12.174)$$

as a function of the transition betwen state $\xi(t)$ and state $\xi(t+1)$. This function can be represented by a matrix $\mathbf{G}$ with entries $g(k,\ell)$, where $1 \leq k, \ell \leq 3^L$.

Example 12.5, continued

For the FIR channel $H(z)$ of order $L = 2$ given by (12.166), the matrix $\mathbf{G}$ with entries $g(k,\ell)$ for $1 \leq k,\ \ell \leq 9$ is given by

$$\mathbf{G} = \begin{bmatrix} -0.3 & 0.5 & 1.3 & x & x & x & x & x & x \\ x & x & x & -0.9 & -0.1 & 0.7 & x & x & x \\ x & x & x & x & x & x & -1.5 & -0.7 & 0.1 \\ -0.2 & 0.6 & 1.4 & x & x & x & x & x & x \\ x & x & x & -0.8 & 0 & 0.8 & x & x & x \\ x & x & x & x & x & x & -1.4 & -0.6 & 0.2 \\ -0.1 & 0.7 & 1.5 & x & x & x & x & x & x \\ x & x & x & -0.7 & 0.1 & 0.9 & x & x & x \\ x & x & x & x & x & x & -1.3 & -0.5 & 0.3 \end{bmatrix}, \qquad (12.175)$$

where the entries corresponding to forbidden state transitions are marked by an x. Note again that the matrix $\mathbf{G}$ is skew-centrosymmetric. □

Let $\mathcal{V}$ denote the set of enlarged state trellis paths of the form

$$\boldsymbol{\xi} = (\xi_0,\ \xi_1,\ \cdots,\ \xi_t,\ \cdots)\,.$$

Note that since we assume $D(t) \equiv 0$ for $t \leq 0$, under the mapping (12.172), the initial state for all trellis paths is

$$\xi_0 = (1 + 3^L)/2\,. \qquad (12.176)$$

Furthermore, since we assume $D(1) \neq 0$, i.e. $D(1) = 1$ or -1, the next state is

$$\xi_1 = (1 + 3^L)/2 - 1 \quad \text{or} \quad \xi_1 = (1 + 3^L)/2 + 1\,. \qquad (12.177)$$

Then, if $\mathcal{V}$ denotes the set of enlarged state trellis paths satisfying (12.176) and (12.177), expression (12.171) for the squared minimum distance $d^2_{\min}$ between paths of the original trellis can be rewritten as the solution of the minimization problem

$$d^2_{\min} = \frac{4}{\sigma^2} \min_{\boldsymbol{\xi} \in \mathcal{V}} \sum_{t=0}^{\infty} (g(\xi_t, \xi_{t+1}))^2\,. \qquad (12.178)$$

for the enlarged state trellis. In this expression the summation is infinite, but it is useful to recognize that if a path returns to state $(1+3^L)/2$ at some time $t_0 > 1$, which corresponds to having

$$D(t_0 - k) = 0$$

for $1 \leq k \leq L$, then the optimum trajectory after t_0 will stay in this state forever, since transitions from state $(1+3^L)/2$ to itself have zero cost. This corresponds to selecting

$$D(t) = 0 \text{ for } t \geq t_0 \,.$$

Thus, the minimization problem (12.178) consists of finding the shortest trellis path (the path with least metric) from state $(1+3^L)/2-1$ to state $(1+3^L)/2$. Note that by symmetry of the mapping (12.172) with respect to the center label $(1+3^L)/2-1$ under sign reversal of the sequence $\{D(t),\, t \geq 0\}$, because of the centrosymmetry of the matrix $\mathbf{G}$, the mirror image with respect to the middle label $(1+3^L)/2$ of the shortest path $\boldsymbol{\xi}$ connecting state $(1+3^L)/2-1$ to state $(1+3^L)/2$ is the shortest path connecting state $(1+3^L)/2+1$ to state $(1+3^L)/2$, and it has the same path metric as $\boldsymbol{\xi}$.

To summarize, expression (12.178) indicates that d^2_{min} can be evaluated by solving a shortest path problem over an oriented graph. The vertices of the graph are the enlarged states $1 \leq k \leq 3^L$. An edge exists between vertices k and ℓ if the dynamics (12.173) allow a transition from $\xi_t = k$ to $\xi_{t+1} = \ell$. The cost of a transition from k to ℓ is $c(k,\ell) = g^2(k,\ell)$ if the transition is allowed, and

$$c(k,\ell) = +\infty$$

if the transition is forbidden. Note that all costs are nonnegative. Several algorithms due to Bellman and Ford or Dijkstra [33], among others, can be employed to solve the shortest path problem from state $(1+3^L)/2-1$ to state $(1+3^L)/2$. Since the Bellman-Ford algorithm relies on Bellman's optimality principle described earlier, we focus on this algorithm.

Bellman-Ford algorithm: Given a fixed end vertex q, the algorithm computes sequentially for increasing values of n the minimum distance $d_k(n)$ from every vertex k to q for paths with n edges or less connecting k to q, starting with $n = 0$. Since the distance of node q to itself is zero, we set $d_q(n) = 0$ for all $n \geq 0$. For nodes k for which there does not exist a path with n edges or less connecting k to q, we set

$$d_k(n) = +\infty \,.$$

Since $1 \leq k \leq V$ where V denotes the total number of vertices of the graph (here $V = 3^L$), for a fixed t, the distances $d_k(n)$ for all vertices k can be represented by a vector

$$\mathbf{d}(n) = \begin{bmatrix} d_1(n) \cdots d_k(n) \cdots d_V(n) \end{bmatrix}^T \,.$$

For $n = 0$, only node q can reach q without any transition, so all entries of $\mathbf{d}(0)$ are equal to $+\infty$, except

$$d_q(0) = 0 \,.$$

Then suppose that the shortest paths with n edges or less have been identified for connecting all nodes $k \in V$ to q, so the vector $\mathbf{d}(n)$ has been evaluated. Then, by Bellman's optimality principle, the shortest path with $n+1$ segments or less connecting k to q must either be the shortest path with n segments connecting k to q, or it must be a path connecting k to another vertex ℓ, followed by the shortest path with n segments connecting ℓ to q. This implies that $d_k(n+1)$ can be expressed as

$$d_k(n+1) = \min(d_k(n), \min_{\ell \in N(k)} (c(k,\ell) + d_\ell(n))) \,. \tag{12.179}$$

In (12.179) the set of neighboring vertices $N(k)$ of k are the vertices ℓ such that there exists an edge connecting k to ℓ (the transition from k to ℓ is allowed). The inner minimization in (12.179) has for purpose to identify the best path connecting k to q with $n+1$ segments. This best path is obtained by comparing the costs $c(k,\ell) + d_\ell(n)$ for all paths obtained by concatenating an edge connecting k to one of its neighbors ℓ followed by the shortest n-segment path connecting ℓ to q. The algorithm (12.179) terminates as soon as the vector $\mathbf{d}(n)$ remains unchanged for one iteration, i.e.,

$$\mathbf{d}(n+1) = \mathbf{d}(n) \,,$$

since the form of the recursion (12.179) implies that in this case $\mathbf{d}(n+m) = \mathbf{d}(n)$ for all positive integers m. The vector $\mathbf{d}$ obtained when the algorithm terminates specifies the minimum distances between each vertex k and terminal vertex q.

We have yet to explain how the shortest path between each vertex k and terminal vertex q is traced. In parallel with recursion (12.179) for each n, we keep track of a vector $\mathbf{s}(n)$ where entry $s_k(n)$ represents the successor vertex of k on the shortest path with n segments connecting k to q. Whenever $d_k(n+1) = d_k(n)$, we set

$$s_k(n+1) = s_k(n) \,, \tag{12.180}$$

which indicates that the successor node is unchanged. But when $d_k(n+1) < d_k(n)$, we select

$$s_k(n+1) = \arg \min_{\ell \in N(k)} (c(k,\ell) + d_\ell(n)) \tag{12.181}$$

as the new successor of k. When the Bellman-Ford algorithm terminates, the final successor vector $\mathbf{s}$ can be used to trace the shortest path from k to q in a manner similar to the trace-back stage of the Viterbi algorithm. Specifically, starting from vertex k, its successor s_k is the k-th entry of $\mathbf{s}$. The next vertex is obtained by reading off the s_k-th entry of $\mathbf{s}$. Proceeding sequentially, the shortest path is obtained when vertex q is reached.

Example 12.5, continued

Consider the evaluation of the squared minimum distance $d^2_{\min}$ for the enlarged state trellis corresponding to the channel $H(z)$ of order $L = 2$ given by (12.166). For this case, the enlarged state trellis depicting the allowed transitions is shown in Fig. 12.17, and the matrix $\mathbf{G}$ is given by (12.175). Recall also that the cost structure is specified by

$$c(k,\ell) = (g(k,\ell))^2$$

if the transition from k to ℓ is allowed and $c(k,\ell) = +\infty$ otherwise. Since $L = 2$, the shortest path problem we need to solve is from state 4 to state 5. We initialize the Bellman-Ford recursion with

$$\mathbf{d}(0) = \begin{bmatrix} \infty & \infty & \infty & \infty & 0 & \infty & \infty & \infty & \infty \end{bmatrix}^T .$$

Then for $n = 1$ we obtainv

$$\mathbf{d}(1) = \begin{bmatrix} \infty \\ (0.1)^2 \\ \infty \\ \infty \\ 0 \\ \infty \\ \infty \\ (0.1)^2 \\ \infty \end{bmatrix} , \quad \mathbf{s}(1) = \begin{bmatrix} x \\ 5 \\ x \\ x \\ 5 \\ x \\ x \\ 5 \\ x \end{bmatrix} ,$$

where x indicates that no successor has been determined yet. Similarly for $n = 2$ and $n = 3$,

$$\mathbf{d}(2) = \begin{bmatrix} (0.5)^2 + (0.1)^2 \\ (0.1)^2 \\ (0.7)^2 + (0.1)^2 \\ (0.6)^2 + (0.1)^2 \\ 0 \\ (0.6)^2 + (0.1)^2 \\ (0.7)^2 + (0.1)^2 \\ (0.1)^2 \\ (0.5)^2 + (0.1)^2 \end{bmatrix} , \quad \mathbf{s}(2) = \begin{bmatrix} 2 \\ 5 \\ 8 \\ 2 \\ 5 \\ 8 \\ 2 \\ 5 \\ 8 \end{bmatrix} ,$$

$$\mathbf{d}(3) = \begin{bmatrix} (0.5)^2 + (0.1)^2 \\ (0.1)^2 \\ (0.1)^2 + (0.5)^2 + (0.1)^2 \\ (0.2)^2 + (0.5)^2 + (0.1)^2 \\ 0 \\ (0.2)^2 + (0.5)^2 + (0.1)^2 \\ (0.1)^2 + (0.5)^2 + (0.1)^2 \\ (0.1)^2 \\ (0.5)^2 + (0.1)^2 \end{bmatrix} , \quad \mathbf{s}(2) = \begin{bmatrix} 2 \\ 5 \\ 9 \\ 1 \\ 5 \\ 9 \\ 1 \\ 5 \\ 8 \end{bmatrix} ,$$

and finally

$$\mathbf{d}(n) = \mathbf{d}(3) \quad , \quad \mathbf{s}(n) = \mathbf{s}(3)$$

for $n \geq 4$. The shortest path $\boldsymbol{\xi}$ leaving state 5 at $t = 0$, going to state 4 at $t = 1$, and then returning to state 5 at $t = 4$ is shown below in Fig. 12.18. It corresponds to the sequence

$$D(1) = D(2) = -1$$

and $D(t) = 0$ for $t \geq 3$. The mirror image $\boldsymbol{\xi}_M$ of path $\boldsymbol{\xi}$ with respect to the center state 5 is also shown in Fig. 12.18. Leaving from state 5 at $t = 0$, it

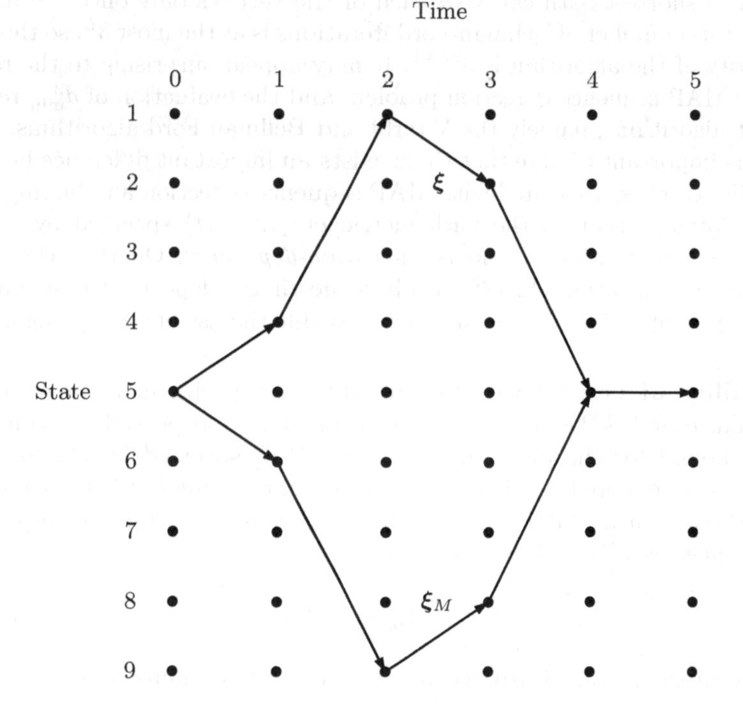

Fig. 12.18. Shortest path $\boldsymbol{\xi}$ from state 4 to state 5. Its mirror image $\boldsymbol{\xi}_M$ is the shortest path from state 6 to state 5.

goes to state 6 at $t = 1$ and then returns to state 5 at $t = 4$. It corresponds to the sign reversed sequence

$$D(1) = D(2) = 1$$

and $D(t) = 0$ for $t \geq 3$.

The squared distance $d^2_{\min}$ is the same for both $\boldsymbol{\xi}$ and its mirror image $\boldsymbol{\xi}_M$. To evaluate $d^2_{\min}$ we observe from (12.178) that $d^2_{\min}$ is obtained by adding to the minimum distance d_4 the cost of the first transition from state 5 to state 4. This gives

$$\begin{aligned} d^2_{\min} &= \frac{4}{\sigma^2}(c(5,4) + d_4) \\ &= \frac{4}{\sigma^2}[(0.8)^2 + (0.2)^2 + (0.5)^2 + (0.1)^2] = 3.76/\sigma^2 \,. \end{aligned}$$

□

Remark: If we consider recursion (12.179) and observe that for the problem we consider a vertex k has only 3 neighbors (corresponding to the 3 different values of $D(t)$), each step of the recursion requires 3×3^L operations. Noting also that a shortest path can visit each of the vertices only once, we deduce that the total number of Bellman-Ford iterations is at the most 3^L, so the total complexity of the algorithm is 3^{2L+1}. It may appear surprising to the reader that the MAP sequence detection problem and the evaluation of $d^2_{\min}$ rely on different algorithms, namely the Viterbi and Bellman-Ford algorithms. However, it is important to note that there exists an important difference between the trellis searches associated with MAP sequence detection and finding $d^2_{\min}$. For the former problem, the path metric $c(x_t, x_{t+1}, t)$ specified by (12.80) depends on observation y_t and is thus *time-dependent.* On the other hand the trellis minimization (12.178) involves the time-independent cost function $c(\xi_t, \xi_{t+1}) = (g(\xi_t, \xi_{t+1}))^2$, which makes possible the use of a faster algorithm.

Probability of error lower bound: To characterize more precisely the performance of MAP sequence equalization, it is also possible to construct a lower bound for the probability of error $P[E]$ specified by (12.102). Let $\{D_0(t), t \geq 1\}$ denote the difference sequence minimizing (12.171). Among all path indices u such that $1 \leq u \leq U$, let $\mathcal{U}_0$ denote the subset corresponding to bit sequences $\{B_u(t),\, t \geq 1\}$ such that

$$B_v(t) = B_u(t) - D_0(t) \tag{12.182}$$

is also a bit sequence. Clearly whenever $D_0(t) = 1$, we must have

$$B_u(t) = 1 \quad , \quad B_v(t) = 0 \,,$$

and when $D_0(t) = -1$, we have

$$B_u(t) = 0 \text{ and } B_v(t) = 1 .$$

On the other hand if $D_0(t) = 0$, $B_u(t)$ can be either 1 or 0, and

$$B_v(t) = B_u(t) .$$

Intuitively, $\mathcal{U}_0$ is the set of indices u such that the set $\mathcal{S}(t, \mathbf{x}_u)$ of simple error events for trellis path $\mathbf{x}_u$ contains a path $\mathbf{x}_v$ located at distance d_{dim} from $\mathbf{x}_u$. Then if the correct path $\mathbf{x}_u$ is such that $u \in \mathcal{U}_0$, an error is commited at time t whenever $\hat{\mathbf{X}}_{\text{MAP}} = \mathbf{x}_v$, so that

$$Q(\frac{d_{\min}}{2}) = P[\hat{\mathbf{X}}_{\text{MAP}} = \mathbf{x}_v | \mathbf{X} = \mathbf{x}_u] \leq P[\hat{X}_t^{\text{MAP}} \neq x_t^u | \mathbf{X} = \mathbf{x}_u] . \quad (12.183)$$

Then, by weighting both sides of inequality (12.183) by the probability p_u of paths in $\mathcal{U}_0$, we find

$$C_L Q(\frac{d_{\min}}{2}) \leq P[E] \quad (12.184)$$

with

$$C_L = \sum_{u \in \mathcal{U}_0} p_u . \quad (12.185)$$

Comparing the lower bound (12.184) with high SNR upper bound (12.168), we find that at high SNR $P[E]$ is bracketed by two terms varying like $Q(d_{\min}/2)$, so it will itself vary like $Q(d_{\min}/2)$.

Comparison of ISI and AWGN channels: The lower and upper bounds (12.184) and (12.168) provide a characterization of the performance of MAP sequence equalization. It is now possible to answer the rather obvious question of whether there exists any advantage in communicating over an ISI channel versus communicating over an additive white Gaussian noise (AWGN) channel of the form

$$Y(t) = AI(t) + V(t) , \quad (12.186)$$

where the symbols $\{I(t), t \geq 1\}$ are independent identically distributed and take values ± 1 with probability 1/2 and $V(t)$ is a zero-mean WGN with variance σ^2. To ensure that the comparison is fair, the same amount of energy needs to be employed to transmit one bit through both channels. Thus in (12.185), the amplitude A satisfies

$$A^2 = E = \sum_{k=0}^{L} h_k^2 \quad (12.187)$$

where E is the energy of the received waveform h_k, $0 \leq k \leq L$ if a single bit $I(1)$ is transmitted through the ISI channel (12.158)–(12.159). Accordingly $A = ||\mathbf{h}||_2$, where $\mathbf{h}$ denotes the vector formed by the channel impulse response. After making this identification, we deduce that the probability of error for transmission over an AWGN channel is

$$P[E] = Q(\frac{d}{2}) \tag{12.188}$$

with

$$d = \frac{2A}{\sigma} = \frac{2||\mathbf{h}||_2}{\sigma}\,. \tag{12.189}$$

For MAP sequence equalization over an ISI channel, the performance is governed by $Q(d_{\min}/2)$. But the minimum of the cost function

$$J(D(\cdot)) = \sum_{t=1}^{\infty} \Big(\sum_{k=0}^{L} h_k D(t-k) \Big)^2 \tag{12.190}$$

is always less than the value of this function for the sequence

$$D_s(1) = 1 \quad , \quad D_s(t) = 0 \text{ otherwise.}\,,$$

which is

$$J(D_s(\cdot)) = \sum_{k=0}^{L} h_k^2 = ||\mathbf{h}||_2^2\,. \tag{12.191}$$

This implies that $d_{\min} < d$ where d is the normalized distance (12.189). Hence, as we might have suspected, it is always preferable to communicate over an AWGN channel. To explain this advantage, observe that the sequence $D_s(\cdot)$ includes only one isolated bit error. An AWGN channel has the feature that all errors are isolated in the sense that there is no propagation of one error to the next. In contrast, for an ISI channel, multiple errors in close proximity may occur in the worst-case sequence $D_0(\cdot)$, so the distance $d_{\min}$ between the two closest paths is lower than the distance d between a path $\mathbf{x}_u$ and neighboring paths containing a single isolated error.

Example 12.5, continued

For the channel $H(z)$ given by (12.166) and transmitted symbols $I(t)$ taking values ± 1, we found that $d_{\min}^2 = 3.76/\sigma^2$. On the other hand,

$$||\mathbf{h}||_2^2 = (0.8)^2 + (0.6)^2 + (0.1)^2 = 1.01\,,$$

so that

$$d^2 = 4\frac{||\mathbf{h}||_2^2}{\sigma^2} = \frac{4.04}{\sigma^2} > d_{\min}^2\,.$$

□

12.5 Bibliographical Notes

The asymptotic decay rates of Section 12.2 for the probability of false alarm and the probability of a miss of a test for completely observed Markov chains were derived by Natarajan [34] (see also [35]), who relied on the method of types for Markov chains [36]. In contrast, the Gartner-Ellis approach that we use here to derive the same results is an adaptation of material appearing in [37]. When the Viterbi algorithm was first introduced, it was expected that it would be applicable primarily to convolutional codes with a short constraint length (a small number of states), but progress in VLSI technology has allowed its implementation in a wide variety of situations. Therefore, most of the research activity centered on the Viterbi algorithm over the last 25 years has focused on its implementation. The upper bound for the probability of error of the Viterbi algorithm was first obtained by Forney [14] for the case of ISI channels. The analysis of [14] contained a minor error which was noted in [38] and later fixed in [24]. What is usually called the Baum-Welch algorithm for HMMs includes really two contributions: the forward and backward algorithm for evaluating the a-posteriori state distribution, and an iterative parameter estimation technique which, for the case of Markov chains, coincides with the EM algorithm. Given that the work of Baum, Welch and their collaborators precedes that of Dempster, Laird and Rubin by several years, they should probably receive some credit for the EM algorithm, even though their version is less general than the one presented in [39]. It is also interesting to note that even though Welch was a co-inventor with Baum of the forward and backward algorithm, his name does not appear on [25], which focuses primarily on the hill climbing property of the iterative parameter estimation method. A detailed account of the development of the Baum-Welch algorithm can be found in Welch's 2003 Shannon Lecture [40].

12.6 Problems

12.1. Consider a binary continuous-phase frequency-shift keyed (CPFSK) signal of the form

$$s(t) = (2E/T)^{1/2} \cos(\omega_c t + \phi(t, \mathbf{I})) , \tag{12.192}$$

where E denotes the bit energy, T the baud period, and ω_c the carrier frequency. The information carrying phase over the n-th bit interval $nT \leq t \leq (n+1)T$ can be expressed as

$$\phi(t, \mathbf{I}) = 2\pi h \sum_{k=-\infty}^{n} I_k q(t - kT) ,$$

where the symbols I_k are independent and take the binary values ± 1 with probability $1/2$ each. The modulation index $h = p/q$ where p and q denote two relatively prime integers, and the phase function $q(t)$ can be represented as

$$q(t) = \int_{-\infty}^{t} g(u)du$$

where $g(t)$ denotes the phase pulse. For CPFSK $g(t)$ is the rectangular 1REC pulse

$$g(t) = \begin{cases} \frac{1}{2T} & 0 \leq t \leq T \\ 0 & \text{otherwise} . \end{cases}$$

Hence

$$q(t) = \begin{cases} 0 & t < 0 \\ \frac{t}{2T} & 0 \leq t \leq T \\ 1/2 & t > T , \end{cases} \tag{12.193}$$

so the phase $\phi(t, \mathbf{I})$ changes linearly by $h\pi I_n$ during the n-th bit interval.

(a) Let

$$X_n \triangleq \frac{q}{\pi}\phi(nT, \mathbf{I}) . \quad \text{mod } (2q)$$

Verify that X_n obeys the recursion

$$X_{n+1} = (X_n + pI_n) \quad \text{mod } (2q) .$$

Explain why X_n forms a Markov chain. Find the number K of states, as well as the number T of transitions from each state.

(b) Let $h = 1/2$, which corresponds to the special case of minimum shift keying (MSK) modulation. Specify the one-step transition probability matrix $\mathbf{P}$ for the MC obtained in part (a) and draw one section of the state trellis.

(c) The signal $s(t)$ is observed in the presence of a zero-mean WGN signal $V(t)$ with variance v, so that the CT observation signal can be expressed as

$$Y(t) = s(t) + V(t) .$$

Assume that the carrier frequency ω_c is much larger than the baud frequency $\omega_b = 2\pi/T$. Then, by extracting the in-phase and quadrature components of $Y(t)$ as indicated in Section 9.2, and sampling them at $t = (n+1)T$, we obtain observations

$$\begin{aligned} Y_{cn} &= E^{1/2}\cos(\pi X_{n+1}/q) + V_{cn} \\ Y_{sn} &= -E^{1/2}\sin(\pi X_{n+1}/q) + V_{sn} , \end{aligned}$$

where V_{cn} and V_{sn} are two independent $N(0, v)$ distributed WGN sequences. Equivalently, if we introduce the complex observations and noise

$$Y_n = Y_{cn} + jY_{sn} \quad , \quad V_n = V_{cn} + jV_{sn} ,$$

the above observations can be expressed in complex form as

$$Y_n = E^{1/2} \exp(-j\pi X_{n+1}/q) + V_n ,$$

where V_n is a complex white $CN(0, 2v)$ distributed sequence. Specify the path metric corresponding to this observation model.

(d) By removing terms that are identical for all paths, show that the MAP sequence detection problem for the MC model of parts (a)–(c) reduces to finding the trellis path $\mathbf{x} = (x_n, 0 \leq n \leq N)$ such that

$$J(\mathbf{x}) = \Re\{\sum_{n=1}^{N} Y_n \exp(j\pi x_{n+1}/q)\} \tag{12.194}$$

is *maximized.*

12.2. Consider the MC model of a CPFSK signal introduced in Problem 12.1.

(a) Show that if $h = p/q$ with p even and if the initial state is known, only half of the states of the MC are visited, so a reduced state MC model can be employed.

(b) For $h = 2/3$, and assuming that the initial state is $X_1 = 0$, obtain the one-step transition probability matrix for the reduced state MC model of part a) and draw a section of the corresponding state trellis.

12.3. Suppose a CPFSK signal is observed in WGN. The variance v of the noise is known. It is not known whether the modulation index of the signal is $H_0 : h = 1/2$ or $H_1 : h = 2/3$. We seek to simultaneously test H_0 against H_1 and find the corresponding MAP state sequence. This is accomplished by joint maximization of the a-posteriori probability $P[\mathbf{x}, H_i|\mathbf{y}]$ over $i = 0, 1$ and $\mathbf{x} \in \mathcal{U}_i$, where $\mathcal{U}_i$ denotes the set of trellis paths under H_i.

(a) Verify that maximizing the a-posteriori probability $P[\mathbf{x}, H_i|\mathbf{y}]$ is equivalent to maximizing

$$\ln f(\mathbf{y}|\mathbf{x}, H_i) + \ln P[\mathbf{x}|H_i] + \ln \pi_i ,$$

where

$$\pi_0 = P[H_0] \text{ and } \pi_1 = P[H_1]$$

denote the a-priori probabilities of the two hypotheses. Next, check that all trajectories $\mathbf{x}$ of trellises $\mathcal{U}_0$ and $\mathcal{U}_1$ have the same probability $P[\mathbf{x}|H_i]$ for $i = 0, 1$.

(b) Use the results of part (a) to derive a joint test between H_0 and H_1 and MAP sequence detector. Show that an implementation of the optimum test and sequence detector consists of running two Viterbi algorithms separately for trellises $\mathcal{U}_0$ and $\mathcal{U}_1$ and the path metric $J(\mathbf{x})$ given by (12.194).

Then if $\hat{\mathbf{x}}_i$ is the MAP path under hypothesis H_i with $i = 0,\ 1$, the optimum MAP test and sequence detector is obtained by maximizing

$$J(\hat{\mathbf{x}}_i) + v \ln \pi_i$$

for $i = 0,\ 1$. Of course, when the two hypotheses are equally likely a-priori, i.e.,

$$\pi_0 = \pi_1 = 1/2\,,$$

the second term can be dropped. Verify that in this case, the optimum test just requires running the Viterbi algorithm on $\mathcal{U}_0$ and $\mathcal{U}_1$ separately and selecting the path with largest overall metric.

12.4. Consider the MC model of a CPFSK signal with modulation index $h = p/q$ derived in Problem 12.1.

(a) Let $\mathbf{x}_u$ and $\mathbf{x}_r$ denote two paths of the MC trellis $\mathcal{U}$. Introduce the path addition modulo $2q$ as

$$\begin{aligned}\mathbf{x}_w &= \mathbf{x}_u \oplus \mathbf{x}_r \\ &= (x_n^u + x_n^r \mod (2q),\ 1 \leq n \leq N+1)\,.\end{aligned}$$

Then if $\mathbf{x}_v$ is in the set $\mathcal{S}(t, \mathbf{x}_u)$ of simple error events for path $\mathbf{x}_u$, verify that for any path $\mathbf{x}_r$, the path $\mathbf{x}_v \oplus \mathbf{x}_r$ is in the set of simple error events for path $\mathbf{x}_u \oplus \mathbf{x}_r$. Show that the path distance d_{uv} is invariant under this transformation.

(b) Given an arbitrary path $\mathbf{x}_u$ corresponding to symbol sequence $\{I_n,\ 1 \leq n \leq N\}$, let $-\mathbf{x}_u$ denote the trellis path corresponding to the sign reversed sequence $\{-I_n,\ 1 \leq n \leq N\}$. Verify that $\mathbf{x}_u \oplus -\mathbf{x}_u$ is the all zero state path. Then conclude from part a) that to evaluate the minimum path distance $d_{\min}$, we only need to find the path $\mathbf{x}_u$ of $\mathcal{S}(t, \mathbf{0})$ which minimizes the distance between signal $\mathbf{s}_u$ and the signal $\mathbf{s}_0$ corresponding to the all zero state path $\mathbf{0}$.

(c) For $h = 1/2$, evaluate the minimum distance $d_{\min}$.

12.5. In duobinary MSK modulation, a signal of the form (12.192) is transmitted, where over interval $nT \leq t \leq (n+1)T$, the phase $\phi(t, \mathbf{I})$ takes the form

$$\phi(t, \mathbf{I}) = \pi \sum_{k=-\infty}^{n} D_k q(t - kT)$$

with

$$D_k = (I_k + I_{k-1})/2\,.$$

In this expression $q(t)$ is the integrated 1REC pulse given by (12.193), and the symbols $\{I_n,\ n \geq 1\}$ are independent and take the binary values ± 1 with probability $1/2$ each.

Observe that in this modulation format, the pseudo-symbol

$$D_n = \begin{cases} 1 & \text{for } I_n = I_{n-1} = 1 \\ -1 & \text{for } I_n = I_{n-1} = -1 \\ 0 & \text{for } I_n \neq I_{n-1} , \end{cases}$$

so the phase changes only over half of the transmission intervals, and accordingly the bandwidth of the transmission signal is smaller than for standard MSK.

(a) Consider the phase

$$\phi(nT, \mathbf{I}) = \frac{\pi}{2} \sum_{k=-\infty}^{n-1} D_k$$

at the end of the $n-1$-th signaling interval. What are its possible values? Show that the pair

$$X_n \triangleq (\phi(nT, \mathbf{I}), I_{n-1})$$

forms a state of the MC. Find the number K of states and the number T of possible transitions out of each state. Draw the state transition diagram, and specify the one-step transition probability matrix $\mathbf{P}$

(b) By following an approach similar to the one outlined in Problem 12.1 for CPFSK, obtain the branch metric for duobinary MSK sequence decoding.

12.6. Consider the MAP sequence equalization problem for observations (12.158)–(12.159), where the transmitted symbols $I(t)$ are independent and take the binary values ± 1 with probability 1/2, $V(t)$ is a zero-mean WGN independent of $I(t)$ with variance σ^2, and the channel

$$H(z) = 1 + az^{-1} + a^2 z^{-2} .$$

(a) Find the normalized minimum distance $d_{\min}(a)$ as a function of a for $0 \leq |a| < 1$. For each a, find the minimizing difference sequence $D(t)$ for $t \geq 1$.

(b) Let

$$||\mathbf{h}||^2 = 1 + a^2 + a^4 .$$

For what values of a is

$$d^2(a) = \frac{4||\mathbf{h}||^2}{\sigma^2} > d^2_{\min}(a) \, ?$$

References

1. G. Ungerboeck, "Trellis-coded modulation with redundant signal sets, Parts I and II," *IEEE Communications Magazine*, vol. 25, pp. 5–21, 1987.
2. C. Douillard, M. Jezequel, C. Berrou, A. Picart, P. Didier, and A. Glavieux, "Iterative correction of intersymbol interference: Turbo equalization," *European Trans. Telecomm.*, vol. 2, pp. 259–263, June 1998.

3. R. Koetter, A. C. Singer, and M. Tüchler, "Turbo equalization," *IEEE Signal Processing Mag.*, vol. 21, pp. 67–80, Jan. 2004.
4. G. Ferrari, G. Colavolpe, and R. Raheli, *Detection Algorithms for Wireless Communications With Applications to Wired and Storage Systems.* Chichester, England: J. Wiley & Sons, 2004.
5. K. Chugg, A. Anastasopoulos, and X. Chen, *Iterative Detection: Adaptivity, Complexity Reduction, and Applications.* Boston: Kluwer Acad. Publ., 2001.
6. L. R. Rabiner, "A tutorial on hidden Markov models and selected applications in speech recognition," *Proc. IEEE*, vol. 77, pp. 257–286, Feb. 1989.
7. L. Rabiner and B.-H. Juang, *Fundamentals of Speech Recognition.* Englewood Cliffs, NJ: Prentice Hall, 1993.
8. J. G. Proakis, *Digital Communications, Fourth Edition.* New York: McGraw-Hill, 2000.
9. R. A. Horn and C. R. Johnson, *Matrix Analysis.* Cambridge, UK: Cambridge Univ. Press, 1985.
10. R. G. Gallager, *Discrete Stochastic Processes.* Boston: Kluwer Acad. Publ., 1996.
11. R. A. Horn and C. R. Johnson, *Topics in Matrix Analysis.* Cambridge, United Kingdom: Cambridge Univ. Press, 1994.
12. S.-I. Amari and H. Nagaoka, *Methods of Information Geometry.* Providence, RI: American Mathematical Soc., 2000.
13. A. J. Viterbi, "Error bounds for convolutional codes and an asymptotically optimum decoding algorithm," *IEEE Trans. Informat. Theory*, vol. 13, pp. 260–269, 1967.
14. J. G. D. Forney, "Maximum-likelihood sequence estimation of digital sequences in the presence of intersymbol interference," *IEEE Trans. Informat. Theory*, vol. 18, pp. 363–378, May 1972.
15. J. G. D. Forney, "The Viterbi algorithm," *Proc. IEEE*, vol. 31, pp. 268–278, Mar. 1973.
16. R. Bellman, "The theory of dynamic programming," *Proc. Nat. Acad. Sci.*, vol. 38, pp. 716–719, 1952.
17. R. Bellman, *Dynamic Programming.* Princeton, NJ: Princeton Univ. Press, 1957. Reprinted by Dover Publ., Mineola, NY, 2003.
18. C. Rader, "Memory management in a Viterbi decoder," *IEEE Trans. Commun.*, vol. 29, pp. 1399–1401, Sept. 1981.
19. R. Cypher and C. B. Shung, "Generalized trace-back technique for survivor memory management in the Viterbi algorithm," *J. VLSI Signal Proc.*, vol. 5, pp. 85–94, 1993.
20. G. Feygin and P. G. Gulak, "Architectural tradeoffs for survivor sequence memory management in Viterbi decoders," *IEEE Trans. Commun.*, vol. 41, pp. 425–429, Mar. 1993.
21. H.-L. Lou, "Implementing the Viterbi algorithm," *IEEE Signal Processing Magazine*, vol. 12, pp. 42–52, Sept. 1995.
22. J. R. Barry, E. A. Lee, and D. G. Messerschmitt, *Digital Communications, Third Edition.* New York: Springer Verlag, 2003.
23. K. Ouahada and H. C. Ferreira, "Viterbi decoding of ternary line codes," in *Proc. 2004 IEEE Internat. Conf. on Communications*, vol. 2, (Paris, France), pp. 667–671, June 2004.

24. S. Verdú, "Maximum likelihood sequence detection for intersymbol interference channels: A new upper bound on error probability," *IEEE Trans. Informat. Theory*, vol. 33, pp. 62–68, Jan. 1987.
25. L. E. Baum, T. Petrie, G. Soules, and N. Weiss, "A maximization technique occurring in the statistical analysis of probabilistic functions of Markov chains," *Annals Mathematical Statistics*, vol. 41, pp. 164–171, Feb. 1970.
26. L. R. Bahl, J. Cocke, F. Jelinek, and J. Raviv, "Optimal decoding of linear codes for minimizing symbol error rates," *IEEE Trans. Informat. Theory*, vol. 20, pp. 284–287, Mar. 1974.
27. C. Berrou and A. Glavieux, "Near optimum error correcting coding and decoding: turbo codes," *IEEE Trans. Commun.*, vol. 44, pp. 1261–127, Oct. 1996.
28. J. Hagenauer and P. Hoeher, "A Viterbi algorithm with soft-decision outputs and its applications," in *Proc. IEEE Globecom Conf.*, (Houston,TX), pp. 793–797, Nov. 1989.
29. M. P. C. Fossorier, F. Burkert, S. Lin, and J. Hagenauer, "On the equivalence between SOVA and max-log-MAP decoding," *IEEE Communications Letters*, vol. 5, pp. 137–139, May 1998.
30. G. Battail, "Pondération des symboles décodés par l'algorithme de Viterbi," *Annales des Telecommunications*, pp. 31–38, Jan. 1987.
31. G. M. Vastula and F. S. Hill, "On optimal detection of band-limited PAM signals with excess bandwidth," *IEEE Trans. Commun.*, vol. 29, pp. 886–890, June 1981.
32. K. M. Chugg and A. Polydoros, "MLSE for an unknown channel – Part I: Optimality considerations," *IEEE Trans. Commun.*, vol. 44, pp. 836–846, July 1996.
33. D. Bertsimas and J. Tsitsiklis, *Introduction to Linear Optimization*. Belmont, MA: Athena Scientific, 1997.
34. S. Natarajan, "Large deviations, hypothesis testing, and source coding for finite Markov chains," *IEEE Trans. Informat. Theory*, vol. 31, pp. 360–365, May 1985.
35. V. Anantharam, "A large deviations approach to error exponents in source coding and hypothesis testing," *IEEE Trans. Informat. Theory*, vol. 36, July 1990.
36. L. D. Davisson, G. Longo, and A. Sgarro, "The error exponent for the noiseless encoding of finite ergodic Markov sources," *IEEE Trans. Informat. Theory*, vol. 27, pp. 431–438, July 1981.
37. F. den Hollander, *Large Deviations*. Providence, RI: American Mathematical Soc., 2000.
38. G. J. Foschini, "Performance bound for maximum-likelihood reception of digital data," *IEEE Trans. Informat. Theory*, vol. 21, pp. 47–50, Jan. 1975.
39. A. P. Dempster, N. M. Laird, and D. B. Rubin, "Maximum likelihood from incomplete data via the EM algorithm," *J. Royal Stat. Society, Series B*, vol. 39, no. 1, pp. 1–38, 1977.
40. L. R. Welch, "The Shannon lecture: Hidden Markov models and the Baum-Welch algorithm," *IEEE Information Theory Soc. Newsletter*, vol. 53, Dec. 2003.

13

Detection of Markov Chains with Unknown Parameters

13.1 Introduction

In the last chapter, for the sequence and pointwise detection of partially observed Markov chains, it was assumed that we knew exactly the Markov chain (MC) and the observation model formed by the one-step transition probability matrix $\mathbf{P}$, the initial probability distribution vector $\boldsymbol{\pi}(1)$ and the output map $h(x_t, x_{t+1})$. In communication applications, the model often includes unknown parameters that need to be estimated, such as timing or phase synchronization parameters. For equalization problems, the channel impulse response $\{h_k, 1 \leq k \leq L\}$ is usually unknown and needs to be estimated. While unknown parameters can be estimated by transmitting known training sequences and then using the estimation techniques developed earlier in the book for known signals with unknown parameters in WGN, such an estimation phase introduces some overhead which lowers the maximum achievable channel thoughput. For wireless communication systems in particular, the link between base station and handset changes rapidly, so training blocks need to be included in each transmission block, leading to a noticeable loss of system efficiency. So instead of estimating the system parameters and detecting the transmitted data in two separate stages, it makes sense to attempt to detect the data and estimate the parameters simultaneously.

In this chapter we discuss joint detection and parameter estimation techniques for Markov chains. The techniques that have been developed over the years for solving this type of problem can be divided roughly in two categories, depending on whether they are GLRT or EM-based. As was shown in Chapter 12, a Markov chain detection problem can be viewed as an M-ary detection problem in a space of high dimensionality, where each hypothesis H_u corresponds to an MC trellis path $\mathbf{x}_u$. From this perspective, if $\boldsymbol{\theta}$ denotes the vector of unknown model parameters, the GLRT requires for each possible trellis path $\mathbf{x}_u$ to evaluate the ML estimate $\hat{\boldsymbol{\theta}}_u$ maximizing the density function $f(\mathbf{y}|\boldsymbol{\theta}, H_u)$ of the observation vector $\mathbf{Y}$ obtained by regrouping

B.C. Levy, *Principles of Signal Detection and Parameter Estimation*,
DOI: 10.1007/978-0-387-76544-0_13, © Springer Science+Business Media, LLC 2008

all observations $Y(t)$ for $1 \leq t \leq N$. Once this has been accomplished, the optimal MC sequence produced by the GLR detector is $\mathbf{x}_{u_0}$ where

$$u_0 = \arg, \max_{1 \leq u \leq U} f(\mathbf{y}|\hat{\boldsymbol{\theta}}_u, H_u) \,. \tag{13.1}$$

This general approach was proposed as early as 1960 by Kailath [1] for the detection of communications signals sent through a random channel with unknown statistics. Unfortunately, the cardinality U of the set $\mathcal{U}$ of the set of MC trajectories is huge, so the computational burden associated with the evaluation of $\hat{\boldsymbol{\theta}}_u$ for each trellis path is overwhelming. A suboptimal, but nevertheless effective, technique consists of evaluating $\hat{\boldsymbol{\theta}}_u$ only for promising paths. However, when $\boldsymbol{\theta}$ is unknown, identifying "promising paths" is a difficult task. If the parameter vector $\boldsymbol{\theta}$ is known, the Viterbi algorithm identifies the optimal survivor path for each possible terminal state k at time t. Conversely, if the optimum survivor path terminating in state k at time t is available, finding the optimal ML parameter vector $\hat{\boldsymbol{\theta}}(k,t)$ corresponding to this trajectory is relatively easy. These two observations form the basis for what has become known as the *per survivor processing* (PSP) joint estimation and detection method for MC trajectories. Although the PSP method was first described by Iltis [2] and Seshadri [3], it was given its name, fully developed, and popularized by Polydoros, Raheli, Chugg, and their collaborators [4–6]. It works as follows: for each terminal state k at time t, it is assumed that a survivor path and its associated parameter vector $\hat{\boldsymbol{\theta}}(k,t)$ have been identified. Then the PSP alternates between path extension and parameter vector update stages. During a path extension stage, the ACS unit of the Viterbi algorithm uses the parameter vector $\hat{\boldsymbol{\theta}}(k,t)$ to evaluate branch metrics for all transitions originating from k at time t. After the new survivor paths terminating in each state at time $t+1$ have been identified, the new path parameter vectors are evaluated. This is usually accomplished adaptively, by using a recursive least-squares or stochastic-gradient iteration to update the parameter vector of the predecessor state of terminal state k.

Thus the PSP algorithm can be viewed as a suboptimal, but yet effective, implementation of the GLRT for the detection of hidden Markov Chain models (HMCMs). Recall that according to the nomenclature employed in this book, when a function $h(X(t), X(t+1))$ of the MC transitions is observed in noise, the MC is said to be partially observed, but if some of the model parameters are unknown, we call it a hidden Markov chain model. This terminology differs from the convention adopted in [7,8] and [9, Chap. 6], where any partially observed Markov model is called an HMM. The goal of this departure from generally accepted terminology is to distinguish between the simple situation where the model is known exactly and the more difficult case where not only the state sequence $\{X(t), 1 \leq t \leq N+1\}$ needs to be detected, but the model parameters need to be estimated at the same time. The GLRT/PSP methodology is not the only way of solving the joint detection and model parameter estimation of Markov chains. Another approach consists of using the EM algorithm, which for the case of Markov chain, is usually called the Baum-Welch reestimation method [7], since it predates the work of Dempster et al. [10]. Recall that the key idea of the EM technique is to augment the available data, here the observations $Y(t)$ for $1 \leq t \leq N$, with some missing data in such a way that the maximization of the log-likelihood function for the complete data is easy to perform. For the case of Markov chains, the missing data corresponds usually to the MC trajectory $X(t)$ for $1 \leq t \leq N$. Then the EM procedure alternates between E-phases during which the log-likelihood function of the complete data is estimated, and M-phases during which the estimated log-likelihood function is maximized to generate an updated parameter vector. The maximization stage is particularly easy if the output map $h(x_t, x_{t+1})$ depends linearly on the parameter vector $\boldsymbol{\theta}$, since in this case it reduces to the solution of a linear least-squares problem. As for the E-phase, the original scheme proposed by Baum and Welch for evaluating the estimated log-likelihood function relies on the forward-backward algorithm. It was shown recently in [11–13] that the computational complexity of the E-phase can be reduced by employing the Viterbi algorithm. Note however that while the Baum-Welch reestimation method evaluates exactly the estimated log-likelihood function of the complete data, the EM Viterbi algorithm (EMVA) of [12,13] relies on an approximation consisting of pruning the set of all trellis paths $\mathcal{U}$ by retaining only the survivor paths based on the current model.

13.2 GLR Detector

13.2.1 Model

Consider a homogeneous Markov chain with one-step transition probability matrix $\mathbf{P}$ and initial probability distribution $\boldsymbol{\pi}(1)$. Let $\mathcal{U}$ denote the space of trellis paths. A path of $\mathcal{U}$ is written as $\mathbf{x}_u = (x_t^u, 1 \leq t \leq N+1)$ with $1 \leq u \leq U$, where U represents the total number of paths. For simplicity, we assume in this section that all trellis paths are equally likely. Let $\boldsymbol{\phi} \in \mathbb{R}^q$ denote an unknown parameter vector. Then if the MC follows trajectory $\mathbf{x}_u$, the observation signal takes the form

$$H_u \ : \ Y(t) = s_u(t, \boldsymbol{\phi}) + V(t) \tag{13.2}$$

for $1 \leq t \leq N$, where the signal

$$s_u(t, \boldsymbol{\phi}) = h(x_t^u, x_{t+1}^u, \boldsymbol{\phi}) \tag{13.3}$$

depends only on the transition from state x_t^u to state x_{t+1}^u. Here $V(t)$ is a WGN with zero-mean and variance v, which is independent of the Markov chain state $X(t)$. The overall parameter vector $\boldsymbol{\theta}$ will either be $\boldsymbol{\phi}$ or will include the variance v, in which case

$$\boldsymbol{\theta} = \begin{bmatrix} \boldsymbol{\phi} \\ v \end{bmatrix} . \tag{13.4}$$

The discussion below assumes that the dimension of $\boldsymbol{\theta}$ is smaller than the number N of observations. If we suppress the index u in (13.3), we see that the signal

$$s(t, \boldsymbol{\phi}) = h(X(t), X(t+1), \boldsymbol{\phi}) \tag{13.5}$$

depends only on the transition from state $X(t)$ to state $X(t+1)$ and on vector $\boldsymbol{\phi}$. We shall examine both the general case where h depends nonlinearly on $\boldsymbol{\phi}$, as well as the linear case where h admits the structure

$$h(X(t), X(t+1), \boldsymbol{\phi}) = \mathbf{a}^T(X(t), X(t+1))\boldsymbol{\phi} . \tag{13.6}$$

For example, for the channel equalization problem examined in Section 12.4, if the channel impulse response h_k with $0 \leq k \leq L$ is unknown, we can select

$$\boldsymbol{\phi} = \begin{bmatrix} h_0 \\ h_1 \\ \vdots \\ h_L \end{bmatrix} , \quad \mathbf{a}(X(t), X(t+1)) = \begin{bmatrix} I(t) \\ I(t-1) \\ \vdots \\ I(t-L) \end{bmatrix} ,$$

where $\{I(t),\, t \geq 1\}$ represents the transmitted symbol sequence. Observe that the state $X(t)$ defined in (12.163) encodes symbols $I(t-k)$ for $1 \leq k \leq L$. Thus $I(t)$ is encoded by $X(t+1)$, and accordingly, all symbols stored by the ISI channel can be recovered from the pair $(X(t),\, X(t+1))$.

Let

$$\mathbf{Y} = \begin{bmatrix} Y(1) \\ \vdots \\ Y(t) \\ \vdots \\ Y(N) \end{bmatrix} , \quad \mathbf{s}_u(\boldsymbol{\phi}) = \begin{bmatrix} s_u(1, \boldsymbol{\phi}) \\ \vdots \\ s_u(t, \boldsymbol{\phi}) \\ \vdots \\ s_u(N, \boldsymbol{\phi}) \end{bmatrix} , \quad \mathbf{V} = \begin{bmatrix} V(1) \\ \vdots \\ V(t) \\ \vdots \\ V(N) \end{bmatrix} \tag{13.7}$$

denote respectively the N-dimensional vectors obtained by vectorizing the observations, signal and noise in model (13.2). The noise vector $\mathbf{V}$ is $N(\mathbf{0}, v\mathbf{I}_N)$ distributed. When the signal $s(t, \boldsymbol{\phi})$ admits the linear parametrization (13.6) in terms of the parameter vector $\boldsymbol{\phi}$, $\mathbf{s}_u(\boldsymbol{\phi})$ can be expressed in matrix form as

$$\mathbf{s}_u(\boldsymbol{\phi}) = \mathbf{A}_u \boldsymbol{\phi} \tag{13.8}$$

where

$$\mathbf{A}_u \triangleq \begin{bmatrix} \mathbf{a}^T(x_1^u, x_2^u) \\ \vdots \\ \mathbf{a}^T(x_t^u, x_{t+1}^u) \\ \vdots \\ \mathbf{a}^T(x_N^u, x_{N+1}^u) \end{bmatrix} . \tag{13.9}$$

13.2.2 GLR Test

Assume that the vector $\boldsymbol{\theta}$ of unknown parameters includes both the vector $\boldsymbol{\phi}$ parametrizing the signal $s(t, \boldsymbol{\phi})$ and the noise variance v, as indicated by (13.4). Then the log-likelihood function for the observation vector $\mathbf{Y}$ under hypothesis H_u is given by

$$L_u(\mathbf{y}|\boldsymbol{\theta}) = \ln f_{\mathbf{Y}}(\mathbf{y}|\boldsymbol{\theta}, H_u)$$
$$= -\frac{1}{2v}||\mathbf{Y} - \mathbf{s}_u(\boldsymbol{\phi})||_2^2 - \frac{N}{2}\ln(2\pi v)\,. \quad (13.10)$$

The implementation of the GLR test requires first the evaluation of the ML estimate

$$\hat{\boldsymbol{\theta}}_u = \begin{bmatrix} \hat{\boldsymbol{\phi}}_u \\ \hat{v}_u \end{bmatrix}$$

of $\boldsymbol{\theta}$ under H_u. This estimate maximizes $L_u(\boldsymbol{\theta})$, so

$$\hat{\boldsymbol{\phi}}_u = \arg\min_{\boldsymbol{\phi}} ||\mathbf{Y} - \mathbf{s}_u(\boldsymbol{\phi})||_2^2\,. \quad (13.11)$$

By setting

$$\frac{\partial}{\partial v} L_u(\mathbf{y}|\boldsymbol{\theta}) = \frac{1}{2v^2}||\mathbf{Y} - \mathbf{s}_u(\boldsymbol{\phi})||_2^2 - \frac{N}{2v} \quad (13.12)$$

equal to zero, and substituting the estimate $\hat{\boldsymbol{\phi}}_u$ gives

$$\hat{v}_u = \frac{||\mathbf{Y} - \mathbf{s}_u(\hat{\boldsymbol{\phi}}_u)||_2^2}{N}. \quad (13.13)$$

Substituting the estimated vector $\hat{\boldsymbol{\theta}}_u$ inside the log-likelihood function gives the GLR statistic

$$\mathcal{L}_u(\mathbf{Y}) = -\frac{N}{2}\ln(||\mathbf{Y} - \mathbf{s}_u(\hat{\boldsymbol{\phi}}_u)||_2^2) + c \quad (13.14)$$

where the constant c is independent of $\mathbf{Y}$ and u. Thus the GLR test can be expressed as

$$u_0 = \arg\min_{1\leq u\leq U} ||\mathbf{Y} - \mathbf{s}_u(\hat{\boldsymbol{\phi}}_u)||_2^2\,, \quad (13.15)$$

which is again a minimum distance rule between observation vector $\mathbf{Y}$ and the U observation signals $\mathbf{s}_u(\hat{\boldsymbol{\phi}}_u)$.

As indicated by (13.11), the ML estimate $\hat{\boldsymbol{\phi}}_u$ minimizes

$$J_u(\boldsymbol{\phi}) = ||\mathbf{Y} - \mathbf{s}_u(\boldsymbol{\phi})||_2^2$$
$$= \sum_{t=1}^{N}(Y(t) - s_u(t, \boldsymbol{\phi}))^2 \quad (13.16)$$

which represents the nonlinear least-squares fit between the observed signal $Y(t)$ and the signal $s_u(t, \phi)$ triggered by the MC trajectory $\mathbf{x}_u$. Then according to (13.15) the GLR test selects the trajectory which achieves the tightest fit.

For the special case where the signal $\mathbf{s}_u(t, \phi)$ admits the linear parametrization (13.6), the minimization of

$$J_u(\phi) = ||\mathbf{Y} - \mathbf{A}_u \phi||_2^2$$
$$= \sum_{t=1}^{N} (Y(t) - \mathbf{a}^T(x_t^u, x_{t+1}^u)\phi)^2 \tag{13.17}$$

becomes a linear least-squares problem which can be solved in closed form. We obtain

$$\hat{\phi}_u = (\mathbf{A}_u^T \mathbf{A}_u)^{-1} \mathbf{A}_u^T \mathbf{Y} \tag{13.18}$$

and

$$\mathbf{s}_u(\hat{\phi}_u) = \mathbf{P}_u \mathbf{Y} , \tag{13.19}$$

where

$$\mathbf{P}_u \triangleq \mathbf{A}_u (\mathbf{A}_u^T \mathbf{A}_u)^{-1} \mathbf{A}_u^T \tag{13.20}$$

denotes the projection matrix onto the subspace $\mathcal{A}_u$ spanned by the columns of $\mathbf{A}_u$. Let

$$\mathbf{P}_u^\perp = \mathbf{I}_N - \mathbf{P}_u \tag{13.21}$$

denote the projection onto the space orthogonal to $\mathcal{M}_u$. Since

$$||\mathbf{Y}||_2^2 = ||\mathbf{P}_u \mathbf{Y}||_2^2 + ||\mathbf{P}_u^\perp \mathbf{Y}||_2^2$$

the GLR test can be expressed as either

$$u_0 = \arg \min_{1 \leq u \leq U} ||\mathbf{P}_u^\perp \mathbf{Y}||_2^2 \tag{13.22}$$

or

$$u_0 = \arg \max_{1 \leq u \leq U} ||\mathbf{P}_u \mathbf{Y}||_2^2 , \tag{13.23}$$

so we can either minimize the norm of the projection of the observation vector $\mathbf{Y}$ onto the space orthogonal to $\mathcal{A}_u$, or maximize the norm of the projection of $\mathbf{Y}$ onto $\mathcal{A}_u$.

Of course, the cardinality U of the set of all trellis paths is usually quite large so the GLR test (13.15), or its simplified version (13.22) in the linear case, is primarily of theoretical interest. To make the test implementable, one must find a way to prune the set $\mathcal{U}$ of trellis paths in such a way that only promising paths are retained.

13.3 Per Survivor Processing

A reduction in the set of trellis paths can be achieved by employing a technique which has become known over the last 15 years as the *per survivor processing* (PSP) method. This method is usually combined with the Viterbi algorithm, so the survivors are the survivor paths terminating in each possible MC terminal state k at time t. If there are K Markov chain states, only K survivor paths (instead of U) need therefore be retained in the implementation of the GLR test. For $1 \leq k \leq K$, let $\hat{\mathbf{x}}(k,t)$ be the survivor path corresponding to terminal state k at time t. Since the parameter vector $\boldsymbol{\phi}$ is unknown, we are confronted right away with the key issue: in order to implement the Viterbi algorithm to find the survivor paths, one must have an estimated $\boldsymbol{\phi}$. But in order to perform a nonlinear least-squares fit for each survivor path, the survivors must be available. In the PSP procedure, this issue is sidestepped by assuming that an initialization or acquisition stage is implemented, so that at time t, a set of survivor paths $\hat{\mathbf{x}}(k,t)$ and their matching ML estimates $\hat{\boldsymbol{\phi}}(k,t)$ have been identified.

13.3.1 Path Extension

The branch metric

$$c(k,\ell,t) = (y_t - h(k,\ell,\hat{\boldsymbol{\phi}}(k,t)))^2 \tag{13.24}$$

is employed for transitions originating from state k at time t. It is obtained by replacing the unknown vector $\boldsymbol{\phi}$ by the estimate $\hat{\boldsymbol{\phi}}(k,t)$ obtained by least-squares fit of the signal triggered by survivor path $\hat{\mathbf{x}}(k,t)$. The survivor paths are extended from time t to $t+1$ by using the usual Viterbi recursion

$$V(\ell,t+1) = \min_{k\,:\,(k,\ell)\in\mathcal{T}} (V(k,t) + c(k,\ell,t)) \, , \tag{13.25}$$

where $\mathcal{T}$ denotes the set of allowed transitions for the Markov chain. The predecessor of state ℓ on the survivor path $\hat{\mathbf{x}}(\ell,t+1)$ terminating in ℓ at time $t+1$ is

$$b(\ell,t+1) = \arg \min_{k\,:\,(k,\ell)\in\mathcal{T}} (V(k,t) + c(k,\ell,t)) \, , \tag{13.26}$$

and the new survivor path terminating in ℓ at $t+1$ is therefore

$$\hat{\mathbf{x}}(\ell,t+1) = (\hat{\mathbf{x}}(b(\ell,t+1),t),\ell) \, . \tag{13.27}$$

13.3.2 Parameter Vector Update

We are then left with the problem of evaluating the new parameter vectors $\hat{\boldsymbol{\phi}}(\ell,t+1)$ for each $1 \leq \ell \leq K$. To do so, it is useful to observe that the path $\hat{\mathbf{x}}(\ell,t+1)$ is obtained by concatenating the transition $(b(\ell,t+1),\ell)$ to the path $\hat{\mathbf{x}}(b(\ell, t+1), t)$. But a parameter vector estimate $\hat{\boldsymbol{\phi}}(b(\ell, t+1), t)$ is already available for the path $\hat{\mathbf{x}}(b(\ell, t+1), t)$, so instead of performing a new least-squares fit for the entire path $\hat{\mathbf{x}}(\ell, t+1)$, we can update the vector $\hat{\boldsymbol{\phi}}(b(\ell, t+1), t)$ obtained by fitting the path up to time t by taking into account the new cost associated to the transition from state $b(\ell, t+1)$ at time t to state ℓ at time $t+1$. Specifically, given a trellis path $\mathbf{x}_u$, let

$$J_t^u(\boldsymbol{\phi}) = \sum_{s=1}^{t-1} c_s^u(\boldsymbol{\phi}) \tag{13.28}$$

with

$$c_s^u(\boldsymbol{\phi}) \triangleq (Y(s) - h(x_s^u, x_{s+1}^u, \boldsymbol{\phi}))^2 \tag{13.29}$$

denote the accumulated squared mismatch up to time t between observations $Y(s)$ and the signal $h(x_s^u, x_{s+1}^u \boldsymbol{\phi})$ triggered by $\mathbf{x}_u$. The accumulated mismatch obeys the recursion

$$J_{t+1}^u(\boldsymbol{\phi}) = J_t^u(\boldsymbol{\phi}) + c_t^u(\boldsymbol{\phi})\,. \tag{13.30}$$

Let u be the index of a trellis path coinciding with $\hat{\mathbf{x}}(\ell, t+1)$ up to time $t+1$. To update the parameter vector $\hat{\boldsymbol{\phi}}(b(\ell, t+1), t)$, we assume that it minimizes *exactly* the accumulated mismatch function $J_t^u(\boldsymbol{\phi})$ for the path up to time t. If

$$\nabla_{\boldsymbol{\phi}} = \begin{bmatrix} \frac{\partial}{\partial \phi_1} \\ \frac{\partial}{\partial \phi_2} \\ \vdots \\ \frac{\partial}{\partial \phi_q} \end{bmatrix}$$

represents the gradient with respect to parameter vector $\boldsymbol{\phi}$, the gradient of the accumulated mismatch function up to time $t+1$ can be expressed as

$$\nabla_{\boldsymbol{\phi}} J_{t+1}^u(\boldsymbol{\phi}) = \nabla_{\boldsymbol{\phi}} J_t^u(\boldsymbol{\phi}) + \nabla_{\boldsymbol{\phi}} c_t^u(\boldsymbol{\phi}) \tag{13.31}$$

with

$$\nabla_{\boldsymbol{\phi}} c_t^u(\boldsymbol{\phi}) = -2(Y(t) - h(x_t^u, x_{t+1}^u, \boldsymbol{\phi}))\nabla_{\boldsymbol{\phi}} h(x_t^u, x_{t+1}^u, \boldsymbol{\phi}))\,. \tag{13.32}$$

But since $\hat{\boldsymbol{\phi}}(b(t+1), t)$ minimizes exactly $J_t^u(\boldsymbol{\phi})$, we have

$$\nabla_{\boldsymbol{\phi}} J_t^u(\hat{\boldsymbol{\phi}}(b(\ell, t+1), t) = 0\,. \tag{13.33}$$

Accordingly, if

$$e(\ell, t) \triangleq Y(t) - h(b(\ell, t+1), \ell, \hat{\boldsymbol{\phi}}(b(\ell, t+1), t)) \tag{13.34}$$

denotes the observation mismatch for transition $(b(\ell, t+1), \ell)$ based on parameter vector estimate $\hat{\boldsymbol{\phi}}(b(\ell, t+1), t)$, expression (13.31) for the gradient of the accumulated squared mismatch along path $\mathbf{x}_u$ until time $t+1$ reduces to

$$\nabla_{\phi} J^u_{t+1}(\hat{\phi}(b(\ell,t+1),t) = \nabla_{\phi} c^u_t(\hat{\phi}(b(\ell,t+1),t)$$
$$= -2e(\ell,t)\nabla_{\phi} h(b(\ell,t+1),\ell,\hat{\phi}(b(\ell,t+1),t)) \, . \tag{13.35}$$

This expression depends only on error $e(\ell,t)$ and gradient function $\nabla_{\phi} h(k,\ell,\phi)$, which is assumed to be available in parametric form.

Then to update the parameter vector estimate along each survivor path, one of the following two approaches can be employed:

Gradient descent: Since we seek to minimize $J^u_{t+1}(\phi)$, to update the parameter vector estimate $\hat{\phi}(b(\ell,t+1),t)$ we can move in a direction opposite to the gradient so as to decrease the value of the objective function. This gives the recursion

$$\hat{\phi}(\ell,t+1) = \hat{\phi}(b(\ell,t+1),t) - \frac{\mu}{2}\nabla_{\phi} J^u_{t+1}(\hat{\phi}(b(\ell,t+1),t)$$
$$= \hat{\phi}(b(\ell,t+1),t) + \mu e(\ell,t)\nabla_{\phi} h(b(\ell,t+1),\ell,\hat{\phi}(b(\ell,t+1),t)) \, , \tag{13.36}$$

where $\mu/2$ denotes the iteration step size.

Newton iteration: Alternatively, suppose that the Hessian matrix

$$\mathbf{H}^u_{t+1}(\phi) = \nabla^T_{\phi}\nabla_{\phi} J^u_{t+1}(\phi)$$
$$= \Big(\frac{\partial^2}{\partial\phi_i\partial\phi_j} J^u_{t+1}(\phi)\Big)_{1\le i,\, j\le q} \tag{13.37}$$

of the accumulated mismatch function is available. Observe that the additive structure (13.28) of the cost function J_t implies a similar structure

$$\mathbf{H}^u_{t+1}(\phi) = \sum_{s=0}^{t} \nabla^T_{\phi}\nabla_{\phi} c^u_s(\phi) \tag{13.38}$$

for the Hessian. By setting the gradient equal to zero in the first-order Taylor series expansion

$$\nabla_{\phi} J^u_{t+1}(\phi) = \nabla_{\phi} J^u_{t+1}(\hat{\phi}(b(\ell,t+1),t))$$
$$+\mathbf{H}^u_{t+1}(\hat{\phi}(b(\ell,t+1),t))(\phi - \hat{\phi}(b(\ell,t+1),t) \, , \tag{13.39}$$

and taking into account expression (13.35) for the gradient of J^u_{t+1}, we obtain the *Newton iteration*

$$\hat{\phi}(\ell,t+1) = \hat{\phi}(b(\ell,t+1),t) \tag{13.40}$$
$$+2e(\ell,t)(\mathbf{H}^u_{t+1})^{-1}(\hat{\phi}(b(\ell,t+1),t))\nabla_{\phi} h(b(\ell,t+1),\ell,\hat{\phi}(b(\ell,t+1),t)) \, .$$

Since the iteration (13.40) requires the evaluation of $(\mathbf{H}^u_{t+1})^{-1}\nabla_{\phi} h$, it has a higher complexity than the gradient descent scheme (13.36), but it converges much faster.

Linear case: When the function $h(k, \ell, \phi)$ admits the linear parametrization (13.6), we have

$$\nabla_{\phi} h(k, \ell, \phi) = \mathbf{a}(k, \ell) \tag{13.41}$$

and

$$\mathbf{H}_{t+1}^{u} = 2 \sum_{s=0}^{t} \mathbf{a}(x_s^u, x_{s+1}^u) \mathbf{a}^T(x_s^u, x_{s+1}^u) \,. \tag{13.42}$$

In this case, the gradient update (13.36) takes the form of a least-mean-squares (LMS) adaptation rule [14]

$$\hat{\phi}(\ell, t+1) = \hat{\phi}(b(\ell, t+1), t) + \mu e(\ell, t) \mathbf{a}(b(\ell, t+1), \ell) \,, \tag{13.43}$$

where the mismatch error for segment $(b(\ell, t+1), \ell)$ can now be expressed as

$$e(\ell, t) = Y(t) - \mathbf{a}^T(b(\ell, t+1), \ell) \hat{\phi}(b(\ell, t+1), t) \,. \tag{13.44}$$

Also, if survivor path $\hat{\mathbf{x}}(\ell, t+1)$ is indexed by u, let

$$\mathbf{P}(\ell, t+1) \triangleq 2(\mathbf{H}_{t+1}^{u})^{-1}$$
$$\mathbf{g}(\ell, t+1) \triangleq \mathbf{P}(\ell, t+1) \mathbf{a}(\ell, t+1) \,.$$

Then the Newton iteration (13.40) reduces to the *recursive least-squares* (RLS) algorithm [14]

$$\hat{\phi}(\ell, t+1) = \hat{\phi}(b(\ell, t+1), t) + \mathbf{g}(\ell, t+1) e(\ell, t) \tag{13.45}$$

where the gain $\mathbf{g}(\ell, t+1)$ satisfies the recursion

$$\mathbf{g}(\ell, t+1) = \frac{\mathbf{P}(b(\ell, t+1), t) \mathbf{a}(b(\ell, t+1), \ell)}{1 + \boldsymbol{a}^T(b(\ell, t+1), \ell) \mathbf{P}(b(\ell, t+1), t) \mathbf{a}(b(\ell, t+1), \ell)} \tag{13.46}$$

with

$$\mathbf{P}(\ell, t+1) = \left[\mathbf{I}_q - \mathbf{g}(\ell, t+1) \boldsymbol{a}^T(b(\ell, t+1), \ell)\right] \mathbf{P}(b(\ell, t+1), t) \,. \tag{13.47}$$

Note that the RLS recursion performs an exact minimization of $J_{t+1}^{u}(\phi)$ in one step.

To illustrate the implementation of the PSP method, we consider two examples.

Example 13.1: Joint timing recovery and equalization

Consider a CT observation signal

$$Y(t) = \sum_{\ell=0}^{L} q(t - \ell T) I_{\ell} + V(t)$$

where the symbols I_k are independent and take the binary values ± 1 with probability 1/2, $q(t)$ is a signaling pulse, T denotes the baud period, and $V(t)$ is a WGN independent of symbols I_k. When this signal is passed through a raised cosine filter $r(t)$ and sampled at times $t = nT + \tau$, where τ denotes an unknown timing offset, if we denote $h(t) = r(t) * q(t)$ and

$$Y_n = (r * Y)(nT + \tau) \quad , \quad V_n = (r * V)(nT + \tau) ,$$

we obtain the DT observation model

$$Y_n = \sum_{\ell=0}^{L} h((n - \ell)T + \tau) I_\ell + V_n \tag{13.48}$$

where V_n is a WGN of intensity v. In this model it is assumed that $h(t)$ and its derivative $\dot{h}(t)$ are known in parametric form. For example, for the equalization of a partial response (PR) channel of Class IV considered in [15],

$$h(t) = p(t) - p(t - 2T)$$

where

$$p(t) = \frac{\sin(\pi t/T)}{\pi t/T}$$

is an ideal zero excess bandwidth Nyquist pulse.

In this case, the unknown parameter vector ϕ reduces to the scalar timing offset parameter τ. Except for the presence of offset parameter τ, the linear equalization problem we consider has the same form as the one examined in Section 12.4, so we can use the same Markov chain model. Thus, we assume that the state $X(n)$ encodes the symbols I_{n-m}, $1 \leq m \leq L$ stored in the channel according to the mapping (12.162). This implies that the current symbol I_n is encoded by $X(n + 1)$, so that the observed signal

$$\sum_{m=0}^{L} h(mT + \tau) I_{n-m} = h(X(n), X(n + 1), \tau) \tag{13.49}$$

can be viewed as a function of the transition between state $X(n)$ and $X(n+1)$. Given this model, we see that to a transition from state k at time n to state ℓ at time $n + 1$ corresponds a unique set of symbols I_{n-m} with $0 \leq m \leq L$.

Then, let $\hat{\mathbf{x}}(k, n)$ and $\hat{\tau}(k, n)$ with $1 \leq k \leq K$ denote the set of survivor paths and corresponding timing offset estimates at time n. The observation mismatch error for the transition from state $b(\ell, n + 1)$ at time n to state ℓ at time $n + 1$ based on offset estimate $\hat{\tau}(b(\ell, t + 1), n)$ is given by

$$e(\ell, n) = Y_n - \sum_{m=0}^{L} h(mT + \hat{\tau}(b(\ell, n + 1), n)) \hat{I}_{n-m} ,$$

where the decoded symbols $\hat{I}_{n-m}$, $0 \leq m \leq L$ are those associated to the transition $(b(\ell, n+1), \ell)$. Furthermore, according to (13.49) the gradient of function $h(b(\ell, n+1), \ell, \tau)$ with respect to τ is

$$\nabla_\tau h(b(\ell, n+1), \ell, \tau) = \sum_{m=0}^{L} \dot{h}(mT + \tau)\hat{I}_{n-m} .$$

The gradient update (13.36) for the timing offset estimate $\hat{\tau}(\ell, n+1)$ can therefore be expressed as

$$\begin{aligned}\hat{\tau}(\ell, n+1) &= \hat{\tau}(b(\ell, n+1), n) \\ &\quad + \mu e(\ell, n) \sum_{m=0}^{L} \dot{h}(mT + \hat{\tau}(b(\ell, n+1), n))\hat{I}_{n-m} . \quad (13.50)\end{aligned}$$

Thus, the PSP timing offset update requires the availability of function $\dot{h}(\cdot)$, and employs the decoded symbol $\hat{I}_{n-m}$ corresponding to the transition from predecessor state $b(\ell, n+1)$ at time n to state ℓ at time $n+1$. □

Example 13.2: Blind equalization

Consider a linear equalization problem of the form considered in 12.4, but where the channel impulse response (CIR) h_k with $0 \leq k \leq L$ is *unknown.* For this type of problem, no training sequence is transmitted, so the equalization problem is called "blind" [16] since the channel needs to be estimated while at the same time detecting the symbol sequence $\{I(t), t \geq 1\}$. Thus the observation model can be expressed as

$$Y(t) = h(X(t), X(t+1), \phi) + V(t)$$

where h admits the structure (13.6) with

$$\mathbf{a}(X(t), X(t+1)) = \begin{bmatrix} I(t) \\ I(t-1) \\ \vdots \\ I(t-L) \end{bmatrix} , \quad \phi = \begin{bmatrix} h_0 \\ h_1 \\ \vdots \\ h_L \end{bmatrix} . \quad (13.51)$$

$V(t)$ denotes here a WGN of intensity v which is independent of symbols $I(t)$ which are independent and take values ± 1 with probability 1/2 each. At the outset it is worth noting that channel vector ϕ and symbol sequence $\{i_t, t \geq 1\}$ produce the same observations as vector $-\phi$ and symbols $\{-i_t, t \geq 1\}$, so the sign of the channel impulse response can never be determined from the given observations. To overcome this difficulty, differential coding [17] can be employed to generate symbols $I(t)$, in which case the transmitted information is encoded by the *sign changes* of sequence $I(t)$. Another approach which is adopted here consists of assuming that the initial $L+1$ transmitted symbols are known, say

$$I(1-k) = -1$$

for $0 \leq k \leq L$. In essence, this amounts to using a mini-training sequence which is not sufficient to acquire the CIR, but yet allows the estimation of its sign. Observe indeed that if the first $L+1$ symbols are -1, then

$$Y(1) = -\sum_{k=0}^{L} h_k + V(1)$$

can be used to acquire the sign of $\sum_{k=0}^{L} h_k$.

Then, since the parametrization (13.6) of the function $h(k, \ell, \phi)$ is linear, the channel estimate $\hat{\phi}(k,t)$ can be updated by using recursion (13.44)–(13.47) where as indicated in (13.50), the vector $\mathbf{a}(b(\ell, t+1), \ell)$ corresponds to the symbols encoded by the transition from state $b(\ell, n+1)$ at time t to state ℓ at time $t+1$. □

In conclusion, the PSP method represents a simplification of the GLRT obtained by pruning in an ad hoc manner the space $\mathcal{U}$ of all trellis paths by retaining only K survivor paths, where K denotes the number of MC states. The paths are extended by using the Viterbi algorithm based on the current survivor paths parameter estimates, and after each path has been extended, the new survivor parameter vectors are calculated by applying standard adaptive filtering recursions. A key aspect of the procedure is its initialization phase. The initialization should exploit available information about the parameter vector ϕ, but even if such information is not available, a fully blind acquisition scheme based on arbitrary initial survivors and arbitrary initial parameter vector estimates can be effective, as was demonstrated in [6].

13.4 EM Detector

The joint sequence detection and parameter estimation for Markov chains can also be performed by employing an EM approach. Let $\mathbf{y} \in \mathbb{R}^N$ denote the vector (13.7) formed by all observations $Y(t)$ with $1 \leq t \leq N$. This vector represents the available data. In the EM approach, this vector is augmented by selecting "missing data," which together with the vector $\mathbf{y}$, forms the complete data. Let $\mathcal{S} = \{1, 2, \ldots, K\}$ denote the set of MC states. For the HMM case, the missing data is the point $\mathbf{x}$ of $\mathcal{S}^{N+1}$ representing the entire state sequence $\{x_t, 1 \leq t \leq N+1\}$, where $x_t \in \mathcal{S}$ for each t. Accordingly, the complete data is formed by

$$\mathbf{z} = \begin{bmatrix} \mathbf{y} \\ \mathbf{x} \end{bmatrix} . \tag{13.52}$$

The EM framework allows the consideration of slighly more general parameter estimation problems than the GLRT/PSP approach. Thus, in addition to assuming that the observed signal component $h(x_t, x_{t+1}, \boldsymbol{\phi})$ of observation

$$Y(t) = h(X(t), X(t+1), \boldsymbol{\phi}) + V(t) \tag{13.53}$$

depends on parameter vector $\boldsymbol{\phi} \in \mathbb{R}^q$, the one-step transition probability matrix $\mathbf{P}$ of the MC may also depend on a parameter vector $\boldsymbol{\xi}$. This dependence is denoted as $\mathbf{P}(\boldsymbol{\xi})$. For reasons that will become clear below, we assume that the the graph $\mathcal{T}$ of $\mathbf{P}(\boldsymbol{\xi})$ does not change as $\boldsymbol{\xi}$ varies. For simplicity, we assume that the vectors $\boldsymbol{\phi}$ and $\boldsymbol{\xi}$ parametrizing output map h and transition matrix $\mathbf{P}$ do not have common components. This assumption is reasonable, since the physical processes modelling the state dynamics and the collection of observations are usually completely different. The dependence of $\mathbf{P}$ on vector $\boldsymbol{\xi}$ allows the estimation of some of the entries of $\mathbf{P}$ if they are not known exactly. This feature can be useful in speech-processing applications, but for communications applications, $\mathbf{P}$ is usually completely known. For example, for the equalization problem considered in Section 12.4, the structure of $\mathbf{P}$ depends first on the observation that the symbols $\{I(t-k), 1 \leq k \leq L\}$ stored in the channel can be described by a shift register, where when t is increased by one, all stored symbols are shifted by one position to make place for the new symbol $I(t)$. Together with the observation that each symbol takes values ± 1 with equal probability, this fixes completely the structure of $\mathbf{P}$. Similarly, the finite state machine which is usually used to model continuous-phase modulated signals [18, Sec. 5.3] is completely fixed by the choice of modulation format. Finally, as in the GLRT case, the variance v of the WGN $V(t)$ may be included in the parameter vector $\boldsymbol{\theta}$ to be estimated. Thus when v is unknown, we estimate

$$\boldsymbol{\theta} \triangleq \begin{bmatrix} \boldsymbol{\xi} \\ \boldsymbol{\phi} \\ v \end{bmatrix} ,$$

whereas when v is known, we select

$$\boldsymbol{\theta} = \begin{bmatrix} \boldsymbol{\xi} \\ \boldsymbol{\phi} \end{bmatrix}$$

as parameter vector.

After data augmentation, the mixed probability density/distribution for the complete data $\mathbf{z}$ can be expressed as

$$f_c(\mathbf{z}|\boldsymbol{\theta}) = f(\mathbf{y}|\mathbf{x}, \boldsymbol{\phi}, v) p(\mathbf{x}|\boldsymbol{\xi}) \tag{13.54}$$

where $f(\mathbf{y}|\mathbf{x}, \boldsymbol{\phi}, v)$ represents the probability density of observation vector $\mathbf{Y}$ given the state path $\mathbf{x}$, and $p(\mathbf{x}|\boldsymbol{\xi})$ is the probability of path $\mathbf{x}$. Given observation model (13.53), the density of $\mathbf{Y}$ knowing the state sequence $\mathbf{x}$ can be expressed as

$$f(\mathbf{y}|\mathbf{x}, \boldsymbol{\phi}, v) = \prod_{t=1}^{N} f(y_t|x_t, x_{t+1}, \boldsymbol{\phi}, v) \quad (13.55)$$

with

$$f(y_t|x_t, x_{t+1}, \boldsymbol{\phi}, v) = \frac{1}{(2\pi v)^{1/2}} \exp\left[-\frac{1}{2v}(y_t - h(x_t, x_{t+1}, \boldsymbol{\phi}))^2\right]. \quad (13.56)$$

Also, if we assume that the initial probability distribution

$$\pi_{x_1} = P[X(1) = x_1]$$

of the MC is known, the probability of path $\mathbf{x} = (x_t,\ 1 \leq t \leq N+1)$ can be expressed as

$$p(\mathbf{x}|\boldsymbol{\xi}) = \pi_{x_1} \prod_{t=1}^{N} p_{x_t x_{t+1}}(\boldsymbol{\xi}). \quad (13.57)$$

Combining (13.54)–(13.57), we find that the log-likelihood function for the complete data $\mathbf{z}$ takes the form

$$\begin{aligned} L_c(\mathbf{z}|\boldsymbol{\theta}) &= \ln f_c(\mathbf{z}|\boldsymbol{\theta}) = \ln f(\mathbf{y}|\mathbf{x}, \boldsymbol{\phi}, v) + \ln p(\mathbf{x}|\boldsymbol{\xi}) \\ &= \sum_{t=1}^{N} \ln f(y_t|x_t, x_{t+1}, \boldsymbol{\phi}, v) + \sum_{t=1}^{N} \ln p_{x_t x_{t+1}}(\boldsymbol{\xi}) + \ln \pi_{x_1} \\ &= -\frac{1}{2v}\sum_{t=1}^{N}(y_t - h(x_t, x_{t+1}, \boldsymbol{\phi}))^2 + \sum_{t=1}^{N} \ln p_{x_t x_{t+1}}(\boldsymbol{\xi}) - \frac{N}{2}\ln(v), \quad (13.58) \end{aligned}$$

where in the last equality we have dropped terms independent of $\mathbf{y}$, $\mathbf{x}$, and $\boldsymbol{\theta}$.

Then, as explained in (11.11)–(11.12), the evaluation of the estimated complete likelihood function

$$Q(\boldsymbol{\theta}|\boldsymbol{\theta}') = \sum_{\mathbf{x} \in \mathcal{S}^{N+1}} L_c(\mathbf{z}|\boldsymbol{\theta}) p_c(\boldsymbol{x}|\mathbf{y}, \boldsymbol{\theta}') \quad (13.59)$$

requires first finding the conditional probability distribution

$$p_c(\mathbf{x}|\mathbf{y}, \boldsymbol{\theta}') = \frac{f_c(\mathbf{z}|\boldsymbol{\theta}')}{f(\mathbf{y}|\boldsymbol{\theta}')} \quad (13.60)$$

of the missing data given observations $\mathbf{Y}$ and the model corresponding to $\boldsymbol{\theta}'$. It turns out that this evaluation can be performed in two different ways. The first method proposed by Baum and Welch [7, 19, 20] uses the forward-backward algorithm to perform an exact evaluation of $Q(\boldsymbol{\theta}|\boldsymbol{\theta}')$. In contrast, a recent approach described in [11–13] employs the Viterbi algorithm to approximate $Q(\boldsymbol{\theta}|\boldsymbol{\theta}')$. In spite of its use of an approximation, this second approach may be preferable to the Baum-Welch reestimation method due to the lower computational complexity of the Viterbi algorithm compared to the forward-backward algorithm.

13.4.1 Forward-backward EM

An important feature of expression (13.58) for the log-likelihood of the complete data is that it depends only on consecutive states $X(t)$ and $X(t+1)$ of the Markov chain. So, instead of evaluating the conditional probability distribution of the entire MC trajectory $\mathbf{X}$ given observation vector $\mathbf{Y}$, we only need to evaluate the joint conditional probability distribution

$$p(k,\ell,t|\mathbf{y},\boldsymbol{\theta}') \stackrel{\triangle}{=} P[X(t)=k, X(t+1)=\ell|\mathbf{Y}=\mathbf{y},\boldsymbol{\theta}'] \tag{13.61}$$

of two consecutive states. To keep things simple, we temporarily suppress the argument $\boldsymbol{\theta}'$. Then, consider the joint hybrid probability distribution/density

$$p_{X(t),X(t+1),\mathbf{Y}}(k,\ell,t;\mathbf{y}) \stackrel{\triangle}{=} p(k,\ell,t|\mathbf{y}) f_{\mathbf{Y}}(\mathbf{y}) \tag{13.62}$$

where $f_{\mathbf{Y}}(\mathbf{y})$ denotes the density of $\mathbf{Y}$. If we employ the partition of $\mathbf{Y}$ into $\mathbf{Y}_{t+1}^-$ and $\mathbf{Y}_{t+1}^+$ components specified by (12.120), the hybrid distribution/density (12.62) can be rewritten as

$$p_{X(t),X(t+1),\mathbf{Y}_{t+1}^-,\mathbf{Y}_{t+1}^+}(k,\ell,t;\mathbf{y}_{t+1}^+,\mathbf{y}_{t+1}^+)\,.$$

By conditioning with respect to $X(t)$, $X(t+1)$ and Y_{t+1}^- and employing the Markov property of $X(t)$ we obtain the factorization

$$\begin{aligned} &p_{X(t),X(t+1),\mathbf{Y}_{t+1}^-,\mathbf{Y}_{t+1}^+}(k,\ell,t;\mathbf{y}_{t+1}^-,\mathbf{Y}_{t+1}^+) = \\ &\quad f_{\mathbf{Y}_{t+1}^+|X(t+1)}(\mathbf{y}_{t+1}^+|\ell,t+1) p_{X(t),X(t+1),\mathbf{Y}_{t+1}^-}(k,\ell,t;\mathbf{y}_{t+1}^-)\,. \end{aligned} \tag{13.63}$$

By successive conditioning and using the Markov property of $X(t)$, we also obtain the factorization

$$\begin{aligned} &p_{X(t),X(t+1),\mathbf{Y}_{t+1}^-}(k,\ell,t;\mathbf{y}_{t+1}^-) = \\ &\quad f_{Y(t)|X(t),X(t+1)}(y_t|k,\ell) p_{X(t+1)|X(t)}(\ell|k) p_{X(t),\mathbf{Y}_t^-}(k,t;\mathbf{y}_t^-)\,. \end{aligned} \tag{13.64}$$

Taking into account definitions (12.123) and (12.124) of $\alpha(k,t)$ and $\beta(k,t)$, we recognize

$$\begin{aligned} \alpha(k,t) &= p_{X(t),\mathbf{Y}_t^-}(k,t;\mathbf{y}_t^-) \\ \beta(\ell,t+1) &= f_{\mathbf{Y}_{t+1}^+|X(t+1)}(\mathbf{y}_{t+1}^+|\ell,t+1))\,, \end{aligned}$$

so substituting (13.64) inside (13.63) gives

$$p_{X(t),X(t+1),\mathbf{Y}}(k,\ell,t;\mathbf{y}) = \alpha(k,t) m_{k\ell}(t) \beta(\ell,t+1)\,, \tag{13.65}$$

where

$$m_{k\ell}(t) = p_{k\ell}\frac{1}{(2\pi v)^{1/2}}\exp\big(-\frac{1}{2v}(y_t-h(k,\ell))^2\big) \tag{13.66}$$

is the (k, ℓ)-th entry of matrix $\mathbf{M}(t)$ introduced in (12.136). By marginalizing the hybrid probability distribution/density (13.65), and taking into account the backward propagation equation (12.133), we find that the density of observation vector $\mathbf{Y}$ can be expressed as

$$f_{\mathbf{Y}}(\mathbf{y}) = \sum_{k=1}^{K} \sum_{\ell=1}^{K} p_{X(t),X(t+1),\mathbf{Y}}(k, \ell, t; \mathbf{y})$$
$$= \sum_{k=1}^{K} \alpha(k,t)\beta(k,t) = \boldsymbol{\alpha}(t)\boldsymbol{\beta}(t) = \gamma\,, \quad (13.67)$$

where, as indicated by (12.145), γ does not depend on t.

Reintroducing argument $\boldsymbol{\theta}'$, the joint conditional distribution of $X(t)$ and $X(t+1)$ given $\mathbf{Y}$ is therefore given by

$$p(k, \ell, t|\mathbf{y}, \boldsymbol{\theta}') = \alpha(k, t, \boldsymbol{\theta}')m_{k\ell}(t, \boldsymbol{\theta}')\beta(\ell, t+1, \boldsymbol{\theta}')/\gamma(\boldsymbol{\theta}')\,. \quad (13.68)$$

Thus, we see that the joint conditional distribution of $X(t)$ and $X(t+1)$ given the entire observation sequence $\mathbf{Y}$ can be evaluated by employing the forward and backward algorithm to evaluate vectors $\boldsymbol{\alpha}(t, \boldsymbol{\theta}')$ and $\boldsymbol{\beta}(t+1, \boldsymbol{\theta}')$ and then using the components of $\mathbf{M}(t, \boldsymbol{\theta}')$ to bind the various entries of these two vectors.

E-Phase: We are now in a position to evaluate the estimated complete data-likelihood function $Q(\boldsymbol{\theta}|\boldsymbol{\theta}')$. By substituting expression (13.68) for the joint conditional density of $X(t)$ and $X(t+1)$ inside (13.59), we find that $Q(\boldsymbol{\theta}|\boldsymbol{\theta}')$ admits the decomposition

$$Q(\boldsymbol{\theta}|\boldsymbol{\theta}') = Q_O(\boldsymbol{\phi}, v|\boldsymbol{\theta}') + Q_T(\boldsymbol{\xi}|\boldsymbol{\theta}')\,. \quad (13.69)$$

In this expression

$$Q_O(\boldsymbol{\phi}, v|\boldsymbol{\theta}') = \sum_{t=1}^{N} \sum_{k=1}^{K} \sum_{\ell=1}^{K} \ln f(y_t|k, \ell, \boldsymbol{\phi}, v)p(k, \ell, t|\mathbf{y}, \boldsymbol{\theta}')$$
$$= -\frac{1}{2v} \sum_{t=1}^{N} \sum_{k=1}^{K} \sum_{\ell=1}^{K} (y_t - h(k, \ell, t))^2 \alpha(k, t, \boldsymbol{\theta}')m_{k\ell}(t, \boldsymbol{\theta}')\beta(\ell, t+1, \boldsymbol{\theta}')/\gamma(\theta')$$
$$-\frac{N}{2}\ln(v)$$
$$= -\frac{1}{2v} \sum_{t=1}^{N} \boldsymbol{\alpha}(t, \boldsymbol{\theta}')(\mathbf{C}(t, \boldsymbol{\phi}) \circ \mathbf{M}(t, \boldsymbol{\theta}'))\boldsymbol{\beta}(t+1, \boldsymbol{\theta}')/\gamma(\boldsymbol{\theta}')$$
$$-\frac{N}{2}\ln(v)\,, \quad (13.70)$$

where $\circ$ denotes the Hadamard product of two matrices of identical dimensions, and $\mathbf{C}(t, \boldsymbol{\phi})$ is the $K \times K$ matrix having for (k, ℓ)-th entry.

$$c_{k\ell}(t, \boldsymbol{\phi}) = (y_t - h(k, \ell, \boldsymbol{\phi}))^2 \tag{13.71}$$

with $1 \leq k, \ell \leq K$. Similarly

$$\begin{aligned} Q_T(\boldsymbol{\xi}|\boldsymbol{\theta}') &= \sum_{t=1}^{N} \sum_{k=1}^{L} \sum_{\ell=1}^{K} \ln p_{k\ell}(\boldsymbol{\xi}) p(k, \ell, t|\mathbf{y}, \boldsymbol{\theta}') \\ &= \sum_{t=1}^{N} \sum_{k=1}^{K} \sum_{\ell=1}^{K} \ln p_{k\ell}(\boldsymbol{\xi}) \alpha(k, t, \boldsymbol{\theta}') m_{k\ell}(t, \boldsymbol{\theta}') \beta(\ell, t+1, \boldsymbol{\theta}') / \gamma(\boldsymbol{\theta}') \\ &= \sum_{t=1}^{N} \boldsymbol{\alpha}(t, \boldsymbol{\theta}') (\mathbf{D}(\boldsymbol{\xi}) \circ \mathbf{M}(t, \boldsymbol{\theta}')) \boldsymbol{\beta}(t+1, \boldsymbol{\theta}') , \end{aligned} \tag{13.72}$$

where the (k, ℓ)-th entry of $K \times K$ matrix $\mathbf{D}(\boldsymbol{\xi})$ is given by

$$d_{k\ell}(\boldsymbol{\xi}) = \ln(p_{k\ell}(\boldsymbol{\xi})) \tag{13.73}$$

for $1 \leq k, \ell \leq K$. At this point it is worth recalling that we are assuming that the graph $\mathcal{T}$ of $\mathbf{P}(\boldsymbol{\xi})$ does not change as $\boldsymbol{\xi}$ varies, where $\mathcal{T}$ has an edge between vertices k and ℓ if $p_{k\ell}(\boldsymbol{\xi}) > 0$. Accordingly, when evaluating the Hadamard product $\mathbf{D}(\boldsymbol{\xi}) \circ \mathbf{M}(t, \boldsymbol{\theta}')$ we set the (k, ℓ)-th entry of the product equal to zero if edge $(k, \ell) \notin \mathcal{T}$, and equal to $d_{k,\ell}(\boldsymbol{\xi}) m_{k\ell}(t, \boldsymbol{\theta}')$ if edge $(k, \ell) \in \mathcal{T}$).

The additive decomposition (13.69) of the estimated complete likelihood function Q into an observation component Q_O depending only on parameters $\boldsymbol{\phi}$ and v, and a state transition component Q_T depending only on $\boldsymbol{\xi}$ is convenient, since it decouples the maximization of $Q(\boldsymbol{\theta}|\boldsymbol{\theta}')$ with respect to $\boldsymbol{\theta}$ in two separate maximizations of functions $Q_O(\boldsymbol{\phi}, v|\boldsymbol{\theta}')$ and $Q_T(\boldsymbol{\xi}|\boldsymbol{\theta}')$ with respect to $(\boldsymbol{\phi}, v)$ and $\boldsymbol{\xi}$, respectively.

M-Phase: Let $(\hat{\boldsymbol{\phi}}, \hat{v})$ denote the pair maximizing $Q_O(\boldsymbol{\phi}, v|\boldsymbol{\theta}')$. Given the form (13.70) of Q_O, instead of maximizing this function over $\boldsymbol{\phi}$, we can minimize

$$W(\boldsymbol{\phi}|\boldsymbol{\theta}') \triangleq \sum_{t=1}^{N} \boldsymbol{\alpha}(t, \boldsymbol{\theta}') (\mathbf{C}(t, \boldsymbol{\phi}) \circ \mathbf{M}(t, \boldsymbol{\theta}')) \boldsymbol{\beta}(t+1, \boldsymbol{\theta}') / \gamma(\boldsymbol{\theta}') . \tag{13.74}$$

Then, if

$$\hat{\boldsymbol{\phi}} = \arg \min_{\boldsymbol{\phi}} W(\boldsymbol{\phi}|\boldsymbol{\theta}') , \tag{13.75}$$

by setting the partial derivative of $Q_O(\boldsymbol{\phi}, v|\boldsymbol{\theta}')$ with respect to v equal to zero, we obtain

$$\hat{v} = \frac{1}{N} W(\hat{\boldsymbol{\phi}}|\boldsymbol{\theta}') . \tag{13.76}$$

The vector $\hat{\boldsymbol{\phi}}$ can be evaluated in closed form in the special case where the output map $h(x_t, x_{t+1}, \boldsymbol{\phi})$ admits the linear parametrization (13.6). In this case, by substituting (13.6) inside the definition (13.74) of $W(\boldsymbol{\phi}|\boldsymbol{\theta}')$, we find that W has the quadratic structure

$$W(\boldsymbol{\phi}|\boldsymbol{\theta}') = \boldsymbol{\phi}^T \boldsymbol{\Omega}(\boldsymbol{\theta}')\boldsymbol{\phi} - 2\boldsymbol{\mu}^T(\boldsymbol{\theta}')\boldsymbol{\phi} + \nu \,, \tag{13.77}$$

with

$$\boldsymbol{\Omega}(\boldsymbol{\theta}') \triangleq \sum_{t=1}^{N}\sum_{k=1}^{K}\sum_{\ell=1}^{K} \alpha(k,t,\boldsymbol{\theta}')\mathbf{a}(k,\ell)\mathbf{a}^T(k,\ell)m_{k\ell}(t,\boldsymbol{\theta}')\beta(\ell,t+1,\boldsymbol{\theta}')/\gamma(\boldsymbol{\theta}') \tag{13.78}$$

$$\boldsymbol{\mu}(\boldsymbol{\theta}') \triangleq \sum_{t=1}^{N} y_t \sum_{k=1}^{K}\sum_{\ell=1}^{K} \alpha(k,t,\boldsymbol{\theta}')\mathbf{a}(k,\ell)m_{k\ell}(t,\boldsymbol{\theta}')\beta(\ell,t+1,\boldsymbol{\theta}')/\gamma(\boldsymbol{\theta}') \tag{13.79}$$

and

$$\nu = \sum_{t=1}^{N} y_t^2 = ||\mathbf{y}||_2^2 \,. \tag{13.80}$$

Accordingly, the vector minimizing $W(\boldsymbol{\phi}|\boldsymbol{\theta}')$ is given by

$$\hat{\boldsymbol{\phi}} = \boldsymbol{\Omega}^{-1}(\boldsymbol{\theta}')\boldsymbol{\mu}(\boldsymbol{\theta}') \,. \tag{13.81}$$

After completing the maximization of $Q_O(\boldsymbol{\phi}, v|\boldsymbol{\theta}')$ with respect to $\boldsymbol{\phi}$ and v, we can now proceed to the maximization of $Q_T(\boldsymbol{\xi}|\boldsymbol{\theta}')$ with respect to $\boldsymbol{\xi}$. The vector $\boldsymbol{\xi}$ includes typically some of the transition probabilities $p_{k\ell}$ of the MC. However, it is worth noting that since the sum of all entries of the k-th row of $\mathbf{P}(\boldsymbol{\xi})$ must equal 1, i.e.,

$$\sum_{\ell=1}^{K} p_{k\ell}(\boldsymbol{\xi}) = 1 \,, \tag{13.82}$$

it is not possible to let one element of the the k-th row of $\mathbf{P}$ vary without varying at least another one.

The maximization of $Q_T(\boldsymbol{\xi}|\boldsymbol{\theta}')$ under constraints (13.81) can be formulated by using the method of Lagrange multipliers. Let λ_k with $1 \leq k \leq K$ denote the multiplier corresponding to constraint (13.82). Then consider the Lagrangian

$$L(\boldsymbol{\xi}, \boldsymbol{\lambda}|\boldsymbol{\theta}') \triangleq Q_T(\boldsymbol{\xi}|\boldsymbol{\theta}') + \sum_{k=1}^{K} \lambda_k (1 - \sum_{\ell=1}^{K} p_{k\ell}(\boldsymbol{\xi})) \,, \tag{13.83}$$

where $\boldsymbol{\lambda} \in \mathbb{R}^K$ is the vector with entries λ_k for $1 \leq k \leq K$. Then suppose that $p_{k\ell}$ is one entry of $\boldsymbol{\xi}$. To maximize L with respect to $p_{k\ell}$, we set

$$\frac{\partial L}{\partial p_{k\ell}} = \frac{1}{p_{k\ell}} \sum_{t=1}^{N} \alpha(k,t,\boldsymbol{\theta}') m_{k\ell}(t,\boldsymbol{\theta}') \beta(\ell,t+1,\boldsymbol{\theta}')/\gamma(\boldsymbol{\theta}') - \lambda_k \tag{13.84}$$

equal to zero. This gives

$$\hat{p}_{k\ell} = \frac{1}{\lambda_k} \sum_{t=1}^{N} \alpha(k,t,\boldsymbol{\theta}') m_{k\ell}(t,\boldsymbol{\theta}') \beta(\ell,t+1,\boldsymbol{\theta}')/\gamma(\boldsymbol{\theta}') \,, \tag{13.85}$$

where the Lagrange multiplier λ_k needs to selected such that constraint (13.82) holds. For example, if all entries $p_{k\ell}$ of row k of $\mathbf{P}$ need to be estimated, we obtain

$$\begin{aligned} 1 &= \sum_{\ell=1}^{K} \hat{p}_{k\ell} \\ &= \frac{1}{\lambda_k \gamma(\boldsymbol{\theta}')} \sum_{t=1}^{N} \alpha(k,t,\boldsymbol{\theta}') \big[\sum_{\ell=1}^{K} m_{k\ell}(t,\boldsymbol{\theta}') \beta(\ell,t+1,\boldsymbol{\theta}') \big] \\ &= \frac{1}{\lambda_k \gamma(\boldsymbol{\theta}')} \sum_{t=1}^{N} \alpha(k,t,\boldsymbol{\theta}') \beta(k,t,\boldsymbol{\theta}') \end{aligned} \tag{13.86}$$

where, in going from the second to the third line, we have taken into account the backward propagation equation (12.133) for β. This yields

$$\lambda_k = \frac{1}{\gamma(\boldsymbol{\theta}')} \sum_{t=1}^{N} \alpha(k,t,\boldsymbol{\theta}') \beta(k,t,\boldsymbol{\theta}') \,. \tag{13.87}$$

When only some of the entries of row $\mathbf{P}$ need to be estimated, and the other ones are fixed, the Lagrange multiplier λ_k is obtained by replacing the estimated $p_{k\ell}$'s by (13.85) in constraint (13.82) while leaving the known $p_{k\ell}$'s equal to their value.

The EM iteration, which in the HMM context is usually called the Baum-Welch re estimation method, can therefore be summarized as follows. Suppose that at the k-th iteration, the current estimated parameter vector is $\boldsymbol{\theta}' = \boldsymbol{\theta}^k$. In the E-phase, the forward backward algorithm is used to evaluate the vectors $\boldsymbol{\alpha}(t,\boldsymbol{\theta}')$ and $\boldsymbol{\beta}(t,\boldsymbol{\theta}')$ for $1 \le t \le N+1$. These quantities are then used in the M-phase to evaluate $\hat{\boldsymbol{\phi}}$, $\hat{v}$ and $\hat{\boldsymbol{\xi}}$ using (13.75) (or (13.81) in the linear case), (13.76), and (13.85). Setting

$$\boldsymbol{\theta}^{k+1} = \begin{bmatrix} \hat{\boldsymbol{\xi}} \\ \hat{\boldsymbol{\theta}} \\ \hat{v} \end{bmatrix} \,, \tag{13.88}$$

we can proceed to the next EM iteration. The above procedure terminates when successive estimates are close to each other. Of course, as is the case for the EM algorithm in general, the iteration converges only to a local maximum of the likelihood function $L(\mathbf{y}|\boldsymbol{\theta})$ of the given data. To ensure that this local maximum is a global maximum, and thus the ML estimate of the parameter vector $\boldsymbol{\theta}$, the initial estimate $\boldsymbol{\theta}^0$ needs to be located sufficiently close to $\hat{\boldsymbol{\theta}}_{\text{ML}}$. In the case when not enough a-priori information is available to ensure a close starting point, it is advisable to apply the EM iteration with multiple starting points and select the parameter estimate achieving the largest value of the likelihood function.

State detection: The above discussion addresses only the estimation of parameter vector $\boldsymbol{\theta}$, but we are also interested in detecting the MC states. Since the forward-backward algorithm is a component of the EM iteration, it is easy to see that the pointwise MAP detection of the MC sequence $(x_t, 1 \leq t \leq N+1)$ is just a byproduct of the EM iteration itself. Let $\boldsymbol{\theta}_*$ denote the parameter vector evaluated at the last EM iteration. Then according to (12.126), the pointwise MAP estimate of the MC state $X(t)$ at time t is

$$\hat{x}_t^{\text{P}} = \arg \max_{1 \leq k \leq K} (\alpha(k, t, \boldsymbol{\theta}_*)\beta(k, t, \boldsymbol{\theta}_*)) , \tag{13.89}$$

so that the E-phase of the last iteration of the EM algorithm evaluates all the quantities needed to perform the pointwise detection of the Markov chain states for $1 \leq t \leq N+1$. Of course, to solve the MAP sequence detection problem, we would need to implement the Viterbi algorithm for the MC model corresponding to $\boldsymbol{\theta}_*$. Instead of performing this additional computation as an "add-on" to the already heavy computational load imposed by the EM algorithm, it is easier to implement the EM Viterbi algorithm described in the next section.

Adaptive implementation: It is worth noting that while the PSP algorithm is inherently adaptive, since the path estimates of the parameter vector are updated as survivor paths get extended, in the form presented above, the EM algorithm requires multiple forward-backward passes over the same data block. This is difficult to accomplish in situations where data needs to be processed in real time. For real-time applications, the EM algorithm can be implemented adaptively, by performing only a small number of iterations on each data block, and then using the final parameter vector estimate for the previous block as initial estimate for the next block. For situations where the parameters are either constant or vary very slowly, this scheme can be highly effective, particularly if it is initialized by a training phase. Instead of using a blockwise adaptive algorithm, it is also possible to implement the EM algorithm for HMMs on line, as described in [21].

13.4.2 EM Viterbi Detector

Let $\mathcal{U}$ denote the set of all MC trellis paths. According to (12.54)–(13.57), for $\mathbf{x}_u \in \mathcal{U}$, the mixed probability density/distribution of observation vector $\mathbf{Y}$

and path $\mathbf{x}_u$ can be written as

$$f_c(\mathbf{y}, \mathbf{x}_u|\boldsymbol{\theta}) = \frac{1}{(2\pi v)^{N/2}} \exp\big(-\frac{1}{2v} J(\mathbf{x}_u|\boldsymbol{\theta})\big) \tag{13.90}$$

with

$$J(\mathbf{x}_u|\boldsymbol{\theta}) = \sum_{t=1}^{N} c(x_t^u, x_{t+1}^u, t, \boldsymbol{\theta}) , \tag{13.91}$$

where

$$c(x_t^u, x_{t+1}^u, t, \boldsymbol{\theta}) = (y_t - h(x_t^u, x_{t+1}^u, \boldsymbol{\phi}))^2 - 2v \ln(p_{x_t^u x_{t+1}^u}(\boldsymbol{\xi})) \tag{13.92}$$

denotes the branch metric of the Viterbi algorithm for the MC model with parameter vector $\boldsymbol{\theta}$. From (13.90)–(13.91), we find that the a-posteriori probability

$$p_u(\theta) = P[\mathbf{X} = \mathbf{x}_u | \mathbf{y}, \theta] \tag{13.93}$$

of path $\mathbf{x}_u$ given observation vector $\mathbf{Y}$ can be expressed as

$$p_u(\boldsymbol{\theta}) = \frac{1}{Z(\boldsymbol{\theta})} \exp\big(-\frac{1}{2v} J(\mathbf{x}_u|\boldsymbol{\theta})\big) \tag{13.94}$$

where

$$Z(\boldsymbol{\theta}) \triangleq \sum_{u=1}^{U} \exp\big(-\frac{1}{2v} J(\mathbf{x}_u|\boldsymbol{\theta})\big) . \tag{13.95}$$

In this expression, $U = |\mathcal{U}|$ denotes the total number of trellis paths. Since the a-posteriori path probability depends exponentially on the overall path metric $J(\mathbf{x}_u|\boldsymbol{\theta})$, paths with lower metric have a much higher probability than other paths, and in fact except for a small number of paths, most paths have a negligible probability.

E-phase: Substituting (13.93) inside expression (13.59) for the complete data estimated likelihood function $Q(\boldsymbol{\theta}|\boldsymbol{\theta}')$ yields the additive decomposition (13.69) where the observation and state transition components are given by

$$Q_O(\boldsymbol{\phi}, v|\boldsymbol{\theta}') = -\frac{1}{2v} \sum_{u=1}^{U} \sum_{t=1}^{N} (y_t - h(x_t^u, x_{t+1}^u, \boldsymbol{\phi}))^2 p_u(\boldsymbol{\theta}') - \frac{N}{2} \ln(v) \tag{13.96}$$

and

$$Q_T(\boldsymbol{\theta}|\boldsymbol{\theta}') = \sum_{u=1}^{U} \sum_{t=1}^{N} \ln(p_{x_t^u x_{t+1}^u}(\boldsymbol{\xi})) p_u(\boldsymbol{\theta}') . \tag{13.97}$$

M-phase: The maximization of the estimated complete data likelihood function $Q(\boldsymbol{\theta}|\boldsymbol{\theta}')$ can be accomplished in the same way as in the Baum-Welch approach. Due to the additive form of decomposition (13.69), the maximization of $Q(\boldsymbol{\theta}|\boldsymbol{\theta}')$ splits into separate maximizations of $Q_O(\boldsymbol{\phi}, v|\boldsymbol{\theta}')$ and $Q_T(\boldsymbol{\xi}|\boldsymbol{\theta}')$

with respect to $(\boldsymbol{\phi}, v)$ and $\boldsymbol{\xi}$, respectively. The pair $(\hat{\boldsymbol{\phi}}, \hat{v})$ maximizing Q_O is given by (13.75)–(13.76), where $W(\boldsymbol{\phi}|\boldsymbol{\theta}')$ is now expressed as

$$W(\boldsymbol{\phi}|\boldsymbol{\theta}') = \sum_{u=1}^{U} \sum_{t=1}^{N} (y_t - h(x_t^u, x_{t+1}^u, \boldsymbol{\phi}))^2 p_u(\boldsymbol{\theta}') \tag{13.98}$$

Comparing this expression with (13.16), we see that

$$W(\boldsymbol{\phi}|\boldsymbol{\theta}') = \sum_{u=1}^{U} J_u(\boldsymbol{\phi}) p_u(\boldsymbol{\theta}') \tag{13.99}$$

can be viewed as a weighted sum of the path mismatch functions $J_u(\boldsymbol{\phi})$ which are used by the GLRT to obtain a path parameter vector estimate $\hat{\boldsymbol{\phi}}_u$ for each trellis path $\mathbf{x}_u$. In (13.99), the weight applied to mismatch function $J_u(\boldsymbol{\phi})$ is the probability of path $\mathbf{x}_u$ based on the current model vector $\boldsymbol{\theta}'$. Thus, instead of performing a nonlinear least-squares fit for each path like the GLR test, the EM algorithm performs a single nonlinear least-squares fit for the composite mismatch function W obtained by weighting the squared mismatch of each path by the path probability.

When the output map $h(x_t, x_{t+1}, \boldsymbol{\phi})$ admits the linear parametrization (13.6), W can be written as

$$W(\boldsymbol{\theta}|\boldsymbol{\phi}) = \sum_{u=1}^{U} ||\mathbf{y} - \mathbf{A}_u \boldsymbol{\phi}||_2^2 p_u(\boldsymbol{\theta}') , \tag{13.100}$$

where the matrix $\mathbf{A}_u$ is defined in (13.9). Since W is a weighted sum of quadratic terms in $\boldsymbol{\phi}$, its minimum can be expressed in closed form as

$$\hat{\boldsymbol{\phi}} = \Big(\sum_{u=1}^{U} \mathbf{A}_u^T \mathbf{A}_u p_u(\boldsymbol{\theta}') \Big)^{-1} \Big(\sum_{u=1}^{U} \mathbf{A}_u^T p_u(\boldsymbol{\theta}') \Big) \mathbf{y} . \tag{13.101}$$

Next, we consider the maximization of $Q_T(\boldsymbol{\xi}|\boldsymbol{\theta}')$ with respect to $\boldsymbol{\xi}$. Let $N_{k\ell}(\mathbf{x}_u)$ be the number of transitions from state k to state ℓ observed in the sequence $\mathbf{x}_u = (x_t^u,\ 1 \leq t \leq N+1)$, and denoted by

$$q_{k\ell}^u \triangleq \frac{N_{k\ell}(\mathbf{x}_u)}{N} \tag{13.102}$$

the corresponding empirical probability distribution over $\mathcal{S}^2$. Let also $N_k(\mathbf{x}_u)$ represent the number of times state k appears in the truncated sequence $(x_t^u,\ 1 \leq t \leq N)$ obtained by dropping the last state from sequence $\mathbf{x}_u$, and write as

$$q_k^u \triangleq \frac{N_k(\mathbf{x}_u)}{N} \tag{13.103}$$

the empirical probability distribution of the MC based on the truncation of $\mathbf{x}_u$. Note that

$$N_k(\mathbf{x}_u) = \sum_{\ell=1}^{K} N_{k\ell}(\mathbf{x}_u) , \tag{13.104}$$

so

$$q_k^u = \sum_{\ell=1}^{K} q_{k\ell}^u \tag{13.105}$$

is just the marginal distribution corresponding to the joint empirical distribution $q_{k\ell}^u$ for two consecutive MC states. Given the above pathwise empirical distributions, let

$$\begin{aligned} \bar{q}_{k\ell}(\boldsymbol{\theta}') &= \sum_{u=1}^{U} q_{k\ell}^u p_u(\boldsymbol{\theta}') \\ \bar{q}_k(\boldsymbol{\theta}') &= \sum_{u=1}^{U} q_k^u p_u(\boldsymbol{\theta}') \end{aligned} \tag{13.106}$$

denote the averaged empirical probability distributions obtained by weighting the empirical distribution corresponding to each path by the path probability $p_u(\boldsymbol{\theta}')$ based on the current parameter vector $\boldsymbol{\theta}'$. Note that identity (13.105) implies that $\bar{q}_k(\boldsymbol{\theta}')$ is the marginal distribution corresponding to $\bar{q}_{k\ell}(\boldsymbol{\theta}')$.

Then, expression (13.97) for $Q_T(\boldsymbol{\xi}|\boldsymbol{\theta}')$ can be rewritten as

$$\begin{aligned} Q_T(\boldsymbol{\xi}|\boldsymbol{\theta}') &= \sum_{k=1}^{K}\sum_{\ell=1}^{K} \Big(\sum_{u=1} N_{k\ell}(\mathbf{x}_u) p_u(\boldsymbol{\theta}')\Big) \ln(p_{k\ell}(\boldsymbol{\xi})) \\ &= N \sum_{k=1}^{K}\sum_{\ell=1}^{K} \bar{q}_{k\ell}(\boldsymbol{\theta}') \ln(p_{k\ell}(\boldsymbol{\xi})) \,. \end{aligned} \tag{13.107}$$

Assume that $p_{k\ell}$ is an entry of $\boldsymbol{\xi}$, so that it needs to be estimated. Substituting (13.103) inside the Lagrangian $L(\boldsymbol{\xi}, \boldsymbol{\lambda}|\boldsymbol{\theta}')$ defined in (13.83) and taking the partial derivative of L with respect to $p_{k\ell}$ gives

$$\frac{\partial L}{\partial p_{k\ell}} = N \frac{\bar{q}_{k\ell}(\boldsymbol{\theta}')}{p_{k\ell}} - \lambda_k \,. \tag{13.108}$$

Setting the derivative equal to zero, we obtain as estimate

$$\hat{p}_{k\ell} = \frac{N}{\lambda_k} \bar{q}_{k\ell}(\boldsymbol{\theta}') , \tag{13.109}$$

where the Lagrange multiplier λ_k is selected such that the sum of the entries of row k of $\mathbf{P}(\boldsymbol{\xi})$ equals 1, as indicated in (13.92). When all entries of row k need to be estimated, we find

$$\lambda_k = N\bar{q}_k(\boldsymbol{\theta}') . \tag{13.110}$$

Thus in this case

$$\hat{p}_{k\ell} = \frac{\bar{q}_{k\ell}(\boldsymbol{\theta}')}{\bar{q}_k(\boldsymbol{\theta}')} \tag{13.111}$$

is just the conditional distribution evaluated from the averaged empirical probability distribution $\bar{q}_{k\ell}(\boldsymbol{\theta}')$.

EM Viterbi approximation: Up to this point it is worth observing that all expressions obtained for $Q_O(\boldsymbol{\phi}, v|\boldsymbol{\theta}')$, $Q_T(\boldsymbol{\xi}|\boldsymbol{\theta}')$, and the variables $\hat{\boldsymbol{\phi}}$, $\hat{v}$ and $\hat{p}_{k\ell}$ maximizing the estimated complete likelihood function are *exact*. However, one manifest weakness of the pathwise-based method for implementing the EM iteration is that the number U of paths is huge, so that the proposed computations are not achievable. The approximation employed in the EM Viterbi algorithm (EMVA) consists of observing that the probability $p_u(\boldsymbol{\theta}')$ of almost all paths is negligible and most of the probability mass is concentrated over a very small number of paths. This suggests truncating the sum by retaining only the Viterbi survivors in all summations. Specifically, let $\hat{x}(i, \boldsymbol{\theta}')$ denote the survivor path terminating in state i at time $N+1$ when the Viterbi algorithm based on model parameter $\boldsymbol{\theta}'$ is applied to observation sequence $\mathbf{y}$ and let

$$V(i, \boldsymbol{\theta}') = J(\hat{\mathbf{x}}(i, \boldsymbol{\theta}')) \tag{13.112}$$

be the corresponding path metric. Then the EMVA assigns probability

$$p_i(\boldsymbol{\theta}') = \frac{1}{Z(\boldsymbol{\theta}')} \exp(-V(i, \boldsymbol{\theta}')) , \tag{13.113}$$

with

$$Z(\boldsymbol{\theta}') = \sum_{i=1}^{K} \exp(-V(i, \boldsymbol{\theta}')) \tag{13.114}$$

to the i-th survivor path. This probability can then be used to obtain approximations for $W(\boldsymbol{\phi}|\boldsymbol{\theta}')$, $Q_T(\boldsymbol{\xi}|\boldsymbol{\theta}')$, and the maximizing variables $\hat{\boldsymbol{\phi}}$, $\hat{v}$ and $\hat{p}_{k\ell}$. In all these approximations, the summation over all trellis paths is replaced by a summation over all survivor paths, where the i-th survivor path probability is given by (13.113).

Thus the EMVA employs a trellis pruning strategy similar to the GLRT since it retains only the Viterbi survivors. However, while the GLRT constructs a model for each survivor, the EMVA iteration generates a single model at each iteration. This is accomplished as follows: Suppose that at the k-th iteration, the current model is $\boldsymbol{\theta}^k$. In the E-phase, the Viterbi algorithm based on $\boldsymbol{\theta}^k$ generates survivors $\hat{\mathbf{x}}(i, \boldsymbol{\theta}^k)$ and their metrics $V(i, \boldsymbol{\theta}^k)$. Note that since we have assumed that the MC graph does not depend on parameter vector $\boldsymbol{\xi}$, the same trellis is used to run the Viterbi algorithm at each iteration. The path metrics $V(i, \boldsymbol{\theta}^k)$ specify path probabilities $p_i(\boldsymbol{\theta}^k)$, which are then used to evaluate $W(\boldsymbol{\phi}|\boldsymbol{\theta}^k)$ and $Q_T(\boldsymbol{\xi}|\boldsymbol{\theta}^k)$ using expressions (13.99) and

(13.107). In the M-phase, $\hat{\phi}$, $\hat{v}$ and $\hat{\boldsymbol{\xi}}$ are obtained by using (13.75) (or (13.101) in the linear case), (13.76), and (13.111) with $\boldsymbol{\theta}' = \boldsymbol{\theta}^k$. The parameter vector $\boldsymbol{\theta}^{k+1}$ for the next iteration is then specified by (13.88).

Note that since the EMVA evaluates the Viterbi survivors at each iteration, the MAP sequence $\hat{\mathbf{x}}_{\text{MAP}}$ is obtained by selecting the survivor path with the smallest metric at the last EM iteration. Provided the EM iteration has been initialized properly, the EM iteration converges to the ML estimate $\hat{\boldsymbol{\theta}}_{\text{ML}}$ of $\boldsymbol{\theta}$ based on observation vector $\mathbf{Y}$, so the detected sequence

$$\hat{\mathbf{x}}_{\text{MAP}} = \arg\max_{\mathbf{x}\in\mathcal{U}} P[\mathbf{x}|\mathbf{y}, \hat{\boldsymbol{\theta}}_{\text{ML}}] \tag{13.115}$$

solves the MAP sequence detection problem for vector $\hat{\boldsymbol{\theta}}_{\text{ML}}$. Since under most circumstances $\hat{\theta}_{\text{ML}}$ converges almost surely to the true vector $\boldsymbol{\theta}$ as the length N of the data block goes to infinity, we conclude that for large N, the EMVA detector will perform almost as well as the Viterbi detector for an MC with known parameters.

Comparison: The key features of the forward-backward EM and EMVA detectors can be summarized as follows:

(a) Both the forward-backward EM detector and EMVA perform the MC state detection as part of the EM iteration. The forward-backward iteration implements a pointwise detector, while the EMVA yields a sequence detector.
(b) The forward-backward EM iteration is exact, whereas the EMVA relies on a truncation of the set $\mathcal{U}$ of trellis paths which retains only the survivor paths of the Viterbi algorithm for the current model.
(c) The forward-backward EM scheme has a higher complexity, since it requires $O(KTN)$ operations per iteration, where T denotes the number of transitions out of each state (for simplicity we assume that T is the same for all states). In contrast, the EMVA has a complexity proportional to KN only per iteration.
(d) Since it implements a pointwise detector, the forward-backward EM iteration evaluates the soft information

$$P[X(t) = k|\mathbf{Y}, \hat{\boldsymbol{\theta}}_{\text{ML}}] \tag{13.116}$$

so it is well suited for iterative detection applications where information needs to be exchanged between an equalizer and an error-correction decoder. By comparison, to generate soft information about each MC state, the EMVA would need to run a soft Viterbi algorithm based on the model obtained at the last iteration. This computation represents only a small added complexity, since it can be accomplished by running the Viterbi algorithm *backwards* and then using the min-log-MAP expression (12.150) to combine the value functions of forward and backward Viterbi runs to generate an estimate of the pointwise a-posteriori probability (13.116).

Remark: Since all trellis paths ultimately merge at a certain depth, for long data blocks, instead of performing a nonlinear least-squares fit for the aggregate function obtained by weighting the nonlinear mismatches of each survivor path by the path probability, we could instead just perform a single nonlinear least-squares fit for the MAP path corresponding to the current parameter $\boldsymbol{\theta}'$. Note that since all survivors merge into the MAP path at a certain depth, for long blocks the mismatch functions for survivor paths differ from the mismatch function for the MAP path only by the contribution of the pre merged segments. If we neglect this relatively small difference, this yields an "EM-like" algorithm consisting of running the Viterbi algorithm for the current model $\boldsymbol{\theta}^k$ and then performing a nonlinear fit of the MAP path to generate the next model vector $\boldsymbol{\theta}^{k+1}$. An algorithm of this type was proposed for blind equalization in [22, 23]. This algorithm can be viewed as performing a joint maximization of the conditional likelihood function $L(\mathbf{y}|\mathbf{x}, \boldsymbol{\theta})$ over $\mathbf{x}$ and $\boldsymbol{\theta}$, where maximizations over $\mathbf{x}$ and $\boldsymbol{\theta}$ are conducted in alternance. Hence, it can be described as a maximization-maximization Viterbi algorithm (MMVA) [13], in contrast to the EM algorithm which maximizes $L(\mathbf{y}|\boldsymbol{\theta})$. A comparison between the EMVA and MMVA algorithms was performed in [13], where it was shown that both algorithms have similar performace for long data blocks, but EMVA outperforms MMVA for short data blocks. This is expected, since for short blocks the lengths of the unmerged parts of the survivor paths cannot be neglected compared to the length of the merged part. Also, the MMVA has a higher bit error rate floor at high SNR.

13.5 Example: Blind Equalization

To illustrate the EM algorithm, consider the blind equalization problem described in Example 13.2. We observe

$$Y(t) = \sum_{k=0}^{L} h_k I(t-k) + V(t) \tag{13.117}$$

for $1 \leq t \leq N$, where the symbols $I(t)$ are indendent and take the values ± 1 with probability $1/2$ each. The initial symbols

$$I(0) = \cdots = I(1-L) = -1 ,$$

and $V(t)$ is a zero-mean WGN with variance v which is independent of symbols $I(t)$. We assume that both the channel impulse response h_k, $0 \leq k \leq L$ and the noise variance v are unknown. We can construct a Markov chain model of the observations by employing state mapping (12.163). Since each symbol $I(t)$ can take two values and we observe a symbol block of length N, the state trellis has $U = 2^N$ paths, each with a-priori probability 2^{-N}. The observations can be written in block form as

$$\mathbf{Y} = \mathbf{A}(\mathbf{I})\boldsymbol{\phi} + \mathbf{V} \tag{13.118}$$

where vectors $\mathbf{Y}$ and $\mathbf{V}$ are defined in (13.7), and since $V(t)$ is WGN, $\mathbf{v}$ is $N(\mathbf{0}, v\mathbf{I}_N)$ distributed. Here

$$\boldsymbol{\phi} = \begin{bmatrix} h_0 \ h_1 \ \dots \ h_L \end{bmatrix}^T \tag{13.119}$$

is the vector representing the channel impulse response (CIR), and $\mathbf{A}(\mathbf{I})$ is the $N \times (L+1)$ Toeplitz matrix

$$\mathbf{A}(\mathbf{I}) = \begin{bmatrix} I(1) & -1 & & -1 \\ I(2) & I(1) & -1 & -1 \\ & \ddots & \ddots & -1 \\ I(L) & I(L-1) & & I(1) \\ \vdots & \vdots & & \vdots \\ I(N) & I(N-1) & & I(N-L) \end{bmatrix} \tag{13.120}$$

where the upper triangle of -1 entries is due to the fact that the initial symbols are assumed to be all -1. Note that if the final L symbols are zero, which is not assumed here, the matrix $\mathbf{A}(\mathbf{I})$ would also include a lower triangle of -1 entries. We recall that a matrix has a Toeplitz structure if its entries are constant along diagonals. This structure just reflects the time-invariance of the channel model (13.117). Note that the matrix $\mathbf{A}(\mathbf{I})$ is completely specified by the symbol sequence $\mathbf{I}$ appearing in its first column. In the following, we let $\mathbf{I}_u = (I_u(t),\ 1 \leq t \leq N)$ denote the symbol sequence corresponding to trellis path $\mathbf{x}_u$ and let $\mathbf{A}_u = A(\mathbf{I}_u)$ represent the corresponding Toeplitz matrix.

Then, let $\boldsymbol{\theta}$ be the parameter vector obtained by appending the noise variance v to vector $\boldsymbol{\phi}$ as shown in (13.4). Given an arbitrary trellis path $\mathbf{x}_u$, the joint probability distribution of observation vector $\mathbf{Y}$ and path $\mathbf{x}_u$ is given by

$$f_c(\mathbf{y}, \mathbf{x}_u|\boldsymbol{\theta}) = \frac{1}{2^N (2\pi v)^{N/2}} \exp\Big(-\frac{1}{2v} J(\mathbf{x}_u|\boldsymbol{\phi})\Big) , \tag{13.121}$$

where

$$J(\mathbf{x}_u|\boldsymbol{\phi}) = ||\mathbf{y} - \mathbf{A}_u \boldsymbol{\phi}||_2^2 . \tag{13.122}$$

The a-posteriori probability $p_u(\boldsymbol{\theta})$ of path $\mathbf{x}_u$ is given by (13.94), except that the path cost function $J(\mathbf{x}_u|\boldsymbol{\phi})$ specified by (13.122) depends on $\boldsymbol{\phi}$ alone.

Then, if the parameter vector available at the beginning of of the k-th EM iteration is $\boldsymbol{\theta}^k$, the complete data estimated log-likelihood function can be expressed as

$$Q(\boldsymbol{\theta}|\boldsymbol{\theta}^k) = -\frac{1}{2v} \sum_{u=1}^{U} ||\mathbf{y} - \mathbf{A}_u \boldsymbol{\phi}||_2^2 p_u(\boldsymbol{\theta}^k) - \frac{N}{2} \ln(v) , \tag{13.123}$$

where we have dropped terms independent of $\boldsymbol{\theta}$. Maximizing this expression over $\boldsymbol{\phi}$ and v gives

$$\boldsymbol{\phi}^{k+1} = \Big(\sum_{u=1}^{U} \mathbf{A}_u^T \mathbf{A}_u p_u(\boldsymbol{\theta}^k)\Big)^{-1} \Big(\sum_{u=1}^{U} \mathbf{A}_u^T p_u(\boldsymbol{\theta}^k)\Big)\mathbf{y} \tag{13.124}$$

and

$$v_{k+1} = \frac{1}{N} \sum_{u=1}^{U} ||\mathbf{y} - \mathbf{A}_u \boldsymbol{\phi}^{k+1}||_2^2 \, p_u(\boldsymbol{\theta}^k) \tag{13.125}$$

as the next iterates for the CIR and noise variance.

13.5.1 Convergence Analysis

To analyze the EM iteration, it is convenient to replace update equation (13.125) by

$$v_{k+1} = \frac{1}{N} \sum_{u=1}^{N} ||\mathbf{y} - \mathbf{A}_u \boldsymbol{\phi}^{k}||_2^2 p_u(\boldsymbol{\theta}^k)\,. \tag{13.126}$$

This simplification is similar to the one employed in (11.31) to analyze the EM iteration for estimating the amplitude of a Gaussian signal in WGN of unknown variance. Because $\boldsymbol{\phi}$ is set equal to the current channel $\boldsymbol{\phi}^k$ while performing the maximization of $Q(\boldsymbol{\theta}|\boldsymbol{\theta}^k)$ over v, the iteration formed by (13.124) and (13.126) corresponds to a coordinate-wise maximization of Q which is usually called the AECM algorithm [24],

Consider the probability density

$$f(\mathbf{y}|\boldsymbol{\theta}) = \sum_{u=1}^{U} f_c(\mathbf{y}, \mathbf{x}_u|\boldsymbol{\theta}) = \frac{Z(\boldsymbol{\theta})}{2^N (2\pi v)^{N/2}} \tag{13.127}$$

of observations $\mathbf{Y}$, where the partition function $Z(\boldsymbol{\theta})$ is given by (13.95). Taking the logarithm of this expression yields the likelihood function

$$L(\mathbf{y}|\boldsymbol{\theta}) = \ln Z(\boldsymbol{\theta}) - \frac{N}{2} \ln(v) + c\,, \tag{13.128}$$

where c is a constant independent of $\boldsymbol{\theta}$. Then the gradient of $L(\mathbf{y}|\boldsymbol{\theta})$ with respect to $\boldsymbol{\phi}$ can be expressed as

$$\begin{aligned}
\nabla_{\boldsymbol{\phi}} L(\mathbf{y}|\boldsymbol{\phi}) &= \frac{1}{Z(\boldsymbol{\theta})} \sum_{u=1}^{U} \frac{1}{v} \mathbf{A}_u^T (\mathbf{y} - \mathbf{A}_u \boldsymbol{\phi}) \exp\Big(-\frac{1}{2v}||\mathbf{y} - \mathbf{A}_u \boldsymbol{\phi}||_2^2\Big) \\
&= \frac{1}{v} \sum_{u=1}^{U} \mathbf{A}_u^T (\mathbf{y} - \mathbf{A}_u \boldsymbol{\phi}) p_u(\boldsymbol{\theta}) \\
&= \frac{1}{v} \Big[\Big(\sum_{u=1}^{U} \mathbf{A}_u^T p_u(\boldsymbol{\theta})\Big)\mathbf{y} - \Big(\sum_{u=1}^{U} \mathbf{A}_u^T \mathbf{A}_u p_u(\boldsymbol{\theta})\Big)\boldsymbol{\phi}\Big]\,.
\end{aligned} \tag{13.129}$$

Similarly, the partial derivative of $L(\mathbf{y}|\boldsymbol{\theta})$ with respect to v can be expressed as

$$\begin{aligned}\frac{\partial}{\partial v}L(\mathbf{y}|\boldsymbol{\theta}) &= -\frac{N}{2v} + \frac{1}{Z(\boldsymbol{\theta})}\sum_{u=1}^{U}\frac{1}{2v^2}||\mathbf{y}-\mathbf{A}_u\boldsymbol{\phi}||_2^2 \exp\Big(-\frac{1}{2v}||\mathbf{y}-\mathbf{A}_u\boldsymbol{\phi}||_2^2\Big) \\ &= -\frac{N}{2v} + \frac{1}{2v^2}\sum_{u=1}^{U}||\mathbf{y}-\mathbf{A}_u\boldsymbol{\phi}||_2^2 p_u(\boldsymbol{\theta}) \,. \end{aligned} \quad (13.130)$$

Evaluating these expressions at $\boldsymbol{\theta} = \boldsymbol{\theta}^k$ and taking into account iterations (13.124)–(13.125) gives

$$\nabla_{\boldsymbol{\phi}}L(\mathbf{y}|\boldsymbol{\theta}^k) = \frac{1}{v_k}\Big(\sum_{u=1}^{U}\mathbf{A}_u^T\mathbf{A}_u p_u(\boldsymbol{\theta}^k)\Big)[\boldsymbol{\phi}^{k+1}-\boldsymbol{\phi}^k] \quad (13.131)$$

and

$$\frac{\partial}{\partial v}L(\mathbf{y}|\boldsymbol{\theta}^k) = \frac{N}{2v_k^2}(v_{k+1}-v_k). \quad (13.132)$$

Thus, if

$$\nabla_{\boldsymbol{\theta}} \triangleq \begin{bmatrix}\nabla_{\boldsymbol{\phi}} \\ \frac{\partial}{\partial v}\end{bmatrix}$$

denotes the gradient with respect to vector $\boldsymbol{\theta}$, the EM iteration (13.124)–(13.125) can be rewritten as

$$\boldsymbol{\theta}^{k+1} = \boldsymbol{\theta}^k + \mathbf{D}(\boldsymbol{\theta}^k)\nabla_{\boldsymbol{\theta}}L(\mathbf{y}|\boldsymbol{\theta}^k) \quad (13.133)$$

with

$$\mathbf{D}(\boldsymbol{\theta}^k) = \frac{1}{N}\begin{bmatrix}\mathbf{W}_k & 0 \\ 0 & 2v_k^2\end{bmatrix}, \quad (13.134)$$

where

$$\mathbf{W}_k^{-1} \triangleq \frac{1}{Nv_k}\sum_{u=1}^{U}\mathbf{A}_u^T\mathbf{A}_u p_u(\boldsymbol{\theta}^k) \,. \quad (13.135)$$

These expressions are the counterpart for the blind equalization problem of identities (11.34)–(11.36) for the EM iteration to find the amplitude of a zero-mean Gaussian process in a WGN of unknown variance. They indicate that the EM iteration for the blind equalization problem can be viewed as a quasi-Newton minimization scheme. Assume there exists a constant C such that $||\mathbf{D}(\boldsymbol{\theta}^k)|| < C$ for all k, where the matrix norm $||\mathbf{D}||$ denotes the maximum singular value of matrix $\mathbf{D}$. Then it is shown in [13] that there exists a capture set for any local maximum of the likelihood function $L(\mathbf{y}|\boldsymbol{\theta})$. Specifically, if $\boldsymbol{\theta}_*$ is a local maximum of $L(\mathbf{y}|\boldsymbol{\theta}_*)$, there exists an open set $\mathcal{S}$ containing $\boldsymbol{\theta}_*$ such that if $\boldsymbol{\theta}^0$ is in $\mathcal{S}$, then $\boldsymbol{\theta}^k$ remains in $\mathcal{S}$ for all k and converges to $\boldsymbol{\theta}_*$ as $k \to \infty$. For the case when $\boldsymbol{\theta}_*$ corresponds to the ML estimate (the global maximum of the likelihood function), this means that if the initial estimate $\boldsymbol{\theta}^0$ is close enough to $\hat{\boldsymbol{\theta}}_{\mathrm{ML}}$, then the iterates $\boldsymbol{\theta}^k$ stay close and converge to $\hat{\boldsymbol{\theta}}_{\mathrm{ML}}$ as k tends to ∞, as desired.

Remark: Consider the $L+1$ dimensional vector

$$\mathbf{I}_t = \begin{bmatrix} I(t) \; I(t-1) \ldots I(t-L) \end{bmatrix}^T$$

formed by the symbol $I(t)$ and its L predecessors. By taking into account the Toeplitz structure of $\mathbf{A}_u = \mathbf{A}(\mathbf{I}_u)$ defined by (13.120), it is easy to verify that

$$\mathbf{W}_k^{-1} = \frac{1}{N v_k} \sum_{t=1}^{N} E[\mathbf{I}_t \mathbf{I}_t^T \mid \mathbf{Y}, \boldsymbol{\theta}_k] \, ,$$

which provides an interpretation of $\mathbf{W}_k^{-1}$ similar to the one appearing on the second line of (11.36) for the Gaussian case.

13.5.2 Convergence Rate

The EM iteration for the blind equalization problem can be expressed as

$$\boldsymbol{\theta}^{k+1} = \mathbf{m}(\boldsymbol{\theta}^k) \, , \tag{13.136}$$

where the nonlinear mapping $\mathbf{m}(\boldsymbol{\theta})$ can be decomposed as

$$\boldsymbol{m}(\boldsymbol{\theta}) = \begin{bmatrix} \mathbf{m}_\phi(\boldsymbol{\theta}) \\ m_v(\boldsymbol{\theta}) \end{bmatrix} \tag{13.137}$$

with

$$\begin{aligned} \mathbf{m}_\phi(\boldsymbol{\theta}) &= \Big(\sum_{u=1}^{U} \mathbf{A}_u \mathbf{A}_u^T p_u(\boldsymbol{\theta}) \Big)^{-1} \Big(\sum_{u=1}^{U} \mathbf{A}_u^T p_u(\boldsymbol{\theta}) \Big) \mathbf{y} \\ &= \Big(\sum_{u=1}^{U} \mathbf{A}_u \mathbf{A}_u^T \tilde{p}_u(\boldsymbol{\theta}) \Big)^{-1} \Big(\sum_{u=1}^{U} \mathbf{A}_u^T \tilde{p}_u(\boldsymbol{\theta}) \Big) \mathbf{y} \, , \end{aligned} \tag{13.138}$$

and

$$\begin{aligned} m_v(\boldsymbol{\theta}) &= \frac{1}{N} \sum_{u=1}^{U} ||\mathbf{y} - \mathbf{A}_u \boldsymbol{\phi}||_2^2 p_u(\boldsymbol{\theta}) \\ &= \frac{1}{N Z(\boldsymbol{\theta})} \sum_{u=1}^{U} ||\mathbf{y} - \mathbf{A}_u \boldsymbol{\phi}||_2^2 \tilde{p}_u(\boldsymbol{\theta}) \, . \end{aligned} \tag{13.139}$$

Here

$$\tilde{p}_u(\boldsymbol{\theta}) = \exp(-\frac{1}{2v} ||\mathbf{y} - \mathbf{A}_u \boldsymbol{\phi}||_2^2) \, , \tag{13.140}$$

denotes the unnormalized a-posteriori probability of path $\mathbf{x}_u$,

$$Z(\boldsymbol{\theta}) = \sum_{u=1}^{U} \tilde{p}_u(\boldsymbol{\theta}) \tag{13.141}$$

represents the normalization constant, and

$$p_u(\boldsymbol{\theta}) = \frac{\tilde{p}_u(\boldsymbol{\theta})}{Z(\boldsymbol{\theta})} \tag{13.142}$$

is the normalized a-posteriori probability.

Then, as explained in Section 11.3.3, the rate of convergence of the EM iteration in the vicinity of a stationary point is specified by the spectral radius of the local iteration matrix

$$\begin{aligned}\mathbf{F}(\boldsymbol{\theta}) = \nabla_{\boldsymbol{\theta}}^T \mathbf{m}(\boldsymbol{\theta}) &= \begin{bmatrix} \nabla_{\boldsymbol{\phi}}^T \mathbf{m}_{\boldsymbol{\phi}}(\boldsymbol{\theta}) & \frac{\partial}{\partial v} \mathbf{m}_{\boldsymbol{\phi}}(\boldsymbol{\theta}) \\ \nabla_{\boldsymbol{\phi}}^T m_v(\boldsymbol{\theta}) & \frac{\partial}{\partial v} m_v(\boldsymbol{\theta}) \end{bmatrix} \\ &= \begin{bmatrix} \mathbf{F}_{11}(\boldsymbol{\theta}) & \mathbf{f}_{12}(\boldsymbol{\theta}) \\ \mathbf{f}_{21}(\boldsymbol{\theta}) & f_{22}(\boldsymbol{\theta}) \end{bmatrix} .\end{aligned} \tag{13.143}$$

By applying $\boldsymbol{\nabla}_{\boldsymbol{\phi}}^T$ to both sides of

$$\Big(\sum_{u=1}^{U} \mathbf{A}_u^T \mathbf{A}_u \tilde{p}_u(\boldsymbol{\theta})\Big) \mathbf{m}(\boldsymbol{\phi}) = \Big(\sum_{u=1}^{U} \mathbf{A}_u^T \tilde{p}_u(\boldsymbol{\theta})\Big) \mathbf{y} \,, \tag{13.144}$$

regrouping terms, and solving for $\nabla_{\boldsymbol{\phi}}^T \mathbf{m}_{\boldsymbol{\phi}}$, we obtain

$$\begin{aligned}\mathbf{F}_{11}(\boldsymbol{\theta}) &= \nabla_{\boldsymbol{\phi}}^T \mathbf{m}_{\boldsymbol{\phi}}(\boldsymbol{\theta}) \\ &= \Big(\sum_{u=1}^{U} \mathbf{A}_u^T \mathbf{A}_u \tilde{p}_u(\boldsymbol{\theta})\Big)^{-1} \sum_{u=1}^{U} \mathbf{A}_u^T (\mathbf{y} - \mathbf{A}_u \mathbf{m}_{\boldsymbol{\phi}}(\boldsymbol{\theta})) \nabla_{\boldsymbol{\phi}}^T \tilde{p}_u(\boldsymbol{\theta}) \,.\end{aligned} \tag{13.145}$$

Proceeding in a similar manner we find

$$\begin{aligned}\mathbf{f}_{12}(\boldsymbol{\theta}) &= \frac{\partial}{\partial v} \mathbf{m}_{\boldsymbol{\phi}}(\boldsymbol{\theta}) \\ &= \Big(\sum_{u=1}^{U} \mathbf{A}_u^T \mathbf{A}_u \tilde{p}_u(\boldsymbol{\theta})\Big)^{-1} \sum_{u=1}^{U} \mathbf{A}_u^T (\mathbf{y} - \mathbf{A}_u \mathbf{m}_{\boldsymbol{\phi}}(\boldsymbol{\theta})) \frac{\partial}{\partial v} \tilde{p}_u(\boldsymbol{\theta}) \,.\end{aligned} \tag{13.146}$$

In these expressions we have

$$\nabla_{\boldsymbol{\phi}}^T \tilde{p}_u(\boldsymbol{\theta}) = \frac{1}{v} (\mathbf{y} - \mathbf{A}_u \boldsymbol{\phi})^T \mathbf{A}_u \tilde{p}_u(\boldsymbol{\theta}) \tag{13.147}$$

$$\frac{\partial}{\partial v} \tilde{p}_u(\boldsymbol{\theta}) = \frac{1}{2v^2} ||(\mathbf{y} - \mathbf{A}_u \boldsymbol{\phi})||_2^2 \tilde{p}_u(\boldsymbol{\theta}) \,. \tag{13.148}$$

Substituting (13.147) and (13.148) inside (13.145) and (13.148) gives

$$\mathbf{F}_{11}(\boldsymbol{\theta}) = \frac{1}{v}\Big(\sum_{u=1}^{U} \mathbf{A}_u^T \mathbf{A}_u p_u(\boldsymbol{\theta})\Big)^{-1} \cdot \sum_{u=1}^{U} \mathbf{A}_u^T (\mathbf{y} - \mathbf{A}_u \mathbf{m}_{\boldsymbol{\phi}}(\boldsymbol{\theta}))(\mathbf{y} - \mathbf{A}_u \boldsymbol{\phi})^T \mathbf{A}_u p_u(\boldsymbol{\theta}) \tag{13.149}$$

and

$$\mathbf{f}_{12}(\boldsymbol{\theta}) = \frac{1}{2v^2}\Big(\sum_{u=1}^{U} \mathbf{A}_u^T \mathbf{A}_u p_u(\boldsymbol{\theta})\Big)^{-1} \cdot \sum_{u=1}^{U} \mathbf{A}_u^T (\mathbf{y} - \mathbf{A}_u \mathbf{m}_{\boldsymbol{\phi}}(\boldsymbol{\theta}))||(\mathbf{y} - \mathbf{A}_u \boldsymbol{\phi})||_2^2 \mathbf{A}_u p_u(\boldsymbol{\theta})\,. \tag{13.150}$$

Similarly we have

$$\begin{aligned}
\mathbf{f}_{21}(\boldsymbol{\theta}) &= \nabla_{\boldsymbol{\phi}}^T m_v(\boldsymbol{\theta}) \\
&= \frac{1}{NZ(\boldsymbol{\theta})}\Big[\sum_{u=1}^{U} ||\mathbf{y} - \mathbf{A}_u \boldsymbol{\phi}||_2^2 \nabla_{\boldsymbol{\phi}}^T \tilde{p}_u(\boldsymbol{\theta}) - 2\sum_{u=1}^{U} (\mathbf{y} - \mathbf{A}_u \boldsymbol{\phi})^T \mathbf{A}_u \tilde{p}_u(\boldsymbol{\theta})\Big] \\
&\quad - \frac{1}{NZ^2(\boldsymbol{\theta})}\Big(\sum_{u=1}^{U} ||\mathbf{y} - \mathbf{A}_u \boldsymbol{\phi}||_2^2 \tilde{p}_u(\boldsymbol{\theta})\Big) \nabla_{\boldsymbol{\phi}}^T Z(\boldsymbol{\theta})\,.
\end{aligned} \tag{13.151}$$

In this expression we have

$$\begin{aligned}
\nabla_{\boldsymbol{\phi}}^T Z(\boldsymbol{\theta}) &= \sum_{u=1}^{U} \nabla_{\boldsymbol{\phi}}^T \tilde{p}_u(\boldsymbol{\theta}) \\
&= \frac{1}{v} \sum_{u=1}^{U} (\mathbf{y} - \mathbf{A}_u \boldsymbol{\phi})^T \mathbf{A}_u \tilde{p}_u(\boldsymbol{\theta})\,.
\end{aligned} \tag{13.152}$$

This gives

$$\begin{aligned}
\mathbf{f}_{21}(\boldsymbol{\theta}) &= \frac{1}{Nv}\Big[\sum_{u=1}^{U} ||\mathbf{y} - \mathbf{A}_u \boldsymbol{\phi}||_2^2 (\mathbf{y} - \mathbf{A}_u \boldsymbol{\phi})^T \mathbf{A}_u p_u(\boldsymbol{\theta}) \\
&\quad - \Big(\sum_{u=1}^{U} ||\mathbf{y} - \mathbf{A}_u \boldsymbol{\phi}||_2^2 p_u(\boldsymbol{\theta})\Big)\Big(\sum_{r=1}^{U} (\mathbf{y} - \mathbf{A}_r \boldsymbol{\phi})^T \mathbf{A}_r p_r(\boldsymbol{\theta})\Big)\Big] \\
&\quad - \frac{2}{N} \sum_{u=1}^{U} (\mathbf{y} - \mathbf{A}_u \boldsymbol{\phi})^T \mathbf{A}_u p_u(\boldsymbol{\theta})\,.
\end{aligned} \tag{13.153}$$

Finally, we have

$$
\begin{aligned}
f_{22}(\boldsymbol{\theta}) &= \frac{\partial}{\partial v} m_v(\boldsymbol{\theta}) \\
&= \frac{1}{NZ(\boldsymbol{\theta})} \sum_{u=1}^{U} ||\mathbf{y} - \mathbf{A}_u \boldsymbol{\phi}||_2^2 \frac{\partial}{\partial v} \tilde{p}_u(\boldsymbol{\theta}) \\
&\quad - \frac{1}{NZ^2(\boldsymbol{\theta})} \Big(\sum_{u=1}^{U} ||\mathbf{y} - \mathbf{A}_u \boldsymbol{\phi}||_2^2 \tilde{p}_u(\boldsymbol{\theta}) \Big) \frac{\partial}{\partial v} Z(\boldsymbol{\theta}) , \qquad (13.154)
\end{aligned}
$$

where

$$
\begin{aligned}
\frac{\partial}{\partial v} Z(\boldsymbol{\theta}) &= \sum_{u=1}^{U} \frac{\partial}{\partial v} \tilde{p}_u(\boldsymbol{\theta}) \\
&= \frac{1}{2v^2} \sum_{u=1}^{U} ||\mathbf{y} - \mathbf{A}_u \boldsymbol{\phi}||_2^2 \tilde{p}_u(\boldsymbol{\theta}) . \qquad (13.155)
\end{aligned}
$$

Substituting (13.155) inside (13.154) we find

$$
\begin{aligned}
f_{22}(\boldsymbol{\theta}) &= \frac{1}{2Nv^2} \Big[\sum_{u=1}^{U} ||\mathbf{y} - \mathbf{A}_u \boldsymbol{\phi}||_2^4 p_u(\boldsymbol{\theta}) - \Big(\sum_{u=1}^{U} ||\mathbf{y} - \mathbf{A}_u \boldsymbol{\phi}||_2^2 p_u(\boldsymbol{\theta}) \Big)^2 \Big] \\
&= \frac{1}{2Nv^2} \sum_{u=1}^{U} ||\mathbf{y} - \mathbf{A}_u \boldsymbol{\phi}||_2^4 p_u(\boldsymbol{\theta}) - \frac{N}{2v^2} m_v^2(\boldsymbol{\theta}) . \qquad (13.156)
\end{aligned}
$$

Up to this point, no assumption has been made about the parameter vector $\boldsymbol{\theta}$. Suppose now that

$$
\boldsymbol{\theta} = \boldsymbol{\theta}_* = \begin{bmatrix} \boldsymbol{\phi}_* \\ v_* \end{bmatrix}
$$

is a stationary point of the EM iteration, so that it satisfies $\boldsymbol{\theta}_* = \mathbf{m}(\boldsymbol{\theta}_*)$. This implies

$$
\boldsymbol{\phi}_* = \mathbf{m}_{\boldsymbol{\phi}}(\boldsymbol{\theta}_*) \quad , \quad v_* = m_v(\boldsymbol{\theta}_*) \qquad (13.157)
$$

and

$$
\sum_{u=1}^{U} (\mathbf{y} - \mathbf{A}_u \boldsymbol{\phi}_*) \mathbf{A}_u^T p_u(\boldsymbol{\theta}_*) = 0 . \qquad (13.158)
$$

Taking these identities into account inside (13.149)–(13.150), (13.153), and (13.156), and denoting

$$
\mathbf{e}_{u\,*} = \mathbf{y} - \mathbf{A}_u \boldsymbol{\phi}_* \quad , \quad J_{u\,*} = ||\mathbf{y} - \mathbf{A}_u \boldsymbol{\phi}_*||_2^2 \quad , \quad p_{u\,*} = p_u(\boldsymbol{\theta}_*) ,
$$

and

$$
\mathbf{W}_* = \Big(\sum_{u=1}^{U} \mathbf{A}_u^T \mathbf{A}_u p_{u\,*} \Big)^{-1} N v_* ,
$$

we find that at a stationary point $\boldsymbol{\theta}_*$ of the EM iteration, the block entries of the local iteration matrix can be expressed as

$$\mathbf{F}_{11}(\boldsymbol{\theta}_*) = \frac{\mathbf{W}_*}{Nv_*^2} \sum_{u=1}^{U} \mathbf{A}_u^T \mathbf{e}_{u*} \mathbf{e}_{u*}^T \mathbf{A}_u p_{u*} \tag{13.159}$$

$$\mathbf{f}_{12}(\boldsymbol{\theta}_*) = \frac{\mathbf{W}_*}{2Nv_*^3} \sum_{u=1}^{U} \mathbf{A}_u^T \mathbf{e}_{u*} J_{u*} p_{u*} \tag{13.160}$$

$$\mathbf{f}_{21}(\boldsymbol{\theta}_*) = \frac{1}{Nv_*} \sum_{u=1}^{U} J_{u*} \mathbf{e}_{u*}^T \mathbf{A}_u p_{u*} \tag{13.161}$$

$$f_{22}(\boldsymbol{\theta}_*) = \frac{1}{2Nv_*^2} \sum_{u=1}^{U} J_{u*}^2 p_{u*} - \frac{N}{2} . \tag{13.162}$$

Note that these entries are all expressed in terms of quantities evaluated by the EMVA, so the local iteration matrix $\mathbf{F}$ can be computed online, with $\boldsymbol{\theta}_*$ replaced by the current estimate $\boldsymbol{\theta}^k$ of the parameter vector. The spectral radius ρ of $\mathbf{F}$ then specifies the local rate of convergence of the EM iteration. So the extra computations required to evaluate $\mathbf{F}$ and its spectral radius allow the determination of the local rate of convergence of the EM iteration, which can be used to decide when to terminate this iteration.

Let $W_*^{1/2}$ denote an artitrary matrix square root of W_*, so that

$$W_* = W_*^{1/2} (W_*^{1/2})^T , \tag{13.163}$$

and consider the matrix

$$T \triangleq \begin{bmatrix} W_*^{1/2} & 0 \\ 0 & 2^{1/2} v_* \end{bmatrix} . \tag{13.164}$$

By observing that the matrix

$$\boldsymbol{\Psi}_* = T^{-1} \mathbf{F}(\boldsymbol{\theta}_*) T \tag{13.165}$$

is symmetric, we can conclude that, although the matrix $\mathbf{F}(\boldsymbol{\theta}_*)$ is not symmetric, all its eigenvalues are real, since it is related to the symmetrix matrix $\boldsymbol{\Phi}_*$ by the similarity transformation T.

Thus, the convergence of the EM iteration for the blind equalization problem can be analyzed at several levels. At a completely global level, the general convergence result of Wu [25] establishes that the EM iterates $\boldsymbol{\theta}^k$ will converge to a local maximum of the log-likelihood function $L(\mathbf{y}|\boldsymbol{\theta})$. However one drawback of this result is that it does not explicitly rule out the possibility that successive iterates might jump around from a region located not too far from one local maximum, to another. The mid-range convergence analysis of the EM iteration presented in Section 13.5.1 establishes that the EM iteration is a quasi-Newton method. Thus under mild conditions, each local maximum of L admits a capture set, so if the initial iterate $\boldsymbol{\theta}^0$ is located inside this capture set, all subsequent iterates will remain in the capture set, thus explicitly ruling out the possibility that successive iterates might jump around from one global maximum to another. Note, however, that the size of capture sets is unknown, so the result of 13.5.1 characterizes the behavior of the EM iteration only within a limited range of a local maximum. Finally, the characterization of the convergence rate of the EM iteration described in Section 13.5.2 only holds in the immediate vicinity of a stationary point of the EM iteration, i.e. a local maximum of $L(\mathbf{y}|\boldsymbol{\theta})$. Thus the different types of convergence analyses presented here provide different and complementary insights concerning the convergence behavior of the EM iteration when applied to blind equalization.

13.6 Bibliographical Notes

The PSP methodology for combining MAP sequence estimation with adaptive Markov chain parameter estimation was first described in detail in [4]. Its blind acquisition characterteristics are examined in [6], and digital communications applications such as detector design for unknown phase channels are discussed in [26]. The Baum-Welch reestimation scheme for estimating the parameters of HMMs is described in [7, 9, 20]. This approach uses the forward-backward algorithm to implement the EM iteration. It is exact, unlike the EM Viterbi algorithm introduced more recently in [11–13], which prunes the state trellis by retaining only the survivor paths. The convergence analysis of the EM iteration presented in Section 13.5 is a simplification for the case of PAM communication systems of the more general results derived in [11–13] for linearly or nonlinearly modulated communication systems. Note that for such systems, the corresponding baseband model is complex, and because of this feature, the convergence analysis of the EM iteration is more involved than for the real channel model considered here.

13.7 Problems

13.1. Consider a CT observation signal

$$Y(t) = E^{1/2} \sum_{m=-\infty}^{\infty} p(t - mT) I_m + V(t)$$

where the transmitted symbols I_m are independent and take binary values ± 1 with probability $1/2$, $V(t)$ is a zero-mean WGN independent of symbols I_m with variance v, and $p(t)$ is a unit energy rectangular pulse

$$p(t) = \begin{cases} \frac{1}{T^{1/2}} & 0 \leq t \leq T \\ 0 & \text{otherwise} \ . \end{cases}$$

In this model E and T denote respectively the energy per bit transmitted and the baud period. This signal is passed through a matched filter with impulse response $p(-t)$ followed by a sampler which samples the signal at times $t = nT + \tau$, where τ denotes an unknown sampling offset such that $0 \leq \tau < T$.

(a) Verify that the DT signal at the output of the sampler can be expressed as

$$Y_n = E^{1/2} \sum_{m=-\infty}^{\infty} h((n-m)T + \tau) I_m + V_n$$

where $h(t) = p(-t) * p(t)$ is the triangular pulse

$$h(t) = \begin{cases} 1 + t/T & -T \leq t \leq 0 \\ 1 - t/T & 0 \leq t \leq T \\ 0 & \text{otherwise} \ , \end{cases}$$

and V_n is a zero-mean DT WGN independent of I_n. What is the variance of V_n?

(b) Conclude from the above expression that the n-th observation can be expressed as

$$Y_n = E^{1/2} \Big[(1 - \frac{\tau}{T}) I_n + \frac{\tau}{T} I_{n+1} \Big] + V_n \ .$$

Thus in the absence of any timing offset, i.e., for $\tau = 0$, the observation Y_n is free of any ISI, but for a nonzero τ, ISI is introduced, making it necessary [27] to employ an MC model for MAP sequence detection. Construct an MC model for the observed signal. How many states are needed? Specify the one-step transition probability matrix of the MC, and draw one section of the MC trellis.

(c) For the model of part (b), observe that the output map

$$h(x_t, x_{t+1}, \tau) = E^{1/2} [I_n + \frac{(I_{n+1} - I_n)}{T} \tau]$$

depends linearly on τ. Use this observation to derive a PSP algorithm in the RLS form (13.45)–(13.47) for updating the estimate $\hat{\tau}(k, t)$ for the survivor path terminating in state k at time t. How many survivor paths are present? Make sure to describe the path extension stage of the PSP algorithm.

13.2. Instead of employing an RLS update in the PSP algorithm of Problem 13.1, derive an LMS update of the form (13.50).

13.3. Obtain an EMV algorithm for the joint sequence detection and timing offset estimation of Problem 13.1. Describe precisely both the E- and M-phases of the algorithm. Use the linear parametrization of $h(x_t, x_{t+1}, \tau)$ to show that the path-weighted mismatch function $W(\tau|\tau')$ given by (13.99) is *quadratic* in τ. Use this expression to derive a closed form expression for updating τ in the M-phase of the EMVA.

13.4. Consider the MAP sequence detection of a CPFSK modulated signal described in Problem 12.1. Let $h = p/q$ denote the modulation index. The Markov chain state X_n is defined as in Problem 12.1. Assume that the amplitude A of the transmitted signal is *unknown*, so after demodulation and sampling of the in-phase and quadrature components of the received signal, the DT observed signal can be expressed in vector form as

$$\begin{bmatrix} Y_{cn} \\ Y_{sn} \end{bmatrix} = A \begin{bmatrix} \cos(\pi X_{n+1}/q) \\ -\sin(\pi X_{n+1}/q) \end{bmatrix} + \begin{bmatrix} V_{cn} \\ V_{sn} \end{bmatrix},$$

where V_{cn} and V_{sn} are two independent zero-mean WGN sequences with variance v. Equivalently, if we introduce the complex sequences

$$Y_n = Y_{cn} + jY_{sn} \quad , \quad V_n = V_{cn} + jV_{sn}$$

the observations can be written as

$$Y_n = A\exp(-j\pi X_{n+1}/q) + V_n \tag{13.166}$$

for $n \geq 1$, where V_n is a $CN(0, 2v)$ distributed white sequence. The noise variance v is assumed known.

(a) Observe that the path metric

$$c(x_{n+1}, A) = |Y_n - A\exp(-j\pi X_{n+1}/q)|^2 \tag{13.167}$$

is quadratic in A. Use this observation to obtain a PSP algorithm in RLS form for the amplitude estimate $\hat{A}(k, n)$ associated to the survivor path $\hat{X}(k, n)$ terminating in state k at stage n of the Viterbi algorithm.

(b) For a known reference trajectory, the observation equation (13.166) is linear in A. Since A is constant, $A_n = A$ admits a trivial state-space model of the form

$$A_{n+1} = A_n \, .$$

Show that the PSP algorithm of part (a) can be viewed as formed by K Kalman filters running in parallel, that generate amplitude estimates $\hat{A}(k, n)$, $1 \leq k \leq K$ for each survivor path.

13.5. Design an EMV algorithm for Problem 13.4 consisting of jointly detecting a CPFSK sequence and estimating amplitude parameter A. Assume that the noise variance v is known. Describe precisely both the E- and M-phases of the algorithm. Use the quadratic form (13.167) of the path metric to derive a closed form expression for updating A in the M-phase of the EMVA.

13.6. Design an EMV algorithm for the joint sequence detection of a CPFSK signal and estimation of the parameter vector

$$\boldsymbol{\theta} = \begin{bmatrix} A \\ v \end{bmatrix}$$

formed by the signal amplitude in (13.166) and noise variance v. Describe precisely the E- and M-phases of the algorithm. In particular, obtain closed-form expressions for updating A and v in the M-phase of the EMVA.

References

1. T. Kailath, "Correlation detection of signals perturbed by a random channel," *IRE Trans. on Informat. Theory*, vol. 6, pp. 361–366, June 1960.
2. R. A. Iltis, "A Bayesian maximum-likelihood sequence estimation algorithm for *a priori* unknown channels and symbol timing," *IEEE J. Selected Areas in Commun.*, vol. 10, pp. 579–588, Apr. 1992.
3. N. Seshadri, "Joint data and channel estimation using blind trellis search techniques," *IEEE Trans. Commun.*, vol. 42, pp. 1000–1011, Feb./Mar./Apr. 1994.
4. R. Raheli, A. Polydoros, and C. Tzou, "Per-survivor-processing: A general approach to MLSE in uncertain environments," *IEEE Trans. Commun.*, vol. 43, pp. 354–364, Feb./Apr. 1995.
5. K. M. Chugg and A. Polydoros, "MLSE for an unknown channel – Part I: Optimality considerations," *IEEE Trans. Commun.*, vol. 44, pp. 836–846, July 1996.
6. K. M. Chugg, "Blind acquisition characteristics of PSP-based sequence detectors," *IEEE J. Selected Areas in Commun.*, vol. 16, pp. 1518–1529, Oct. 1998.
7. L. R. Rabiner, "A tutorial on hidden Markov models and selected applications in speech recognition," *Proc. IEEE*, vol. 77, pp. 257–286, Feb. 1989.
8. Y. Ephraim and N. Merhav, "Hidden Markov processes," *IEEE Trans. Informat. Theory*, vol. 48, pp. 1518–1569, June 2002.
9. L. Rabiner and B.-H. Juang, *Fundamentals of Speech Recognition*. Englewood Cliffs, NJ: Prentice Hall, 1993.
10. A. P. Dempster, N. M. Laird, and D. B. Rubin, "Maximum likelihood from incomplete data via the EM algorithm," *J. Royal Stat. Society, Series B*, vol. 39, no. 1, pp. 1–38, 1977.
11. H. Nguyen, *The Expectation-Maximization Viterbi algorithm for blind channel identification and equalization.* PhD thesis, Dept. of Electrical and Computer Engineering, Univ. California, Davis, Aug. 2003.
12. H. Nguyen and B. C. Levy, "Blind and semi-blind equalization of CPM signals with the EMV algorithm," *IEEE Trans. Signal Proc.*, vol. 51, pp. 2650–2664, Oct. 2003.
13. H. Nguyen and B. C. Levy, "The expectation-maximization Viterbi algorithm for blind adaptive channel equalization," *IEEE Trans. Commun.*, vol. 53, pp. 1671–1678, Oct. 2005.
14. A. Sayed, *Fundamentals of Adaptive Filtering.* New York: Wiley Interscience/IEEE Press, 2003.

15. P. Kovintavewat, J. R. Barry, M. F. Erden, and E. Kurtas, "Per-survivor timing recovery for uncoded partial response channels," in *Proc. 2004 IEEE Internat. Conf. on Communications*, vol. 5, (Paris, France), pp. 2715–2719, June 2004.
16. Z. Ding and Y. Li, *Blind Equalization and Identification.* New York, NY: Marcel Dekker, 2001.
17. I. W. J. Weber, "Differential encoding for multiple amplitude and phase-shift-keying systems," *IEEE Trans. Commun.*, vol. 26, pp. 385–391, May 1978.
18. J. G. Proakis, *Digital Communications, Fourth Edition.* New York: McGraw-Hill, 2000.
19. L. E. Baum, T. Petrie, G. Soules, and N. Weiss, "A maximization technique occurring in the statistical analysis of probabilistic functions of Markov chains," *Annals Mathematical Statistics*, vol. 41, pp. 164–171, Feb. 1970.
20. T. K. Moon, "The Expectation-Maximization algorithm," *IEEE Signal Proc. Magazine*, pp. 47–60, Nov. 1996.
21. V. Krishnamurthy and J. B. Moore, "On-line estimation of hidden Markov model parameters based on the Kullback-Leibler information measure," *IEEE Trans. Signal Proc.*, vol. 41, Aug. 1993.
22. M. Feder and J. Catipovic, "Algorithms for joint channel estimation and data recovery– applications to equalization in underwater acoustics," *IEEE J. Ocean Eng.*, vol. 16, pp. 42–55, Jan. 1991.
23. K. H. Chang, W. S. Yuan, and C. N. Georghiades, "Block-by-block channel and sequence estimation for ISI/fading channels," in *Signal Processing in Telecommunications: Proceedings of the 7th Thyrrhenian Workshop on Digital Communications, Viareggio, Italy,* sept. 10–14, 1995 (E. Biglieri and M. Luise, eds.), pp. 153–170, Berlin, Germany: Springer Verlag, 1996.
24. X.-L. Meng and D. van Dyk, "The EM algorithm – an old folk-song sung to a fast new tune," *J. Royal Stat. Soc., Series B*, vol. 59, no. 3, pp. 511–567, 1997.
25. C. F. J. Wu, "On the convergence properties of the EM algorithm," *Annals Statistics*, vol. 11, pp. 95–103, 1983.
26. G. Ferrari, G. Colavolpe, and R. Raheli, *Detection Algorithms for Wireless Communications With Applications to Wired and Storage Systems.* Chichester, England: J. Wiley & Sons, 2004.
27. C. N. Georghiades, "Optimum delay and sequence estimation from incomplete data," *IEEE Trans. Informat. Theory*, vol. 36, pp. 202–208, Jan. 1990.

Index